Introduction to Communication Systems

Showcasing the essential principles behind modern communication systems, this accessible undergraduate textbook provides a solid introduction to the foundations of communication theory.

- Carefully selected topics introduce students to the most important and fundamental concepts, giving them a focused, in-depth understanding of core material, and preparing them for more advanced study.
- Abstract concepts are introduced "just in time" and reinforced by nearly 200 end-of-chapter exercises, alongside numerous MATLAB code fragments, software problems, and practical lab exercises, firmly linking the underlying theory to real-world problems, and providing additional hands-on experience.
- An accessible lecture-style organisation makes it easy for students to navigate to key passages, and quickly identify the most relevant material.

Containing material suitable for a one- or two-semester course, and accompanied online by a password-protected solutions manual and supporting instructor resources, this is the perfect introductory textbook for undergraduate students studying electrical and computer engineering.

Upamanyu Madhow is Professor of Electrical and Computer Engineering at the University of California, Santa Barbara. His research interests broadly span communications, signal processing and networking, with current emphasis on next-generation wireless and bio-inspired approaches to networking and inference. He is a recipient of the NSF CAREER Award and the IEEE Marconi prize paper award in wireless communication, author of the graduate-level textbook *Fundamentals of Digital Communication* (2008), and a Fellow of the IEEE.

"Madhow does it again: *Introduction to Communication Systems* is an accessible yet rigorous new text that does for undergraduates what his digital communications book did for graduate students. It provides a superior treatment of not only the fundamentals of analog and digital communication, but also the theoretical underpinnings needed to understand them, including frequency domain analysis and probability. The book is unusual in that it also includes newer topics of pressing current relevance like multiple antenna communication and OFDM. I strongly recommend this book for faculty teaching senior level courses on communication systems."

Jeffrey G. Andrews
The University of Texas at Austin

"This is an excellent undergraduate text on analog and digital communications. It covers everything from classic analog techniques to recent wireless systems. Students will enjoy the inclusion of advanced topics such as channel coding and MIMO."

David Love
Purdue University

"*Introduction to Communication Systems* by Madhow is truly unique in the vast landscape of introductory books on communication systems. From the basics of signal processing, probability, and communications, to the advanced topics of coding, multipath mitigation, and multiple antenna systems, the book deftly interweaves abstract theory, design principles, and applications in a highly effective and insightful manner. This masterfully-written book will play a key role in teaching and inspiring the next generation of communication system engineers."

Andrea Goldsmith
Stanford University

"This is the textbook on communications I have wanted for a while. Crisply written, it forms the basis for an ideal two semester sequence. It nicely balances rigor, concepts and practice."

Saoura Dasgupta
University of Iowa

"This is a unique introduction to the basic principles of communication system design, with a remarkable combination of rigour and accessibility. The MATLAB exercises are expertly weaved together with theoretical principles making it an excellent textbook for training undergraduate communication systems engineers."

Suhas Diggavi
University of California, Los Angeles

"This is a valuable addition to the current set of textbooks on communication systems. It is comprehensive, and offers a modern perspective shaped by the author's research that has pushed the state of the art. The software labs enhance the practicality of the text and serve to illustrate more advanced material in an accessible way."

Michael Honig
Northwestern University

Introduction to Communication Systems

Upamanyu Madhow
University of California, Santa Barbara

Shaftesbury Road, Cambridge CB2 8EA, United Kingdom

One Liberty Plaza, 20th Floor, New York, NY 10006, USA

477 Williamstown Road, Port Melbourne, VIC 3207, Australia

314–321, 3rd Floor, Plot 3, Splendor Forum, Jasola District Centre, New Delhi – 110025, India

103 Penang Road, #05–06/07, Visioncrest Commercial, Singapore 238467

Cambridge University Press is part of Cambridge University Press & Assessment,
a department of the University of Cambridge.

We share the University's mission to contribute to society through the pursuit of
education, learning and research at the highest international levels of excellence.

www.cambridge.org
Information on this title: www.cambridge.org/9781107022775

First published 2014

A catalogue record for this publication is available from the British Library

Library of Congress Cataloging-in-Publication data
Madhow, Upamanyu.
Introduction to communication systems / Upamanyu Madhow, University of
California, Santa Barbara.
pages cm
Includes bibliographical references.
ISBN 978-1-107-02277-5
1. Telecommunication. I. Title.
TK5101.M273 2014
621.382–dc23
2014004974

ISBN 978-1-107-02277-5 Hardback

Additional resources for this publication at www.cambridge.org/commnsystems

To my family and students

Contents

Preface

Progress in telecommunications over the past two decades has been nothing short of revolutionary, with communications taken for granted in modern society to the same extent as electricity. There is therefore a persistent need for engineers who are well-versed in the principles of communication systems. These principles apply to communication between points in space, as well as communication between points in time (i.e., storage). Digital systems are fast replacing analog systems in both domains. This book has been written in response to the following core question: what is the basic material that an undergraduate student with an interest in communications should learn, in order to be well prepared for either industry or graduate school? For example, some institutions teach only digital communication, assuming that analog communication is dead or dying. Is that the right approach? From a purely pedagogical viewpoint, there are critical questions related to mathematical preparation: how much mathematics must a student learn to become well-versed in system design, what should be assumed as background, and at what point should the mathematics that is not in the background be introduced? Classically, students learn probability and random processes, and then tackle communication. This does not quite work today: students increasingly (and, I believe, rightly) question the applicability of the material they learn, and are less interested in abstraction for its own sake. On the other hand, I have found from my own teaching experience that students get truly excited about abstract concepts when they discover their power in applications, and it is possible to provide the means for such discovery using software packages such as MATLAB. Thus, we have the opportunity to get a new generation of students excited about this field: by covering abstractions "just in time" to shed light on engineering design, and by reinforcing concepts immediately using software experiments in addition to conventional pen-and-paper problem solving, we can remove the lag between learning and application, and ensure that the concepts stick.

This textbook represents my attempt to act upon the preceding observations, and is an outgrowth of my lectures for a two-course undergraduate elective sequence on communication at UCSB, which is often also taken by some beginning graduate students. Thus, it can be used as the basis for a two-course sequence in communication systems, or a single course on digital communication, at the undergraduate or beginning graduate level. The book also provides a review or introduction to communication systems for practitioners, easing the path to study of more advanced graduate texts and the research literature. The prerequisite is a course on signals and systems, together with an introductory course on probability. The required material on random processes is included in the text.

A student who masters the material here should be well-prepared for either graduate school or the telecommunications industry. The student should leave with an understanding of baseband and passband signals and channels, modulation formats appropriate for these channels, random processes and noise, a systematic framework for optimum demodulation based on signal-space concepts, performance analysis and power–bandwidth tradeoffs for common modulation schemes, a hint of the power of information theory and channel coding, and an introduction to communication techniques for dispersive channels and multiple antenna systems. Given the significant ongoing research and development activity in wireless communication, and the fact that an understanding of wireless link design provides a sound background for approaching other communication links, material enabling hands-on discovery of key concepts for wireless system design is distributed throughout the textbook.

I should add that I firmly believe that the utility of this material goes well beyond communications, important as that field is. Communications systems design merges concepts from signals and systems, probability and random processes, and statistical inference. Given the broad applicability of these concepts, a background in communications is of value in a large variety of areas requiring "systems thinking," as I discuss briefly at the end of Chapter 1.

The goal of the lecture-style exposition in this book is to clearly articulate a selection of concepts that I deem *fundamental* to communication system design, rather than to provide comprehensive coverage. "Just in time" coverage is provided by organizing and limiting the material so that we get to core concepts and applications as quickly as possible, and by sometimes asking the reader to operate with partial information (which is, of course, standard operating procedure in the real world of engineering design). However, the topics that we do cover are covered in sufficient detail to enable the student to solve nontrivial problems and to obtain hands-on involvement via software labs. Descriptive material that can easily be looked up online is omitted.

Organization

- Chapter 1 provides a perspective on communication systems, including a discussion of the transition from analog to digital communication and how it colors the selection of material in this text.
- Chapter 2 provides a review of signals and systems (biased towards communications applications), and then discusses the complex-baseband representation of passband signals and systems, emphasizing its critical role in modeling, design, and implementation. A software lab on modeling and undoing phase offsets in complex baseband, while providing a sneak preview of digital modulation, is included. Chapter 2 also includes a section on wireless-channel modeling in complex baseband using ray tracing, reinforced by a software lab that applies these ideas to simulate link time variations for a lamppost-based broadband wireless network.
- Chapter 3 covers analog communication techniques that remain relevant even as the world goes digital, including superheterodyne reception and phase-locked loops. Legacy analog modulation techniques are discussed to illustrate core concepts, as well as in

recognition of the fact that suboptimal analog techniques such as envelope detection and limiter–discriminator detection may have to be resurrected as we push the limits of digital communication in terms of speed and power consumption. Software labs reinforce and extend concepts in amplitude and angle modulation.

- Chapter 4 discusses digital modulation, including linear modulation using constellations such as pulse amplitude modulation (PAM), quadrature amplitude modulation (QAM), and phase-shift keying (PSK), and orthogonal modulation and its variants. The chapter includes discussion of the number of degrees of freedom available on a bandlimited channel, the Nyquist criterion for avoidance of intersymbol interference, and typical choices of Nyquist and square-root Nyquist signaling pulses. We also provide a sneak preview of power–bandwidth tradeoffs (with detailed discussion postponed until the effect of noise has been modeled in Chapters 5 and 6). A software lab providing a hands-on feel for Nyquist signaling is included in this chapter.

The material in Chapters 2 through 4 requires only a background in signals and systems.

- Chapter 5 provides a review of basic probability and random variables, and then introduces random processes. This chapter provides detailed discussion of Gaussian random variables, vectors and processes; this is essential for modeling noise in communication systems. Examples giving a preview of receiver operations in communication systems, and computation of performance measures such as error probability and signal-to-noise ratio (SNR), are provided. A discussion of the circular symmetry of white noise, and noise analysis of analog modulation techniques, are placed in an appendix, since this is material that is often skipped in modern courses on communication systems.
- Chapter 6 covers classical material on optimum demodulation for M-ary signaling in the presence of additive white Gaussian noise (AWGN). The background on Gaussian random variables, vectors, and processes developed in Chapter 5 is applied to derive optimal receivers, and to analyze their performance. After discussing error probability computation as a function of SNR, we are able to combine the materials in Chapters 4 and 6 for a detailed discussion of power–bandwidth tradeoffs. Chapter 6 concludes with an introduction to link-budget analysis, which provides guidelines on the choice of physical link parameters such as transmit and receive antenna gains, and distance between transmitter and receiver, using what we know about the dependence of error probability as a function of SNR. This chapter includes a software lab that builds on the Nyquist signaling lab in Chapter 4 by investigating the effect of noise. It also includes another software lab simulating performance over a time-varying wireless channel, examining the effects of fading and diversity, and introduces the concept of differential demodulation for avoidance of explicit channel tracking.

Chapters 2 through 6 provide a systematic lecture-style exposition of what I consider core concepts in communication at an undergraduate level.

- Chapter 7 provides a glimpse of information theory and coding whose goal is to stimulate the reader to explore further using more advanced resources such as graduate courses and textbooks. It shows the critical role of channel coding, provides an initial exposure to information-theoretic performance benchmarks, and discusses belief propagation in detail, reinforcing the basic concepts through a software lab.

- Chapter 8 provides a first exposure to the more advanced topics of communication over dispersive channels, and to multiple antenna systems, often termed space–time communication, or multiple-input, multiple-output (MIMO) communication. These topics are grouped together because they use similar signal processing tools. We emphasize lab-style "discovery" in this chapter using three software labs, one on adaptive linear equalization for single-carrier modulation, one on basic orthogonal frequency-division multiplexing (OFDM) transceiver operations, and one on MIMO signal processing for space–time coding and spatial multiplexing. The goal is for students to acquire hands-on insight that should motivate them to undertake a deeper and more systematic investigation.
- Finally, the epilogue contains speculation on future directions in communications research and technology. The goal is to provide a high-level perspective on where mastery of the introductory material in this textbook could lead, and to argue that the innovations which this field has already seen set the stage for many exciting developments to come.

The role of software. Software problems and labs are integrated into the text, with "code fragments" implementing core functionalities provided in the text. While code can be provided online, separate from the text (and, indeed, sample code is made available online for instructors), code fragments are integrated into the text for two reasons. First, they enable readers to immediately see the software realization of a key concept as they read the text. Second, I feel that students learn more by putting in the work of writing their own code, building on these code fragments if they wish, rather than using code that is easily available online. The particular software that we use is MATLAB, because of its widespread availability, and because of its importance in design and performance evaluation both in academia and in industry. However, the code fragments can also be viewed as "pseudocode," and can be easily implemented using other software packages or languages. Block-based packages such as Simulink (which builds upon MATLAB) are avoided here, because the use of software here is pedagogical rather than aimed at, say, designing a complete system by putting together subsystems as one might do in industry.

Suggestions for using this book

I view Chapter 2 (complex baseband), Chapter 4 (digital modulation), and Chapter 6 (optimum demodulation) as core material that *must* be studied to understand the concepts underlying modern communication systems. Chapter 6 relies on the probability and random processes material in Chapter 5, especially the material on jointly Gaussian random variables and white Gaussian noise (WGN), but the remaining material in Chapter 5 can be skipped or covered selectively, depending on the students' background. Chapter 3 (analog communication techniques) is designed such that it can be completely skipped if one wishes to focus solely on digital communication. Finally, Chapter 7 and Chapter 8 contain glimpses of advanced material that can be sampled according to the instructor's discretion. The qualitative discussion in the epilogue is meant to provide the student with perspective, and is not intended for formal coverage in the classroom.

In my own teaching at UCSB, this material forms the basis for a two-course sequence, with Chapters 2–4 covered in the first course, and Chapters 5 and 6 covered in the second course, with the dispersive channels portion of Chapter 8 providing the basis for the labs in the second course. The content of these courses is constantly being revised, and it is expected that the material on channel coding and MIMO may displace some of the existing material in the future. UCSB is on a quarter system, hence the coverage is fast-paced, and many topics are omitted or skimmed. There is ample material here for a two-semester undergraduate course sequence. For a one-semester course, one possible organization is to cover Chapter 2 (focusing on the complex envelope), Chapter 4, a selection of Chapter 5, Chapter 6, and, if time permits, Chapter 7.

The slides accompanying the book are intended not to provide comprehensive coverage of the material, but rather to provide an example of selections from the material to be covered in the classroom. I must comment in particular on Chapter 5. While much of the book follows the format in which I lecture, Chapter 5 is structured as a reference on probability, random variables, and random processes that the instructor must pick and choose from, depending on the background of the students in the class. The particular choices I make in my own lectures on this material are reflected in the slides for this chapter.

Acknowledgements

This book is an outgrowth of lecture notes for an undergraduate elective course sequence in communications at UCSB, and I am grateful to the succession of students who have used, and provided encouraging comments on, the evolution of the course sequence and the notes. I would also like to acknowledge faculty in the communications area at UCSB who were kind enough to give me a "lock" on these courses over the past few years, as I was developing this textbook.

The first priority in a research university is to run a vibrant research program, hence I must acknowledge the extraordinarily capable graduate students in my research group over the years during which this textbook was developed. They have done superb research with minimal supervision from me, and the strength of their peer interactions and collaborations is what gave me the mental space, and time, needed to write this textbook. Current and former group members who have directly helped with aspects of this book include Andrew Irish, Babak Mamandipoor, Dinesh Ramasamy, Maryam Eslami Rasekh, Sumit Singh, Sriram Venkateswaran, and Aseem Wadhwa.

I gratefully acknowledge the funding agencies which have provided support for our research group in recent years, including the National Science Foundation (NSF), the Army Research Office (ARO), the Defense Advanced Research Projects Agency (DARPA), and Systems on Nanoscale Information Fabrics (SONIC), a center supported by DARPA and Microelectronics Advanced Research Corporation (MARCO). One of the primary advantages of a research university is that undergraduate education is influenced, and kept up to date, by cutting-edge research. This textbook embodies this paradigm both in its approach (an emphasis on what one can *do* with what one learns) and in its content (emphasis of concepts that are fundamental background for research in the area).

I thank Phil Meyler and his colleagues at Cambridge University Press for encouraging me to initiate this project, and for their blend of patience and persistence in getting me to see it through despite a host of other commitments. I also thank the anonymous reviewers of the book proposal and sample chapters sent to Cambridge several years back for their encouragement and constructive comments. I am also grateful to a number of faculty colleagues who have given encouragement, helpful suggestions, and pointers to alternative pedagogical approaches: Professor Soura Dasgupta (University of Iowa), Professor Jerry Gibson (UCSB), Professor Gerhard Kramer (Technische Universität München, Munich), Professor Phil Schniter (Ohio State University), and Professor Venu Veeravalli (University of Illinois at Urbana-Champaign).

Finally, I thank the anonymous reviewers of the almost-final manuscript for their helpful comments.

Finally, I would like to thank my wife and children for always being the most enjoyable and interesting people to spend time with. Recharging my batteries in their company, and that of our many pets, is what provides me with the energy needed for an active professional life.

1 Introduction

This textbook provides an introduction to the conceptual underpinnings of communication technologies. Most of us directly experience such technologies daily: browsing (and audio/video streaming from) the Internet, sending/receiving emails, watching television, or carrying out a phone conversation. Many of these experiences occur on mobile devices that we carry around with us, so that we are always connected to the cyberworld of modern communication systems. In addition, there is a huge amount of machine-to-machine communication that we do not directly experience, but which is indispensable for the operation of modern society. This includes, for example, signaling between routers on the Internet, or between processors and memories on any computing device.

We define *communication* as the process of *information transfer across space or time.* Communication across space is something we have an intuitive understanding of: for example, radio waves carry our phone conversation between our cell phone and the nearest base station, and coaxial cables (or optical fiber, or radio waves from a satellite) deliver television from a remote location to our home. However, a moment's thought shows that that communication across time, or storage of information, is also an everyday experience, given our use of storage media such as compact discs (CDs), digital video discs (DVDs), hard drives, and memory sticks. In all of these instances, the key steps in the operation of a communication link are as follows:

(a) insertion of information into a signal, termed the *transmitted signal,* compatible with the physical medium of interest;
(b) propagation of the signal through the physical medium (termed the *channel*) in space or time; and
(c) extraction of information from the signal (termed the *received signal*) obtained after propagation through the medium.

In this book, we study the fundamentals of modeling and design for these steps.

Chapter plan

In Section 1.1, we provide a high-level description of analog and digital communication systems, and discuss why digital communication is the inevitable design choice in modern systems. In Section 1.2, we briefly provide a technological perspective on recent developments in communication. We do not attempt to provide a comprehensive discussion of the fascinating history of communication: thanks to the advances in communication that brought us the Internet, it is easy to look it up online! A discussion of the scope of this textbook is provided in Section 1.3.

1.1 Analog or digital?

Even without defining information formally, we intuitively understand that speech, audio, and video signals contain information. We use the term *message signals* for such signals, since these are the messages we wish to convey over a communication system. In their original form – both during generation and consumption – these message signals are *analog*: they are continuous-time signals, with the signal values also lying in a continuum. When someone plays the violin, an analog acoustic signal is generated (often translated to an analog electrical signal using a microphone). Even when this music is recorded onto a digital storage medium such as a CD (using the digital communication framework outlined in Section 1.1.2), when we ultimately listen to the CD being played on an audio system, we hear an analog acoustic signal. The transmitted signals corresponding to physical communication media are also analog. For example, in both wireless and optical communication, we employ electromagnetic waves, which correspond to continuous-time electric and magnetic fields taking values in a continuum.

1.1.1 Analog communication

Given the analog nature of both the message signal and the communication medium, a natural design choice is to map the analog message signal (e.g., an audio signal, translated from the acoustic to the electrical domain using a microphone) to an analog transmitted signal (e.g., a radio wave carrying the audio signal) that is compatible with the physical medium over which we wish to communicate (e.g., broadcasting audio over the air from an FM radio station). This approach to communication system design, depicted in Figure 1.1, is termed *analog communication.* Early communication systems were all analog: examples include AM (amplitude modulation) and FM (frequency modulation) radio, analog television, first-generation cellular-phone technology (based on FM), vinyl records, audio cassettes, and VHS or Betamax videocassettes

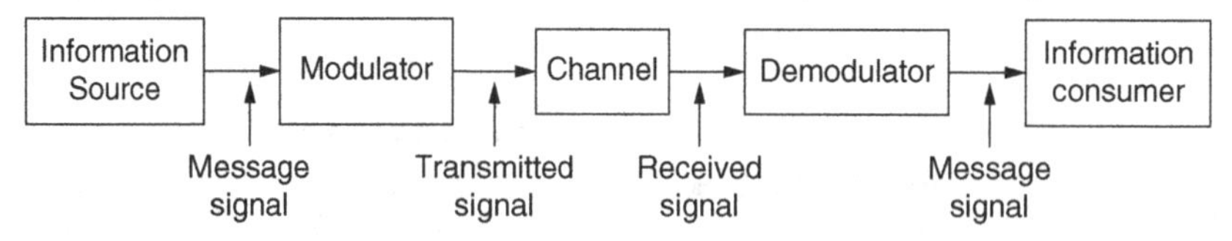

Figure 1.1 A block diagram for an analog communication system. The modulator transforms the message signal into the transmitted signal. The channel distorts and adds noise to the transmitted signal. The demodulator extracts an estimate of the message signal from the received signal arriving from the channel.

While analog communication might seem like the most natural option, it is in fact obsolete. Cellular-phone technologies from the second generation onwards are digital; vinyl records and audio cassettes have been supplanted by CDs, and videocassettes by DVDs. Broadcast technologies such as radio and television are often slower to upgrade because of economic and political factors, but digital broadcast radio and television technologies are either replacing or sidestepping (e.g., via satellite) analog FM/AM radio and television broadcast. Let us now define what we mean by digital communication, before discussing the reasons for the inexorable trend away from analog and towards digital communication.

1.1.2 Digital communication

The conceptual basis for digital communication was established in 1948 by Claude Shannon, when he founded the field of information theory. There are two main threads to this theory.

- **Source coding and compression.** Any information-bearing signal can be represented efficiently, to within a desired accuracy of reproduction, by a digital signal (i.e., a discrete-time signal taking values from a discrete set), which in its simplest form is just a sequence of binary digits (zeros or ones), or *bits.* This is true irrespective of whether the information source is text, speech, audio, or video. Techniques for performing the mapping from the original source signal to a bit sequence are generically termed *source coding.* They often involve *compression,* or removal of redundancy, in a manner that exploits the properties of the source signal (e.g., the heavy spatial correlation among adjacent pixels in an image can be exploited to represent it more efficiently than a pixel-by-pixel representation).
- **Digital information transfer.** Once the source encoding has been done, our communication task reduces to reliably transferring the bit sequence at the output of the source encoder across space or time, without worrying about the original source and the sophisticated tricks that have been used to encode it. The performance of any communication system depends on the relative strengths of the signal and noise or interference, and the distortions imposed by the channel. Shannon showed that, once we have fixed these operational parameters for any communication channel, there exists a maximum possible rate of reliable communication, termed the *channel capacity.* Thus, given the information bits at the output of the source encoder, in principle, we can transmit them reliably over a given link as long as the information rate is smaller than the channel capacity, and we cannot transmit them reliably if the information rate is larger than the channel capacity. This sharp transition between reliable and unreliable communication differs fundamentally from analog communication, where the quality of the reproduced source signal typically degrades gradually as the channel conditions get worse.

A block diagram for a typical digital communication system based on these two threads is shown in Figure 1.2. We now briefly describe the role of each component, together with simplified examples of its function.

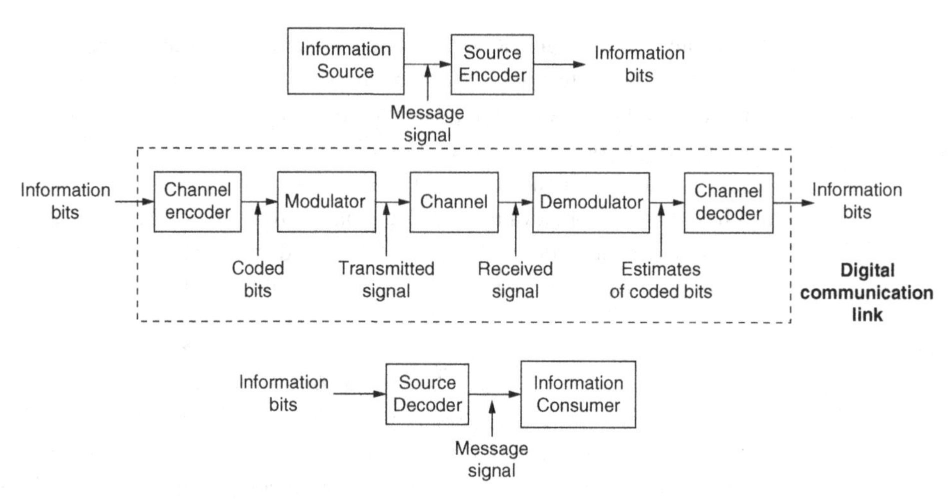

Figure 1.2 Components of a digital communication system.

Source encoder As already discussed, the source encoder converts the message signal into a sequence of information bits. The information bit rate depends on the nature of the message signal (e.g., speech, audio, video) and the application requirements. Even when we fix the class of message signals, the choice of source encoder is heavily dependent on the setting. For example, video signals are heavily compressed when they are sent over a cellular link to a mobile device, but are lightly compressed when sent to a high-definition television (HDTV) set. A cellular link can support a much smaller bit rate than, say, the cable connecting a DVD player to an HDTV set, and a smaller mobile display device requires lower resolution than a large HDTV screen. In general, the source encoder must be chosen such that the bit rate it generates can be supported by the digital communication link we wish to transfer information over. Other than this, source coding can be decoupled entirely from link design (we comment further on this a bit later).

Example. A laptop display may have resolution 1024×768 pixels. For a grayscale digital image, the intensity for each pixel might be represented by 8 bits. Multiplying by the number of pixels gives us about 6.3 million bits, or about 0.8 Mbyte (a byte equals 8 bits). However, for a typical image, the intensities for neighboring pixels are heavily correlated, which can be exploited for significantly reducing the number of bits required to represent the image, without noticeably distorting it. For example, one could take a two-dimensional Fourier transform, which concentrates most of the information in the image at lower frequencies, and then discard many of the high-frequency coefficients. There are other possible transforms one could use, and also several more processing stages, but the bottom line is that, for natural images, state-of-the-art image-compression algorithms can provide $10\times$ compression (i.e., reduction in the number of bits relative to the original uncompressed digital image) with hardly any perceptual degradation. Far more aggressive compression ratios are possible if we are willing to tolerate more distortion. For video, in addition to the spatial correlation exploited for image compression, we can also exploit temporal correlation across successive frames.

Channel encoder The channel encoder adds redundancy to the information bits obtained from the source encoder, in order to facilitate error recovery after transmission over the channel. It might appear that we are putting in too much work, adding redundancy just after the source encoder has removed it. However, the redundancy added by the channel encoder is tailored to the channel over which information transfer is to occur, whereas the redundancy in the original message signal is beyond our control, so that it would be inefficient to keep it when we transmit the signal over the channel.

Example. The noise and distortion introduced by the channel can cause errors in the bits we send over it. Consider the following abstraction for a channel: we can send a string of bits (zeros or ones) over it, and the channel randomly flips each bit with probability 0.01 (i.e., the channel has a 1% error rate). If we cannot tolerate this error rate, we could repeat each bit that we wish to send three times, and use a majority rule to decide on its value. Now, we only make an error if two or more of the three bits are flipped by the channel. It is left as an exercise to calculate that an error now happens with probability approximately 0.0003 (i.e., the error rate has gone down to 0.03%). That is, we have improved performance by introducing redundancy. Of course, there are far more sophisticated and efficient techniques for introducing redundancy than the simple repetition strategy just described; see Chapter 7.

Modulator The modulator maps the coded bits at the output of the channel encoder to a transmitted signal to be sent over the channel. For example, we may insist that the transmitted signal fit within a given frequency band and adhere to stringent power constraints in a wireless system, where interference between users and between co-existing systems is a major concern. Unlicensed WiFi transmissions typically occupy 20–40 MHz of bandwidth in the 2.4- or 5-GHz bands. Transmissions in fourth-generation cellular systems may often occupy bandwidths ranging from 1 to 20 MHz at frequencies ranging from 700 MHz to 3 GHz. While these signal bandwidths are being increased in an effort to increase data rates (e.g., up to 160 GHz for emerging WiFi standards, and up to 100 MHz for emerging cellular standards), and new frequency bands are being actively explored (see the epilogue for more discussion), the transmitted signal still needs to be shaped to fit within certain spectral constraints.

Example. Suppose that we send bit value 0 by transmitting the signal $s(t)$, and bit value 1 by transmitting $-s(t)$. Even for this simple example, we must design the signal $s(t)$ so it fits within spectral constraints (e.g., two different users may use two different segments of spectrum to avoid interfering with each other), and we must figure out how to prevent successive bits of the same user from interfering with each other. For wireless communication, these signals are voltages generated by circuits coupled to antennas, and are ultimately emitted as electromagnetic waves from the antennas.

The channel encoder and modulator are typically jointly designed, keeping in mind the anticipated channel conditions, and the result is termed a *coded modulator.*

Channel The channel distorts and adds noise, and possibly interference, to the transmitted signal. Much of our success in developing communication technologies has resulted from being able to optimize communication strategies based on accurate mathematical models for the channel. Such models are typically statistical, and are developed with significant effort using a combination of measurement and computation. The physical characteristics of the communication medium vary widely, and hence so do the channel models. Wireline channels are typically well modeled as linear and time-invariant, while optical-fiber channels exhibit nonlinearities. Wireless mobile channels are particularly challenging because of the time variations caused by mobility, and due to the potential for interference due to the broadcast nature of the medium. The link design also depends on system-level characteristics, such as whether or not the transmitter has feedback regarding the channel, and what strategy is used to manage interference.

Example. Consider communication between a cellular base station and a mobile device. The electromagnetic waves emitted by the base station can reach the mobile's antennas through multiple paths, including bounces off streets and building surfaces. The received signal at the mobile can be modeled as multiple copies of the transmitted signal with different gains and delays. These gains and delays change due to mobility, but the rate of change is often slow compared with the data rate, hence, over short intervals, we can get away with modeling the channel as a linear time-invariant system that the transmitted signal goes through before arriving at the receiver.

Demodulator The demodulator processes the signal received from the channel to produce bit estimates to be fed to the channel decoder. It typically performs a number of signal-processing tasks, such as synchronization of phase, frequency, and timing, and compensating for distortions induced by the channel.

Example. Consider the simplest possible channel model, where the channel just adds noise to the transmitted signal. In our earlier example of sending $\pm s(t)$ to send 0 or 1, the demodulator must guess, based on the noisy received signal, which of these two options is true. It might make a hard decision (e.g., guess that 0 was sent), or hedge its bets, and make a soft decision, saying, for example, that it is 80% sure that the transmitted bit is a zero. There are many other aspects of demodulation that we have swept under the rug: for example, before making any decisions, the demodulator has to perform functions such as synchronization (making sure that the receiver's notion of time and frequency is consistent with the transmitter's) and equalization (compensating for the distortions due to the channel).

Channel decoder The channel decoder processes the imperfect bit estimates provided by the demodulator, and exploits the controlled redundancy introduced by the channel encoder to estimate the information bits.

Example. The channel decoder takes the guesses from the demodulator and uses the redundancies in the channel code to clean up the decisions. In our simple example of repeating every bit three times, it might use a majority rule to make its final decision if the demodulator is putting out hard decisions. For soft decisions, it might use more sophisticated combining rules with improved performance.

While we have described the demodulator and decoder as operating separately and in sequence for simplicity, there can be significant benefits from iterative information exchange between the two. In addition, for certain coded modulation strategies in which channel coding and modulation are tightly coupled, the demodulator and channel decoder may be integrated into a single entity.

Source decoder The source decoder processes the estimated information bits at the output of the channel decoder to obtain an estimate of the message. The message format may, but need not, be the same as that of the original message input to the source encoder: for example, the source encoder may translate speech to text before encoding into bits, and the source decoder may output a text message to the end user.

Example. For the example of a digital image considered earlier, the compressed image can be translated back to a pixel-by-pixel representation by taking the inverse spatial Fourier transform of the coefficients that survived the compression.

We are now ready to compare analog and digital communication, and discuss why the trend towards digital is inevitable.

1.1.3 Why digital?

On comparing the block diagrams for analog and digital communication in Figures 1.1 and 1.2, respectively, we see that the digital communication system involves far more processing. However, this is not an obstacle for modern transceiver design, due to the exponential increase in the computational power of low-cost silicon integrated circuits. Digital communication has the following key advantages.

Optimality For a point-to-point link, it is optimal to separately optimize source coding and channel coding, as long as we do not mind the delay and processing incurred in doing so. Owing to this *source–channel separation principle,* we can leverage the best available source codes and the best available channel codes in designing a digital communication system, independently of each other. Efficient source encoders must be highly specialized. For example, state-of-the-art speech encoders, video-compression algorithms, and text-compression algorithms are very different from each other, and are each the result of significant effort over many years by a large community of researchers. However, once source encoding has been performed, the coded modulation scheme used over the communication link can be engineered to transmit the information bits reliably, regardless of what kind of source they correspond to, with the bit rate limited only by the channel and transceiver characteristics. Thus, the design of a digital communication link is *source-independent* and *channel-optimized.* In contrast, the waveform transmitted in an analog communication system depends on the message signal, which is beyond the control of the link designer, hence we do not have the freedom to optimize link performance over all possible communication schemes. This is not just a theoretical observation: in practice, huge performance gains are obtained from switching from analog to digital communication.

Scalability While Figure 1.2 shows a single digital communication link between source encoder and decoder, under the source–channel-separation principle, there is nothing preventing us from inserting additional links, putting the source encoder and decoder at the end points. This is because digital communication allows *ideal regeneration* of the information bits, hence every time we add a link, we can focus on communicating reliably over that particular link. (Of course, information bits do not always get through reliably, hence we typically add error-recovery mechanisms such as retransmission, at the level of an individual link or "end-to-end" over a sequence of links between the information source and sink.) Another consequence of the source–channel-separation principle is that, since information bits are transported without interpretation, the same link can be used to carry multiple kinds of messages. A particularly useful approach is to chop the information bits up into discrete chunks, or *packets*, which can then be processed independently on each link. These properties of digital communication are critical for enabling massively scalable, general-purpose, *communication networks* such as the Internet. Such networks can have large numbers of digital communication links, possibly with different characteristics, independently engineered to provide "bit pipes" that can support data rates. Messages of various kinds, after source encoding, are reduced to packets, and these packets are switched along different paths along the network, depending on the identities of the source and destination nodes, and the loads on different links in the network. None of this would be possible with analog communication: link performance in an analog communication system depends on message properties, and successive links incur noise accumulation, which limits the number of links which can be cascaded.

The preceding makes it clear that source–channel separation, and the associated bit-pipe abstraction, is crucial in the formation and growth of modern communication networks. However, there are some important caveats that are worth noting. Joint source–channel design can provide better performance in some settings, especially when there are constraints on delay or complexity, or if multiple users are being supported simultaneously on a given communication medium. In practice, this means that "local" violations of the separation principle (e.g., over a wireless last hop in a communication network) may be a useful design trick. Similarly, the bit-pipe abstraction used by network designers is too simplistic for the design of wireless networks at the edge of the Internet: physical properties of the wireless channel such as interference, multipath propagation, and mobility must be taken into account in network engineering.

1.1.4 Why analog design remains important

While we are interested in transporting bits in digital communication, the physical link over which these bits are sent is analog. Thus, analog and mixed-signal (digital/analog) design continue to play a crucial role in modern digital communication systems. Analog design of digital-to-analog converters, mixers, amplifiers, and antennas is required in order to translate bits to physical waveforms to be emitted by the transmitter. At the receiver, analog design of antennas, amplifiers, mixers, and analog-to-digital converters is required in order to translate the physical received waveforms to digital (discrete-valued, discrete-time) signals that are amenable to the digital signal processing that is at the core of modern transceivers. Analog circuit design for communications is therefore a thriving field in its own right, which this textbook makes no attempt to cover. However, the material in Chapter 3 on analog communication techniques is intended to introduce digital communication system designers to some of the high-level issues addressed by analog circuit designers.

The goal is to establish enough of a common language to facilitate interaction between system and circuit designers. While much of digital communication system design can be carried out by abstracting out the intervening analog design (as done in Chapters 4 through 8), closer interaction between system and circuit designers becomes increasingly important as we push the limits of communication systems, as briefly indicated in the epilogue.

1.2 A technology perspective

We now discuss some technology trends and concepts that have driven the astonishing growth in communication systems in the past two decades, and that are expected to impact future developments in this area. Our discussion is structured in terms of big technology "stories."

Technology story 1: the Internet Some of the key ingredients that contributed to its growth and the essential role it plays in our lives are as follows.

- Any kind of message can be chopped up into packets and routed across the network, using an Internet Protocol (IP) that is simple to implement in software.
- Advances in optical-fiber communication and high-speed digital hardware enable a super-fast "core" of routers connected by very high-speed, long-range links, that enable worldwide coverage;
- The World Wide Web, or web, makes it easy to organize information into interlinked hypertext documents, which can be browsed from anywhere in the world.
- The digitization of content (audio, video, books) means that ultimately "all" information is expected to be available on the web.
- Search engines enable us to efficiently search for this information.
- Connectivity applications such as email, teleconferencing, videoconferencing and online social networks have become indispensable in our daily lives.

Technology story 2: wireless Cellular mobile networks are everywhere, and are based on the breakthrough concept that ubiquitous tetherless connectivity can be provided by breaking the world into cells, with "spatial reuse" of precious spectrum resources in cells that are "far enough" apart. Base stations serve mobiles in their cells, and hand them off to adjacent base stations when the mobile moves to another cell. While cellular networks were invented to support voice calls for mobile users, today's mobile devices (e.g., "smart phones" and tablet computers) are actually powerful computers with displays large enough for users to consume video on the go. Thus, cellular networks must now support seamless access to the Internet. The billions of mobile devices in use easily outnumber desktop and laptop computers, so that the most important parts of the Internet today are arguably the cellular networks at its edge. Mobile service providers are having great difficulty keeping up with the increase in demand resulting from this convergence of cellular and Internet; by some estimates, the capacity of cellular networks must be scaled up by several orders of magnitude, at least in densely populated urban areas! As discussed in the epilogue, a major challenge for the communication researcher and technologist, therefore, is to come up with the breakthroughs required to deliver such capacity gains.

Another major success in wireless is WiFi, a catchy term for a class of standardized wireless local-area network (WLAN) technologies based on the IEEE 802.11 family of standards. Currently, WiFi networks use unlicensed spectrum in the 2.4- and 5-GHz bands, and have come into widespread use in both residential and commercial environments. WiFi transceivers are now incorporated into almost every computer and mobile device. One way of alleviating the cellular capacity crunch that was just mentioned is to offload Internet access to the nearest WiFi network. Of course, since different WiFi networks are often controlled by different entities, seamless switching between cellular and WiFi is not always possible.

It is instructive to devote some thought to the contrast between cellular and WiFi technologies. Cellular transceivers and networks are far more tightly engineered. They employ spectrum that mobile operators pay a great deal of money to license, hence it is critical to use this spectrum efficiently. Furthermore, cellular networks must provide robust wide-area coverage in the face of rapid mobility (e.g., automobiles at highway speeds). In contrast, WiFi uses unlicensed (i.e., free!) spectrum, must provide only local coverage, and typically handles much slower mobility (e.g., pedestrian motion through a home or building). As a result, WiFi can be more loosely engineered than cellular. It is interesting to note that, despite the deployment of many uncoordinated WiFi networks in an unlicensed setting, WiFi typically provides acceptable performance, partly because the relatively large amount of unlicensed spectrum (especially in the 5-GHz band) allows channel switching on encountering excessive interference, and partly because of naturally occurring spatial reuse (WiFi networks that are "far enough" from each other do not interfere with each other). Of course, in densely populated urban environments with many independently deployed WiFi networks, the performance can deteriorate significantly, a phenomenon sometimes referred to as a tragedy of the commons (individually selfish behavior leading to poor utilization of a shared resource). As we briefly discuss in the epilogue, both the cellular and the WiFi design paradigms need to evolve to meet our future needs.

Technology story 3: Moore's law Moore's "law" is actually an empirical observation attributed to Gordon Moore, one of the founders of Intel Corporation. It can be paraphrased as saying that the density of transistors in an integrated circuit, and hence the amount of computation per unit cost, can be expected to increase exponentially over time. This observation has become a self-fulfilling prophecy, because it has been taken up by the semiconductor industry as a growth benchmark driving their technology roadmap. While the growth in density implied by Moore's law may be slowing down somewhat, it has already had a spectacular impact on the communications industry by drastically lowering the cost and increasing the speed of digital computation. By converting analog signals to the digital domain as soon as possible, advanced transceiver algorithms can be implemented in digital signal processing (DSP) using low-cost integrated circuits, so that research breakthroughs in coding and modulation can be quickly transitioned into products. This leads to economies of scale that have been critical to the growth of mass-market products in both wireless (e.g., cellular and WiFi) and wireline (e.g., cable modems and DSL) communication.

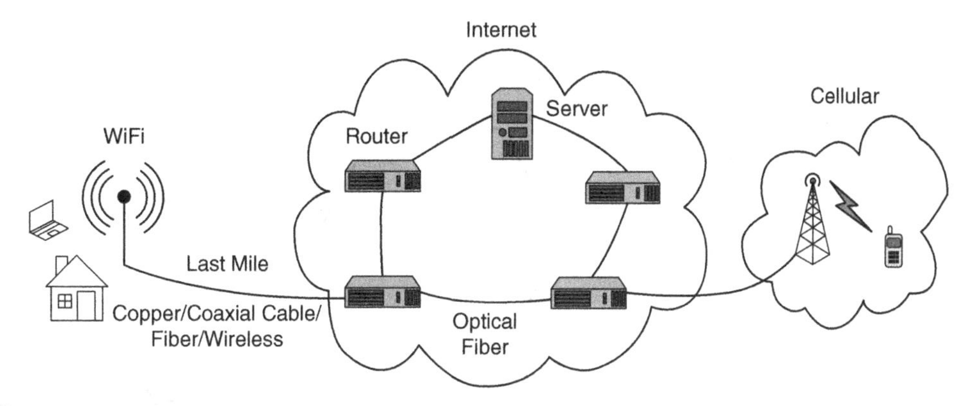

Figure 1.3 The Internet has a core of routers and servers connected by high-speed fiber links, with wireless networks hanging off the edge (figure courtesy of Aseem Wadhwa).

How do these stories come together? The sketch in Figure 1.3 highlights key building blocks of the Internet today. The *core* of the network consists of powerful routers that direct packets of data from an incoming edge to an outgoing edge, and *servers* (often housed in large *data centers*) that serve up content requested by *clients* such as personal computers and mobile devices. The elements in the core network are connected by high-speed optical fiber. Wireless can be viewed as hanging off the edge of the Internet. Wide-area cellular networks may have worldwide coverage, but each base station is typically connected by a high-speed link to the wired Internet. WiFi networks are wireless local-area networks, typically deployed indoors (but potentially also providing outdoor coverage for low-mobility scenarios) in homes and office buildings, connected to the Internet via *last-mile* links, which might run over copper wires (a legacy of wired telephony, with transceivers typically upgraded to support broadband Internet access) or coaxial cable (originally deployed to deliver cable television, but now also providing broadband Internet access). Some areas have been upgraded to optical fiber to the curb or even to the home, while some others might be remote enough to require wireless last-mile solutions.

Zooming in now on cellular networks, Figure 1.4 shows three adjacent cells in a cellular network with hexagonal cells. A working definition of a cell is that it is the area around a base station where the signal strength is higher than that from other base stations. Of course, under realistic propagation conditions, cells are never hexagonal, but the concept of spatial reuse still holds: the interference between distant cells can be neglected, hence they can use the same communication resources. For example, in Figure 1.4, we might decide to use three different frequency bands in the three cells shown, but might then reuse these bands in other cells. Figure 1.4 also shows that a user may be simultaneously in range of multiple base stations when near cell boundaries. Crossing these boundaries may result in a *handoff* from one base station to another. In addition, near cell boundaries, a mobile device may be in communication with multiple base stations simultaneously, a concept known as *soft handoff*.

It is useful for a communication system designer to be aware of the preceding "big picture" of technology trends and network architectures in order to understand how to

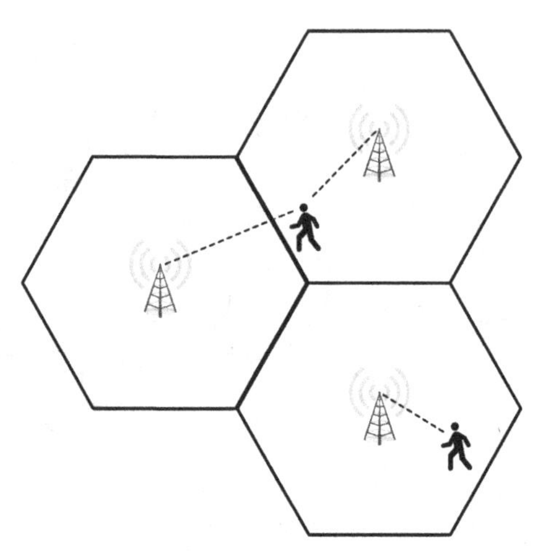

Figure 1.4 A segment of a cellular network with idealized hexagonal shapes (figure courtesy of Aseem Wadhwa).

direct his or her talents as these systems continue to evolve (the epilogue contains more detailed speculation regarding this evolution). However, the first order of business is to acquire the *fundamentals* required to get going in this field. These are quite simply stated: a communication system designer must be comfortable with mathematical modeling (in order to understand the state of the art, as well as to devise new models as required), and with devising and evaluating signal-processing algorithms based on these models. The goal of this textbook is to provide a first exposure to such a technical background.

1.3 The scope of this textbook

Referring to the block diagram of a digital communication system in Figure 1.2, our focus in this textbook is to provide an introduction to the design of a digital communication link as shown inside the dashed box. While we are primarily interested in digital communication, circuit designers implementing such systems must deal with analog waveforms, hence we believe that a rudimentary background in analog communication techniques, as provided in this book, is useful for the communication system designer. We do not discuss source encoding and decoding in this book; these topics are highly specialized and technical, and doing them justice requires an entire textbook of its own at the graduate level. A detailed outline of the book is provided in the preface, hence we restrict ourselves here to summarizing the roles of the various chapters:

Chapter 2 introduces the signal-processing background required for DSP-centric implementations of communication transceivers;

Chapter 3 provides just enough background on analog communication techniques (this can be skipped to focus exclusively on digital communication);

Chapter 4 discusses digital modulation techniques;

Chapter 5 provides the probability background required for receiver design, including noise modeling;

Chapter 6 discusses design and performance analysis of demodulators in digital communication systems for idealized link models;
Chapter 7 provides an initial exposure to channel coding techniques and benchmarks;
Chapter 8 provides an introduction to approaches for handling channel dispersion, and to multiple antenna communication; and the
Epilogue discusses emerging trends shaping research and development in communications.

Chapters 2, 4, and 6 are core material that must be mastered (much of Chapter 5 is also core material, but some readers may already have enough probability background that they can skip, or skim, it). Chapter 3 is highly recommended for communication system designers with an interest in radio-frequency circuit design, since it highlights, at a high level, some of the ideas and issues that come up there. Chapters 7 and 8 are independent of each other, and contain more advanced material that might not always fit within an undergraduate curriculum. They contain "hands-on" introductions to these topics via code fragments and software labs that should encourage the reader to explore further.

1.4 Why study communication systems?

Before launching into our formal study, it makes sense to ask why the material in this textbook is worth studying. There are several obvious answers to this question. The indispensable role of communications in modern life, and the success of the communications industry, implies that a solid understanding of this material constitutes a valuable skill set. The vibrant future of communications (see the epilogue) ensures the continuing value of this skill set for many decades to come. However, there is also an indirect, and perhaps more fundamental, answer to this question. The design of communication systems today represents a triumph of mathematical modeling and statistical signal processing. Detailed, hands-on experience building confidence in such techniques is therefore excellent preparation for tackling more complex systems for which complete mathematical models might not be available, as the author has discovered in his own research. Examples of such systems include the Internet itself, online social networks running on the Internet, financial systems exhibiting a complex web of interdependences, as well as signal processing, inference, and machine-learning techniques for the huge volumes of data ("big data") being generated in a host of other applications.

1.5 Concept summary

The goal of this chapter is to provide an intellectual framework and motivation for the rest of this textbook. Some of the key concepts are as follows.

- Communication refers to information transfer across either space or time, where the latter refers to storage media.

- Signals carrying information and signals that can be sent over a communication medium are both inherently analog (i.e., continuous-time, continuous-valued).
- Analog communication corresponds to transforming an analog message waveform directly into an analog transmitted waveform at the transmitter, and undoing this transformation at the receiver.
- Digital communication corresponds to first reducing message waveforms to information bits, and then transporting these bits over the communication channel.
- Digital communication requires the following steps: source encoding and decoding, modulation and demodulation, channel encoding and decoding.
- While digital communication requires more processing steps than analog communication, it has the advantages of optimality and scalability, hence there is an unstoppable trend from analog to digital.
- The growth in communication has been driven by major technology stories including the Internet, wireless, and Moore's law.
- Key components of the communication system designer's toolbox are mathematical modeling and signal processing.

1.6 Notes

There are many textbooks on communication systems at both undergraduate and graduate level. Undergraduate texts include Haykin [1], Proakis and Salehi [2], Pursley [3], and Ziemer and Tranter [4]. Graduate texts, which typically focus on digital communication, include Barry, Lee, and Messerschmitt [5], Benedetto and Biglieri [6], Madhow [7], and Proakis and Salehi [8]. The first coherent exposition of the modern theory of communication receiver design is in the classical (graduate level) textbook by Wozencraft and Jacobs [9]. Other important classical graduate-level texts are Viterbi and Omura [10] and Blahut [11]. More specialized references (e.g., on signal processing, information theory, channel coding, wireless communication) are mentioned in later chapters. In addition to these textbooks, an overview of many important topics can be found in the recently updated mobile communications handbook [12] edited by Gibson.

This book is intended to be accessible to readers who have never been exposed to communication systems before. It has some overlap with more advanced graduate texts (e.g., Chapters 2, 4, 5, and 6 here overlap heavily with Chapters 2 and 3 in the author's own graduate text [7]), and provides the technical background and motivation required to easily access these more advanced texts. Of course, the best way to continue building expertise in the field is by actually working in it. Research and development in this field requires study of the research literature, of more specialized texts (e.g., on information theory, channel coding, synchronization), and of commercial standards. The Institute for Electrical and Electronics Engineers (IEEE) is responsible for publication of many conference proceedings and journals in communications: major conferences include that IEEE Global Telecommunications Conference (Globecom) and the IEEE International Communications Conference (ICC), major journals and magazines include *IEEE Communications*

Magazine, the *IEEE Transactions on Communications*, and the *IEEE Journal on Selected Areas in Communications*. Closely related fields such as information theory and signal processing have their own conferences, journals, and magazines. Major conferences include the IEEE International Symposium on Information Theory (ISIT) and IEEE International Conference on Acoustics, Speech and Signal Processing (ICASSP); journals include the *IEEE Transactions on Information Theory* and the *IEEE Transactions on Signal Processing*. The IEEE also publishes a number of standards online, such as the IEEE 802 family of standards for local-area networks.

A useful resource for learning source coding and data compression, which are not discussed in this text, is the textbook by Sayood [13]. Textbooks on core concepts in communication networks include Bertsekas and Gallager [14], Kumar, Manjunath, and Kuri [15], and Walrand and Varaiya [16].

2 Signals and systems

A communication link involves several stages of signal manipulation: the transmitter transforms the message into a signal that can be sent over a communication channel; the channel distorts the signal and adds noise to it; and the receiver processes the noisy received signal to extract the message. Thus, communication systems design must be based on a sound understanding of signals, and the systems that shape them. In this chapter, we discuss concepts and terminology from signals and systems, with a focus on how we plan to apply them in our discussion of communication systems. Much of this chapter is a review of concepts with which the reader might already be familiar from prior exposure to signals and systems. However, special attention should be paid to the discussion of baseband and passband signals and systems (Sections 2.7 and 2.8). This material, which is crucial for our purpose, is typically not emphasized in a first course on signals and systems. Additional material on the geometric relationship between signals is covered in later chapters, when we discuss digital communication.

Chapter plan

After a review of complex numbers and complex arithmetic in Section 2.1, we provide some examples of useful signals in Section 2.2. We then discuss LTI systems and convolution in Section 2.3. This is followed by Fourier series (Section 2.4) and the Fourier transform (Section 2.5). These sections (Sections 2.1 through Section 2.5) correspond to a review of material that is part of the assumed background for the core content of this textbook. However, even readers familiar with the material are encouraged to skim through it quickly in order to gain familiarity with the notation. This gets us to the point where we can classify signals and systems based on the frequency band they occupy. Specifically, we discuss baseband and passband signals and systems in Sections 2.7 and 2.8. Messages are typically baseband, while signals sent over channels (especially radio channels) are typically passband. We discuss methods for going from baseband to passband and back. We specifically emphasize the fact that a real-valued passband signal is equivalent (in a mathematically convenient and physically meaningful sense) to a complex-valued baseband signal, called the *complex-baseband representation*, or *complex envelope*, of the passband signal. We note that the information carried by a passband signal resides in its complex envelope, so that modulation (or the process of encoding messages in waveforms that can be sent over physical channels) consists of mapping information into a complex envelope, and then converting this complex envelope into a passband signal. We discuss the physical significance of the rectangular form of the complex envelope, which corresponds to the *in-phase (I)* and *quadrature (Q)* components of the passband signal, and that of the polar form of the complex envelope, which corresponds to the *envelope* and *phase* of the passband signal. We conclude by discussing the role of complex baseband in transceiver implementations, and by illustrating its use for wireless channel modeling.

Software

The software labs in this chapter introduce the use of MATLAB for signal processing. They provide practice in writing MATLAB code from scratch (i.e., without using prepackaged routines or Simulink) for simple computations. Software Lab 2.1 is an introduction to the use of MATLAB for typical operations of interest to us, and illustrates how we approximate continuous-time operations in discrete time. Software Lab 2.2 shows how to model and undo the effects of carrier-phase offsets in complex baseband. Software Lab 2.3 develops complex-baseband models for wireless multipath channels, and explores the phenomenon of signal fading due to constructive and destructive interference between the paths.

2.1 Complex numbers

A complex number z can be written as $z = x + jy$, where x and y are real numbers, and $j = \sqrt{-1}$. We say that $x = \text{Re}(z)$ is the real part of z and $y = \text{Im}(z)$ is the imaginary part of z. As depicted in Figure 2.1, it is often advantageous to interpret the complex number z as a two-dimensional real vector, which can be represented in rectangular form as $(x, y) = (\text{Re}(z), \text{Im}(z))$, or in polar form (r, θ) as

$$\begin{aligned} r &= |z| = \sqrt{x^2 + y^2} \\ \theta &= \angle z = \tan^{-1}(y/x) \end{aligned} \tag{2.1}$$

We can go back from polar form to rectangular form as follows:

$$x = r\cos\theta, \quad y = r\sin\theta \tag{2.2}$$

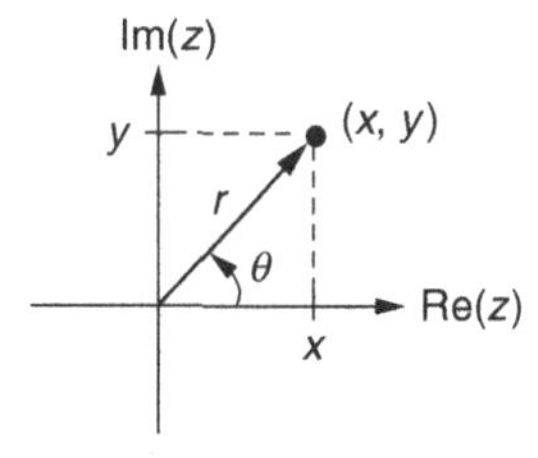

Figure 2.1 A complex number z represented in the two-dimensional real plane.

Complex conjugation For a complex number $z = x + jy = re^{j\theta}$, its complex conjugate

$$z^* = x - jy = re^{-j\theta} \tag{2.3}$$

That is,

$$\begin{aligned} \mathrm{Re}(z^*) = \mathrm{Re}(z), \quad \mathrm{Im}(z^*) = -\mathrm{Im}(z) \\ |z^*| = |z|, \quad \angle z^* = -\angle z \end{aligned} \tag{2.4}$$

The real and imaginary parts of a complex number z can be written in terms of z and z^* as follows:

$$\mathrm{Re}(z) = \frac{z + z^*}{2}, \quad \mathrm{Im}(z) = \frac{z - z^*}{2j} \tag{2.5}$$

Euler's formula This formula is of fundamental importance in complex analysis, and relates the rectangular and polar forms of a complex number:

$$e^{j\theta} = \cos\theta + j\sin\theta \tag{2.6}$$

The complex conjugate of $e^{j\theta}$ is given by

$$e^{-j\theta} = \left(e^{j\theta}\right)^* = \cos\theta - j\sin\theta$$

We can express cosines and sines in terms of $e^{j\theta}$ and its complex conjugate as follows:

$$\mathrm{Re}(e^{j\theta}) = \frac{e^{j\theta} + e^{-j\theta}}{2} = \cos\theta, \quad \mathrm{Im}(e^{j\theta}) = \frac{e^{j\theta} - e^{-j\theta}}{2j} = \sin\theta \tag{2.7}$$

On applying Euler's formula to (2.1), we can write

$$z = x + jy = r\cos\theta + jr\sin\theta = re^{j\theta} \tag{2.8}$$

Being able to go back and forth between the rectangular and polar forms of a complex number is useful. For example, it is easier to add in the rectangular form, but it is easier to multiply in the polar form.

Complex addition For two complex numbers $z_1 = x_1 + jy_1$ and $z_2 = x_2 + jy_2$,

$$z_1 + z_2 = (x_1 + x_2) + j(y_1 + y_2) \tag{2.9}$$

That is,

$$\mathrm{Re}(z_1 + z_2) = \mathrm{Re}(z_1) + \mathrm{Re}(z_2), \quad \mathrm{Im}(z_1 + z_2) = \mathrm{Im}(z_1) + \mathrm{Im}(z_2) \tag{2.10}$$

Complex multiplication (rectangular form) For two complex numbers $z_1 = x_1 + jy_1$ and $z_2 = x_2 + jy_2$,

$$z_1 z_2 = (x_1 x_2 - y_1 y_2) + j(y_1 x_2 + x_1 y_2) \tag{2.11}$$

This follows simply by multiplying out, and setting $j^2 = -1$. We have

$$\mathrm{Re}(z_1 z_2) = \mathrm{Re}(z_1)\mathrm{Re}(z_2) - \mathrm{Im}(z_1)\mathrm{Im}(z_2), \quad \mathrm{Im}(z_1 z_2) = \mathrm{Im}(z_1)\mathrm{Re}(z_2) + \mathrm{Re}(z_1)\mathrm{Im}(z_2) \tag{2.12}$$

Note that, using the rectangular form, a single complex multiplication requires four real multiplications.

Complex multiplication (polar form) Complex multiplication is easier when the numbers are expressed in polar form. For $z_1 = r_1 e^{j\theta_1}$ and $z_2 = r_2 e^{j\theta_2}$, we have

$$z_1 z_2 = r_1 r_2 e^{j(\theta_1+\theta_2)} \tag{2.13}$$

That is,

$$|z_1 z_2| = |z_1||z_2|, \quad \angle z_1 z_2 = \angle z_1 + \angle z_2 \tag{2.14}$$

Division For two complex numbers $z_1 = x_1 + jy_1 = r_1 e^{j\theta_1}$ and $z_2 = x_2 + jy_2 = r_2 e^{j\theta_2}$ (with $z_2 \neq 0$, i.e., $r_2 > 0$), it is easiest to express the result of division in polar form:

$$z_1/z_2 = (r_1/r_2) e^{j(\theta_1-\theta_2)} \tag{2.15}$$

That is,

$$|z_1/z_2| = |z_1|/|z_2|, \quad \angle z_1/z_2 = \angle z_1 - \angle z_2 \tag{2.16}$$

In order to divide using rectangular form, it is convenient to multiply the numerator and the denominator by z_2^*, which gives

$$z_1/z_2 = z_1 z_2^*/(z_2 z_2^*) = z_1 z_2^*/|z_2|^2 = \frac{(x_1 + jy_1)(x_2 - jy_2)}{x_2^2 + y_2^2}$$

On multiplying out as usual, we get

$$z_1/z_2 = \frac{(x_1 x_2 + y_1 y_2) + j(-x_1 y_2 + y_1 x_2)}{x_2^2 + y_2^2} \tag{2.17}$$

Example 2.1.1 (Computations with complex numbers) Consider the complex numbers $z_1 = 1+j$ and $z_2 = 2e^{-j\pi/6}$. Find $z_1 + z_2$, $z_1 z_2$, and z_1/z_2. Also specify z_1^* and z_2^*.

For complex addition, it is convenient to express both numbers in rectangular form. Thus,

$$z_2 = 2(\cos(-\pi/6) + j\sin(-\pi/6)) = \sqrt{3} - j$$

and

$$z_1 + z_2 = (1+j) + \left(\sqrt{3} - j\right) = \sqrt{3} + 1$$

For complex multiplication and division, it is convenient to express both numbers in polar form. We obtain $z_1 = \sqrt{2} e^{j\pi/4}$ by applying (2.1). Now, from (2.11), we have

$$z_1 z_2 = \sqrt{2} e^{j\pi/4} 2 e^{-j\pi/6} = 2\sqrt{2} e^{j(\pi/4 - \pi/6)} = 2\sqrt{2} e^{j\pi/12}$$

Similarly,

$$z_1/z_2 = \frac{\sqrt{2} e^{j\pi/4}}{2e^{-j\pi/6}} = \frac{1}{\sqrt{2}} e^{j(\pi/4+\pi/6)} = \frac{1}{\sqrt{2}} e^{j5\pi/12}$$

Multiplication using the rectangular forms of the complex numbers yields the following:

$$z_1 z_2 = (1+j)\left(\sqrt{3} - j\right) = \sqrt{3} - j + \sqrt{3}j + 1 = \left(\sqrt{3} + 1\right) + j\left(\sqrt{3} - 1\right)$$

Note that $z_1^* = 1 - j = \sqrt{2}e^{-j\pi/4}$ and $z_2^* = 2e^{j\pi/6} = \sqrt{3} + j$. Division using rectangular forms gives

$$z_1/z_2 = z_1 z_2^*/|z_2|^2 = (1+j)\left(\sqrt{3}+j\right)\Big/ 2^2 = \frac{\sqrt{3}-1}{4} + j\frac{\sqrt{3}+1}{4}$$

No need to memorize trigonometric identities any more Once we can do computations using complex numbers, we can use Euler's formula to quickly derive well-known trigonometric identities involving sines and cosines. For example,

$$\cos(\theta_1 + \theta_2) = \mathrm{Re}(e^{j(\theta_1+\theta_2)})$$

But

$$\begin{aligned} e^{j(\theta_1+\theta_2)} &= e^{j\theta_1}e^{j\theta_2} = (\cos\theta_1 + j\sin\theta_1)(\cos\theta_2 + j\sin\theta_2) \\ &= (\cos\theta_1\cos\theta_2 - \sin\theta_1\sin\theta_2) + j(\cos\theta_1\sin\theta_2 + \sin\theta_1\cos\theta_2) \end{aligned}$$

Taking the real part, we can read off the identity

$$\cos(\theta_1 + \theta_2) = \cos\theta_1\cos\theta_2 - \sin\theta_1\sin\theta_2 \tag{2.18}$$

Moreover, taking the imaginary part, we can read off

$$\sin(\theta_1 + \theta_2) = \cos\theta_1\sin\theta_2 + \sin\theta_1\cos\theta_2 \tag{2.19}$$

2.2 Signals

Signal A signal $s(t)$ is a function of time (or some other independent variable, such as frequency, or spatial coordinates) that has an interesting physical interpretation. For example, it is generated by a transmitter, or processed by a receiver. While physically realizable signals such as those sent over a wire or over the air must take real values, we shall see that it is extremely useful (and physically meaningful) to consider *a pair* of real-valued signals, interpreted as the real and imaginary parts of a complex-valued signal. Thus, in general, we allow signals to take complex values.

Discrete versus continuous time We generically use the notation $x(t)$ to denote continuous-time signals (t taking real values), and $x[n]$ to denote discrete-time signals (n taking integer values). A continuous-time signal $x(t)$ sampled at rate T_s produces discrete time samples $x(nT_s + t_0)$ (t_0 is an arbitrary offset), which we often denote as a discrete-time signal $x[n]$. While signals sent over a physical communication channel are inherently continuous-time, implementations both at the transmitter and at the receiver make heavy use of discrete-time implementations on digitized samples corresponding to the analog continuous-time waveforms of interest.

We now introduce some signals that recur often in this text.

Sinusoid This is a periodic function of time of the form

$$s(t) = A\cos(2\pi f_0 t + \theta) \tag{2.20}$$

where $A > 0$ is the amplitude, f_0 is the frequency, and $\theta \in [0, 2\pi]$ is the phase. By setting $\theta = 0$, we obtain a pure cosine $A\cos(2\pi f_c t)$, and by setting $\theta = -\pi/2$, we obtain a pure sine $A\sin(2\pi f_c t)$. In general, using (2.18), we can rewrite (2.20) as

$$s(t) = A_c \cos(2\pi f_0 t) - A_s \sin(2\pi f_0 t) \tag{2.21}$$

where $A_c = A\cos\theta$ and $A_s = A\sin\theta$ are real numbers. Using Euler's formula, we can write

$$Ae^{j\theta} = A_c + jA_s \tag{2.22}$$

Thus, the parameters of a sinusoid at frequency f_0 can be represented by the complex number in (2.22), with (2.20) using the polar form, and (2.21) the rectangular form, of this number. Note that $A = \sqrt{A_c^2 + A_s^2}$ and $\theta = \tan^{-1}(A_s/A_c)$.

Clearly, sinusoids with known amplitude, phase, and frequency are perfectly predictable, and hence cannot carry any information. As we shall see, information can be transmitted by making the complex number $Ae^{j\theta} = A_c + jA_s$ associated with the parameters of the sinusoid vary in a way that depends on the message to be conveyed. Of course, once this has been done, the resulting signal will no longer be a pure sinusoid, and part of the work of the communication system designer is to decide what shape such a signal should take in the frequency domain.

We now define complex exponentials, which play a key role in understanding signals and systems in the frequency domain.

Complex exponential A complex exponential at a frequency f_0 is defined as

$$s(t) = Ae^{j(2\pi f_0 t + \theta)} = \alpha e^{j2\pi f_0 t} \tag{2.23}$$

where $A > 0$ is the amplitude, f_0 is the frequency, $\theta \in [0, 2\pi]$ is the phase, and $\alpha = Ae^{j\theta}$ is a complex number that contains both the amplitude and the phase information. Let us now make three observations. First, note the ease with which we handle amplitude and phase for complex exponentials: they simply combine into a complex number that factors out of the complex exponential. Second, by Euler's formula,

$$\mathrm{Re}(Ae^{j(2\pi f_0 t + \theta)}) = A\cos(2\pi f_0 t + \theta)$$

so that real-valued sinusoids are "contained in" complex exponentials. Third, as we shall soon see, the set of complex exponentials $\{e^{j2\pi f t}\}$, where f takes values in $(-\infty, \infty)$, forms a "basis" for a large class of signals (basically, for all signals that are of interest to us), and the Fourier transform of a signal is simply its expansion with respect to this basis. Such observations are key to why complex exponentials play such an important role in signals and systems in general, and in communication systems in particular.

The delta, or impulse, function Another signal that plays a crucial role in signals and systems is the delta function, or the unit impulse, which we denote by $\delta(t)$. Physically, we can think of it as a narrow, tall pulse with unit area: examples are shown in Figure 2.2. Mathematically, we can think of it as a limit of such pulses as the pulse width shrinks (and hence the

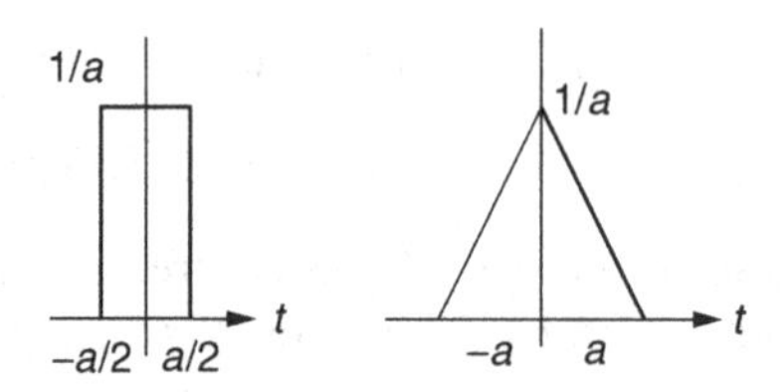

Figure 2.2 The impulse function may be viewed as a limit of tall thin pulses ($a \rightarrow 0$ in the examples shown in the figure).

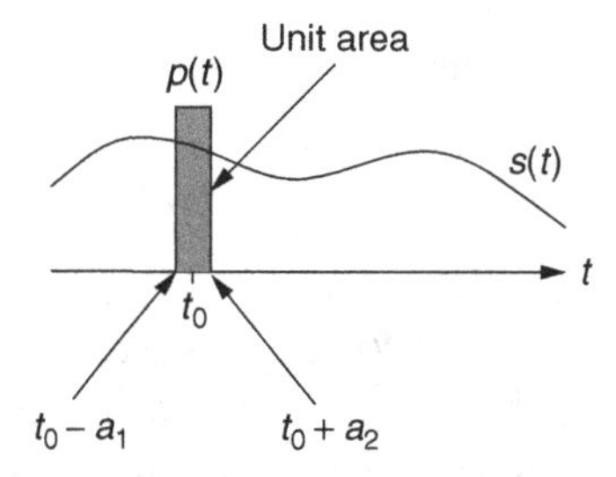

Figure 2.3 Multiplying a signal by a tall thin pulse to select its value at t_0.

pulse height goes to infinity). Such a limit is not physically realizable, but it serves a very useful purpose in terms of understanding the structure of physically realizable signals. To see this, consider a signal $s(t)$ that varies smoothly, and multiply it by a tall, thin pulse of unit area, centered at time t_0, as shown in Figure 2.3. If we now integrate the product, we obtain

$$\int_{-\infty}^{\infty} s(t)p(t)dt = \int_{t_0-a_1}^{t_0+a_2} s(t)p(t)dt \approx s(t_0) \int_{t_0-a_1}^{t_0+a_1} p(t)dt = s(t_0)$$

That is, the preceding operation "selects" the value of the signal at time t_0. On taking the limit of the tall thin pulse as its width $a_1 + a_2 \rightarrow 0$, we get a translated version of the delta function, namely $\delta(t - t_0)$. Note that the exact shape of the pulse does not matter in the preceding argument. The delta function is therefore *defined* by means of the following sifting property: for any "smooth" function $s(t)$, we have

$$\int_{-\infty}^{\infty} s(t)\delta(t - t_0)dt = s(t_0) \quad \textbf{sifting property of the impulse} \tag{2.24}$$

Thus, the delta function is defined mathematically by the way it acts on other signals, rather than as a signal by itself. However, it is also important to keep in mind its intuitive interpretation as (the limit of) a tall, thin, pulse of unit area.

The following function is useful for expressing signals compactly.

Indicator function We use I_A to denote the indicator function of a set A, defined as

$$I_A(x) = \begin{cases} 1, & x \in A \\ 0, & \text{otherwise} \end{cases}$$

The indicator function of an interval is a rectangular pulse, as shown in Figure 2.4.

The indicator function can also be used to compactly express more complex signals, as shown in the examples in Figure 2.5.

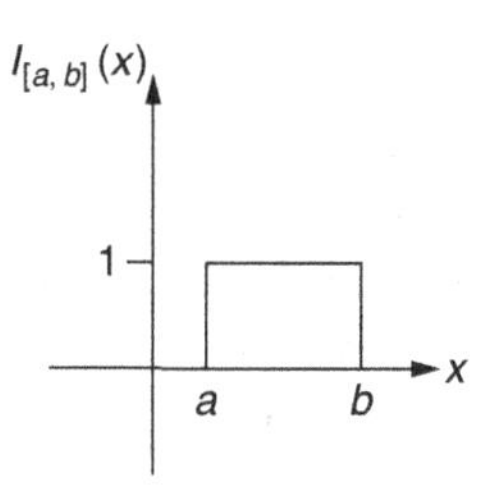

Figure 2.4 The indicator function of an interval is a rectangular pulse.

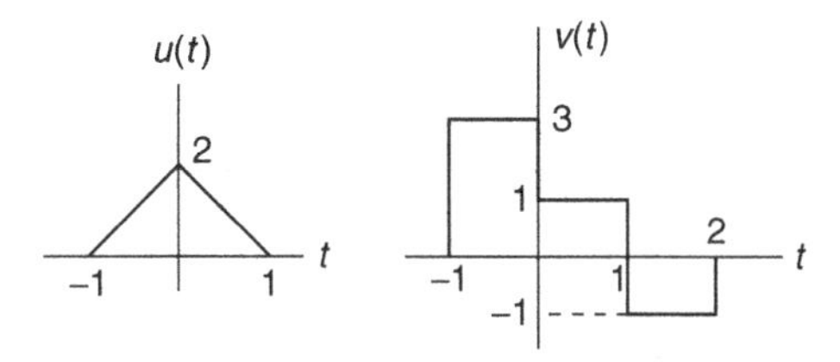

Figure 2.5 The functions $u(t) = 2(1 - |t|)I_{[-1,1]}(t)$ and $v(t) = 3I_{[-1,0]}(t) + I_{[0,1]}(t) - I_{[1,2]}(t)$ can be written compactly in terms of indicator functions.

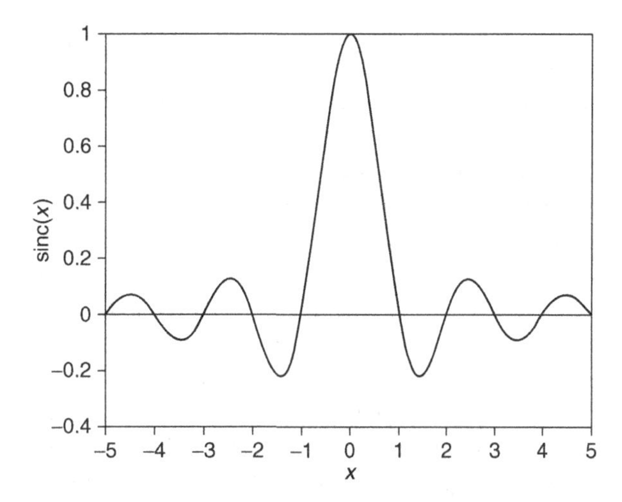

Figure 2.6 The sinc function.

Sinc function The sinc function, plotted in Figure 2.6, is defined as

$$\text{sinc}(x) = \frac{\sin(\pi x)}{\pi x}$$

where the value at $x = 0$ is defined in the limit as $x \to 0$ to be $\text{sinc}(0) = 1$. Since $|\sin(\pi x)| \leq 1$, we have that $|\text{sinc}(x)| \leq 1/(\pi|x|)$, with equality if and only if x is an odd multiple of $1/2$. That is, the sinc function exhibits a sinusoidal variation, with an envelope that decays as $1/|x|$.

The analogy between signals and vectors Even though signals can be complicated functions of time that live in an infinite-dimensional space, the mathematics for manipulating them is very similar to that for manipulating finite-dimensional vectors, with sums replaced by integrals. A key building block of communication theory is the relative geometry of the signals used, which is governed by the inner products between signals. Inner products for continuous-time signals can be defined in a manner exactly analogous to the corresponding definitions in finite-dimensional vector space.

Inner product The inner product for two $m \times 1$ complex vectors $\mathbf{s} = (s[1], \ldots, s[m])^{\mathrm{T}}$ and $\mathbf{r} = (r[1], \ldots, r[m])^{\mathrm{T}}$ is given by

$$\langle \mathbf{s}, \mathbf{r} \rangle = \sum_{i=1}^{m} s[i] r^*[i] = \mathbf{r}^{\mathrm{H}} \mathbf{s} \tag{2.25}$$

Similarly, we define the inner product of two (possibly complex-valued) signals $s(t)$ and $r(t)$ as follows:

$$\langle s, r \rangle = \int_{-\infty}^{\infty} s(t) r^*(t) dt \tag{2.26}$$

The inner product obeys the following linearity properties:

$$\langle a_1 s_1 + a_2 s_2, r \rangle = a_1 \langle s_1, r \rangle + a_2 \langle s_2, r \rangle$$
$$\langle s, a_1 r_1 + a_2 r_2 \rangle = a_1^* \langle s, r_1 \rangle + a_2^* \langle s, r_2 \rangle$$

where a_1 and a_2 are complex-valued constants, and s, s_1, s_2, r, r_1, and r_2 are signals (or vectors). The complex conjugation when we pull out constants from the second argument of the inner product is something that we need to maintain awareness of when computing inner products for complex-valued signals.

Energy and norm The *energy* E_s of a signal s is defined as its inner product with itself:

$$E_s = ||s||^2 = \langle s, s \rangle = \int_{-\infty}^{\infty} |s(t)|^2 \, dt \tag{2.27}$$

where $||s||$ denotes the *norm* of s. If the energy of s is zero, then s must be zero "almost everywhere" (e.g., $s(t)$ cannot be nonzero over any interval, no matter how small its length). For continuous-time signals, we take this to be equivalent to being zero everywhere. With this understanding, $||s|| = 0$ implies that s is zero, which is a property that is true for norms in finite-dimensional vector spaces.

Example 2.2.1 (Energy computations) Consider $s(t) = 2I_{[0,T]} + jI_{[T/2,2T]}$. On writing it out in more detail, we have

$$s(t) = \begin{cases} 2, & 0 \le t < T/2 \\ 2+j, & T/2 \le t < T \\ j, & T \le t < 2T \end{cases}$$

so that its energy is given by

$$||s||^2 = \int_0^{T/2} 2^2 \, dt + \int_{T/2}^{T} |2+j|^2 \, dt + \int_T^{2T} |j|^2 \, dt = 4(T/2) + 5(T/2) + T = 11T/2$$

As another example, consider $s(t) = e^{-3|t|+j2\pi t}$, for which the energy is given by

$$||s||^2 = \int_{-\infty}^{\infty} |e^{-3|t|+j2\pi t}|^2\, dt = \int_{-\infty}^{\infty} e^{-6|t|}\, dt = 2\int_{0}^{\infty} e^{-6t}\, dt = 1/3$$

Note that the complex phase term $j2\pi t$ does not affect the energy, since it goes away when we take the magnitude.

Power The power of a signal $s(t)$ is defined as the time average of its energy computed over a large time interval:

$$P_s = \lim_{T_o \to \infty} \frac{1}{T_o} \int_{-T_o/2}^{T_o/2} |s(t)|^2\, dt \tag{2.28}$$

Finite-energy signals, of course, have zero power.

We see from (2.28) that power is defined as a time average. It is useful to introduce a compact notation for time averages.

Time average For a function $g(t)$, define the time average as

$$\overline{g} = \lim_{T_o \to \infty} \frac{1}{T_o} \int_{-T_o/2}^{T_o/2} g(t) dt \tag{2.29}$$

That is, we compute the time average over an observation interval of length T_o, and then let the observation interval get large. We can now rewrite the power computation in (2.28) in this notation as follows.

Power The power of a signal $s(t)$ is defined as

$$P_s = \overline{|s(t)|^2} \tag{2.30}$$

Another time average of interest is the DC value of a signal.

DC value The DC value of $s(t)$ is defined as $\overline{s(t)}$.

Let us compute these quantities for the simple example of a complex exponential, $s(t) = Ae^{j(2\pi f_0 t+\theta)}$, where $A > 0$ is the amplitude, $\theta \in [0, 2\pi]$ is the phase, and f_0 is a real-valued frequency. Since $|s(t)|^2 \equiv A^2$ for all t, we get the same value when we average it. Thus, the power is given by $P_s = \overline{s^2(t)} = A^2$. For nonzero frequency f_0, it is intuitively clear that all the power in s is concentrated away from DC, since $s(t) = Ae^{j(2\pi f_0 t+\theta)} \leftrightarrow S(f) = Ae^{j\theta}\delta(f - f_0)$. We therefore see that the DC value is zero. While this is a convincing intuitive argument, it is instructive to prove this starting from the definition (2.29).

Proving that a complex exponential has zero DC value For $s(t) = Ae^{j(2\pi f_0 t+\theta)}$, the integral over its period (of length $1/f_0$) is zero. As shown in Figure 2.7, the length L of any interval I can be written as $L = K/f_0 + \ell$, where K is a nonnegative integer and $0 \le \ell < 1/f_0$ is the length of the remaining interval I_r. Since the integral over an integer number of periods is zero, we have

$$\int_I s(t) dt = \int_{I_r} s(t) dt$$

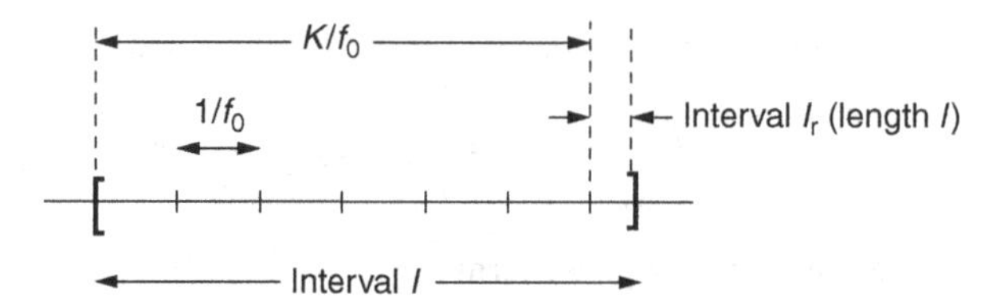

Figure 2.7 The interval I for computing the time average of a periodic function with period $1/f_0$ can be decomposed into an integer number K of periods, with the remaining interval I_r of length $\ell < 1/f_0$.

Thus,

$$\left|\int_I s(t)dt\right| = \left|\int_{I_r} s(t)dt\right| \leq \ell \max_t |s(t)| = A\ell < \frac{A}{f_0}$$

since $|s(t)| = A$. We therefore obtain

$$\left|\int_{-T_o/2}^{T_o/2} s(t)dt\right| \leq A/f_0$$

which yields that the DC value $\bar{s} = 0$, since

$$|\bar{s}| = \left|\lim_{T_o \to \infty} \frac{1}{T_o} \int_{-T_o/2}^{T_o/2} s(t)dt\right| \leq \lim_{T_o \to \infty} \frac{A}{f_0 T_o} = 0$$

Essentially the same argument implies that, in general, the time average of a periodic signal equals the average over a single period. We use this fact without further comment henceforth.

Power and DC value of a sinusoid For a real-valued sinusoid $s(t) = A\cos(2\pi f_0 t + \theta)$, we can use the results derived for complex exponentials above. Using Euler's identity, a real-valued sinusoid at f_0 is a sum of complex exponentials at $\pm f_0$:

$$s(t) = \frac{A}{2}e^{j(2\pi f_0 t+\theta)} + \frac{A}{2}e^{-j(2\pi f_0 t+\theta)}$$

Since each complex exponential has zero DC value, we obtain

$$\bar{s} = 0$$

That is, the DC value of any real-valued sinusoid is zero. Then

$$P_s = \overline{s^2(t)} = \overline{A^2\cos^2(2\pi f_0 t + \theta)} = \overline{\frac{A^2}{2} + \frac{A^2}{2}\cos(4\pi f_0 t + 2\theta)} = \frac{A^2}{2}$$

since the DC value of the sinusoid at $2f_0$ is zero.

2.3 Linear time-invariant systems

System A system takes as input one or more signals, and produces as output one or more signals. A system is specified once we characterize its input–output relationship; that is, if we can determine the output, or response, $y(t)$, corresponding to any possible input $x(t)$ in a given class of signals of interest.

Our primary focus here is on *linear time-invariant (LTI)* systems, which provide good models for filters at the transmitter and receiver, as well as for the distortion induced by a variety of channels. We shall see that the input–output relationship is particularly easy to characterize for such systems.

Linear system Let $x_1(t)$ and $x_2(t)$ denote arbitrary input signals, and let $y_1(t)$ and $y_2(t)$ denote the corresponding system outputs, respectively. Then, for arbitrary scalars a_1 and a_2, the response of the system to input $a_1x_1(t) + a_2x_2(t)$ is $a_1y_1(t) + a_2y_2(t)$.

Time-invariant system Let $y(t)$ denote the system response to an input $x(t)$. Then the system response to a time-shifted version of the input, $x_1(t) = x(t - t_0)$, is $y_1(t) = y(t - t_0)$. That is, a time shift in the input causes an identical time shift in the output.

Example 2.3.1 (Examples of linear systems) It can (and should) be checked that the following systems are linear. These examples show that linear systems may, but need not, be time-invariant:

$$y(t) = 2x(t-1) - jx(t-2) \quad \textit{time-invariant}$$

$$y(t) = (3-2j)x(1-t) \quad \textit{time-varying}$$

$$y(t) = x(t)\cos(100\pi t) - x(t-1)\sin(100\pi t) \quad \textit{time-varying}$$

$$y(t) = \int_{t-1}^{t+1} x(\tau)d\tau \quad \textit{time-invariant}$$

Example 2.3.2 (Examples of time-invariant systems) It can (and should) be checked that the following systems are time-invariant. These examples show that time-invariant systems may, but need not, be linear:

$$y(t) = e^{2x(t-1)} \quad \textit{nonlinear}$$

$$y(t) = \int_{-\infty}^{t} x(\tau)e^{-(t-\tau)}\,d\tau \quad \textit{linear}$$

$$y(t) = \int_{t-1}^{t+1} x^2(\tau)d\tau \quad \textit{nonlinear}$$

Linear time-invariant system A linear time-invariant (LTI) system is (unsurprisingly) defined to be a system that is both linear and time-invariant. What is surprising, however, is how powerful the LTI property is in terms of dictating what the input–output relationship must look like. Specifically, if we know the *impulse response* of an LTI system (i.e., the output signal when the input signal is the delta function), then we can compute the system response for *any* input signal. Before deriving and stating this result, we illustrate the LTI property using an example; see Figure 2.8. Suppose that the response of an LTI system to the rectangular pulse $p_1(t) = I_{[-1/2,1/2]}(t)$ is given by the trapezoidal waveform $h_1(t)$.

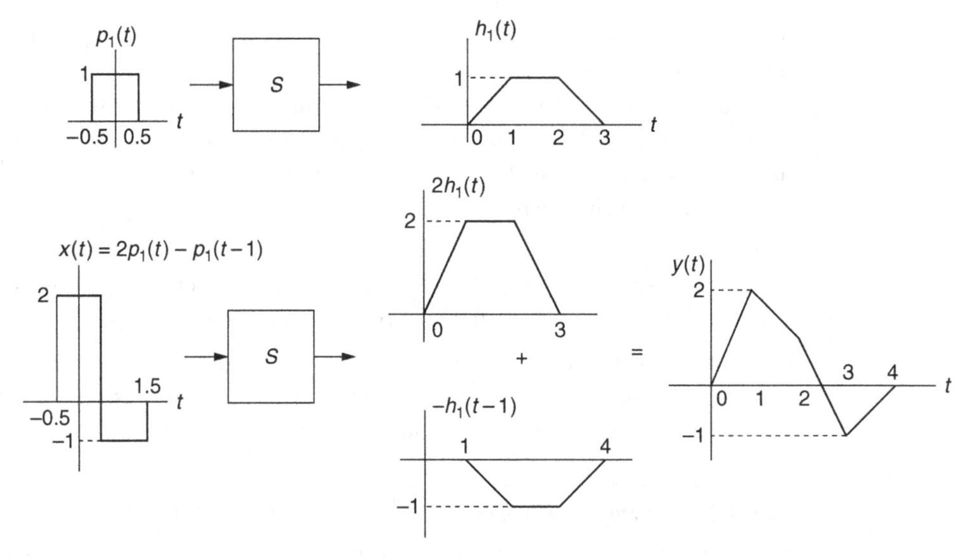

Figure 2.8 Given that the response of an LTI system $\mathcal{S}$ to the pulse $p_1(t)$ is $h_1(t)$, we can use the LTI property to infer that the response to $x(t) = 2p_1(t) - p_1(t-1)$ is $y(t) = 2h_1(t) - h_1(t-1)$.

We can now compute the system response to any linear combination of time shifts of the pulse $p(t)$, as illustrated by the example in Figure 2.8. More generally, using the LTI property, we infer that the response to an input signal of the form $x(t) = \sum_i a_i p_1(t - t_i)$ is $y(t) = \sum_i a_i h_1(t - t_i)$.

Can we extend the preceding idea to compute the system response to arbitrary input signals? The answer is yes: if we know the system response to thinner and thinner pulses, then we can approximate arbitrary signals better and better using linear combinations of shifts of these pulses. Consider $p_\Delta(t) = (1/\Delta)I_{[-\Delta/2,\Delta/2]}(t)$, where $\Delta > 0$ is getting smaller and smaller. Note that we have normalized the area of the pulse to unity, so that the limit of $p_\Delta(t)$ as $\Delta \to 0$ is the delta function. Figure 2.9 shows how to approximate a smooth input signal as a linear combination of shifts of $p_\Delta(t)$. That is, for Δ small, we have

$$x(t) \approx x_\Delta(t) = \sum_{k=-\infty}^{\infty} x(k\Delta)\Delta p_\Delta(t - k\Delta) \tag{2.31}$$

If the system response to $p_\Delta(t)$ is $h_\Delta(t)$, then we can use the LTI property to compute the response $y_\Delta(t)$ to $x_\Delta(t)$, and use this to approximate the response $y(t)$ to the input $x(t)$, as follows:

$$y(t) \approx y_\Delta(t) = \sum_{k=-\infty}^{\infty} x(k\Delta)\Delta h_\Delta(t - k\Delta) \tag{2.32}$$

As $\Delta \to 0$, the sums above tend to integrals, and the pulse $p_\Delta(t)$ tends to the delta function $\delta(t)$. The approximation to the input signal in Equation (2.31) becomes exact, with the sum tending to an integral:

$$\lim_{\Delta \to 0} x_\Delta(t) = x(t) = \int_{-\infty}^{\infty} x(\tau)\delta(t - \tau)d\tau$$

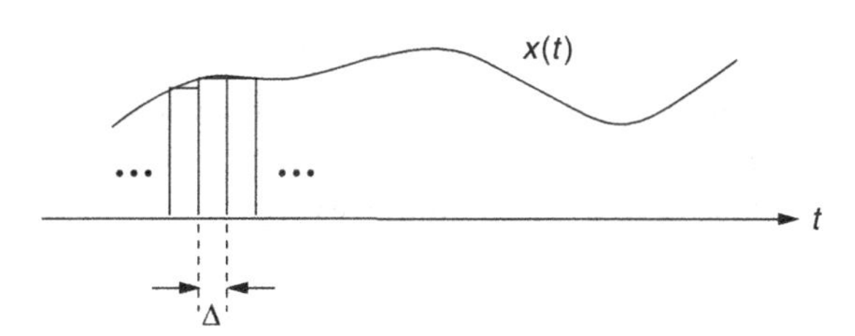

Figure 2.9 A smooth signal can be approximated as a linear combination of shifts of tall thin pulses.

replacing the discrete time shifts $k\Delta$ by the continuous variable τ, the discrete increment Δ by the infinitesimal $d\tau$, and the sum by an integral. This is just a restatement of the sifting property of the impulse. That is, an arbitrary input signal can be expressed as a linear combination of time-shifted versions of the delta function, where we now consider a continuum of time shifts.

In similar fashion, the approximation to the output signal in (2.32) becomes exact, with the sum reducing to the following *convolution* integral:

$$\lim_{\Delta \to 0} y_\Delta(t) = y(t) = \int_{-\infty}^{\infty} x(\tau)h(t-\tau)d\tau \tag{2.33}$$

where $h(t)$ denotes the *impulse response* of the LTI system.

Convolution and its computation The convolution $v(t)$ of two signals $u_1(t)$ and $u_2(t)$ is given by

$$v(t) = (u_1 * u_2)(t) = \int_{-\infty}^{\infty} u_1(\tau)u_2(t-\tau)d\tau = \int_{-\infty}^{\infty} u_1(t-\tau)u_2(\tau)d\tau \tag{2.34}$$

Note that τ is a dummy variable that is integrated out in order to determine the value of the signal $v(t)$ at each possible time t. The roles of u_1 and u_2 in the integral can be exchanged. This can be proved using a change of variables, replacing $t-\tau$ by τ. We often drop the time variable, and write $v = u_1 * u_2 = u_2 * u_1$.

An LTI system is completely characterized by its impulse response As derived in (2.33), the output y of an LTI system is the convolution of the input signal u and the system impulse response h. That is, $y = u * h$. From (2.34), we realize that the roles of the signal and the system can be exchanged: that is, we would get the same output y if a signal h were sent through a system with impulse response u.

Flip and slide Consider the expression for the convolution in (2.34):

$$v(t) = \int_{-\infty}^{\infty} u_1(\tau)u_2(t-\tau)d\tau$$

Fix a value of time t at which we wish to evaluate v. In order to compute $v(t)$, we must multiply two functions of a "dummy variable" τ and then integrate over τ. In particular, $s_2(\tau) = u_2(-\tau)$ is the signal $u_2(\tau)$ flipped around the origin, so that $u_2(t-\tau) = u_2(-(\tau-t)) = s_2(\tau-t)$ is $s_2(\tau)$ translated to the right by t (if $t < 0$, translation to the right by

t actually corresponds to translation to the left by $|t|$). In short, the mechanics of computing the convolution involves flipping and sliding one of the signals, multiplying by the other signal, and integrating. Pictures are extremely helpful when doing such computations by hand, as illustrated by the following example.

Example 2.3.3 (Convolving rectangular pulses) Consider the rectangular pulses $u_1(t) = I_{[5,11]}(t)$ and $u_2(t) = I_{[1,3]}(t)$. We wish to compute the convolution

$$v(t) = (u_1 * u_2)(t) = \int_{-\infty}^{\infty} u_1(\tau)u_2(t-\tau)d\tau$$

We now draw pictures of the signals involved in these "flip and slide" computations in order to figure out the limits of integration for different ranges of t. Figure 2.10 shows that there are five different ranges of interest, and yields the following results.

(a) For $t < 6$, $u_1(\tau)u_2(t-\tau) \equiv 0$, so that $v(t) = 0$.

(b) For $6 < t < 8$, $u_1(\tau)u_2(t-\tau) = 1$ for $5 < \tau < t-1$, so that

$$v(t) = \int_5^{t-1} d\tau = t - 6$$

(c) For $8 < t < 12$, $u_1(\tau)u_2(t-\tau) = 1$ for $t-3 < \tau < t-1$, so that

$$v(t) = \int_{t-3}^{t-1} d\tau = 2$$

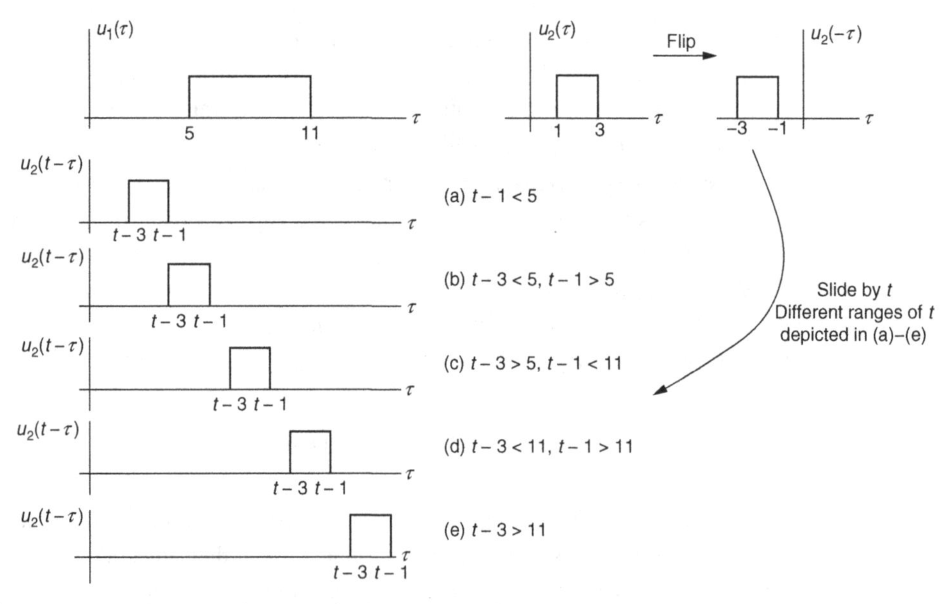

Figure 2.10 Illustrating the "flip and slide" operation for the convolution of two rectangular pulses.

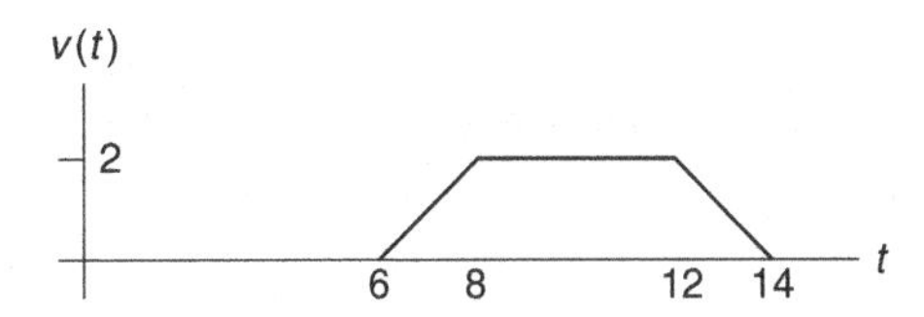

Figure 2.11 The convolution of the two rectangular pulses in Example 2.3.3 results in a trapezoidal pulse.

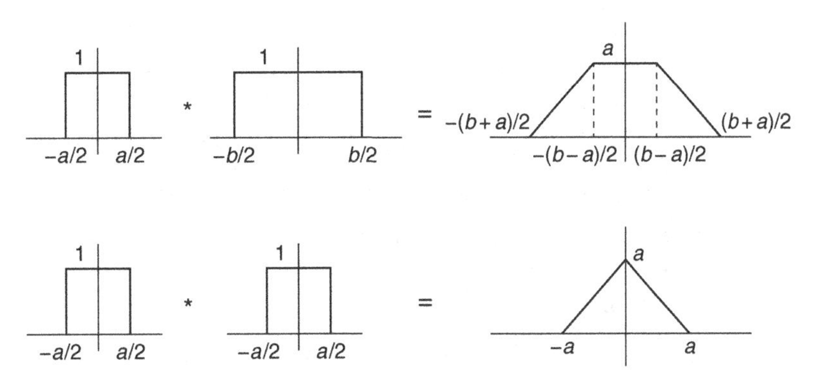

Figure 2.12 Convolution of two rectangular pulses as a function of the pulse durations. The trapezoidal pulse reduces to a triangular pulse for equal pulse durations.

(d) For $12 < t < 14$, $u_1(\tau)u_2(t-\tau) = 1$ for $t-3 < \tau < 11$, so that

$$v(t) = \int_{t-3}^{11} d\tau = 11 - (t-3) = 14 - t$$

(e) For $t > 14$, $u_1(\tau)u_2(t-\tau) \equiv 0$, so that $v(t) = 0$.

The result of the convolution is the trapezoidal pulse sketched in Figure 2.11.

It is useful to record the general form of the convolution between two rectangular pulses of the form $I_{[-a/2,a/2]}(t)$ and $I_{[-b/2,b/2]}(t)$, where we take $a \leq b$ without loss of generality. The result is a trapezoidal pulse, which reduces to a triangular pulse for $a = b$, as shown in Figure 2.12. Once we know this, using the LTI property, we can infer the convolution of any signals that can be expressed as linear combinations of shifts of rectangular pulses.

Occasional notational sloppiness can be useful As the preceding example shows, a convolution computation as in (2.34) requires a careful distinction between the variable t at which the convolution is being evaluated and the dummy variable τ. This is why we make sure that the dummy variable does not appear in our notation $(s * r)(t)$ for the convolution between signals $s(t)$ and $r(t)$. However, it is sometimes convenient to abuse notation and use the notation $s(t) * r(t)$ instead, as long we remain aware of what we are doing. For example, this enables us to compactly state the following linear time-invariance (LTI) property:

$$(a_1 s_1(t-t_1) + a_2 s_2(t-t_2)) * r(t) = a_1(s_1 * r)(t-t_1) + a_2(s_2 * r)(t-t_2)$$

for any complex gains a_1 and a_2, and any time offsets t_1 and t_2.

Example 2.3.4 (Modeling a multipath channel) We can get a delayed version of a signal by convolving it with a delayed impulse as follows:

$$y_1(t) = u(t) * \delta(t - t_1) = u(t - t_1) \tag{2.35}$$

To see this, compute

$$y_1(t) = \int u(\tau)\delta(t - \tau - t_1)d\tau = \int u(\tau)\delta(\tau - (t - t_1))d\tau = u(t - t_1)$$

where we first use the fact that the delta function is even, and then use its sifting property.

Equation (2.35) immediately tells us how to model *multipath channels,* in which multiply scattered versions of a transmitted signal $u(t)$ combine to give a received signal $y(t)$ that is a superposition of delayed versions of the transmitted signal, as illustrated in Figure 2.13:

$$y(t) = \alpha_1 u(t - \tau_1) + \cdots + \alpha_m u(t - \tau_m)$$

(plus noise, which we have not talked about yet). From (2.35), we see that we can write

$$\begin{aligned} y(t) &= \alpha_1 u(t) * \delta(t - \tau_1) + \cdots + \alpha_m u(t) * \delta(t - \tau_m) \\ &= u(t) * (\alpha_1 \delta(t - \tau_1) + \cdots + \alpha_m \delta(t - \tau_m)) \end{aligned}$$

That is, we can model the received signal as a convolution of the transmitted signal with a channel impulse response that is a linear combination of time-shifted impulses:

$$h(t) = \alpha_1 \delta(t - \tau_1) + \cdots + \alpha_m \delta(t - \tau_m) \tag{2.36}$$

Figure 2.14 illustrates how a rectangular pulse spreads as it goes through a multipath channel with impulse response $h(t) = \delta(t - 1) - 0.5\delta(t - 1.5) + 0.5\delta(t - 3.5)$. While the gains $\{\alpha_k\}$ in this example are real-valued, as we shall soon see (in Section 2.8), we need to allow both the signal $u(t)$ and the gains $\{\alpha_k\}$ to take complex values in order to model, for example, signals carrying information over radio channels.

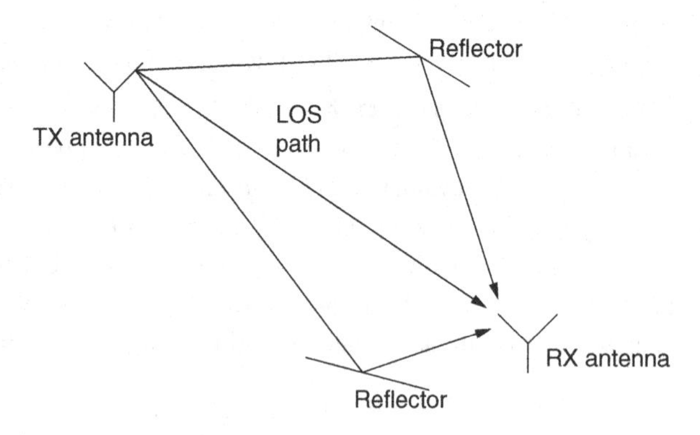

Figure 2.13 Multipath channels typical of wireless communication can include line-of-sight (LOS) and reflected paths.

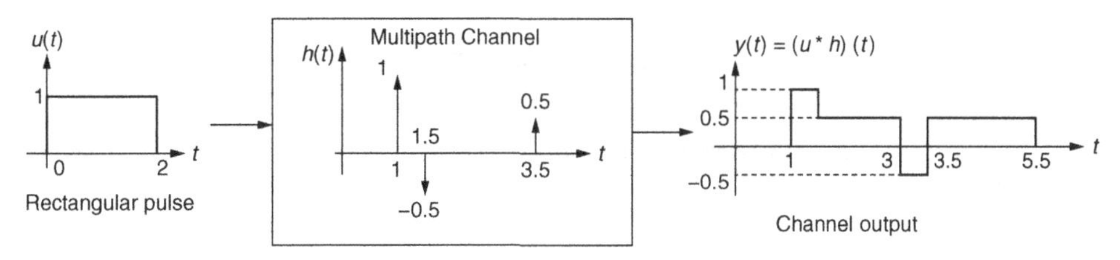

Figure 2.14 A rectangular pulse through a multipath channel.

Complex exponential through an LTI system In order to understand LTI systems in the frequency domain, let us consider what happens to a complex exponential $u(t) = e^{j2\pi f_0 t}$ when it goes through an LTI system with impulse response $h(t)$. The output is given by

$$y(t) = (u * h)(t) = \int_{-\infty}^{\infty} h(\tau) e^{j2\pi f_0 (t-\tau)} \, d\tau$$
$$= e^{j2\pi f_0 t} \int_{-\infty}^{\infty} h(\tau) e^{-j2\pi f_0 \tau} \, d\tau = H(f_0) e^{j2\pi f_0 t} \tag{2.37}$$

where

$$H(f_0) = \int_{-\infty}^{\infty} h(\tau) e^{-j2\pi f_0 \tau} \, d\tau$$

is the Fourier transform of h evaluated at the frequency f_0. We shall discuss the Fourier transform and its properties in more detail shortly.

Complex exponentials are eigenfunctions of LTI systems Recall that an eigenvector of a matrix $\mathbf{H}$ is any vector $\mathbf{x}$ that satisfies $\mathbf{Hx} = \lambda \mathbf{x}$. That is, the matrix leaves its eigenvectors unchanged except for a scale factor λ, which is the eigenvalue associated with that eigenvector. In an entirely analogous fashion, we see that the complex exponential signal $e^{j2\pi f_0 t}$ is an *eigenfunction* of the LTI system with impulse response h, with eigenvalue $H(f_0)$. See Figure 2.15. Since we have not constrained h, we conclude that complex exponentials are eigenfunctions of *any* LTI system. We shall soon see, when we discuss Fourier transforms, that this eigenfunction property allows us to characterize LTI systems in the frequency-domain, which in turn enables powerful frequency domain design and analysis tools.

2.3.1 Discrete-time convolution

DSP-based implementations of convolutions are inherently discrete-time operations. For two discrete-time sequences $\{u_1[n]\}$ and $\{u_2[n]\}$, their convolution $y = u_1 * u_2$ is defined analogously to continuous-time convolution, replacing integration by summation:

$$y[n] = \sum_k u_1[k] u_2[n-k] \tag{2.38}$$

MATLAB implements this using the "conv" function. This can be interpreted as u_1 being the input to a system with impulse response u_2, where a discrete time impulse is simply a one, followed by all zeros.

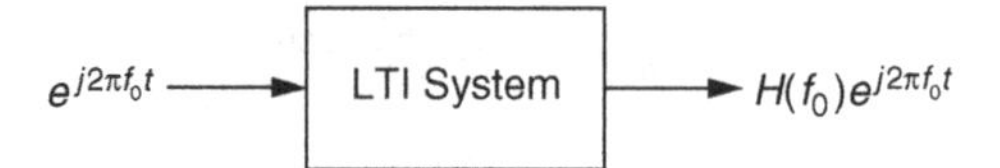

Figure 2.15 Complex exponentials are eigenfunctions of LTI systems.

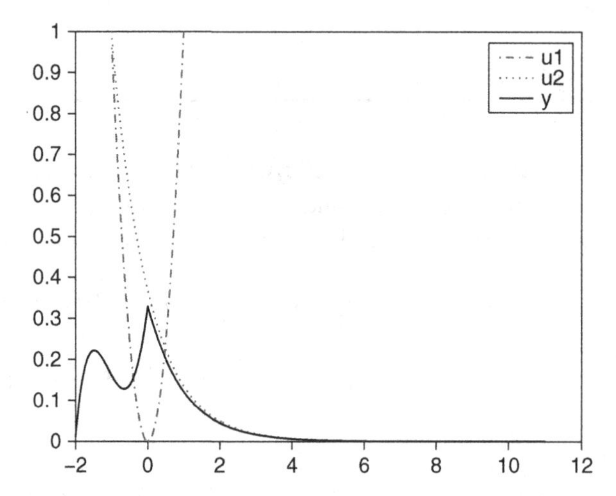

Figure 2.16 Two signals and their continuous-time convolution, computed in discrete time using Code Fragment 2.3.1.

Continuous-time convolution between $u_1(t)$ and $u_2(t)$ can be approximated using discrete-time convolutions between the corresponding sampled signals. For example, for samples taken at rate $1/T_s$, the infinitesimal dt is replaced by the sampling interval T_s as follows:

$$y(t) = (u_1 * u_2)(t) = \int u_1(\tau)u_2(t-\tau)d\tau \approx \sum_k u_1(kT_s)u_2(t-kT_s)T_s$$

Evaluating at a sampling time $t = nT_s$, we have

$$y(nT_s) = T_s \sum_k u_1(kT_s)u_2(nT_s - kT_s)$$

Letting $x[n] = x(nT_s)$ denote the discrete-time waveform corresponding to the nth sample for each of the preceding waveforms, we have

$$y(nT_s) = y[n] \approx T_s \sum_k u_1[k]u_2[n-k] = T_s(u_1 * u_2)[n] \tag{2.39}$$

which shows us how to implement continuous-time convolution using discrete-time operations.

The following MATLAB code provides an example of a continuous-time convolution approximated numerically using discrete-time convolution, and then plotted against the original continuous-time index t, as shown in Figure 2.16 (cosmetic touches not included in the code below). The two waveforms convolved are $u_1(t) = t^2 I_{[-1,1]}(t)$ and $u_2(t) = e^{-(t+1)}I_{[-1,\infty)}$ (the latter is truncated in our discrete-time implementation).

Code Fragment 2.3.1 (Discrete-time computation of continuous time convolution)

```
dt = 0.01; %sampling interval T_s
%%FIRST SIGNAL
u1start = -1; u1end = 1; %start and end times for first signal
t1 = u1start:dt:u1end; %sampling times for first signal
u1 = t1.^2; %discrete-time version of first signal
%%SECOND SIGNAL (exponential truncated when it gets small)
u2start = -1; u2end = 10;
t2 = u2start:dt:u2end;
u2 = exp(-(t2 + 1));
%%APPROXIMATION OF CONTINUOUS-TIME CONVOLUTION
y = dt * conv(u1,u2);
%%PLOT OF SIGNALS AND THEIR CONVOLUTION
ystart = u1start + u2start; %start time for convolution output
time_axis = ystart:dt:ystart + dt*(length(y) - 1);
%%PLOT u1, u2, and y
plot(t1,u1,'r-.');
hold on;
plot(t2,u2,'r:');
plot(time_axis,y);
legend('u1','u2','y','Location','NorthEast');
hold off;
```

2.3.2 Multi-rate systems

While continuous-time signals can be converted to discrete-time signals by sampling "fast enough," it is often required that we operate at multiple sampling rates. For example, in digital communication, we may send a string of symbols $\{b[n]\}$ (think of these as taking values $+1$ or -1 for now) by *modulating* them onto shifted versions of a pulse $p(t)$ as follows:

$$u(t) = \sum_n b[n]p(t - nT) \tag{2.40}$$

where $1/T$ is the rate at which symbols are generated (termed the *symbol rate*). In order to represent the analog pulse $p(t)$ as discrete-time samples, we may sample it at rate $1/T_s$, typically chosen to be an integer multiple of the symbol rate, so that $T = mT_s$, where m is a positive integer. Typical values employed in transmitter DSP modules might be $m = 4$ or $m = 8$. Thus, the system in which we are interested is multi-rate: waveforms are sampled at rate $1/T_s = m/T$, but the input is at rate $1/T$. Set $u[k] = u(kT_s)$ and $p[k] = p(kT_s)$ as the discrete-time signals corresponding to samples of the transmitted waveform $u(t)$ and the pulse $p(t)$, respectively. We can write the sampled version of (2.40) as

$$u[k] = \sum_n b[n]p(kT_s - nT) = \sum_n b[n]p[k - nm] \tag{2.41}$$

The preceding almost has the form of a discrete-time convolution, but the key difference is that the successive symbols $\{b[n]\}$ are spaced by time T, which corresponds to $m > 1$ samples at the sampling rate $1/T_s$. Thus, in order to implement this system using

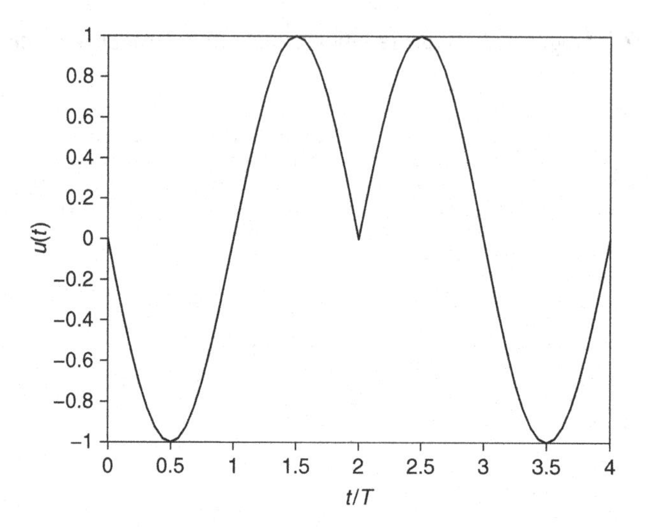

Figure 2.17 The digitally modulated waveform obtained using Code Fragment 2.3.2.

convolution at rate $1/T_s$, we must space out the input symbols by inserting $m-1$ zeros between successive symbols $b[n]$, thus converting a rate $1/T$ signal to a rate $1/T_s = m/T$ signal. This process is termed *upsampling*. While the upsampling function is available in certain MATLAB toolboxes, we provide a self-contained code fragment below that illustrates its use for digital modulation, and plots the waveform obtained for the symbol sequence $-1, +1, +1, -1$. See Figure 2.17. The modulating pulse, is a sine pulse: $p(t) = \sin(\pi t/T) I_{[0,T]}$, and our convention is to set $T = 1$ without loss of generality (or, equivalently, to replace t by t/T). We set the *oversampling factor* $M = 16$ in order to obtain smooth plots, even though typical implementations in communication transmitters may use smaller values.

Code Fragment 2.3.2 (Upsampling for digital modulation)

```
m = 16; %sampling rate as multiple of symbol rate
%discrete-time representation of sine pulse
time_p = 0:1/m:1; %sampling times over duration of pulse
p = sin(pi * time_p); %samples of the pulse
%symbols to be modulated
symbols = [-1;1;1;-1];
%UPSAMPLE BY m
nsymbols = length(symbols); %length of original symbol sequence
nsymbols_upsampled = 1 + (nsymbols - 1) * m; %length of upsampled symbol sequence
symbols_upsampled = zeros(nsymbols_upsampled,1);
symbols_upsampled(1:m:nsymbols_upsampled) = symbols; %insert symbols with
    spacing M
%GENERATE MODULATED SIGNAL BY DISCRETE-TIME CONVOLUTION
u = conv(symbols_upsampled,p);
%PLOT MODULATED SIGNAL
time_u = 0:1/m:(length(u) - 1)/m; %unit of time = symbol time T
plot(time_u,u);
xlabel('t/T');
```

2.4 Fourier series

Fourier series represent periodic signals in terms of sinusoids or complex exponentials. A signal $u(t)$ is periodic with period T if $u(t+T) = u(t)$ for all t. Note that, if u is periodic with period T, then it is also periodic with period nT, where n is any positive integer. The smallest time interval for which $u(t)$ is periodic is termed the fundamental period. Let us denote this by T_0, and define the corresponding fundamental frequency $f_0 = 1/T_0$ (measured in hertz if T_0 is measured in seconds). It is easy to show that, if $u(t)$ is periodic with period T, then T must be an integer multiple of T_0. In the following, we often simply refer to the fundamental period as "period."

Using mathematical machinery beyond our current scope, it can be shown that any periodic signal with period T_0 (subject to mild technical conditions) can be expressed as a linear combination of complex exponentials

$$\psi_m(t) = e^{j2\pi m f_0 t} = e^{j2\pi mt/T_0}, \quad m = 0, \pm 1, \pm 2, \ldots$$

whose frequencies are integer multiples of the fundamental frequency f_0. That is, we can write

$$u(t) = \sum_{n=-\infty}^{\infty} u_n \psi_n(t) = \sum_{n=-\infty}^{\infty} u_n e^{j2\pi n f_0 t} \tag{2.42}$$

The coefficients $\{u_n\}$ are in general complex-valued, and are called the Fourier series for $u(t)$. They can be computed as follows:

$$u_k = \frac{1}{T_0} \int_{T_0} u(t) e^{-j2\pi k f_0 t}\, dt \tag{2.43}$$

where $\int_{T_0}$ denotes an integral over any interval of length T_0.

Let us now derive (2.43). For m a nonzero integer, consider an arbitrary interval of length T_0, of the form $[D, D+T_0]$, where the offset D is free to take on any real value. Then, for any nonzero integer $m \neq 0$, we have

$$\begin{aligned}\int_D^{D+T_0} e^{j2\pi m f_0 t}\, dt &= \left. \frac{e^{j2\pi m f_0 t}}{j2\pi m f_0} \right|_D^{D+T_0} \\ &= \frac{e^{j2\pi f_0 m D} - e^{j(2\pi m f_0 D + 2\pi m)}}{j2\pi m f_0} = 0\end{aligned} \tag{2.44}$$

since $e^{j2\pi m} = 1$. Thus, when we multiply both sides of (2.42) by $e^{-j2\pi k f_0 t}$ and integrate over a period, all terms corresponding to $n \neq k$ drop out by virtue of (2.44), and we are left only with the $n = k$ term:

$$\begin{aligned}\int_D^{D+T_0} u(t) e^{-j2\pi k f_0 t}\, dt &= \int_D^{D+T_0} \left[\sum_{n=-\infty}^{\infty} u_n e^{j2\pi n f_0 t} \right] e^{-j2\pi k f_0 t}\, dt \\ &= u_k \int_D^{D+T_0} e^{j2\pi k f_0 t} e^{-j2\pi k f_0 t}\, dt + \sum_{n \neq k} u_n \int_D^{D+T_0} e^{j2\pi (n-k) f_0 t}\, dt \\ &= u_k T_0 + 0\end{aligned}$$

which proves (2.43).

We denote the Fourier-series relationship in (2.42) and (2.43) as $u(t) \leftrightarrow \{u_n\}$. It is useful to keep in mind the geometric meaning of this relationship. The space of periodic signals with period $T_0 = 1/f_0$ can be thought of in the same way as the finite-dimensional vector spaces we are familiar with, except that the inner product between two periodic signals is given by

$$\langle u, v \rangle_{T_0} = \int_{T_0} u(t) v^*(t) dt$$

The energy over a period for a signal u is given by $||u||_{T_0}^2 = \langle u, u \rangle_{T_0}$, where $||u||_{T_0}$ denotes the norm computed over a period. We have assumed that the Fourier basis $\{\psi_n(t)\}$ spans this vector space, and have computed the Fourier series after showing that the basis is orthogonal,

$$\langle \psi_n, \psi_m \rangle_{T_0} = 0, \; n \neq m$$

and of equal energy,

$$||\psi_n||_{T_0}^2 = \langle \psi_n, \psi_n \rangle_{T_0} = T_0$$

The computation of the expression for the Fourier series $\{u_k\}$ can be rewritten in these vector-space terms as follows. A periodic signal $u(t)$ can be expanded in terms of the Fourier basis as

$$u(t) = \sum_{n=-\infty}^{\infty} u_n \psi_n(t) \tag{2.45}$$

Using the orthogonality of the basis functions, we have

$$\langle u, \psi_k \rangle_{T_0} = \sum_n u_n \langle \psi_n, \psi_k \rangle_{T_0} = u_k ||\psi_k||^2$$

That is,

$$u_k = \frac{\langle u, \psi_k \rangle_{T_0}}{||\psi_k||^2} = \frac{\langle u, \psi_k \rangle_{T_0}}{T_0} \tag{2.46}$$

In general, the Fourier series of an arbitrary periodic signal may have an infinite number of terms. In practice, one might truncate the Fourier series at a finite number of terms, with the number of terms required to provide a good approximation to the signal depending on the nature of the signal.

Example 2.4.1 (Fourier series of a square wave) Consider the periodic waveform $u(t)$ as shown in Figure 2.18. For $k = 0$, we get the DC value $u_0 = (A_{\max} + A_{\min})/2$. For $k \neq 0$, we have, using (2.43), that

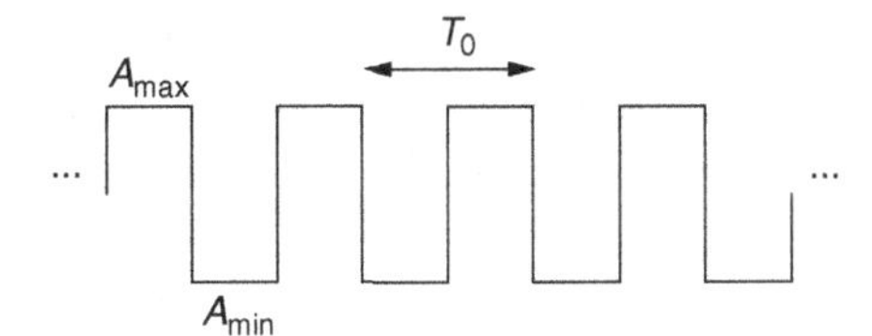

Figure 2.18 A square wave with period T_0.

$$
\begin{aligned}
u_k &= \frac{1}{T_0}\int_{-T_0/2}^{0} A_{\min} e^{-j2\pi kt/T_0}\, dt + \frac{1}{T_0}\int_{0}^{T_0/2} A_{\max} e^{-j2\pi kt/T_0}\, dt \\
&= \frac{A_{\min}}{T_0}\frac{e^{-j2\pi kt/T_0}}{-j2\pi k/T_0}\Big|_{-T_0/2}^{0} + \frac{A_{\max}}{T_0}\frac{e^{-j2\pi kt/T_0}}{-j2\pi k/T_0}\Big|_{0}^{T_0/2} \\
&= \frac{A_{\min}\left(1 - e^{j\pi k}\right) + A_{\max}\left(e^{-j\pi k} - 1\right)}{-j2\pi k}
\end{aligned}
$$

For k even, $e^{j\pi k} = e^{-j\pi k} = 1$, which yields $u_k = 0$. That is, there are no even harmonics. For k odd, $e^{j\pi k} = e^{-j\pi k} = -1$, which yields $u_k = (A_{\max} - A_{\min})/(j\pi k)$. We therefore obtain

$$
u_k = \begin{cases} 0, & k \text{ even} \\ (A_{\max} - A_{\min})/(j\pi k), & k \text{ odd} \end{cases}
$$

On combining the terms for positive and negative k, we obtain

$$
u(t) = \frac{A_{\max} + A_{\min}}{2} + \sum_{k \text{ odd}} \frac{2(A_{\max} - A_{\min})}{\pi k} \sin(2\pi kt/T_0)
$$

Example 2.4.2 (Fourier series of an impulse train) Even though the delta function is not physically realizable, the Fourier series of an impulse train, as shown in Figure 2.19, turns out to be extremely useful in theoretical development and in computations. Specifically, consider

$$
u(t) = \sum_{n=-\infty}^{\infty} \delta(t - nT_0)
$$

By integrating over an interval of length T_0 centered around the origin, we obtain

$$
u_k = \frac{1}{T_0}\int_{-T_0/2}^{T_0/2} u(t) e^{-j2\pi kf_0 t}\, dt = \frac{1}{T_0}\int_{-T_0/2}^{T_0/2} \delta(t) e^{-j2\pi kf_0 t}\, dt = \frac{1}{T_0}
$$

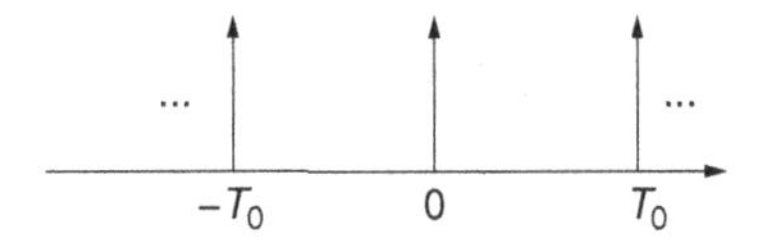

Figure 2.19 An impulse train of period T_0.

using the sifting property of the impulse. That is, the delta function has equal frequency content at all harmonics. This is yet another manifestation of the physical unrealizability of the impulse: for well-behaved signals, the Fourier series should decay as the frequency increases.

While we have considered signals that are periodic functions of time, the concept of Fourier series applies to periodic functions in general, whatever the physical interpretation of the argument of the function. In particular, as we shall see when we discuss the effect of time-domain sampling in the context of digital communication, the time-domain samples of a waveform can be interpreted as the Fourier series for a particular periodic function of frequency.

2.4.1 Fourier-series properties and applications

We now state some Fourier-series properties that are helpful both for computation and for developing intuition. The derivations are omitted, since they follow in a straightforward manner from (2.42) and (2.43), and are included in any standard text on signals and systems. In the following, $u(t)$ and $v(t)$ denote periodic waveforms of period T_0 and Fourier series $\{u_k\}$ and $\{v_k\}$, respectively.

Linearity For arbitrary complex numbers α and β,

$$\alpha u(t) + \beta v(t) \leftrightarrow \{\alpha u_k + \beta v_k\}$$

Time delay corresponds to linear phase in the frequency domain We have

$$u(t-d) \leftrightarrow \{u_k e^{-j2\pi k f_0 d} = u_k e^{-j2\pi kd/T_0}\}$$

The Fourier series of a real-valued signal is conjugate symmetric If $u(t)$ is real-valued, then $u_k = u_{-k}^*$.

Harmonic structure of real-valued periodic signals While both the Fourier-series coefficients and the complex exponential basis functions are complex-valued, for real-valued $u(t)$, the linear combination on the right-hand side of (2.42) must be real-valued. In particular, as we show below, the terms corresponding to u_k and u_{-k} ($k \geq 1$) combine together into a real-valued sinusoid which we term the kth harmonic. Specifically, writing $u_k = A_k e^{j\phi_k}$ in polar form, we invoke the conjugate symmetry of the Fourier series for real-valued $u(t)$ to infer that $u_{-k} = u_k^* = A_k e^{-j\phi_k}$. The Fourier series can therefore be written as

$$u(t) = u_0 + \sum_{k=1}^{\infty} u_k e^{j2\pi k f_0 t} + u_{-k} e^{-j2\pi k f_0 t} = u_0 + \sum_{k=1}^{\infty} \left(A_k e^{j\phi_k} e^{j2\pi k f_0 t} + A_k e^{-j\phi_k} e^{-j2\pi k f_0 t} \right)$$

This yields the following Fourier series in terms of real-valued sinusoids:

$$u(t) = u_0 + \sum_{k=1}^{\infty} 2A_k \cos(2\pi k f_0 t + \phi_k) = u_0 + \sum_{k=1}^{\infty} 2|u_k| \cos(2\pi k f_0 t + \angle u_k) \quad (2.47)$$

Differentiation amplifies higher frequencies We have

$$x(t) = \frac{d}{dt}u(t) \leftrightarrow x_k = j2\pi k f_0 u_k \tag{2.48}$$

Note that differentiation kills the DC term, i.e., $x_0 = 0$. However, the information at all other frequencies is preserved. That is, if we know $\{x_k\}$ then we can recover $\{u_k, k \neq 0\}$ as follows:

$$u_k = \frac{x_k}{j2\pi f_0 k}, \; k \neq 0 \tag{2.49}$$

This is a useful property, since differentiation often makes Fourier series easier to compute.

Example 2.4.1 redone (using differentiation to simplify Fourier series computation) Differentiating the square wave in Figure 2.18 gives us two interleaved impulse trains, one corresponding to the upward edges of the rectangular pulses and the other to the downward edges of the rectangular pulses, as shown in Figure 2.20:

$$x(t) = \frac{d}{dt}u(t) = (A_{\max} - A_{\min})\left(\sum_k \delta(t - kT_0) - \sum_k \delta(t - kT_0 - T_0/2)\right)$$

Compared with the impulse train in Example 2.4.2, the first impulse train above is offset by 0, while the second is offset by $T_0/2$ (and inverted). We can therefore infer their Fourier series using the time-delay property, and add them up by linearity, to obtain

$$x_k = \frac{A_{\max} - A_{\min}}{T_0} - \frac{A_{\max} - A_{\min}}{T_0} e^{-j2\pi f_0 k T_0/2} = \frac{A_{\max} - A_{\min}}{T_0}(1 - e^{-j\pi k}), \; k \neq 0$$

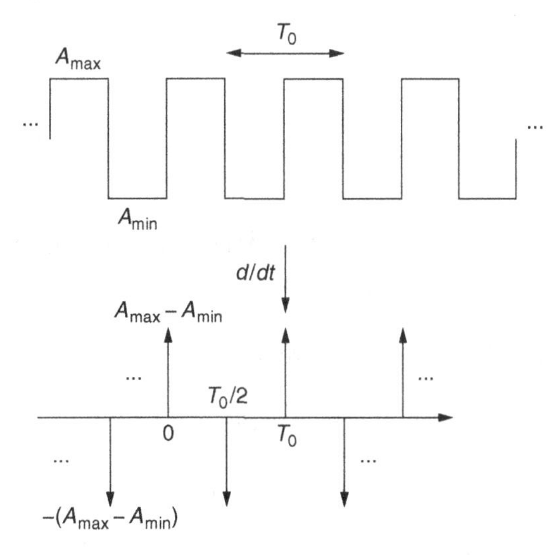

Figure 2.20 The derivative of a square wave is two interleaved impulse trains.

Using the differentiation property, we can therefore infer that

$$u_k = \frac{x_k}{j2\pi f_0 k} = \frac{A_{max} - A_{min}}{-j2\pi k f_0 T_0}(1 - e^{-j\pi k})$$

which gives us the same result as before. Note that the DC term u_0 cannot be obtained using this approach, since it vanishes upon differentiation. But it is easy to compute, since it is just the average value of $u(t)$, which can be seen to be $u_0 = (A_{max} + A_{min})/2$ by inspection.

In addition to simplifying computation for waveforms that can be described (or approximated) as polynomial functions of time (so that enough differentiation ultimately reduces them to impulse trains), the differentiation method explicitly reveals how the harmonic structure (i.e., the strength and location of the harmonics) of a periodic waveform is related to its transitions in the time domain. Once we understand the harmonic structure, we can shape it by appropriate filtering. For example, if we wish to generate a sinusoid of frequency 300 MHz using a digital circuit capable of generating symmetric square waves of frequency 100 MHz, we can choose a filter to isolate the third harmonic. However, we cannot generate a sinusoid of frequency 200 MHz (unless we make the square wave suitably asymmetric), since the even harmonics do not exist for a symmetric square wave (i.e., a square wave whose high and low durations are the same).

Parseval's identity (periodic inner product/power can be computed in either the time or the frequency domain) Using the orthogonality of complex exponentials over a period, it can be shown that

$$\langle u, v \rangle_{T_0} = \int_{T_0} u(t)v^*(t)dt = T_0 \sum_{k=-\infty}^{\infty} u_k v_k^* \tag{2.50}$$

On setting $v = u$, and dividing both sides by T_0, the preceding specializes to an expression for the signal power (which can be computed for a periodic signal by averaging over a period):

$$\frac{1}{T_0} \int_{T_0} |u(t)|^2 \, dt = \sum_{k=-\infty}^{\infty} |u_k|^2 \tag{2.51}$$

2.5 The Fourier transform

We define the Fourier transform $U(f)$ for an aperiodic, finite-energy waveform $u(t)$ as

$$U(f) = \int_{-\infty}^{\infty} u(t)e^{-j2\pi ft} \, dt, \quad -\infty < f < \infty \quad \textbf{Fourier transform} \tag{2.52}$$

The inverse Fourier transform is given by

$$u(t) = \int_{-\infty}^{\infty} U(f)e^{j2\pi ft} \, df, \quad -\infty < t < \infty \quad \textbf{inverse Fourier transform} \tag{2.53}$$

The inverse Fourier transform tells us that any finite-energy signal can be written as a linear combination of a continuum of complex exponentials, with the coefficients of the linear combination given by the Fourier transform $U(f)$.

Notation We call a signal and its Fourier transform a Fourier-transform pair, and denote them as $u(t) \leftrightarrow U(f)$. We also denote the Fourier-transform operation by $\mathcal{F}$, so that $U(f) = \mathcal{F}(u(t))$.

Example 2.5.1 (Rectangular pulse and sinc function form a Fourier-transform pair) Consider the rectangular pulse $u(t) = I_{[-T/2,T/2]}(t)$ of duration T. Its Fourier transform is given by

$$\begin{aligned} U(f) &= \int_{-\infty}^{\infty} u(t)e^{-j2\pi ft}\, dt = \int_{-T/2}^{T/2} e^{-j2\pi ft}\, dt \\ &= \left. \frac{e^{-j2\pi ft}}{-j2\pi f} \right|_{-T/2}^{T/2} = \frac{e^{-j\pi fT} - e^{j\pi fT}}{-j2\pi f} \\ &= \frac{\sin(\pi fT)}{\pi f} = T\,\mathrm{sinc}(fT) \end{aligned}$$

We denote this as

$$I_{[-T/2,T/2]}(t) \leftrightarrow T\,\mathrm{sinc}(fT)$$

Duality Given the similarity in form of the Fourier transform (2.52) and inverse Fourier transform (2.53), we can see that the roles of time and frequency can be switched simply by negating one of the arguments. In particular, suppose that $u(t) \leftrightarrow U(f)$. Define the time-domain signal $s(t) = U(t)$, replacing f by t. Then the Fourier transform of $s(t)$ is given by $S(f) = u(-f)$, replacing t by $-f$. Since negating the argument corresponds to reflection around the origin, we can simply switch time and frequency for signals that are symmetric around the origin. By applying duality to Example 2.5.1, we infer that a signal that is ideally bandlimited in frequency corresponds to a sinc function in time:

$$I_{[-W/2,W/2]}(f) \leftrightarrow W\,\mathrm{sinc}(Wt)$$

Relation to Fourier series The Fourier transform can be obtained by taking the limit of the Fourier series as the period gets large, with $T_0 \to \infty$ and $f_0 \to 0$ (think of an aperiodic signal as periodic with infinite period). We do not provide details, but sketch the process of taking this limit: $T_0 u_k$ tends to $U(f)$, where $f = kf_0$, and the Fourier-series sum in (2.42) becomes the inverse Fourier-transform integral in (2.53), with f_0 becoming df. Not surprisingly, therefore, the Fourier transform exhibits properties entirely analogous to those for Fourier series, as we shall see shortly. However, the Fourier transform applies to a broader class of signals, and we can take advantage of time–frequency duality more easily, because both time and frequency are now continuous-valued variables.

Application to infinite-energy signals In engineering applications, we routinely apply the Fourier transform and the inverse Fourier transform to infinite energy signals, even though the derivation of the Fourier transform as the limit of a Fourier series is based on the assumption that the signal has finite energy. While infinite-energy signals are not physically realizable, they are useful approximations of finite-energy signals, often simplifying mathematical manipulations. For example, instead of considering a sinusoid over a large time interval, we can consider a sinusoid of infinite duration. As we shall see, this leads to an impulsive function in the frequency domain. As another example, delta functions in the time domain are useful in modeling the impulse response of wireless multipath channels. Basically, once we are willing to work with impulses, we can use the Fourier transform on a very broad class of signals.

Example 2.5.2 (The delta function and the constant function form a Fourier-transform pair) For $u(t) = \delta(t)$, we have

$$U(f) = \int_{-\infty}^{\infty} \delta(t) e^{-j2\pi f t}\, dt = 1$$

for all f. That is,

$$\delta(t) \leftrightarrow I_{(-\infty,\infty)}(f)$$

Example 2.5.3 (Complex exponentials in the time domain correspond to impulses in the frequency domain) Let us show this using the inverse Fourier transform. For a frequency-domain impulse at f_0, $U(f) = \delta(f - f_0)$, the inverse Fourier transform is given by

$$u(t) = \int_{-\infty}^{\infty} \delta(f - f_0) e^{j2\pi f t}\, df = e^{j2\pi f_0 t}$$

using the sifting property of the impulse. That is,

$$e^{j2\pi f_0 t} \leftrightarrow \delta(f - f_0)$$

Once we embrace frequency-domain impulses, we can fold Fourier series into Fourier transforms as follows.

Example 2.5.4 (Fourier series expressed in terms of Fourier transforms) We know that a periodic signal $u(t)$ with period T_0 can be written as

$$u(t) = \sum_{n=-\infty}^{\infty} u_n e^{j2\pi n f_0 t}$$

where $f_0 = 1/T_0$ is the fundamental frequency and $\{u_n\}$ are the Fourier-series coefficients. Using Example 2.5.3 to take the Fourier transform of both sides, we obtain

$$U(f) = \sum_{n=-\infty}^{\infty} u_n \delta(f - nf_0)$$

Thus, the Fourier transform of a periodic signal is constituted of impulses at the harmonics, with coefficients given by the Fourier series.

Now that we have seen both the Fourier series and the Fourier transform, it is worth commenting on the following frequently asked questions.

What do negative frequencies mean? Why do we need them? Consider a real-valued sinusoid $A\cos(2\pi f_0 t + \theta)$, where $f_0 > 0$. If we now replace f_0 by $-f_0$, we obtain $A\cos(-2\pi f_0 t + \theta) = A\cos(2\pi f_0 t - \theta)$, using the fact that cosine is an even function. Thus, we do not need negative frequencies when working with real-valued sinusoids. However, unlike complex exponentials, real-valued sinusoids are not eigenfunctions of LTI systems: we can pass a cosine through an LTI system and get a sine, for example. Thus, once we decide to work with a basis formed by complex exponentials, we do need both positive and negative frequencies in order to describe all signals of interest. For example, a real-valued sinusoid can be written in terms of complex exponentials as

$$A\cos(2\pi f_0 t + \theta) = \frac{A}{2}\left(e^{j(2\pi f_0 t + \theta)} + e^{-j(2\pi f_0 t + \theta)}\right) = \frac{A}{2} e^{j\theta} e^{j2\pi f_0 t} + \frac{A}{2} e^{-j\theta} e^{-j2\pi f_0 t}$$

so that we need complex exponentials at both $+f_0$ and $-f_0$ to describe a real-valued sinusoid at frequency f_0. Of course, the coefficients multiplying these two complex exponentials are not arbitrary: they are complex conjugates of each other. More generally, as we have already seen, such conjugate symmetry holds both for Fourier series and for Fourier transforms of real-valued signals. We can therefore state the following.

(a) We do need both positive and negative frequencies to form a complete basis using complex exponentials.

(b) For real-valued (i.e., physically realizable) signals, the expansion in terms of a complex exponential basis, whether it is the Fourier series or the Fourier transform, exhibits conjugate symmetry. Hence, we need to know only the Fourier series or the Fourier transform of a real-valued signal for positive frequencies.

2.5.1 Fourier-transform properties

We now state some key properties of the Fourier transform. In the following, $u(t)$ and $v(t)$ denote signals with Fourier transforms $U(f)$ and $V(f)$, respectively.

Linearity For arbitrary complex numbers α and β,

$$\alpha u(t) + \beta v(t) \leftrightarrow \alpha U(f) + \beta V(f)$$

Time delay corresponds to linear phase in the frequency domain We have

$$u(t - t_0) \leftrightarrow U(f) e^{-j2\pi f t_0}$$

Frequency shift corresponds to modulation by a complex exponential We have

$$U(f - f_0) \leftrightarrow u(t)e^{j2\pi f_0 t}$$

The Fourier transform of a real-valued signal is conjugate symmetric If $u(t)$ is real-valued, then $U(f) = U^*(-f)$.

Differentiation in the time domain amplifies higher frequencies We have

$$x(t) = \frac{d}{dt}u(t) \leftrightarrow X(f) = j2\pi f U(f)$$

As for Fourier series, differentiation kills the DC term, i.e., $X(0) = 0$. However, the information at all other frequencies is preserved. Thus, if we know $X(f)$ then we can recover $U(f)$ for $f \neq 0$ as follows:

$$U(f) = \frac{X(f)}{j2\pi f}, \; f \neq 0 \tag{2.54}$$

This specifies the Fourier transform almost everywhere (except at DC: $f = 0$). If $U(f)$ is finite everywhere, then we do not need to worry about its value at a particular point, and can leave $U(0)$ unspecified, or define it as the limit of (2.54) as $f \to 0$ (if this limit does not exist, we can set $U(0)$ to be the left limit, or the right limit, or any number in between). In short, we can simply adopt (2.54) as the expression for $U(f)$ for all f, when $U(0)$ is finite. However, the DC term does matter when $u(t)$ has a nonzero average value, in which case we get an impulse at DC. The average value of $u(t)$ is given by

$$\bar{u} = \lim_{T \to \infty} \frac{1}{T} \int_{-T/2}^{T/2} u(t)dt$$

and has a Fourier transform given by $\bar{u}(t) \equiv \bar{u} \leftrightarrow \bar{u}\delta(f)$. Thus, we can write the overall Fourier transform as

$$U(f) = \frac{X(f)}{j2\pi f} + \bar{u}\delta(f) \tag{2.55}$$

We illustrate this via the following example.

Example 2.5.5 (Fourier transform of a step function) Let us use differentiation to compute the Fourier transform of the unit step function

$$u(t) = \begin{cases} 0, & t < 0 \\ 1, & t \geq 0 \end{cases}$$

Its DC value is given by

$$\bar{u} = 1/2$$

and its derivative is the delta function (see Figure 2.21):

$$x(t) = \frac{d}{dt}u(t) = \delta(t) \leftrightarrow X(f) \equiv 1$$

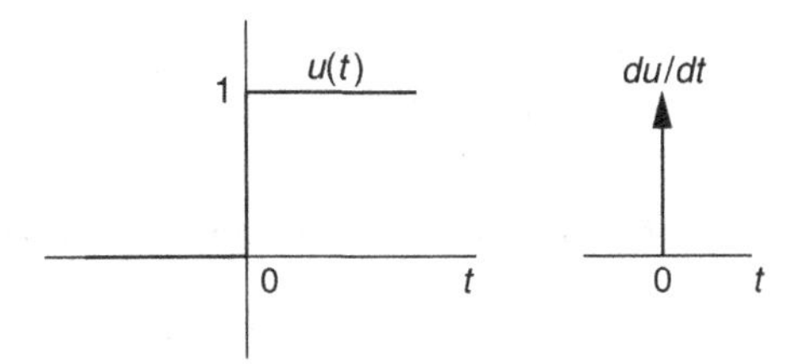

Figure 2.21 The unit step function and its derivative, the delta function.

On applying (2.55), we obtain that the Fourier transform of the unit step function is given by

$$U(f) = \frac{1}{j2\pi f} + \frac{1}{2}\delta(f)$$

Parseval's identity (the inner product/energy can be computed in either the time or the frequency domain) We have

$$\langle u, v\rangle = \int_{-\infty}^{\infty} u(t)v^*(t)dt = \int_{-\infty}^{\infty} U(f)V^*(f)df$$

On setting $v = u$, we get an expression for the energy of a signal:

$$||u||^2 = \int_{-\infty}^{\infty} |u(t)|^2\, dt = \int_{-\infty}^{\infty} |U(f)|^2\, df$$

Next, we discuss the significance of the Fourier transform in understanding the effect of LTI systems.

Transfer function for an LTI system The transfer function $H(f)$ of an LTI system is defined to be the Fourier transform of its impulse response $h(t)$. That is, $H(f) = \mathcal{F}(h(t))$. We now discuss its significance.

From (2.37), we know that, when the input to an LTI system is the complex exponential $e^{j2\pi f_0 t}$, the output is given by $H(f_0)e^{j2\pi f_0 t}$. From the inverse Fourier transform (2.53), we know that any input $u(t)$ can be expressed as a linear combination of complex exponentials. Thus, the corresponding response, which we know is given by $y(t) = (u * h)(t)$, must be a linear combination of the responses to these complex exponentials. Thus, we have

$$y(t) = \int_{-\infty}^{\infty} U(f)H(f)e^{j2\pi ft}\, df$$

We recognize that the preceding function is in the form of an inverse Fourier transform, and read off $Y(f) = U(f)H(f)$. That is, the Fourier transform of the output is simply the product of the Fourier transform of the input and the system transfer function. This is because complex exponentials at different frequencies propagate through an LTI system without mixing with each other, with a complex exponential at frequency f passing through with a scaling of $H(f)$.

Of course, we have also derived an expression for $y(t)$ in terms of a convolution of the input signal with the system impulse response: $y(t) = (u * h)(t)$. We can now infer the following key property.

Convolution in the time domain corresponds to multiplication in the frequency domain We have

$$y(t) = (u * h)(t) \leftrightarrow Y(f) = U(f)H(f) \tag{2.56}$$

We can also infer the following dual property, either by using duality or by directly deriving it from first principles.

Multiplication in the time domain corresponds to convolution in the frequency domain We have

$$y(t) = u(t)v(t) \leftrightarrow Y(f) = (U * V)(f) \tag{2.57}$$

LTI system response to real-valued sinusoidal signals For a sinusoidal input $u(t) = \cos(2\pi f_0 t$ $k + \theta)$, the response of an LTI system with real-valued impulse response h is given by

$$y(t) = (u * h)(t) = |H(f_0)| \cos(2\pi f_0 t + \theta + \underline{/H(f_0)})$$

This can be inferred from what we know about the response for complex exponentials, thanks to Euler's formula. Specifically, we have

$$u(t) = \frac{1}{2}\left(e^{j(2\pi f_0 t+\theta)} + e^{-j(2\pi f_0 t+\theta)}\right) = \frac{1}{2}e^{j\theta}e^{j2\pi f_0 t} + \frac{1}{2}e^{-j\theta}e^{-j2\pi f_0 t}$$

When u goes through an LTI system with transfer function $H(f)$, the output is given by

$$y(t) = \frac{1}{2}e^{j\theta}H(f_0)e^{j2\pi f_0 t} + \frac{1}{2}e^{-j\theta}H(-f_0)e^{-j2\pi f_0 t}$$

If the system is physically realizable, the impulse response $h(t)$ is real-valued, and the transfer function is conjugate symmetric. Thus, if $H(f_0) = Ge^{j\phi}$ $(G \geq 0)$, then $H(-f_0) = H^*(f_0) = Ge^{-j\phi}$. On substituting, we obtain

$$y(t) = \frac{G}{2}e^{j(2\pi f_0 t+\theta+\phi)} + \frac{G}{2}e^{-j(2\pi f_0 t+\theta+\phi)} = G\cos(2\pi f_0 t + \theta + \phi)$$

This yields the well-known result that the sinusoid gets scaled by the magnitude of the transfer function $G = |H(f_0)|$, and gets phase shifted by the phase of the transfer function $\phi = \underline{/H(f_0)}$.

Example 2.5.6 (Delay spread, coherence bandwidth, and fading for a multipath channel) The transfer function of a multipath channel as in (2.36) is given by

$$H(f) = \alpha_1 e^{-j2\pi f \tau_1} + \cdots + \alpha_m e^{-j2\pi f \tau_m} \tag{2.58}$$

Thus, the channel transfer function is a linear combination of *complex exponentials in the frequency domain.* As with any sinusoids, these can interfere constructively or destructively, leading to significant fluctuations in $H(f)$ as f varies. For wireless channels, this phenomenon is called *frequency-selective fading.* Let us examine the structure of the fading a little further. Suppose, without loss of generality, that the delays are in increasing order (i.e., $\tau_1 < \tau_2 < \ldots < \tau_m$). We can then rewrite the transfer function as

$$H(f) = e^{-j2\pi f\tau_1} \sum_{k=1}^{m} \alpha_k e^{-j2\pi f(\tau_k - \tau_1)}$$

The first term, $e^{-j2\pi f\tau_1}$, corresponds simply to a pure delay τ_1 (seen by all frequencies), and can be dropped (taking τ_1 as our time origin, without loss of generality), so that the transfer function can be rewritten as

$$H(f) = \alpha_1 + \sum_{k=2}^{m} \alpha_k e^{-j2\pi f(\tau_k - \tau_1)} \tag{2.59}$$

The period of the kth sinusoid above ($k \geq 2$) is $1/(\tau_k - \tau_1)$, so that the smallest period, and hence the fastest fluctuations as a function of f, occurs because of the largest delay difference $\tau_d = \tau_m - \tau_1$, which we call the channel *delay spread.* Thus, for a frequency interval that is significantly smaller than $1/\tau_d$, the variation of $|H(f)|$ over the interval is small. We define the channel *coherence bandwidth* as the inverse of the delay spread, i.e., as $B_c = 1/(\tau_m - \tau_1) = 1/\tau_d$ (this definition is not unique, but, in general, the coherence bandwidth is defined to be inversely proportional to some appropriately defined measure of the channel delay spread). As we have noted, $H(f)$ can be well modeled as constant over intervals significantly smaller than the coherence bandwidth.

Let us apply this to the example in Figure 2.14, where we have a multipath channel with impulse response $h(t) = \delta(t-1) - 0.5\delta(t-1.5) + 0.5\delta(t-3.5)$. On dropping the first delay as before, we have

$$H(f) = 1 - 0.5e^{-j\pi f} + 0.5e^{-j5\pi f}$$

For concreteness, suppose that time is measured in *microseconds* (typical numbers for an outdoor wireless cellular link), so that frequency is measured in MHz. The delay spread is 2.5 μs, hence the coherence bandwidth is 400 kHz. We therefore ballpark the size of the frequency interval over which $H(f)$ can be approximated as constant to about 40 kHz (i.e., of size 10% of the coherence bandwidth). Note that this is a very fuzzy estimate: if the larger delays occur with smaller relative amplitudes, as is typical, then they have a smaller effect on $H(f)$, and we could potentially approximate $H(f)$ as constant over a larger fraction of the coherence bandwidth. Figure 2.22 depicts the fluctuations in $H(f)$ first on a linear scale and then on a log scale. A plot of the transfer-function magnitude is shown in Figure 2.22(a). This is the amplitude gain on a linear scale, and shows significant variations as a function of f (while we do not show it here, zooming in to 40-kHz bands shows relatively small fluctuations). The amount of fluctuation becomes even more apparent on a log scale. Interpreting the gain at the smallest delay ($\alpha_1 = 1$ in our case) as that of a nominal channel, the *fading gain* is defined as the power gain relative to this nominal, and is given by $20\log_{10}(|H(f)|/|\alpha_1|)$ in decibels (dB). This is shown in Figure 2.22(b). Note that the fading gain can dip below -18 dB in our example, which we term a fade of *depth* 18 dB. If we are using a "narrowband" signal that has a bandwidth that is small compared with the coherence bandwidth, and happen to get hit by such a fade,

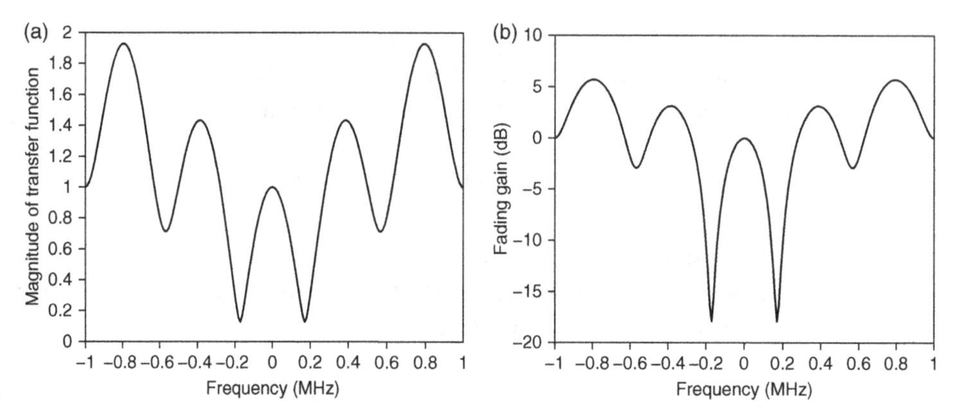

Figure 2.22 Multipath propagation causes severe frequency-selective fading: (a) transfer-function magnitude (linear scale) and (b) frequency-selective fading (dB).

then we can expect much poorer performance than nominal. To combat this, one must use *diversity*. For example, a "wideband" signal whose bandwidth is larger than the coherence bandwidth provides frequency diversity, while, if we are constrained to use narrowband signals, we may need to introduce other forms of diversity (e.g., antenna diversity as in Software Lab 2.3).

2.5.2 Numerical computation using DFT

In many practical settings, we do not have nice analytical expressions for the Fourier or inverse Fourier transforms, and must resort to numerical computation, typically using the discrete Fourier transform (DFT). The DFT of a discrete time sequence $\{u[n], n = 0, \ldots, N-1\}$ of length N is given by

$$U[m] = \sum_{n=0}^{N-1} u[n] e^{-j2\pi mn/N}, \quad m = 0, 1, \ldots, N-1. \tag{2.60}$$

MATLAB is good at doing DFTs. When N is a power of 2, the DFT can be computed very efficiently, and this procedure is called a fast Fourier transform (FFT). Comparing (2.60) with the Fourier-transform expression

$$U(f) = \int_{-\infty}^{\infty} u(t) e^{-j2\pi ft}\, dt \tag{2.61}$$

we can view the sum in the DFT (2.60) as an approximation for the integral in (2.61) under the right set of conditions. Let us first assume that $u(t) = 0$ for $t < 0$: any waveform that can be truncated so that most of its energy falls within a finite interval can be shifted so that this is true. Next, suppose that we sample the waveform with spacing t_s to get

$$u[n] = u(nt_s)$$

Now, suppose we want to compute the Fourier transform $U(f)$ for $f = mf_s$, where f_s is the desired frequency resolution. We can approximate the integral for the Fourier transform by a sum, using t_s-spaced time samples as follows:

$$U(mf_s) = \int_{-\infty}^{\infty} u(t)e^{-j2\pi mf_s t}\, dt \approx \sum_n u(nt_s)e^{-j2\pi mf_s nt_s} t_s$$

(dt in the integral is replaced by the sample spacing t_s). Since $u[n] = u(nt_s)$, the approximation can be computed using the DFT formula (2.60) as follows:

$$U(mf_s) \approx t_s U[m] \tag{2.62}$$

as long as $f_s t_s = 1/N$. That is, using a DFT of length N, we can get a frequency granularity of $f_s = 1/(Nt_s)$. This implies that, if we choose the time samples close together (in order to represent $u(t)$ accurately), then we must also use a large N to get a desired frequency granularity. Often this means that we must pad the time-domain samples with zeros.

Another important observation is that, while the DFT in (2.60) ranges over $m = 0, \ldots, N-1$, it actually computes the Fourier transform for both positive and negative frequencies. Noting that $e^{j2\pi mn/N} = e^{j2\pi(-N+m)n/N}$, we realize that the DFT values for $m = N/2, \ldots, N-1$ correspond to the Fourier transform evaluated at frequencies $(m-N)f_s = -(N/2)f_s, \ldots, -f_s$. The DFT values for $m = 0, \ldots, N/2-1$ correspond to the Fourier transform evaluated at frequencies $0, f_s, \ldots, (N/2-1)f_s$. Thus, we should swap the left and right halves of the DFT output in order to represent positive and negative frequencies, with DC falling in the middle. MATLAB actually has a function, fftshift, that does this.

Note that the DFT (2.60) is periodic with period N, so that the Fourier-transform approximation (2.62) is periodic with period $Nf_s = 1/t_s$. We typically limit the range of frequencies over which we use the DFT to compute the Fourier transform to the fundamental period $(-1/(2t_s), 1/(2t_s))$. This is consistent with the sampling theorem, which says that the sampling rate $1/t_s$ must be at least as large as the size of the frequency band of interest. (The sampling theorem is reviewed in Chapter 4, when we discuss digital modulation.)

Example 2.5.7 (DFT-based Fourier transform computation) Suppose that we want to compute the Fourier transform of the sine pulse $u(t) = \sin(\pi t)I_{[0,1]}(t)$. The Fourier transform for this can be computed analytically (see Problem 2.9) to be

$$U(f) = \frac{2\cos(\pi f)}{\pi(1-4f^2)} e^{-j\pi f} \tag{2.63}$$

Note that $U(f)$ has a $0/0$ form at $f = 1/2$, but, using l'Hôpital's rule, we can show that $U(1/2) \neq 0$. Thus, the first zeros of $U(f)$ are at $f = \pm 3/2$. This is a timelimited pulse and hence cannot be bandlimited, but $U(f)$ decays as $1/f^2$ for f large, so we can capture most of the energy of the pulse within a suitably chosen finite frequency interval. Let us use the DFT to compute $U(f)$ over $f \in (-8, 8)$. This means that we set $1/(2t_s) = 8$, or $t_s = 1/16$, which yields about 16 samples over the interval $[0, 1]$ over which the signal $u(t)$

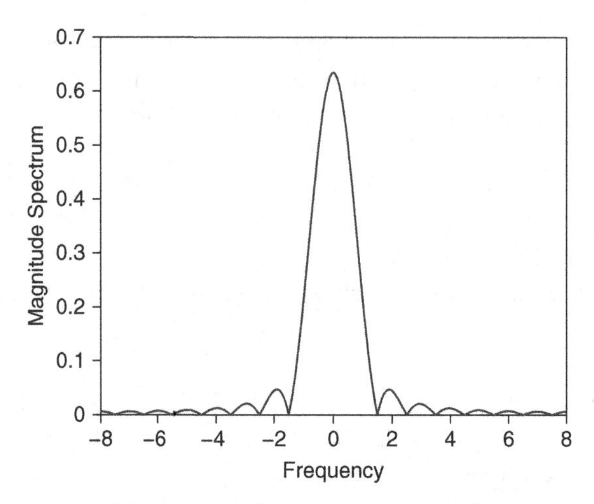

Figure 2.23 A plot of the magnitude spectrum of the sine pulse in Example 2.5.7 obtained numerically using the DFT.

has support. Suppose now that we want the frequency granularity to be at least $f_s = 1/160$. Then we must use a DFT with $N \geq 1/(t_s f_s) = 2560 = N_{\min}$. In order to efficiently compute the DFT using the FFT, we choose $N = 4096$, the next power of 2 at least as large as $N_{\min}$. Code Fragment 2.5.1 performs and plots this DFT. The resulting plot (with cosmetic touches not included in the code below) is displayed in Figure 2.23. It is useful to compare this with a plot obtained from the analytical formula (2.63), and we leave that as an exercise.

Code Fragment 2.5.1 (Numerical computation of Fourier transform using FFT)

```
ts = 1/16; %sampling interval
time_interval = 0:ts:1; %sampling time instants
%time-domain signal evaluated at sampling instants
signal_timedomain = sin(pi * time_interval); %sinusoidal pulse in our example
fs_desired = 1/160; %desired frequency granularity
Nmin = ceil(1/(fs_desired * ts)); %minimum-length DFT for desired frequency granularity
%for efficient computation, choose FFT size to be power of 2
Nfft = 2^(nextpow2(Nmin)) %FFT size = the next power of 2 at least as big as Nmin
%Alternatively, one could also use DFT size equal to the minimum length
%Nfft = Nmin;
%note: fft function in Matlab is just the DFT when Nfft is not a power of 2
%frequency-domain signal computed using DFT
%fft function of size Nfft automatically zeropads as needed
signal_freqdomain = ts * fft(signal_timedomain,Nfft);
%fftshift function shifts DC to center of spectrum
signal_freqdomain_centered = fftshift(signal_freqdomain);
fs = 1/(Nfft * ts); %actual frequency resolution attained
%set of frequencies for which Fourier transform has been computed using DFT
freqs = ((1:Nfft) - 1 - Nfft/2) * fs;
%plot the magnitude spectrum
plot(freqs,abs(signal_freqdomain_centered));
xlabel('Frequency');
ylabel('Magnitude Spectrum');
```

2.6 Energy spectral density and bandwidth

Communication channels have frequency-dependent characteristics, hence it is useful to appropriately shape the frequency-domain characteristics of the signals sent over them. Furthermore, for wireless communication systems, frequency spectrum is a particularly precious commodity, since wireless is a broadcast medium to be shared by multiple signals. It is therefore important to quantify the frequency occupancy of communication signals. We provide a first exposure to these concepts here via the notion of energy spectral density for finite-energy signals. These ideas are extended to finite-power signals, for which we can define the analogous concept of power spectral density, in Chapter 4, "just in time" for our discussion of the spectral occupancy of digitally modulated signals. Once we know the energy or power spectral density of a signal, we shall see that there are several possible definitions of bandwidth, which is a measure of the size of the frequency interval occupied by the signal.

Energy spectral density The energy spectral density $E_u(f)$ of a signal $u(t)$ can be defined operationally as shown in Figure 2.24. Pass the signal $u(t)$ through an ideal narrowband filter with transfer function as follows:

$$H_{f^*}(f) = \begin{cases} 1, & f^* - \Delta f/2 < f < f^* + \Delta f/2 \\ 0, & \text{else} \end{cases}$$

The energy spectral density $E_u(f^*)$ is defined to be the energy at the output of the filter, divided by the width Δf (in the limit as $\Delta f \to 0$). That is, the energy at the output of the filter is approximately $E_u(f^*)\Delta f$. But the Fourier transform of the filter output is

$$Y(f) = U(f)H(f) = \begin{cases} U(f), & f^* - \Delta f/2 < f < f^* + \Delta f/2 \\ 0, & \text{else} \end{cases}$$

By Parseval's identity, the energy at the output of the filter is

$$\int_{-\infty}^{\infty} |Y(f)|^2 \, df = \int_{f^*-\Delta f/2}^{f^*+\Delta f/2} |U(f)|^2 \, df \approx |U(f^*)|^2 \, \Delta f$$

assuming that $U(f)$ varies smoothly and Δf is small enough. We can now infer that the energy spectral density is simply the magnitude squared of the Fourier transform:

$$E_u(f) = |U(f)|^2 \tag{2.64}$$

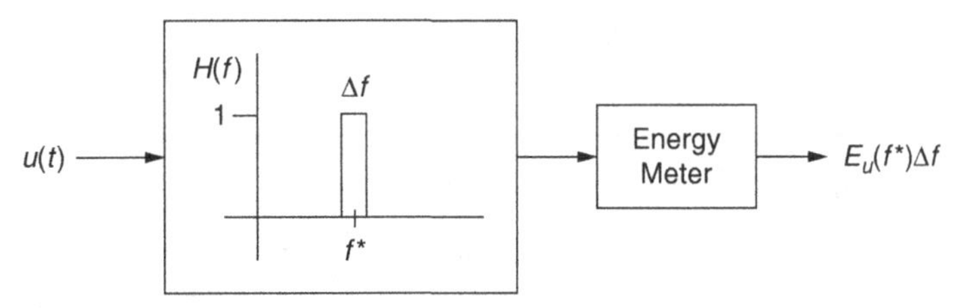

Figure 2.24 Operational definition of energy spectral density.

The integral of the energy spectral density equals the signal energy, which is consistent with Parseval's identity.

The inverse Fourier transform of the energy spectral density has a nice intuitive interpretation. Noting that $|U(f)|^2 = U(f)U^*(f)$ and $U^*(f) \leftrightarrow u^*(-t)$, let us define $u_{MF}(t) = u^*(-t)$ as (the impulse response of) the *matched filter* for $u(t)$, where the reasons for this term will be clarified later. Then

$$|U(f)|^2 = U(f)U^*(f) \leftrightarrow (u * u_{MF})(\tau) = \int u(t)u_{MF}(\tau - t)dt$$
$$= \int u(t)u^*(t - \tau)dt \tag{2.65}$$

where t is a dummy variable for the integration, and the convolution is evaluated at the time variable τ, which denotes the delay between the two versions of u being correlated: the extreme right-hand side is simply the correlation of u with itself (after complex conjugation), evaluated at different delays τ. We call this the *autocorrelation function* of the signal u. We have therefore shown the following: *for a finite-energy signal, the energy spectral density and the autocorrelation function form a Fourier-transform pair.*

Bandwidth The bandwidth of a signal $u(t)$ is loosely defined to be the size of the band of frequencies occupied by $U(f)$. The definition is "loose" because the concept of occupancy can vary, depending on the application, since signals are seldom strictly bandlimited. One possibility is to consider the band over which $|U(f)|^2$ is within some fraction of its peak value (setting the fraction equal to $\frac{1}{2}$ corresponds to the 3-dB bandwidth). Alternatively, we might be interested in the energy-containment bandwidth, which is the size of the smallest band which contains a specified fraction of the signal energy (for a finite-power signal, we define analogously the power-containment bandwidth).

Only positive frequencies count when computing bandwidth for physical (real-valued) signals For a physically realizable (i.e., real-valued) signal, the bandwidth is defined as its occupancy of positive frequencies, because conjugate symmetry implies that the information at negative frequencies is redundant.

While physically realizable time-domain signals are real-valued, we shall soon introduce complex-valued signals that have a useful physical interpretation, in the sense that they have a well-defined mapping to physically realizable signals. Conjugate symmetry in the frequency domain does not hold for complex-valued time-domain signals, with different information contained in positive and negative frequencies in general. Thus, the bandwidth for a complex-valued signal is defined as the size of the frequency band it occupies over both positive and negative frequencies. The justification for this convention becomes apparent later in this chapter.

Example 2.6.1 (Some bandwidth computations)

(a) Consider $u(t) = \text{sinc}(2t)$, where the unit of time is microseconds. Then the unit of frequency is MHz, and $U(f) = \frac{1}{2}I_{[-1,1]}(f)$ is strictly bandlimited with a bandwidth of 2 MHz.

(b) Now, consider the timelimited waveform $u(t) = I_{[2,4]}(t)$, where the unit of time is microseconds. Then $U(f) = 2\ \text{sinc}(2f)e^{-j6\pi f}$, which is not bandlimited. The 99% energy-containment bandwidth W is defined by the equation

$$\int_{-W}^{W} |U(f)|^2\, df = 0.99 \int_{-\infty}^{\infty} |U(f)|^2\, df = 0.99 \int_{-\infty}^{\infty} |u(t)|^2\, dt = 0.99 \int_{2}^{4} 1^2\, dt = 1.98$$

where we use Parseval's identity to simplify computation for timelimited waveforms. Using the fact that $|U(f)|$ is even, we obtain that

$$1.98 = 2\int_{0}^{W} |U(f)|^2\, df = 2\int_{0}^{W} 4\ \text{sinc}^2(2f) df$$

We can now solve numerically to obtain $W \approx 5.1$ MHz.

2.7 Baseband and passband signals

Baseband A signal $u(t)$ is said to be baseband if the signal energy is concentrated in a band around DC, and

$$U(f) \approx 0, \quad |f| > W \tag{2.66}$$

for some $W > 0$. Similarly, a channel modeled as a linear time-invariant system is said to be baseband if its transfer function $H(f)$ has support concentrated around DC, and satisfies (2.66).

A signal $u(t)$ is said to be *passband* if its energy is concentrated in a band away from DC, with

$$U(f) \approx 0, \quad |f \pm f_c| > W \tag{2.67}$$

where $f_c > W > 0$. A channel modeled as a linear time-invariant system is said to be passband if its transfer function $H(f)$ satisfies (2.67).

Examples of baseband and passband signals are shown in Figures 2.25 and 2.26, respectively. Physically realizable signals must be real-valued in the time domain, which means that their Fourier transforms, which can be complex-valued, must be conjugate symmetric: $U(-f) = U^*(f)$. As discussed earlier, the bandwidth B for a real-valued signal $u(t)$ is the size of the frequency interval (counting only positive frequencies) occupied by $U(f)$.

Information sources typically emit baseband signals. For example, an analog audio signal has significant frequency content ranging from DC to around 20 kHz. A digital signal in which zeros and ones are represented by pulses is also a baseband signal, with the frequency content governed by the shape of the pulse (as we shall see in more detail in Chapter 4). Even when the pulse is timelimited, and hence not strictly bandlimited, most of the energy is concentrated in a band around DC.

Wired channels (e.g., telephone lines, USB connectors) are typically modeled as baseband: the attenuation over the wire increases with frequency, so that it makes sense to design the transmitted signal to utilize a frequency band around DC. An example of

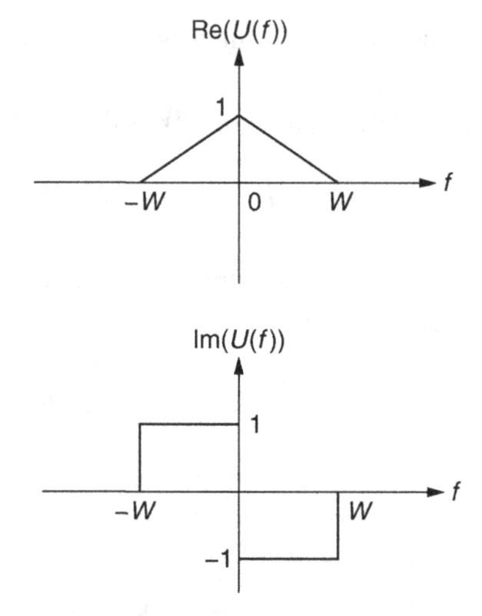

Figure 2.25 An example of the spectrum $U(f)$ for a real-valued baseband signal. The bandwidth of the signal is W.

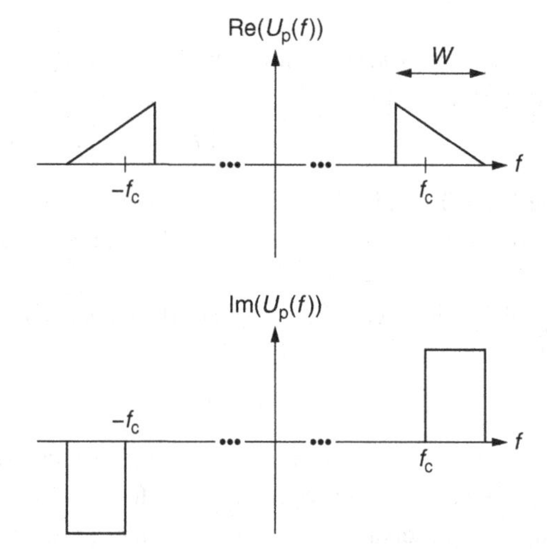

Figure 2.26 An example of the spectrum $U(f)$ for a real-valued passband signal. The bandwidth of the signal is W. The figure shows an arbitrarily chosen frequency f_c within the band in which $U(f)$ is nonzero. Typically, f_c is much larger than the signal bandwidth W.

passband communication over a wire is a digital subscriber line (DSL), where high-speed data transmission using frequencies above 25 kHz co-exists with voice transmission in the band 0–4 kHz. The design and use of passband signals for communication is particularly important for wireless communication, in which the transmitted signals must fit within frequency bands dictated by regulatory agencies, such as the Federal Communication Commission (FCC) in the United States. For example, an amplitude-modulation (AM) radio signal typically occupies a frequency interval of length 10 kHz somewhere in the 540–1600-kHz band allocated for AM radio. Thus, the baseband audio message signal must be transformed into a passband signal before it can be sent over the passband channel spanning the desired band. As another example, a transmitted signal in a WiFi network may be designed to fit within a 20-MHz frequency interval in the 2.4-GHz unlicensed band, so that digital messages to be sent over WiFi must be encoded onto passband signals occupying the designated spectral band.

2.8 The structure of a passband signal

In order to employ a passband channel for communication, we need to understand how to design a passband transmitted signal to carry information, and how to recover this information from a passband received signal. We also need to understand how the transmitted signal is affected by a passband channel.

2.8.1 Time-domain relationships

Let us start by considering a real-valued baseband message signal $m(t)$ of bandwidth W, to be sent over a passband channel centered around f_c. As illustrated in Figure 2.27, we can translate the message to passband simply by multiplying it by a sinusoid at f_c:

$$u_p(t) = m(t)\cos(2\pi f_c t) \leftrightarrow U_p(f) = \frac{1}{2}(M(f - f_c) + M(f + f_c))$$

We use the term *carrier frequency* for f_c, and the term *carrier* for a sinusoid at the carrier frequency, since the modulated sinusoid is "carrying" the message information over a passband channel. Instead of a cosine, we could also use a sine:

$$v_p(t) = m(t)\sin(2\pi f_c t) \leftrightarrow V_p(f) = \frac{1}{2j}(M(f - f_c) - M(f + f_c))$$

Note that $|U_p(f)|$ and $|V_p(f)|$ have frequency content in a band around f_c, and are passband signals (i.e., living in a band not containing DC) as long as $f_c > W$.

I and Q components If we use both cosine and sine carriers, we can construct a passband signal of the form

$$u_p(t) = u_c(t)\cos(2\pi f_c t) - u_s(t)\sin(2\pi f_c t) \tag{2.68}$$

where u_c and u_s are real baseband signals of bandwidth at most W, with $f_c > W$. The signal $u_c(t)$ is called the in-phase (or I) component, and $u_s(t)$ is called the quadrature (or Q)

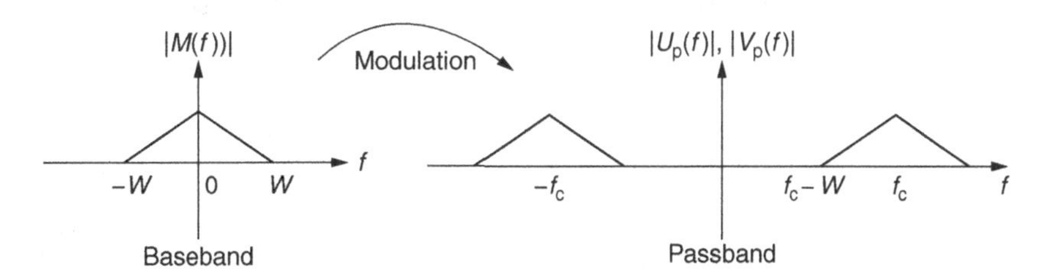

Figure 2.27 A baseband message of bandwidth W is translated to passband by multiplying by a sinusoid at frequency f_c, as long as $f_c > W$.

component. The negative sign for the Q term is a standard convention. Since the sinusoidal terms are entirely predictable once we specify f_c, all information in the passband signal u_p must be contained in the I and Q components. Modulation for a passband channel therefore corresponds to choosing a method of encoding information into the I and Q components of the transmitted signal, while demodulation corresponds to extracting this information from the received passband signal. In order to accomplish modulation and demodulation, we must be able to *upconvert* from baseband to passband, and *downconvert* from passband to baseband, as follows.

Upconversion and downconversion Equation (2.68) immediately tells us how to *upconvert* from baseband to passband. To *downconvert* from passband to baseband, consider

$$\begin{aligned} 2u_p(t)\cos(2\pi f_c t) &= 2u_c(t)\cos^2(2\pi f_c t) - 2u_s(t)\sin(2\pi f_c t)\cos(2\pi f_c t) \\ &= u_c(t) + u_c(t)\cos(4\pi f_c t) - u_s(t)\sin(4\pi f_c t) \end{aligned}$$

The first term on the extreme right-hand side is the I component $u_c(t)$, a baseband signal. The second and third terms are passband signals at $2f_c$, which we can get rid of by lowpass filtering. Similarly, we can obtain the Q component $u_s(t)$ by lowpass filtering $-2u_p(t)\sin(2\pi f_c t)$. Block diagrams for upconversion and downconversion are depicted in Figure 2.28. The implementation of these operations could, in practice, be done in multiple stages, and requires careful analog circuit design.

We now dig deeper into the structure of a passband signal. First, can we choose the I and Q components freely, independently of each other? The answer is yes: the I and Q components provide two parallel, orthogonal "channels" for encoding information, as we show next.

Orthogonality of I and Q channels The passband waveform $a_p(t) = u_c(t)\cos(2\pi f_c t)$, corresponding to the I component, and the passband waveform $b_p(t) = u_s(t)\sin(2\pi f_c t)$, corresponding to the Q component, are orthogonal. That is,

$$\langle a_p, b_p \rangle = 0 \tag{2.69}$$

Let

$$x(t) = a_p(t)b_p(t) = u_c(t)u_s(t)\cos(2\pi f_c t)\sin(2\pi f_c t) = \frac{1}{2}u_c(t)u_s(t)\sin(4\pi f_c t)$$

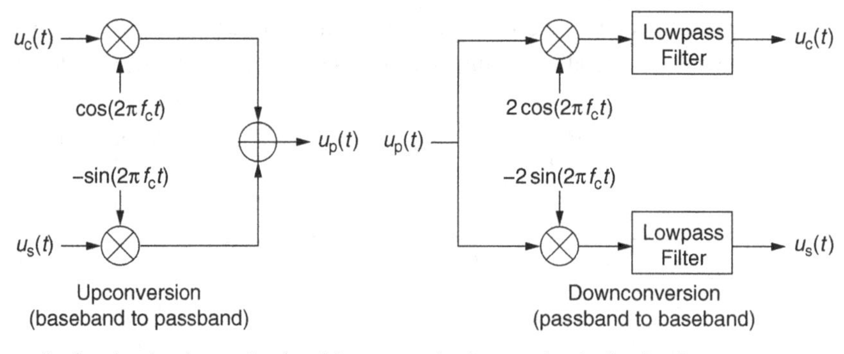

Figure 2.28 Upconversion from baseband to passband, and downconversion from passband to baseband.

We prove the desired result by showing that $x(t)$ is a passband signal at $2f_c$, so that its DC component is zero. That is,

$$\int_{-\infty}^{\infty} x(t)dt = X(0) = 0$$

which is the desired result. To show this, note that

$$p(t) = \frac{1}{2}u_c(t)u_s(t) \leftrightarrow \frac{1}{2}(U_c * U_s)(f)$$

is a baseband signal: if $U_c(f)$ is baseband with bandwidth W_1 and $U_s(f)$ is baseband with bandwidth W_2, then their convolution has bandwidth at most $W_1 + W_2$. In order for a_p to be passband, we must have $f_c > W_1$, and in order for b_p to be passband, we must have $f_c > W_2$. Thus, $2f_c > W_1 + W_2$, which means that $x(t) = p(t)\sin(4\pi f_c t)$ is passband around $2f_c$, and is therefore zero at DC. This completes the derivation.

Example 2.8.1 (Passband signal) The signal

$$u_p(t) = I_{[0,1]}(t)\cos(300\pi t) - (1 - |t|)I_{[-1,1]}(t)\sin(300\pi t)$$

is a passband signal with I component $u_c(t) = I_{[0,1]}(t)$ and Q component $u_s(t) = (1 - |t|)I_{[-1,1]}(t)$. This example illustrates that we do not require strict bandwidth limitations in our definitions of passband and baseband: the I and Q components are timelimited, and hence cannot be bandlimited. However, they are termed baseband signals because most of their energy lies in baseband. Similarly, $u_p(t)$ is termed a passband signal, since most of its frequency content lies in a small band around 150 Hz.

Envelope and phase Since a passband signal u_p is equivalent to a pair of real-valued baseband waveforms (u_c, u_s), passband modulation is often called *two-dimensional modulation.* The representation (2.68) in terms of I and Q components corresponds to thinking of this two-dimensional waveform in rectangular coordinates (the "cosine axis" and the "sine axis"). We can also represent the passband waveform using polar coordinates. Consider the rectangular–polar transformation

$$e(t) = \sqrt{u_c^2(t) + u_s^2(t)}, \quad \theta(t) = \tan^{-1}\left(\frac{u_s(t)}{u_c(t)}\right)$$

where $e(t) \geq 0$ is termed the *envelope* and $\theta(t)$ is the *phase*. This corresponds to $u_c(t) = e(t)\cos\theta(t)$ and $u_s(t) = e(t)\sin\theta(t)$. On substituting into (2.68), we obtain

$$u_p(t) = e(t)\cos\theta(t)\cos(2\pi f_c t) - e(t)\sin\theta(t)\sin(2\pi f_c t) = e(t)\cos(2\pi f_c t + \theta(t)) \quad (2.70)$$

This provides an alternate representation of the passband signal in terms of baseband envelope and phase signals.

Complex envelope To obtain a third representation of a passband signal, we note that a two-dimensional point can also be mapped to a complex number; see Section 2.1. We define the *complex envelope* $u(t)$ of the passband signal $u_p(t)$ in (2.68) and (2.70) as follows:

$$u(t) = u_c(t) + ju_s(t) = e(t)e^{j\theta(t)} \tag{2.71}$$

We can now express the passband signal in terms of its complex envelope. From (2.70), we see that

$$u_p(t) = e(t)\text{Re}\left(e^{j(2\pi f_c t + \theta(t))}\right) = \text{Re}\left(e(t)e^{j(2\pi f_c t + \theta(t))}\right) = \text{Re}\left(e(t)e^{j\theta(t)}e^{j2\pi f_c t}\right)$$

This leads to our third representation of a passband signal:

$$u_p(t) = \text{Re}\left(u(t)e^{j2\pi f_c t}\right) \tag{2.72}$$

While we have obtained (2.72) using the polar representation (2.70), we should also check that it is consistent with the rectangular representation (2.68), by writing out the real and imaginary parts of the complex waveforms above as follows:

$$\begin{aligned} u(t)e^{j2\pi f_c t} &= (u_c(t) + ju_s(t))(\cos(2\pi f_c t) + j\sin(2\pi f_c t)) \\ &= (u_c(t)\cos(2\pi f_c t) - u_s(t)\sin(2\pi f_c t)) + j(u_s(t)\cos(2\pi f_c t) + u_c(t)\sin(2\pi f_c t)) \end{aligned} \tag{2.73}$$

On taking the real part, we obtain the expression (2.68) for $u_p(t)$.

The relationship among the three time-domain representations of a passband signal in terms of its complex envelope is depicted in Figure 2.29. We now specify the corresponding frequency-domain relationship.

Information resides in complex baseband The complex-baseband representation corresponds to subtracting out the rapid, but predictable, phase variation due to the fixed reference frequency f_c, and then considering the much slower amplitude and phase variations induced by baseband modulation. Since the phase variation due to f_c is predictable, it cannot convey any information. Thus, all the information in a passband signal is contained in its complex envelope.

Choice of frequency/phase reference is arbitrary We can define the complex-baseband representation of a passband signal using an arbitrary frequency reference f_c (and can also vary the phase reference), as long as we satisfy $f_c > W$, where W is the bandwidth. We may often wish to transform the complex-baseband representations for two different references.

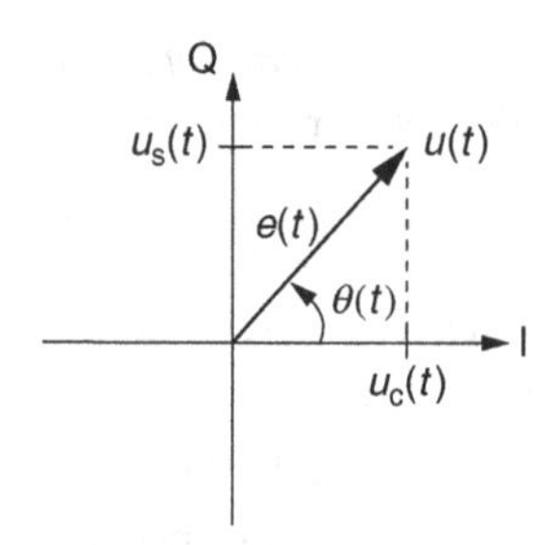

Figure 2.29 The geometry of the complex envelope.

For example, we can write

$$\begin{aligned} u_p(t) &= u_{c1}(t)\cos(2\pi f_1 t + \theta_1) - u_{s1}(t)\sin(2\pi f_1 t + \theta_1) \\ &= u_{c2}(t)\cos(2\pi f_2 t + \theta_2) - u_{s2}(t)\sin(2\pi f_2 t + \theta_2) \end{aligned}$$

We can express this more compactly in terms of the complex envelopes $u_1 = u_{c1} + ju_{s1}$ and $u_2 = u_{c2} + ju_{s2}$:

$$u_p(t) = \text{Re}\left(u_1(t)e^{j(2\pi f_1 t + \theta_1)}\right) = \text{Re}\left(u_2(t)e^{j(2\pi f_2 t + \theta_2)}\right) \tag{2.74}$$

We can now find the relationship between these complex envelopes by transforming the exponential term for one reference to the other:

$$u_p(t) = \text{Re}\left(u_1(t)e^{j(2\pi f_1 t + \theta_1)}\right) = \text{Re}\left([u_1(t)e^{j(2\pi(f_1 - f_2)t + \theta_1 - \theta_2)}]e^{j(2\pi f_2 t + \theta_2)}\right) \tag{2.75}$$

By comparing with the extreme right-hand sides of (2.74) and (2.75), we can read off that

$$u_2(t) = u_1(t)e^{j(2\pi(f_1 - f_2)t + \theta_1 - \theta_2)}$$

While we derived this result using algebraic manipulations, it has the following intuitive interpretation: if the instantaneous phase $2\pi f_i t + \theta_i$ of the reference is ahead/behind, then the complex envelope must be correspondingly retarded/advanced, so that the instantaneous phase of the overall passband signal stays the same. We illustrate this via some examples below.

Example 2.8.2 (Change of reference frequency/phase) Consider the passband signal $u_p(t) = I_{[-1,1]}(t)\cos(400\pi t)$.

(a) Find the output when $u_p(t)\cos(401\pi t)$ is passed through a lowpass filter.
(b) Find the output when $u_p(t)\sin(400\pi t - \pi/4)$ is passed through a lowpass filter.

Solution

From Figure 2.28, we recognize that both (a) and (b) correspond to downconversion operations with different frequency and phase references. Thus, by converting the complex envelope with respect to the appropriate reference, we can read off the answers.

(a) Letting $u_1 = u_{c1} + ju_{s1}$ denote the complex envelope with respect to the reference $e^{j401\pi t}$, we recognize that the output of the LPF is $u_{c1}/2$. The passband signal can be written as

$$u_p(t) = I_{[-1,1]}(t)\cos(400\pi t) = \text{Re}\left(I_{[-1,1]}(t)e^{j400\pi t}\right)$$

We can now massage it to read off the complex envelope for the new reference:

$$u_p(t) = \text{Re}\left(I_{[-1,1]}(t)e^{-j\pi t}e^{j401\pi t}\right)$$

from which we see that $u_1(t) = I_{[-1,1]}(t)e^{-j\pi t} = I_{[-1,1]}(t)(\cos(\pi t) - j\sin(\pi t))$. Taking real and imaginary parts, we obtain $u_{c1}(t) = I_{[-1,1]}(t)\cos(\pi t)$ and $u_{s1}(t) = -I_{[-1,1]}(t)\sin(\pi t)$, respectively. Thus, the LPF output is $\frac{1}{2}I_{[-1,1]}(t)\cos(\pi t)$.

(b) On letting $u_2 = u_{c2} + ju_{s2}$ denote the complex envelope with respect to the reference $e^{j(400\pi t - \pi/4)}$, we recognize that the output of the LPF is $-u_{s2}/2$. We can convert to the new reference as before:

$$u_p(t) = \text{Re}\left(I_{[-1,1]}(t)e^{j\pi/4}e^{j(400\pi t - \pi/4)}\right)$$

which gives the complex envelope $u_2 = I_{[-1,1]}(t)e^{j\pi/4} = I_{[-1,1]}(t)(\cos(\pi/4) + j\sin(\pi/4))$. On taking the real and imaginary parts, we obtain $u_{c2}(t) = I_{[-1,1]}(t)\cos(\pi/4)$ and $u_{s2}(t) = I_{[-1,1]}(t)\sin(\pi/4)$, respectively. Thus, the LPF output is given by $-u_{s2}/2 = -\frac{1}{2}I_{[-1,1]}(t)\sin(\pi/4) = -(1/(2\sqrt{2}))I_{[-1,1]}(t)$.

From a practical point of view, keeping track of frequency/phase references becomes important for the task of synchronization. For example, the carrier frequency used by the transmitter for upconversion might not be exactly equal to that used by the receiver for downconversion. Thus, the receiver must compensate for the phase rotation incurred by the complex envelope at the output of the downconverter, as illustrated by the following example.

Example 2.8.3 (Modeling and compensating for frequency/phase offsets in complex baseband) Consider the passband signal u_p (2.68), with complex baseband representation $u = u_c + ju_s$. Now, consider a phase-shifted version of the passband signal

$$\tilde{u}_p(t) = u_c(t)\cos(2\pi f_c t + \theta(t)) - u_s(t)\sin(2\pi f_c t + \theta(t))$$

where $\theta(t)$ may vary slowly with time. For example, the situation with a carrier-frequency offset Δf and a phase offset γ corresponds to $\theta(t) = 2\pi\Delta f\, t + \gamma$. Suppose, now, that the signal is downconverted as in Figure 2.28, where we take the phase reference as that of the receiver's local oscillator (LO). How do the I and Q components depend on the phase offset of the received signal relative to the LO? The easiest way to answer this is to find the complex envelope of $\tilde{u}_p$ with respect to f_c. To do this, we write $\tilde{u}_p$ in the standard form (2.70) as follows:

$$\tilde{u}_p(t) = \text{Re}\left(u(t)e^{j(2\pi f_c t + \theta(t))}\right)$$

By comparing with the desired form

$$\tilde{u}_p(t) = \text{Re}(\tilde{u}(t)e^{j2\pi f_c t})$$

we can read off

$$\tilde{u}(t) = u(t)e^{j\theta(t)} \tag{2.76}$$

Equation (2.76) relates the complex envelopes before and after a phase offset. We can expand out this "polar form" representation to obtain the corresponding relationship between the I and Q components. Suppressing the time dependence from the notation, we can rewrite (2.76) as

$$\tilde{u}_c + j\tilde{u}_s = (u_c + ju_s)(\cos\theta + j\sin\theta)$$

using Euler's formula. By equating real and imaginary parts on both sides, we obtain

$$\begin{aligned} \tilde{u}_c &= u_c \cos\theta - u_s \sin\theta \\ \tilde{u}_s &= u_c \sin\theta + u_s \cos\theta \end{aligned} \tag{2.77}$$

The phase offset therefore results in the I and Q components being mixed together at the output of the downconverter. Thus, for a coherent receiver to recover the original I and Q components u_c and u_s, we must account for the (possibly time-varying) phase offset $\theta(t)$. In particular, if we have an estimate of the phase offset, then we can undo it by inverting the relationship in (2.76):

$$u(t) = \tilde{u}(t)e^{-j\theta(t)} \tag{2.78}$$

which can be written out in terms of real-valued operations as follows:

$$\begin{aligned} u_c &= \tilde{u}_c \cos\theta + \tilde{u}_s \sin\theta \\ u_s &= -\tilde{u}_c \sin\theta + \tilde{u}_s \cos\theta \end{aligned} \tag{2.79}$$

The preceding computations provide a typical example of the advantage of working in complex baseband. Relationships between passband signals can be compactly represented in complex baseband, as in (2.76) and (2.78). For signal processing using real-valued arithmetic, these complex-baseband relationships can be expanded out to obtain relationships involving real-valued quantities, as in (2.77) and (2.79). See Software Lab 2.2 for an example of such computations.

2.8.2 Frequency-domain relationships

Consider an arbitrary complex-valued baseband waveform $u(t)$ whose frequency content is contained in $[-W, W]$, and suppose that $f_c > W$. We want to show that

$$u_p(t) = \text{Re}\left(u(t)e^{j2\pi f_c t}\right) \tag{2.80}$$

is a real-valued passband signal whose frequency is concentrated around $\pm f_c$, away from DC. Let

$$c(t) = u(t)e^{j2\pi f_c t} \leftrightarrow C(f) = U(f - f_c) \tag{2.81}$$

That is, $C(f)$ is the complex envelope $U(f)$, shifted to the right by f_c. Since $U(f)$ has frequency content in $[-W, W]$, $C(f)$ has frequency content around $[f_c - W, f_c + W]$. Since $f_c - W > 0$, this band does not include DC. Now,

$$u_p(t) = \text{Re}(c(t)) = \frac{1}{2}(c(t) + c^*(t)) \leftrightarrow U_p(f) = \frac{1}{2}(C(f) + C^*(-f))$$

Since $C^*(-f)$ is the complex conjugated version of $C(f)$, flipped around the origin, it has frequency content in the band of negative frequencies $[-f_c - W, -f_c + W]$ around $-f_c$, which does not include DC because $-f_c + W < 0$. Thus, we have shown that $u_p(t)$ is a passband signal. It is real-valued by virtue of its construction using the time-domain equation (2.80), which involves taking the real part. But we can also double check for consistency in the frequency domain: $U_p(f)$ is conjugate symmetric, since its positive-frequency component is $C(f)$, and its negative-frequency component is $C^*(-f)$. On substituting $C(f)$ by $U(f - f_c)$, we obtain the passband spectrum in terms of the complex baseband spectrum:

$$U_p(f) = \frac{1}{2}(U(f - f_c) + U^*(-f - f_c)) \tag{2.82}$$

So far, we have seen how to construct a real-valued passband signal given a complex-valued baseband signal. To go in reverse, we must answer the following questions: do the equivalent representations (2.68), (2.70), (2.72), and (2.82) hold for any passband signal, and, if so, how do we find the spectrum of the complex envelope given the spectrum of the passband signal? To answer these questions, we simply trace back the steps we used to arrive at (2.82). Given the spectrum $U_p(f)$ for a real-valued passband signal $u_p(t)$, we construct $C(f)$ as a scaled version of $U_p^+(f) = U_p(f)I_{[0,\infty)}(f)$, the positive-frequency part of $U_p(f)$, as follows:

$$C(f) = 2U_p^+(f) = \begin{cases} 2U_p(f), & f > 0 \\ 0, & f < 0 \end{cases}$$

This means that $U_p(f) = \frac{1}{2}C(f)$ for positive frequencies. By virtue of the conjugate symmetry of $U_p(f)$, the negative-frequency component must be $\frac{1}{2}C^*(-f)$, so that $U_p(f) = \frac{1}{2}C(f) + \frac{1}{2}C^*(-f)$. In the time domain, this corresponds to

$$u_p(t) = \frac{1}{2}c(t) + \frac{1}{2}c^*(t) = \mathrm{Re}(c(t)) \tag{2.83}$$

Now, let us define the complex envelope as follows:

$$u(t) = c(t)e^{-j2\pi f_c t} \leftrightarrow U(f) = C(f + f_c)$$

Since $c(t) = u(t)e^{j2\pi f_c t}$, we obtain the desired relationship (2.68) on substituting into (2.83). Since $C(f)$ has frequency content in a band around f_c, $U(f)$, which is obtained by shifting $C(f)$ to the left by f_c, is indeed a baseband signal with frequency content in a band around DC.

Frequency-domain expressions for I and Q components If we are given the time-domain complex envelope, we can read off the I and Q components as the real and imaginary parts:

$$u_c(t) = \mathrm{Re}(u(t)) = (u(t) + u^*(t))/2$$
$$u_s(t) = \mathrm{Im}(u(t)) = (u(t) - u^*(t))/(2j)$$

On taking Fourier transforms, we obtain

$$U_c(f) = \tfrac{1}{2}(U(f) + U^*(-f))$$
$$U_s(f) = (1/(2j))(U(f) - U^*(-f))$$

Figure 2.30 shows the relation among the passband signal $U_p(f)$, its scaled version $C(f)$ restricted to positive frequencies, and the complex-baseband signal $U(f)$. As this example emphasizes, all of these spectra can, in general, be complex-valued. Equation (2.80) corresponds to starting with an arbitrary baseband signal $U(f)$, as in the bottom of Figure 2.30, and constructing $C(f)$ as depicted in the middle of the figure. We then use $C(f)$ to construct a conjugate-symmetric passband signal $U_p(f)$, proceeding from the middle of the figure to

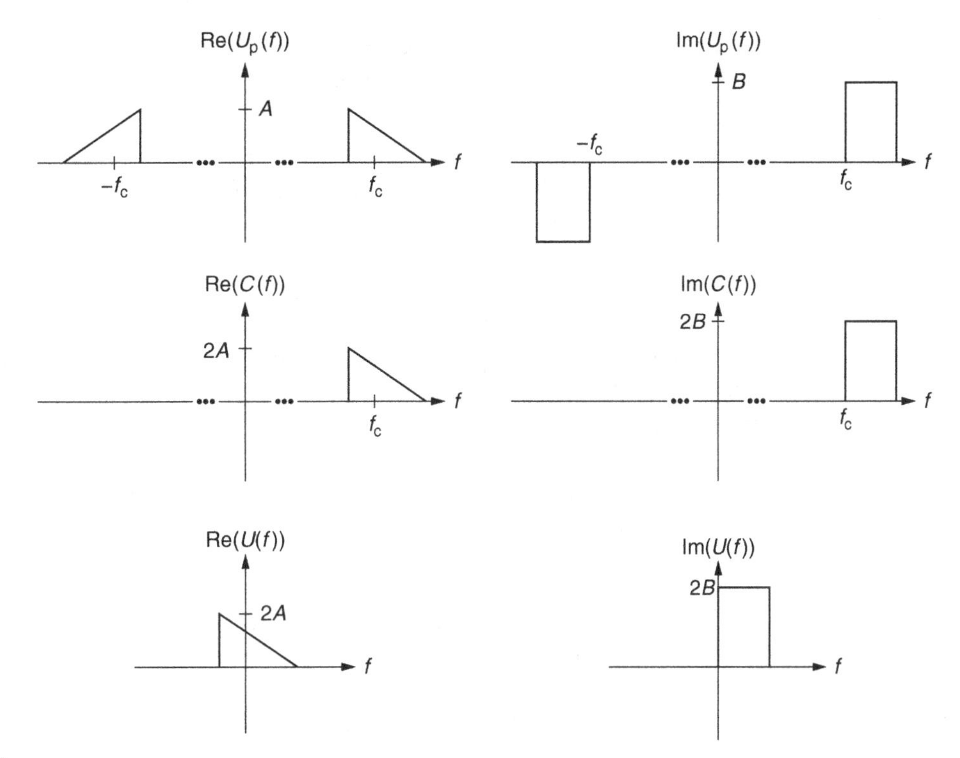

Figure 2.30 The frequency-domain relationship between a real-valued passband signal and its complex envelope. The figure shows the spectrum $U_p(f)$ of the passband signal, its scaled restriction to positive frequencies $C(f)$, and the spectrum $U(f)$ of the complex envelope.

the top. This example also shows that $U(f)$ does not, in general, obey conjugate symmetry, so that the baseband signal $u(t)$ is, in general, complex-valued. However, by construction, $U_p(f)$ is conjugate symmetric, and hence the passband signal $u_p(t)$ is real-valued.

Example 2.8.4 Let $v_p(t)$ denote a real-valued passband signal, with Fourier transform $V_p(f)$ specified as follows for negative frequencies:

$$V_p(f) = \begin{cases} -(f+99) & -101 \leq f \leq -99 \\ 0 & f < -101 \text{ or } -99 < f \leq 0 \end{cases}$$

(a) Sketch $V_p(f)$ for both positive and negative frequencies.
(b) Without explicitly taking the inverse Fourier transform, can you say whether $v_p(t) = v_p(-t)$ or not?
(c) Find and sketch $V_c(f)$ and $V_s(f)$, the Fourier transforms of the I and Q components with respect to a reference frequency $f_c = 99$. Do this without going to the time domain.
(d) Find an explicit time-domain expression for the output when $v_p(t)\cos(200\pi t)$ is passed through an ideal lowpass filter of bandwidth 4.

(e) Find an explicit time-domain expression for the output when $v_p(t)\sin(202\pi t)$ is passed through an ideal lowpass filter of bandwidth 4.

Solution

(a) Since $v_p(t)$ is real-valued, we have $V_p(f) = V_p^*(-f)$. Since the spectrum is also given to be real-valued for $f \leq 0$, we have $V_p^*(-f) = V_p(-f)$. The spectrum is sketched in Figure 2.31.

(b) Yes, $v_p(t) = v_p(-t)$. Since $v_p(t)$ is real-valued, we have $v_p(-t) = v_p^*(-t) \leftrightarrow V_p^*(f)$. But $V_p^*(f) = V_p(f)$, since the spectrum is real-valued.

(c) The spectrum of the complex envelope and the I and Q components are shown in Figure 2.32. The complex envelope is obtained as $V(f) = 2V_p^+(f+f_c)$, while the I and Q components satisfy

$$V_c(f) = \frac{V(f) + V^*(-f)}{2}, \quad V_s(f) = \frac{V(f) - V^*(-f)}{2j}$$

In our case, $V_c(f) = |f| I_{[-2,2]}(f)$ and $jV_s(f) = f I_{[-2,2]}(f)$ are real-valued, and are plotted in Figure 2.32.

(d) The output of the LPF is $v_c(t)/2$, where v_c is the I component with respect to $f_c = 100$. In Figure 2.33, we construct the complex envelope and the I component as in (c), except that the reference frequency is different. Clearly, the boxcar spectrum corresponds to $v_c(t) = 4\,\text{sinc}(2t)$, so that the output is $2\,\text{sinc}(2t)$.

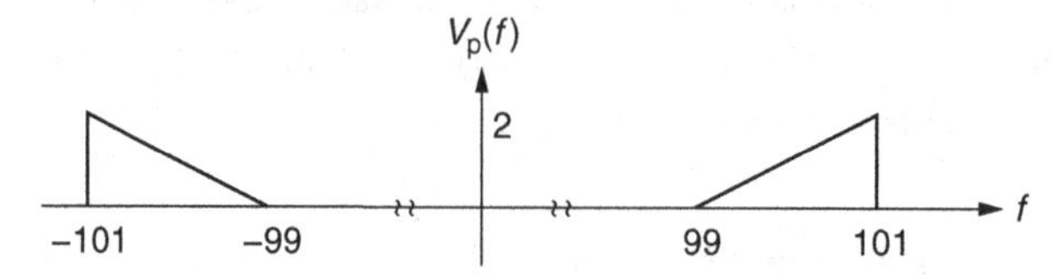

Figure 2.31 A sketch of the passband spectrum for Example 2.8.4.

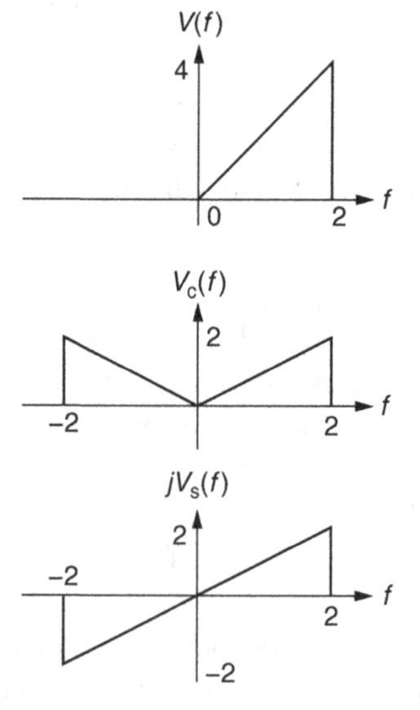

Figure 2.32 A sketch of I and Q spectra in Example 2.8.4(c), taking the reference frequency as $f_c = 99$.

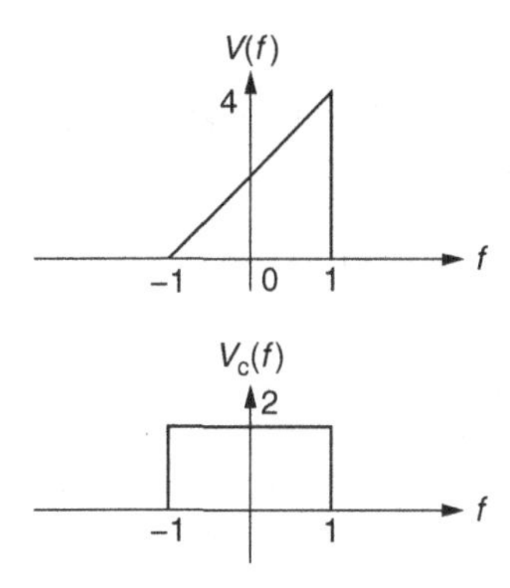

Figure 2.33 Finding the I component in Example 2.8.4(d), taking the reference frequency as $f_c = 100$.

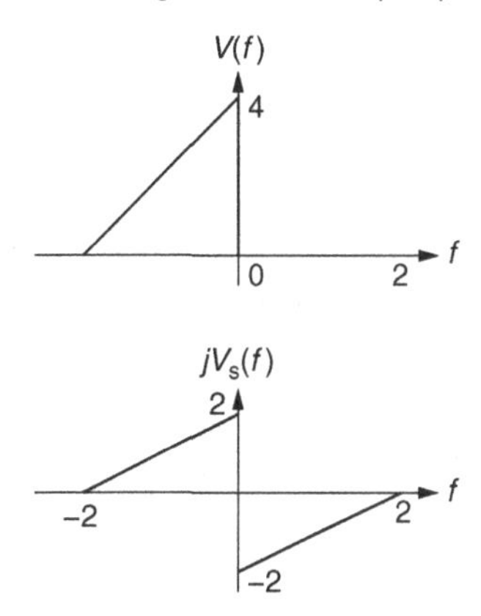

Figure 2.34 Finding the Q component in Example 2.8.4(e), taking the reference frequency as $f_c = 101$.

(e) The output of the LPF is $-v_s(t)/2$, where v_s is the I component with respect to $f_c = 101$. In Figure 2.34, we construct the complex envelope and the Q component as in (c), except that the reference frequency is different. We now have to take the inverse Fourier transform, which is a little painful if we do it from scratch. Instead, let us differentiate to see that

$$j\frac{dV_s(f)}{df} = I_{[-2,2]}(f) - 4\delta(f) \leftrightarrow 4\,\mathrm{sinc}(4t) - 4$$

But $dV_s(f)/df \leftrightarrow -j2\pi t v_s(t)$, so that $jdV_s(f)/df \leftrightarrow 2\pi t v_s(t)$. We therefore obtain that $2\pi t v_s(t) = 4\,\mathrm{sinc}(4t) - 4$, or $v_s(t) = 2(\mathrm{sinc}(4t) - 1)/(\pi t)$. Thus, the output of the LPF is $-v_s(t)/2$, or $(1 - \mathrm{sinc}(4t))/(\pi t)$.

2.8.3 The complex-baseband equivalent of passband filtering

We now state another result that is extremely relevant to transceiver operations; namely, any passband filter can be implemented in complex baseband. This result applies to filtering operations that we desire to perform at the transmitter (e.g., to conform to spectral masks), at the receiver (e.g., to filter out noise), and to a broad class of channels modeled as linear filters. Suppose that a passband signal $u_p(t) = u_c(t)\cos(2\pi f_c t) - u_s(t)\sin(2\pi f_c t)$ is passed through a passband filter with impulse response $h_p(t) = h_c(t)\cos(2\pi f_c t) - h_s(t)\sin(2\pi f_c t)$ to get an output $y_p(t) = (u_p * h_p)(t)$. In the frequency domain, $Y_p(f) = H_p(f)U_p(f)$, so that the output $y_p(t)$ is also passband, and can be written as $y_p(t) = y_c(t)\cos(2\pi f_c t) - y_s(t)\sin(2\pi f_c t)$. How are the I and Q components of the output related to those of the input and the filter impulse response? We now show that a compact answer is given in terms of complex envelopes: the complex envelope y is the convolution of the complex envelopes of the input and the impulse response, up to a scale factor. Let y, u, and h denote the complex envelopes for y_p, u_p and h_p, respectively, with respect to a common frequency reference f_c. Since real-valued passband signals are completely characterized by their spectra for positive frequencies, the passband filtering equation $Y_p(f) = U_p(f)H_p(f)$ can be separately (and redundantly) written out for positive and negative frequencies, because the waveforms are conjugate symmetric around the origin, and there is no energy around $f = 0$. Thus, focusing on the positive-frequency segments $Y^+(f) = Y_p(f)I_{\{f>0\}}$, $U^+(f) = U_p(f)I_{\{f>0\}}$, and $H^+(f) = H_p(f)I_{\{f>0\}}$, we have $Y^+(f) = U^+(f)H^+(f)$, from which we conclude that the complex envelope of y is given by

$$Y(f) = 2Y^+(f + f_c) = 2U^+(f + f_c)H^+(f + f_c) = \frac{1}{2}U(f)H(f)$$

Figure 2.35 depicts the relationship between the passband and complex-baseband waveforms in the frequency domain, and supplies a pictorial proof of the preceding relationship. We now restate this important result in the time domain:

$$y(t) = \frac{1}{2}(u * h)(t) \tag{2.84}$$

A practical consequence of this is that any desired passband filtering function can be realized in complex baseband. As shown in Figure 2.36, this requires four real baseband filters: on writing out the real and imaginary parts of (2.84), we obtain

$$y_c = \frac{1}{2}(u_c * h_c - u_s * h_s), \quad y_s = \frac{1}{2}(u_s * h_c + u_c * h_s) \tag{2.85}$$

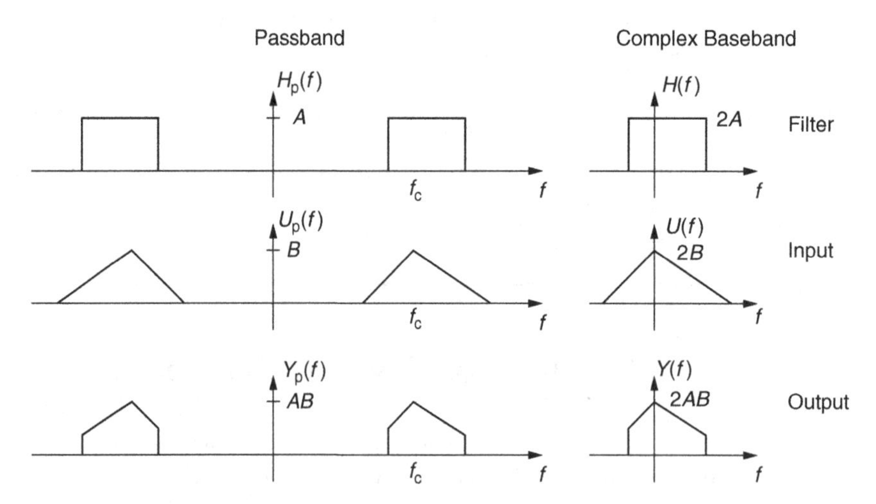

Figure 2.35 The relationship between passband filtering and its complex baseband analog.

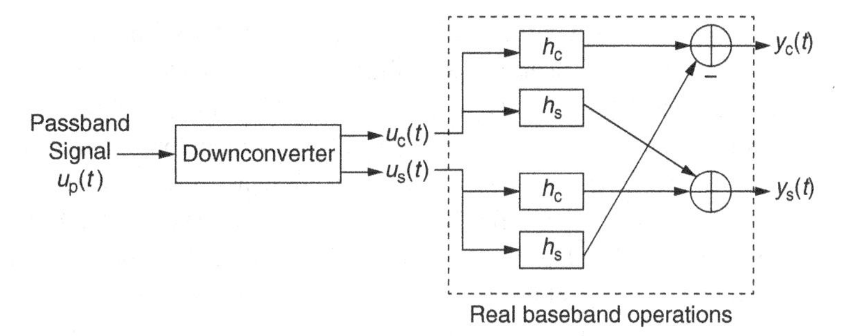

Figure 2.36 Complex-baseband realization of a passband filter. The constant scale factors of $\frac{1}{2}$ have been omitted.

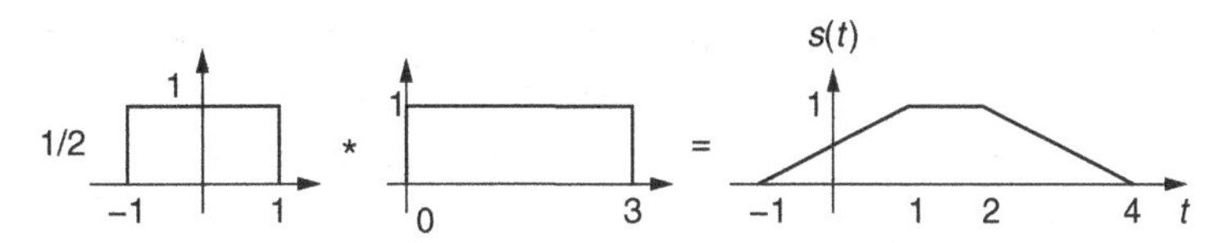

Figure 2.37 The convolution of two boxes for Example 2.8.5.

Example 2.8.5 The passband signal $u(t) = I_{[-1,1]}(t)\cos(100\pi t)$ is passed through the passband filter $h(t) = I_{[0,3]}(t)\sin(100\pi t)$. Find an explicit time domain expression for the filter output.

Solution

We need to find the convolution $y_p(t)$ of the signal $u_p(t) = I_{[-1,1]}(t)\cos(100\pi t)$ with the impulse response $h_p(t) = I_{[0,3]}(t)\sin(100\pi t)$, where we have inserted the subscript to explicitly denote that the signals are passband. The corresponding relationship in complex baseband is $y = (1/2)u * h$. Taking a reference frequency $f_c = 50$, we can read off the complex envelopes $u(t) = I_{[-1,1]}(t)$ and $h(t) = -jI_{[0,3]}(t)$, so that

$$y = (-j/2)I_{[-1,1]}(t) * I_{[0,3]}(t)$$

Let $s(t) = (1/2)I_{[-1,1]}(t) * I_{[0,3]}(t)$ denote the trapezoid obtained by convolving the two boxes, as shown in Figure 2.37. Then

$$y(t) = -js(t)$$

That is, $y_c = 0$ and $y_s = -s(t)$, so that $y_p(t) = s(t)\sin(100\pi t)$.

2.8.4 General comments on complex baseband

Remark 2.8.1 (Complex baseband in transceiver implementations) Given the equivalence of passband and complex baseband, and the fact that key operations such as linear filtering can be performed in complex baseband, it is understandable why, in typical modern passband transceivers, most of the intelligence is moved to baseband processing. For moderate bandwidths at which analog-to-digital and digital-to-analog conversion can be accomplished inexpensively, baseband operations can be efficiently performed in DSP. These digital algorithms are independent of the passband over which communication eventually occurs, and are amenable to a variety of low-cost implementations, including very-large-scale integrated circuits (VLSI), field programmable gate arrays (FPGA), and general-purpose DSP engines. On the other hand, analog components such as local oscillators, power amplifiers, and low-noise amplifiers must be optimized for the bands of interest, and are often bulky. Thus, the trend in modern transceivers is to accomplish as much as possible using baseband DSP algorithms. For example, complicated filters shaping the transmitted waveform to a spectral mask dictated by the FCC can be achieved with baseband DSP algorithms, allowing the use of relatively sloppy analog filters at passband. Another example is the elimination of analog phase-locked loops for carrier synchronization in many modern receivers; the receiver instead employs a fixed analog local oscillator for downconversion, followed by a digital phase-locked loop, or a one-shot carrier-frequency/phase estimate, implemented in complex baseband.

Energy and power The energy of a passband signal equals that of its complex envelope, up to a scale factor that depends on the particular convention we adopt. In particular, for the convention in (2.68), we have

$$||u_p||^2 = \frac{1}{2}\left(||u_c||^2 + ||u_s||^2\right) = \frac{1}{2}||u||^2 \tag{2.86}$$

That is, the energy equals the sum of the energies of the I and Q components, up to a scalar constant. The same relationship holds for the powers of finite-power passband signals and their complex envelopes, since power is computed as a time average of energy. To show (2.86), consider

$$\begin{aligned}||u_p||^2 &= \int (u_c(t)\cos(2\pi f_c t) - u_s(t)\sin(2\pi f_c t)^2)dt \\ &= \int u_c^2(t)\cos^2(2\pi f_c t)dt + \int u_s^2(t)\sin^2(2\pi f_c t)dt \\ &\quad - 2\int u_c(t)\cos(2\pi f_c t)u_s(t)\sin(2\pi f_c t)dt\end{aligned}$$

The I–Q cross term drops out due to I–Q orthogonality, so that we are left with the I–I and Q–Q terms, as follows:

$$||u_p||^2 = \int u_c^2(t)\cos^2(2\pi f_c t)dt + \int u_s^2(t)\sin^2(2\pi f_c t)dt$$

Now, $\cos^2(2\pi f_c t) = \frac{1}{2} + \frac{1}{2}\cos(4\pi f_c t)$ and $\sin^2(2\pi f_c t) = \frac{1}{2} - \frac{1}{2}\cos(4\pi f_c t)$. We therefore obtain

$$||u_p||^2 = \frac{1}{2}\int u_c^2(t)dt + \frac{1}{2}\int u_s^2(t)dt + \frac{1}{2}\int u_c^2(t)\cos(4\pi f_c t)dt - \frac{1}{2}\int u_s^2(t)\cos(4\pi f_c t)dt$$

The last two terms are zero, since they are equal to the DC components of passband waveforms centered around $2f_c$, arguing in exactly the same fashion as in our derivation of I–Q orthogonality. This gives the desired result (2.86).

Correlation between two signals The correlation, or inner product, of two real-valued passband signals u_p and v_p is defined as

$$\langle u_p, v_p \rangle = \int_{-\infty}^{\infty} u_p(t) v_p(t) dt$$

Using exactly the same reasoning as above, we can show that

$$\langle u_p, v_p \rangle = \frac{1}{2} (\langle u_c, v_c \rangle + \langle u_s, v_s \rangle) \tag{2.87}$$

That is, we can implement a passband correlation by first downconverting, and then employing baseband operations: correlating I against I, and Q against Q, and then summing the results. It is also worth noting how this is related to the complex-baseband inner product, which is defined as

$$\begin{aligned} \langle u, v \rangle &= \int_{-\infty}^{\infty} u(t) v^*(t) dt = \int_{-\infty}^{\infty} (u_c(t) + j u_s(t)) (v_c(t) - j v_s(t)) \\ &= (\langle u_c, v_c \rangle + \langle u_s, v_s \rangle) + j(\langle u_s, v_c \rangle - \langle u_c, v_s \rangle) \end{aligned} \tag{2.88}$$

On comparing with (2.87), we obtain that

$$\langle u_p, v_p \rangle = \frac{1}{2} \operatorname{Re}(\langle u, v \rangle)$$

That is, the passband inner product is the real part of the complex-baseband inner product (up to a scale factor). Does the imaginary part of the complex-baseband inner product have any meaning? Indeed it does: it becomes important when there is phase uncertainty in the downconversion operation, which causes the I and Q components to leak into each other. However, we postpone discussion of such issues to later chapters.

2.9 Wireless-channel modeling in complex baseband

We now provide a glimpse of wireless-channel modeling using complex baseband. There are two key differences between wireless and wireline communication. The first, which is what we focus on now, is multipath propagation due to reflections off of scatterers adding up at the receiver. This addition can be constructive or destructive (as we saw in Example 2.5.6), and is sensitive to small changes in the relative location of the transmitter and receiver that produce changes in the relative delays of the various paths. The resulting fluctuations in signal strength are termed *fading*. The second key feature of wireless, which we explore in a different wireless module, is interference: wireless is a broadcast medium, hence the receiver can also hear transmissions other than the one it is interested in. We now explore the effects of multipath fading for some simple scenarios. While we just made up the example impulse response in Example 2.5.6, we now consider more detailed, but still simplified, models of the propagation environment and the associated channel models.

Consider a passband transmitted signal at carrier frequency, of the form

$$u_p(t) = u_c(t)\cos(2\pi f_c t) - u_s(t)\sin(2\pi f_c t) = e(t)\cos(2\pi f_c t + \theta(t))$$

where

$$u(t) = u_c(t) + ju_s(t) = e(t)e^{j\theta(t)}$$

is the complex-baseband representation, or complex envelope. In order to model the propagation of this signal through a multipath environment, let us consider its propagation through a path of length r. The propagation attenuates the field by a factor of $1/r$, and introduces a delay of $\tau(r) = r/c$, where c denotes the speed of light. Suppressing the dependence of τ on r, the received signal is given by

$$v_p(t) = \frac{A}{r}e(t-\tau)\cos(2\pi f_c(t-\tau) + \theta(t-\tau) + \phi)$$

where we consider relative values (across paths) for the constants A and ϕ. The complex envelope of $v_p(t)$ with respect to the reference $e^{j2\pi f_c t}$ is given by

$$v(t) = \frac{A}{r}u(t-\tau)e^{-j(2\pi f_c \tau + \phi)} \tag{2.89}$$

For example, we may take $A = 1$ and $\phi = 0$ for a direct, or line-of-sight (LOS), path from transmitter to receiver, which we may take as a reference. Figure 2.38 shows the geometry for a reflected path corresponding to a single bounce, relative to the LOS path. Following standard terminology, θ_i denotes the angle of incidence, and $\theta_g = \pi/2 - \theta_i$ is the *grazing angle*. The change in relative amplitude and phase due to the reflection depends on the carrier frequency, the reflector material, the angle of incidence, and the polarization with respect to the orientation of the reflector surface. Since we do not wish to get

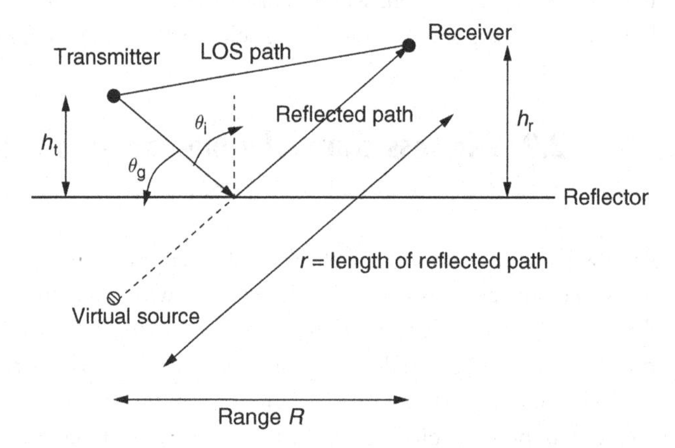

Figure 2.38 Ray tracing for a single-bounce path. We can reflect the transmitter around the reflector to create a virtual source. The line between the virtual source and the receiver tells us where the ray will hit the reflector, following the law of reflection that the angles of incidence and reflection must be equal. The length of the line equals the length of the reflected ray to be plugged into (2.92).

into the underlying electromagnetics, we consider simplified models of relative amplitude and phase. In particular, we note that for grazing incidence ($\theta_g \approx 0$) we have $A \approx 1$ and $\phi \approx \pi$.

On generalizing (2.89) to multiple paths of length $r_1, r_2, \ldots$, the complex envelope of the received signal is given by

$$v(t) = \sum_i \frac{A_i}{r_i} u(t - \tau_i) e^{-j(2\pi f_c \tau_i + \phi_i)} \tag{2.90}$$

where $\tau_i = r_i/c$, and A_i and ϕ_i depend on the reflector characteristic and incidence angle for the ith ray. This corresponds to the complex-baseband channel impulse response

$$h(t) = \sum_i \frac{A_i}{r_i} e^{-j(2\pi f_c \tau_i + \phi_i)} \delta(t - \tau_i) \tag{2.91}$$

This is in exact correspondence to our original multipath model (2.36), with $\alpha_i = (A_i/r_i) e^{-j(2\pi f_c \tau_i + \phi_i)}$. The corresponding frequency-domain response is given by

$$H(f) = \sum_i \frac{A_i}{r_i} e^{-j(2\pi f_c \tau_i + \phi_i)} e^{-j2\pi f \tau_i} \tag{2.92}$$

Since we are modeling in complex baseband, f takes values around DC, with $f = 0$ corresponding to the passband reference frequency f_c.

Channel delay spread and coherence bandwidth We have already introduced these concepts in Example 2.5.6, but reiterate them here. Let $\tau_{\min}$ and $\tau_{\max}$ denote the minimum and maximum of the delays $\{\tau_i\}$. The difference $\tau_d = \tau_{\max} - \tau_{\min}$ is called the *channel delay spread.* The reciprocal of the delay spread is termed the *channel coherence bandwidth,* $B_c = 1/\tau_d$. A baseband signal of bandwidth W is said to be *narrowband* if $W\tau_d = W/B_c \ll 1$, or, equivalently, if its bandwidth is significantly smaller than the channel coherence bandwidth.

We can now infer that, for a narrowband signal around the reference frequency, the received complex-baseband signal equals a delayed version of the transmitted signal, scaled by the complex channel gain

$$h = H(0) = \sum_i \frac{A_i}{r_i} e^{-j(2\pi f_c \tau_i + \phi_i)} \tag{2.93}$$

Example 2.9.1 (Two-ray model) Suppose our propagation environment consists of the LOS ray and the single reflected ray shown in Figure 2.38. Then we have two rays, with $r_1 = \sqrt{R^2 + (h_r - h_t)^2}$ and $r_2 = \sqrt{R^2 + (h_r + h_t)^2}$. The corresponding delays are $\tau_i = r_i/c$, $i = 1, 2$, where c denotes the speed of propagation. The grazing angle is given by $\theta_g = \tan^{-1}((h_t + h_r)/R)$. Setting $A_1 = 1$ and $\phi_1 = 0$, once we specify A_2 and ϕ_2 for the reflected path, we can specify the complex-baseband channel. Numerical examples are explored in Problem 2.21 and in Software Lab 2.2.

2.10 Concept summary

In addition to a review of basic signals and systems concepts such as convolution and Fourier transforms, the main focus of this chapter is to develop the complex-baseband representation of passband signals, and to emphasize its crucial role in modeling and implementation of communication systems.

Review

- Euler's formula: $e^{j\theta} = \cos\theta + j\sin\theta$
- Important signals: delta function (sifting property), indicator function, complex exponential, sinusoid, sinc
- Signals analogous to vectors: inner product, energy, and norm
- LTI systems: impulse response, convolution, complex exponentials as eigenfunctions, multipath channel modeling
- Fourier series: complex exponentials or sinusoids as basis for periodic signals, conjugate symmetry for real-valued signals, Parseval's identity, use of differentiation to simplify computation
- Fourier transform: standard pairs (sinc and boxcar, impulse and constant), effect of time delay and frequency shift, conjugate symmetry for real-valued signals, Parseval's identity, use of differentiation to simplify computation, numerical computation using DFT
- Bandwidth: for physical signals, given by occupancy of positive frequencies; energy spectral density equals magnitude squared of Fourier transform; computation of fractional energy-containment bandwidth from energy spectral density

Complex-baseband representation

- Complex envelope of passband signal: rectangular form (I and Q components), polar form (envelope and phase), upconversion and downconversion, orthogonality of I and Q components (under ideal synchronization), frequency-domain relationship between passband signal and its complex envelope
- Passband filtering can be accomplished in complex baseband
- Passband inner product and energy in terms of complex-baseband quantities

Modeling in complex baseband

- Frequency and phase offsets: rotating phasor multiplying complex envelope, derotation to undo offsets
- Wireless multipath channel: impulse response modeled as sum of impulses with complex-valued coefficients, ray tracing, delay spread, and coherence bandwidth

2.11 Notes

A detailed treatment of the material reviewed in Sections 2.1–2.5 can be found in basic textbooks on signals and systems such as Oppenheim, Willsky, and Nawab [17] and Lathi [18].

The MATLAB code fragments and software labs presented at various points in this textbook provide a glimpse of the use of DSP in communication. However, for a background in core DSP algorithms, we refer the reader to textbooks such as Oppenheim and Schafer [19] and Mitra [20].

2.12 Problems

LTI systems and convolution

Problem 2.1 A system with input $x(t)$ has output given by

$$y(t) = \int_{-\infty}^{t} e^{u-t} x(u) du$$

(a) Show that the system is LTI and find its impulse response.
(b) Find the transfer function $H(f)$ and plot $|H(f)|$.
(c) If the input $x(t) = 2\,\mathrm{sinc}(2t)$, find the energy of the output.

Problem 2.2 Find and sketch $y = x_1 * x_2$ for the following.

(a) $x_1(t) = e^{-t} I_{[0,\infty)}(t)$, $x_2(t) = x_1(-t)$.
(b) $x_1(t) = I_{[0,2]}(t) - 3I_{[1,4]}(t)$, $x_2(t) = I_{[0,1]}(t)$.

Hint. In (b), you can use the LTI property and the known result in Figure 2.12 on the convolution of two boxes.

Fourier series

Problem 2.3 A digital circuit generates the following periodic waveform with period 0.5:

$$u(t) = \begin{cases} 1, & 0 \le t < 0.1 \\ 0, & 1 \le t < 0.5 \end{cases}$$

where **the unit of time is microseconds** throughout this problem.

(a) Find the complex exponential Fourier series for du/dt.
(b) Find the complex exponential Fourier series for $u(t)$, using the results of (a).
(c) Find an explicit time-domain expression for the output when $u(t)$ is passed through an ideal lowpass filter of bandwidth 100 kHz.
(d) Repeat (c) when the filter bandwidth is increased to 300 kHz.

(e) Find an explicit time-domain expression for the output when $u(t)$ is passed through a filter with impulse response $h_2(t) = \text{sinc}(t)\cos(8\pi t)$.

(f) Can you generate a sinusoidal waveform of frequency 1 MHz by appropriately filtering $u(t)$? If so, specify in detail how you would do it.

Fourier transform and bandwidth

Problem 2.4 Find and sketch the Fourier transforms for the following signals.

(a) $u(t) = (1 - |t|)I_{[-1,1]}(t)$.
(b) $v(t) = \text{sinc}(2t)\text{sinc}(4t)$.
(c) $s(t) = v(t)\cos(200\pi t)$.
(d) Classify each of the signals in (a)–(c) as baseband or passband.

Problem 2.5 Use Parseval's identity to compute the following integrals.

(a) $\int_{-\infty}^{\infty} \text{sinc}^2(2t)dt$.
(b) $\int_0^{\infty} \text{sinc}(t)\text{sinc}(2t)dt$.

Problem 2.6

(a) For $u(t) = \text{sinc}(t)\text{sinc}(2t)$, where t is in microseconds, find and plot the magnitude spectrum $|U(f)|$, carefully labeling the units of frequency on the x axis.

(b) Now, consider $s(t) = u(t)\cos(200\pi t)$. Plot the magnitude spectrum $|S(f)|$, again labeling the units of frequency and carefully showing the frequency intervals over which the spectrum is nonzero.

Problem 2.7 The signal $s(t) = \text{sinc}(4t)$ is passed through a filter with impulse response $h(t) = \text{sinc}^2 t \cos(4\pi t)$ to obtain output $y(t)$. Find and sketch the Fourier transform $Y(f)$ of the output (sketch the real and imaginary parts separately if the spectrum is complex-valued).

Problem 2.8 Consider the tent signal $s(t) = (1 - |t|)I_{[-1,1]}(t)$.

(a) Find and sketch the Fourier transform $S(f)$.

(b) Compute the 99% energy-containment bandwidth in kHz, assuming that the unit of time is milliseconds.

Problem 2.9 Consider the cosine pulse

$$p(t) = \cos(\pi t)I_{[-1/2,1/2]}(t)$$

(a) Show that the Fourier transform of this pulse is given by

$$P(f) = \frac{2\cos(\pi f)}{\pi(1 - 4f^2)}$$

(b) Use this result to derive the formula (2.63) for the sine pulse in Example 2.5.7.

Problem 2.10 (Numerical computation of the Fourier transform) Modify Code Fragment 2.5.1 for Example 2.5.7 to numerically compute the Fourier transform of the tent function in Problem 2.8. Display the magnitude spectra of the DFT-based numerically computed Fourier transform and the analytically computed Fourier transform (from Problem 2.8) in the same plot, over the frequency interval $[-10, 10]$. Comment on the accuracy of the DFT-based computation.

Introducing the matched filter

Problem 2.11 For a signal $s(t)$, the *matched filter* is defined as a filter with impulse response $h(t) = s_{\mathrm{MF}}(t) = s^*(-t)$ (we allow signals to be complex-valued, since we want to handle complex-baseband signals as well as physical real-valued signals).

(a) Sketch the matched-filter impulse response for $s(t) = I_{[1,3]}(t)$.
(b) Find and sketch the convolution $y(t) = (s * s_{\mathrm{MF}})(t)$. This is the output when the signal is passed through its matched filter. Where does the peak of the output occur?
(c) Is is true that $Y(f) \geq 0$ for all f?

Problem 2.12 Repeat Problem 2.11 for $s(t) = I_{[1,3]}(t) - 2I_{[2,5]}(t)$.

Introducing delay spread and coherence bandwidth

Problem 2.13 A wireless channel has impulse response given by $h(t) = 2\delta(t-0.1)+j\delta(t-0.64) - 0.8\delta(t-2.2)$, where the unit of time is in microseconds.

(a) What is the delay spread and coherence bandwidth?
(b) Plot the magnitude and phase of the channel transfer function $H(f)$ over the interval $[-2B_{\mathrm{c}}, 2B_{\mathrm{c}}]$, where B_{c} denotes the coherence bandwidth computed in (a). Comment on how the phase behaves when $|H(f)|$ is small.
(c) Express $|H(f)|$ in dB, taking 0 dB as the gain of a nominal channel $h_{\mathrm{nom}}(t) = 2\delta(t-0.1)$ corresponding to the first ray alone. What are the fading depths that you see with respect to this nominal?

Define the average channel power gain over a band $[-W/2, W/2]$ as

$$\bar{G}(W) = \frac{1}{W}\int_{-W/2}^{W/2} |H(f)|^2 \, df$$

This is a simplified measure of how increasing the signal bandwidth W can help compensate for frequency-selective fading: we hope that, as W gets large, we can average out fluctuations in $|H(f)|$.

(d) Plot $\bar{G}(W)$ as a function of W/B_{c}, and comment on how large the bandwidth needs to be (as a multiple of B_{c}) to provide "enough averaging."

Complex envelope of passband signals

Problem 2.14 Consider a passband signal of the form

$$u_p(t) = a(t)\cos(200\pi t)$$

where $a(t) = \text{sinc}(2t)$, and where the unit of time is in microseconds.

(a) What is the frequency band occupied by $u_p(t)$?
(b) The signal $u_p(t)\cos(199\pi t)$ is passed through a lowpass filter to obtain an output $b(t)$. Give an explicit expression for $b(t)$, and sketch $B(f)$ (if $B(f)$ is complex-valued, sketch its real and imaginary parts separately).
(c) The signal $u_p(t)\sin(199\pi t)$ is passed through a lowpass filter to obtain an output $c(t)$. Give an explicit expression for $c(t)$, and sketch $C(f)$ (if $C(f)$ is complex-valued, sketch its real and imaginary parts separately).
(d) Can you reconstruct $a(t)$ from simple real-valued operations performed on $b(t)$ and $c(t)$? If so, sketch a block diagram for the operations required. If not, say why not.

Problem 2.15 Consider the signal $s(t) = I_{[-1,1]}(t)\cos(400\pi t)$.

(a) Find and sketch the baseband signal $u(t)$ that results when $s(t)$ is downconverted as shown in the upper branch of Figure 2.39.
(b) The signal $s(t)$ is passed through the bandpass filter with impulse response $h(t) = I_{[0,1]}(t)\sin(400\pi t + \pi/4)$. Find and sketch the baseband signal $v(t)$ that results when the filter output $y(t) = (s * h)(t)$ is downconverted as shown in the lower branch of Figure 2.39.

Problem 2.16 Consider the signals given by $u_1(t) = I_{[0,1]}(t)\cos(100\pi t)$ and $u_2(t) = I_{[0,1]}(t)\sin(100\pi t)$.

(a) Find the numerical value of the inner product $\int_{-\infty}^{\infty} u_1(t)u_2(t)dt$.
(b) Find an explicit time-domain expression for the convolution $y(t) = (u_1 * u_2)(t)$.
(c) Sketch the magnitude spectrum $|Y(f)|$ for the convolution in (b).

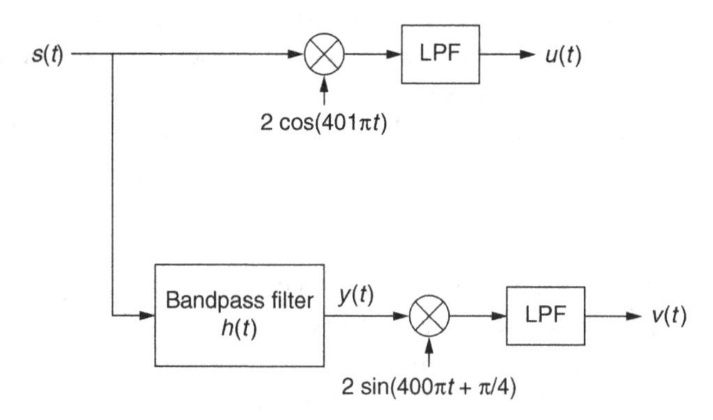

Figure 2.39 Operations involved in Problem 2.15.

Problem 2.17 Consider a real-valued passband signal $v_p(t)$ whose Fourier transform for positive frequencies is given by

$$\mathrm{Re}(V_p(f)) = \begin{cases} 2, & 30 \leq f \leq 32 \\ 0, & 0 \leq f < 30 \\ 0, & 32 < f < \infty \end{cases}$$

$$\mathrm{Im}(V_p(f)) = \begin{cases} 1 - |f - 32|, & 31 \leq f \leq 33 \\ 0, & 0 \leq f < 31 \\ 0, & 33 < f < \infty \end{cases}$$

(a) Sketch the real and imaginary parts of $V_p(f)$ for both positive and negative frequencies.

(b) Specify, both in the time domain and in the frequency domain, the waveform you get when you pass $v_p(t)\cos(60\pi t)$ through a lowpass filter.

Problem 2.18 The passband signal $u(t) = I_{[-1,1]}(t)\cos(100\pi t)$ is passed through the passband filter $h(t) = I_{[0,3]}(t)\sin(100\pi t)$. Find an explicit time-domain expression for the filter output.

Problem 2.19 Consider the passband signal $u_p(t) = \mathrm{sinc}(t)\cos(20\pi t)$, where the unit of time is in microseconds.

(a) Use MATLAB to plot the signal (plot over a large enough time interval so as to include "most" of the signal energy). Label the units on the time axis.

Remark Since you will be plotting a discretized version, the sampling rate you should choose should be large enough that the carrier waveform looks reasonably smooth (e.g., a rate of at least 10 times the carrier frequency).

(b) Write a MATLAB program to implement a simple downconverter as follows. Pass $x(t) = 2u_p(t)\cos(20\pi t)$ through a lowpass filter that consists of computing a sliding window average over a window of 1 microsecond. That is, the LPF output is given by $y(t) = \int_{t-1}^{t} x(\tau)d\tau$. Plot the output and comment on whether it is what you expect to see.

Problem 2.20 Consider the following two passband signals:

$$u_p(t) = \mathrm{sinc}(2t)\cos(100\pi t)$$

and

$$v_p(t) = \mathrm{sinc}(t)\sin(101\pi t + \pi/4)$$

(a) Find the complex envelopes $u(t)$ and $v(t)$ for u_p and v_p, respectively, with respect to the frequency reference $f_c = 50$.

(b) What is the bandwidth of $u_p(t)$? What is the bandwidth of $v_p(t)$?

(c) Find the inner product $\langle u_p, v_p \rangle$, using the result in (a).

(d) Find the convolution $y_p(t) = (u_p * v_p)(t)$, using the result in (a).

Wireless-channel modeling

Problem 2.21 Consider the two-ray wireless channel model in Example 2.9.1.

(a) Show that, as long as the range $R \gg h_t, h_r$, the delay spread is well approximated as

$$\tau_d \approx \frac{2h_t h_r}{Rc}$$

where c denotes the propagation speed. We assume free-space propagation with $c = 3 \times 10^8$ m/s.

(b) Compare the approximation in (a) with the actual value of the delay spread for $R = 200$ m, $h_t = 2$ m, and $h_r = 10$ m (e.g., modeling an outdoor link with LOS and a single ground bounce).

(c) What is the coherence bandwidth for the numerical example in (b)?

(d) Redo (b) and (c) for $R = 10$ m and $h_t = h_r = 2$ m (e.g., a model for an indoor link modeling LOS plus a single wall bounce).

Problem 2.22 Consider $R = 200$ m, $h_t = 2$ m, and $h_r = 10$ m in the two-ray wireless channel model in Example 2.9.1. Assume $A_1 = 1$ and $\phi_1 = 0$, set $A_2 = 0.95$ and $\phi_2 = \pi$, and assume that the carrier frequency is 5 GHz.

(a) Specify the channel impulse response, normalizing the LOS path with respect to unit gain and zero delay. Make sure you specify the unit of time being used.

(b) Plot the magnitude and phase of the channel transfer function over $[-3B_c, 3B_c]$, where B_c denotes the channel coherence bandwidth.

(c) Plot the frequency-selective fading gain in dB over $[-3B_c, 3B_c]$, using a LOS channel as nominal. Comment on the fading depth.

(d) As in Problem 2.13, compute the frequency-averaged power gain $\bar{G}(W)$ and plot it as a function of W/B_c.

How much bandwidth is needed to average out the effects of frequency-selective fading?

Software Lab 2.1: signals and systems computations using MATLAB

Lab objectives The goal of this lab is to gain familiarity with computations and plots with MATLAB, and to reinforce key concepts in signals and systems. The questions are chosen to illustrate how we can emulate continuous time operations using the discrete time framework provided by MATLAB.

Reading Sections 2.2, 2.3, 2.5 (basic material on signals and systems).

Laboratory assignment

Functions and plots

(1.1) (a) Write a MATLAB function *signalx* that evaluates the following signal at an arbitrary set of points:

$$x(t) = \begin{cases} 2e^{t+2}, & -3 \le t \le -1 \\ 2e^{-t}\cos(2\pi t), & -1 \le t \le 4 \\ 0, & \text{else} \end{cases}$$

That is, given an input vector of time points, the function should give an output vector with the values of x evaluated at those time points. For time points falling outside $[-3, 4]$, the function should return the value zero.

(b) Use the function *signalx* to plot $x(t)$ versus t, for $-6 \le t \le 6$. To do this, create a vector of sampling times spaced closely enough to get a smooth plot. Generate a corresponding vector using *signalx*. Then plot one against the other.

(c) Use the function *signalx* to plot $x(t-3)$ versus t.

(d) Use the function *signalx* to plot $x(3-t)$ versus t.

(e) Use the function *signalx* to plot $x(2t)$ versus t.

Convolution

(1.2) (a) Write a MATLAB function *contconv* that computes an approximation to continuous-time convolution as follows.

Inputs Vectors $\mathbf{x}_1$ and $\mathbf{x}_2$ representing samples of two signals to be convolved. Scalars t_1, t_2, and dt, representing the starting time for the samples of $\mathbf{x}_1$, the starting time for the samples in $\mathbf{x}_2$, and the spacing of the samples.

Outputs Vectors $\mathbf{y}$ and $\mathbf{t}$, corresponding to the samples of the convolution output and the sampling times.

(b) Check that your function works by using it to convolve two boxes, $3I_{[-2,-1]}$ and $4I_{[1,3]}$, to get a trapezoid (e.g., using the following code fragment):

```
dt=0.01; %sample spacing
s1 = -2:dt:-1; %sampling times over the interval [-2,-1]
s2 = 1:dt:3; %sampling times over the interval [1,3]
x1 = 3 * ones(length(s1),1); %samples for first box
x2 = 4 * ones(length(s2),1); %samples for second box
[y,t] = contconv(x1,x2,s1(1),s2(1),dt);
figure(1);
plot(t,y);
```

Check that the trapezoid you get spans the correct interval (based on the analytical answer) and has the correct scaling.

Matched filter

(1.3) (a) Consider the signal $u(t) = 2I_{[1,3]}(t) - 3I_{[2,4]}(t)$. Plot $u(t)$ and its matched filter $u_{MF}(t) = u(-t)$ on the same plot.

(b) Use the function *contconv* to convolve $u(t)$ and $u_{MF}(t)$. Plot the result of the convolution. Where is the peak of the signal?

(c) Now, consider a complex-valued signal $s(t) = u(t) + jv(t)$, where $v(t) = I_{[-1,2]}(t) + 2I_{[0,1]}(t)$. The matched filter is given by $s_{MF}(t) = s^*(-t)$. Plot the real parts of $s(t)$ and $s_{MF}(t)$ on one plot, and the imaginary parts on another.

(d) Use the function *contconv* to convolve $s(t)$ and $s_{MF}(t)$. Plot the real part, the imaginary part, and the magnitude of the output. Do you see a peak?

(e) Now, use the function *contconv* to convolve $s_1(t) = s(t - t_0)e^{j\theta}$ and $s_{MF}(t)$, for $t_0 = 2$ and $\theta = \pi/4$. Plot the real part, the imaginary part, and the magnitude of the output. Do you see a peak?

(f) If you did not know t_0 and θ, could you estimate them from the output of the convolution in (e)? Try out some ideas and report on the results.

Fourier transform

The following MATLAB function is a modification of Code Fragment 2.5.1.

```
function [X,f,df] = contFT(x,tstart,dt,df_desired)
%Use Matlab DFT for approximate computation of continuous time Fourier
%transform
%INPUTS
%x = vector of time-domain samples, assumed uniformly spaced
%tstart = time at which first sample is taken
%dt = spacing between samples
%df_desired = desired frequency resolution
%OUTPUTS
%X = vector of samples of Fourier transform
%f = corresponding vector of frequencies at which samples are obtained
%df = frequency resolution attained (redundant -- already available from
%difference of consecutive entries of f)
%%%%%%%%%
%minimum FFT size determined by desired frequency resolution or length of x
Nmin = max(ceil(1/(df_desired * dt)),length(x));
%choose FFT size to be the next power of 2
Nfft = 2^(nextpow2(Nmin))
%compute Fourier transform, centering around DC
X = dt * fftshift(fft(x,Nfft));
%achieved frequency resolution
df = 1/(Nfft * dt)
%range of frequencies covered
f = ((0:Nfft - 1) - Nfft/2) * df; %same as f = -1/(2 * dt):df:1/(2 * dt) - df
%phase shift associated with start time
X = X.* exp(-j * 2 * pi * f * tstart);
end
```

(1.4) (a) Use the function *contFT* to compute the Fourier transform of $s(t) = 3\,\text{sinc}(2t - 3)$, where the unit of time is a microsecond, and the signal is sampled at the rate of 16 MHz, and truncated to the range $[-8, 8]$ microseconds. We wish to attain a frequency resolution of 1 kHz or better. Plot the magnitude of the Fourier transform versus frequency, making sure you specify the units on the frequency axis. Check that the plot conforms to your expectations.

(b) Plot the phase of the Fourier transform obtained in (a) versus frequency (again, make sure that the units on the frequency axis are specified). What is the range of frequencies over which the phase plot has meaning?

Matched filter in the frequency domain

(1.5) (a) Consider the signal $s(t)$ in 3(c). Assuming that the unit of time is a millisecond and the desired frequency resolution is 1 Hz, use the function *contFT* to compute and plot $|S(f)|$.

(b) Use the function *contFT* to compute and plot the magnitude of the Fourier transform of the convolution $s * s_{\mathrm{MF}}$ numerically computed in 3(d). Also plot for comparison $|S(f)|^2$, using the output of 5(a). The two plots should match.

(c) Plot the phase of the Fourier transform of $s * s_{\mathrm{MF}}$ obtained in 5(b). Comment on whether the plot matches your expectations.

Lab report

- Discuss the results you obtain, answer any specific questions that are asked, and print out the most useful plots to support your answers.
- Append your programs to the report. Make sure you include comments within them in enough detail, so that they are easy to understand. In addition to the functions you are asked to write, label the code fragments used for each assigned segment (1 through 5) separately.
- Write a paragraph about any questions or confusions that you may have experienced with this lab.

Software Lab 2.2: modeling carrier-phase uncertainty

Lab objectives The goal of this lab is to explore modeling and receiver operations in complex baseband, In particular, we model and undo the effect of carrier-phase mismatch between the receiver LO and the incoming carrier.

Reading Section 2.8 (complex-baseband basics).

Laboratory assignment

Consider a pair of independently modulated signals, $u_{\mathrm{c}}(t) = \sum_{n=1}^{N} b_{\mathrm{c}}[n]p(t-n)$ and $u_{\mathrm{s}}(t) = \sum_{n=1}^{N} b_{\mathrm{s}}[n]p(t-n)$, where the symbols $b_{\mathrm{c}}[n]$ and $b_{\mathrm{s}}[n]$ are chosen with equal probability to be +1 and -1, and $p(t) = I_{[0,1]}(t)$ is a rectangular pulse. Let $N = 100$.

(2.1) Use MATLAB to plot a typical realization of $u_{\mathrm{c}}(t)$ and $u_{\mathrm{s}}(t)$ over 10 symbols. Make sure you sample fast enough for the plot to look reasonably "nice."

(2.2) Upconvert the baseband waveform $u_c(t)$ to get

$$u_{p,1}(t) = u_c(t)\cos(40\pi t)$$

This is a so-called binary phase-shift keyed (BPSK) signal, since the changes in phase are due to the changes in the signs of the transmitted symbols. Plot the passband signal $u_{p,1}(t)$ over four symbols (you will need to sample at a multiple of the carrier frequency for the plot to look nice, which means you might have to go back and increase the sampling rate beyond what was required for the baseband plots to look nice).

(1.3) Now, add in the Q component to obtain the passband signal

$$u_p(t) = u_c(t)\cos(40\pi t) - u_s(t)\sin(40\pi t)$$

Plot the resulting quaternary phase-shift keyed (QPSK) signal $u_p(t)$ over four symbols.

(1.4) Downconvert $u_p(t)$ by passing $2u_p(t)\cos(40\pi t + \theta)$ and $2u_p(t)\sin(40\pi t + \theta)$ through crude lowpass filters with impulse response $h(t) = I_{[0,0.25]}(t)$. Denote the resulting I and Q components by $v_c(t)$ and $v_s(t)$, respectively. Plot v_c and v_s for $\theta = 0$ over 10 symbols. How do they compare with u_c and u_s? Can you read off the corresponding bits $b_c[n]$ and $b_s[n]$ by eyeballing the plots for v_c and v_s?

(1.5) Plot v_c and v_s for $\theta = \pi/4$. How do they compare with u_c and u_s? Can you read off the corresponding bits $b_c[n]$ and $b_s[n]$ by eyeballing the plots for v_c and v_s?

(1.6) Figure out how to recover u_c and u_s from v_c and v_s if a genie tells you the value of θ (we are looking for an approximate reconstruction – the LPFs used in downconversion are non-ideal, and the original waveforms are not exactly bandlimited). Check whether your method for undoing the phase offset works for $\theta = \pi/4$, the scenario in (1.5). Plot the resulting reconstructions $\tilde{u}_c$ and $\tilde{u}_s$, and compare them with the original I and Q components. Can you read off the corresponding bits $b_c[n]$ and $b_s[n]$ by eyeballing the plots for $\tilde{u}_c$ and $\tilde{u}_s$?

Lab report

- Answer all questions and print out the most useful plots to support your answers.
- Write a paragraph about any questions or confusions that you may have experienced with this lab.

Software Lab 2.3: modeling a lamppost-based broadband network

Lab objectives The goal of this lab is to illustrate how wireless multipath channels can be modeled in complex baseband.

Reading The background for this lab is provided in Section 2.9, which discusses wireless-channel modeling. This material should be reviewed prior to doing the lab.

Laboratory assignment

Consider a lamppost-based network supplying broadband access using unlicensed spectrum at 5 GHz. Figure 2.40 shows two kinds of links: lamppost-to-lamppost for backhaul, and lamppost-to-mobile for access, where we show nominal values of antenna heights and distances. We explore simple channel models for each case, consisting only of the direct path and the ground reflection. For simplicity, assume throughout that $A_1 = 1$, $\phi_1 = 0$ for the direct path, and $A_2 = 0.98$, $\phi_2 = \pi$ for the ground reflection (we assume a phase shift of π for the reflected ray even though it might not be at grazing incidence, especially for the lamppost-to-mobile link).

(2.1) Find the delay spread and coherence bandwidth for the lamppost-to-lamppost link. If the message signal has bandwidth 20 MHz, is it "narrowband" with respect to this channel?

(2.2) Repeat item (2.1) for the lamppost-to-car link when the car is 100 m away from each lamppost.

Fading and diversity for the backhaul link

First, let us explore the sensitivity of the lamppost-to-lamppost link to variations in range and height. Fix the height of the transmitter on lamppost 1 at 10 m. Vary the height of the receiver on lamppost 2 from 9.5 to 10.5 m.

(2.3) Letting h_{nom} denote the nominal channel gain between two lampposts if you consider only the direct path and h the net complex gain including the reflected path, plot the normalized power gain in dB, $20\log_{10}(|h|/|h_{\text{nom}}|)$, as a function of the variation in

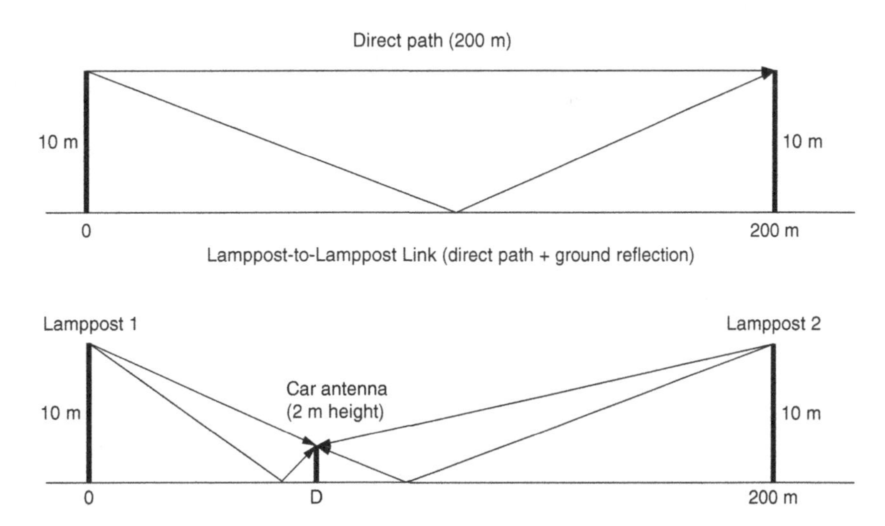

Figure 2.40 Links in a lamppost-based network.

the receiver height. Comment on the sensitivity of channel quality to variations in the receiver height.

(2.4) By modeling the variations in receiver height as coming from a uniform distribution over [9.5, 10.5], find the probability that the normalized power gain is smaller than -20 dB (i.e., that we have a fade in signal power of 20 dB or worse).

(2.5) Now, suppose that the transmitter has two antennas, vertically spaced by 25 cm, with the lower one at a height of 10 m. Let h_1 and h_2 denote the channels from the two antennas to the receiver. Let h_{nom} be defined as in (2.3) above. Plot the normalized power gains in dB, $20\log_{10}(|h_i|/|h_{\text{nom}}|)$, $i = 1, 2$. Comment on whether or not both gains dip or peak at the same time.

(2.6) Plot $20\log_{10}(\max(|h_1|, |h_2|)/|h_{\text{nom}}|)$, which is the normalized power gain you would get if you switched to the transmit antenna which has the better channel. This strategy is termed *switched diversity*.

(2.7) Find the probability that the normalized power gain of the switched diversity scheme is smaller than -20 dB.

(2.8) Comment on whether, and to what extent, diversity helped in combating fading.

Fading on the access link

Consider the access channel from lamppost 1 to the car. Let $h_{\text{nom}}(D)$ denote the nominal channel gain from the lamppost to the car, *ignoring the ground reflection.* Taking into account the ground reflection, let the channel gain be denoted as $h(D)$. Here D is the distance of the car from the bottom of lamppost 1, as shown in Figure 2.40.

(2.9) Plot $|h_{\text{nom}}|$ and $|h|$ as a function of D on a dB scale (an amplitude α is expressed on the dB scale as $20\log_{10}\alpha$). Comment on the "long-term" variation due to range, and the "short-term" variation due to multipath fading.

Lab report

- Answer all questions and print out the most useful plots to support your answers.
- Write a paragraph about any questions or confusions that you may have experienced with this lab.

3 Analog communication techniques

Modulation is the process of encoding information into a signal that can be transmitted (or recorded) over a channel of interest. In analog modulation, a baseband message signal, such as speech, audio, or video, is directly transformed into a signal that can be transmitted over a designated channel, typically a passband radio-frequency (RF) channel. Digital modulation differs from this only in the following additional step: bits are encoded into baseband message signals, which are then transformed into passband signals to be transmitted. Thus, despite the relentless transition from digital to analog modulation, many of the techniques developed for analog communication systems remain important for the digital communication systems designer, and our goal in this chapter is to study an important subset of these techniques, using legacy analog communication systems as examples to reinforce concepts.

From Chapter 2, we know that a passband signal carries information in its complex envelope, and that the complex envelope can be represented either in terms of I and Q components or in terms of envelope and phase. We study two broad classes of techniques: **amplitude modulation**, in which the analog message signal appears directly in the I and/or Q components; and **angle modulation**, in which the analog message signal appears directly in the phase or in the instantaneous frequency (i.e., in the derivative of the phase) of the transmitted signal. Examples of analog communication in space include AM radio, FM radio, and broadcast television, as well as a variety of specialized radios. Examples of analog communication in time (i.e., for storage) include audiocassettes and VHS videotapes.

The analog-centric techniques covered in this chapter include envelope detection, superheterodyne reception, limiter discriminators, and phase-locked loops. At a high level, these techniques tell us how to go from baseband message signals to passband transmitted signals, and back from passband received signals to baseband message signals. For analog communication, this is enough, since we consider continuous-time message signals that are directly transformed to passband through amplitude or angle modulation. For digital communication, we need to also figure out how to decode the encoded bits from the received passband signal, typically after downconversion to baseband; this is a subject discussed in later chapters. However, between encoding at the transmitter and decoding at the receiver, analog communication techniques are relevant: for example, we need to decide between direct and superheterodyne architectures for upconversion and downconversion, and tailor our frequency planning appropriately; we may use a phase-locked loop (PLL) to synthesize the local oscillator frequencies at the transmitter and receiver; and the basic techniques for mapping baseband signals to passband remain the same (amplitude and/or angle modulation). In addition, while many classical analog processing functionalities are replaced by digital signal processing in modern digital communication transceivers, when we push the limits of digital communication systems, in terms of lowering power consumption or increasing data rates, it is often necessary to fall back on analog-centric, or hybrid digital–analog, techniques. This is because the analog-to-digital conversion required for digital transceiver implementations may often be too costly or power-hungry for ultra-high-speed, or ultra-low-power, implementations.

Chapter plan

After a quick discussion of terminology and notation in Section 3.1, we discuss various forms of amplitude modulation in Section 3.2, including bandwidth requirements and the tradeoffs between power efficiency and simplicity of demodulation. We discuss angle modulation in Section 3.3, including the relation between phase and frequency modulation, the bandwidth of angle-modulated signals, and simple suboptimal demodulation strategies.

The superheterodyne up/downconversion architecture is discussed in Section 3.4, and the design considerations are illustrated via the example of analog AM radio. The PLL is discussed in Section 3.5, including discussion of applications such as frequency synthesis and FM demodulation, linearized modeling and analysis, and a glimpse of the insights provided by nonlinear models. Finally, as a historical note, we discuss some legacy analog communication systems in Section 3.6, mainly to highlight some of the creative design choices that were made in times when sophisticated digital signal-processing techniques were not available. This last section can be skipped if the reader's interest is limited to learning analog-centric techniques for digital communication system design.

Software

Software Lab 3.1 reinforces concepts in amplitude modulation, and shows how envelope detection, used for analog amplitude modulation, actually remains relevant for downconversion for systems where we are pushing the limits in terms of carrier frequency (e.g., coherent optical communication). Angle modulation is explored further in Software Lab 3.2, which includes an introduction to a digital communication technique based on angle modulation.

3.1 Terminology and notation

Message signal In the remainder of this chapter, the analog baseband message signal is denoted by $m(t)$. Depending on convenience of exposition, we shall think of this message as being of either finite power or finite energy. In practice, any message we would encounter would have finite energy when we consider a finite time interval. However, when modeling transmissions over long time intervals, it is useful to think of messages as finite-power signals spanning an infinite time interval. On the other hand, when discussing the effect of the message spectrum on the spectrum of the transmitted signal, it may be convenient to consider a finite energy message signal. Since we consider physical message signals, the time-domain signal is real-valued, so that its Fourier transform (defined for a finite-energy signal) is conjugate symmetric: $M(f) = M^*(-f)$. For a finite-power (infinite-energy) message, recall from Chapter 2 that the power is defined as a time average in the limit of an infinite observation interval, as follows:

$$\overline{m^2} = \lim_{T_o \to \infty} \frac{1}{T_o} \int_0^{T_o} m^2(t)dt$$

Similarly, the DC value is defined as

$$\overline{m} = \lim_{T_o \to \infty} \frac{1}{T_o} \int_0^{T_o} m(t)dt$$

We typically assume that the DC value of the message is zero: $\overline{m} = 0$.

A simple example, shown in Figure 3.1, that we shall use often is a finite-power sinusoidal message signal, $m(t) = A_m \cos(2\pi f_m t)$, whose spectrum consists of impulses at $\pm f_m$: $M(f) = (A_m/2)(\delta(f - f_m) + \delta(f + f_m))$. For this message, $\overline{m} = 0$ and $\overline{m^2} = A_m^2/2$.

Transmitted signal When the signal transmitted over the channel is a passband signal, it can be written as (see Chapter 2)

$$u_p(t) = u_c(t)\cos(2\pi f_c t) - u_s(t)\sin(2\pi f_c t) = e(t)\cos(2\pi f_c t + \theta(t))$$

where f_c is a carrier frequency, $u_c(t)$ is the I component, $u_s(t)$ is the Q component, $e(t) \geq 0$ is the envelope, and $\theta(t)$ is the phase. Modulation consists of encoding the message in $u_c(t)$ and $u_s(t)$, or, equivalently, in $e(t)$ and $\theta(t)$. In most of the analog amplitude modulation schemes considered, the message modulates the I component (with the Q component occasionally playing a "supporting role") as discussed in Section 3.2. The exception is quadrature amplitude modulation, in which both I and Q components carry separate

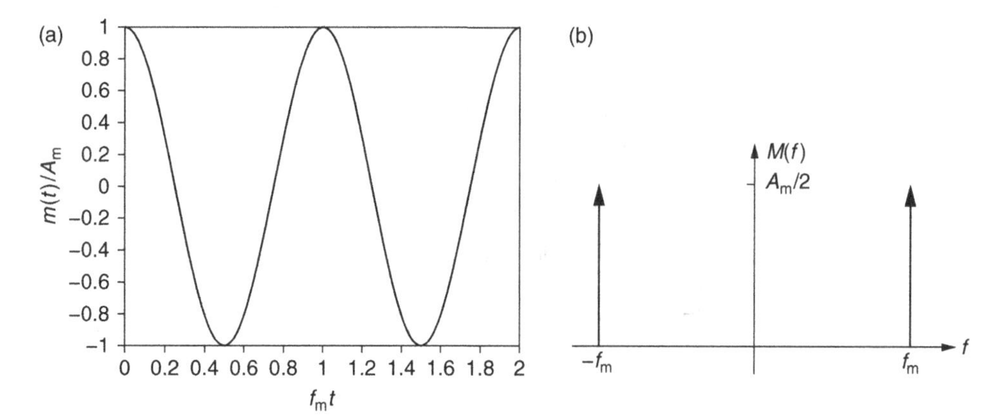

Figure 3.1 A sinusoidal message (a) and its spectrum (b).

messages. In phase and frequency modulation, or angle modulation, the message directly modulates the phase $\theta(t)$ or its derivative, keeping the envelope $e(t)$ unchanged.

3.2 Amplitude modulation

We now discuss a number of variants of amplitude modulation, in which the baseband message signal modulates the amplitude of a sinusoidal carrier whose frequency falls in the passband over which we wish to communicate.

3.2.1 Double-sideband (DSB) suppressed carrier (SC)

Here, the message m modulates the I component of the passband transmitted signal u as follows:

$$u_{\rm DSB}(t) = Am(t)\cos(2\pi f_{\rm c}t) \tag{3.1}$$

On taking Fourier transforms, we have

$$U_{\rm DSB}(f) = \frac{A}{2}(M(f - f_{\rm c}) + M(f + f_{\rm c})) \tag{3.2}$$

The time-domain and frequency-domain DSB signals for a sinusoidal message are shown in Figure 3.2.

As another example, consider the finite-energy message whose spectrum is shown in Figure 3.3. Since the time-domain message $m(t)$ is real-valued, its spectrum exhibits conjugate symmetry (we have chosen a complex-valued message spectrum to emphasize the latter property). The message bandwidth is denoted by B. The bandwidth of the DSB-SC signal is $2B$, which is twice the message bandwidth. This indicates that we are being redundant in our use of spectrum. To see this, consider the upper sideband (USB) and lower sideband (LSB) depicted in Figure 3.4. The shape of the signal in the USB (i.e.,

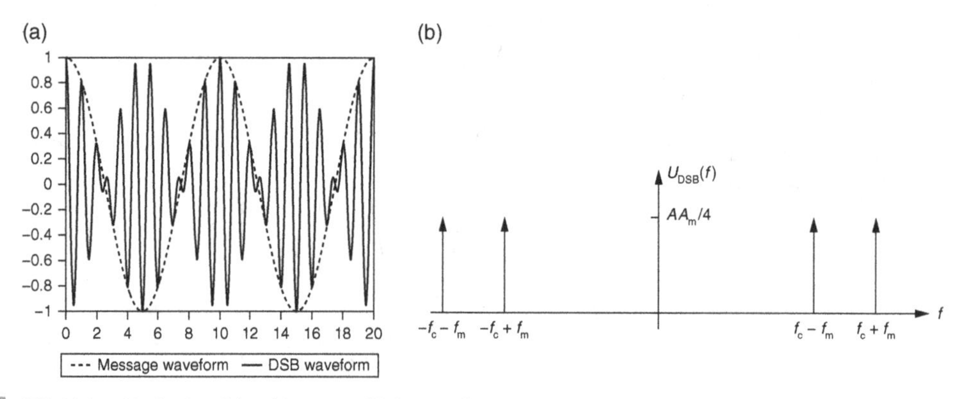

Figure 3.2 DSB-SC signal in the time (a) and frequency (b) domains for the sinusoidal message $m(t) = A_{\rm m}\cos(2\pi f_{\rm m}t)$ of Figure 3.1.

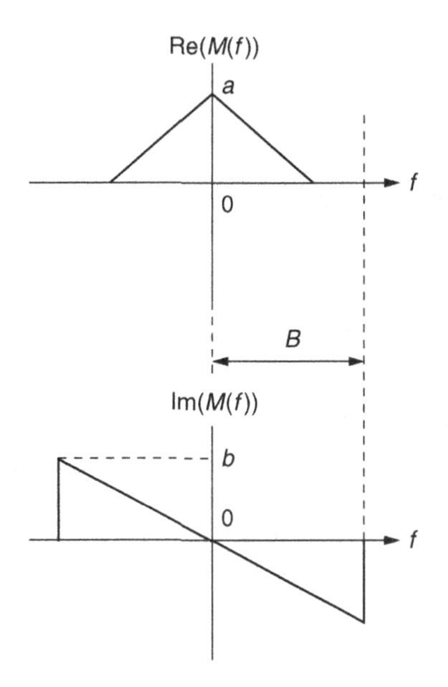

Figure 3.3 An example message spectrum.

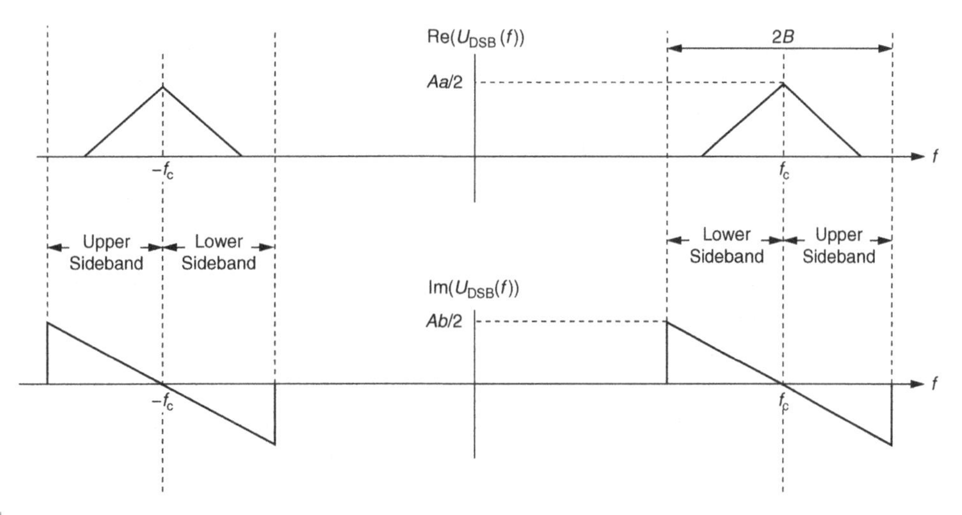

Figure 3.4 The spectrum of the passband DSB-SC signal for the example message in Figure 3.3.

$U_p(f)$ for $f_c < f \leq f_c + B$) is the same as that of the message for positive frequencies (i.e., $M(f), f > 0$). The shape of the signal in the LSB (i.e., $U_p(f)$ for $f_c - B \leq f < f_c$) is the same as that of the message for negative frequencies (i.e., $M(f), f < 0$). Since $m(t)$ is real-valued, we have $M(-f) = M^*(f)$, so that we can reconstruct the message if we know its content at either positive or negative frequencies. Thus, the USB and LSB of $u(t)$ *each* contain enough information to reconstruct the message. The term DSB refers to the fact that we are sending both sidebands. Doing this, of course, is wasteful of spectrum. This motivates single-sideband (SSB) and vestigial-sideband (VSB) modulation, which are discussed a little later.

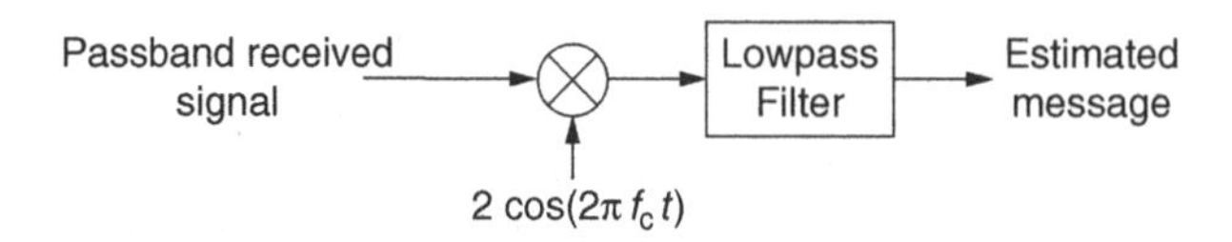

Figure 3.5 Coherent demodulation for AM.

The term *suppressed carrier* is employed because, for a message with no DC component, we see from (3.2) that the transmitted signal does not have a discrete component at the carrier frequency (i.e., $U_p(f)$ does not have impulses at $\pm f_c$).

Demodulation of DSB-SC Since the message is contained in the I component, demodulation consists of extracting the I component of the received signal, which we know how to do from Chapter 2: multiply the received signal by the cosine of the carrier, and pass it through a lowpass filter. Ignoring noise, the received signal is given by

$$y_p(t) = Am(t)\cos(2\pi f_c t + \theta_r) \tag{3.3}$$

where θ_r is the phase of the received carrier relative to the local copy of the carrier produced by the receiver's local oscillator (LO), and A is the received amplitude, taking into account the propagation channel from the transmitter to the receiver. The demodulator is shown in Figure 3.5. In order for this demodulator to work well, we must have θ_r as close to zero as possible; that is, the carrier produced by the LO must be *coherent* with the received carrier. To see the effect of phase mismatch, let us compute the demodulator output for arbitrary θ_r. Using the trigonometric identity $2\cos\theta_1 \cos\theta_2 = \cos(\theta_1 - \theta_2) + \cos(\theta_1 + \theta_2)$, we have

$$2y_p(t)\cos(2\pi f_c t) = Am(t)\cos(2\pi f_c t + \theta_r)\cos(2\pi f_c t) = Am(t)\cos\theta_r + Am(t)\cos(4\pi f_c t + \theta_r)$$

We recognize the second term on the extreme right-hand side as being a passband signal at $2f_c$ (since it is a baseband message multiplied by a carrier whose frequency exceeds the message bandwidth). It is therefore rejected by the lowpass filter. The first term is a baseband signal proportional to the message, which appears unchanged at the output of the LPF (except possibly for scaling), as long as the LPF response has been designed to be flat over the message bandwidth. The output of the demodulator is therefore given by

$$\hat{m}(t) = Am(t)\cos\theta_r \tag{3.4}$$

We can also infer this using the complex-baseband representation, which is what we prefer to employ instead of unwieldy trigonometric identities. The coherent demodulator in Figure 3.5 extracts the I component relative to the receiver's LO. The received signal can be written as

$$y_p(t) = Am(t)\cos(2\pi f_c t + \theta_r) = \text{Re}\left(Am(t)e^{j(2\pi f_c t + \theta_r)}\right) = \text{Re}\left(Am(t)e^{j\theta_r}e^{j2\pi f_c t}\right)$$

from which we can read off the complex envelope $y(t) = Am(t)e^{j\theta_r}$. The real part $y_c(t) = Am(t)\cos\theta_r$ is the I component extracted by the demodulator.

The demodulator output (3.4) is proportional to the message, which is what we want, but the proportionality constant varies with the phase of the received carrier relative to the LO. In particular, the signal gets significantly attenuated as the phase mismatch increases, and gets completely wiped out for $\theta_r = \pi/2$. Note that, if the carrier frequency of the LO is not synchronized with that of the received carrier (say with frequency offset Δf), then $\theta_r(t) = 2\pi \Delta f\, t + \phi$ is a time-varying phase that takes all values in $[0, 2\pi)$, which leads to time-varying signal degradation in amplitude, as well as unwanted sign changes. Thus, for coherent demodulation to be successful, we must drive Δf to zero, and make ϕ as small as possible; that is, we must synchronize with respect to the received carrier. One possible approach is to use feedback-based techniques such as the PLL, discussed later in this chapter.

3.2.2 Conventional AM

In conventional AM, we add a large carrier component to a DSB-SC signal, so that the passband transmitted signal is of the form

$$u_{AM}(t) = Am(t)\cos(2\pi f_c t) + A_c \cos(2\pi f_c t) \tag{3.5}$$

On taking the Fourier transform, we have

$$U_{AM}(f) = \frac{A}{2}(M(f - f_c) + M(f + f_c)) + \frac{A_c}{2}(\delta(f - f_c) + \delta(f + f_c))$$

which means that, in addition to the USB and LSB due to the message modulation, we also have impulses at $\pm f_c$ due to the unmodulated carrier. Figure 3.6 shows the resulting spectrum.

The key concept behind conventional AM is that, by making A_c large enough, the message can be demodulated using a simple envelope detector. Large A_c corresponds to expending transmitter power on sending an unmodulated carrier that carries no message information, in order to simplify the receiver. This tradeoff makes sense in a broadcast context, where one powerful transmitter may be sending information to a large number of low-cost receivers, and is the design approach that has been adopted for broadcast AM radio. A more detailed discussion follows.

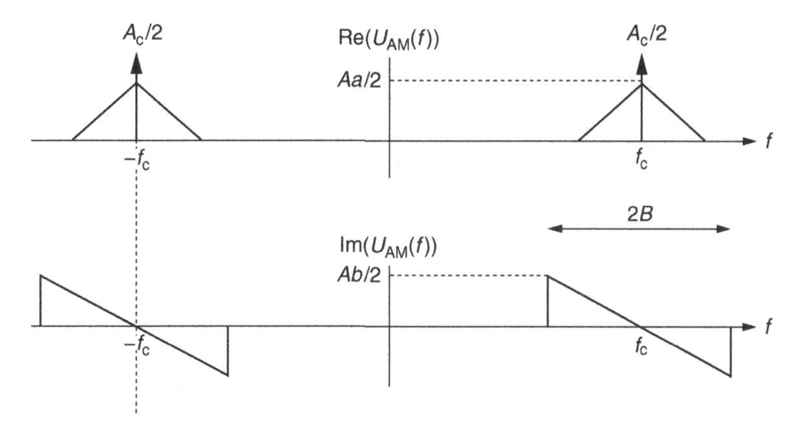

Figure 3.6 The spectrum of a conventional AM signal for the example message in Figure 3.3.

The envelope of the AM signal in (3.5) is given by

$$e(t) = |Am(t) + A_c|$$

If the term inside the magnitude operation is always nonnegative, we have $e(t)=Am(t) + A_c$. In this case, we can read off the message signal directly from the envelope, using AC coupling to get rid of the DC offset due to the second term. For this to happen, we must have

$$Am(t) + A_c \geq 0 \ \text{ for all } t \quad \Longleftrightarrow A \min_t m(t) + A_c > 0 \tag{3.6}$$

Let $\min_t m(t) = -M_0$, where $M_0 = |\min_t m(t)|$. (Note that the minimum value of the message must be negative if the message has zero DC value.) Equation (3.6) reduces to $-AM_0 + A_c \geq 0$, or $A_c \geq AM_0$. Let us define the *modulation index* a_{mod} as the ratio of the size of the biggest negative incursion due to the message term to the size of the unmodulated carrier term:

$$a_{\mathrm{mod}} = \frac{AM_0}{A_c} = \frac{A|\min_t m(t)|}{A_c}$$

The condition (3.6) for accurately recovering the message using envelope detection can now be rewritten as

$$a_{\mathrm{mod}} \leq 1 \tag{3.7}$$

It is also convenient to define a normalized version of the message as follows:

$$m_n(t) = \frac{m(t)}{M_0} = \frac{m(t)}{|\min_t m(t)|} \tag{3.8}$$

which satisfies

$$\min_t m_n(t) = \frac{\min_t m(t)}{M_0} = -1$$

It is easy to see that the AM signal (3.5) can be rewritten as

$$u_{\mathrm{AM}}(t) = A_c(1 + a_{\mathrm{mod}} m_n(t))\cos(2\pi f_c t) \tag{3.9}$$

which clearly brings out the role of the modulation index in ensuring that envelope detection works.

Figure 3.7 illustrates the impact of the modulation index on the viability of envelope detection, where the message signal is the sinusoidal message in Figure 3.1. For $a_{\mathrm{mod}} = 0.5$ and $a_{\mathrm{mod}} = 1$, we see that the envelope equals a scaled and DC-shifted version of the message. For $a_{\mathrm{mod}} = 1.5$, we see that the envelope no longer follows the shape of the message.

Demodulation of conventional AM Ignoring noise, the received signal is given by

$$y_p(t) = B(1 + a_{\mathrm{mod}} m_n(t))\cos(2\pi f_c t + \theta_r) \tag{3.10}$$

where θ_r is a phase offset that is unknown *a priori,* if we do not perform carrier synchronization. However, as long as $a_{\mathrm{mod}} \leq 1$, we can recover the message without knowing θ_r using envelope detection, since the envelope is still just a scaled and DC-shifted version of the message. Of course, the message can also be recovered by coherent detection, since the

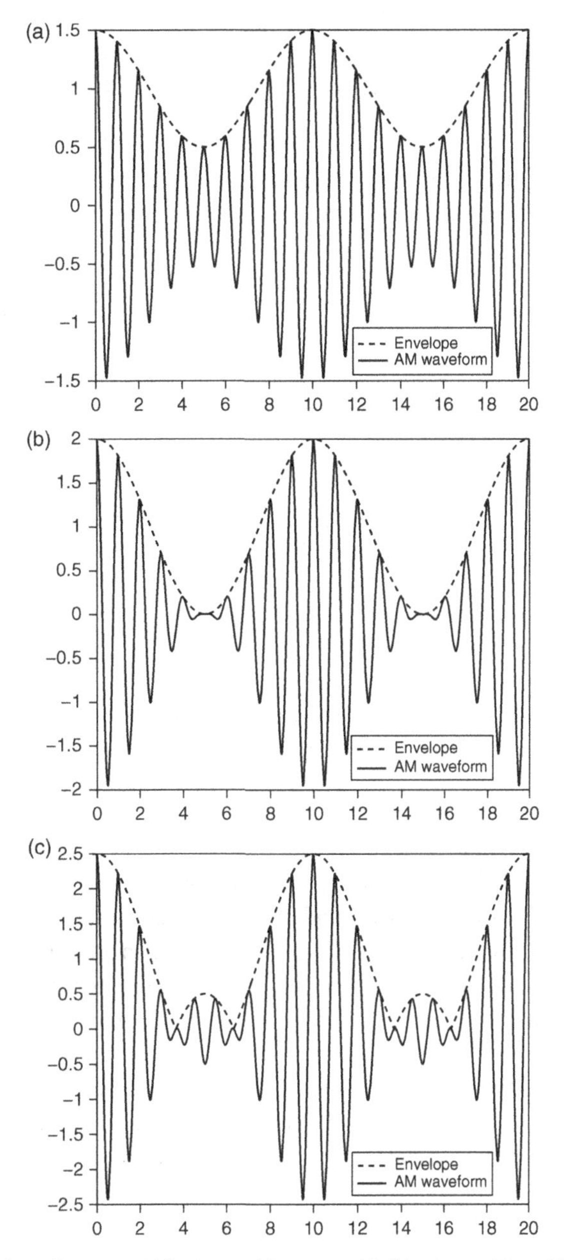

Figure 3.7 Time-domain AM waveforms for a sinusoidal message: (a) $a_{\mathrm{mod}} = 0.5$, (b) $a_{\mathrm{mod}} = 1.0$, and (c) $a_{\mathrm{mod}} = 1.5$. The envelope no longer follows the message for modulation index a_{mod} larger than one.

I component of the received carrier equals a scaled and DC-shifted version of the message. However, by doing envelope detection instead, we can avoid carrier synchronization, thus reducing receiver complexity drastically. An envelope detector is shown in Figure 3.8, and an example (where the envelope is a straight line) showing how it works is depicted in Figure 3.9. The diode (we assume that it is ideal) conducts in only the forward direction,

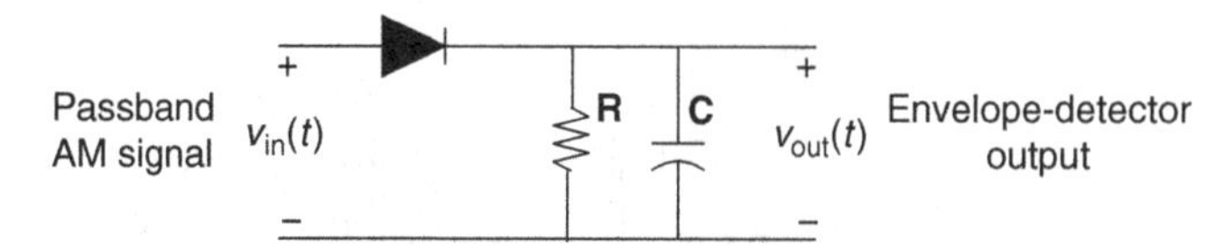

Figure 3.8 Envelope-detector demodulation of AM. The envelope-detector output is typically passed through a DC blocking capacitance (not shown) to eliminate the DC offset due to the carrier component of the AM signal.

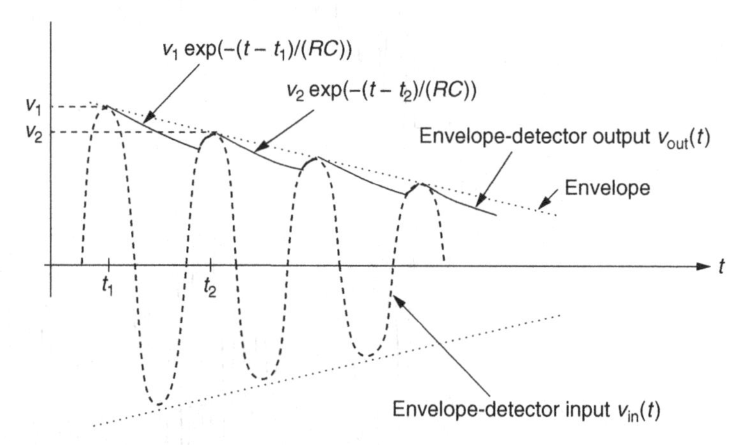

Figure 3.9 The relation between the envelope-detector output $v_{out}(t)$ (shown in bold) and input $v_{in}(t)$ (shown as a dashed line). The output closely follows the envelope (shown as a dotted line).

when the input voltage $v_{in}(t)$ of the passband signal is larger than the output voltage $v_{out}(t)$ across the RC filter. When this happens, the output voltage becomes equal to the input voltage instantaneously (under the idealization that the diode has zero resistance). In this regime, we have $v_{out}(t) = v_{in}(t)$. When the input voltage is smaller than the output voltage, the diode does not conduct, and the capacitor starts discharging through the resistor with time constant RC. As shown in Figure 3.9, in this regime, starting at time t_1, we have $v(t) = v_1 e^{-(t-t_1)/(RC)}$, where $v_1 = v(t_1)$, as shown in Figure 3.9.

Roughly speaking, the capacitor gets charged at each carrier peak, and discharges between peaks. The time interval between successive charging episodes is therefore approximately equal to $1/f_c$, the time between successive carrier peaks. The factor by which the output voltage is reduced during this period due to capacitor discharge is $\exp(-1/(f_c RC))$. This must be close to one in order for the voltage to follow the envelope, rather than the variations in the sinusoidal carrier. That is, we must have $f_c RC \gg 1$. On the other hand, the decay in the envelope-detector output must be fast enough (i.e., the RC time constant must be small enough) that it can follow changes in the envelope. Since the time constant for envelope variations is inversely proportional to the message bandwidth B, we must have $RC \ll 1/B$. On combining these two conditions for envelope detection to work well, we have

$$\frac{1}{f_c} \ll RC \ll \frac{1}{B} \tag{3.11}$$

This of course requires that $f_c \gg B$ (the carrier frequency much larger than the message bandwidth), which is typically satisfied in practice. For example, the carrier frequencies in broadcast AM radio are over 500 kHz, whereas the message bandwidth is limited to 5 kHz. Applying (3.11), the RC time constant for an envelope detector should be chosen so that

$$2\ \mu\text{s} \ll RC \ll 200\ \mu\text{s}$$

In this case, a good choice of parameters would be $RC = 20\ \mu$s, for example, with $R = 50\ \Omega$ (ohms), and $C = 400$ nF (nanofarads).

Software Lab 3.1 introduces a different application of envelope detection. Adding a strong carrier component *at the receiver,* followed by envelope detection, provides an alternative approach to downconversion that avoids the use of mixers, which are difficult to implement at very high carrier frequencies (e.g., for coherent optical communication).

Power efficiency of conventional AM The price we pay for the receiver simplicity of conventional AM is power inefficiency: in (3.5) the unmodulated carrier $A_c \cos(2\pi f_c t)$ is not carrying any information regarding the message. We now compute the power efficiency η_{AM}, which is defined as the ratio of the transmitted power due to the message-bearing term $Am(t)\cos(2\pi f_c t)$ to the total power of $u_{AM}(t)$. In order to express the result in terms of the modulation index, let us use the expression (3.9):

$$\begin{aligned}\overline{u_{AM}^2(t)} &= \overline{A_c^2(1 + a_{mod}m_n(t))^2 \cos^2(2\pi f_c t)} \\ &= \frac{A_c^2}{2}\overline{(1 + a_{mod}m_n(t))^2} + \frac{A_c^2}{2}\overline{(1 + a_{mod}m_n(t))^2 \cos(4\pi f_c t)}\end{aligned}$$

The second term on the right-hand side is the DC value of a passband signal at $2f_c$, which is zero. Upon expanding out the first term, we have

$$\overline{u_{AM}^2(t)} = \frac{A_c^2}{2}\left(1 + a_{mod}^2\overline{m_n^2} + 2a_{mod}\overline{m_n}\right) = \frac{A_c^2}{2}\left(1 + a_{mod}^2\overline{m_n^2}\right) \tag{3.12}$$

assuming that the message has zero DC value. The power of the message-bearing term can be similarly computed as

$$\overline{(A_c a_{mod} m_n(t))^2 \cos^2(2\pi f_c t)} = \frac{A_c^2}{2} a_{mod}^2 \overline{m_n^2}$$

so that the power efficiency is given by

$$\eta_{AM} = \frac{a_{mod}^2\overline{m_n^2}}{1 + a_{mod}^2\overline{m_n^2}} \tag{3.13}$$

Noting that m_n is normalized so that its most negative value is -1, for messages that have comparable positive and negative excursions around zero, we expect $|m_n(t)| \leq 1$, and hence average power $\overline{m_n^2} \leq 1$ (typical values are much smaller than one). Since $a_{mod} \leq 1$ for envelope detection to work, the power efficiency of conventional AM is at best 50%. For a sinusoidal message, for example, it is easy to see that $\overline{m_n^2} = 1/2$, so the power efficiency is at most 33%. For speech signals, which have significantly higher peak-to-average ratio, the power efficiency is even smaller.

Example 3.2.1 (AM power efficiency computation) The message $m(t) = 2\sin(2000\pi t) - 3\cos(4000\pi t)$ is used in an AM system with a modulation index of 70% and carrier frequency of 580 kHz. What is the power efficiency? If the net transmitted power is 10 W (watts), find the magnitude spectrum of the transmitted signal.

We need to find $M_0 = |\min_t m(t)|$ in order to determine the normalized form $m_n(t) = m(t)/M_0$. To simplify the notation, let $x = 2000\pi t$, and minimize $g(x) = 2\sin x - 3\cos(2x)$. Since g is periodic with period 2π, we can minimize it numerically over a period. However, we can perform the minimization analytically in this case. By differentiating g, we obtain

$$g'(x) = 2\cos x + 6\sin(2x) = 0$$

This gives

$$2\cos x + 12\sin x\cos x = 2\cos x(1 + 6\sin x) = 0$$

There are two solutions, namely $\cos x = 0$ and $\sin x = -\frac{1}{6}$. The first solution gives $\cos(2x) = 2\cos^2 x - 1 = -1$ and $\sin x = \pm 1$, which gives $g(x) = 1, 5$. The second solution gives $\cos(2x) = 1 - 2\sin^2 x = 1 - 2/36 = 17/18$, which gives $g(x) = 2(-1/6) -3(17/18) = -19/6$. We therefore obtain

$$M_0 = |\min_t m(t)| = 19/6$$

This gives

$$m_n(t) = \frac{m(t)}{M_0} = \frac{12}{19}\sin(1000\pi t) - \frac{18}{19}\cos(2000\pi t)$$

This gives

$$\overline{m_n^2} = (12/19)^2(1/2) + (18/19)^2(1/2) = 0.65$$

On substituting into (3.13), setting $a_{\text{mod}} = 0.7$, we obtain a power efficiency $\eta_{\text{AM}} = 0.24$, or 24%.

To figure out the spectrum of the transmitted signal, we must find A_c in the formula (3.9). The power of the transmitted signal is given by (3.12) to be

$$10 = \frac{A_c^2}{2}\left(1 + a_{\text{mod}}^2\overline{m_n^2}\right) = \frac{A_c^2}{2}\left(1 + (0.7^2)(0.65)\right)$$

which yields $A_c \approx 3.9$. The overall AM signal is given by

$$u_{\text{AM}}(t) = A_c(1 + a_{\text{mod}}m_n(t))\cos(2\pi f_c t) = A_c(1 + a_1\sin(2\pi f_1 t) + a_2\cos(4\pi f_1 t))\cos(2\pi f_c t)$$

where $a_1 = 0.7(12/19) = 0.44$, $a_2 = 0.7(-18/19) = -0.66$, $f_1 = 1$ kHz, and $f_c = 580$ kHz. The magnitude spectrum is given by

$$\begin{aligned}|U_{\text{AM}}(f)| &= (A_c/2)(\delta(f - f_c) + \delta(f + f_c)) \\ &\quad + (A_c|a_1|/4)(\delta(f - f_c - f_1) + \delta(f - f_c + f_1) + \delta(f + f_c + f_1) + \delta(f + f_c - f_1)) \\ &\quad + (A_c|a_2|/4)(\delta(f - f_c - 2f_1) + \delta(f - f_c + 2f_1) + \delta(f + f_c + 2f_1) \\ &\quad + \delta(f + f_c - 2f_1))\end{aligned}$$

with the numerical values shown in Figure 3.10.

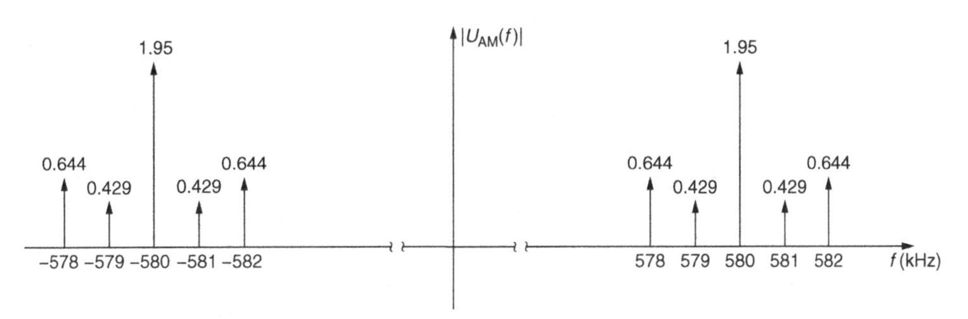

Figure 3.10 The magnitude spectrum for the AM waveform in Example 3.2.1.

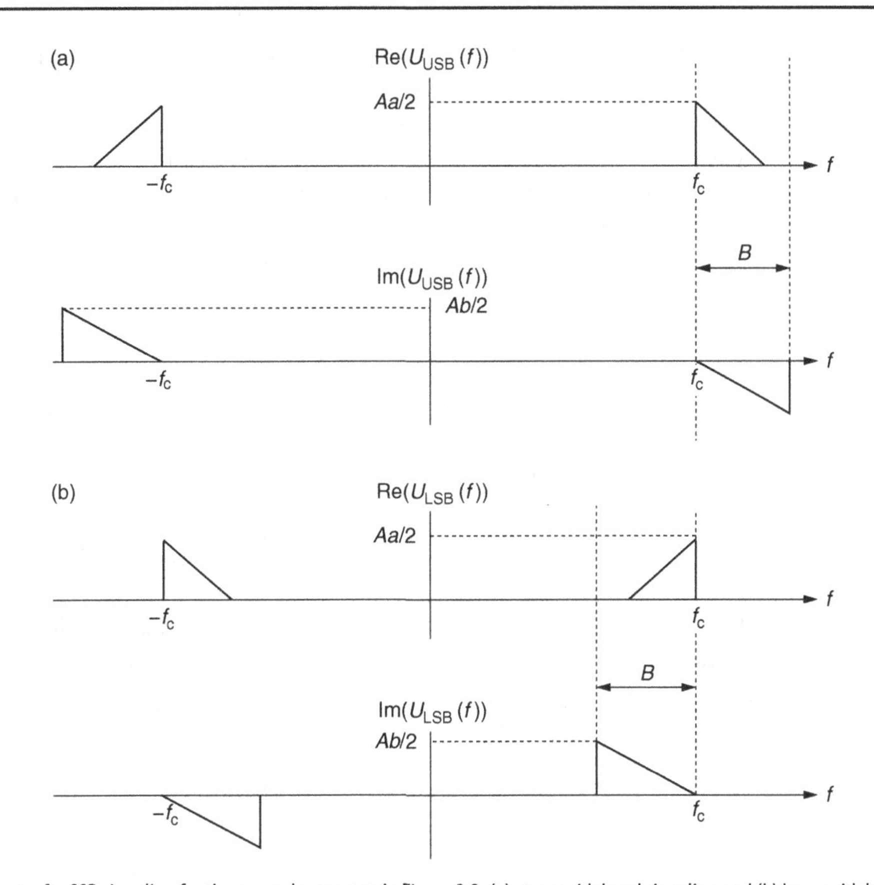

Figure 3.11 Spectra for SSB signaling for the example message in Figure 3.3: (a) upper-sideband signaling and (b) lower-sideband signaling.

3.2.3 Single-sideband modulation (SSB)

In SSB modulation, we send either the upper sideband or the lower sideband of a DSB-SC signal. For the running example, the spectra of the passband USB and LSB signals are shown in Figure 3.11.

From our discussion of DSB-SC, we know that each sideband provides enough information to reconstruct the message. But how do we physically reconstruct the message from an SSB signal? To see this, consider the USB signal depicted in Figure 3.11(a). We can reconstruct the baseband message if we can move the component near $+f_c$ to the left by f_c, and the component near $-f_c$ to the right by f_c; that is, if we move the passband components in towards the origin. These two frequency translations can be accomplished by multiplying the USB signal by $2\cos(2\pi f_c t) = e^{j2\pi f_c t} + e^{-j2\pi f_c t}$, as shown in Figure 3.5, which creates the desired message signal at baseband, as well as undesired frequency components at $\pm 2f_c$, which can be rejected by a lowpass filter. It can be checked that the same argument applies to LSB signals as well.

It follows from the preceding discussion that SSB signals can be demodulated in exactly the same fashion as DSB-SC, using the coherent demodulator depicted in Figure 3.5. Since this demodulator simply extracts the I component of the passband signal, the I component of the SSB signal must be the message. In order to understand the structure of an SSB signal, it remains to identify the Q component. This is most easily done by considering the complex envelope of the passband transmitted signal. Consider again the example USB signal in Figure 3.11(a). The spectrum $U(f)$ of its complex envelope relative to f_c is shown in Figure 3.12. Now, the spectra of I and Q components can be inferred as follows:

$$U_c(f) = \frac{U(f) + U^*(-f)}{2}, \quad U_s(f) = \frac{U(f) - U^*(-f)}{2j}$$

By applying these equations, we get I and Q components as shown in Figure 3.13.

Thus, up to scaling, the I component $U_c(f) = M(f)$, and the Q component is a transformation of the message given by

$$U_s(f) = \begin{cases} -jM(f), & f > 0 \\ jM(f), & f < 0 \end{cases} = M(f)(-j\,\mathrm{sgn}(f)) \tag{3.14}$$

That is, the Q component is a filtered version of the message, where the filter transfer function is $H(f) = -j\,\mathrm{sgn}(f)$. This transformation is called the *Hilbert transform.*

Hilbert transform The Hilbert transform of a signal $x(t)$ is denoted by $\check{x}(t)$, and is specified in the frequency domain as

$$\check{X}(f) = (-j\,\mathrm{sgn}(f))X(f)$$

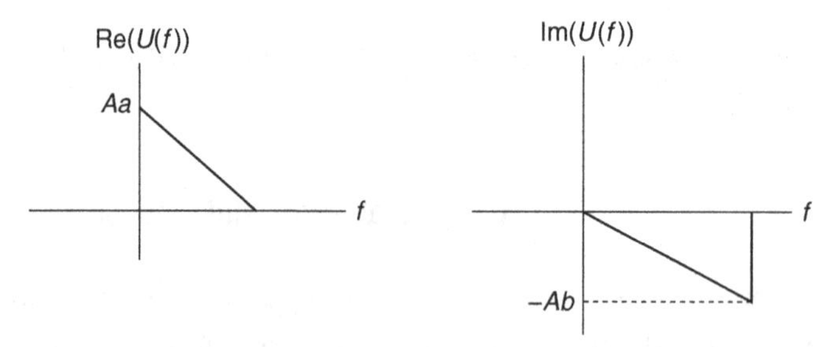

Figure 3.12 Complex envelope for the USB signal in Figure 3.11(a).

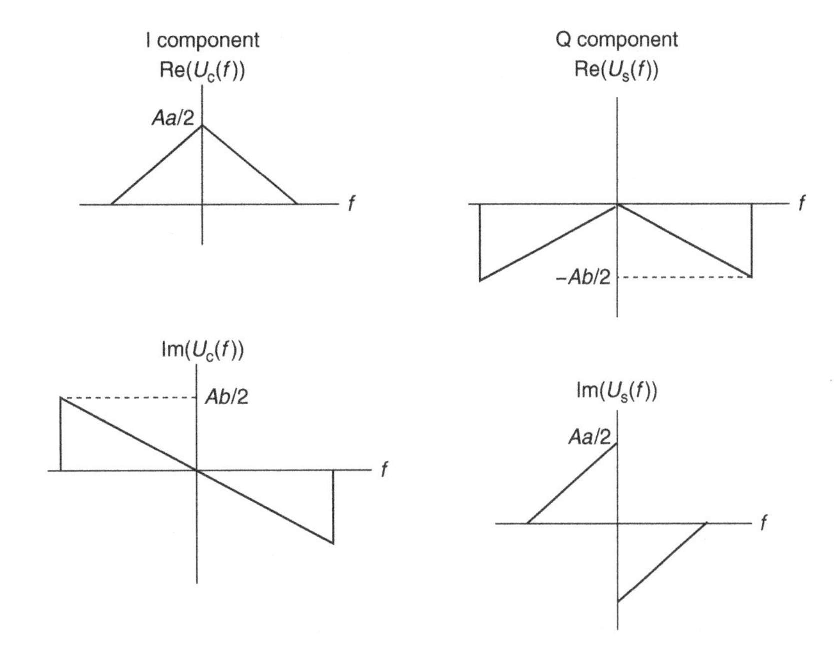

Figure 3.13 I and Q components for the USB signal in Figure 3.11(a).

This corresponds to passing u through a filter with transfer function

$$H(f) = -j\,\mathrm{sgn}(f) \leftrightarrow h(t) = \frac{1}{\pi t}$$

where the derivation of the impulse response is left as an exercise.

Figure 3.14 shows the spectrum of the Hilbert transform of the example message in Figure 3.3. We see that it is the same (up to scaling) as the Q component of the USB signal, shown in Figure 3.13.

Physical interpretation of the Hilbert transform If $x(t)$ is real-valued, then so is its Hilbert transform $\check{x}(t)$. Thus, the Fourier transforms $X(f)$ and $\check{X}(f)$ must both satisfy conjugate symmetry, and we need to discuss only what happens at positive frequencies. For $f > 0$, we have $\check{X}(f) = -j\,\mathrm{sgn}(f)X(f) = -jX(f) = e^{-j\pi/2}X(f)$. That is, the Hilbert transform simply imposes a $\pi/2$ phase lag at all (positive) frequencies, leaving the magnitude of the Fourier transform unchanged.

Example 3.2.2 (Hilbert transform of a sinusoid) From the preceding argument, a sinusoid $s(t) = \cos(2\pi f_0 t + \phi)$ has Hilbert transform $\check{s}(t) = \cos(2\pi f_0 t + \phi - \pi/2) = \sin(2\pi f_0 t + \phi)$. We can also do this the hard way, as follows:

$$\begin{aligned} s(t) = \cos(2\pi f_0 t + \phi) &= \tfrac{1}{2}\left(e^{j(2\pi f_0 t+\phi)} + e^{-j(2\pi f_0 t+\phi)}\right) \\ \leftrightarrow S(f) &= \tfrac{1}{2}\left(e^{j\phi}\delta(f - f_0) + e^{-j\phi}\delta(f + f_0)\right) \end{aligned}$$

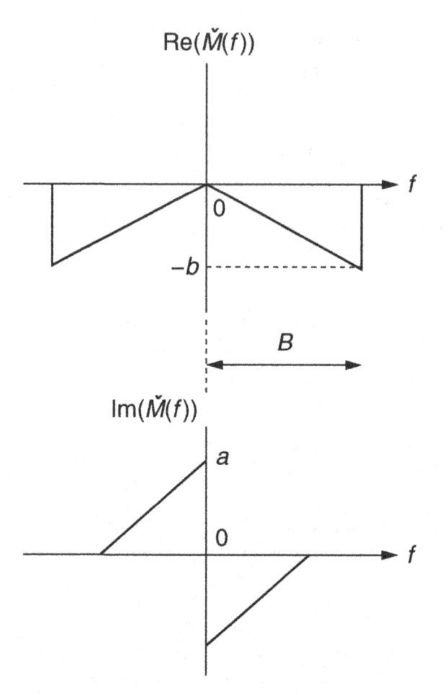

Figure 3.14 The spectrum of the Hilbert transform of the example message in Figure 3.3.

Thus,

$$\check{S}(f) = -j\,\mathrm{sgn}(f)S(f) = \tfrac{1}{2}\left(e^{j\phi}(-j)\delta(f-f_0) + e^{-j\phi}(j)\delta(f+f_0)\right)$$
$$\leftrightarrow \check{s}(t) = \tfrac{1}{2}\left(e^{j\phi}(-j)e^{j2\pi f_0 t} + e^{-j\phi}(j)e^{j2\pi f_0 t}\right)$$

which simplifies to

$$\check{s}(t) = \frac{1}{2j}\left(e^{j(2\pi f_0 t+\phi)} - e^{-j(2\pi f_0 t+\phi)}\right) = \sin(2\pi f_0 t + \phi)$$

Equation (3.14) shows that the Q component of the USB signal is $\check{m}(t)$, the Hilbert transform of the message. Thus, the passband USB signal can be written as

$$u_{\mathrm{USB}}(t) = m(t)\cos(2\pi f_c t) - \check{m}(t)\sin(2\pi f_c t) \tag{3.15}$$

Similarly, we can show that the Q component of an LSB signal is $-\check{m}(t)$, so the passband LSB signal is given by

$$u_{\mathrm{LSB}}(t) = m(t)\cos(2\pi f_c t) + \check{m}(t)\sin(2\pi f_c t) \tag{3.16}$$

SSB modulation Conceptually, an SSB signal can be generated by filtering out one of the sidebands of a DSB-SC signal. However, it is difficult to implement the required sharp cutoff at f_c, especially if we wish to preserve the information contained at the boundary of the two sidebands, which corresponds to the message information near DC. Thus, an implementation of SSB based on sharp bandpass filters runs into trouble when the message has significant frequency content near DC. The representations in (3.15) and (3.16) provide an alternative approach to generating SSB signals, as shown in Figure 3.15. We have emphasized the role of 90° phase lags in generating the I and Q components, as well as the LO signals used for upconversion.

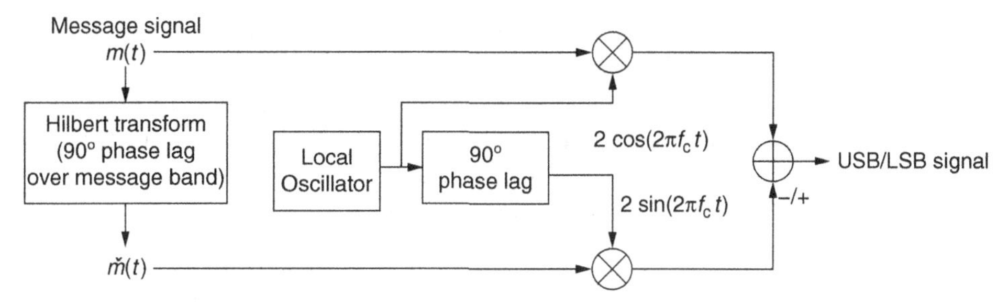

Figure 3.15 SSB modulation using the Hilbert transform of the message.

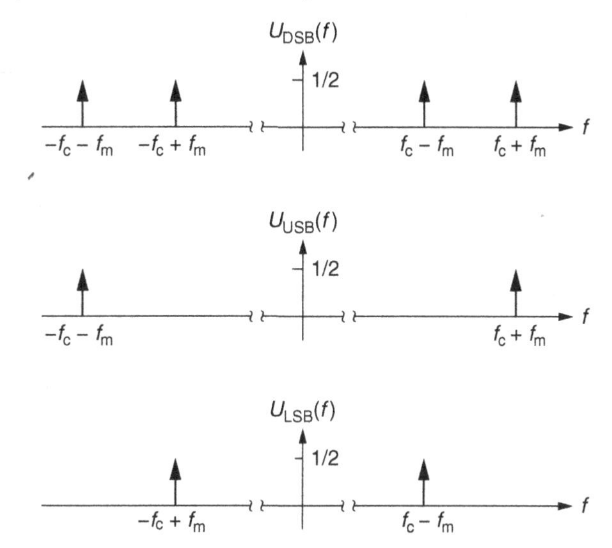

Figure 3.16 DSB and SSB spectra for a sinusoidal message.

Example 3.2.3 (SSB waveforms for a sinusoidal message) For a sinusoidal message $m(t) = \cos(2\pi f_m t)$, we have $\check{m}(t) = \sin(2\pi f_m t)$ from Example 3.2.2. Consider the DSB signal

$$u_{\text{DSB}}(t) = 2\cos(2\pi f_m t)\cos(2\pi f_c t)$$

where we have normalized the signal power to one: $\overline{u_{\text{DSB}}^2} = 1$. The DSB, USB, and SSB spectra are shown in Figure 3.16. From the SSB spectra shown, we can immediately write down the following time-domain expressions:

$$u_{\text{USB}}(t) = \cos(2\pi(f_c + f_m)t) = \cos(2\pi f_m t)\cos(2\pi f_c t) - \sin(2\pi f_m t)\sin(2\pi f_c t)$$

$$u_{\text{LSB}}(t) = \cos(2\pi(f_c - f_m)t) = \cos(2\pi f_m t)\cos(2\pi f_c t) + \sin(2\pi f_m t)\sin(2\pi f_c t)$$

The preceding equations are consistent with (3.15) and (3.16). For both the USB and LSB signals, the I component equals the message: $u_c(t) = m(t) = \cos(2\pi f_m t)$. The Q component for the USB signal is $u_s(t) = \check{m}(t) = \sin(2\pi f_m t)$, and the Q component for the LSB signal is $u_s(t) = -\check{m}(t) = -\sin(2\pi f_m t)$.

SSB demodulation We know now that the message can be recovered from an SSB signal by extracting its I component using a coherent demodulator as in Figure 3.5. The difficulty of coherent demodulation lies in the requirement for carrier synchronization, and we have discussed the adverse impact of imperfect synchronization for DSB-SC signals. We now show that the performance degradation is even more significant for SSB signals. Consider a USB received signal of the form (ignoring scale factors)

$$y_p(t) = m(t)\cos(2\pi f_c t + \theta_r) - \check{m}(t)\sin(2\pi f_c t + \theta_r) \tag{3.17}$$

where θ_r is the phase offset with respect to the receiver LO. The complex envelope with respect to the receiver LO is given by

$$y(t) = (m(t) + j\check{m}(t))e^{j\theta_r} = (m(t) + j\check{m}(t))(\cos\theta_r + j\sin\theta_r)$$

On taking the real part, we obtain that the I component extracted by the coherent demodulator is

$$y_c(t) = m(t)\cos\theta_r - \check{m}(t)\sin\theta_r$$

Thus, as the phase error θ_r increases, we get not only an attenuation in the first term corresponding to the desired message (as in DSB), but also interference due to the second term from the Hilbert transform of the message. Thus, for coherent demodulation, accurate carrier synchronization is even more crucial for SSB than for DSB.

Noncoherent demodulation is also possible for SSB if we add a strong carrier term, as in conventional AM. Specifically, for a received signal given by

$$y_p(t) = (A + m(t))\cos(2\pi f_c t + \theta_r) \pm \check{m}(t)\sin(2\pi f_c t + \theta_r)$$

the envelope is given by

$$e(t) = \sqrt{(A + m(t))^2 + \check{m}^2(t)} \approx A + m(t) \tag{3.18}$$

if $|A + m(t)| \gg |\check{m}(t)|$. Subject to the approximation in (3.18), an envelope detector works just as in conventional AM.

3.2.4 Vestigial-sideband (VSB) modulation

VSB is similar to SSB, in that it also tries to reduce the transmitted bandwidth relative to DSB, and the transmitted signal is a filtered version of the DSB signal. The idea is to mainly transmit one of the two sidebands, but to leave a *vestige* of the other sideband in order to ease the filtering requirements. The passband filter used to shape the DSB signal in this fashion is chosen so that the I component of the transmitted signal equals the message. To see this, consider the DSB-SC signal

$$2m(t)\cos(2\pi f_c t) \leftrightarrow M(f - f_c) + M(f + f_c)$$

This is filtered by a passband VSB filter with transfer function $H_p(f)$, as shown in Figure 3.17, to obtain the transmitted signal with spectrum

$$U_{\text{VSB}}(f) = H_p(f)(M(f - f_c) + M(f + f_c)) \tag{3.19}$$

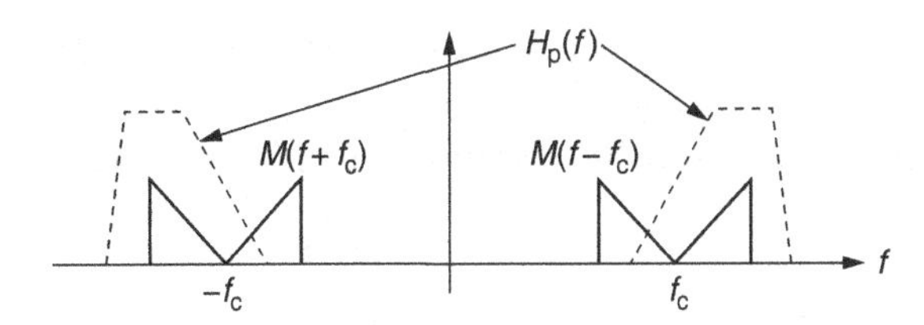

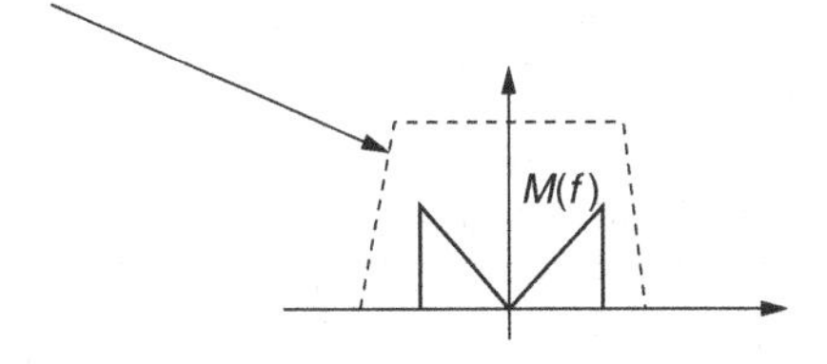

Figure 3.17 Relevant passband and baseband spectra for VSB.

A coherent demodulator extracting the I component passes $2u_{VSB}(t)\cos(2\pi f_c t)$ through a lowpass filter. But

$$2u_{VSB}(t)\cos(2\pi f_c t) \leftrightarrow U_{VSB}(f - f_c) + U_{VSB}(f + f_c)$$

which equals (substituting from (3.19)

$$H_p(f - f_c)(M(f - 2f_c) + M(f)) + H_p(f + f_c)(M(f) + M(f + 2f_c)) \tag{3.20}$$

The $2f_c$ term, $H_p(f - f_c)M(f - 2f_c) + H_p(f + f_c)M(f + 2f_c)$, is filtered out by the lowpass filter. The output of the LPF consists of the lowpass terms in (3.20), which equal the I component, and are given by

$$M(f)(H_p(f - f_c) + H_p(f + f_c))$$

In order for this to equal (a scaled version of) the desired message, we must have

$$H_p(f + f_c) + H_p(f - f_c) = \text{ constant}, \quad |f| < W \tag{3.21}$$

as shown in the example in Figure 3.17. To understand what this implies about the structure of the passband VSB filter, note that the filter impulse response can be written as $h_p(t) = h_c(t)\cos(2\pi f_c t) - h_s(t)\sin(2\pi f_c t)$, where $h_c(t)$ is obtained by passing $2h_p(t)\cos(2\pi f_c t)$ through a lowpass filter. But $2h_p(t)\cos(2\pi f_c t) \leftrightarrow H_p(f - f_c) + H_p(f + f_c)$. Thus, the Fourier transform involved in (3.21) is precisely the lowpass restriction of $2h_p(t)\cos(2\pi f_c t)$, i.e., it is $H_c(f)$. Thus, the correct demodulation condition for VSB in (3.21) is equivalent to requiring that $H_c(f)$ be constant over the message band. Further discussion of the structure of VSB signals is provided via problems.

As with SSB, if we add a strong carrier component to the VSB signal, we can demodulate it noncoherently using an envelope detector, again at the cost of some distortion arising from the presence of the Q component.

3.2.5 Quadrature amplitude modulation

The transmitted signal in quadrature amplitude modulation (QAM) is of the form

$$u_{QAM}(t) = m_c(t)\cos(2\pi f_c t) - m_s(t)\sin(2\pi f_c t)$$

where $m_c(t)$ and $m_s(t)$ are separate messages (unlike for SSB and VSB, where the Q component is a transformation of the message carried by the I component). In other words, a complex-valued message $m = m_c(t) + jm_s(t)$ is encoded in the complex envelope of the passband transmitted signal. QAM is extensively employed in digital communication, as we shall see in later chapters.

Demodulation is achieved using a coherent receiver that extracts both the I component and the Q component, as shown in Figure 3.18. If the received signal has a phase offset θ relative to the receiver's LO, then we get both attenuation in the desired message and interference from the undesired message, as follows. Ignoring noise and scale factors, the reconstructed complex-baseband message is given by

$$\hat{m}(t) = \hat{m}_c(t) + j\hat{m}_s(t) = (m_c(t) + jm_s(t))e^{j\theta(t)} = m(t)e^{j\theta(t)}$$

from which we conclude that

$$\hat{m}_c(t) = m_c(t)\cos\theta(t) - m_s(t)\sin\theta(t)$$
$$\hat{m}_s(t) = m_s(t)\cos\theta(t) + m_c(t)\sin\theta(t)$$

Thus, accurate carrier synchronization ($\theta(t)$ as close to zero as possible) is important for QAM demodulation to function properly.

3.2.6 Concept synthesis for AM

Here is a worked problem that synthesizes a few of the concepts we have discussed for AM.

Example 3.2.4 The signal $m(t) = 2\cos(20\pi t) - \cos(40\pi t)$, where the **unit of time is milliseconds**, is amplitude modulated using a carrier frequency f_c of 600 kHz. The AM signal is given by

$$x(t) = 5\cos(2\pi f_c t) + m(t)\cos(2\pi f_c t)$$

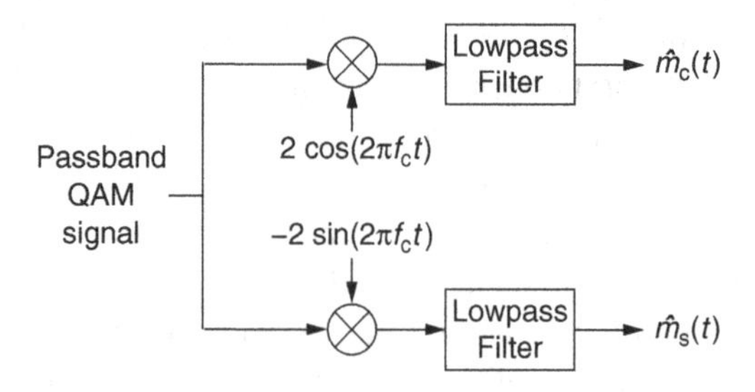

Figure 3.18 Demodulation for quadrature amplitude modulation.

(a) Sketch the magnitude spectrum of x. What is its bandwidth?

(b) What is the modulation index?

(c) The AM signal is passed through an ideal highpass filter with cutoff frequency 595 kHz (i.e., the filter passes all frequencies above 595 kHz, and cuts off all frequencies below 595 kHz). Find an explicit time-domain expression for the Q component of the filter output with respect to a 600-kHz frequency reference.

Solution

(a) The message spectrum $M(f) = \delta(f-10) + \delta(f+10) - \frac{1}{2}\delta(f-20) - \frac{1}{2}\delta(f+20)$. The spectrum of the AM signal is given by

$$X(f) = \frac{5}{2}\delta(f-f_c) + \frac{5}{2}\delta(f+f_c) + \frac{1}{2}M(f-f_c) + \frac{1}{2}M(f+f_c)$$

These spectra are sketched in Figure 3.19.

(b) The modulation index $a_{\mathrm{mod}} = M_0/A_c$, where $-M_0 = \min_t m(t)$. To simplify the notation, let us minimize $g(x) = 2\cos x - \cos(2x)$. We can actually do this by inspection: for $x = \pi$, $\cos x = -1$ and $\cos(2x) = 1$, so that $\min_x g(x) = -3$. Alternatively, we could set the derivative to zero: $g'(x) = -2\sin x + 2\sin(2x) = -2\sin x + 4\sin x\cos x = 2\sin x(-1 + 2\cos x)$ is satisfied if $\sin x = 0$ (i.e., $\cos x = \pm 1$) or $\cos x = \frac{1}{2}$. We can check that the first solution with $\cos x = -1$ minimizes $g(x)$. Thus, we obtain $M_0 = 3$ and hence $a_{\mathrm{mod}} = M_0/A_c = 3/5$ or 60%.

(c) From Figure 3.19, it is clear that a highpass filter with cutoff at 595 kHz selects the USB signal plus the carrier. The passband output has the spectrum shown in Figure 3.20(a), and the complex envelope with respect to 600 kHz is shown in Figure 3.20(b). On taking the inverse Fourier transform, the time-domain complex envelope is given by

$$\tilde{y}(t) = 5 + e^{j20\pi t} - \frac{1}{2}e^{j40\pi t}$$

We can now find the Q component to be

$$y_s(t) = \mathrm{Im}(\tilde{y}(t)) = \sin(20\pi t) - \frac{1}{2}\sin(40\pi t)$$

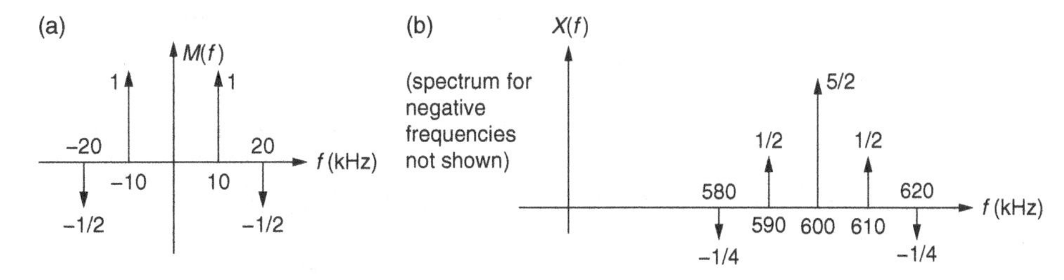

Figure 3.19 The spectrum of the message (a) and the corresponding AM signal (b) in Example 3.2.4. The axes are not to scale.

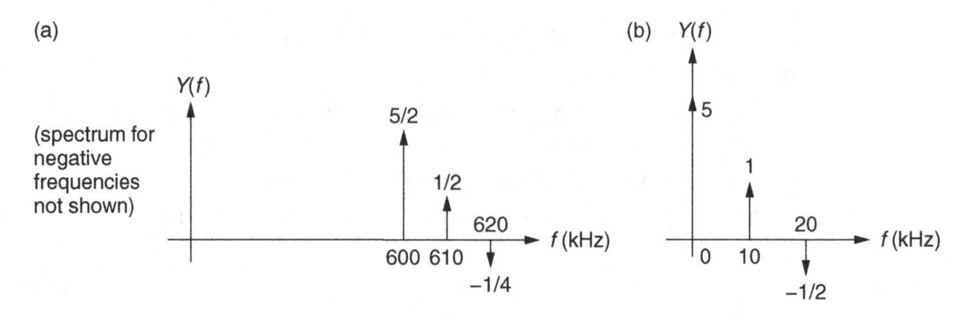

Figure 3.20 The passband output of a bandpass filter (a) and its complex envelope (b) with respect to a 600-kHz reference, for Example 3.2.4. The axes are not to scale.

where t is in milliseconds. Another approach is to recognize that the Q component is the Q component of the USB signal, which is known to be the Hilbert transform of the message. Yet another approach is to find the Q component in the frequency domain using $jY_s(f) = \left(\tilde{Y}(f) - \tilde{Y}^*(f)\right)/2$ and then take the inverse Fourier transform. In this particular example, the first approach is probably the simplest.

3.3 Angle modulation

We know that a passband signal can be represented as $e(t)\cos(2\pi f_c t + \theta(t))$, where $e(t)$ is the envelope, and $\theta(t)$ is the phase. Let us define the instantaneous frequency offset relative to the carrier as

$$f(t) = \frac{1}{2\pi}\frac{d\theta(t)}{dt}$$

In frequency modulation (FM) and phase modulation (PM), we encode information into the phase $\theta(t)$, with the envelope remaining constant. The transmitted signal is given by

$$u(t) = A_c \cos(2\pi f_c t + \theta(t)), \quad \textbf{angle modulation} \text{ (information carried in } \theta)$$

For a message $m(t)$, we have

$$\theta(t) = k_p m(t), \quad \textbf{phase modulation} \tag{3.22}$$

and

$$\frac{1}{2\pi}\frac{d\theta(t)}{dt} = f(t) = k_f m(t), \quad \textbf{frequency modulation} \tag{3.23}$$

where k_p and k_f are constants. Upon integrating (3.23), the phase of the FM waveform is given by

$$\theta(t) = \theta(0) + 2\pi k_f \int_0^t m(\tau)d\tau \tag{3.24}$$

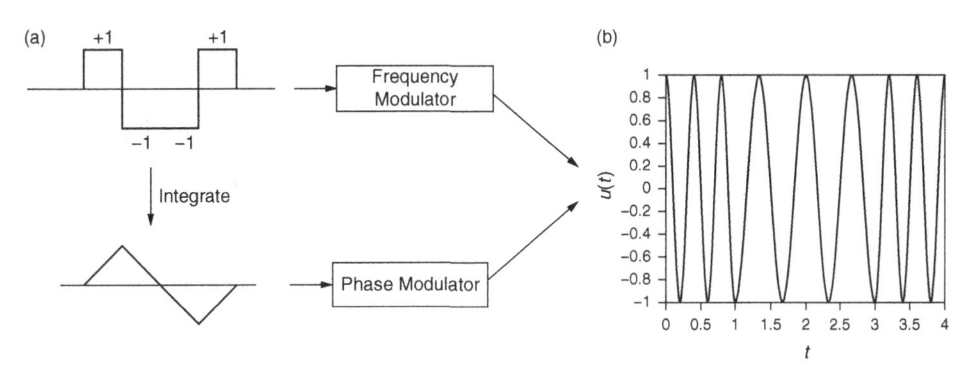

Figure 3.21 The equivalence of FM and PM: (a) messages used for angle modulation and (b) the angle-modulated signal.

On comparing (3.24) with (3.22), we see that FM is equivalent to PM with the integral of the message. Similarly, for differentiable messages, PM can be interpreted as FM, with the input to the FM modulator being the derivative of the message. Figure 3.21 provides an example illustrating this relationship; this is actually a digital modulation scheme called continuous phase modulation, as we shall see when we study digital communication. In this example, the digital message $+1, -1, -1, +1$ is the input to an FM modulator: the instantaneous frequency switches from $f_c + k_f$ (for one time unit) to $f_c - k_f$ (for two time units) and then back to $f_c + k_f$ again. The same waveform is produced when we feed the integral of the message into a PM modulator, as shown in Figure 3.21.

When the digital message of Figure 3.21 is input to a *phase* modulator, then we get a modulated waveform with phase discontinuities when the message changes sign. This is in contrast to the output in Figure 3.21, where the phase is continuous. That is, if we compare FM and PM for the same message, we infer that FM waveforms should have less abrupt phase transitions due to the smoothing resulting from integration: compare the expressions for the phases of the modulated signals in (3.22) and (3.24) for the same message $m(t)$. Thus, for a given level of message variations, we expect FM to have smaller bandwidth. FM is therefore preferred to PM for analog modulation, where the communication system designer does not have control over the properties of the message signal (e.g., the system designer cannot require the message to be smooth). For this reason, and also given the basic equivalence of the two formats, we restrict the discussion in the remainder of this section to FM for the most part. PM, however, is extensively employed in digital communication, where the system designer has significant flexibility in shaping the message signal. In this context, we use the term phase-shift keying (PSK) to denote the discrete nature of the information encoded in the message. Figure 3.22 is actually a simple example of PSK, although in practice, the phase of the modulated signal is shaped to be smoother in order to improve bandwidth efficiency.

Frequency deviation and modulation index The maximum deviation in instantaneous frequency due to a message $m(t)$ is given by

$$\Delta f_{\max} = k_f \max_t |m(t)|$$

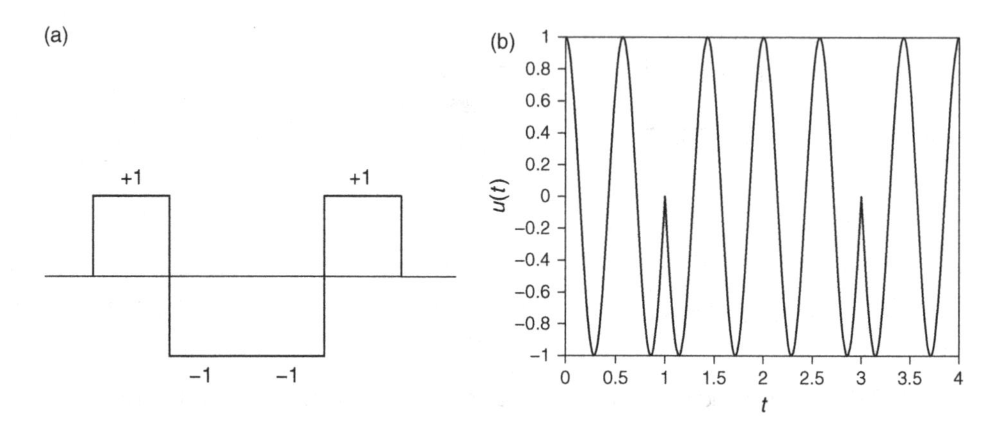

Figure 3.22 Phase discontinuities in a PM signal due to sharp message transitions: (a) digital input to the phase modulator and (b) the phase-shift keyed signal.

If the bandwidth of the message is B, the modulation index is defined as

$$\beta = \frac{\Delta f_{\max}}{B} = \frac{k_f \max_t |m(t)|}{B}$$

We use the term *narrowband FM* if $\beta < 1$ (typically much smaller than one), and the term *wideband FM* if $\beta > 1$. We discuss the bandwidth occupancy of FM signals in more detail a little later, but note for now that the bandwidth of narrowband FM signals is dominated by that of the message, while the bandwidth of wideband FM signals is dominated by the frequency deviation.

Consider the FM signal corresponding to a sinusoidal message $m(t) = A_m \cos(2\pi f_m t)$. The phase deviation due to this message is given by

$$\theta(t) = 2\pi k_f \int_0^t A_m \cos(2\pi f_m \tau) d\tau = \frac{A_m k_f}{f_m} \sin(2\pi f_m t)$$

Since the maximum frequency deviation $\Delta f_{\max} = A_m k_f$ and the message bandwidth $B = f_m$, the modulation index is given by $\beta = A_m k_f / f_m$, so the phase deviation can be written as

$$\theta(t) = \beta \sin(2\pi f_m t) \tag{3.25}$$

Modulation An FM modulator, by definition, is a voltage-controlled oscillator (VCO), whose output is a sinusoidal wave whose instantaneous frequency offset from a reference frequency is proportional to the input signal. VCO implementations are often based on the use of varactor diodes, which provide voltage-controlled capacitance, in *LC*-tuned circuits. This is termed *direct* FM modulation, in that the output of the VCO produces a passband signal with the desired frequency deviation as a function of the message. The VCO output may be at the desired carrier frequency, or at an intermediate frequency. In the latter scenario, it must be upconverted further to the carrier frequency, but this operation does not change the frequency modulation. Direct FM modulation may be employed for both narrowband and wideband modulation.

An alternative approach to wideband modulation is to first generate a narrowband FM signal (typically using a phase modulator), and then multiply the frequency (often over multiple stages) using nonlinearities, thus increasing the frequency deviation as well as the carrier frequency. This method, which is termed *indirect* FM modulation, is of historical importance, but is not used in present-day FM systems because direct modulation for wideband FM is now feasible and cost-effective.

Demodulation Many different approaches to FM demodulation have evolved over the past century. Here we discuss two important classes of demodulators: the limiter–discriminator demodulator in Section 3.3.1, and the phase-locked loop in Section 3.5.

3.3.1 Limiter–discriminator demodulation

The task of an FM demodulator is to convert frequency variations in the passband received signal into amplitude variations, thus recovering an estimate of the message. Ideally, therefore, an FM demodulator would produce the derivative of the phase of the received signal; this is termed a *discriminator,* as shown in Figure 3.23. While an ideal FM signal as in (3.26) does not have amplitude fluctuations, noise and channel distortions might create such fluctuations, leading to unwanted contributions to the discriminator output. In practice, therefore, as shown in Figure 3.23, the discriminator is typically preceded by a *limiter*, which removes amplitude fluctuations due to noise and channel distortions that lead to unwanted contributions to the discriminator output. This is achieved by passing the modulated sinusoidal waveform through a hardlimiter, which generates a square wave, and then selecting the right harmonic using a bandpass filter tuned to the carrier frequency. The overall structure is termed a *limiter–discriminator.*

Ideal limiter–discriminator Following the limiter, we have an FM signal of the form

$$y_p(t) = A\cos(2\pi f_c t + \theta(t))$$

where $\theta(t)$ may include contributions due to channel and noise impairments (to be discussed later), as well as the angle modulation due to the message. An ideal discriminator now produces the output $d\theta(t)/dt$ (where we ignore scaling factors).

A crude realization of a discriminator, which converts fluctuations in frequency to fluctuations in envelope, is shown in Figure 3.24. On taking the derivative of the FM signal

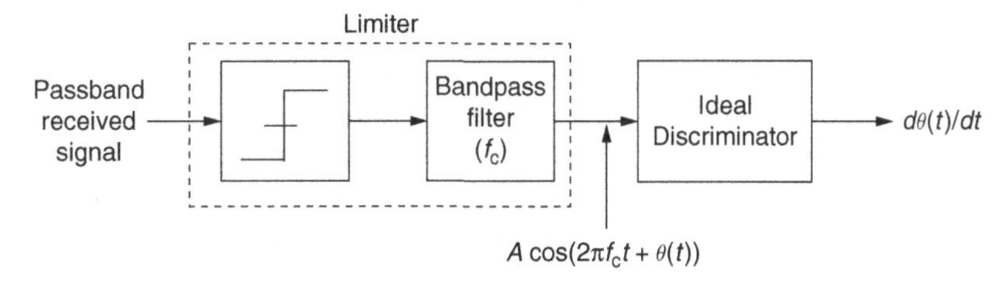

Figure 3.23 Limiter–discriminator demodulation of FM.

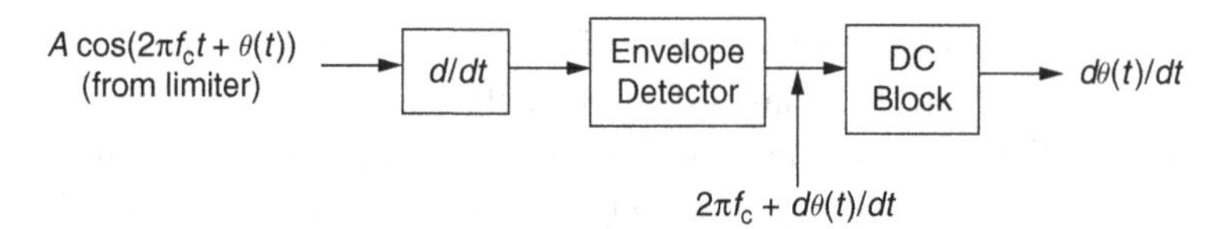

Figure 3.24 A crude discriminator based on differentiation and envelope detection.

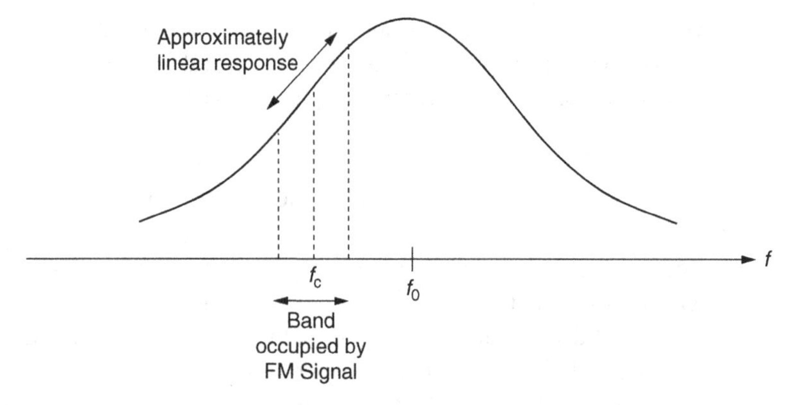

Figure 3.25 A slope detector using a tuned circuit offset from resonance.

$$u_{FM}(t) = A_c \cos\left(2\pi f_c t + 2\pi k_f \int_0^t m(\tau)d\tau + \theta_0\right) \tag{3.26}$$

we have

$$v(t) = \frac{du_{FM}(t)}{dt} = -A_c(2\pi f_c + 2\pi k_f m(t))\sin\left(2\pi f_c t + 2\pi k_f \int_0^t m(\tau)d\tau + \theta_0\right)$$

The envelope of $v(t)$ is $2\pi A_c|f_c + k_f m(t)|$. On noting that $k_f m(t)$ is the instantaneous frequency deviation from the carrier, whose magnitude is much smaller than f_c for a properly designed system, we realize that $f_c + k_f m(t) > 0$ for all t. Thus, the envelope equals $2\pi A_c(f_c + k_f m(t))$, so that passing the discriminator output through an envelope detector yields a scaled and DC-shifted version of the message. Using AC coupling to reject the DC term, we obtain a scaled version of the message $m(t)$, just as in conventional AM.

The discriminator as described above corresponds to the frequency-domain transfer function $H(f) = j2\pi f$, and can therefore be approximated (up to DC offsets) by transfer functions that are approximately linear over the FM band of interest. An example of such a *slope detector* is given in Figure 3.25, where the carrier frequency f_c is chosen at an offset from the resonance frequency f_0 of a tuned circuit.

One problem with the simple discriminator and its approximations is that the envelope-detector output has a significant DC component: when we get rid of this using AC coupling, we also attenuate low-frequency components near DC. This limitation can be overcome by employing circuits that rely on the approximately linear variations in amplitude and phase of tuned circuits around resonance to synthesize approximations to an ideal discriminator whose output is the derivative of the phase. These include the Foster–Seely detector and the ratio detector. Circuit-level details of such implementations are beyond our scope.

3.3.2 FM spectrum

We first consider a naive but useful estimate of FM bandwidth termed *Carson's rule.* We then show that the spectral properties of FM are actually quite complicated, even for a simple sinusoidal message, and outline methods for obtaining more detailed bandwidth estimates.

Consider an angle-modulated signal, $u_p(t) = A_c \cos(2\pi f_c t + \theta(t))$, where $\theta(t)$ contains the message information. For a baseband message $m(t)$ of bandwidth B, the phase $\theta(t)$ for PM is also a baseband signal with the same bandwidth. The phase $\theta(t)$ for FM is the integral of the message. Since integration smooths out the time-domain signal, or equivalently, attenuates higher frequencies, $\theta(t)$ is a baseband signal with bandwidth at most B. We therefore loosely think of $\theta(t)$ as having a bandwidth equal to B, the message bandwidth, for the remainder of this section.

The complex envelope of u_p with respect to f_c is given by

$$u(t) = A_c e^{j\theta(t)} = A_c \cos\theta(t) + jA_c \sin\theta(t)$$

Now, if $|\theta(t)|$ is small, as is the case for narrowband angle modulation, then $\cos\theta(t) \approx 1$ and $\sin\theta(t) \approx \theta(t)$, so the complex envelope is approximately given by

$$u(t) \approx A_c + jA_c\theta(t)$$

Thus, the passband signal is approximately given by

$$u_p(t) \approx A_c \cos(2\pi f_c t) - \theta(t) A_c \sin(2\pi f_c t)$$

Thus, the I component has a large unmodulated carrier contribution as in conventional AM, but the message information is now in the Q component instead of in the I component, as in AM. The Fourier transform is given by

$$U_p(f) = \frac{A_c}{2}(\delta(f - f_c) + \delta(f + f_c)) - \frac{A_c}{2j}(\Theta(f - f_c) - \Theta(f + f_c))$$

where $\Theta(f)$ denotes the Fourier transform of $\theta(t)$. The magnitude spectrum is therefore given by

$$|U_p(f)| = \frac{A_c}{2}(\delta(f - f_c) + \delta(f + f_c)) + \frac{A_c}{2}(|\Theta(f - f_c)| + |\Theta(f + f_c)|) \qquad (3.27)$$

Thus, the bandwidth of a *narrowband* FM signal is $2B$, or twice the message bandwidth, just as in AM. For example, narrowband angle modulation with a sinusoidal message $m(t) = \cos(2\pi f_m t)$ occupies a bandwidth of $2f_m$: $\theta(t) = (k_f/f_m)\sin(2\pi f_m t)$ for FM, and $\theta(t) = k_p \cos(2\pi f_m t)$ for PM.

For wideband FM, we would expect the bandwidth to be dominated by the frequency deviation $k_f m(t)$. For messages that have positive and negative peaks of similar size, the frequency deviation ranges between $-\Delta f_{max}$ and Δf_{max}, where $\Delta f_{max} = k_f \max_t |m(t)|$. In this case, we expect the bandwidth to be dominated by the instantaneous deviations around the carrier frequency, which spans an interval of length $2\,\Delta f_{max}$.

Carson's rule This is an estimate for the bandwidth of a general FM signal, which is based on simply adding up the estimates from our separate discussion of narrowband and wideband modulation:

$$B_{\mathrm{FM}} \approx 2B + 2\,\Delta f_{\max} = 2B(\beta + 1), \quad \textbf{Carson's rule} \tag{3.28}$$

where $\beta = \Delta f_{\max}/B$ is the modulation index, which is also called the FM deviation ratio, as defined earlier.

FM spectrum for a sinusoidal message In order to get more detailed insight into what the spectrum of an FM signal looks like, let us now consider the example of a sinusoidal message, for which the phase deviation is given by $\theta(t) = \beta \sin(2\pi f_{\mathrm{m}} t)$, from (3.25). The complex envelope of the FM signal with respect to f_{c} is given by

$$u(t) = e^{j\theta(t)} = e^{j\beta \sin(2\pi f_{\mathrm{m}} t)}$$

Since the sinusoid in the exponent is periodic with period $1/f_{\mathrm{m}}$, so is $u(t)$. It can therefore be expanded into a Fourier series of the form

$$u(t) = \sum_{n=-\infty}^{\infty} u[n] e^{j2\pi n f_{\mathrm{m}} t}$$

where the Fourier coefficients $\{u[n]\}$ are given by

$$u[n] = f_{\mathrm{m}} \int_{-1/(2f_{\mathrm{m}})}^{1/(2f_{\mathrm{m}})} u(t) e^{-j2\pi n f_{\mathrm{m}} t}\, dt = f_{\mathrm{m}} \int_{-1/(2f_{\mathrm{m}})}^{1/(2f_{\mathrm{m}})} e^{j\beta \sin(2\pi f_{\mathrm{m}} t)} e^{-j2\pi n f_{\mathrm{m}} t}\, dt$$

Using the change of variables $2\pi f_{\mathrm{m}} t = x$, we have

$$u[n] = \frac{1}{2\pi} \int_{-\pi}^{\pi} e^{j(\beta \sin x - nx)}\, dx = J_n(\beta)$$

where $J_n(\cdot)$ is the Bessel function of the first kind of order n. While the integrand above is complex-valued, the integral is real-valued. To see this, use Euler's formula:

$$e^{j(\beta \sin x - nx)} = \cos(\beta \sin x - nx) + j \sin(\beta \sin x - nx)$$

Since $\beta \sin x - nx$ and the sine function are both odd, the imaginary term $\sin(\beta \sin x - nx)$ above is an odd function, and integrates out to zero over $[-\pi, \pi]$. The real part is even, hence the integral over $[-\pi, \pi]$ is simply twice that over $[0, \pi]$. We summarize as follows:

$$u[n] = J_n(\beta) = \frac{1}{2\pi} \int_{-\pi}^{\pi} e^{j(\beta \sin x - nx)}\, dx = \frac{1}{\pi} \int_0^{\pi} \cos(\beta \sin x - nx) dx \tag{3.29}$$

Bessel functions are available in mathematical software packages such as MATLAB and Mathematica. Figure 3.26 shows some Bessel-function plots. Some properties of Bessel-functions worth noting are as follows.

- For n integer, $J_n(\beta) = (-1)^n J_{-n}(\beta) = (-1)^n J_n(-\beta)$.
- For fixed β, $J_n(\beta)$ tends to zero fast as n gets large, so the complex envelope is well approximated by a finite number of Fourier-series components. In particular, a good approximation is that $J_n(\beta)$ is small for $|n| > \beta + 1$. This leads to an approximation for

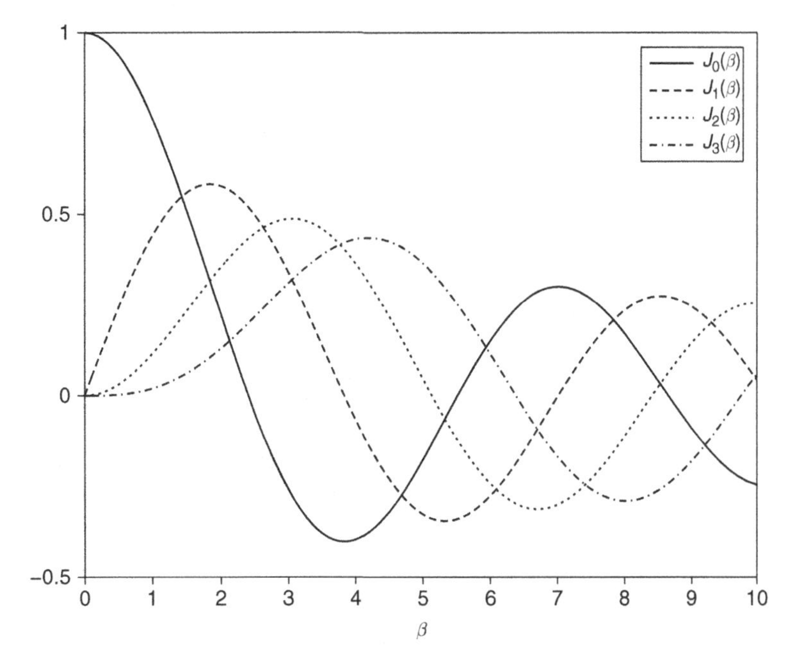

Figure 3.26 Bessel functions of the first kind, $J_n(\beta)$ versus β, for $n = 0, 1, 2, 3$.

the bandwidth of the FM signal given by $2(\beta + 1)f_{\mathrm{m}}$, which is consistent with Carson's rule.

- For fixed n, $J_n(\beta)$ vanishes for specific values of β, a fact that can be used for spectral shaping.

To summarize, the complex envelope of an FM signal modulated by a sinusoidal message can be written as

$$u(t) = e^{j\beta \sin(2\pi f_{\mathrm{m}} t)} = \sum_{n=-\infty}^{\infty} J_n(\beta) e^{j2\pi n f_{\mathrm{m}} t} \tag{3.30}$$

The corresponding spectrum is given by

$$U(f) = \sum_{n=-\infty}^{\infty} J_n(\beta) \delta(f - n f_{\mathrm{m}}) \tag{3.31}$$

Noting that $|J_{-n}(\beta)| = |J_n(\beta)|$, the complex envelope has discrete frequency components at $\pm n f_{\mathrm{m}}$ of strength $|J_n(\beta)|$: these correspond to frequency components at $f_{\mathrm{c}} \pm n f_{\mathrm{m}}$ in the passband FM signal.

Fractional power containment bandwidth By virtue of Parseval's identity for Fourier series, the power of the complex envelope is given by

$$1 = |u(t)|^2 = \overline{|u(t)|^2} = \sum_{n=-\infty}^{\infty} J_n^2(\beta) = J_0^2(\beta) + 2\sum_{n=1}^{\infty} J_n^2(\beta)$$

We can compute the fractional power-containment bandwidth as $2Kf_m$, where $K \geq 1$ is the smallest integer such that

$$J_0^2(\beta) + 2\sum_{n=1}^{K} J_n^2(\beta) \geq \alpha$$

where α is the desired fraction of power within the band (e.g., $\alpha = 0.99$ for the 99% power-containment bandwidth). For integer values of $\beta = 1, \ldots, 10$, we find that $K = \beta + 1$ provides a good approximation to the 99% power-containment bandwidth, which is again consistent with Carson's formula.

3.3.3 Concept synthesis for FM

The following worked problem brings together some of the concepts we have discussed regarding FM.

Example 3.3.1 The signal $a(t)$ shown in Figure 3.27 is fed to a VCO with a quiescent frequency of 5 MHz and a frequency deviation of 25 kHz/mV. Denote the output of the VCO by $y(t)$.

(a) Provide an estimate of the bandwidth of y. Clearly state the assumptions that you make.
(b) The signal $y(t)$ is passed through an ideal bandpass filter of bandwidth 5 kHz, centered at 5.005 MHz. Provide the simplest possible expression for the power at the filter output (if you can give a numerical answer, do so).

Solution

(a) The VCO output is an FM signal with

$$\Delta f_{\max} = k_f \max_t m(t) = 25 \text{ kHz/mV} \times 2 \text{ mV} = 50 \text{ kHz}$$

The message is periodic with period 100 microseconds, hence its fundamental frequency is 10 kHz. Approximating its bandwidth by its first harmonic, we have $B \approx 10$ kHz. Using Carson's formula, we can approximate the bandwidth of the FM signal at the VCO output as

$$B_{\text{FM}} \approx 2\,\Delta f_{\max} + 2B \approx 120 \text{ kHz}$$

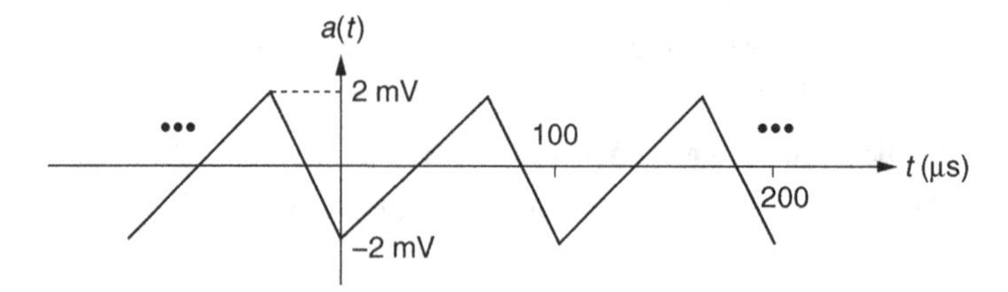

Figure 3.27 Input to the VCO in Example 3.3.1.

(b) The complex envelope of the VCO output is given by $e^{j\theta(t)}$, where

$$\theta(t) = 2\pi k_{\rm f} \int m(\tau) d\tau$$

For periodic messages with zero DC value (as is the case for $m(t)$ here), $\theta(t)$, and hence $e^{j\theta(t)}$, has the same period as the message. We can therefore express the complex envelope as a Fourier series with complex exponentials at frequencies $nf_{\rm m}$, where $f_{\rm m} =$ 10 kHz is the fundamental frequency for the message, and where n takes integer values. Thus, the FM signal has discrete components at $f_{\rm c} + nf_{\rm m}$, where $f_{\rm c} = 5$ MHz in this example. A bandpass filter at 5.005 MHz with bandwidth 5 kHz does not capture any of these components, since it spans the interval [5.0025, 5.0075] MHz, whereas the nearest Fourier components are at 5 MHz and 5.01 MHz. Thus, the power at the output of the bandpass filter is zero.

3.4 The superheterodyne receiver

The receiver in a radio communication system must *downconvert* the passband received signal down to baseband in order to recover the message. At the turn of the twentieth century, it was difficult to produce amplification at frequencies beyond a few MHz using the vacuum-tube technology of that time. However, higher carrier frequencies are desirable because of the larger available bandwidths and the smaller antennas required. The invention of the *superheterodyne*, or *superhet*, receiver was motivated by these considerations. Basically, the idea is to use sloppy design for front-end filtering of the received radio-frequency (RF) signal, and for translating it to a lower *intermediate frequency (IF)*. The IF signal is then processed using carefully designed filters and amplifiers. Subsequently, the IF signal can be converted to baseband in a number of different ways: for example, using an envelope detector for AM radio, a phase-locked loop or discriminator for FM radio, and a coherent quadrature demodulator for digital cellular telephone receivers (Figure 3.28).

While the original motivation for the superheterodyne receiver is no longer strictly applicable (modern analog electronics is capable of providing amplification at the carrier frequencies in commercial use), it is still true that gain is easier to provide at lower

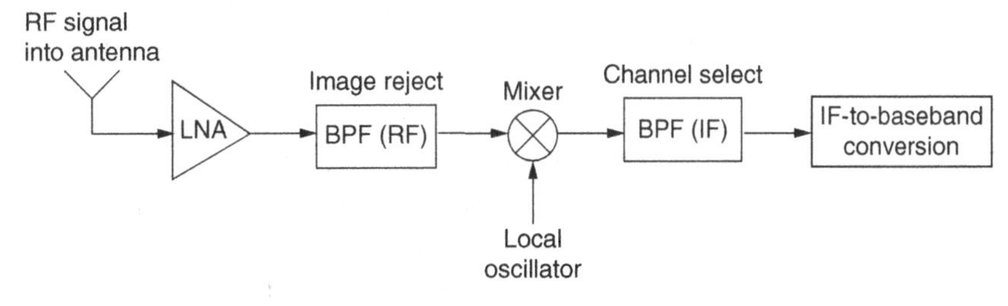

Figure 3.28 A generic block diagram for a superhet receiver.

frequencies than at higher frequencies. Furthermore, it becomes possible to closely optimize the processing at a fixed IF (in terms of amplifier and filter design), while permitting a tunable RF front end with more relaxed specifications, which is important for the design of radios that operate over a wide range of carrier frequencies. For example, the superhet architecture is commonly employed for AM and FM broadcast radio receivers, where the RF front end tunes to the desired station, translating the received signal to a fixed IF. Radio receivers built with discrete components often take advantage of the widespread availability of inexpensive filters at certain commonly used IF frequencies, such as 455 kHz (used for AM radio) and 10.7 MHz (used for FM radio). As carrier frequencies scale up to the GHz range (as is the case for modern digital cellular and wireless local-area network transceivers), circuit components shrink with the carrier wavelength, and it becomes possible to implement RF amplifiers and filters using integrated circuits. In such settings, a *direct-conversion* architecture, in which the passband signal is directly translated to baseband, becomes increasingly attractive, as discussed later in this section.

The key element in frequency translation is a mixer, which multiplies two input signals. For our present purpose, one of these inputs is a passband received signal $A\cos(2\pi f_{RF}t+\theta)$, where the envelope $A(t)$ and phase $\theta(t)$ are baseband signals that contain message information. The second input is a *local oscillator (LO)* signal, which is a locally generated sinusoid $\cos(2\pi f_{LO}t)$ (we set the LO phase to zero without loss of generality, effectively adopting it as our phase reference). The output of the mixer is therefore given by

$$A\cos(2\pi f_{RF}t+\theta)\cos(2\pi f_{LO}t) = \frac{A}{2}\cos(2\pi(f_{RF}-f_{LO})t+\theta) + \frac{A}{2}\cos(2\pi(f_{RF}+f_{LO})t+\theta)$$

Thus, there are two frequency components at the output of the mixer, $f_{RF}+f_{LO}$ and $|f_{RF}-f_{LO}|$ (remember that we need to talk only about positive frequencies when discussing physically realizable signals, due to the conjugate symmetry of the Fourier transform of real-valued time signals). In the superhet receiver, we set one of these as our IF, typically the difference frequency: $f_{IF} = |f_{RF}-f_{LO}|$.

For a given RF and a fixed IF, we therefore have two choices of LO frequency when $f_{IF} = |f_{RF}-f_{LO}|$: $f_{LO} = f_{RF}-f_{IF}$ and $f_{LO} = f_{RF}+f_{IF}$. To continue the discussion, let us consider the example of AM broadcast radio, which operates over the band from 540 to 1600 kHz, with 10 kHz spacing between the carrier frequencies for different stations. The audio message signal is limited to 5-kHz bandwidth, modulated using conventional AM to obtain an RF signal of bandwidth 10 kHz. Figure 3.29 shows a block diagram for the superhet architecture commonly used in AM receivers. The RF bandpass filter must be tuned to the carrier frequency for the desired station, and, at the same time, the LO frequency into the mixer must be chosen so that the difference frequency equals the IF frequency of 455 kHz. If $f_{LO} = f_{RF}+f_{IF}$, then the LO frequency ranges from 995 to 2055 kHz, corresponding to an approximately two-fold variation in tuning range. If $f_{LO} = f_{RF}-f_{IF}$, then the LO frequency ranges from 85 to 1145 kHz, corresponding to a more than 13-fold variation in tuning range. The first choice is therefore preferred, because it is easier to implement a tunable oscillator over a smaller tuning range.

Having fixed the LO frequency, we have a desired signal at $f_{RF} = f_{LO}-f_{IF}$ that leads to a component at IF, and potentially an undesired *image frequency* at $f_{IM} = f_{LO}+f_{IF} =$

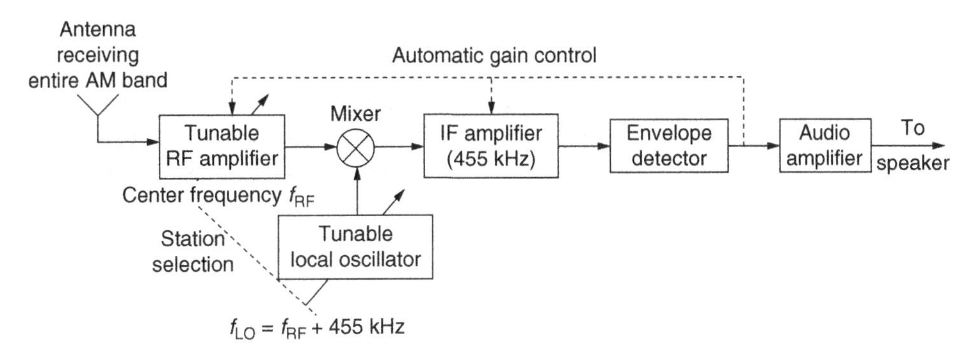

Figure 3.29 A superhet AM receiver.

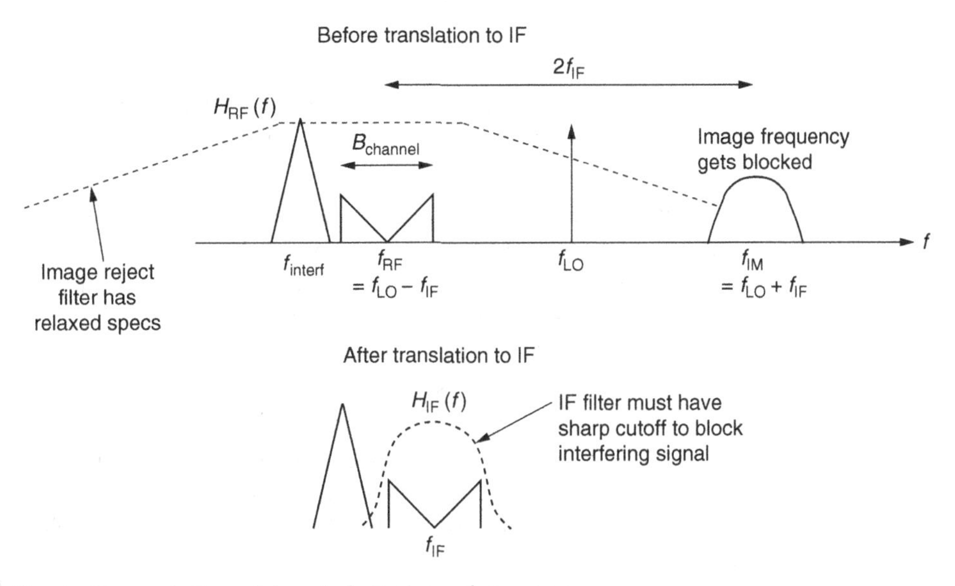

Figure 3.30 The role of image rejection and channel selection in superhet receivers.

$f_{RF} + 2f_{IF}$ that also leads to a component at IF. The job of the RF bandpass filter is to block this image frequency. Thus, the filter must let in the desired signal at f_{RF} (so that its bandwidth must be larger than 10 kHz), but severely attenuate the image frequency which is 910 kHz away from the center frequency. It is therefore termed an *image-reject* filter. We see that, for the AM broadcast radio application, a superhet architecture allows us to design the tunable image-reject filter to somewhat relaxed specifications. However, the image-reject filter does let in not only the signal from the desired station, but also those from adjacent stations. It is the job of the IF filter, which is tuned to the fixed frequency of 455 kHz, to filter out these adjacent stations. For this purpose, we use a highly selective filter at IF with a bandwidth of 10 kHz. Figure 3.30 illustrates these design considerations more generally.

Receivers for FM broadcast radio also commonly use a superhet architecture. The FM broadcast band ranges from 88 to 108 MHz, with a carrier-frequency separation of 200 kHz between adjacent stations. The IF is chosen at 10.7 MHz, so that the LO is tuned from 98.7 to 118.7 MHz for the choice $f_{LO} = f_{RF} + f_{IF}$. The RF filter specifications remain relaxed: it has to let in the desired signal of bandwidth 200 kHz, while rejecting an image frequency that is $2f_{IF} = 21.4$ MHz away from its center frequency. We discuss the structure of the FM broadcast signal, particularly the way in which stereo FM is transmitted, in more detail in Section 3.6.

Roughly indexing the difficulty of implementing a filter by the ratio of its center frequency to its bandwidth, or its *Q factor,* with high Q being more difficult to implement, we have the following fundamental tradeoff for superhet receivers. If we use a large IF, then the Q needed for the image reject filter is smaller. On the other hand, the Q needed for the IF filter to reject an interfering signal whose frequency is near that of the desired signal becomes higher. In modern digital communication applications, superheterodyne reception with multiple IF stages may be used in order to work around this tradeoff, in order to achieve the desired gain for the signal of interest and to attenuate sufficiently interference from other signals, while achieving an adequate degree of image rejection. Image rejection can be enhanced by employing appropriately designed image-reject mixer architectures.

Direct conversion receivers With the trend towards increasing monolithic integration of digital communication transceivers for applications such as cellular telephony and wireless local area networks, the superhet architecture is often being supplanted by *direct-conversion* (or *zero IF*) receivers, in which the passband received signal is directly converted down to baseband using a quadrature mixer at the RF carrier frequency. In this case, the desired signal is its own image, which removes the necessity for image rejection. Moreover, interfering signals can be filtered out at baseband, often using sophisticated digital signal processing after analog-to-digital conversion (ADC), provided that there is enough dynamic range in the circuitry to prevent a strong interferer from swamping the desired signal prior to the ADC. In contrast, the high-Q bandpass filters required for image rejection and interference suppression in the superhet design must often be implemented off-chip using, for example, surface-acoustic-wave (SAW) devices, which are bulky and costly. Thus, direct conversion is in some sense the "obvious" thing to do, except that, historically, people were unable to make it work, leading to the superhet architecture serving as the default design throughout most of the twentieth century.

A key problem with direct conversion is that LO leakage into the RF input of the mixer causes self-mixing, leading to a DC offset. While a DC offset can be calibrated out, the main problem is that it can saturate the amplifiers following the mixer, thus swamping out the contribution of the weaker received signal. Note that the DC offset due to LO leakage is not a problem with a superhet architecture, since the DC term gets filtered out by the passband IF filter. Other problems with direct conversion include $1/f$ noise and susceptibility to second-order nonlinearities, but discussion of these issues is beyond our current scope. However, since the 1990s, integrated-circuit designers have managed to overcome these and other obstacles, and direct-conversion receivers have become the norm for monolithic implementations of modern digital communication transceivers. These include cellular systems in various licensed bands ranging from 900 MHz to 2 GHz, and WLANs in the 2.4-GHz and 5-GHz unlicensed bands.

The insatiable demand for communication bandwidth virtually assures us that we will seek to exploit frequency bands well beyond 5 GHz, and circuit designers will be making informed choices between the superhet and direct conversion architectures for radios at these higher frequencies. For example, the 60-GHz band in the United States has 7 GHz of unlicensed spectrum; this band is susceptible to oxygen absorption, and is ideally suited for short-range (e.g. over a distance of 10–500 m) communication both indoors and outdoors. Similarly, the 71–76-GHz and 81–86-GHz bands, which avoid oxygen-absorption loss, are available for semi-unlicensed point-to-point "last-mile" links. Just as with cellular and WLAN applications in lower-frequency bands, we expect that proliferation of applications using these "millimeter (mm)-wave" bands would require low-cost integrated-circuit transceiver implementations. Given the trends at lower frequencies, one is tempted to conjecture that initial circuit designs might be based on superhet architectures, with direct-conversion receivers becoming subsequently more popular as designers become more comfortable with working at these higher frequencies. It is interesting to note that the design experience at lower carrier frequencies does not go to waste; for example, direct-conversion receivers at, say, 5 GHz, can serve as the IF stage for superhet receivers for mm-wave communication.

3.5 The phase-locked loop

The phase-locked loop (PLL) is an effective FM demodulator, but also has a far broader range of applications, including frequency synthesis and synchronization. We therefore treat it separately from our coverage of FM. The PLL provides a canonical example of the use of feedback for estimation and synchronization in communication systems, a principle that is employed in variants such as the Costas loop and the delay-locked loop.

The key idea behind the PLL, depicted in Figure 3.31, is as follows: we would like to lock on to the phase of the input to the PLL. We compare the phase of the input with that of the output of a voltage-controlled oscillator (VCO) using a phase detector. The difference between the phases drives the input of the VCO. If the VCO output is ahead of the PLL input in phase, then we would like to retard the VCO output phase. If the VCO output is behind the PLL input in phase, we would like to advance the VCO output phase. This is done by using the phase difference to control the VCO input. Typically, rather than using the output of the phase detector directly for this purpose, we smooth it out using a loop filter in order to reduce the effect of noise.

Mixer as phase detector The classical analog realization of the PLL is based on using a mixer (i.e., a multiplier) as a phase detector. To see how this works, consider the product of two sinusoids whose phases we are trying to align:

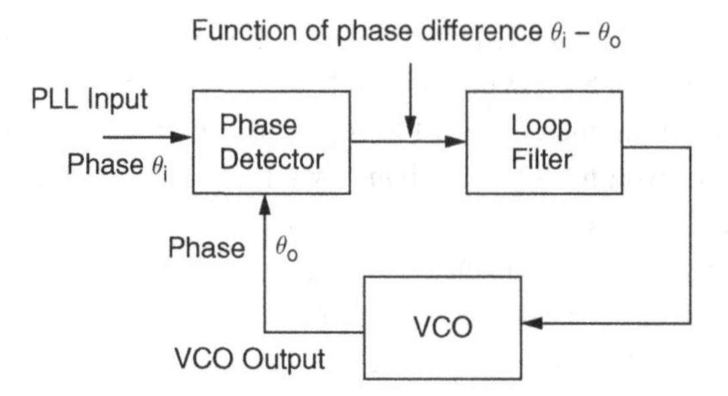

Figure 3.31 The PLL block diagram.

$$\cos(2\pi f_c t + \theta_1)\cos(2\pi f_c t + \theta_2) = \frac{1}{2}\cos(\theta_1 - \theta_2) + \frac{1}{2}\cos(4\pi f_c t + \theta_1 + \theta_2)$$

The second term on the right-hand side is a passband signal at $2f_c$, which can be filtered out by a lowpass filter. The first term contains the phase difference $\theta_1 - \theta_2$, and is to be used to drive the VCO so that we eventually match the phases. Thus, the first term should be small when we are near a phase match. Since the driving term is the *cosine* of the phase difference, the phase-match condition is $\theta_1 - \theta_2 = \pi/2$. That is, using a mixer as our phase detector means that, when the PLL is locked, the phase at the VCO output is offset by 90° from the phase of the PLL input. Now that we know this, we adopt a more convenient notation, changing variables to define a phase difference whose value at the desired matched state is zero rather than $\pi/2$. Let the PLL input be denoted by $A_c\cos(2\pi f_c + \theta_i(t))$, and let the VCO output be denoted by $A_v\cos(2\pi f_c + \theta_o(t) + \pi/2) = -A_v\sin(2\pi f_c + \theta_o(t))$. The output of the mixer is now given by

$$-A_c A_v \cos(2\pi f_c + \theta_i(t))\sin(2\pi f_c + \theta_o(t))$$

$$= (A_c A_v/2)\sin(\theta_i(t) - \theta_o(t)) - (A_c A_v/2)\sin(4\pi f_c t + \theta_i(t) + \theta_o(t))$$

The second term on the right-hand side is a passband signal at $2f_c$, which can be filtered out as before. The first term is the desired driving term, and, with the change of notation, we note that the desired state, when the driving term is zero, corresponds to $\theta_i = \theta_o$. The mixer-based realization of the PLL is shown in Figure 3.32.

The instantaneous frequency of the VCO is proportional to its input. Thus the phase of the VCO output, $-\sin(2\pi f_c t + \theta_o(t))$, is given by

$$\theta_o(t) = K_v \int_0^t x(\tau)d\tau$$

ignoring integration constants. On taking Laplace transforms, we have $\Theta_o(s) = K_v X(s)/s$. The reference frequency f_c is chosen as the *quiescent frequency* of the VCO, which is the frequency it would produce when its input voltage is zero.

Mixed-signal phase detectors Modern hardware realizations of the PLL, particularly for applications involving digital waveforms (e.g., a clock signal), often realize the phase detector using digital logic. The most rudimentary of these is an exclusive-or (XOR) gate, as shown in Figure 3.33. For the scenario depicted in that figure, we see that the average

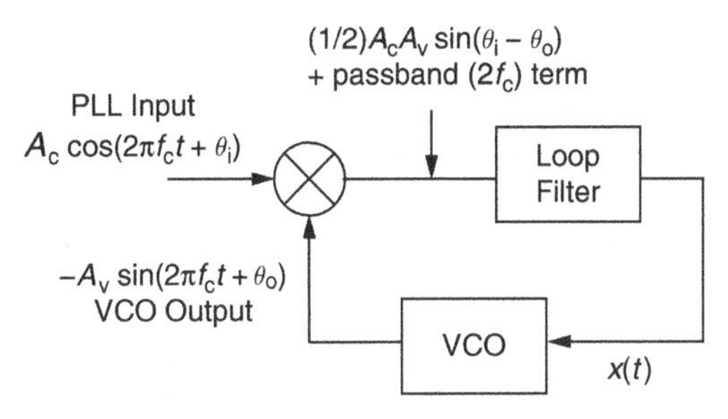

Figure 3.32 PLL realization using a mixer as the phase detector.

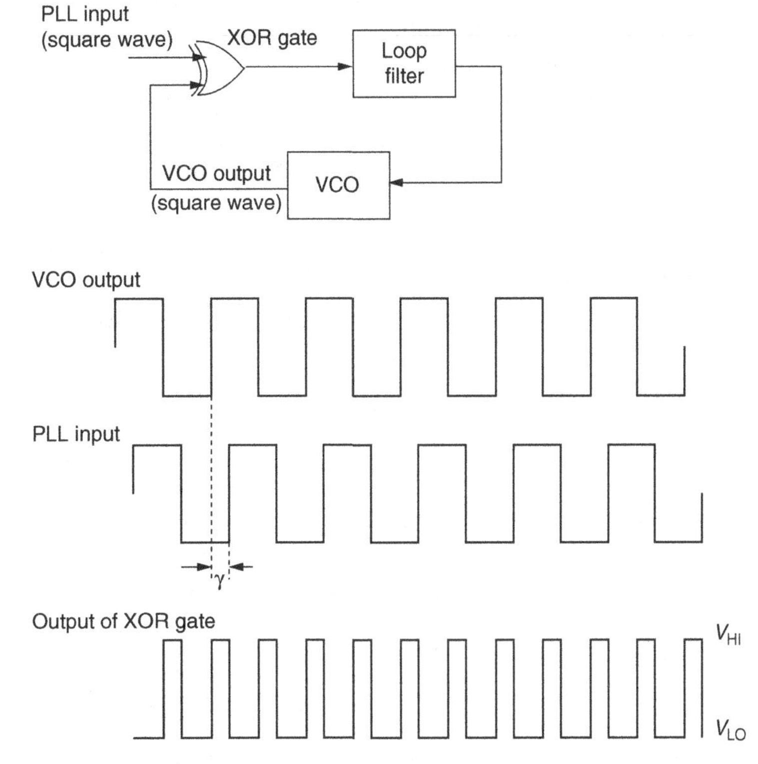

Figure 3.33 PLL realization using an XOR gate as the phase detector.

value of the output of the XOR gate is linearly related to the phase offset γ. On normalizing a period of the square wave to length 2π, this DC value V' is related to γ as shown in Figure 3.34(a). Note that, for zero phase offset, we have $V' = V_{HI}$, and that the response is symmetric around $\gamma = 0$. In order to get a linear phase-detector response going through the origin, we translate this curve along both axes: we define $V = V' - (V_{LO} + V_{HI})/2$ as a centered response, and we define the phase offset $\theta = \gamma - \pi/2$. Thus, the lock condition ($\theta = 0$) corresponds to the square waves being 90° out of phase. This translation gives us the phase response shown in Figure 3.34(b), which looks like a triangular version of the sinusoidal response for the mixer-based phase detector.

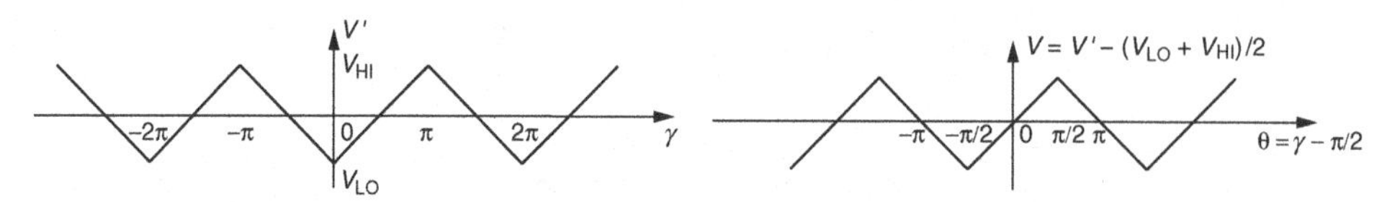

Figure 3.34 The response for the XOR phase detector: (a) the DC value of the output of the XOR gate and (b) the XOR phase detector output after axes translation.

The simple XOR-based phase detector has the disadvantage of requiring that the waveforms have 50% duty cycle. In practice, more sophisticated phase detectors, often based on edge detection, are used. These include "phase-frequency detectors" that directly provide information on frequency differences, which is useful for rapid locking. While discussion of the many phase detector variants employed in hardware design is beyond our scope, references for further study are provided at the end of this chapter.

3.5.1 PLL applications

Before trying to understand how a PLL works in more detail, let us discuss how we would use it, assuming that it has been properly designed. That is, suppose we can design a system such as that depicted in Figure 3.32, such that $\theta_o(t) \approx \theta_i(t)$. What would we do with such a system?

PLL as FM demodulator If the PLL input is an FM signal, its phase is given by

$$\theta_i(t) = 2\pi k_f \int_0^t m(\tau)d\tau$$

The VCO output phase is given by

$$\theta_o(t) = K_v \int_0^t x(\tau)d\tau$$

If $\theta_o \approx \theta_i$, then $d\theta_o/dt \approx d\theta_i/dt$, so that

$$K_v x(t) \approx 2\pi k_f m(t)$$

That is, the VCO input is approximately equal to a scaled version of the message. Thus, the PLL is an FM demodulator, where the FM signal is the input to the PLL, and the demodulator output is the VCO input, as shown in Figure 3.35.

PLL as frequency synthesizer The PLL is often used to synthesize the local oscillators used in communication transmitters and receivers. In a typical scenario, we might have a crystal oscillator that provides an accurate frequency reference at a relatively low frequency, say 40 MHz. We wish to use this to derive an accurate frequency reference at a higher frequency, say 1 GHz, which might be the local oscillator used at an IF or RF stage in the transceiver. We have a VCO that can produce frequencies around 1 GHz (but is not calibrated to produce the exact value of the desired frequency), and we wish to use it to obtain a frequency f_0 that is exactly K times the crystal frequency f_{crystal}. This can be achieved by adding a frequency divider into the PLL loop, as shown in Figure 3.36. Such frequency dividers

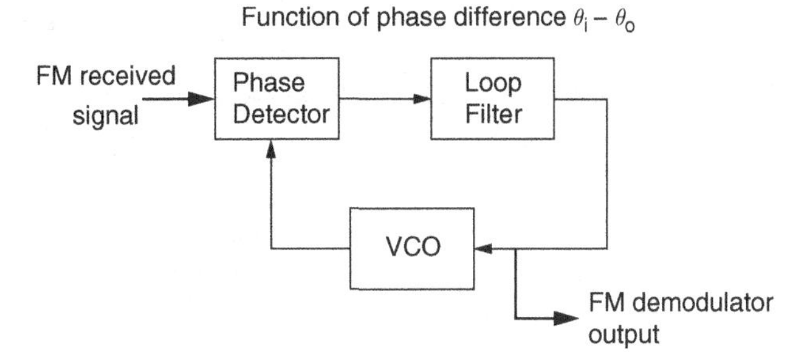

Figure 3.35 The PLL is an FM demodulator.

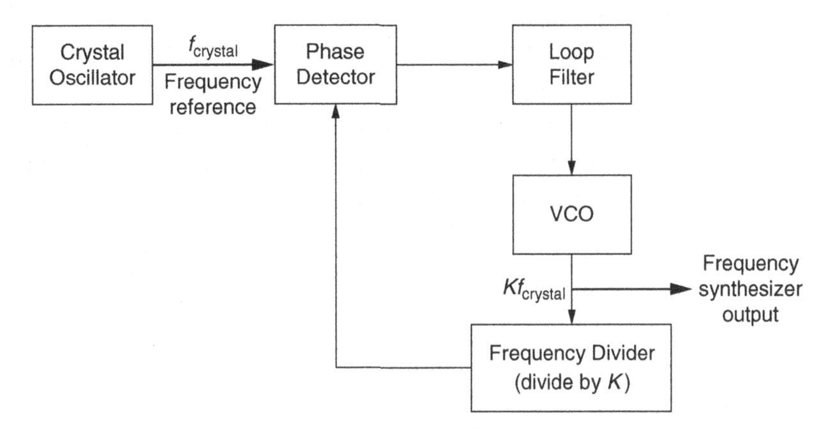

Figure 3.36 Frequency synthesis using a PLL by inserting a frequency divider into the loop.

can be implemented digitally by appropriately skipping pulses. Many variants of this basic concept are possible, such as using multiple frequency dividers, frequency multipliers, or multiple interacting loops.

All of these applications rely on the basic property that the VCO output phase successfully tracks some reference phase using the feedback in the loop. Let us now try to get some insight into how this happens, and into the impact of various parameters on the PLL's performance.

3.5.2 A mathematical model for the PLL

The mixer-based PLL in Figure 3.32 can be modeled as shown in Figure 3.37, where $\theta_i(t)$ is the input phase, and $\theta_o(t)$ is the output phase. It is also useful to define the corresponding instantaneous frequencies (or, rather, frequency deviations from the VCO quiescent frequency f_c):

$$f_i(t) = \frac{1}{2\pi}\frac{d\theta_i(t)}{dt}, \quad f_o(t) = \frac{1}{2\pi}\frac{d\theta_o(t)}{dt}$$

The phase and frequency errors are defined as

$$\theta_e(t) = \theta_i(t) - \theta_o(t), \quad f_e(t) = f_i(t) - f_o(t)$$

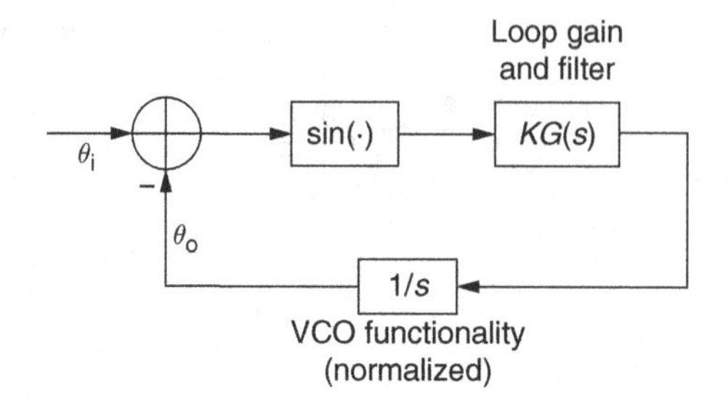

Figure 3.37 A nonlinear model for mixer-based PLL.

In deriving this model, we can ignore the passband term at $2f_c$, which will get rejected by the integration operation due to the VCO, as well as by the loop filter (if a nontrivial lowpass loop filter is employed). From Figure 3.32, the sine of the phase difference is amplified by $\frac{1}{2}A_c A_v$ due to the amplitudes of the PLL input and VCO output. This is passed through the loop filter, which has transfer function $G(s)$, and then through the VCO, which has a transfer function K_v/s. The loop gain K shown in Figure 3.37 is set to be the product $K = \frac{1}{2}A_c A_v K_v$ (in addition, the loop gain also includes additional amplification or attenuation in the loop that is not accounted for in the transfer function $G(s)$).

The model in Figure 3.37 is difficult to analyze because of the $\sin(\cdot)$ nonlinearity after the phase difference operation. One way to avoid this difficulty is to linearize the model by simply dropping the nonlinearity. The motivation is that, when the input and output phases are close, as is the case when the PLL is in tracking mode, then

$$\sin(\theta_i - \theta_o) \approx \theta_i - \theta_o$$

By applying this approximation, we obtain the linearized model of Figure 3.38. Note that, for the XOR-based response shown in Figure 3.34(b), the response is exactly linear for $|\theta| \leq \pi/2$.

3.5.3 PLL analysis

Under the linearized model, the PLL becomes an LTI system whose analysis is conveniently performed using the Laplace transform. From Figure 3.38, we see that

$$(\Theta_i(s) - \Theta_o(s))KG(s)/s = \Theta_o(s)$$

from which we infer the input–output relationship

$$H(s) = \frac{\Theta_o(s)}{\Theta_i(s)} = \frac{KG(s)}{s + KG(s)} \tag{3.32}$$

It is also useful to express the phase error θ_e in terms of the input θ_i, as follows:

$$H_e(s) = \frac{\Theta_e(s)}{\Theta_i(s)} = \frac{\Theta_i(s) - \Theta_o(s)}{\Theta_i(s)} = 1 - H(s) = \frac{s}{s + KG(s)} \tag{3.33}$$

For this LTI model, the same transfer functions also govern the relationships between the input and output instantaneous frequencies: since $F_i(s) = (s/(2\pi))\Theta_i(s)$ and $F_o(s) = (s/(2\pi))\Theta_o(s)$, we obtain $F_o(s)/F_i(s) = \Theta_o(s)/\Theta_i(s)$. Thus, we have

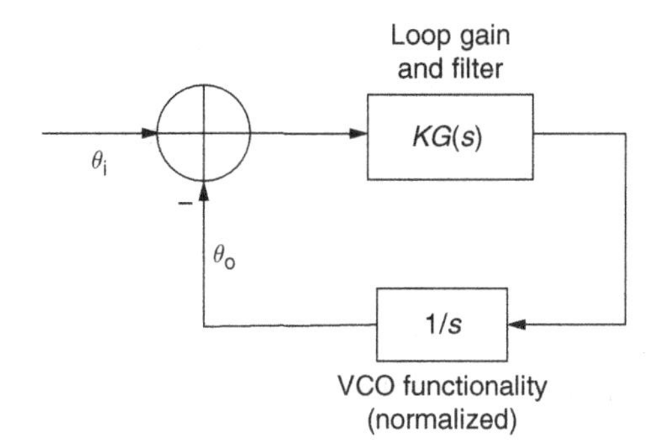

Figure 3.38 The linearized PLL model.

$$\frac{F_o(s)}{F_i(s)} = H(s) = \frac{KG(s)}{s + KG(s)} \tag{3.34}$$

$$\frac{F_i(s) - F_o(s)}{F_i(s)} = H_e(s) = \frac{s}{s + KG(s)} \tag{3.35}$$

First-order PLL When we have a trivial loop filter, $G(s) = 1$, we obtain the first-order response

$$H(s) = \frac{K}{s+K}, \quad H_e(s) = \frac{s}{s+K}$$

which is a stable response for loop gain $K > 0$, with a single pole at $s = -K$. It is interesting to see what happens when the input phase is a step function, $\theta_i(t) = \Delta\theta \, I_{[0,\infty)}(t)$, or $\Theta_i(s) = \Delta\theta/s$. We obtain

$$\Theta_o(s) = H(s)\Theta_i(s) = \frac{K \, \Delta\theta}{s(s+K)} = \frac{\Delta\theta}{s} - \frac{\Delta\theta}{s+K}$$

On taking the inverse Laplace transform, we obtain

$$\theta_o(t) = \Delta\theta(1 - e^{-Kt})I_{[0,\infty)}(t)$$

so that $\theta_o(t) \to \Delta\theta$ as $t \to \infty$. Thus, the first-order PLL can track a sudden change in phase, with the output phase converging to the input phase exponentially fast. The residual phase error is zero. Note that we could also have inferred this quickly from the final-value theorem, without taking the inverse Laplace transform:

$$\lim_{t\to\infty} \theta_e(t) = \lim_{s\to 0} s\Theta_e(s) = \lim_{s\to 0} sH_e(s)\Theta_i(s) \tag{3.36}$$

On specializing to the setting of interest, we obtain

$$\lim_{t\to\infty} \theta_e(t) = \lim_{s\to 0} s\frac{s}{s+K}\frac{\Delta\theta_0}{s} = 0$$

We now examine the response of the first-order PLL to a frequency step Δf, such that the instantaneous input frequency is $f_i(t) = \Delta f \, I_{[0,\infty)}(t)$. The corresponding Laplace transform is $F_i(s) = \Delta f/s$. The input phase is the integral of the instantaneous frequency:

$$\theta_i(t) = 2\pi \int_0^t f_i(\tau)d\tau$$

The Laplace transform of the input phase is therefore given by

$$\Theta_i(s) = 2\pi F(s)/s = \frac{2\pi\ \Delta f}{s^2}$$

Given that the input–output relationships are identical for frequency and phase, we can reuse the computations we did for the phase-step input, replacing phase by frequency, to conclude that $f_o(t) = \Delta f(1 - e^{-Kt})I_{[0,\infty)}(t) \to \Delta f$ as $t \to \infty$, so that the steady-state frequency error is zero. The corresponding output phase trajectory is left as an exercise, but we can use the final-value theorem to compute the limiting value of the phase error:

$$\lim_{t\to\infty} \theta_e(t) = \lim_{s\to 0} s \frac{s}{s+K} \frac{2\pi\ \Delta f}{s^2} = \frac{2\pi\ \Delta f}{K}$$

Thus, the first-order PLL can adapt its frequency to track a step frequency change, but there is a nonzero steady-state phase error. This can be fixed by increasing the order of the PLL, as we now show below.

Second-order PLL We now introduce a loop filter that feeds back both the phase error and the integral of the phase error to the VCO input (in control-theory terminology, we are using "proportional plus integral" feedback). That is, $G(s) = 1 + a/s$, where $a > 0$. This yields the second-order response

$$H(s) = \frac{KG(s)}{s + KG(s)} = \frac{K(s+a)}{s^2 + Ks + Ka}$$

$$H_e(s) = \frac{s}{s + KG(s)} = \frac{s^2}{s^2 + Ks + Ka}$$

The poles of the response are at $s = (-K \pm \sqrt{K^2 - 4Ka})/2$. It is easy to check that the response is stable (i.e., the poles are in the left half plane) for $K > 0$. The poles are conjugate symmetric with an imaginary component if $K^2 - 4Ka < 0$, or $K < 4a$; otherwise they are both real-valued. Note that the phase error due to a step frequency response does go to zero. This is easily seen by invoking the final-value theorem (3.36):

$$\lim_{t\to\infty} \theta_e(t) = \lim_{s\to 0} s \frac{s^2}{s^2 + Ks + Ka} \frac{2\pi\ \Delta f}{s^2} = 0$$

Thus, the second-order PLL has zero steady-state frequency and phase errors when responding to a constant-frequency offset.

We have seen now that the first-order PLL can handle step phase changes, and the second-order PLL can handle step frequency changes, while driving the steady-state phase error to zero. This pattern continues as we keep increasing the order of the PLL: for example, a third-order PLL can handle a linear frequency ramp, which corresponds to $\Theta_i(s)$ being proportional to $1/s^3$.

Linearized analysis provides quick insight into the complexity of the phase/frequency variations that the PLL can track, as a function of the choice of loop filter and loop gain. We now take another look at the first-order PLL, accounting for the $\sin(\cdot)$ nonlinearity in Figure 3.37, in order to provide a glimpse of the approach used for handling the nonlinear differential equations involved, and to compare the results with the linearized analysis.

Nonlinear model for the first-order PLL Let us try to express the phase error θ_e in terms of the input phase for a first-order PLL, with $G(s) = 1$. The model of Figure 3.37 can be expressed in the time domain as

$$\int_0^t K \sin(\theta_e(\tau))d\tau = \theta_o(t) = \theta_i(t) - \theta_e(t)$$

By differentiating with respect to t, we obtain

$$K \sin\theta_e = \frac{d\theta_i}{dt} - \frac{d\theta_e}{dt} \tag{3.37}$$

(Both θ_e and θ_i are functions of t, but we suppress the dependence for the sake of notational simplicity.)

Let us now specialize to the specific example of a step frequency input, for which

$$\frac{d\theta_i}{dt} = 2\pi\ \Delta f$$

On plugging this into (3.37) and rearranging, we get

$$\frac{d\theta_e}{dt} = 2\pi\ \Delta f - K \sin\theta_e \tag{3.38}$$

We cannot solve the nonlinear differential equation (3.38) for θ_e analytically, but we can get useful insight from a "phase-plane plot" of $d\theta_e/dt$ against θ_e, as shown in Figure 3.39. Since $\sin\theta_e \leq 1$, we have $d\theta_e/dt \geq 2\pi\ \Delta f - K$, so that, if $\Delta f > K/(2\pi)$, then $d\theta_e/dt > 0$ for all t. Thus, for large enough frequency offset, the loop never locks. On the other hand, if $\Delta f < K/(2\pi)$, then the loop does lock. In this case, starting from an initial error, say

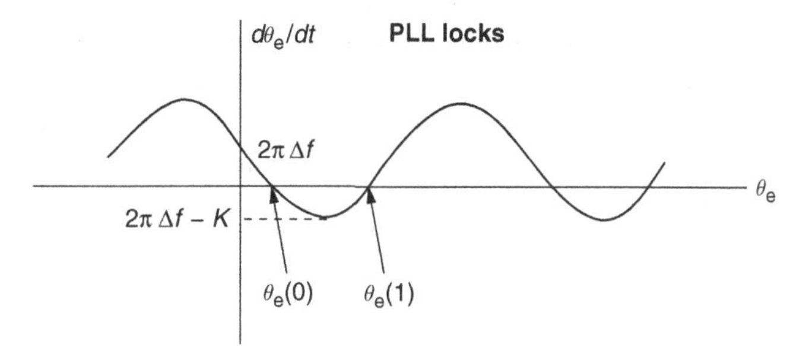

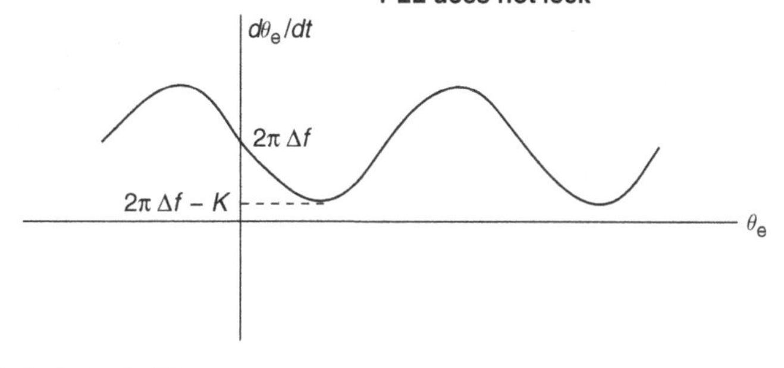

Figure 3.39 Phase-plane plots for first-order PLL.

$\theta_e(0)$, the phase error follows the trajectory to the right (if the derivative is positive) or left (if the derivative is negative) until it hits a point at which $d\theta_e/dt = 0$. From (3.38), this happens when

$$\sin\theta_e = \frac{2\pi\ \Delta f}{K} \tag{3.39}$$

Owing to the periodicity of the sine function, if θ is a solution to the preceding equation, so is $\theta + 2\pi$. Thus, if the equation has a solution, there must be at least one solution in the basic interval $[-\pi, \pi]$. Moreover, since $\sin\theta = \sin(\pi - \theta)$, if θ is a solution, so is $\pi - \theta$, so that there are actually two solutions in $[-\pi, \pi]$. Let us denote by $\theta_e(0) = \sin^{-1}(2\pi\ \Delta f/K)$ the solution that lies in the interval $[-\pi/2, \pi/2]$. This forms a stable equilibrium: from (3.38), we see that the derivative is negative for a phase error slightly above $\theta_e(0)$, and is positive as the phase error becomes slightly below $\theta_e(0)$, so that the phase error is driven back to $\theta_e(0)$ in each case. Using exactly the same argument, we see that the points $\theta_e(0) + 2n\pi$ are also stable equilibria, where n takes integer values. However, another solution to (3.39) is $\theta_e(1) = \pi - \theta(0)$, and translations of it by 2π. It is easy to see that this is an unstable equilibrium: when there is a slight perturbation, the sign of the derivative is such that it drives the phase error away from $\theta_e(1)$. In general, $\theta_e(1) + 2n\pi$ are unstable equilibria, where n takes integer values. Thus, if the frequency offset is within the "pull-in range" $K/(2\pi)$ of the first-order PLL, then the steady-state phase offset (modulo 2π) is $\theta_e(0) = \sin^{-1}(2\pi\ \Delta f/K)$, which, for small values of $2\pi\ \Delta f/K$, is approximately equal to the value $2\pi\ \Delta f/K$ predicted by the linearized analysis.

Linear versus nonlinear model Roughly speaking, the nonlinear model (which we simply simulate when phase-plane plots get too complicated) tells us when the PLL locks, while the linearized analysis provides accurate estimates when the PLL does lock. The linearized model also tells us something about scenarios when the PLL does not lock: when the phase error blows up for the linearized model, it indicates that the PLL will perform poorly. This is because the linearized model holds under the assumption that the phase error is small; if the phase error under this optimistic assumption turns out not to be small, then our initial assumption must have been wrong, and the actual phase error must also be large.

The following worked problem illustrates the application of linearized PLL analysis.

Example 3.5.1 Consider the PLL shown in Figure 3.40, assumed to be locked at time zero.

(a) Suppose that the input phase jumps by $e = 2.72$ radians at time zero (set the phase just before the jump to zero, without loss of generality). How long does it take for the difference between the PLL input phase and the VCO output phase to shrink to 1 radian? (Make sure you specify the unit of time that you use.)

(b) Find the limiting value of the phase error (in radians) if the frequency jumps by 1 kHz just after time zero.

Solution

Let $\theta_e(t) = \theta_i(t) - \theta_o(t)$ denote the phase error. In the s domain, it is related to the input phase as follows:

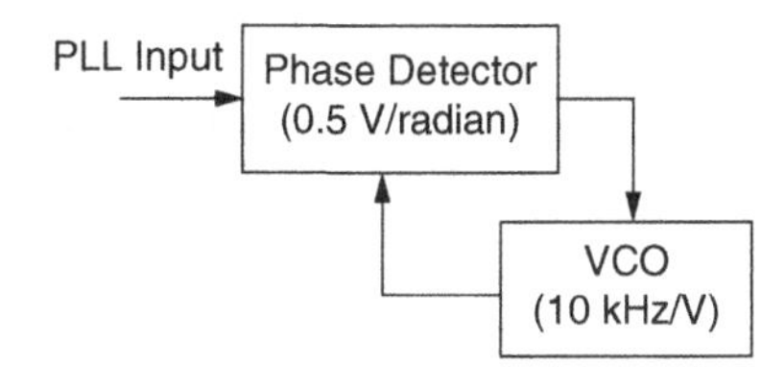

Figure 3.40 The PLL for Example 3.5.1.

$$\Theta_{\text{i}}(s) - \frac{K}{s}\Theta_{\text{e}}(s) = \Theta_{\text{e}}(s)$$

so that

$$\frac{\Theta_{\text{e}}(s)}{\Theta_{\text{i}}(s)} = \frac{s}{s+K}$$

(a) For a phase jump of e radians at time zero, we have $\Theta_{\text{i}}(s) = e/s$, which yields

$$\Theta_{\text{e}}(s) = \Theta_{\text{i}}(s)\frac{s}{s+K} = \frac{e}{s+K}$$

On going to the time domain, we have

$$\theta_{\text{e}}(t) = ee^{-Kt} = e^{1-Kt}$$

so that $\theta_{\text{e}}(t) = 1$ for $1 - Kt = 0$, or $t = 1/K = \frac{1}{5}$ milliseconds.

(b) For a frequency jump of Δf, the Laplace transform of the input phase is given by

$$\Theta_{\text{i}}(s) = \frac{2\pi\ \Delta f}{s^2}$$

so that the phase error is given by

$$\Theta_{\text{e}}(s) = \Theta_{\text{i}}(s)\frac{s}{s+K} = \frac{2\pi\ \Delta f}{s(s+K)}$$

Using the final-value theorem, we have

$$\lim_{t\to\infty} \theta_{\text{e}}(t) = \lim_{s\to 0} s\Theta_{\text{e}}(s) = \frac{2\pi\ \Delta f}{K}$$

For $\Delta f = 1$ kHz and $K = 5$ kHz/radian, this yields a phase error of $2\pi/5$ radians, or $72°$.

3.6 Some analog communication systems

Some of the analog communication systems that we encounter (or at least, used to encounter) in our daily lives include broadcast radio and television. We have already discussed AM radio in the context of the superhet receiver. We now briefly discuss FM radio and television. Our goal is to highlight design concepts, and the role played in these systems by the various modulation formats we have studied, rather than to provide a detailed technical description. Other commonly encountered examples of analog communication that we do not discuss include analog storage media (audiotapes and videotapes), analog wireline telephony, analog cellular telephony, amateur ham radio, and wireless microphones.

3.6.1 FM radio

FM mono radio employs a peak frequency deviation of 75 kHz, with the baseband audio message signal bandlimited to 15 kHz; this corresponds to a modulation index of 5. Using Carson's formula, the bandwidth of the FM radio signal can be estimated as 180 kHz. The separation between adjacent radio stations is 200 kHz. An FM stereo broadcast transmits two audio channels, "left" and "right," in a manner that is backwards compatible with mono broadcast, in that a standard mono receiver can extract the sum of the left and right channels, while remaining oblivious to whether the broadcast signal is mono or stereo. The structure of the baseband signal into the FM modulator is shown in Figure 3.41. The sum of the left and right channels, or the $L+R$ signal, occupies a band from 30 Hz to 15 kHz. The difference, or the $L-R$ signal (which also has a bandwidth of 15 kHz), is modulated using DSB-SC, using a carrier frequency of 38 kHz, and hence occupies a band from 23 kHz to 53 kHz. A pilot tone at 19 kHz, at half the carrier frequency for the DSB signal, is provided to enable coherent demodulation of the DSB-SC signal. The spacing between adjacent FM stereo broadcast stations is still 200 kHz, which makes it a somewhat tight fit (if we apply Carson's formula with a maximum frequency deviation of 75 kHz, we obtain an RF bandwidth of 256 kHz).

The format of the baseband signal in Figure 3.41 (in particular, the DSB-SC modulation of the difference signal) seems rather contrived, but the corresponding modulator can be implemented quite simply, as sketched in Figure 3.42: we simply switch between the left and right channel audio signals using a 38-kHz clock. As we show in one of the problems, this directly yields the $L+R$ signal, plus the DSB-SC modulated $L-R$ signal. It remains to add in the 19-kHz pilot before feeding the composite baseband signal to the FM modulator.

The receiver employs an FM demodulator to obtain an estimate of the baseband transmitted signal. The $L+R$ signal is obtained by bandlimiting the output of the FM demodulator to 15 kHz using a lowpass filter; this is what an oblivious mono receiver would do. A stereo receiver, in addition, processes the output of the FM demodulator in the band from 15 kHz to 53 kHz. It extracts the 19-kHz pilot tone, doubles its frequency to obtain a coherent carrier reference, and uses that to demodulate the $L-R$ signal sent using DSB-SC. It then

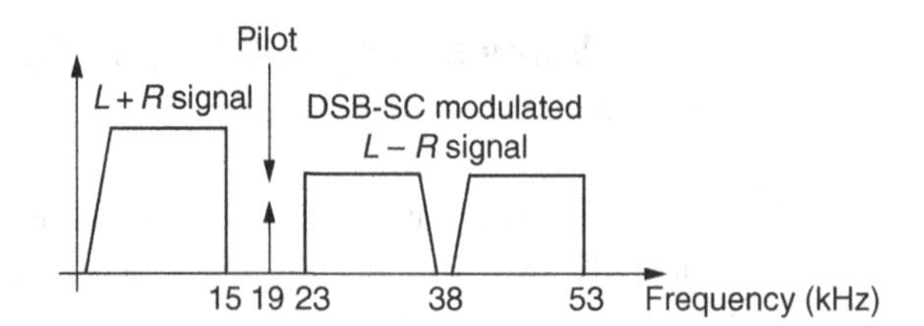

Figure 3.41 The spectrum of baseband input to an FM modulator for FM stereo broadcast.

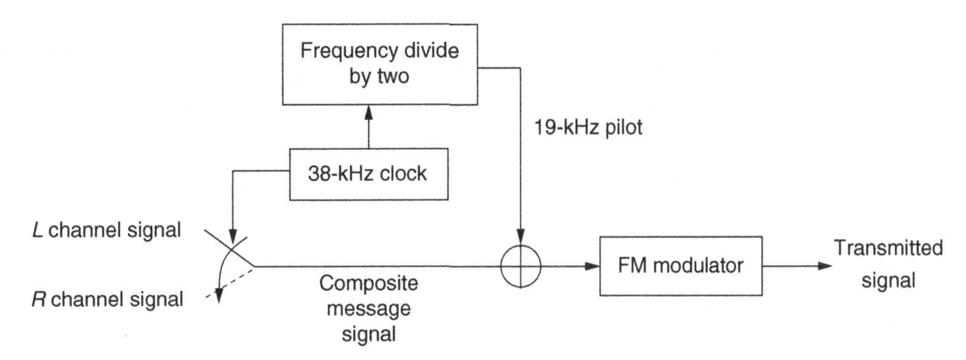

Figure 3.42 A block diagram of a simple FM stereo transmitter.

obtains the left and right channels by adding the $L + R$ and $L - R$ signals and subtracting one of them from the other, respectively.

3.6.2 Analog broadcast TV

While analog broadcast TV is obsolete, and is being replaced by digital TV as we speak, we discuss it briefly here to highlight a few features. First, it illustrates an application of several modulation schemes: VSB (for intensity information), quadrature modulation (for the color information), and FM (for audio information). Second, it is an interesting example of how the embedding of different kinds of information in analog form must account for the characteristics of the information source (video) and destination (a cathode-ray-tube (CRT) TV monitor). This customized, and rather painful, design process is in contrast to the generality and conceptual simplification provided by the source-channel-separation principle in digital communication (as mentioned in Chapter 1). Indeed, from Chapter 4 onwards, where we restrict our attention to digital communication, we do not need to discuss source characteristics.

We first need a quick discussion of CRT TV monitors. An electron beam impinging on a fluorescent screen is used to emit the light that we perceive as the image on the TV. The electron beam is "raster scanned" in horizontal lines moving down the screen, with its horizontal and vertical location controlled by two magnetic fields created by voltages, as shown in Figure 3.43. We rely on the persistence of human vision to piece together these discrete scans into a continuous image in space and time. Black-and-white TV monitors use a phosphor (or fluorescent material) that emits white light when struck by electrons. Color TV monitors use three kinds of phosphors, typically arranged as dots on the screen, which emit red, green, and blue light, respectively, when struck by electrons. Three electron beams are used, one for each color. The intensity of the emitted light is controlled by the intensity of the electron beam. For historical reasons, the scan rate is chosen to be equal to the frequency of the AC power (otherwise, for the power supplies used at the time, rolling bars would appear on the TV screen). In the United States, this means that the scan rate is set at 60 Hz (the frequency of the AC mains).

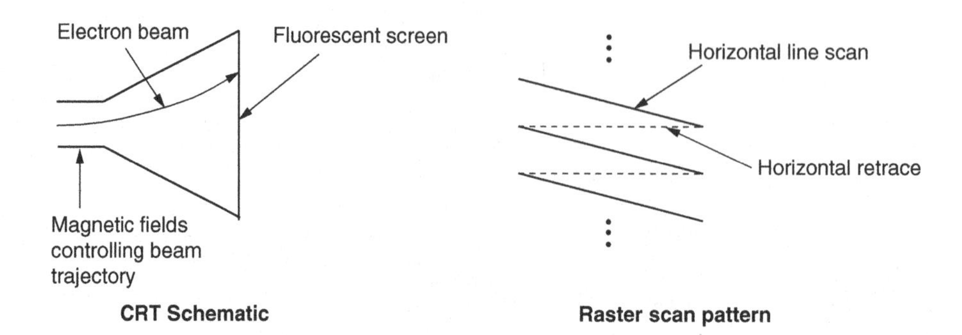

Figure 3.43 Implementing raster scan in a CRT monitor requires magnetic fields controlled by sawtooth waveforms.

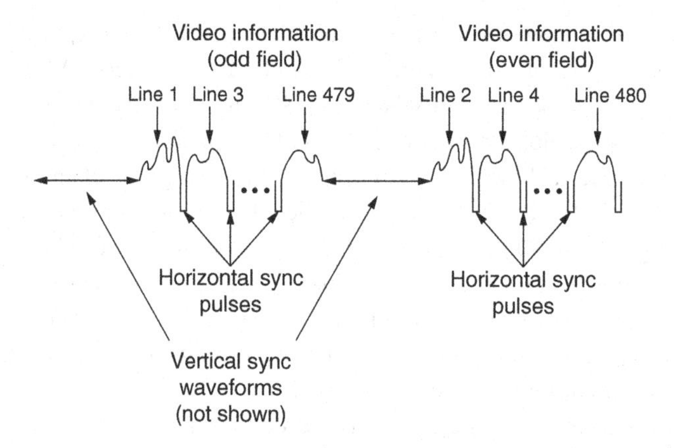

Figure 3.44 The structure of a black-and-white composite video signal (numbers apply to the NTSC standard).

In order to enable the TV receiver to control the operation of the CRT monitor, the received signal must contain not only intensity and color information, but also the timing information required to correctly implement the raster scan. Figure 3.44 shows the format of the *composite video* signal containing this information. In order to reduce flicker (again a historical legacy, since older CRT monitors could not maintain intensities for long enough if the time between refreshes is too long), the CRT screen is painted in two rounds for each image (or frame): first the odd lines (comprising the odd field) are scanned, then the even lines (comprising the even field) are scanned. For the NTSC standard, this is done at a rate of 60 fields per second, or 30 frames per second. A horizontal sync pulse is inserted between each line. A more complex vertical synchronization waveform is inserted between each field; this enables vertical synchronization (as well as other functionalities that we do not discuss here). The receiver can extract the horizontal and vertical timing information from the composite video signal, and generate the sawtooth waveforms required for controlling the electron beam (one of the first widespread commercial applications of the PLL was for this purpose). For the NTSC standard, the composite video signal spans 525 lines, about 486 of which are actually painted (counting both the even and odd fields). The remaining 39 lines accommodate the vertical synchronization waveforms.

The bandwidth of the baseband video signal can be roughly estimated as follows. Assuming about 480 lines, with about 640 pixels per line (for an aspect ratio of 4:3), we have about 300,000 pixels, refreshed at the rate of 30 times per second. Thus, our overall sampling rate is about 9 Msamples/second. This can accurately represent a signal of bandwidth 4.5 MHz. For a 6 MHz TV channel bandwidth, DSB and wideband FM are therefore out of the question, and VSB was chosen to modulate the composite video signal. However, the careful shaping of the spectrum required for VSB is not carried out at the transmitter, because this would require the design of high-power electronics with tight specifications. Instead, the transmitter uses a simple filter, while the receiver, which deals with a low-power signal, accomplishes the VSB shaping requirement in (3.21). Audio modulation is done using FM in a band adjacent to the one carrying the video signal.

While the signaling for black-and-white TV is essentially the same for all existing analog TV standards, the insertion of color differs among standards such as NTSC, PAL, and SECAM. We do not go into details here, but, taking NTSC as an example, we note that the frequency-domain characteristics of the black-and-white composite video signal are exploited in rather a clever way to insert color information. The black-and-white signal exhibits a clustering of power around the Fourier-series components corresponding to the horizontal scan rate, with the power decaying around the higher-order harmonics. The color-modulated signal uses the same band as the black-and-white signal, but is inserted between two such harmonics, so as to minimize the mutual interference between the intensity information and the color information. The color information is encoded in two baseband signals, which are modulated on to the I and Q components using QAM. Synchronization information that permits coherent recovery of the color subcarrier for quadrature demodulation is embedded in the vertical synchronization waveform.

3.7 Concept summary

This chapter provides an introduction to analog communication techniques, focusing mainly on concepts that remain relevant in the age of digital communication.

Amplitude modulation

- Double-sideband (DSB) suppressed-carrier (SC) modulation refers to translation of a real baseband message to passband by mixing (multiplying) it by a sinusoid at the carrier frequency. The bandwidth of a DSB-SC signal is twice that of the baseband message.
- For DSB-SC, the message is the I component of the passband waveform, and therefore can be recovered by standard downconversion. However, this requires carrier synchronization, and is vulnerable to synchronization errors.

- Conventional AM refers to a DSB-SC signal plus a strong carrier component. An AM signal with modulation index smaller than one preserves the shape of the message in the envelope, since the latter equals the message plus a DC offset. The message can therefore be demodulated without carrier synchronization using envelope detection.
- The power efficiency of conventional AM can be quite small for messages with large dynamic ranges, but it does result in simple, low-cost receivers. This is the tradeoff that designers have traditionally made for broadcast systems such as AM radio in which a single powerful transmitter serves a large number of receivers. (However, such tradeoffs are somewhat obsolete today, since advances in integrated-circuit implementations of communication receivers imply that receivers can now be both sophisticated and low-cost.)
- Single-sideband (SSB) modulation corresponds to sending only the upper or lower sideband of a DSB signal. The passband SSB signal therefore has the same bandwidth as the original message.
- The I component of an SSB signal is the message, while the Q component is its Hilbert transform (with sign depending on which sideband is sent). Thus, an SSB signal can be constructed at baseband and then upconverted, rather than employing stringent filtering at passband to isolate a sideband.
- The Hilbert transform of a message corresponds to imposing a phase lag of 90° at all (positive) frequencies.
- The message can be recovered from an SSB signal using synchronous downconversion, but this is vulnerable to carrier-synchronization errors. Another option is to add a strong carrier component at the transmitter and to employ envelope detection.
- Vestigial-sideband modulation (VSB) may be viewed as a generalization of SSB, in which a DSB signal is filtered so as to let through one sideband and a vestige of the other. For an appropriately designed VSB filter, the I component of the VSB signal is the message, while the Q component is a filtered version of the message whose form depends on the VSB filter. Demodulation can be performed as for an SSB signal.
- Quadrature amplitude modulation (QAM) corresponds to sending different messages on both the I and Q components of a passband signal. It requires synchronous demodulation. QAM is a popular design approach for digital communication, and its use for this purpose is discussed in Chapter 4 and beyond.

Angle modulation

- Angle modulation refers to the baseband message being embedded in the frequency/phase of a passband waveform.
- Frequency modulation may be interpreted as phase modulation with the integral of the message. This smoothing operation leads to better spectral properties, hence FM is preferred for analog communication, where the message waveform is given by nature. However, phase modulation is often used for digital communication, where the designer can carefully shape the transmitted waveforms.

- The bandwidth of an FM waveform depends on both the maximum frequency deviation and the message bandwidth, and can be approximated using Carson's formula.
- For periodic messages, the complex envelope of an FM waveform is periodic. Fourier series can therefore be used to characterize the spectrum of the complex envelope, and hence the FM waveform.
- A simple-to-understand, suboptimal demodulator for FM is the limiter–discriminator, which consists of differentiation followed by envelope detection. However, feedback-based techniques such as the PLL provide superior performance for analog communication, while demodulation techniques exploiting the structure of the message are preferred for digital communication.

Superheterodyne receiver

- The superhet receiver achieves downconversion in multiple stages, mixing the passband received waveform at RF down to another passband waveform at IF by beating it against a local oscillator offset from the desired carrier frequency by f_{IF}. This is followed by demodulation to baseband by any of a variety of techniques, including coherent downconversion and envelope detection.
- For operation over multiple bands, the RF filter and the LO are tunable. The specifications on the RF, or image reject, filter can be fairly relaxed, since its key function is to reject the image frequency (which is separated from the desired band by $2f_{IF}$).
- The image-reject filter typically lets in bands adjacent to the desired frequency, hence the IF filter must have sharp cutoffs in order to suppress "adjacent-channel interference" from these bands. The fact that the IF is fixed facilitates designing to such tight specifications.

Phase-locked loop

- The PLL enables phase/frequency tracking of a desired passband waveform using feedback. The phase difference between the desired waveform and a locally generated copy, obtained using a phase detector, is filtered using a loop filter, and is fed back to drive the VCO generating the local copy.
- The PLL can be used for a variety of functions, including FM demodulation and frequency synthesis (the latter is its most important application in the digital age).
- Classically, phase detectors were implemented using mixers, but modern implementations often use mixed-signal (digital and analog) approaches.
- The phase detector is nonlinear, but PLLs are often analyzed using linearized models characterized by the transfer function between the input phase and the output phase. The order of the PLL is the degree of the denominator in the Laplace transform of the transfer function.
- First-order PLLs (trivial loop filters) can track phase jumps but not frequency jumps. Second order PLLs (in which the loop filter provides proportional plus integral feedback) can track both phase and frequency jumps.

3.8 Notes

A towering figure in the history of analog communication techniques was Edwin Howard Armstrong, who invented the regenerative circuit (for feedback-based amplification), the superhet receiver, and FM. Mentioning Armstrong gives us the opportunity for a quick discussion on the evolution of design philosophy from analog to digital communication. From what we know about Armstrong, he was a clear thinker who would use systematic experimentation rather than trial and error, along with physical intuition, to arrive at his inventions. However, he distrusted results obtained only using mathematics because of the potential for hidden, and potentially flawed, assumptions in mathematical models. While some amount of skepticism of this nature is warranted, it is worth noting that digital communication would not exist today if it were not for mathematical abstractions of the physical world. Indeed, as mentioned in Chapter 7, Claude Shannon created information theory as a mathematical framework promising the existence of reliable and efficient communication systems in 1948, but it took many decades of effort by communication system designers to build practical systems approaching information-theoretic limits. Shannon's promise, which was based purely on idealized mathematical models (which Armstrong would perhaps not have approved of) was essential in motivating this effort. Furthermore, as we gain more confidence in the accuracy of our mathematical models, they play a bigger role in design, as we shall see when we transition from the piecemeal ideas in this chapter to the more systematic framework for digital communication in forthcoming chapters. Specifically, the digital communication system designer today employs sophisticated and accurate mathematical models for communication channels (which have been developed using Armstrong's blend of physical intuition and experimentation) to establish systematic principles for practical transceiver design and implementation that approach the performance limits promised by Shannon's theoretical framework.

Since our treatment of analog communication techniques here emphasizes those that remain relevant in the digital age, we refer readers interested in a deeper look at analog communication techniques to the excellent treatment in Ziemer and Tranter [4]. For classic treatments of the PLL, see Gardner [21] and Viterbi [22] (the latter provides analysis for a nonlinear PLL model). More recent books include those by Best [23] and Razavi [24].

3.9 Problems

Amplitude modulation

Problem 3.1 Figure 3.45 shows a signal obtained after amplitude modulation by a sinusoidal message. The carrier frequency is difficult to determine from the figure, and is not needed for answering the questions below.

(a) Find the modulation index.
(b) Find the signal power.
(c) Find the bandwidth of the AM signal.

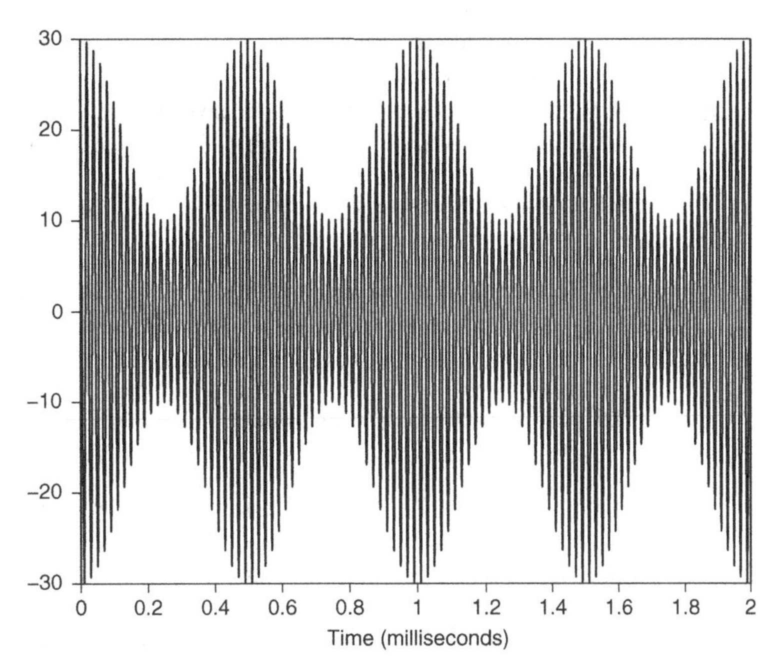

Figure 3.45 The amplitude-modulated signal for Problem 3.1.

Problem 3.2 Consider a message signal $m(t) = 2\cos(2\pi t + \pi/4)$.

(a) Sketch the spectrum $U(f)$ of the DSB-SC signal $u_p(t) = 8m(t)\cos(400\pi t)$. What is the power of u?
(b) Carefully sketch the output of an ideal envelope detector with input u_p. On the same plot, sketch the message signal $m(t)$.
(c) Let $v_p(t)$ denote the waveform obtained by high-pass filtering the signal $u(t)$ so as to let through only frequencies above 200 Hz. Find $v_c(t)$ and $v_s(t)$ such that we can write

$$v_p(t) = v_c(t)\cos(400\pi t) - v_s(t)\sin(400\pi t)$$

and sketch the envelope of v.

Problem 3.3 A message to be transmitted using AM is given by

$$m(t) = 3\cos(2\pi t) + 4\sin(6\pi t)$$

where **the unit of time is milliseconds.** It is to be sent using a carrier frequency of 600 kHz.

(a) What is the message bandwidth? Sketch its magnitude spectrum, clearly specifying the units used on the frequency axis.
(b) Find an expression for the normalized message $m_n(t)$.
(c) For a modulation index of 50%, write an explicit time-domain expression for the AM signal.
(d) What is the power efficiency of the AM signal?

(e) Sketch the magnitude spectrum for the AM signal, again clearly specifying the units used on the frequency axis.
(f) The AM signal is to be detected using an envelope detector (as shown in Figure 3.8), with $R = 50$ ohms. What is a good range of choices for the capacitance C?

Problem 3.4 Consider a message signal $m(t) = \cos(2\pi f_m t + \phi)$ and a corresponding DSB-SC signal $u_p(t) = Am(t)\cos(2\pi f_c t)$, where $f_c > f_m$.

(a) Sketch the spectra of the corresponding LSB and USB signals (if the spectrum is complex-valued, sketch the real and imaginary parts separately).
(b) Find explicit time-domain expressions for the LSB and USB signals.

Problem 3.5 One way of avoiding the use of a mixer in generating AM is to pass $x(t) = m(t) + \alpha\cos(2\pi f_c t)$ through a memoryless nonlinearity and then a bandpass filter.

(a) Suppose that $M(f) = (1 - |f|/10)I_{[-10,10]}$ (the frequency is in units of kHz) and f_c is 900 kHz. For a nonlinearity $f(x) = \beta x^2 + x$, sketch the magnitude spectrum at the output of the nonlinearity when the input is $x(t)$, carefully labeling the frequency axis.
(b) For the specific settings in (a), characterize the bandpass filter (BPF) you should use at the output of the nonlinearity so as to generate an AM signal carrying the message $m(t)$. That is, describe the set of the frequencies which the BPF must reject, and those which it must pass.

Problem 3.6 Consider a DSB signal corresponding to the message $m(t) = \text{sinc}(2t)$ and a carrier frequency f_c that is 100 times larger than the message bandwidth, where **the unit of time is milliseconds.**

(a) Sketch the magnitude spectrum of the DSB signal $10m(t)\cos(2\pi f_c t)$, specifying the units on the frequency axis.
(b) Specify a time-domain expression for the corresponding LSB signal.
(c) Now, suppose that the DSB signal is passed through a bandpass filter whose transfer function is given by

$$H_p(f) = \left(f - f_c + \frac{1}{2}\right) I_{[f_c-1/2,\, f_c+1/2]} + I_{[f_c+1/2,\, f_c+3/2]}, \quad f > 0$$

Sketch the magnitude spectrum of the corresponding VSB signal.
(d) Find a time-domain expression for the VSB signal of the form

$$u_c(t)\cos(2\pi f_c t) - u_s(t)\sin(2\pi f_c t)$$

carefully specifying u_c and u_s, the I and Q components.

Problem 3.7 Figure 3.46 shows a block diagram of Weaver's SSB modulator, which works if we choose f_1, f_2, and the bandwidth of the lowpass filter (LPF) appropriately. Let us work through these choices for a waveform of the form $m(t) = A_L\cos(2\pi f_L t + \phi_L) + A_H\cos(2\pi f_H t + \phi_H)$, where $f_H > f_L$ (the design choices we obtain will work for any message whose spectrum lies in the band $[f_L, f_H]$).

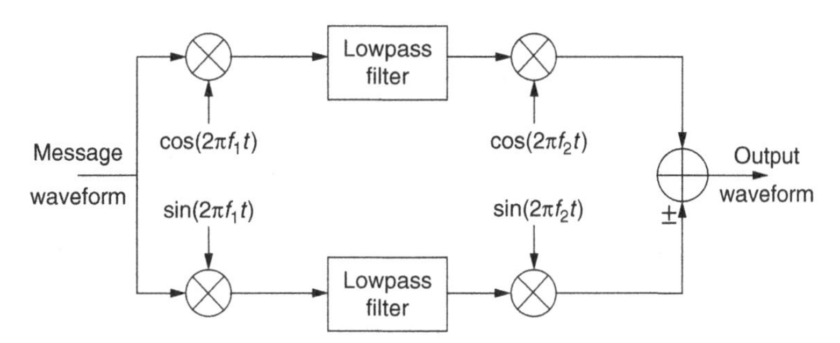

Figure 3.46 The block diagram of Weaver's SSB modulator for Problem 3.7.

(a) For $f_1 = (f_L + f_H)/2$ (i.e., choosing the first LO frequency to be in the middle of the message band), find the time-domain waveforms at the outputs of the upper and lower branches after the first mixer.

(b) Choose the bandwidth of the LPF to be $W = (f_H + 2f_L)/2$ (assume that the LPF is ideal). Find the time-domain waveforms at the outputs of the upper and lower branches after the LPF.

(c) Now, assuming that $f_2 \gg f_H$, find a time-domain expression for the output waveform, assuming that the upper and lower branches are added together. Is this an LSB or USB waveform? What is the carrier frequency?

(d) Repeat (c) when the lower branch is subtracted from the upper branch.

Remark Weaver's modulator does not require bandpass filters with sharp cutoffs, unlike the direct approach to generating SSB waveforms by filtering DSB-SC waveforms. It is also simpler than the Hilbert-transform method (the latter requires implementation of a $\pi/2$ phase shift over the entire message band).

Problem 3.8 Consider the AM signal $u_p(t) = 2(10 + \cos(2\pi f_m t))\cos(2\pi f_c t)$, where the message frequency f_m is 1 MHz and the carrier frequency f_c is 885 MHz.

(a) Suppose that we use superheterodyne reception with an IF of 10.7 MHz, and envelope detection after the IF filter. Envelope detection is accomplished as in Figure 3.8, using a diode and an RC circuit. What would be a good choice of C if $R = 100$ ohms?

(b) The AM signal $u_p(t)$ is passed through the bandpass filter with transfer function $H_p(f)$ depicted (for positive frequencies) in Figure 3.47. Find the I and Q components of the filter output with respect to a reference frequency f_c of 885 MHz. Does the filter output represent a form of modulation you are familiar with?

Problem 3.9 Consider a message signal $m(t)$ with spectrum $M(f) = I_{[-2,2]}(f)$.

(a) Sketch the spectrum of the DSB-SC signal $u_{\text{DSB-SC}} = 10m(t)\cos(300\pi t)$. What are the power and bandwidth of u?

(b) The signal in (a) is passed through an envelope detector. Sketch the output, and comment on how it is related to the message.

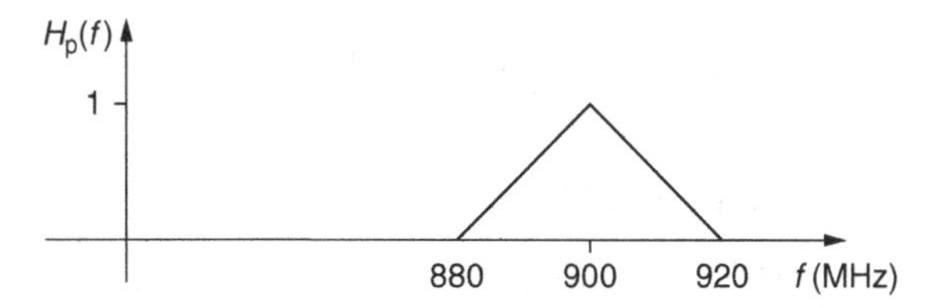

Figure 3.47 The bandpass filter for Problem 3.8.

(c) What is the smallest value of A such that the message can be recovered without distortion from the AM signal $u_{AM} = (A + m(t))\cos(300\pi t)$ by envelope detection?

(d) Give a time-domain expression of the form

$$u_p(t) = u_c(t)\cos(300\pi t) - u_s(t)\sin(300\pi t)$$

obtained by high-pass filtering the DSB signal in (a) so as to let through only frequencies above 150 Hz.

(e) Consider a VSB signal constructed by passing the signal in (a) through a passband filter with transfer function for positive frequencies specified by

$$H_p(f) = \begin{cases} f - 149, & 149 \leq f \leq 151 \\ 2, & f \geq 151 \end{cases}$$

(you should be able to sketch $H_p(f)$ for both positive and negative frequencies). Find a time-domain expression for the VSB signal of the form

$$u_p(t) = u_c(t)\cos(300\pi t) - u_s(t)\sin(300\pi t)$$

Problem 3.10 Consider Figure 3.17 depicting VSB spectra. Suppose that the passband VSB filter $H_p(f)$ is specified (for positive frequencies) as follows:

$$H_p(f) = \begin{cases} 1, & 101 \leq f < 102 \\ \frac{1}{2}(f - 99), & 99 \leq f \leq 101 \\ 0, & \text{else} \end{cases}$$

(a) Sketch the passband transfer function $H_p(f)$ for both positive and negative frequencies.

(b) Sketch the spectrum of the complex envelope $H(f)$, taking $f_c = 100$ as a reference.

(c) Sketch the spectra (show the real and imaginary parts separately) of the I and Q components of the impulse response of the passband filter.

(d) Consider a message signal of the form $m(t) = 4\,\text{sinc}(4t) - 2\cos(2\pi t)$. Sketch the spectrum of the DSB signal that results when the message is modulated by a carrier at $f_c = 100$.

(e) Now, suppose that the DSB signal in (d) is passed through the VSB filter in (a)–(c). Sketch the spectra of the I and Q components of the resulting VSB signal, showing the real and imaginary parts separately.

(f) Find a time-domain expression for the Q component.

Problem 3.11 Consider the periodic signal $m(t) = \sum_{n=-\infty}^{\infty} p(t - 2n)$, where $p(t) = tI_{[-1,1]}(t)$.

(a) Sketch the AM signal $x(t) = (4 + m(t))\cos(100\pi t)$.
(b) What is the power efficiency?

Problem 3.12 Find an explicit time-domain expression for the Hilbert transform of $m(t) = \text{sinc}(2t)$.

Superheterodyne reception

Problem 3.13 A dual-band radio operates at 900 MHz and 1.8 GHz. The channel spacing in each band is 1 MHz. We wish to design a superheterodyne receiver with an IF of 250 MHz. The LO is built using a frequency synthesizer that is tunable from 1.9 to 2.25 GHz, and frequency divider circuits if needed (assume that you can only implement frequency division by an integer).

(a) How would you design a superhet receiver to receive a passband signal restricted to the band 1800–1801 MHz? Specify the characteristics of the RF and IF filters, and how you would choose and synthesize the LO frequency.
(b) Repeat (a) for when the signal to be received lies in the band 900–901 MHz.

Angle modulation

Problem 3.14 Figure 3.48 shows, as a function of time, the phase deviation of a bandpass FM signal modulated by a sinusoidal message.

(a) Find the modulation index (assume that it is an integer multiple of π for your estimate).
(b) Find the message bandwidth.
(c) Estimate the bandwidth of the FM signal using Carson's formula.

Problem 3.15 The input $m(t)$ to an FM modulator with $k_f = 1$ has Fourier transform

$$M(f) = \begin{cases} j2\pi f, & |f| < 1 \\ 0, & \text{else} \end{cases}$$

The output of the FM modulator is given by

$$u(t) = A\cos(2\pi f_c t + \phi(t))$$

where f_c is the carrier frequency.

(a) Find an explicit time-domain expression for $\phi(t)$ and carefully sketch $\phi(t)$ as a function of time.
(b) Find the magnitude of the instantaneous frequency deviation from the carrier at time $t = \frac{1}{4}$.
(c) Using the result from (b) as an approximation for the maximum frequency deviation, estimate the bandwidth of $u(t)$.

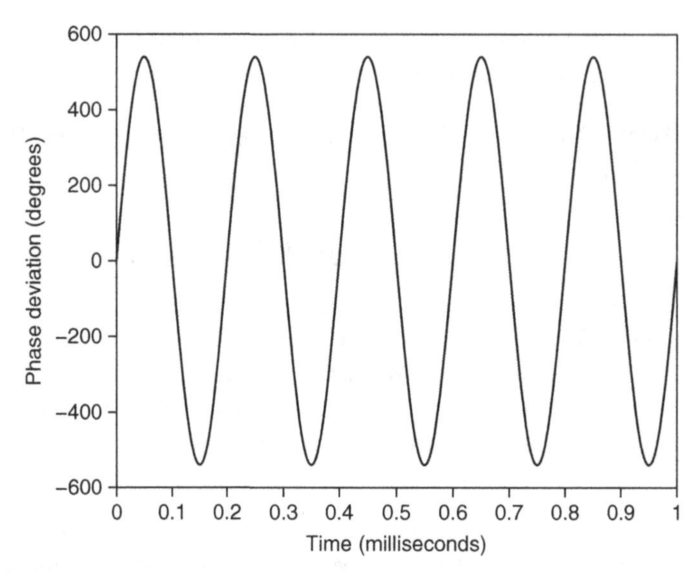

Figure 3.48 The phase deviation of the FM signal for Problem 3.14.

Problem 3.16 Let $p(t) = I_{[-1/2,1/2]}(t)$ denote a rectangular pulse of unit duration. Construct the signal

$$m(t) = \sum_{n=-\infty}^{\infty} (-1)^n p(t-n)$$

The signal $m(t)$ is input to an FM modulator, whose output is given by

$$u(t) = 20\cos(2\pi f_c t + \phi(t))$$

where

$$\phi(t) = 20\pi \int_{-\infty}^{t} m(\tau)d\tau + a$$

and a is chosen such that $\phi(0) = 0$.

(a) Carefully sketch both $m(t)$ and $\phi(t)$ as a function of time.

(b) Approximating the bandwidth of $m(t)$ as $W \approx 2$, estimate the bandwidth of $u(t)$ using Carson's formula.

(c) Suppose that a very narrow ideal BPF (with bandwidth less than 0.1) is placed at $f_c + \alpha$. For which (if any) of the following choices of α will you get nonzero power at the output of the BPF: (i) $\alpha = 0.5$, (ii) $\alpha = 0.75$, and (iii) $\alpha = 1$.

Problem 3.17 Let $u(t) = 20\cos(2000\pi t + \phi(t))$ denote an angle modulated signal.

(a) For $\phi(t) = 0.1\cos(2\pi t)$, what is the approximate bandwidth of u?

(b) Let $y(t) = u^{12}(t)$. Specify the frequency bands spanned by $y(t)$. In particular, specify the output when y is passed through

  (i) a BPF centered at 12 kHz (using Carson's formula, determine the bandwidth of the BPF required to recover most of the information in ϕ from the output);

(ii) an ideal LPF of bandwidth 200 Hz; and

(iii) a BPF of bandwidth 100 Hz centered at 11 kHz.

(c) For $\phi(t) = 2\sum_n s(t-2n)$, where $s(t) = (1-|t|)I_{[-1,1]}$,

(i) sketch the instantaneous frequency deviation from the carrier frequency of 1 kHz, and

(ii) show that we can write

$$u(t) = \sum_n c_n \cos(2000\pi t + n\alpha t)$$

Specify α, and write down an explicit integral expression for c_n.

Problem 3.18 Consider the set-up of Problem 3.16, taking the unit of time in milliseconds for concreteness. You do not need the value of f_c, but you can take it to be 1 MHz.

(a) Numerically (e.g., using MATLAB) compute the Fourier-series expansion for the complex envelope of the FM waveform, in the same manner as was done for a sinusoidal message. Report the magnitudes of the Fourier-series coefficients for the first five harmonics.

(b) Find the 90%, 95%, and 99% power-containment bandwidths. Compare your answers with the estimate from Carson's formula obtained in Problem 3.16(b).

Problem 3.19 A VCO with a quiescent frequency of 1 GHz, with a frequency sweep of 2 MHz/mV, produces an angle-modulated signal whose phase deviation $\theta(t)$ from a carrier frequency f_c of 1 GHz is shown in Figure 3.49.

(a) Sketch the input $m(t)$ to the VCO, carefully labeling the voltage and time axes.

(b) Estimate the bandwidth of the angle-modulated signal at the VCO output. You may approximate the bandwidth of a periodic signal by that of its first harmonic.

Uncategorized problems

Problem 3.20 The signal $m(t) = 2\cos(20\pi t) - \cos(40\pi t)$, where the **unit of time is milliseconds** and **the unit of amplitude is millivolts (mV)**, is fed to a VCO with a quiescent frequency of 5 MHz and a frequency deviation of 100 kHz/mV. Denote the output of the VCO by $y(t)$.

(a) Provide an estimate of the bandwidth of y.

(b) The signal $y(t)$ is passed through an ideal bandpass filter of bandwidth 5 kHz, centered at 5.005 MHz. Describe in detail how you would compute the power at the filter output (if you can compute the power in closed form, do so).

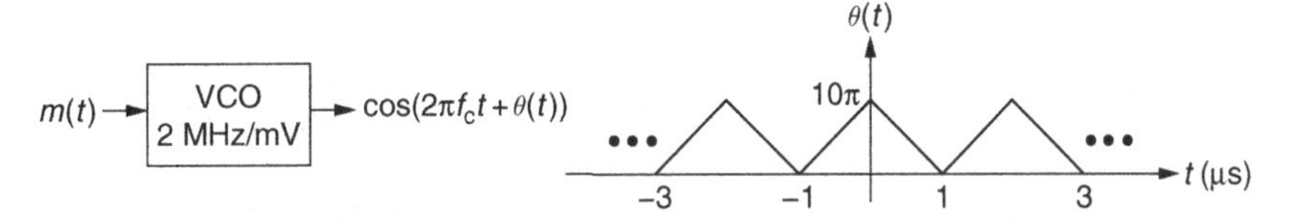

Figure 3.49 The set-up for Problem 3.19.

Problem 3.21 Consider the AM signal $u_p(t) = (A + m(t))\cos(400\pi t)$ (t in ms) with message signal $m(t)$ as in Figure 3.50, where A is 10 mV.

(a) If the AM signal is demodulated using an envelope detector with an RC filter, how should you choose C if $R = 500$ ohms? Try to ensure that the first harmonic (i.e., the fundamental) and the third harmonic of the message are reproduced with minimal distortion.
(b) Now, consider an attempt at synchronous demodulation, where the AM signal is downconverted using a 201-kHz LO, as shown in Figure 3.51. Find and sketch the I and Q components, $u_c(t)$ and $u_s(t)$, for $0 \le t \le 2$ (t in ms).
(c) Describe how you would recover the original message $m(t)$ from the downconverter outputs $u_c(t)$ and $u_s(t)$, drawing block diagrams as needed.

Problem 3.22 The square-wave message signal $m(t)$ in Figure 3.50 is input to a VCO with quiescent frequency 200 kHz and frequency deviation 1 kHz/mV. Denote the output of the VCO by $u_p(t)$.

(a) Sketch the I and Q components of the FM signal (with respect to a frequency reference of 200 kHz and a phase reference chosen such that the phase is zero at time zero) over the time interval $0 \le t \le 2$ (t in ms), clearly labeling the axes.
(b) In order to extract the I and Q components using a standard downconverter (mix with an LO and then a lowpass filter), how would you choose the bandwidth of the LPFs used at the mixer outputs?

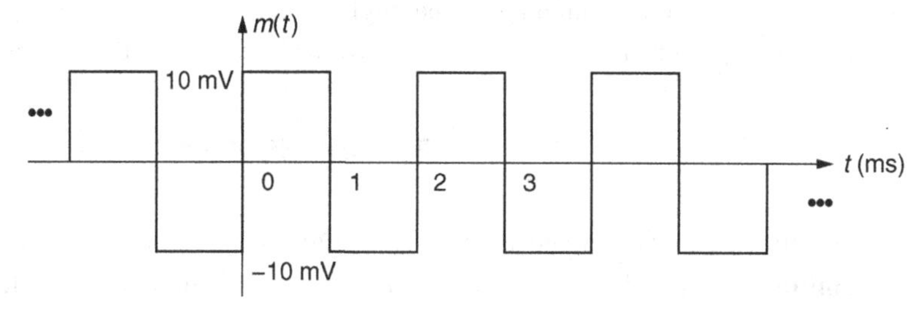

Figure 3.50 The message signal for Problems 3.21 and 3.22.

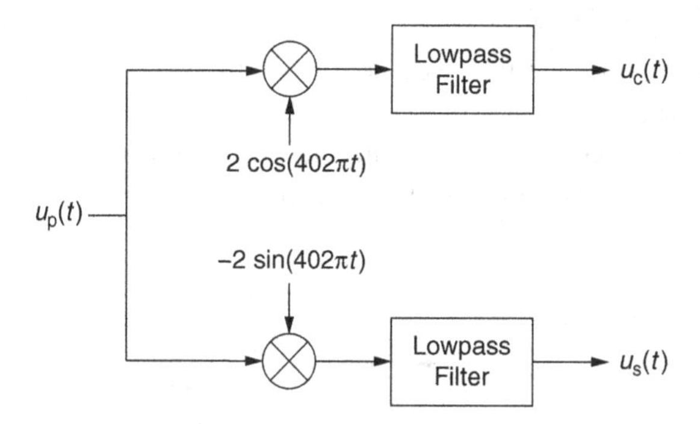

Figure 3.51 Downconversion using a 201-kHz LO (t in ms in the figure) for parts (b) and (c) of Problem 3.21.

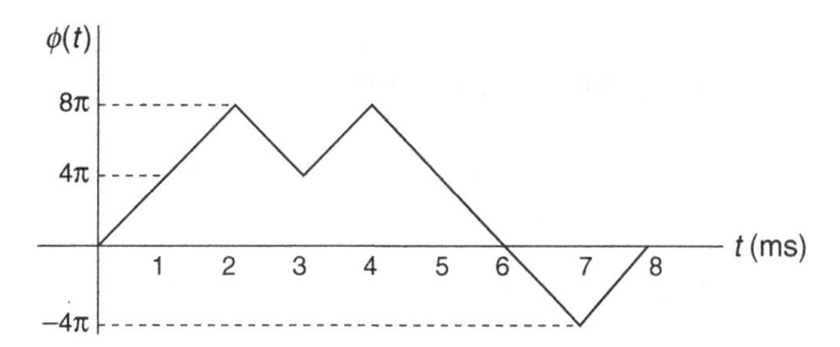

Figure 3.52 The phase evolution in Problem 3.23.

Problem 3.23 The output of an FM modulator is the bandpass signal $y(t) = 10\cos(300\pi t + \phi(t))$, where the **unit of time is milliseconds**, and the phase $\phi(t)$ is as sketched in Figure 3.52.

(a) Supposing that $y(t)$ is the output of a VCO with frequency deviation 1 kHz/mV and quiescent frequency 149 kHz, find and sketch the input to the VCO.
(b) Use Carson's formula to estimate the bandwidth of $y(t)$, clearly stating the approximations that you make.

Phase-locked loop

Set-up for PLL problems For the next few problems on PLL modeling and analysis, consider the linearized model in Figure 3.38, with the following notation: loop filter $G(s)$, loop gain K, and VCO modeled as $1/s$. Recall from your background on signals and systems that a second-order system of the form $1/(s^2 + 2\zeta\omega_n s + \omega_n^2)$ is said to have natural frequency ω_n (in radians/second) and damping factor ζ.

Problem 3.24 Let $H(s)$ denote the gain from the PLL input to the output of the VCO. Let $H_e(s)$ denote the gain from the PLL input to the input to the loop filter. Let $H_m(s)$ denote the gain from the PLL input to the VCO input.

(a) Write down the formulas for $H(s)$, $H_e(s)$, and $H_m(s)$, in terms of K and $G(s)$.
(b) Which is the relevant transfer function if the PLL is being used for FM demodulation?
(c) Which is the relevant transfer function if the PLL is being used for carrier-phase tracking?
(d) For $G(s) = (s + 8)/s$ and $K = 2$, write down expressions for $H(s)$, $H_e(s)$, and $H_m(s)$. What are the natural frequency and the damping factor?

Problem 3.25 Suppose the PLL input exhibits a frequency jump of 1 kHz.

(a) How would you choose the loop gain K for a first-order PLL ($G(s) = 1$) to ensure a steady-state error of at most 5 degrees?

(b) How would you choose the parameters a and K for a second-order PLL ($G(s) = (s+a)/s$) to have a natural frequency of 1.414 kHz and a damping factor of $1/\sqrt{2}$. Specify the units for a and K.

(c) For the parameter choices in (b), find and roughly sketch the phase error as a function of time for a frequency jump of 1 kHz.

Problem 3.26 Suppose that $G(s) = (s+16)/(s+4)$ and $K = 4$.

(a) Find the transfer function $\Theta_o(s)/\Theta_i(s)$.

(b) Suppose that the PLL is used for FM demodulation, with the input to the PLL being an FM signal with instantaneous frequency deviation of the FM signal $(10/\pi)m(t)$, where the message $m(t) = 2\cos t + \sin(2t)$. Using the linearized model for the PLL, find a time-domain expression for the estimated message provided by the PLL-based demodulator.

Hint. What happens to a sinusoid of frequency ω passing through a linear system with transfer function $H(s)$?

Problem 3.27 Consider the PLL depicted in Figure 3.53, with input phase $\phi(t)$. The output signal of interest to us here is $v(t)$, the VCO input. The parameter for the loop filter $G(s)$ is given by $a = 1000\pi$ radians/s.

(a) Assume that the PLL is locked at time 0, and suppose that $\phi(t) = 1000\pi t I_{\{t>0\}}$. Find the limiting value of $v(t)$.

(b) Now, suppose that $\phi(t) = 4\pi \sin(1000\pi t)$. Find an approximate expression for $v(t)$. For full credit, simplify as much as possible.

(c) For part (b), estimate the bandwidth of the passband signal at the PLL input.

Quiz on analog communication systems

Problem 3.28 Answer the following questions regarding commercial analog communication systems (some of which may no longer exist in your neighborhood).

(a) **(True or false?)** The modulation format for analog cellular telephony was conventional AM.

(b) **(Multiple choice)** FM was used in analog TV as follows:
   (i) to modulate the video signal
   (ii) to modulate the audio signal
   (iii) FM was not used in analog TV systems.

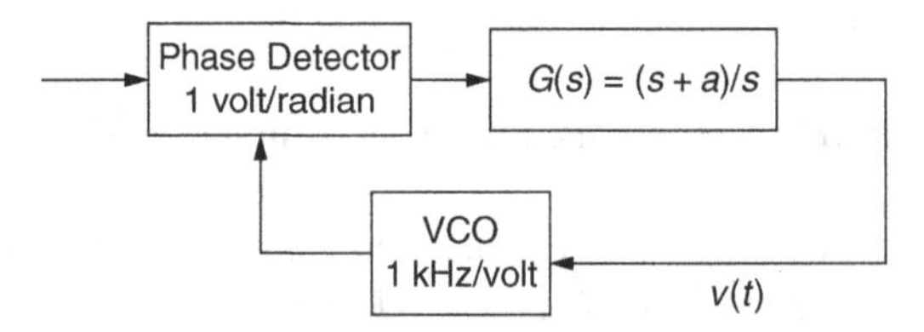

Figure 3.53 The system for Problem 3.27.

(c) A superheterodyne receiver for AM radio employs an intermediate frequency (IF) of 455 kHz, and has stations spaced at 10 kHz. Comment briefly on each of the following statements.
   (i) The AM band is small enough that the problem of image frequencies does not occur.
   (ii) A bandwidth of 20 kHz for the RF front end is a good choice.
   (iii) A bandwidth of 20 kHz for the IF filter is a good choice.

Software Lab 3.1: amplitude modulation and envelope detection

Lab objectives The goal of this lab is to illustrate amplitude modulation and envelope detection using digitally modulated messages. In addition to using envelope detection to demodulate conventional amplitude modulation, we also illustrate how it can be used for I/Q downconversion when quadrature mixers are not available.

Reading Section 3.2 (amplitude modulation).

Laboratory assignment

(1.1) Generate a message signal $m(t)$ using binary digital modulation with a sine pulse by modifying Code Fragment 2.3.2 to use random bits. Set the symbol time T to 1 millisecond. Take a waveform segment spanning n_s symbols and take its Fourier transform by modifying Code Fragment 2.5.1 (or using the code from Lab 2.2), choosing the length of the FFT (and hence n_s) so as to get a frequency resolution of 1 Hz. Plot the magnitude squared of the Fourier transform, divided by $n_s T$, the length of the observation interval. This is an estimate of the *power spectral density (PSD)* $S_m(f)$, which is formally defined later, in Chapters 4 and 5.

(1.2) Repeat the PSD estimation in (1.1) above over multiple runs and average the estimates, choosing a number of runs large enough to get a smooth estimate of the PSD. Eyeball the PSD to estimate the bandwidth of the signal (the units should be consistent with our assumption of $T = 1$ ms).

(1.3) Now, generate the DSB signal $u(t) = m(t)\cos(2\pi f_c t)$, where $f_c = 10/T$. Choose the sampling rate for generating discrete time samples as $4f_c$. Plot the DSB signal over four symbols.

(1.4) Estimate the PSD $S_u(f)$ of the DSB signal generated in (1.3) by choosing a large enough number of symbols as in (1.1), and averaging over several runs as in (1.2). What is the relationship with the PSD obtained in (1.2)?

(1.5) Repeat (1.3) and (1.4) for the AM signal $u(t) = (A_c + m(t))\cos(2\pi f_c t)$, where $f_c = 10/T$ and A_c is chosen to have the smallest possible value that allows envelope detection. As before, choose the sampling rate for generating discrete-time samples as $4f_c$. Do you run into difficulty when computing the PSD? Explain.

(1.6) Starting with an AM signal as in (1.5), implement an envelope detector as follows.

(a) Pass $u(t)$ through an idealized diode to obtain $u_+(t) = u(t)I_{u(t)\geq 0}$.

(b) Pass $u_+(t)$ through an RC filter with impulse response $h(t) = e^{-t/(HC)}I_{t\geq 0}$. You can use the *contconv* function in Lab 2.2 for this purpose. Choose the value of RC based on the design rule of thumb discussed in Chapter 3.

(c) Implement a DC block simply by subtracting out the empirical mean from the output of (b).

Plot the output of the envelope detector, along with the original message $m(t)$.

(1.7) Repeat (1.6) for different values of RC (both too large and too small), and comment on how the resulting message estimate is affected by the value of RC.

Envelope-detector-based I/Q downconversion

We know from Chapter 2 that a passband signal can be downconverted to complex baseband by mixing with the cosine and sine of the carrier and then lowpass filtering. However, implementing mixers might not be easy at really high carrier frequencies (e.g., for coherent optical communication). Envelope detection, after adding strong locally generated carrier components to the received waveform, provides an alternative approach for downconversion that may be easier to implement in such scenarios. Consider the QAM received signal

$$v(t) = v_c(t)\cos(2\pi f_c t) - v_s(t)\sin(2\pi f_c t) \qquad (3.40)$$

where v_c and v_s are real baseband messages. The receiver's local oscillator generates $A_c \cos(2\pi f_c t + \theta)$ and $A_c \sin(2\pi f_c t + \theta)$, where θ is the offset between the carrier reference used at the transmitter to generate the QAM signal and the local copy of the carrier at the receiver. Instead of mixing the local oscillator outputs against $v(t)$, we add them to $v(t)$ and then perform envelope detection, as described in (1.6). That is, we perform the following operations.

- Pass $v(t) + A_c \cos(2\pi f_c t + \theta)$ through an envelope detector to obtain $\tilde{v}_c(t)$.
- Pass $v(t) + A_c \sin(2\pi f_c t + \theta)$ through an envelope detector to obtain $\tilde{v}_s(t)$.

(1.8) For A_c large enough, can you find simple relationships between $\tilde{v}_c(t)$, $\tilde{v}_s(t)$ and the original messages $v_c(t)$, $v_s(t)$? Is there a simple relationship between the complex-baseband waveforms $v(t) = v_c(t) + jv_s(t)$ and $\tilde{v}(t) = \tilde{v}_c(t) + j\tilde{v}_s(t)$?

(1.9) Generate v_c and v_s as in (1.1), using different sequences of random bits, and generate $v(t)$ as in (3.40), setting $f_c = 10/T$ as in (1.3). Implement the preceding envelope-detection operations, first setting $\theta = 0$. Plot $\tilde{v}_c(t)$ and $\tilde{v}_s(t)$ for A_c "large enough." Also plot for reference $v_c(t)$ and $v_s(t)$. Comment on whether the results conform to your expectations from (1.7). How small can you make A_c while still getting "good" results?

(1.10) Repeat (1.9) for $\theta = \pi/4$.

(1.11) Assuming that the receiver knows θ, can you recover estimates of $v_c(t)$ and $v_s(t)$ from $\tilde{v}_c(t)$ and $\tilde{v}_s(t)$? Implement these operations and show plots of the estimates and the original waveforms.

Lab report

- Answer all questions and print out the most useful plots to support your answers.
- Write a paragraph about any questions or confusions that you may have experienced with this lab.

Software Lab 3.2: frequency-modulation basics

Lab objectives The goal of this lab to explore the characteristics of frequency-modulated signals using digitally modulated messages, and to explore demodulation using differentiation in baseband.

Reading Section 3.3 (angle modulation).

Laboratory assignment

Consider the message signal $m(t) = \sum_n b[n]p(t - nT)$, where $b[n]$ are chosen from $\{-1, +1\}$, and T is the symbol interval. You can generate such a signal by modifying Code Fragment 2.3.2. Define a passband FM signal modulated by the message by

$$s_p(t) = \cos(2\pi f_c t + \theta(t))$$

where f_c is the carrier frequency, and the phase

$$\theta(t) = 2\pi k_f \int_0^t m(\tau)d\tau$$

(assume $t \geq 0$). The complex envelope of s_p with respect to the reference $2\pi f_c t$ is given by $s(t) = e^{j\theta(t)}$. We consider a digital message signal of the form $m(t) = \sum_n b[n]p(t - nT)$, where $b[n]$ are chosen independently and with equal probability from $\{-1, +1\}$.

The following code fragment generates the FM waveform for a rectangular pulse $I_{[0,1]}$.

```
oversampling_factor = 16;
%for a pulse with amplitude one, the maximum frequency deviation is given by kf
kf = 4;
%increase the oversampling factor if kf (and hence frequency deviation,
%and hence bw of FM signal) is large
oversampling_factor = ceil(max(kf,1) * oversampling_factor);
ts = 1/oversampling_factor; %sampling time
nsamples = ceil(1/ts);
pulse = ones(nsamples,1); %rectangular pulse
nsymbols = 10;
symbols = zeros(nsymbols,1);
%random symbol sequence
symbols = sign(rand(nsymbols,1) - 0.5);
%generate digitally modulated message
nsymbols_upsampled = 1 + (nsymbols - 1) * nsamples;
symbols_upsampled = zeros(nsymbols_upsampled,1);
symbols_upsampled(1:nsamples:nsymbols_upsampled) = symbols;
message = conv(symbols_upsampled,pulse);
%FM signal phase obtained by integrating the message
```

```
theta = 2 * pi * kf * ts * cumsum(message);
cenvelope = exp(j * theta);
L = length(cenvelope);
time = (0:L - 1) * ts;
Icomponent = real(cenvelope);
Qcomponent = imag(cenvelope);
%plot I component
plot(time,Icomponent);
```

(2.1) By modifying and enhancing the preceding code fragment as needed, plot the I and Q components of the complex envelope for a random sequence of bits as a function of time for $k_f = 0.25$. Also plot $\theta(t)/\pi$ versus t. How big are the changes in $\theta(t)$ corresponding to a given message bit $b[n]$? Do you notice a pattern in how the I and Q components depend on the message bits $\{b[n]\}$?

Remark The special case of $k_f = 1/4$ is a digital modulation scheme known as minimum-shift keying (MSK). It can be viewed as FM modulation using a digital message, but the plots of the I and Q components should indicate that MSK can also be interpreted as the I and Q components each being amplitude modulated by a different set of bits, with an offset between the I and Q components.

(2.2) Now redo (2.1) for $k_f = 4$. The patterns in the I and Q components are much harder to see now. We typically do not use such wideband FM for digital modulation, but may use it for analog messages.

(2.3) For a complex-baseband waveform $y(t) = y_c(t) + jy_s(t) = e(t)e^{j\theta(t)}$, we know that

$$\theta(t) = \tan^{-1}\left(\frac{y_s(t)}{y_c(t)}\right)$$

Show that

$$\frac{d}{dt}\theta(t) = \frac{y_c(t)y_s'(t) - y_s(t)y_c'(t)}{y_c^2(t) + y_s^2(t)} \tag{3.41}$$

For an FM signal, the message can be estimated as $(1/(2\pi k_f))d\theta(t)/dt$, with the derivative computed using a highpass filter. Thus, this can be viewed as a baseband version of the limiter–discriminator demodulator for FM. It can be implemented using the following code fragment:

```
%baseband discriminator
%differencing operation approximates derivative
Iderivative = [0;diff(Icomponent)]/ts;
Qderivative = [0;diff(Qcomponent)]/ts;
message_estimate = (1/(2 * pi * kf)) * (Icomponent.* Qderivative
                   - Qcomponent.* Iderivative).
```

(2.4) Apply the preceding approach to the noiseless FM signals generated in (2.1) and (2.2) above. Plot the estimated message and the original message on the same plot, and comment on whether you are getting a good estimate.

(2.5) Add an arbitrary phase to the complex envelope:

```
phi = 2 * pi * rand; %phase uniform over [0,2 pi]
cenvelope = cenvelope * exp(j * phi);
%now apply baseband discriminator
```

Redo (2.4). What happens to the estimated message? Are you still getting a good estimate of the original message?

(2.6) Now, add a frequency offset as well as a phase offset to the complex envelope:

```
phi = 2 * pi * rand; %phase uniform over [0,2 pi]
df = 0.3;
cenvelope = cenvelope.* exp(j * (2 * pi * df * time + phi));
%now apply baseband discriminator
```

Redo (2.4). What happens to the estimated message? Are you still getting a good estimate of the original message? If you are not quite getting the original message back, what can you do to fix the situation?

Remark You should find that this crude differentiation technique does work for low noise (we are considering *zero* noise). However, it is rather fragile when noise is inserted. We do not explore this in this lab, but you are welcome to try adding Gaussian noise samples to the I and Q components to see how the discriminator performs for different values of noise variance. At the very least, one would need to lowpass filter the message estimate obtained above to average out noise, but it is far better to use feedback-based techniques such as the PLL for general FM demodulation, or, for digital messages, to use demodulation techniques that use the structure of the message. We do not discuss such techniques in this lab.

(2.7) We now explore the spectral properties of FM. For the complex envelope $s(t) = e^{j\theta(t)}$, compute the Fourier transform numerically, choosing the length of the FFT (and hence n_s) so as to get a frequency resolution of 0.1. You can modify Code Fragment 2.5.1 or reuse code from Lab 2.2. Compute the power spectral density (PSD), defined as the magnitude squared of the Fourier transform divided by the interval over which you are computing it, and then averaged over multiple runs. Plot the PSD for $k_f = 1/4$ and $k_f = 4$. You can modify the following code fragment as needed.

```
nsymbols = 1000;
symbols = zeros(nsymbols,1);
nruns = 1000;
fs = 0.1;
Nmin = ceil(1/(fs_desired * ts)); %minimum length DFT for desired
    frequency granularity
message_length = 1 + (nsymbols - 1) * nsamples + length(pulse) - 1;
Nmin = max(message_length,Nmin);
%for efficient computation, choose FFT size to be power of 2
Nfft = 2^(nextpow2(Nmin)) %FFT size = the next power of 2 at least as
  big as Nmin
psd = zeros(Nfft,1);
for runs = 1:nruns,
%random symbol sequence
symbols = sign(rand(nsymbols,1) - 0.5);
```

```
nsymbols_upsampled = 1 + (nsymbols - 1) * nsamples;
symbols_upsampled = zeros(nsymbols_upsampled,1);
symbols_upsampled(1:nsamples:nsymbols_upsampled) = symbols;
message = conv(symbols_upsampled,pulse);
%FM signal phase
theta = 2 * pi * kf * ts * cumsum(message);
cenvelope = exp(j * theta);
time = (0:length(cenvelope) - 1) * ts;
%frequency-domain signal computed using DFT
cenvelope_freq = ts * fft(cenvelope,Nfft);
%FFT of size Nfft, automatically zeropads as needed
cenvelope_freq_centered = fftshift(cenvelope_freq); %shifts DC to
  center of spectrum
psd = psd + abs(cenvelope_freq_centered).^2;
end
psd = psd/(nruns * nsymbols);
fs = 1/(Nfft * ts) %actual frequency resolution attained
%set of frequencies for which Fourier transform has been computed using DFT
freqs = ((1:Nfft) - 1 - Nfft/2) * fs;
%plot the PSD
plot(freqs,psd);
```

(2.8) Plot the PSD for $k_f = \frac{1}{4}$ and $k_f = 4$. Are your results consistent with Carson's formula?

(2.9) Redo (2.8), replacing the rectangular pulse by a sine pulse: $p(t) = \sin(\pi t) I_{[0,1]}(t)$. Are your results consistent with Carson's formula? Compare the spectrum occupancy in (2.8) with that in this case, commenting on the roles of k_f and $p(t)$.

```
pulse = transpose(sin(pi * (0:ts:1)));
```

(2.10) Now, let us increase the dynamic range of the message by replacing the bits by numbers drawn from a Gaussian distribution with the same variance:

```
symbols = randn(nsymbols,1);
```

Compute and plot the PSD for a sinusoidal pulse, and compare the spectral occupancy with that in (2.9).

(2.11) Assuming that the unit of time is 1 ms, estimate the bandwidth of the FM signals whose PSDs you plotted in (2.9). You can either eyeball it, or estimate the length of the interval over which 95% of the signal power is contained.

Lab report

- Answer all questions and print out the most useful plots to support your answers.
- Write a paragraph about any questions or confusions that you may have experienced with this lab.

4 Digital modulation

Digital modulation is the process of translating bits to analog waveforms that can be sent over a physical channel. Figure 4.1 shows an example of a baseband digitally modulated waveform, where bits that take values in $\{0, 1\}$ are mapped to symbols in $\{+1, -1\}$, which are then used to modulate translates of a rectangular pulse, where the translation corresponding to successive symbols is the symbol interval T. The modulated waveform can be represented as a sequence of symbols (taking values ± 1 in the example) multiplying translates of a pulse (rectangular in the example). This is an example of a widely used form of digital modulation termed *linear modulation*, where the transmitted signal depends linearly on the symbols to be sent. Our treatment of linear modulation in this chapter generalizes this example in several ways. The modulated signal in Figure 4.1 is a baseband signal, but what if we are constrained to use a passband channel (e.g., a wireless cellular system operating at 900 MHz)? One way to handle this to simply translate this baseband waveform to passband by upconversion; that is, send $u_{\mathrm{p}}(t) = u(t)\cos(2\pi f_{\mathrm{c}} t)$, where the carrier frequency f_{c} lies in the desired frequency band. However, what if the frequency occupancy of the passband signal is strictly constrained? (Such constraints are often the result of guidelines from standards or regulatory bodies, and serve to limit interference between users operating in adjacent channels.) Clearly, the timelimited modulation pulse used in Figure 4.1 spreads out significantly in frequency. We must therefore learn to work with modulation pulses that are better constrained in frequency. We may also wish to send information on both the I components and the Q components. Finally, we may wish to pack in more bits per symbol; for example, we could send two bits per symbol by using four levels, say $\{\pm 1, \pm 3\}$.

Chapter plan

In Section 4.1, we develop an understanding of the structure of linearly modulated signals, using the binary modulation in Figure 4.1 to lead into variants of this example corresponding to different signaling constellations that can be used for baseband and passband channels. In Section 4.2, we discuss how to quantify the bandwidth of linearly modulated signals by computing the power spectral density. With these basic insights in place, we turn in Section 4.3 to a discussion of modulation for bandlimited channels, treating signaling over baseband and passband channels in a unified framework using the complex-baseband representation. We note, invoking Nyquist's sampling theorem to determine the degrees of freedom offered by bandlimited channels, that linear modulation with a bandlimited

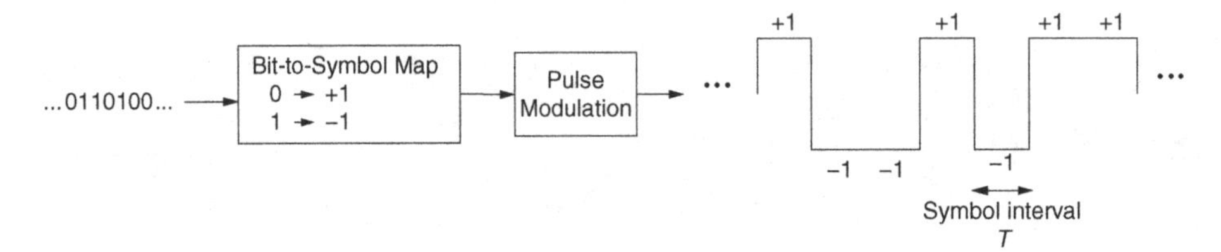

Figure 4.1 Running example: binary antipodal signaling using a timelimited pulse.

modulation pulse can be used to fill all of these degrees of freedom. We discuss how to design bandlimited modulation pulses employing the Nyquist criterion for avoidance of intersymbol interference (ISI). These concepts are reinforced by Software Lab 8.1, which provides a hands-on demonstration of Nyquist signaling in the absence of noise. Finally, we discuss orthogonal and biorthogonal modulation in Section 4.4.

Software

Over the course of this and later chapters, we develop a simulation framework for simulating linear modulation over noisy dispersive channels. Software Lab 4.1 in this chapter is a first step in this direction. Appendix 4.B provides guidance for developing the software for this lab.

4.1 Signal constellations

The linearly modulated signal depicted in Figure 4.1 can be written in the following general form:

$$u(t) = \sum_n b[n]p(t - nT) \tag{4.1}$$

where $\{b[n]\}$ is a sequence of *symbols* and $p(t)$ is the *modulating pulse*. The symbols take values in $\{-1, +1\}$ in our example, and the modulating pulse is a rectangular timelimited pulse. As we proceed through this chapter, we shall see that linear modulation as in (4.1) is far more generally applicable, in terms of the set of possible values taken by the symbol sequence, as well as the choice of modulating pulse.

The modulated waveform (4.1) is a baseband waveform. While it is timelimited in our example, and hence cannot be strictly bandlimited, it is approximately bandlimited to a band around DC. Now, if we are given a passband channel over which to send the information encoded in this waveform, one easy approach is to send the passband signal

$$u_p(t) = u(t)\cos(2\pi f_c t) \tag{4.2}$$

where f_c is the carrier frequency. That is, the modulated baseband signal is sent as the I component of the passband signal. To see what happens to the passband signal as a consequence of the modulation, we plot it in Figure 4.2. For the nth symbol interval

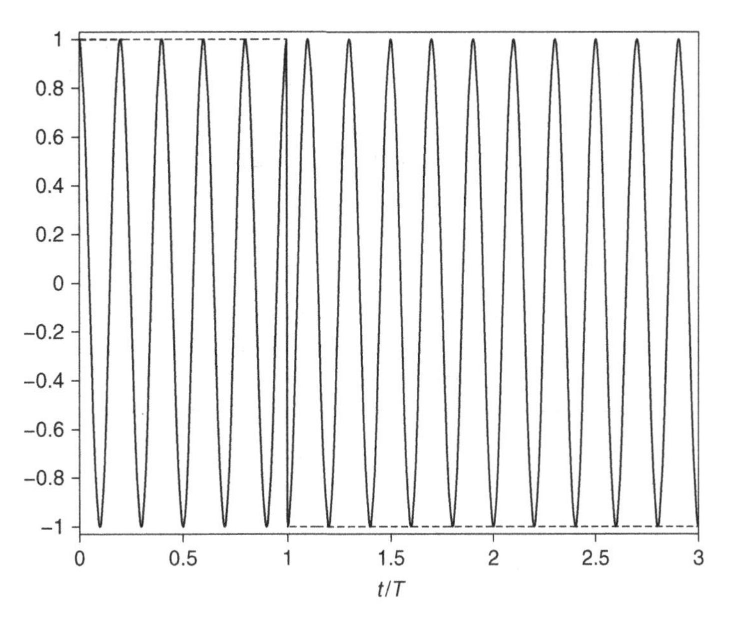

Figure 4.2 BPSK illustrated for $f_c = 4/T$ and the symbol sequence $+1, -1, -1$. The solid line corresponds to the passband signal $u_p(t)$, and the dashed line to the baseband signal $u(t)$. Note that, due to the change in sign between the first and second symbols, there is a phase discontinuity of π at $t = T$.

$nT \leq t < (n+1)T$, we have $u_p(t) = \cos(2\pi f_c t)$ if $b[n] = +1$, and $u_p(t) = -\cos(2\pi f_c t) = \cos(2\pi f_c t + \pi)$ if $b[n] = -1$. Thus, binary antipodal modulation switches the phase of the carrier between two values 0 and π, which is why it is termed binary phase-shift keying (BPSK) when applied to a passband channel:

We know from Chapter 2 that any passband signal can be represented in terms of two real-valued baseband waveforms, the I and Q components,

$$u_p(t) = u_c(t)\cos(2\pi f_c t) - u_s(t)\sin(2\pi f_c t)$$

The complex envelope of $u_p(t)$ is given by $u(t) = u_c(t) + ju_s(t)$. For BPSK, the I component is modulated using binary antipodal signaling, while the Q component is not used, so that $u(t) = u_c(t)$. However, noting that the two signals, $u_c(t)\cos(2\pi f_c t)$ and $u_s(t)\sin(2\pi f_c t)$ are orthogonal regardless of the choice of u_c and u_s, we realize that we can modulate both I and Q components independently, without affecting their orthogonality. In this case, we have

$$u_c(t) = \sum_n b_c[n]p(t - nT), \quad u_s(t) = \sum_n b_s[n]p(t - nT)$$

The complex envelope is given by

$$u(t) = u_c(t) + ju_s(t) = \sum_n (b_c[n] + jb_s[n])p(t - nT) = \sum_n b[n]p(t - nT) \tag{4.3}$$

where $\{b[n] = b_c[n] + jb_s[n]\}$ are complex-valued symbols.

Let us see what happens to the passband signal when both $b_c[n]$ and $b_s[n]$ take values in $\{\pm 1\}$ (i.e., $b[n] = b_c[n] + jb_s[n]$ takes values in $\{\pm 1 \pm j\}$). For the nth symbol interval $nT \leq t < (n+1)T$:

$$u_p(t) = \cos(2\pi f_c t) - \sin(2\pi f_c t) = \sqrt{2}\cos(2\pi f_c t + \pi/4) \text{ if } b_c[n] = +1, b_s[n] = +1;$$
$$u_p(t) = \cos(2\pi f_c t) + \sin(2\pi f_c t) = \sqrt{2}\cos(2\pi f_c t - \pi/4) \text{ if } b_c[n] = +1, b_s[n] = -1;$$
$$u_p(t) = -\cos(2\pi f_c t) - \sin(2\pi f_c t) = \sqrt{2}\cos(2\pi f_c t + 3\pi/4) \text{ if } b_c[n] = -1, b_s[n] = +1;$$
$$u_p(t) = -\cos(2\pi f_c t) + \sin(2\pi f_c t) = \sqrt{2}\cos(2\pi f_c t - 3\pi/4) \text{ if } b_c[n] = -1, b_s[n] = -1.$$

Thus, the modulation causes the passband signal to switch its phase among four possibilities, $\{\pm\pi/4, \pm 3\pi/4\}$, as illustrated in Figure 4.3, which is why we call it quadrature phase-shift keying (QPSK).

Equivalently, we could have seen this from the complex envelope. Note that the QPSK symbols can be written as $b[n] = \sqrt{2}e^{j\theta[n]}$, where $\theta[n] \in \{\pm\pi/4, \pm 3\pi/4\}$. Thus, over the nth symbol, we have

$$u_p(t) = \text{Re}\left(b[n]e^{j2\pi f_c t}\right) = \text{Re}\left(\sqrt{2}e^{j\theta[n]}e^{j2\pi f_c t}\right) = \sqrt{2}\cos(2\pi f_c t + \theta[n]), \quad nT \leq t < (n+1)T$$

This indicates that it is actually easier to figure out what is happening to the passband signal by working with the complex envelope. We therefore work in the complex-baseband domain for the remainder of this chapter.

In general, the complex envelope for a linearly modulated signal is given by (4.1), where $b[n] = b_c[n] + jb_s[n] = r[n]e^{j\theta[n]}$ can be complex-valued. We can view this as $b_c[n]$

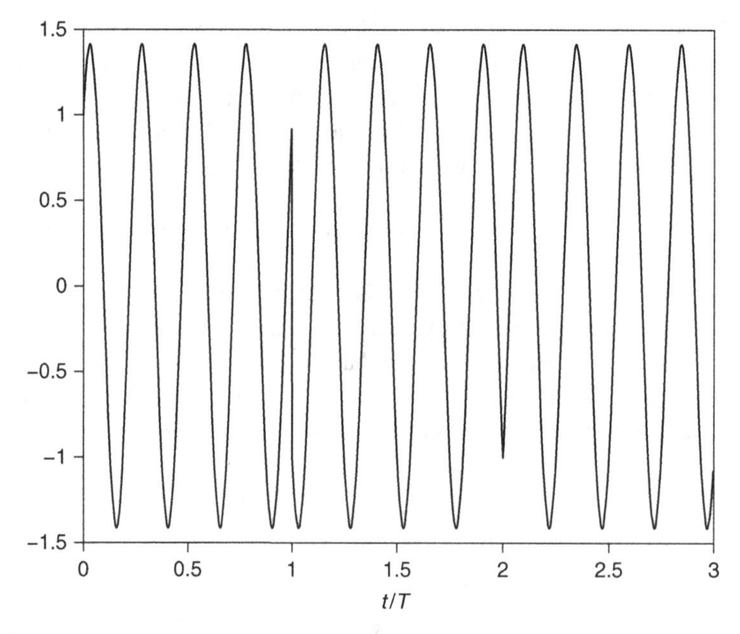

Figure 4.3 QPSK illustrated for $f_c = 4/T$, with symbol sequences $\{b_c[n]\} = \{+1, -1, -1\}$ and $\{b_s[n]\} = \{-1, +1, -1\}$. The phase of the passband signal is $-\pi/4$ in the first symbol interval, switching to $3\pi/4$ in the second, and to $-3\pi/4$ in the third.

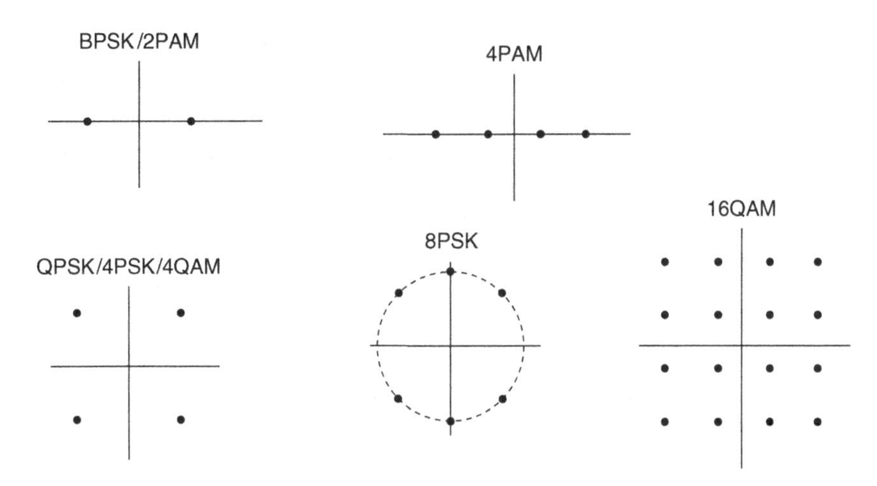

Figure 4.4 Some commonly used constellations. Note that 2PAM and 4PAM can be used over both baseband and passband channels, while the two-dimensional constellations QPSK, 8PSK, and 16QAM are for use over passband channels.

modulating the I component and $b_s[n]$ modulating the Q component, or as scaling the envelope by $r[n]$ and switching the phase by $\theta[n]$. The set of values that each symbol can take is called the signaling *alphabet*, or *constellation*. We can plot the constellation in a two-dimensional plot, with the x axis denoting the real part $b_c[n]$ (corresponding to the I component) and the y axis denoting the imaginary part $b_s[n]$ (corresponding to the Q component). Indeed, this is why linear modulation over passband channels is also termed *two-dimensional* modulation. Note that this provides a unified description of constellations that can be used over both baseband and passband channels: for physical baseband channels, we simply constrain $b[n] = b_c[n]$ to be real-valued, setting $b_s[n] = 0$.

Figure 4.4 shows some common constellations. Pulse amplitude modulation (PAM) corresponds to using multiple amplitude levels along the I component (setting the Q component to zero). This is often used for signaling over physical baseband channels. Using PAM along both I and Q axes corresponds to quadrature amplitude modulation (QAM). If the constellation points lie on a circle, they affect only the phase of the carrier: such signaling schemes are termed phase-shift keying (PSK). When naming a modulation scheme, we usually indicate the number of points in the constellations. BPSK and QPSK are special: BPSK (or 2PSK) can also be classified as 2PAM, while QPSK (or 4PSK) can also be classified as 4QAM.

Each symbol in a constellation of size M can be uniquely mapped to $\log_2 M$ bits. For a symbol rate of $1/T$ symbols per unit time, the *bit rate* is therefore $(\log_2 M)/T$ bits per unit time. Since the transmitted bits often contain redundancy due to a channel code employed for error correction or detection, the *information rate* is typically smaller than the bit rate. The choice of constellation for a particular application depends on considerations such as power–bandwidth tradeoffs and implementation complexity. We shall discuss these issues once we have developed more background.

4.2 Bandwidth occupancy

Bandwidth is a precious commodity, hence it is important to quantify the frequency occupancy of communication signals. To this end, consider the complex envelope of a linearly modulated signal (the two-sided bandwidth of this complex envelope equals the physical bandwidth of the corresponding passband signal), which has the form given in (4.1): $u(t) = \sum_n b[n]p(t - nT)$. The complex-valued symbol sequence $\{b[n]\}$ is modeled as random. Modeling the sequence as random at the transmitter makes sense because the latter does not control the information being sent (e.g., it depends on the specific computer file or digital audio signal being sent). Since this information is mapped to the symbols in some fashion, it follows that the symbols themselves are also random rather than deterministic. Modeling the symbols as random at the receiver makes even more sense, since the receiver by definition does not know the symbol sequence (otherwise there would be no need to transmit). However, for characterizing the bandwidth occupancy of the digitally modulated signal u, we do not compute statistics across different possible realizations of the symbol sequence $\{b[n]\}$. Rather, we define the quantities of interest in terms of averages across time, treating $u(t)$ as a finite-power signal that can be modeled as deterministic once the symbol sequence $\{b[n]\}$ is fixed. (We discuss concepts of statistical averaging across realizations later, when we discuss random processes in Chapter 5.)

We introduce the concept of PSD in Section 4.2.1. In Section 4.2.2, we state our main result on the PSD of digitally modulated signals, and discuss how to compute bandwidth once we know the PSD.

4.2.1 Power spectral density

We now introduce the important concept of power spectral density (PSD), which specifies how the power in a signal is distributed in different frequency bands.

Power spectral density The power spectral density (PSD), $S_x(f)$, for a finite-power signal $x(t)$ is defined through the conceptual measurement depicted in Figure 4.5. Pass $x(t)$ through an ideal narrowband filter with transfer function

$$H_\nu(f) = \begin{cases} 1, & \nu - \Delta f/2 < f < \nu + \Delta f/2 \\ 0, & \text{else} \end{cases}$$

The PSD evaluated at ν, $S_x(\nu)$, is defined as the measured power at the filter output, divided by the filter width Δf (in the limit as $\Delta f \to 0$).

Example (PSD of complex exponentials) Let us now find the PSD of $x(t) = Ae^{j(2\pi f_0 t + \theta)}$. Since the frequency content of x is concentrated at f_0, the power meter in Figure 4.5 will have zero output for $\nu \neq f_0$ (as $\Delta f \to 0$, f_0 falls outside the filter bandwidth for any such f_0). Thus, $S_x(f) = 0$ for $f \neq f_0$. On the other hand, for $\nu = f_0$, the output of the power meter is the entire power of x, which is

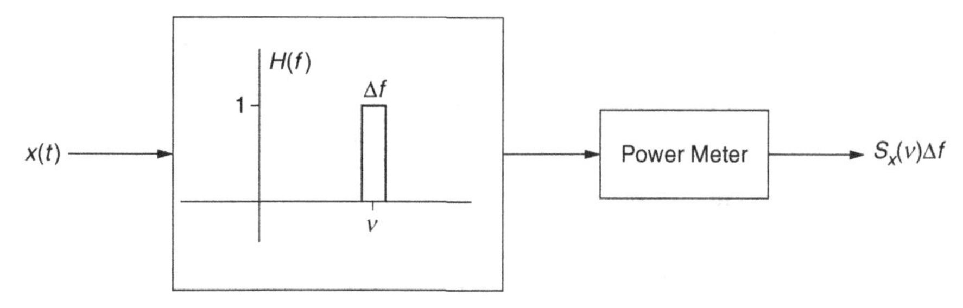

Figure 4.5 The operational definition of PSD.

$$P_x = A^2 = \int_{f_0-\Delta f/2}^{f_0+\Delta f/2} S_x(f) df$$

We conclude that the PSD is $S_x(f) = A^2\delta(f - f_0)$. Extending this reasoning to a sum of complex exponentials, we have

$$\text{PSD of } \sum_i A_i e^{j(2\pi f_i t+\theta_i)} = \sum_i A_i^2 \delta(f - f_i)$$

where f_i are distinct frequencies (positive or negative), and A_i and θ_i are the amplitude and phase, respectively, of the ith complex exponential. Thus, for a real-valued sinusoid, we obtain

$$S_x(f) = \frac{1}{4}\delta(f-f_0)+\frac{1}{4}\delta(f+f_0), \quad \text{for } x(t) = \cos(2\pi f_0 t+\theta) = \frac{1}{2}e^{j(2\pi f_0 t+\theta)}+\frac{1}{2}e^{-j(2\pi f_0 t+\theta)} \tag{4.4}$$

Periodogram-based PSD estimation One way to carry out the conceptual measurement in Figure 4.5 is to limit $x(t)$ to a finite observation interval, compute its Fourier transform and hence its energy spectral density (which is the magnitude square of the Fourier transform), and then divide by the length of the observation interval. The PSD is obtained by letting the observation interval get large. Specifically, define the time-windowed version of x as

$$x_{T_o}(t) = x(t) I_{[-T_o/2,T_o/2]}(t) \tag{4.5}$$

where T_o is the length of the observation interval. Since T_o is finite and $x(t)$ has finite power, $x_{T_o}(t)$ has finite energy, and we can compute its Fourier transform

$$X_{T_o}(f) = \mathcal{F}(x_{T_o})$$

The energy spectral density of x_{T_o} is given by $|X_{T_o}(f)|^2$. By averaging this over the observation interval, we obtain the estimated PSD

$$\hat{S}_x(f) = \frac{|X_{T_o}(f)|^2}{T_o} \tag{4.6}$$

The estimate in (4.6), which is termed a *periodogram,* can typically be obtained by taking the DFT of a sampled version of the time-windowed signal; the time interval T_o must be large enough to give the desired frequency resolution, while the sampling rate must be large enough to capture the variations in $x(t)$. The estimated PSDs obtained over multiple observation intervals can then be averaged further to get smoother estimates.

Formally, we can define the PSD in the limit of large time windows as follows:

$$S_x(f) = \lim_{T_o \to \infty} \frac{|X_{T_o}(f)|^2}{T_o} \tag{4.7}$$

Units for PSD Power per unit frequency has the same units as power multiplied by time, or energy. Thus, the PSD is expressed in units of watts/hertz, or joules.

Power in terms of PSD The power P_x of a finite-power signal x is given by integrating its PSD:

$$P_x = \int_{-\infty}^{\infty} S_x(f) df \tag{4.8}$$

4.2.2 The PSD of a linearly modulated signal

We are now ready to state our result on the PSD of a linearly modulated signal $u(t) = \sum_n b[n]p(t - nT)$. While we derive a more general result in Appendix 4.A, our result here applies to the following important special case:

(a) the symbols have *zero DC value*: $\lim_{N\to\infty}(1/(2N+1)) \sum_{n=-N}^{N} b[n] = 0$; and
(b) the symbols are *uncorrelated*: $\lim_{N\to\infty}(1/(2N)) \sum_{n=-N}^{N} b[n]b^*[n-k] = 0$ for $k \neq 0$.

Theorem 4.2.1 (PSD of a linearly modulated signal) *Consider a linearly modulated signal*

$$u(t) = \sum_n b[n]p(t - nT)$$

where the symbol sequence $\{b[n]\}$ is zero mean and uncorrelated with average symbol energy

$$\sigma_b^2 = \overline{|b[n]|^2} = \lim_{N\to\infty} \frac{1}{2N+1} \sum_{n=-N}^{N} |b[n]|^2$$

Then the PSD is given by

$$S_u(f) = \frac{|P(f)|^2}{T} \sigma_b^2 \tag{4.9}$$

and the power of the modulated signal is

$$P_u = \frac{\sigma_b^2 ||p||^2}{T} \tag{4.10}$$

where $||p||^2$ denotes the energy of the modulating pulse.

See Appendix 4.A for a proof of (4.9), which follows from specializing a more general expression.

The expression for the power follows from integrating the PSD:

$$P_u = \int_{-\infty}^{\infty} S_u(f)df = \frac{\sigma_b^2}{T}\int_{-\infty}^{\infty} |P(f)|^2\, df = \frac{\sigma_b^2}{T}\int_{-\infty}^{\infty} |p(t)|^2\, dt = \frac{\sigma_b^2 ||p||^2}{T}$$

where we have used Parseval's identity.

An intuitive interpretation of this theorem is as follows. Every T time units, we send a pulse of the form $b[n]p(t - nT)$ with average energy spectral density $\sigma_b^2|P(f)|^2$, so the PSD is obtained by dividing this by T. The same reasoning applies to the expression for power: every T time units, we send a pulse $b[n]p(t - nT)$ with average energy $\sigma_b^2||p||^2$, so the power is obtained by dividing by T. The preceding intuition does not apply when successive symbols are correlated, in which case we get the more complicated expression (4.32) for the PSD in Appendix 4.A.

Once we know the PSD, we can define the bandwidth of u in a number of ways.

3-dB bandwidth For symmetric $S_u(f)$ with a maximum at $f = 0$, the *3-dB bandwidth* $B_{3\text{dB}}$ is defined by $S_u(B_{3\text{dB}}/2) = S_{\text{u}}(-B_{3\text{dB}}/2) = \frac{1}{2}S_u(0)$. That is, the 3-dB bandwidth is the size of the interval between the points at which the PSD is 3 dB, or a factor of $\frac{1}{2}$, smaller than its maximum value.

Fractional power-containment bandwidth This is the size of the smallest interval that contains a given fraction of the power. For example, for symmetric $S_{\text{u}}(f)$, the 99% fractional power-containment bandwidth B is defined by

$$\int_{-B/2}^{B/2} S_u(f)df = 0.99P_u = 0.99\int_{-\infty}^{\infty} S_u(f)df$$

(replace 0.99 in the preceding equation by any desired fraction γ to get the corresponding γ power-containment bandwidth).

Time/frequency normalization Before we discuss examples in detail, let us simplify our task by making a simple observation on time and frequency scaling. Suppose we have a linearly modulated system operating at a symbol rate of $1/T$, as in (4.1). We can think of it as a normalized system operating at a symbol rate of one, where the unit of time is T. This implies that the unit of frequency is $1/T$. In terms of these new units, we can write the linearly modulated signal as

$$u_1(t) = \sum_n b[n]p_1(t - n)$$

where $p_1(t)$ is the modulation pulse for the normalized system. For example, for a rectangular pulse timelimited to the symbol interval, we have $p_1(t) = I_{[0,1]}(t)$. Suppose now that the bandwidth of the normalized system (computed using any definition that we please) is B_1. Since the unit of frequency is $1/T$, the bandwidth in the original system is B_1/T. Thus, in terms of determining the frequency occupancy, we can work, without loss of generality, with the normalized system. In the original system, what we are really doing is working with the normalized time t/T and the normalized frequency fT.

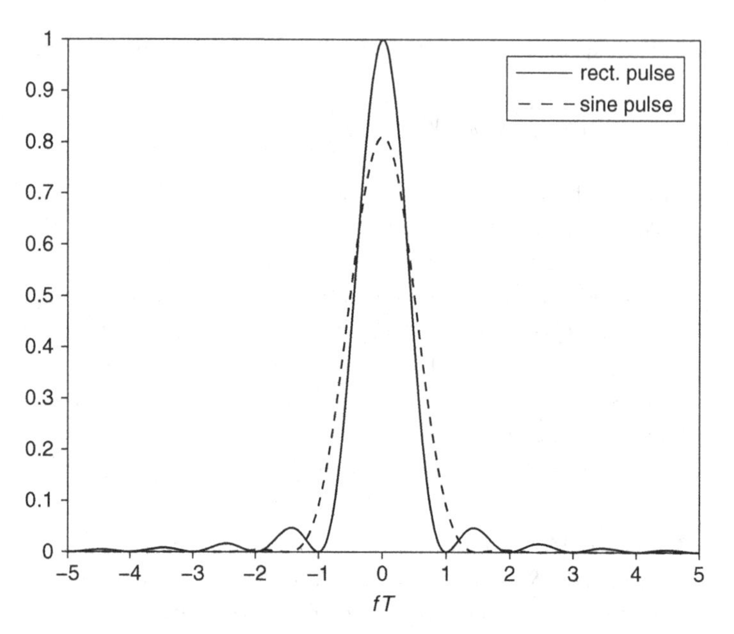

Figure 4.6 The PSD corresponding to rectangular (rect.) and sine timelimited pulses. The main lobe of the PSD is broader for the sine pulse, but its 99% power-containment bandwidth is much smaller.

Rectangular pulse Without loss of generality, consider a normalized system with $p_1(t) = I_{[0,1]}(t)$, for which $P_1(f) = \text{sinc}(f)e^{-j\pi f}$. For $\{b[n]\}$ i.i.d., taking values ± 1 with equal probability, we have $\sigma_b^2 = 1$. On applying (4.9), we obtain

$$S_{u_1}(f) = \sigma_b^2 \, \text{sinc}^2 f \tag{4.11}$$

By integrating, or applying (4.10), we obtain $P_u = \sigma_b^2$. The scale factor of σ_b^2 is not important, since it drops out for any definition of bandwidth. We therefore set it to $\sigma_b^2 = 1$. The PSD for the rectangular pulse, along with that for a sine pulse, to be introduced shortly, is plotted in Figure 4.6.

Note that the PSD for the rectangular pulse has much fatter tails, which does not bode well for its bandwidth efficiency. For a fractional power-containment bandwidth with fraction γ, we have the equation

$$\int_{-B_1/2}^{B_1/2} \text{sinc}^2 f \, df = \gamma \int_{-\infty}^{\infty} \text{sinc}^2 f \, df = \gamma \int_0^1 1^2 \, dt = \gamma$$

using Parseval's identity. We therefore obtain, using the symmetry of the PSD, that the bandwidth is the numerical solution to the equation

$$\int_0^{B_1/2} \text{sinc}^2 f \, df = \gamma/2 \tag{4.12}$$

For example, for $\gamma = 0.99$, we obtain $B_1 = 10.2$, while for $\gamma = 0.9$, we obtain $B_1 = 0.85$. Thus, if we wish to be strict about power containment (e.g., in order to limit adjacent-channel interference in wireless systems), the rectangular timelimited pulse is a very poor choice. On the other hand, in systems where interference and regulation are not significant issues (e.g., low-cost wired systems), this pulse may be a good choice because of its ease of implementation using digital logic.

Example 4.2.1 (Bandwidth computation) Consider a passband system operating at a carrier frequency of 2.4 GHz at a bit rate of 20 Mbps. A rectangular modulation pulse timelimited to the symbol interval is employed.

(a) Find the 99% and 90% power-containment bandwidths if the constellation used is 16-QAM.
(b) Find the 99% and 90% power-containment bandwidths if the constellation used is QPSK.

Solution

(a) The 16-QAM system sends 4 bits/symbol, so that the symbol rate $1/T$ equals $(20\ \text{Mbits/s})/(4\ \text{bits/symbol}) = 5$ Msymbols/s. Since the 99% power-containment bandwidth for the normalized system is $B_1 = 10.2$, the required bandwidth is $B_1/T = 51$ MHz. Since the 90% power-containment for the normalized system is $B_1 = 0.85$, the required bandwidth B_1/T equals 4.25 MHz.
(b) The QPSK system sends 2 bits/symbol, so that the symbol rate is 10 Msymbols/s. The bandwidths required are therefore double those in (a): the 99% power-containment bandwidth is 102 MHz, while the 90% power-containment bandwidth is 8.5 MHz.

Clearly, when the criterion for defining bandwidth is the same, then 16-QAM consumes half the bandwidth compared with QPSK for a fixed bit rate. However, it is interesting to note that, for the rectangular timelimited pulse, a QPSK system where we are sloppy about power leakage (90% power-containment bandwidth of 8.5 MHz) can require far less bandwidth than a system using a more bandwidth-efficient 16-QAM constellation where we are strict about power leakage (99% power-containment bandwidth of 51 MHz). This extreme variation of bandwidth when we tweak definitions slightly is because of the poor frequency-domain containment of the rectangular timelimited pulse. Thus, if we are serious about limiting frequency occupancy, we need to think about more sophisticated designs for the modulation pulse.

Smoothing out the rectangular pulse A useful alternative to using the rectangular pulse, while still keeping the modulating pulse timelimited to a symbol interval, is the sine pulse, which for the normalized system equals

$$p_1(t) = \sqrt{2}\sin(\pi t) I_{[0,1]}(t)$$

Since the sine pulse does not have the sharp edges of the rectangular pulse in the time domain, we expect it to be more compact in the frequency domain. Note that we have normalized the pulse to have unit energy, as we did for the normalized rectangular pulse.

This implies that the power of the modulated signal is the same in the two cases, so that we can compare PSDs under the constraint that the area under the PSDs remains constant. On setting $\sigma_b^2 = 1$ and using (4.9), we obtain (see Problem 4.1)

$$S_{u_1}(f) = |P_1(f)|^2 = \frac{8}{\pi^2} \frac{\cos^2(\pi f)}{\left(1 - 4f^2\right)^2} \tag{4.13}$$

Proceeding as we did for obtaining (4.12), the fractional power-containment bandwidth for fraction γ is given by the formula

$$\int_0^{B_1/2} \frac{8}{\pi^2} \frac{\cos^2(\pi f)}{\left(1 - 4f^2\right)^2} \, df = \gamma/2 \tag{4.14}$$

For $\gamma = 0.99$, we obtain $B_1 = 1.2$, which is an order-of-magnitude improvement over the corresponding value of $B_1 = 10.2$ for the rectangular pulse.

While the sine pulse has better frequency-domain containment than the rectangular pulse, it is still not suitable for *strictly* bandlimited channels. We discuss pulse design for such channels next.

4.3 Design for bandlimited channels

Suppose that you are told to design your digital communication system so that the transmitted signal fits between 2.39 and 2.41 GHz; that is, you are given a passband channel of bandwidth 20 MHz at a carrier frequency of 2.4 GHz. Any signal that you transmit over this band has a complex envelope with respect to 2.4 GHz that occupies a band from −10 MHz to 10 MHz. Similarly, the passband channel (modeled as an LTI system) has an impulse response whose complex envelope is bandlimited from −10 MHz to 10 MHz. In general, for a passband channel or signal of bandwidth W, with an appropriate choice of reference frequency, we have a corresponding complex-baseband signal spanning the band $[-W/2, W/2]$. Thus, we restrict our design to the complex-baseband domain, with the understanding that the designs can be translated to passband channels by upconversion of the I and Q components at the transmitter, and downconversion at the receiver. Also, note that the designs specialize to physical baseband channels if we restrict the baseband signals to being real-valued.

4.3.1 Nyquist's sampling theorem and the sinc pulse

Our first step in understanding communication system design for such a bandlimited channel is to understand the structure of bandlimited signals. To this end, suppose that the signal $s(t)$ is bandlimited to $[-W/2, W/2]$. We can now invoke Nyquist's sampling theorem (the proof of which is postponed to Section 4.5) to express the signal in terms of its samples at rate W.

Theorem 4.3.1 (Nyquist's sampling theorem) *Any signal $s(t)$ bandlimited to $[-W/2, W/2]$ can be described completely by its samples $\{s(n/W)\}$ at rate W. The signal $s(t)$ can be recovered from its samples using the following interpolation formula:*

$$s(t) = \sum_{n=-\infty}^{\infty} s\left(\frac{n}{W}\right) p\left(t - \frac{n}{W}\right) \tag{4.15}$$

where $p(t) = \mathrm{sinc}(Wt)$.

Degrees of freedom What does the sampling theorem tell us about digital modulation? The interpolation formula (4.15) tells us that we can interpret $s(t)$ as a linearly modulated signal with symbol sequence equal to the samples $\{s(n/W)\}$, symbol rate $1/T$ equal to the bandwidth W, and modulation pulse given by $p(t) = \mathrm{sinc}(Wt) \leftrightarrow P(f) = (1/W)I_{[-W/2,W/2]}(f)$. Thus, linear modulation with the sinc pulse is able to exploit all of the "degrees of freedom" available in a bandlimited channel.

Signal space If we signal over an observation interval of length T_0 using linear modulation according to the interpolation formula (4.15), then we have approximately WT_0 complex-valued samples. Thus, while the signals we send are continuous-time signals, which, in general, lie in an infinite-dimensional space, the set of possible signals we can send in a finite observation interval of length T_0 lives in a complex-valued vector space of *finite* dimension WT_0, or, equivalently, a real-valued vector space of dimension $2WT_0$. Such geometric views of communication signals as vectors, often termed *signal-space concepts,* are particularly useful in design and analysis, as we explore in more detail in Chapter 6.

The concept of Nyquist signaling Since the sinc pulse is not timelimited to a symbol interval, in principle, the symbols could interfere with each other. The time-domain signal corresponding to a bandlimited modulation pulse such as the sinc spans an interval significantly larger than the symbol interval (in theory, the interval is infinitely large, but we always truncate the waveform in implementations). This means that successive pulses corresponding to successive symbols that are spaced by the symbol interval (i.e., $b[n]p(t - nT)$ as we increment n) overlap with, and therefore can interfere with, each other. Figure 4.7 shows the sinc pulse modulated by three bits, +1, −1, +1. While the pulses corresponding to the three symbols do overlap, notice that, by sampling at $t = 0$, $t = T$, and $t = 2T$, we can recover the three symbols because exactly one of the pulses is nonzero at each of these times. That is, at sampling times spaced by integer multiples of the symbol time T, there is no *intersymbol interference.* We call such a pulse *Nyquist* for signaling at rate $1/T$, and we discuss other examples of such pulses soon. Designing pulses based on the Nyquist criterion allows us the freedom to expand the modulation pulses in time beyond the symbol interval (thus enabling better containment in the frequency domain), while ensuring that there is no ISI at appropriately chosen sampling times despite the significant overlap between successive pulses.

The problem with sinc Are we done then? Should we just use linear modulation with a sinc pulse when confronted with a bandlimited channel? Unfortunately, the answer is no: just as the rectangular timelimited pulse decays too slowly in frequency, the rectangular bandlimited pulse, corresponding to the sinc pulse in the time domain, decays too slowly in time. Let us see what happens as a consequence. Figure 4.8 shows a plot of the modulated

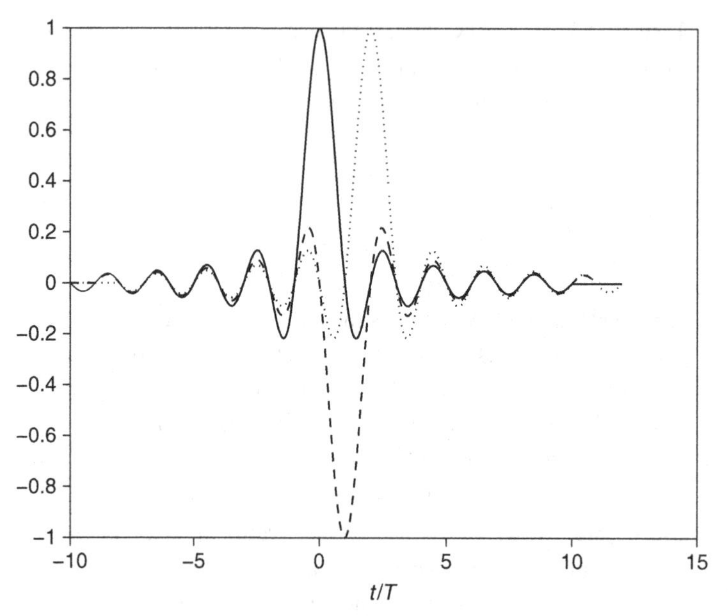

Figure 4.7 Three successive sinc pulses (each pulse is truncated to a length of 10 symbol intervals on each side) modulated by $+1$, -1, $+1$. The actual transmitted signal is the sum of these pulses (not shown). Note that, while the pulses overlap, the samples at $t = 0, T, 2T$ are equal to the transmitted bits because only one pulse is nonzero at these times.

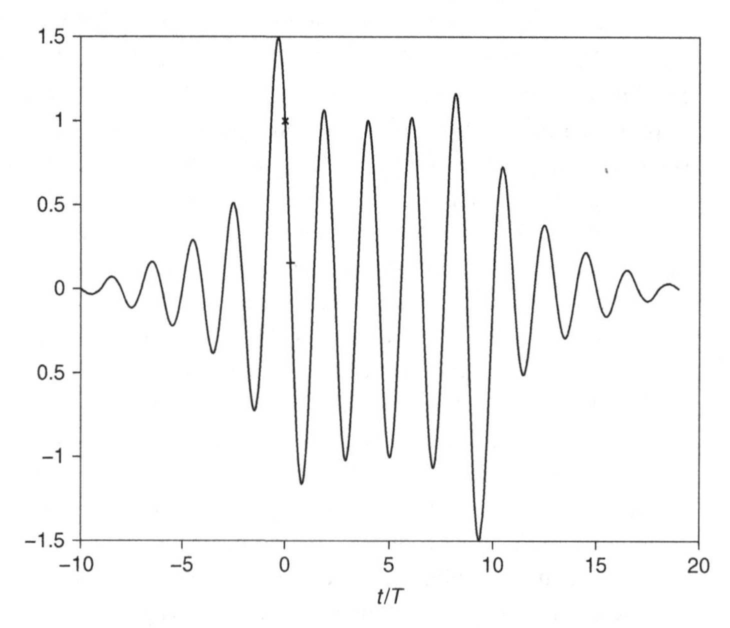

Figure 4.8 The baseband signal for 10 BPSK symbols of alternating signs, modulated using the sinc pulse. The first symbol is $+1$, and the sample at time $t = 0$, marked with "x," equals $+1$, as desired (no ISI). However, if the sampling time is off by $0.25T$, the sample value, marked by "+," becomes much smaller because of ISI. While it still has the right sign, the ISI causes it to have significantly smaller noise immunity. See Problem 4.14 for an example in which the ISI due to timing mismatch actually causes the sign to flip.

waveform for a bit sequence of alternating sign. At the correct sampling times, there is no ISI. However, if we consider a timing error of $0.25T$, the ISI causes the sample value to drop drastically, making the system more vulnerable to noise. What is happening is that, when there is a small sampling offset, we can make the ISI add up to a large value by choosing the interfering symbols so that their contributions all have signs opposite to that of the desired symbol at the sampling time. Since the sinc pulse decays as $1/t$, the ISI created for a given symbol by an interfering symbol that is n symbol intervals away decays as $1/n$, so that, in the worst case, the contributions from the interfering symbols roughly have the form $\sum_n 1/n$, a series that is known to diverge. Thus, in theory, if we do not truncate the sinc pulse, we can make the ISI arbitrarily large when there is a small timing offset. In practice, we do truncate the modulation pulse, so that we see ISI only from a finite number of symbols. However, even when we do truncate, as we see from Figure 4.8, the slow decay of the sinc pulse means that the ISI adds up quickly, and significantly reduces the margin of error when noise is introduced into the system.

While using the sinc pulse might not be a good idea in practice, the idea of using bandwidth-efficient Nyquist pulses is a good one, and we now develop it further.

4.3.2 The Nyquist criterion for ISI avoidance

Nyquist signaling Consider a linearly modulated signal

$$u(t) = \sum_n b[n]p(t - nT)$$

We say that the pulse $p(t)$ is Nyquist (or satisfies the Nyquist criterion) for signaling at rate $1/T$ if the symbol-spaced samples of the modulated signal are equal to the symbols (or a fixed scalar multiple of the symbols); that is, $u(kT) = b[k]$ for all k. That is, there is no ISI at appropriately chosen sampling times spaced by the symbol interval.

In the time domain, it is quite easy to see what is required to satisfy the Nyquist criterion. The samples satisfy $u(kT) = \sum_n b[n]p(kT - nT) = b[k]$ (or a scalar multiple of $b[k]$) for all k if and only if $p(0) = 1$ (or some nonzero constant) and $p(mT) = 0$ for all integers $m \neq 0$. However, for design of bandwidth efficient pulses, it is important to characterize the Nyquist criterion in the frequency domain. This is given by the following theorem.

Theorem 4.3.2 (Nyquist criterion for ISI avoidance) *The pulse $p(t) \leftrightarrow P(f)$ is Nyquist for signaling at rate $1/T$ if*

$$p(mT) = \delta_{m0} = \begin{cases} 1, & m = 0 \\ 0, & m \neq 0 \end{cases} \tag{4.16}$$

or, equivalently,

$$\frac{1}{T} \sum_{k=-\infty}^{\infty} P\left(f + \frac{k}{T}\right) = 1 \quad \text{for all } f \tag{4.17}$$

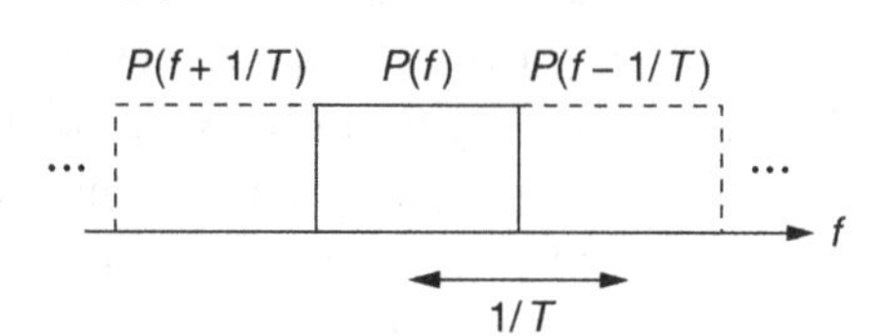

Figure 4.9 The minimum-bandwidth Nyquist pulse is a sinc.

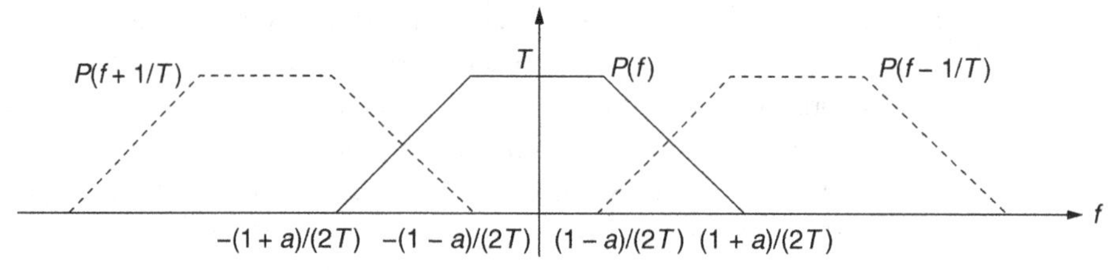

Figure 4.10 A trapezoidal pulse that is Nyquist at rate $1/T$. The (fractional) excess bandwidth is a.

The proof of this theorem is given in Section 4.5, where we show that both the Nyquist sampling theorem, Theorem 4.3.1, and the preceding theorem are based on the same mathematical result, namely that the samples of a time-domain signal have a one-to-one mapping with the sum of translated (or *aliased*) versions of its Fourier transform.

In this section, we explore the design implications of Theorem 4.3.2. In the frequency domain, the translates of $P(f)$ by integer multiples of $1/T$ must add up to a constant. As illustrated by Figure 4.9, the minimum-bandwidth pulse for which this happens is the ideal bandlimited pulse over an interval of length $1/T$.

Minimum-bandwidth Nyquist pulse The minimum-bandwidth Nyquist pulse is

$$P(f) = \begin{cases} T, & |f| \leq 1/(2T) \\ 0, & \text{else} \end{cases}$$

corresponding to the time-domain pulse

$$p(t) = \text{sinc}(t/T)$$

As we have already discussed, the sinc pulse is not a good choice in practice because of its slow decay in time. To speed up the decay in time, we must expand in the frequency domain, while conforming to the Nyquist criterion. The trapezoidal pulse depicted in Figure 4.10 is an example of such a pulse.

The role of excess bandwidth We have noted earlier that the problem with the sinc pulse arises because of its $1/t$ decay and the divergence of the harmonic series $\sum_{n=1}^{\infty} 1/n$, which implies that the worst-case contribution from "distant" interfering symbols at a given sampling instant can blow up. Using the same reasoning, however, a pulse $p(t)$ decaying as $1/t^b$ for $b > 1$ should work, since the series $\sum_{n=1}^{\infty} 1/n^b$ does converge for $b > 1$. A faster time decay requires a slower decay in frequency. Thus, we need *excess bandwidth,* beyond the minimum bandwidth dictated by the Nyquist criterion, to fix the problems associated with the sinc pulse. The (fractional) excess bandwidth for a linear modulation scheme is defined to be the fraction of bandwidth over the minimum required for ISI avoidance at a given symbol rate. In particular, Figure 4.10 shows that a trapezoidal pulse (in the frequency domain) can be Nyquist for suitably chosen parameters, since the translates $\{P(f + k/T)\}$ as shown in the figure add up to a constant. Since trapezoidal $P(f)$ is the convolution of two boxes in the frequency domain, the time-domain pulse $p(t)$ is the product of two sinc functions, as worked out in the example below. Since each sinc decays as $1/t$, the product decays as $1/t^2$, which implies that the worst-case ISI with timing mismatch is indeed bounded.

Example 4.3.1 Consider the trapezoidal pulse of excess bandwidth a shown in Figure 4.10.

(a) Find an explicit expression for the time-domain pulse $p(t)$.

(b) What is the bandwidth required for a passband system using this pulse operating at 120 Mbps using 64QAM, with an excess bandwidth of 25%?

Solution

(a) It is easy to check that the trapezoid is a convolution of two boxes as follows (we assume $0 < a \leq 1$):

$$P(f) = \frac{T^2}{a} I_{[-1/(2T),1/(2T)]}(f) * I_{[-a/(2T),a/(2T)]}(f)$$

Taking inverse Fourier transforms, we obtain

$$p(t) = \frac{T^2}{a}\left(\frac{1}{T}\,\mathrm{sinc}(t/T)\right)\left(\frac{a}{T}\,\mathrm{sinc}(at/T)\right) = \mathrm{sinc}(t/T)\mathrm{sinc}(at/T) \quad (4.18)$$

The presence of the first sinc provides the zeros required by the time-domain Nyquist criterion: $p(mT) = 0$ for nonzero integers $m \neq 0$. The presence of a second sinc yields a $1/t^2$ decay, providing robustness against timing mismatch.

(b) Since $64 = 2^6$, the use of 64QAM corresponds to sending 6 bits/symbol, so the symbol rate is $120/6 = 20$ Msymbols/s. The minimum bandwidth required is therefore 20 MHz, so that 25% excess bandwidth corresponds to a bandwidth of $20 \times 1.25 = 25$ MHz.

Raised-cosine pulse Replacing the straight line of the trapezoid with a smoother cosine-shaped curve in the frequency domain gives us the raised-cosine pulse, which has a faster, $1/t^3$, decay in the time domain:

$$P(f) = \begin{cases} T, & |f| \leq (1-a)/(2T) \\ (T/2)[1 + \cos((|f| - (1-a)/(2T))\pi T/a)], & (1-a)/(2T) \leq |f| \leq (1+a)/(2T) \\ 0, & |f| > (1+a)/(2T) \end{cases}$$

where a is the fractional excess bandwidth, which is typically chosen in the range where $0 \leq a < 1$. As shown in Problem 4.11, the time-domain pulse $s(t)$ is given by

$$p(t) = \text{sinc}\left(\frac{t}{T}\right)\frac{\cos(\pi at/T)}{1-(2at/T)^2}$$

This pulse inherits the Nyquist property of the sinc pulse, while having an additional multiplicative factor that gives an overall $1/t^3$ decay with time. The faster time decay compared with the sinc pulse is evident from a comparison of Figures 4.11(b) and 4.12(b).

We now comment on some interesting properties of Nyquist pulses.

For both the trapezoidal and the raised-cosine waveforms, the time-domain pulse has a sinc(at) term that provides zeros at integer multiples of $T = 1/a$. This means that the pulse is Nyquist at rate $1/T = a$. In other words, a time-domain factor that provides "zeros at rate a" (i.e., spaced by $1/a$) enables Nyquist signaling at rate a. A pulse that is trapezoidal

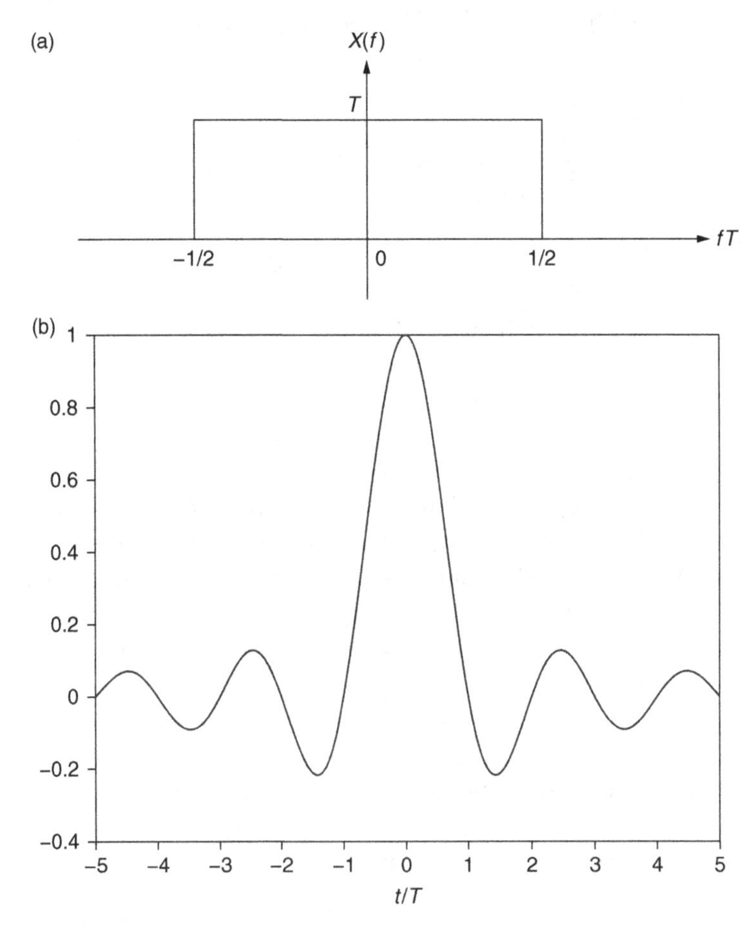

Figure 4.11 A sinc pulse for minimum bandwidth ISI-free signaling at rate $1/T$. (a) Frequency-domain boxcar. (b) Time-domain sinc pulse. Both time and frequency axes are normalized to be dimensionless.

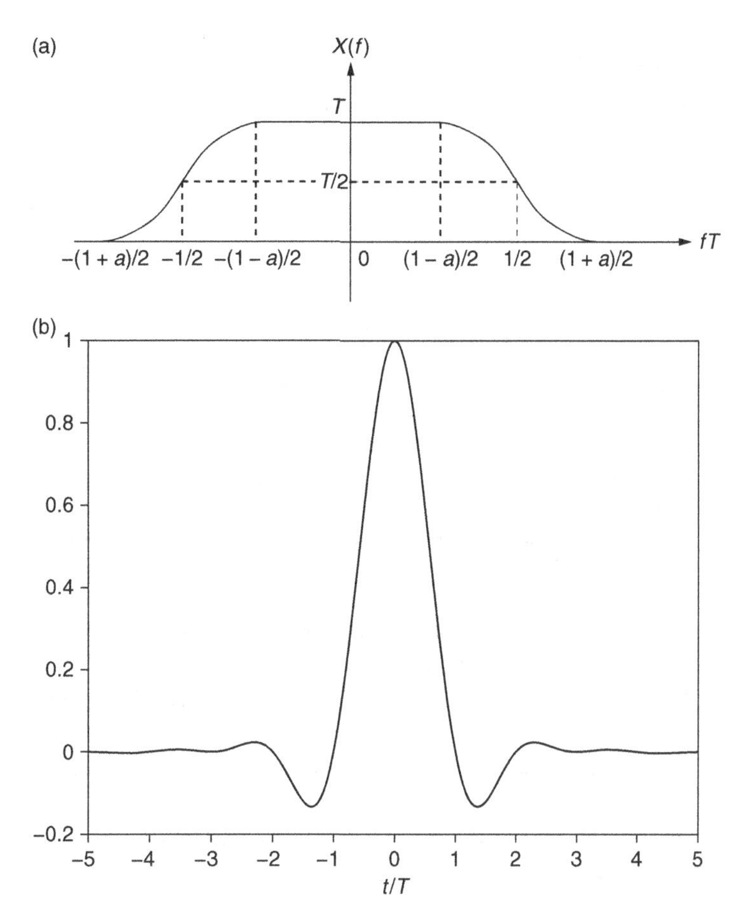

Figure 4.12 A raised-cosine pulse for minimum bandwidth ISI-free signaling at rate $1/T$, with excess bandwidth a. (a) Frequency-domain raised cosine. (b) Time-domain pulse (excess bandwidth $a = 0.5$). Both time and frequency axes are normalized to be dimensionless.

has a time-domain pulse of the form $\text{sinc}(at)\text{sinc}(bt)$, which provides zeros at rate a as well as at rate b. Thus, this pulse is Nyquist at rate a and at rate b.

It is also interesting to note that, once we have zeros at integer multiples of T, we also have zeros at integer multiples of KT, where K is any positive integer. In other words, if a pulse is Nyquist at rate $1/T$, then it is also Nyquist at integer submultiples of this rate; that is, it is Nyquist for all rates of the form $1/(KT)$, for K a positive integer. Thus, a factor $\text{sinc}(at)$ in the pulse guarantees the Nyquist property for all rates a/K.

Of course, we are typically interested only in the highest rate for a given bandwidth, but it is interesting to play with the preceding observations, as we do in the following example involving a trapezoidal pulse.

Example 4.3.2 Consider passband linear modulation using the bandlimited pulse shown in Figure 4.13. Answer the following true/false questions, clearly stating your reasoning.

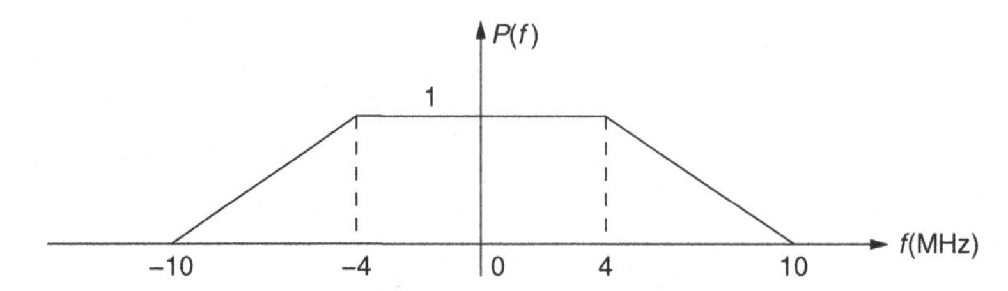

Figure 4.13 The pulse for Example 4.3.2.

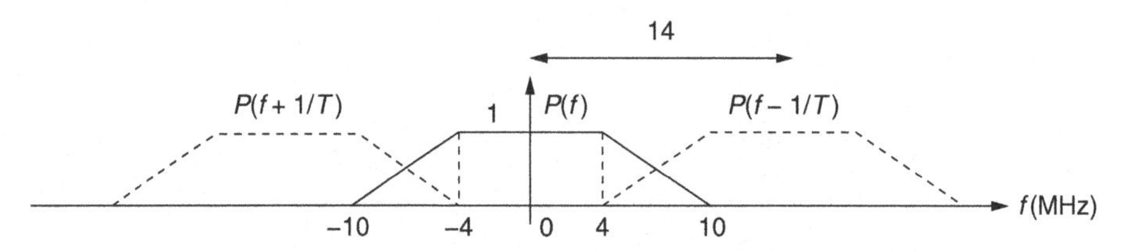

Figure 4.14 The frequency-domain Nyquist criterion is satisfied for $1/T = 14$ Msymbols/s in Example 4.3.2(a).

(a) *True or false:* the pulse $p(t)$ can be used for Nyquist signaling at a bit rate of 56 Mbps using a 16QAM constellation.

(b) *True or false:* the pulse $p(t)$ can be used for Nyquist signaling at a bit rate of 21 Mbps using an 8PSK constellation.

(c) *True or false:* the pulse $p(t)$ can be used for Nyquist signaling at a bit rate of 18 Mbps using an 8PSK constellation.

(d) *True or false:* the pulse $p(t)$ can be used for Nyquist signaling at a bit rate of 25 Mbps using a QPSK constellation.

Solution

(a) The symbol rate is

$$\frac{1}{T} = \frac{56 \text{ Mbps}}{4 \text{ bits/symbol}} = 14 \text{ Msymbols/s}$$

From Figure 4.14, we see that for this rate, the frequency-domain Nyquist criterion is satisfied: $\sum_k P(f + k/T)$ is constant. Alternatively, we know that the frequency-domain trapezoid corresponds to $p(t) = \text{sinc}(at)\text{sinc}(bt)$ in the time domain, where $(a - b)/2 = 4$ and $(a + b)/2 = 10$. On solving, we obtain $a = 14$ MHz and $b = 6$ MHz. Thus, the time-domain pulse provides zeros at rates 14 MHz and 6 MHz; hence it is indeed Nyquist at rate 14 Msymbols/s. The statement is therefore *true.*

(b) The symbol rate is $1/T = 21$ Mbps/(3 bits/symbol) $= 7$ Msymbols/s. Since the pulse is Nyquist at 14 Msymbols/s, it is also Nyquist at $14/2 = 7$ Msymbols/s. The statement is therefore *true.*

(c) The symbol rate is $1/T = 18$ Mbps/(3 bits/symbol) $= 6$ Msymbols/s. As shown in (a), the pulse has a $\text{sinc}(6t)$ term that provides zeros at rate 6 MHz; hence the statement is *true.*

(d) The symbol rate is $1/T = 25$ Mbps/(2 bits/symbol) $= 12.5$ Msymbols/s. This is not an integer submultiple of either 14 MHz or 6 MHz, the rates at which zeros are provided by the two sinc factors. Thus, the Nyquist property does not hold, and the statement is *false*.

4.3.3 Bandwidth efficiency

We define the *bandwidth efficiency* of linear modulation with an M-ary alphabet as

$$\eta_{\mathrm{B}} = \log_2 M \text{ bits/symbol}$$

The Nyquist criterion for ISI avoidance says that the minimum bandwidth required for ISI-free transmission using linear modulation equals the symbol rate, using the sinc as the modulation pulse. For such an idealized system, we can think of η_{B} as bits/second per hertz, since the symbol rate equals the bandwidth. Thus, knowing the bit rate R_{b} and the bandwidth efficiency η_{B} of the modulation scheme, we can determine the symbol rate, and hence the minimum required bandwidth, B_{min}, as follows:

$$B_{\mathrm{min}} = \frac{R_{\mathrm{b}}}{\eta_{\mathrm{B}}}$$

This bandwidth would then be expanded by the excess bandwidth used in the modulating pulse. However, this is not included in our definition of bandwidth efficiency, because excess bandwidth is a highly variable quantity dictated by a variety of implementation considerations. Once we have decided on the fractional excess bandwidth a, the actual bandwidth required is

$$B = (1 + a)B_{\mathrm{min}} = (1 + a)\frac{R_{\mathrm{b}}}{\eta_{\mathrm{B}}}$$

4.3.4 Power–bandwidth tradeoffs: a sneak preview

Clearly, we can increase bandwidth efficiency simply by increasing M, the constellation size. For example, the bandwidth efficiency of QPSK is 2 bits/symbol, while that of 16QAM is 4 bits/symbol. What stops us from increasing the constellation size, and hence bandwidth efficiency, indefinitely is noise, and the fact that we cannot use arbitrarily large transmit power (which is typically limited by cost or physical and regulatory constraints) to overcome it. Noise in digital communication systems must be modeled statistically, hence rigorous discussion of a formal model and its design consequences is postponed to Chapters 5 and 6. However, that does not prevent us from giving a handwaving sneak preview of the bottom line here. *Note that this subsection is meant as a teaser: it can be safely skipped, since these issues are covered in detail in Chapter 6.*

Intuitively speaking, the effect of noise is to perturb constellation points from the nominal locations shown in Figure 4.4, which leads to the possibility of making an error in deciding which point was transmitted. For a given noise "strength" (which determines how

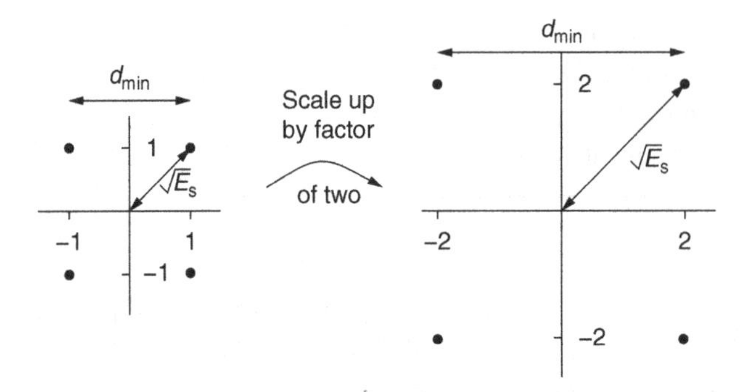

Figure 4.15 Scaling of minimum distance and energy per symbol.

much movement the noise can produce), the closer the constellation points, the more likely such errors. In particular, as we shall see in Chapter 6, the minimum distance between constellation points, termed $d_{\rm min}$, provides a good measure of how vulnerable we are to noise. For a given constellation shape, we can increase $d_{\rm min}$ simply by scaling up the constellation, as shown in Figure 4.15, but this comes with a corresponding increase in energy expenditure. To quantify this, define the *energy per symbol* $E_{\rm s}$ for a constellation as the average of the squared Euclidean distances of the points from the origin. For an M-ary constellation, each symbol carries $\log_2 M$ bits of information, and we can define the average *energy per bit* $E_{\rm b}$ as $E_{\rm b} = E_{\rm s}/\log_2 M$. Specifically, $d_{\rm min}$ increases from 2 to 4 by scaling as shown in Figure 4.15. Correspondingly, $E_{\rm s} = 2$ and $E_{\rm b} = 1$ are increased to $E_{\rm s} = 8$ and $E_{\rm b} = 4$ in Figure 4.15(b). Thus, doubling the minimum distance in Figure 4.15 leads to a four-fold increase in $E_{\rm s}$ and $E_{\rm b}$. However, the quantity $d_{\rm min}^2/E_{\rm b}$ does not change due to scaling; it depends only on the relative geometry of the constellation points. We therefore adopt this scale-invariant measure as our notion of *power efficiency* for a constellation:

$$\eta_{\rm P} = \frac{d_{\rm min}^2}{E_{\rm b}} \tag{4.19}$$

Since this quantity is scale-invariant, we can choose any convenient scaling in computing it: for QPSK, choosing the scaling on the left in Figure 4.15, we have $d_{\rm min} = 2$, $E_{\rm s} = 2$, and $E_{\rm b} = 1$, giving $\eta_{\rm P} = 4$.

It is important to understand how these quantities relate to physical link parameters. For a given bit rate $R_{\rm b}$ and received power $P_{\rm RX}$, the energy per bit is given by $E_{\rm b} = P_{\rm RX}/R_{\rm b}$. It is worth verifying that the units make sense: the numerator has units of watts, or joules/second, while the denominator has units of bits/second, so $E_{\rm b}$ has units of joules/bit. We shall see in Chapter 6 that the reliability of communication is determined by the power efficiency $\eta_{\rm P}$ (a scale-invariant quantity that is a function of the constellation shape) and the dimensionless signal-to-noise ratio (SNR) measure $E_{\rm b}/N_0$, where N_0 is the noise power spectral density, which has units of W/Hz, or J. Specifically, the reliability can be approximately characterized by the product $\eta_{\rm P} E_{\rm b}/N_0$, so that, for a given desired reliability, the required energy per bit (and hence power) scales inversely with the power efficiency for a fixed bit rate. Communication link designers use such concepts as the basis for forming a "link budget" that can be used to choose link parameters such as transmit power, antenna gains, and range.

Even on the basis of these rather sketchy and oversimplified arguments, we can draw quick conclusions on the power–bandwidth tradeoffs in using different constellations, as shown in the following example.

Example 4.3.3 We wish to design a passband communication system operating at a bit rate of 40 Mbps.

(a) What is the bandwidth required if we employ QPSK, with an excess bandwidth of 25%.
(b) What if we now employ 16QAM, again with excess bandwidth 25%.
(c) Suppose that the QPSK system in (a) attains a desired reliability when the transmit power is 50 mW. Give an estimate of the transmit power needed for the 16QAM system in (b) to attain a similar reliability.
(d) How does the bandwidth and transmit power required change for the QPSK system if we increase the bit rate to 80 Mbps.
(e) How does the bandwidth and transmit power required change for the QPSK system if we increase the bit rate to 80 Mbps.

Solution

(a) The bandwidth efficiency of QPSK is 2 bits/symbol, hence the minimum bandwidth required is 20 MHz. For excess bandwidth of 25%, the bandwidth required is 25 MHz.
(b) The bandwidth efficiency of 16QAM is 4 bits/symbol, hence, reasoning as in (a), the bandwidth required is 12.5 MHz.
(c) We wish to set $\eta_P \, E_b/N_0$ to be equal for the two systems in order to keep the reliability roughly the same. Assuming that the noise PSD N_0 is the same for both systems, the required E_b scales as $1/\eta_P$. Since the bit rates R_b for the two systems are equal, the required received power $P = E_b R_b$ (and hence the required transmit power, assuming that the received power scales linearly with the transmit power) also scales as $1/\eta_P$. We already know that $\eta_P = 4$ for QPSK. It remains to find η_P for 16QAM, which is shown in Problem 4.15 to equal $8/5$. We therefore conclude that the transmit power for the 16QAM system can be estimated as

$$P_T(16QAM) = P_T(QPSK) \frac{\eta_P(QPSK)}{\eta_P(16QAM)}$$

which evaluates to 125 mW.
(d) For fixed bandwidth efficiency, the required bandwidth scales linearly with the bit rate; hence the new bandwidth required is 50 MHz. In order to maintain a given reliability, we must maintain the same value of $\eta_P E_b/N_0$ as in (c). The power efficiency η_P is unchanged, since we are using the same constellation. Assuming that the noise PSD N_0 is unchanged, the required energy per bit E_b is unchanged; hence the transmit power

must scale up linearly with the bit rate R_b. Thus, the power required using QPSK is now 100 mW.

(e) Arguing as in (d), we require a bandwidth of 25 MHz and a power of 250 mW for 16QAM, using the results in (b) and (c).

4.3.5 The Nyquist criterion at the link level

Figure 4.16 shows a block diagram for a link using linear modulation, with the entire model expressed in complex baseband. The symbols $\{b[n]\}$ are passed through the transmit filter to obtain the waveform $\sum_n b[n]g_{TX}(t - nT)$. This then goes through the channel filter $g_C(t)$, and then the receive filter $g_{RX}(t)$. Thus, at the output of the receive filter, we have the linearly modulated signal $\sum_n b[n]p(t-nT)$, where $p(t) = (g_{TX} * g_C * g_{RX})(t)$ is the cascade of the transmit, channel, and receive filters. We would like the pulse $p(t)$ to be Nyquist at rate $1/T$, so that, in the absence of noise, the symbol rate samples at the output of the receive filter equal the transmitted symbols. Of course, in practice, we do not have control over the channel; hence we often assume an ideal channel, and design such that the cascade of the transmit and receive filters, given by $(g_{TX} * g_{RX})(t)G_{TX}(f)G_{RX}(f)$, is Nyquist. One possible choice is to set G_{TX} to be a Nyquist pulse, and G_{RX} to be a wideband filter whose response is flat over the band of interest. Another choice that is even more popular is to set $G_{TX}(f)$ and $G_{RX}(f)$ to be square roots of a Nyquist pulse. In particular, the square-root raised-cosine (SRRC) pulse is often used in practice.

A framework for software simulations of linear modulated systems with raised-cosine and SRRC pulses, including MATLAB code fragments, is given in Appendix 4.B, and provides a foundation for Software Lab 4.1.

Square-root Nyquist pulses and their time-domain interpretation A pulse $g(t) \leftrightarrow G(f)$ is defined to be square-root Nyquist at rate $1/T$ if $|G(f)|^2$ is Nyquist at rate $1/T$. Note that $P(f) = |G(f)|^2 \leftrightarrow p(t) = (g * g_{MF})(t)$, where $g_{MF}(t) = g^*(-t)$. The time-domain Nyquist condition is given by

$$p(mT) = (g * g_{MF})(mT) = \int g(t)g^*(t - mT)dt = \delta_{m0} \tag{4.20}$$

That is, a square-root Nyquist pulse has an autocorrelation function that vanishes at nonzero integer multiples of T. In other words, the waveforms $\{g(t-kT), k = 0, \pm 1, \pm 2, \ldots\}$ are orthonormal, and can be used to provide a basis for constructing more complex waveforms, as we shall see in Section 4.3.6.

Food for thought *True or false?* Any pulse timelimited to $[0, T]$ is square-root Nyquist at rate $1/T$.

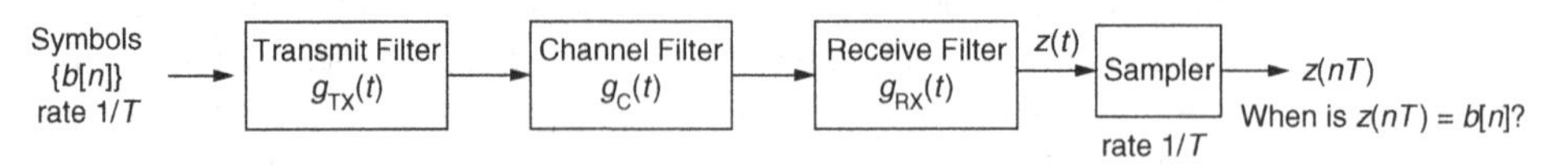

Figure 4.16 The Nyquist criterion at the link level.

4.3.6 Linear modulation as a building block

Linear modulation can be used as a building block for constructing more sophisticated waveforms, using discrete-time sequences modulated by square-root Nyquist pulses. Thus, one symbol would be made up of multiple "chips," linearly modulated by a square-root Nyquist "chip waveform." Specifically, suppose that $\psi(t)$ is square-root Nyquist at a chip rate $1/T_c$. N chips make up one symbol, so that the symbol rate is $1/T_s = 1/(NT_c)$, and a symbol waveform is given by linearly modulating a code vector $\mathbf{s} = (s[0], \ldots, s[N-1])$ consisting of N chips, as follows:

$$s(t) = \sum_{k=0}^{N} s[k]\psi(t - kT_c)$$

Since $\{\psi(t - kT_c)\}$ are orthonormal (see (4.20)), we have simply expressed the code vector in a continuous-time basis. Thus, the continuous-time inner product between two symbol waveforms (which determines their geometric relationships and their performance in terms of noise, as we shall see in the next chapter) is equal to the discrete-time inner product between the corresponding code vectors. Specifically, suppose that $s_1(t)$ and $s_2(t)$ are two symbol waveforms corresponding to code vectors $\mathbf{s}_1$ and $\mathbf{s}_2$, respectively. Then their inner product satisfies

$$\langle s_1, s_2 \rangle = \sum_{k=0}^{N-1}\sum_{l=0}^{N-1} s_1[k]s_2^*[l] \int \psi(t - kT_c)\psi^*(t - lT_c)dt = \sum_{k=0}^{N-1} s_1[k]s_2^*[k] = \langle \mathbf{s}_1, \mathbf{s}_2 \rangle$$

where we have used the orthonormality of the translates $\{\psi(t - kT_c)\}$. This means that we can design discrete-time code vectors to have certain desired properties, and then linearly modulate square-root Nyquist chip waveforms to get symbol waveforms that have the same desired properties. For example, if $\mathbf{s}_1$ and $\mathbf{s}_2$ are orthogonal, then so are $s_1(t)$ and $s_2(t)$; we use this in the next section when we discuss orthogonal modulation.

Examples of square-root Nyquist chip waveforms include a rectangular pulse timelimited to an interval of length T_c, as well as bandlimited pulses such as the square-root raised cosine. From Theorem 4.2.1, we see that the PSD of the modulated waveform is proportional to $|\Psi(f)|^2$ (it is typically a good approximation to assume that the chips $\{s[k]\}$ are uncorrelated). That is, the bandwidth occupancy is determined by that of the chip waveform ψ.

4.4 Orthogonal and biorthogonal modulation

While linear modulation with larger and larger constellations is a means of increasing bandwidth efficiency, we shall see that orthogonal modulation with larger and larger constellations is a means of increasing the power efficiency (at the cost of making the bandwidth efficiency smaller). Consider first M-ary frequency-shift keying (FSK), a classical form of orthogonal modulation in which one of M sinusoidal tones, which are successively spaced by Δf, is transmitted every T units of time, where $1/T$ is the symbol rate. Thus, the bit rate is $(\log_2 M)/T$, and, for a typical symbol interval, the transmitted passband signal is chosen from one of M possibilities:

$$u_{p,k}(t) = \cos(2\pi(f_0 + k\ \Delta f)t),\ 0 \le t \le T,\ \ k = 0, 1, \ldots, M-1$$

where we typically have $f_0 \gg 1/T$. On taking f_0 as a reference, the corresponding complex baseband waveforms are

$$u_k(t) = \exp(j2\pi k\ \Delta f\ t),\ 0 \le t \le T,\ \ k = 0, 1, \ldots, M-1$$

Let us now understand how the tones should be chosen in order to ensure orthogonality. Recall that the passband and complex-baseband inner products are related as follows:

$$\langle u_{p,k}, u_{p,l} \rangle = \frac{1}{2} \operatorname{Re}\langle u_k, u_l \rangle$$

so we can develop criteria for orthogonality working in complex baseband. Setting $k = l$, we see that

$$||u_k||^2 = T$$

For two adjacent tones, $l = k + 1$, we leave it as an exercise to show that

$$\operatorname{Re}\langle u_k, u_{k+1} \rangle = \frac{\sin(2\pi\ \Delta f\ T)}{2\pi\ \Delta f}$$

We see that the minimum value of Δf for which the preceding quantity is zero is given by $2\pi\ \Delta f\ T = \pi$, or $\Delta f = 1/(2T)$.

Thus, from the point of view of the receiver, a tone spacing of $1/(2T)$ ensures that, when there is an incoming wave at the kth tone, then correlating against the kth tone will give a large output, but correlating against the $(k+1)$th tone will give zero output (in the absence of noise). However, this assumes a *coherent* system in which the tones we are correlating against are synchronized in phase with the incoming wave. What happens if they are 90° out of phase? Then correlation of the kth tone with itself yields

$$\int_0^T \cos(2\pi(f_0 + k\ \Delta f)t) \cos\left(2\pi(f_0 + k\ \Delta f)t + \frac{\pi}{2}\right) dt = 0$$

(by orthogonality of the cosine and sine), so that the output we desire to be large is actually zero! Robustness to such variations can be obtained by employing *noncoherent* reception, which we describe next.

Noncoherent reception Let us develop the concept of noncoherent reception in generality, because it is a concept that is useful in many settings, not just for orthogonal modulation. Suppose that we transmit a passband waveform, and wish to detect it at the receiver by correlating it against the receiver's copy of the waveform. However, the receiver's local oscillator might not be synchronized in phase with the phase of the incoming wave. Let us denote the receiver's copy of the signal as

$$u_p(t) = u_c(t)\cos(2\pi f_c t) - u_s(t)\sin(2\pi f_c t)$$

and the incoming passband signal as

$$y_p(t) = y_c(t)\cos(2\pi f_c t) - y_s(t)\sin(2\pi f_c t) = u_c(t)\cos(2\pi f_c t + \theta) - u_s(t)\sin(2\pi f_c t + \theta)$$

Using the receiver's local oscillator as a reference, the complex envelope of the receiver's copy is $u(t) = u_c + ju_s(t)$, while that of the incoming wave is $y(t) = u(t)e^{j\theta}$. Thus, the inner product

$$\langle y_p, u_p \rangle = \frac{1}{2}\operatorname{Re}\langle y, u \rangle = \frac{1}{2}\operatorname{Re}\langle ue^{j\theta}, u \rangle = \frac{1}{2}\operatorname{Re}\left(||u||^2 e^{j\theta}\right) = \frac{||u||^2}{2}\cos\theta$$

Thus, the output of the correlator is degraded by the factor $\cos\theta$, and can actually become zero, as we have already observed, if the phase offset $\theta = \pi/2$. In order to get around this problem, let us look at the complex-baseband inner product again:

$$\langle y, u \rangle = \langle ue^{j\theta}, u \rangle = e^{j\theta}||u||^2$$

We could ensure that this output remains large regardless of the value of θ if we took its *magnitude,* rather than the real part. Thus, noncoherent reception corresponds to computing $|\langle y, u \rangle|$ or $|\langle y, u \rangle|^2$. Let us unwrap the complex inner product to see what this entails:

$$\begin{aligned}\langle y, u \rangle = \int y(t)u^*(t)dt &= \int (y_c(t) + jy_s(t))(u_c(t) - ju_s(t))dt \\ &= (\langle y_c, u_c \rangle + \langle y_s, u_s \rangle) + j(\langle y_s, u_c \rangle - \langle y_c, u_s \rangle)\end{aligned}$$

Thus, the noncoherent receiver computes the quantity

$$|\langle y, u \rangle|^2 = (\langle y_c, u_c \rangle + \langle y_s, u_s \rangle)^2 + (\langle y_s, u_c \rangle - \langle y_c, u_s \rangle)^2$$

In contrast, the coherent receiver computes

$$\operatorname{Re}\langle y, u \rangle = \langle y_c, u_c \rangle + \langle y_s, u_s \rangle$$

That is, when the receiver LO is synchronized to the phase of the incoming wave, we can correlate the I component of the received waveform with the I component of the receiver's copy, and similarly correlate the Q components, and sum them up. However, in the presence of phase asynchrony, the I and Q components get mixed up, and we must compute the magnitude of the complex inner product to recover all the energy of the incoming wave. Figure 4.17 shows the receiver operations corresponding to coherent and noncoherent reception.

Back to FSK Going back to FSK, if we now use noncoherent reception, then, in order to ensure that we get a zero output (in the absence of noise) when receiving the kth tone with a noncoherent receiver for the $(k+1)$th tone, we must ensure that

$$|\langle u_k, u_{k+1} \rangle| = 0$$

We leave it as an exercise (Problem 4.18) to show that the minimum tone spacing for noncoherent FSK is $1/T$, which is double that required for orthogonality in coherent FSK. The bandwidth for coherent M-ary FSK is approximately $M/(2T)$, which corresponds to a time–bandwidth product of approximately $M/2$. This corresponds to a complex vector space of dimension $M/2$, or a real vector space of dimension M, in which we can fit

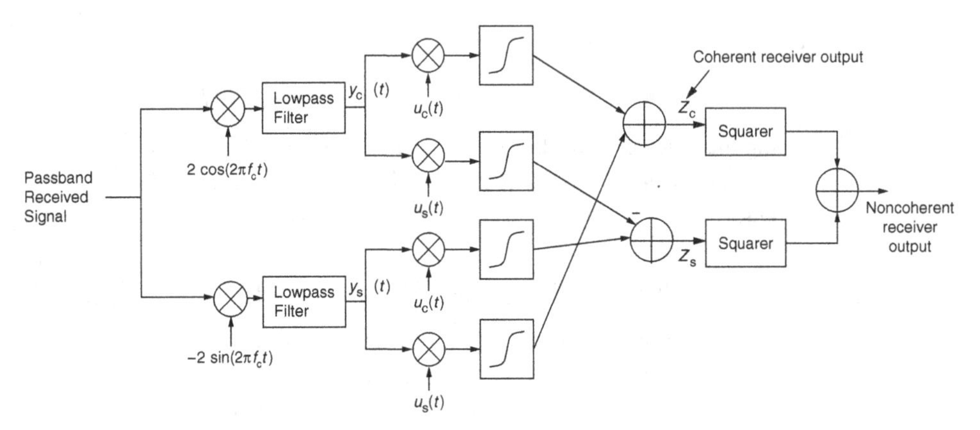

Figure 4.17 The structure of coherent and noncoherent receivers.

M orthogonal signals. On the other hand, M-ary noncoherent signaling requires M complex dimensions, since the complex-baseband signals must remain orthogonal even under multiplication by complex-valued scalars.

Summarizing the concept of orthogonality To summarize, when we say "orthogonal" modulation, we must specify whether we mean coherent or noncoherent reception, because the concept of orthogonality is different in the two cases. For a signal set $\{s_k(t)\}$, orthogonality requires that, for $k \neq l$, we have

$$\begin{aligned} \operatorname{Re}(\langle s_k, s_l \rangle) &= 0 \ \textbf{coherent orthogonality criterion} \\ \langle s_k, s_l \rangle &= 0 \ \textbf{noncoherent orthogonality criterion} \end{aligned} \tag{4.21}$$

Bandwidth efficiency We conclude from the example of orthogonal FSK that the bandwidth efficiency of orthogonal signaling is $\eta_{\mathrm{B}} = (\log_2(2M))/M$ bits/complex dimension for coherent systems, and $\eta_{\mathrm{B}} = (\log_2 M)/M$ bits/complex dimension for noncoherent systems. This is a general observation that holds for any realization of orthogonal signaling. In a signal space of complex dimension D (and hence real dimension $2D$), we can fit $2D$ signals satisfying the coherent orthogonality criterion, but only D signals satisfying the noncoherent orthogonality criterion. As M gets large, the bandwidth efficiency tends to zero. In compensation, as we shall see in Chapter 6, the power efficiency of orthogonal signaling for large M is the "best possible."

Orthogonal Walsh–Hadamard codes

Section 4.3.6 shows how to map vectors to waveforms while preserving inner products, by using linear modulation with a square-root Nyquist chip waveform. Applying this construction, the problem of designing orthogonal waveforms $\{s_i\}$ now reduces to designing orthogonal code vectors $\{\mathbf{s}_i\}$. Walsh–Hadamard codes are a standard construction employed for this purpose, and can be constructed recursively as follows: at the nth stage, we generate 2^n orthogonal vectors, using the 2^{n-1} vectors constructed in the $(n-1)$th stage. Let $\mathbf{H}_n$ denote a matrix whose rows are 2^n orthogonal codes obtained after the nth stage, with $H_0 = (1)$. Then

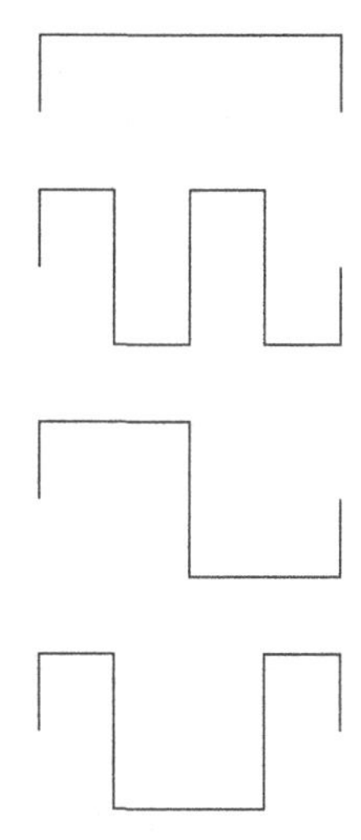

Figure 4.18 Walsh–Hadamard codes for 4-ary orthogonal modulation.

$$\mathbf{H}_n = \begin{pmatrix} \mathbf{H}_{n-1} & \mathbf{H}_{n-1} \\ \mathbf{H}_{n-1} & -\mathbf{H}_{n-1} \end{pmatrix}$$

We therefore get

$$H_1 = \begin{pmatrix} 1 & 1 \\ 1 & -1 \end{pmatrix}, \quad H_2 = \begin{pmatrix} 1 & 1 & 1 & 1 \\ 1 & -1 & 1 & -1 \\ 1 & 1 & -1 & -1 \\ 1 & -1 & -1 & 1 \end{pmatrix}, \quad \text{etc.}$$

Figure 4.18 depicts the waveforms corresponding to the 4-ary signal set in H_2 using a rectangular timelimited chip waveform to go from sequences to signals, as described in Section 4.3.6.

The signals $\{s_i\}$ obtained above can be used for noncoherent orthogonal signaling, since they satisfy the orthogonality criterion $\langle s_i, s_j \rangle = 0$ for $i \neq j$. However, just as for FSK, we can fit twice as many signals into the same number of degrees of freedom if we used the weaker notion of orthogonality required for coherent signaling, namely $\mathrm{Re}(\langle s_i, s_j \rangle) = 0$ for $i \neq j$. It is easy to check that, for M-ary Walsh–Hadamard signals $\{s_i, i = 1, \ldots, M\}$, we can get $2M$ orthogonal signals for coherent signaling: $\{s_i, js_i, i = 1, \ldots, M\}$. This construction corresponds to independently modulating the I and Q components with a Walsh–Hadamard code; that is, using passband waveforms $s_i(t)\cos(2\pi f_c t)$ and $-s_i(t)\sin(2\pi f_c t)$ (the negative sign is only to conform to our convention for I and Q, and can be dropped, which corresponds to replacing js_i by $-js_i$ in complex baseband), $i = 1, \ldots, M$.

Biorthogonal modulation

Given an orthogonal signal set, a biorthogonal signal set of twice the size can be obtained by including a negated copy of each signal. Since signals s and $-s$ cannot be distinguished in a noncoherent system, biorthogonal signaling is applicable to coherent systems. Thus, for an M-ary Walsh–Hadamard signal set $\{s_i\}$ with M signals obeying the noncoherent orthogonality criterion, we can construct a coherent orthogonal signal set $\{s_i, js_i\}$ of size $2M$, and hence a biorthogonal signal set of size $4M$, e.g., $\{s_i, js_i, -s_i, -js_i\}$. These correspond to the $4M$ passband waveforms $\pm s_i(t)\cos(2\pi f_c t)$ and $\pm s_i(t)\sin(2\pi f_c t)$, $i = 1, \ldots, M$.

4.5 Proofs of the Nyquist theorems

We have used Nyquist's sampling theorem, Theorem 4.3.1, to argue that linear modulation using the sinc pulse is able to use all the degrees of freedom in a bandlimited channel. On the other hand, Nyquist's criterion for ISI avoidance, Theorem 4.3.2, tells us, roughly speaking, that we must have enough degrees of freedom in order to avoid ISI (and that the sinc pulse provides the minimum number of such degrees of freedom). As it turns out, both theorems are based on the same mathematical relationship between samples in the time domain and aliased spectra in the frequency domain, stated in the following theorem.

Theorem 4.5.1 (Sampling and aliasing) *Consider a signal $s(t)$, sampled at rate $1/T_s$. Let $S(f)$ denote the spectrum of $s(t)$, and let*

$$B(f) = \frac{1}{T_s} \sum_{k=-\infty}^{\infty} S\left(f + \frac{k}{T_s}\right) \tag{4.22}$$

denote the sum of translates of the spectrum. Then the following observations hold:

(a) $B(f)$ is periodic with period $1/T_s$.

(b) The samples $\{s(nT_s)\}$ are the Fourier series for $B(f)$, satisfying

$$s(nT_s) = T_s \int_{-1/(2T_s)}^{1/(2T_s)} B(f) e^{j2\pi f n T_s} \, df \tag{4.23}$$

and

$$B(f) = \sum_{n=-\infty}^{\infty} s(nT_s) e^{-j2\pi f n T_s} \tag{4.24}$$

Remark Note that the signs of the exponents for the frequency-domain Fourier series in the theorem are reversed from the convention in the usual time-domain Fourier series (analogous to the reversal of the sign of the exponent for the inverse Fourier transform compared with the Fourier transform).

Proof of Theorem 4.5.1 The periodicity of $B(f)$ follows by its very construction. To prove (b), apply the the inverse Fourier transform to obtain

$$s(nT_s) = \int_{-\infty}^{\infty} S(f) e^{j2\pi f n T_s} \, df$$

We now write the integral as an infinite sum of integrals over segments of length $1/T$,

$$s(nT_s) = \sum_{k=-\infty}^{\infty} \int_{(k-\frac{1}{2})/T_s}^{(k+\frac{1}{2})/T_s} S(f)e^{j2\pi fnT_s}\, df$$

In the integral over the kth segment, make the substitution $\nu = f - k/T_s$ and rewrite it as

$$\int_{-1/(2T_s)}^{1/(2T_s)} S\left(\nu + \frac{k}{T_s}\right) e^{j2\pi(\nu+k/T_s)nT_s}\, d\nu = \int_{-1/(2T_s)}^{1/(2T_s)} S\left(\nu + \frac{k}{T_s}\right) e^{j2\pi\nu nT_s}\, d\nu$$

Now that the limits of all segments and the complex exponential in the integrand are the same (i.e., independent of k), we can move the summation inside to obtain

$$\begin{aligned} s(nT_s) &= \int_{-1/(2T_s)}^{1/(2T_s)} \left(\sum_{k=-\infty}^{\infty} S\left(\nu + \frac{k}{T_s}\right)\right) e^{j2\pi\nu nT_s}\, d\nu \\ &= T_s \int_{-1/(2T_s)}^{1/(2T_s)} B(\nu)e^{j2\pi\nu nT_s}\, d\nu \end{aligned}$$

proving (4.23). We can now recognize that this is just the formula for the Fourier-series coefficients of $B(f)$, from which (4.24) follows. □

Inferring Nyquist's sampling theorem from Theorem 4.5.1 Suppose that $s(t)$ is bandlimited to $[-W/2, W/2]$. The samples of $s(t)$ at rate $1/T_s$ can be used to reconstruct $B(f)$, since they are the Fourier series for $B(f)$. But $S(f)$ can be recovered from $B(f)$ if and only if the translates $S(f - k/T_s)$ do not overlap, as shown in Figure 4.19. This happens if and only if $1/T_s \geq W$. Once this condition is satisfied, $(1/T_s)S(f)$ can be recovered from $B(f)$ by passing it through an ideal bandlimited filter $H(f) = I_{[-W/2,W/2]}(f)$. We therefore obtain that

$$\frac{1}{T_s}S(f) = B(f)H(f) = \sum_{n=-\infty}^{\infty} s(nT_s)e^{-j2\pi fnT_s} I_{[-W/2,W/2]}(f) \tag{4.25}$$

Noting that $I_{[-W/2,W/2]}(f) \leftrightarrow W\,\text{sinc}(Wt)$, we have

$$e^{-j2\pi fnT_s} I_{[-W/2,W/2]}(f) \leftrightarrow W\,\text{sinc}(W(t - nT_s))$$

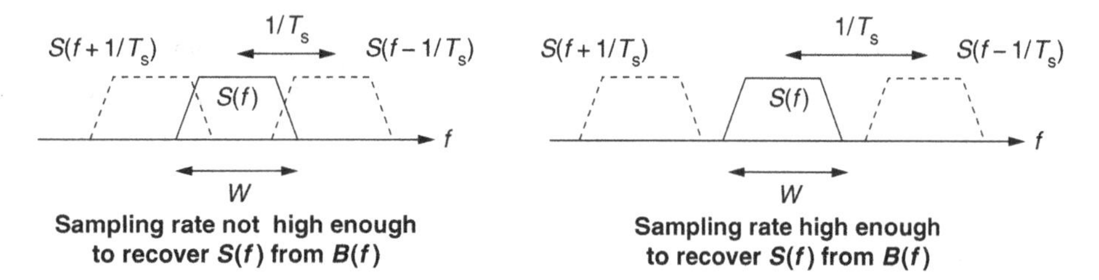

Figure 4.19 Recovering a signal from its samples requires a high enough sampling rate for translates of the spectrum not to overlap.

Taking inverse Fourier transforms, we get the interpolation formula

$$\frac{1}{T_s}s(t) = \sum_{n=-\infty}^{\infty} s(nT_s)W \operatorname{sinc}(W(t - nT_s))$$

which reduces to (4.15) for $1/T_s = W$. This completes the proof of the sampling theorem, Theorem 4.3.1. □

Inferring Nyquist's criterion for ISI avoidance from Theorem 4.5.1 A Nyquist pulse $p(t)$ at rate $1/T$ must satisfy $p(nT) = \delta_{n0}$. On applying Theorem 4.5.1 with $s(t) = p(t)$ and $T_s = T$, it follows immediately from (4.24) that $p(nT) = \delta_{n0}$ (i.e., the time-domain Nyquist criterion holds) if and only if

$$B(f) = \frac{1}{T}\sum_{k=-\infty}^{\infty} P\left(f + \frac{k}{T_s}\right) = 1$$

In other words, if the Fourier series has only a DC term, then the periodic waveform to which it corresponds must be constant. □

4.6 Concept summary

This chapter provides an introduction to how bits can be translated into information-carrying signals that satisfy certain constraints (e.g., fitting within a given frequency band). We focus on linear modulation over passband channels.

Modulation basics

- Information bits can be encoded into two-dimensional (complex-valued) constellations, which can be modulated onto baseband pulses to produce a complex-baseband waveform. Constellations may carry information in both amplitude and phase (e.g., QAM) or in phase only (e.g., PSK). This modulated waveform can then be upconverted to the appropriate frequency band for passband signaling.
- The PSD of a linearly modulated waveform using pulse $p(t)$ is proportional to $|P(f)|^2$, so that the choice of modulating pulse is critical for determining bandwidth occupancy. Fractional power containment provides a useful notion of bandwidth.
- Time-limited pulses with sharp edges have large bandwidth, but this can be reduced by smoothing out the edges (e.g., by replacing a rectangular pulse by a trapezoidal pulse or by a sinusoidal pulse).

Degrees of freedom

- Nyquist's sampling theorem says that a signal bandlimited over $[-W/2, W/2]$ is completely characterized by its samples at rate W (or higher). By applying this to the complex envelope of a passband signal of bandwidth W, we infer that a passband channel of bandwidth W provides W complex-valued degrees of freedom per unit time for carrying information.

- The (time-domain) sinc pulse, which corresponds to a frequency-domain boxcar, allows us to utilize all degrees of freedom in a bandlimited channel, but it decays too slowly, at rate $1/t$, for practical use: it can lead to unbounded signal amplitude and, in the presence of timing mismatch, unbounded ISI.

ISI avoidance

- The Nyquist criterion for ISI avoidance requires that the end-to-end signaling pulse vanish at nonzero integer multiples of the symbol time. In the frequency domain, this corresponds to aliased versions of the pulse summing to a constant.
- The sinc pulse is the minimum-bandwidth Nyquist pulse, but decays too slowly with time. It can be replaced, at the expense of some excess bandwidth, by pulses with less sharp transitions in the frequency domain to obtain faster decay in time. The raised-cosine pulse is a popular choice, giving a $1/t^3$ decay.
- If the receive filter is matched to the transmit filter, each has to be a *square-root Nyquist* pulse, with their cascade being Nyquist. The SRRC is a popular choice.

Power–bandwidth tradeoffs

- For an M-ary constellation, the bandwidth efficiency is $\log_2 M$ bits per symbol, so larger constellations are more bandwidth-efficient.
- The power efficiency for a constellation is well characterized by the scale-invariant quantity $d_{\min}^2/E_b$. Large constellations are typically less power-efficient.

Beyond linear modulation

- Linear modulation using square-root Nyquist pulses can be used to translate signal design from discrete time to continuous time while preserving geometric relationships such as inner products. This is because, if $\psi(t)$ is square-root Nyquist at rate $1/T_c$, then $\{\psi(t - kT_c)\}$, its translates by integer multiples of T_c, form an orthonormal basis.
- Orthogonal modulation can be used with either coherent or noncoherent reception, but the concept of orthogonality is more stringent (eating up more degrees of freedom) for noncoherent orthogonal signaling. Waveforms for orthogonal modulation can be constructed in a variety of ways, including FSK and Walsh–Hadamard sequences modulated onto square-root Nyquist pulses. Biorthogonal signaling doubles the signaling alphabet for coherent orthogonal signaling by adding the negative of each signal to the constellation.

Sampling and aliasing

- Time-domain sampling corresponds to frequency-domain aliasing. Specifically, the samples of a waveform $x(t)$ at rate $1/T$ are the Fourier series for the periodic frequency-domain waveform $(1/T)\sum_k X(f - k/T)$ obtained by summing the frequency-domain waveform and its aliases $X(f - k/T)$ (k integer).
- The Nyquist sampling theorem corresponds to requiring that the aliased copies are far enough apart (i.e., the sampling rate is high enough) that we can recover the original frequency-domain waveform by filtering the sum of the aliased waveforms.
- The Nyquist criterion for interference avoidance requires that the samples of the signaling pulse form a discrete delta function, or that the corresponding sum of the aliased waveforms is a constant.

4.7 Notes

While we use linear modulation in the time domain for our introduction to modulation, an alternative frequency-domain approach is to divide the available bandwidth into thin slices, or subcarriers, and to transmit symbols in parallel on each subcarrier. Such a strategy is termed orthogonal frequency-division multiplexing (OFDM) or multicarrier modulation, and we discuss it in more detail in Chapter 8. OFDM is also termed multicarrier modulation, while the time-domain linear modulation schemes covered here are classified as single-carrier modulation. In addition to the degrees of freedom provided by time and frequency, additional *spatial* degrees of freedom can be obtained by employing multiple antennas at the transmitter and receiver, and we provide a glimpse of such multiple-input, multiple-output (MIMO) techniques in Chapter 8.

While the basic linear modulation strategies discussed here, in either single-carrier or multicarrier modulation formats, are employed in many existing and emerging communication systems, it is worth mentioning a number of other strategies in which modulation with memory is used to shape the transmitted waveform in various ways, including insertion of spectral nulls (e.g., line codes, which are often used for baseband wireline transmission), avoidance of long runs of zeros and ones that can disrupt synchronization (e.g., runlength-constrained codes, which are often used for magnetic recording channels), controlling variations in the signal envelope (e.g., constant phase modulation), and controlling ISI (e.g., partial response signaling). Memory can also be inserted in the manner that bits are encoded into symbols (e.g., differential encoding for alleviating the need to track a time-varying channel), without changing the basic linear modulation format. The preceding discussion, while not containing enough detail to convey the underlying concepts, is meant to provide keywords to facilitate further exploration, with more advanced communication theory texts such as [5, 7, 8] serving as a good starting point.

4.8 Problems

Timelimited pulses

Problem 4.1 (Sine pulse) Consider the sine pulse pulse $p(t) = \sin(\pi t) I_{[0,1]}(t)$.

(a) Show that its Fourier transform is given by

$$P(f) = \frac{2\cos(\pi f)e^{-j\pi f}}{\pi(1-4f^2)}$$

(b) Consider the linearly modulated signal $u(t) = \sum_n b[n]p(t-n)$, where $b[n]$ are independently chosen to take values in a QPSK constellation (each point is chosen with equal probability), and the time is in units of microseconds. Find the 95% power-containment bandwidth (specify the units).

Problem 4.2 Consider the pulse

$$p(t) = \begin{cases} t/a, & 0 \le t \le a \\ 1, & a \le t \le 1-a \\ (1-t)/a, & 1-a \le t \le 1 \\ 0, & \text{else} \end{cases}$$

where $0 \le a \le \frac{1}{2}$.

(a) Sketch $p(t)$ and find its Fourier transform $P(f)$.
(b) Consider the linearly modulated signal $u(t) = \sum_n b[n]p(t-n)$, where $b[n]$ take values independently and with equal probability in a 4PAM alphabet $\{\pm 1, \pm 3\}$. Find an expression for the PSD of u as a function of the pulse-shape parameter a.
(c) Numerically estimate the 95% fractional power-containment bandwidth for u and plot it as a function of $0 \le a \le \frac{1}{2}$. For concreteness, assume that the unit of time is 100 picoseconds and specify the units of bandwidth in your plot.

Basic concepts in Nyquist signaling

Problem 4.3 Consider a pulse $s(t) = \mathrm{sinc}(at)\mathrm{sinc}(bt)$, where $a \ge b$.

(a) Sketch the frequency-domain response $S(f)$ of the pulse.
(b) Suppose that the pulse is to be used over an ideal real-baseband channel with one-sided bandwidth 400 Hz. Choose a and b so that the pulse is Nyquist for 4PAM signaling at 1200 bits/s and exactly fills the channel bandwidth.
(c) Now, suppose that the pulse is to be used over a passband channel spanning the frequency range 2.4–2.42 GHz. Assuming that we use 64QAM signaling at 60 Mbits/s, choose a and b so that the pulse is Nyquist and exactly fills the channel bandwidth.
(d) Sketch an argument showing that the magnitude of the transmitted waveform in the preceding settings is always finite.

Problem 4.4 Consider the pulse $p(t)$ whose Fourier transform satisfies

$$P(f) = \begin{cases} 1, & 0 \leq |f| \leq A \\ (B - |f|)/(B - A), & A \leq |f| \leq B \\ 0, & \text{else} \end{cases}$$

where $A = 250$ kHz and $B = 1.25$ MHz.

(a) **(True or false?)** The pulse $p(t)$ can be used for Nyquist signaling at rate 3 Mbps using an 8PSK constellation.
(b) **(True or false?)** The pulse $p(t)$ can be used for Nyquist signaling at rate 4.5 Mbps using an 8PSK constellation.

Problem 4.5 Consider the pulse

$$p(t) = \begin{cases} 1 - |t|/T, & 0 \leq |t| \leq T \\ 0, & \text{else} \end{cases}$$

Let $P(f)$ denote the Fourier transform of $p(t)$.

(a) **(True or false?)** The pulse $p(t)$ is Nyquist at rate $1/T$.
(b) **(True or false?)** The pulse $p(t)$ is square root Nyquist at rate $1/T$. (i.e., $|P(f)|^2$ is Nyquist at rate $1/T$).

Problem 4.6 Consider Nyquist signaling at 80 Mbps using a 16QAM constellation with 50% excess bandwidth. The signaling pulse has the spectrum shown in Figure 4.20.

(a) Find the values of a and b in Figure 4.20, making sure you specify the units.
(b) **(True or false?)** The pulse is also Nyquist for signaling at 20 Mbps using QPSK. (Justify your answer.)

Problem 4.7 Consider linear modulation with a signaling pulse $p(t) = \text{sinc}(at)\text{sinc}(bt)$, where a and b are to be determined.

(a) How should a and b be chosen so that $p(t)$ is Nyquist with 50% excess bandwidth for a data rate of 40 Mbps using 16QAM? Specify the occupied bandwidth.

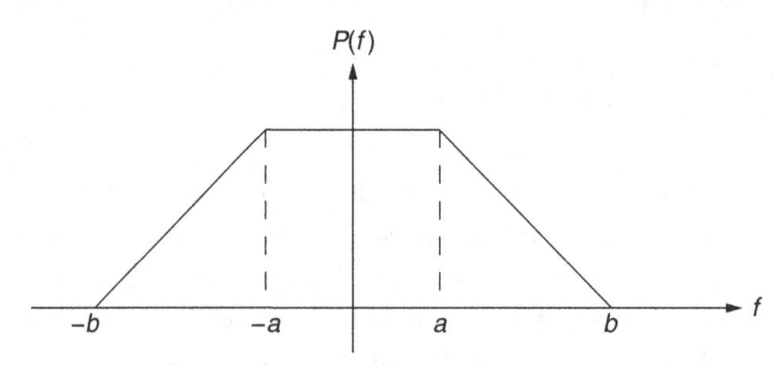

Figure 4.20 The signaling pulse for Problem 4.6.

(b) How should a and b be chosen so that $p(t)$ can be used for Nyquist signaling both for a 16QAM system with data rate 40 Mbps, and for an 8PSK system with data rate 18 Mbps? Specify the occupied bandwidth.

Problem 4.8 Consider a passband communication link operating at a bit rate of 16 Mbps using a 256QAM constellation.

(a) What must we set the unit of time as, so that $p(t) = \sin(\pi t) I_{[0,1]}(t)$ is square-root Nyquist for the system of interest, while occupying the smallest possible bandwidth?
(b) What must we set the unit of time as, so that $p(t) = \text{sinc}(t)\text{sinc}(2t)$ is Nyquist for the system of interest, while occupying the smallest possible bandwidth?

Problem 4.9 Consider passband linear modulation with a pulse of the form $p(t) = \text{sinc}(3t)\text{sinc}(2t)$, where the unit of time is *microseconds*.

(a) Sketch the spectrum $P(f)$ versus f. Make sure you specify the units on the f axis.
(b) What is the largest achievable *bit rate* for Nyquist signaling using $p(t)$ if we employ a 16QAM constellation? What is the fractional excess bandwidth for this bit rate?
(c) **(True or False?)** The pulse $p(t)$ can be used for Nyquist signaling at a bit rate of 4 Mbps using a QPSK constellation.

Problem 4.10 (True or False?) Any pulse timelimited to duration T is square-root Nyquist (up to scaling) at rate $1/T$.

Problem 4.11 (Raised-cosine pulse) In this problem, we derive the time domain response of the frequency-domain raised cosine pulse. Let $R(f) = I_{[-1/2,1/2]}(f)$ denote an ideal boxcar transfer function, and let $C(f) = (\pi/(2a))\cos((\pi/a)f)I_{[-a/2,a/2]}$ denote a cosine transfer function.

(a) Sketch $R(f)$ and $C(f)$, assuming that $0 < a < 1$.
(b) Show that the frequency-domain raised-cosine pulse can be written as

$$S(f) = (R * C)(f)$$

(c) Find the time-domain pulse $s(t) = r(t)c(t)$. Where are the zeros of $s(t)$? Conclude that $s(t/T)$ is Nyquist at rate $1/T$.
(d) Sketch an argument that shows that, if the pulse $s(t/T)$ is used for BPSK signaling at rate $1/T$, then the magnitude of the transmitted waveform is always finite.

Software experiments with Nyquist and square-root Nyquist pulses

Problem 4.12 (Software exercise for the raised cosine pulse) Code Fragment 4.B.1 in the appendix implements a discrete-time truncated raised-cosine pulse.

(a) Run the code fragment for 25%, 50%, and 100% excess bandwidths and plot the time-domain waveforms versus normalized time t/T over the interval $[-5T, 5T]$, sampling fast enough (e.g., at rate $32/T$ or higher) to obtain smooth curves. Comment on the effect of varying the excess bandwidth on these waveforms.

(b) For excess bandwidth of 50%, numerically explore the effect of time-domain truncation on frequency-domain spillage. Specifically, compute the Fourier transform for two cases: truncation to $[-2T, 2T]$ and truncation to $[-5T, 5T]$, using the DFT as described in Code Fragment 2.5.1 to obtain a frequency resolution at least as good as $1/(64T)$. Plot these Fourier transforms against the normalized frequency fT, and comment on how much of an increase in bandwidth, if any, you see due to truncation in the two cases.

(c) Numerically compute the 95% bandwidth of the two pulses in (b), and compare it with the nominal bandwidth without truncation.

Problem 4.13 (Software exercise for the SRRC pulse)

(a) Write a function for generating a sampled SRRC pulse, analogous to Code Fragment 4.B.1, where you can specify the sampling rate, the excess bandwidth, and the truncation length. The time domain expression for the SRRC pulse is given by (4.45) in the appendix.

Remark The zero in the denominator can be handled by either analytical or numerical implementation of l'Hôpital's rule. See the comments in Code Fragment 4.B.1.

(b) Plot the SRRC pulses versus normalized time t/T, for excess bandwidths of 25%, 50%, and 100%. Comment on the effect of varying excess bandwidth on these waveforms.

Effect of timing errors

Problem 4.14 (Effect of timing errors) Consider digital modulation at rate $1/T$ using the sinc pulse $s(t) = \text{sinc}(2Wt)$, with transmitted waveform

$$y(t) = \sum_{n=1}^{100} b_n s(t - (n-1)T)$$

where $1/T$ is the symbol rate and $\{b_n\}$ is the bit stream being sent (assume that each b_n takes one of the values ± 1 with equal probability). The receiver makes bit decisions based on the samples $r_n = y((n-1)T)$, $n = 1, \ldots, 100$.

(a) For what value of T (as a function of W) is $r_n = b_n$, $n = 1, \ldots, 100$?

Remark In this case, we simply use the sign of the nth sample r_n as an estimate of b_n.

(b) For the choice of T as in (a), suppose that the receiver sampling times are off by $0.25T$. That is, the nth sample is given by $r_n = y((n-1)T + 0.25T)$, $n = 1, \ldots, 100$. In this case, we do have ISI of different degrees of severity, depending on the bit pattern. Consider the following bit pattern:

$$b_n = \begin{cases} (-1)^{n-1}, & 1 \leq n \leq 49 \\ (-1)^n, & 50 \leq n \leq 100 \end{cases}$$

Numerically evaluate the 50th sample r_{50}. Does it have the same sign as the 50th bit b_{50}?

Remark The preceding bit pattern creates the worst possible ISI for the 50th bit. Since the sinc pulse dies off slowly with time, the ISI contributions due to the 99 other bits to the 50th sample sum up to a number larger in magnitude, and opposite in sign, relative to the contribution due to b_{50}. A decision on b_{50} based on the sign of r_{50} would therefore be wrong. This sensitivity to timing error is why the sinc pulse is seldom used in practice.

(c) Now, consider the digitally modulated signal in (a) with the pulse $s(t) = \text{sinc}(2Wt)\text{sinc}(Wt)$. For ideal sampling as in (a), what are the two values of T such that $r_n = b_n$?

(d) For the smaller of the two values of T found in (c) (which corresponds to faster signaling, since the symbol rate is $1/T$), repeat the computation in (b). That is, find r_{50} and compare its sign with that of b_{50} for the bit pattern in (b).

(e) Find and sketch the frequency response of the pulse in (c). What is the excess bandwidth relative to the pulse in (a), assuming Nyquist signaling at the same symbol rate?

(f) Discuss the impact of the excess bandwidth on the severity of the ISI due to timing mismatch.

Power–bandwidth tradeoffs

Problem 4.15 (Power efficiency of 16QAM) In this problem, we sketch the computation of the power efficiency for the 16QAM constellation shown in Figure 4.21.

(a) Note that the minimum distance for the particular scaling chosen in the figure is $d_{\min} = 2$.

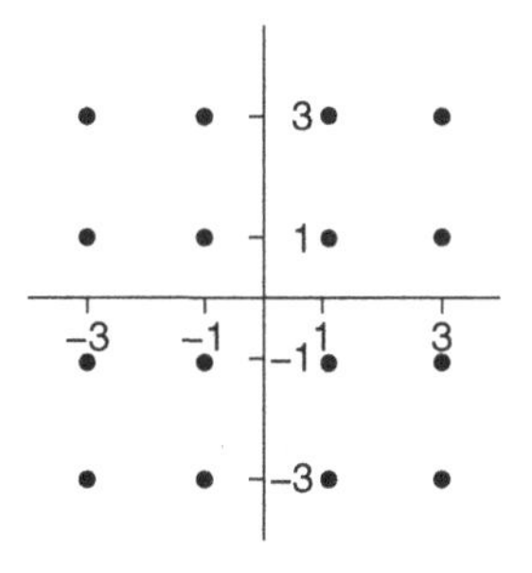

Figure 4.21 The 16QAM constellation with scaling chosen for convenient computation of the power efficiency.

(b) Show that the constellation points divide into three categories based on their distance from the origin, corresponding to squared distances, or energies, of $1^2 + 1^2$, $1^2 + 3^2$, and $3^2 + 3^2$. Averaging over these energies (weighting by the number of points in each category), show that the average energy per symbol is $E_s = 10$.

(c) Using (a) and (b), and accounting for the number of bits/symbol, show that the power efficiency is given by $\eta_P = d_{\min}^2/E_b = \frac{8}{5}$.

Problem 4.16 (Power–bandwidth tradeoffs) A 16QAM system transmits at 50 Mbps using an excess bandwidth of 50%. The transmit power is 100 mW.

(a) Assuming that the carrier frequency is 5.2 GHz, specify the frequency interval occupied by the passband modulated signal.

(b) Using the same frequency band in (a), how fast could you signal using QPSK with the same excess bandwidth?

(c) Estimate the transmit power needed in the QPSK system, assuming the same range and reliability requirements as in the 16QAM system.

Minimum-shift keying

Problem 4.17 (OQPSK and MSK) Linear modulation with a bandlimited pulse can perform poorly over nonlinear passband channels. For example, the output of a passband hardlimiter (which is a good model for power amplifiers operating in a saturated regime) has a constant envelope, but a PSK signal employing a bandlimited pulse has an envelope that passes through zero during a 180° phase transition, as shown in Figure 4.22. One way to alleviate this problem is to not allow 180° phase transitions. Offset QPSK (OQPSK) is one example of such a scheme, where the transmitted signal is given by

$$s(t) = \sum_{n=-\infty}^{\infty} b_c[n]p(t - nT) + jb_s[n]p\left(t - nT - \frac{T}{2}\right) \tag{4.26}$$

where $\{b_c[n]\}$ and $b_s[n]$ are ± 1 BPSK symbols modulating the I and Q channels, with the I and Q signals being staggered by half a symbol interval. This leads to phase transitions

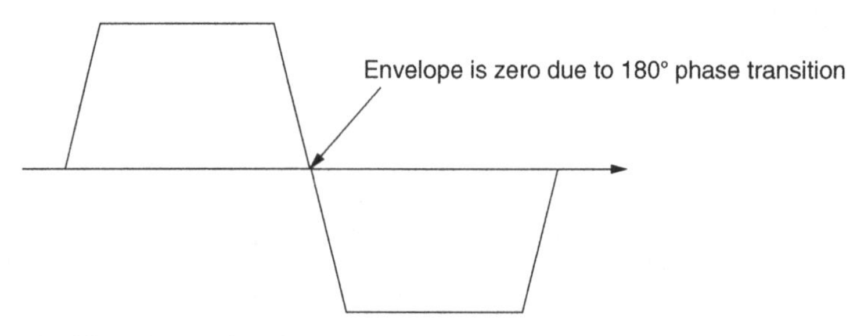

Figure 4.22 The envelope of a PSK signal passes through zero during a 180° phase transition, and gets distorted over a nonlinear channel.

of at most 90° at integer multiples of the *bit time* $T_b = T/2$. Minimum-shift keying (MSK) is a special case of OQPSK with timelimited modulating pulse

$$p(t) = \sqrt{2}\sin\left(\frac{\pi t}{T}\right)I_{[0,T]}(t) \tag{4.27}$$

(a) Sketch the I and Q waveforms for a typical MSK signal, clearly showing the timing relationship between the waveforms.
(b) Show that the MSK waveform has a constant envelope (an extremely desirable property for nonlinear channels).
(c) Find an analytical expression for the PSD of an MSK signal, assuming that all bits sent are i.i.d., taking values ± 1 with equal probability. Plot the PSD versus normalized frequency fT.
(d) Find the 99% power-containment normalized bandwidth of MSK. Compare this with the minimum Nyquist bandwidth and with the 99% power-containment bandwidth of OQPSK using a rectangular pulse.
(e) Recognize that Figure 4.6 gives the PSD for OQPSK and MSK, and reproduce this figure, normalizing the area under the PSD curve to be the same for both modulation formats.

Orthogonal signaling

Problem 4.18 (FSK tone spacing) Consider two real-valued passband pulses of the form

$$s_0(t) = \cos(2\pi f_0 t + \phi_0), \quad 0 \le t \le T$$
$$s_1(t) = \cos(2\pi f_1 t + \phi_1), \quad 0 \le t \le T$$

where $f_1 > f_0 \gg 1/T$. The pulses are said to be *orthogonal* if $\langle s_0, s_1 \rangle = \int_0^T s_0(t)s_1(t)\, dt = 0$.

(a) If $\phi_0 = \phi_1 = 0$, show that the minimum frequency separation such that the pulses are orthogonal is $f_1 - f_0 = 1/(2T)$.
(b) If ϕ_0 and ϕ_1 are arbitrary phases, show that the minimum separation for the pulses to be orthogonal regardless of ϕ_0, ϕ_1 is $f_1 - f_0 = 1/T$.

Remark The results of this problem can be used to determine the bandwidth requirements for coherent and noncoherent FSK, respectively.

Problem 4.19 (Walsh–Hadamard codes)

(a) Specify the Walsh–Hadamard codes for 8-ary orthogonal signaling with noncoherent reception.
(b) Plot the baseband waveforms corresponding to sending these codes using a square-root raised cosine pulse with excess bandwidth of 50%.
(c) What is the fractional increase in bandwidth efficiency if we use these eight waveforms as building blocks for biorthogonal signaling with coherent reception?

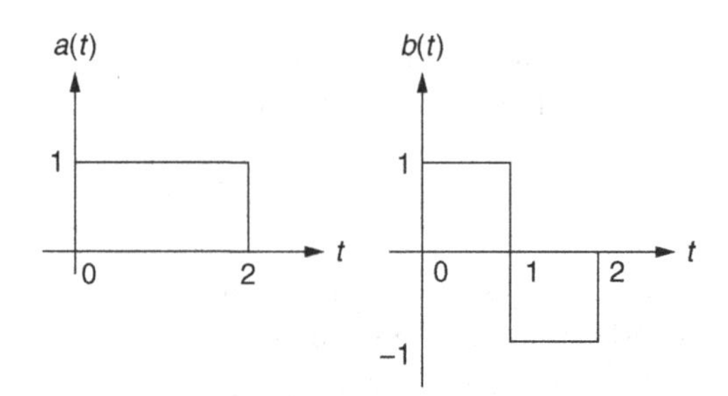

Figure 4.23 Baseband signals for Problem 4.20.

Problem 4.20 The two orthogonal baseband signals shown in Figure 4.23 are used as building blocks for constructing passband signals as follows:

$$\begin{aligned} u_p(t) &= a(t)\cos(2\pi f_c t) - b(t)\sin(2\pi f_c t) \\ v_p(t) &= b(t)\cos(2\pi f_c t) - a(t)\sin(2\pi f_c t) \\ w_p(t) &= b(t)\cos(2\pi f_c t) + a(t)\sin(2\pi f_c t) \\ x_p(t) &= a(t)\cos(2\pi f_c t) + b(t)\sin(2\pi f_c t) \end{aligned}$$

where $f_c \gg 1$.

(a) **(True or false?)** The signal set can be used for 4-ary orthogonal modulation with coherent demodulation.

(b) **(True or false?)** The signal set can be used for 4-ary orthogonal modulation with noncoherent demodulation.

Bandwidth occupancy as a function of modulation format

Problem 4.21 We wish to send at a rate of 10 Mbits/s over a passband channel. Assuming that an excess bandwidth of 50% is used, how much bandwidth is needed for each of the following schemes: QPSK, 64QAM, and 64-ary noncoherent orthogonal modulation using a Walsh–Hadamard code.

Problem 4.22 Consider 64-ary orthogonal signaling using Walsh–Hadamard codes. Assuming that the chip pulse is square-root raised cosine with excess bandwidth 25%, what is the bandwidth required for sending data at 20 kbps over a passband channel assuming (a) coherent reception and (b) noncoherent reception.

Software Lab 4.1: linear modulation over a noiseless ideal channel

Lab objectives This is the first of a sequence of software labs that gradually develop a reasonably complete MATLAB simulator for a linearly modulated system. The follow-on labs are Software Lab 6.1, in Chapter 6, and Software Lab 8.1, in Chapter 8.

Reading Sections 4.3 (signaling over bandlimited channels) and 4.B (simulations with bandlimited pulses), together with the following background.

Background

Figure 4.24 shows block diagrams corresponding to a typical DSP-centric realization of a communication transceiver employing linear modulation. In the labs, we model the core components of such a system using the complex-baseband representation, as shown in Figure 4.25. Given the equivalence of passband and complex baseband, we are only skipping the modeling of finite-precision effects due to digital-to-analog conversion (DAC) and analog-to-digital conversion (ADC). These effects can easily be incorporated into MATLAB models such as those we develop, but are beyond our current scope.

A few points worth noting about the model of Figure 4.25 are as follows.

Choice of transmit filter. The PSD of the transmitted signal is proportional to $|G_{TX}(f)|^2$ (see Chapter 4). The choice of transmit filter is made on the basis of spectral

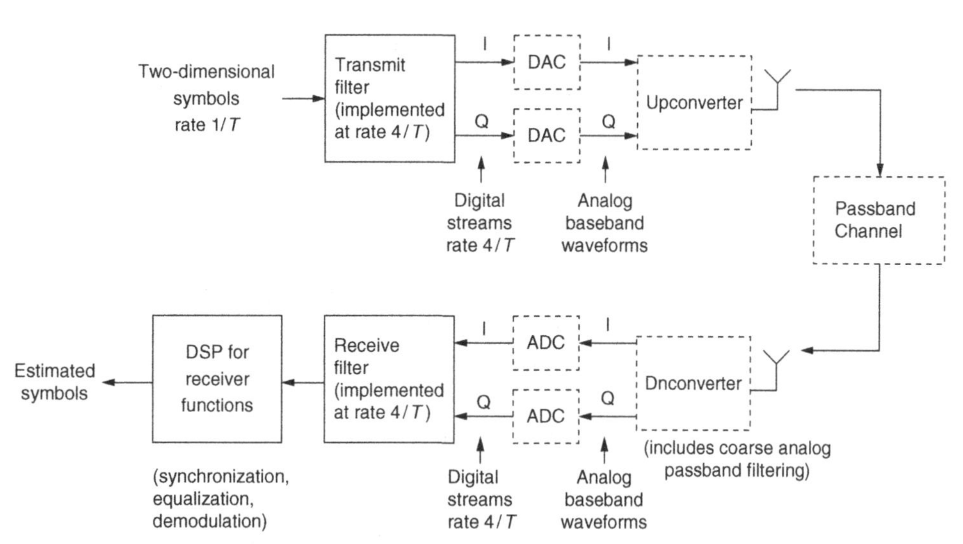

Figure 4.24 A typical DSP-centric transceiver realization. Our model does not include the blocks shown within dashed lines. Finite-precision effects such as DAC and ADC are not considered. The upconversion and downconversion operations are not modeled. The passband channel is modeled as an LTI system in complex baseband.

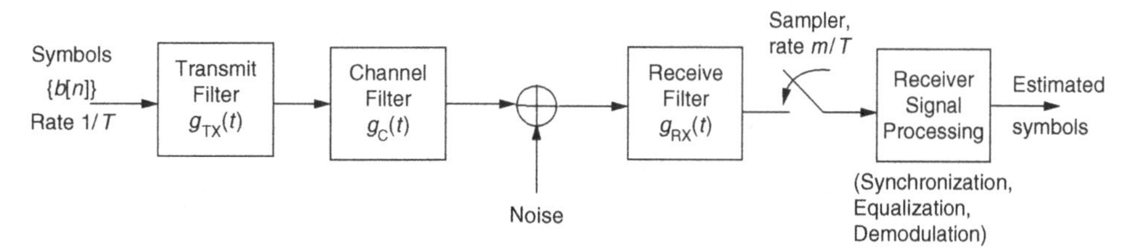

Figure 4.25 The block diagram of a linearly modulated system, modeled in complex baseband.

constraints, as well as considerations such as sensitivity to receiver timing errors and intersymbol interference. Typically, the bandwidth employed is of the order of $1/T$.

Channel model. We typically model the channel as an linear time-invariant (LTI) system. For certain applications, such as wireless communications, the channel may be modeled as slowly time-varying.

Noise model. Noise is introduced in a later lab (in Chapter 6).

Receive filter and sampler. The optimal choice of receive filter is actually a filter matched to the cascade of the transmit filter and the channel. In this case, there is no information loss in sampling the output of the receive filter at the symbol rate $1/T$. Often, however, we use a suboptimal choice of receive filter (e.g., a wideband filter flat over the signal band, or a filter matched to the transmit filter). In this case, it is typically advantageous to sample faster than the symbol rate. In general, we assume that the sampler operates at rate m/T, where m is a positive integer. The output of the sampler is then processed, typically using digital signal processing (DSP), to perform receiver functions such as synchronization, equalization, and demodulation.

The simulation of a linearly modulated system typically involves the following steps.

Step 1. Generating random symbols to be sent

We restrict attention in this lab to binary phase-shift keying (BPSK). That is, the symbols $\{b_n\}$ in Figure 4.25 take values ± 1.

Step 2. Implementing the transmit, channel, and receive filters

Since the bandwidth of these filters is of the order of $1/T$, they can be accurately implemented in DSP by using FIR filters operating on samples at a rate that is a suitable multiple of $1/T$. The default choice of sampling rate in the labs is $4/T$, unless specified otherwise. If the filter is specified in continuous time, then, typically, one simply samples the impulse response at rate $4/T$, taking a large enough filter length to capture most of the energy in the impulse response. Code Fragment 4.B.1 in the appendix illustrates generating a discrete-time filter corresponding to a truncated raised-cosine pulse.

Step 3. Sending the symbols through the filters

To send symbols at rate $1/T$ through filters implemented at rate $4/T$, it is necessary to upsample the symbols before convolving them with the filter impulse response determined in Step 2. Code Fragment 4.B.2 in the appendix illustrates this for a raised-cosine pulse.

Step 4. Adding noise

Typically, we add white Gaussian noise (the model for this will be specified in a later lab) at the input to the receive filter.

Step 5. Processing at the receive filter output

If there is no intersymbol interference (ISI), the processing simply consists of sampling at rate $1/T$ to get decision statistics for the symbols of interest. For BPSK, you might simply take the sign of the decision statistic to make your bit decision.

If the ISI is significant, then channel equalization (discussed in a later lab) is required prior to making symbol decisions.

Laboratory assignment

(1.1) Write a MATLAB function analogous to Code Fragment 4.B.1 to generate an SRRC pulse (i.e., do Problem 4.13(a)) where you can specify the truncation length and the excess bandwidth.

(1.2) Set the transmit filter to an SRRC pulse with excess bandwidth 22%, sampled at rate $4/T$ and truncated to $[-5T, 5T]$. Plot the impulse response of the transmit filter versus t/T.

If you have difficulty generating the SRRC pulse, use the following code fragment to generate the transmit filter:

Code Fragment 4.8.1 (Explicit specification of transmit filter)

```
%first specify half of the filter
hhalf = [-0.025288315;-0.034167931;-0.035752323;-0.016733702;0.021602514;
0.064938487;0.091002137;0.081894974;0.037071157;-0.021998074;-0.060716277;
-0.051178658;0.007874526;0.084368728;0.126869306;0.094528345;-0.012839661;
-0.143477028;-0.211829088;-0.140513128;0.094601918;0.441387140;0.785875640;
1.0];
transmit_filter = [hhalf;flipud(hhalf)];
```

(1.3) Using the DFT (as in Code Fragment 2.5.1 for Example 2.5.4), compute the magnitude of the transfer function of the transmit filter versus the normalized frequency fT (make sure the resolution in frequency is good enough to get a smooth plot, e.g., at least as good as $1/(64T)$). By eyeballing the plot, check whether the normalized bandwidth (i.e., the bandwidth as a multiple of $1/T$) is well predicted by the nominal excess bandwidth.

(1.4) Use the transmit filter in Code Fragment 4.B.2, which implements upsampling and allows sending a programmable number of symbols through the system. Set the receive filter to be the matched filter corresponding to the transmit filter, and plot the response at the output of the receive filter to a single symbol. Is the cascade of the transmit and receive filters Nyquist at rate $1/T$?

(1.5) Generate 100 random bits $\{a[n]\}$ taking values in $\{0, 1\}$, and map them to symbols $\{b[n]\}$ taking values in $\{-1, +1\}$, with 0 mapped to $+1$ and 1 to -1.

(1.6) Send the 100 symbols $\{b[n]\}$ through the system. What is the length of the corresponding output of the transmit filter? What is the length of the corresponding output of the receive filter? Plot separately the input to the receive filter, and the output of the receive filter versus time, with one unit of time on the x axis equal to the symbol time T.

(1.7) Do the best job you can in recovering the transmitted bits $\{a[n]\}$ by directly sampling the **input** to the receive filter, and add lines in the MATLAB code for implementing your idea. That is, select a set of 100 samples, and estimate the 100 transmitted bits from the sign of these samples. (What sampling delay and spacing would you use?) Estimate the probability of error (note that no noise has been added).

(1.8) Do the best job you can in recovering the transmitted bits by directly sampling the **output** of the receive filter, and add lines in the MATLAB code for implementing your idea. That is, select a set of 100 samples, and estimate the 100 transmitted bits from the sign of these samples. (What sampling delay and spacing would you use?) Estimate the probability of error. Also estimate the probability of error if you chose an incorrect delay, offset from the correct delay by $T/2$.

(1.9) Suppose that the receiver LO used for downconversion is ahead in frequency and phase relative to the incoming wave by $\Delta f = 1/(40T)$ and a phase of $\pi/2$. Modify your complex-baseband model to include the effects of the carrier phase and frequency offset. When you now sample at the "correct" delay as determined in (1.8), construct a scatter plot of the complex-valued samples $\{y[n], n = 1, \ldots, 100\}$ that you obtain. Can you make correct decisions by taking the sign of the real part of the samples, as in (1.8)?

(1.10) Now consider a *differentially encoded* system in which we send $\{a[n], n = 1, \ldots, 99\}$, where $a[n] \in \{0, 1\}$, by sending the following ± 1 bits: $b[1] = +1$, and, for $n = 2, \ldots, 100$,

$$b[n] = \begin{cases} b[n-1], & a[n] = 0 \\ -b[n-1], & a[n] = 1 \end{cases}$$

Devise estimates for the bits $\{a[n]\}$ from the samples $\{y[n]\}$ in (1.9), and estimate the probability of error.
Hint. What does $y[n]y^*[n-1]$ look like?

Lab report Your lab report should answer the preceding questions in order, and should document the reasoning you used and the difficulties you encountered. Comment on whether you get better error probability in (1.7) or (1.8), and specify why this is so.

Appendix 4.A Power spectral density of a linearly modulated signal

We wish to compute the PSD of a linearly modulated signal of the form

$$u(t) = \sum_n b[n]p(t - nT)$$

While we model the complex-valued symbol sequence $\{b[n]\}$ as random, we do not need to invoke concepts from probability and random processes to compute the PSD, but can simply model time-averaged quantities for the symbol sequence. For example, the DC value, which is typically designed to be zero, is defined by

$$\overline{b[n]} = \lim_{N\to\infty} \frac{1}{2N+1} \sum_{n=-N}^{N} b[n] \tag{4.28}$$

We also define the time-averaged autocorrelation function $R_b[k] = \overline{b[n]b^*[n-k]}$ for the symbol sequence as the following limit:

$$R_b[k] = \lim_{N\to\infty} \frac{1}{2N} \sum_{n=-N}^{N} b[n]b^*[n-k] \tag{4.29}$$

Note that we are being deliberately sloppy about the limits of summation in n on the right-hand side to avoid messy notation. Actually, since $-N \le m = n - k \le N$, we have the constraint $-N + k \le n \le N + k$ in addition to the constraint $-N \le n \le N$. Thus, the summation in n should depend on the delay k at which we are evaluating the autocorrelation function, going from $n = -N$ to $n = N + k$ for $k < 0$, and from $n = -N + k$ to $n = N$ for $k \ge 0$. However, we ignore these edge effects, since they become negligible when we let N get large while keeping k fixed.

We now compute the time-averaged PSD. As described in Section 4.2.1, the steps for computing the PSD for a finite-power signal $u(t)$ are as follows:

(a) timelimit to a finite observation interval of length T_o to get a finite-energy signal $u_{T_o}(t)$;
(b) compute the Fourier transform $U_{T_o}(f)$, and hence obtain the energy spectral density $\left|U_{T_o}(f)\right|^2$;
(c) estimate the PSD as $\hat{S}_u(f) = \left|U_{T_o}(f)\right|^2 / T_o$, and take the limit $T_0 \to \infty$ to obtain $S_u(f)$.

Consider the observation interval $[-NT, NT]$, which fits roughly $2N$ symbols. In general, the modulation pulse $p(t)$ need not be timelimited to the symbol duration T. However, we can neglect the edge effects caused by this, since we eventually take the limit as the observation interval gets large. Thus, we can write

$$u_{T_o}(t) \approx \sum_{n=-N}^{N} b[n]p(t - nT)$$

On taking the Fourier transform, we obtain

$$U_{T_o}(f) = \sum_{n=-N}^{N} b[n]P(f)e^{-j2\pi fnT}$$

The energy spectral density is therefore given by

$$\left|U_{T_o}(f)\right|^2 = U_{T_o}(f)U^*_{T_o}(f) = \sum_{n=-N}^{N} b[n]P(f)e^{-j2\pi fnT} \sum_{m=-N}^{N} b^*[m]P^*(f)e^{j2\pi fmT}$$

where we need to use two different dummy variables, n and m, for the summations corresponding to $U_{T_o}(f)$ and $U^*_{T_o}(f)$, respectively. Thus,

$$\left|U_{T_o}(f)\right|^2 = |P(f)|^2 \sum_{m=-N}^{N} \sum_{n=-N}^{N} b[n]b^*[m]e^{-j2\pi(m-n)fT}$$

and the PSD is estimated as

$$\hat{S}_u(f) = \frac{|U_{T_o}(f)|^2}{2NT} = \frac{|P(f)|^2}{T} \left\{ \frac{1}{2N} \sum_{m=-N}^{N} \sum_{n=-N}^{N} b[n]b^*[m]e^{-j2\pi f(n-m)T} \right\} \tag{4.30}$$

Thus, the PSD factors into two components: the first is a term $|P(f)|^2/T$ that depends only on the spectrum of the modulation pulse $p(t)$, while the second term (in curly brackets) depends only on the symbol sequence $\{b[n]\}$. Let us now work on simplifying the latter. By grouping terms of the form $m = n - k$ for each fixed k, we can rewrite this term as

$$\frac{1}{2N} \sum_{m=-N}^{N} \sum_{n=-N}^{N} b[n]b^*[m]e^{-j2\pi f(n-m)T} = \sum_k \frac{1}{2N} \sum_{n=-N}^{N} b[n]b^*[n-k]e^{-j2\pi fkT} \tag{4.31}$$

From (4.29), we see that taking the limit $N \to \infty$ in (4.31) yields $\sum_k R_b[k]e^{-j2\pi fkT}$. On substituting this into (4.30), we obtain that the PSD is given by

$$S_u(f) = \frac{|P(f)|^2}{T} \sum_k R_b[k]e^{-j2\pi fkT} \tag{4.32}$$

Thus, we see that the PSD depends both on the modulating pulse $p(t)$ and on the properties of the symbol sequence $\{b[n]\}$. We shall explore how the dependence on the symbol sequence can be exploited for shaping the spectrum in the problems. However, for most systems, the symbol sequence can be modeled as uncorrelated and of zero mean, In this case, $R_b[k] = 0$ for $k \neq 0$. Specializing to this important setting yields Theorem 4.2.1.

Appendix 4.B Simulation resource: bandlimited pulses and upsampling

The discussion in this appendix should be helpful for Software Lab 4.1. In order to simulate a linearly modulated system, we must specify the transmit and receive filters, typically chosen so that their cascade is Nyquist at the symbol rate. As mentioned earlier, there are two popular choices. One choice is to set the transmit filter to a Nyquist pulse, and the receive filter to a wideband pulse that has a response that is roughly flat over the band of interest. Another is to set the transmit and receive filters to be square roots (in the frequency domain) of a Nyquist pulse. We discuss software implementations of both choices here.

Consider the raised-cosine pulse, which is the most common choice for bandlimited Nyquist pulses. On setting the symbol rate $1/T = 1$ without loss of generality (this is equivalent to expressing all results in terms of t/T or fT), this pulse is given by

$$P(f) = \begin{cases} 1, & 0 \le |f| \le (1-a)/2 \\ \frac{1}{2}[1 + \cos((\pi/a)(|f| - (1-a)/2))], & (1-a)/2 \le |f| \le (1+a)/2 \\ 0, & \text{else} \end{cases} \tag{4.33}$$

The corresponding time-domain pulse is given by

$$p(t) = \text{sinc}(t)\frac{\cos(\pi at)}{1 - 4a^2t^2} \tag{4.34}$$

where $0 \leq a \leq 1$ denotes the excess bandwidth. When generating a sampled version of this pulse, we must account for the zero in the denominator at $t = \pm 1/(2a)$. An example MATLAB function for generating a sampled version of the raised cosine pulse is provided below. Note that the code must account for the zero in the denominator at $t = \pm 1/(2a)$. It is left as an exercise to show, using l'Hôpital's rule, that the 0/0 form taken by $\cos(\pi at)/(1 - 4a^2t^2)$ at these times evaluates to $\pi/4$.

Code Fragment 4.B.1 (Sampled raised-cosine pulse)

```
%time-domain pulse for raised cosine, together with time vector to plot it against
%oversampling factor = how much faster than the symbol rate we sample at
%length = where to truncate response (multiple of symbol time) on each side of peak
%a = excess bandwidth
function [rc,time_axis] = raised_cosine(a,m,length)
     length_os = floor(length * m); %number of samples on each side of peak
     %time vector (in units of symbol interval) on one side of the peak
     z = cumsum(ones(length_os,1))/m;
     A = sin(pi * z)./(pi * z); %term 1
     B = cos(pi * a * z); %term 2
     C = 1 - (2 * a * z).^2; %term 3
     zerotest = m/(2 * a); %location of zero in denominator
     %check whether any sample coincides with zero location
     if (zerotest == floor(zerotest)),
       B(zerotest) = pi * a;
       C(zerotest) = 4 * a;
       alternative is to perturb around the sample
       %(find l'Hôpital limit numerically)
       %B(zerotest) = cos(pi * a * (z(zerotest) + 0.001));
       %C(zerotest) = 1 - (2 * a * (z(zerotest) + 0.001))^2;
end
D = (A.* B)./C; %response to one side of peak
rc = [flipud(D);1;D]; %add in peak and other side
time_axis = [flipud(-z);0;z];
```

This can, for example, be used to generate a plot of the raised cosine pulse, as follows, where we would typically oversample by a large factor (e.g., $m = 32$) in order to get a smooth plot:

```
%%plot time-domain raised-cosine pulse
a = 0.5; % desired excess bandwidth
m = 32; %oversample by a lot to get smooth plot
length = 10; % where to truncate the time-domain response
             %(one-sided, multiple of symbol time)
[rc,time] = raised_cosine(a,m,length);
plot(time,rc);
```

The code for the raised-cosine function can also be used to generate the coefficients of a discrete-time transmit filter. Here, the oversampling factor would be dictated by our DSP-centric implementation, and would usually be far less than what is required for a smooth plot: the digital-to-analog converter would perform the interpolation required to provide a smooth analog waveform for upconversion. A typical choice is $m = 4$, as in the MATLAB code below for generating a noiseless BPSK modulated signal.

Upsampling As noted in our preview of digital modulation in Section 2.3.2, the symbols come in every T seconds, while the samples of the transmit filter are spaced by T/m. For example, the nth symbol contributes $b[n]p(t - nT)$ to the transmit filter output, and the $(n+1)$st symbol contributes $b[n+1]p(t-(n+1)T)$. Since $p(t-nT)$ and $p(t-(n+1)T)$ are offset by T, they must be offset by m samples when sampling at a rate of m/T. Thus, if the symbols are input to a transmit filter whose discrete-time impulse response is expressed at sampling rate m/T, then successive symbols at the input to the filter must be spaced by m samples. That is, in order to get the output as a convolution of the symbols with the transmit filter expressed at rate m/T, we must insert $m - 1$ zeros between successive symbols to convert them to a sampling rate of m/T.

For completeness, we reproduce part of the upsampling Code Fragment 2.3.2 below in implementing a raised-cosine transmit filter.

Code Fragment 4.B.2 (Sampled transmitter output)

```
oversampling_factor = 4;
m = oversampling_factor;
%parameters for sampled raised-cosine pulse
a = 0.5;
length = 10; % (truncated outside [-length * T,length * T])
%raised-cosine transmit filter (time vector set to a dummy variable which is not
%used)
[transmit_filter,dummy] = raised_cosine(a,m,length);
%NUMBER OF SYMBOLS
nsymbols = 100;
%BPSK SYMBOL GENERATION
symbols = sign(rand(nsymbols,1) -.5);
%UPSAMPLE BY m
nsymbols_upsampled = 1 + (nsymbols - 1) * m; %length of upsampled symbol sequence
symbols_upsampled = zeros(nsymbols_upsampled,1); %initialize
symbols_upsampled(1:m:nsymbols_upsampled) = symbols;
%insert symbols with spacing m
%NOISELESS MODULATED SIGNAL
tx_output = conv(symbols_upsampled,transmit_filter);
```

Let us now discuss the implementation of an alternative transmit filter, the square-root raised cosine (SRRC). The frequency-domain SRRC pulse is given by $G(f) = \sqrt{P(f)}$, where $P(f)$ is as in (4.33). We now need to find a sampled version of the time-domain pulse $g(t)$ in order to implement linear modulation as above. While this could be done numerically by sampling the frequency-domain pulse and computing an inverse DFT, we can also find an analytical formula for $g(t)$, as follows. Given the practical importance of the SRRC pulse, we provide the formula and sketch its derivation. Noting that $1 + \cos(2\theta) = 2\cos^2\theta$, we can rewrite the frequency-domain expression (4.33) for the raised-cosine pulse as

$$P(f) = \begin{cases} 1, & 0 \le |f| \le (1-a)/2 \\ \cos^2((\pi/(2a))(|f| - (1-a)/2)), & (1-a)/2 \le |f| \le (1+a)/2 \\ 0, & \text{else} \end{cases} \tag{4.35}$$

We can now take the square root to get an analytical expression for the SRRC pulse in the frequency domain as follows:

$$G(f) = \begin{cases} 1, & 0 \leq |f| \leq (1-a)/2 \\ \cos((\pi/2a))(|f| - (1-a)/2)), & (1-a)/2 \leq |f| \leq (1+a)/2 \\ 0, & \text{else} \end{cases} \quad (4.36)$$

frequency-domain SRRC pulse

Finding the time-domain SRRC pulse is now a matter of computing the inverse Fourier transform. Since it is also an interesting exercise in utilizing Fourier-transform properties, we sketch the derivation. First, we break up the frequency-domain pulse into segments whose inverse Fourier transforms are well known. Setting $b = (1-a)/2$, we have

$$G(f) = G_1(f) + G_2(f) \quad (4.37)$$

where

$$G_1(f) = I_{[-b,b]}(f) \leftrightarrow g_1(t) = 2b\ \text{sinc}(2bt) = \frac{\sin(2\pi bt)}{\pi t} = \frac{\sin(\pi(1-a)t)}{\pi t} \quad (4.38)$$

and

$$G_2(f) = U(f-b) + U(-f-b) \quad (4.39)$$

with

$$U(f) = \cos\left(\frac{\pi}{2a}f\right) I_{[0,a]}(f) = \frac{1}{2}\left(e^{j(\pi/(2a))f} + e^{-j(\pi/(2a))f}\right) I_{[0,a]}(f) \quad (4.40)$$

To evaluate $g_2(t)$, note first that

$$I_{[0,a]}(f) \leftrightarrow = a\ \text{sinc}(at)e^{j\pi at} \quad (4.41)$$

Multiplication by $e^{j(\pi/(2a))f}$ in the frequency domain corresponds to a leftward time shift by $1/(4a)$, while multiplication by $e^{-j(\pi/(2a))f}$ corresponds to a rightward time shift by $1/(4a)$. From (4.40) and (4.41), we therefore obtain that

$$\begin{aligned} U(f) &= \cos\left(\frac{\pi}{2a}f\right) I_{[0,a]}(f) \leftrightarrow u(t) \\ &= \frac{a}{2}\ \text{sinc}\left(a\left(t + \frac{1}{4a}\right)\right) e^{j\pi a(t+1/(4a))} + \frac{a}{2}\ \text{sinc}\left(a\left(t - \frac{1}{4a}\right)\right) e^{j\pi a(t-1/(4a))} \end{aligned}$$

By simplifying, we obtain that

$$u(t) = \frac{a}{\pi}\frac{2e^{j2\pi at} - j8at}{1 - 16a^2t^2} \quad (4.42)$$

Now,

$$G_2(f) = U(f-b) + U(-f-b) \leftrightarrow g_2(t) = u(t)e^{j2\pi bt} + u^*(t)e^{-j2\pi bt} = \text{Re}\left(2u(t)e^{j2\pi bt}\right) \quad (4.43)$$

On plugging this into (4.42), and substituting the value of $b = (1-a)/2$, we obtain upon simplification that

$$2u(t)e^{j2\pi bt} = \frac{2a}{\pi}\frac{2e^{j\pi(1+a)t} - j8ate^{j\pi(1-a)t}}{1 - 16a^2t^2}$$

On taking the real part, we obtain

$$g_2(t) = \frac{1}{\pi} \frac{4a\cos(\pi(1+a)t) + 16a^2 t \sin(\pi(1-a)t)}{1 - 16a^2t^2} \tag{4.44}$$

By combining (4.38) and (4.44) and simplifying, we obtain the following expression for the SRRC pulse $g(t) = g_1(t) + g_2(t)$:

$$g(t) = \frac{4a\cos(\pi(1+a)t) + (\sin(\pi(1-a)t))/t}{\pi(1 - 16a^2t^2)} \quad \textbf{time-domain SRRC pulse} \tag{4.45}$$

We leave it as an exercise to write MATLAB code to generate a sampled version of the SRRC pulse (analogous to Code Fragment 4.B.1), taking into account the zeros in the denominator. This can then be used to generate a noiseless transmit waveform as in Code Fragment 4.B.2 simply by replacing the transmit filter by an SRRC pulse.

5 Probability and random processes

Probability theory is fundamental to communication system design, especially for digital communication. Not only are there uncontrolled sources of uncertainty such as noise, interference, and other channel impairments that are amenable only to statistical modeling, but also the very notion of information underlying digital communication is based on uncertainty. In particular, the receiver in a communication system does not know *a priori* what the transmitter is sending (otherwise the transmission would be pointless), hence the receiver designer must employ statistical models for the transmitted signal. In this chapter, we review basic concepts of probability and random variables with examples motivated by communications applications. We also introduce the concept of random processes, which are used to model both signals and noise in communication systems.

Chapter plan

The goal of this chapter is to develop the statistical modeling tools required in later chapters. *For readers who are already comfortable with probability and random processes, the shortest path to Chapter 6 is to review the material on Gaussian random variables in Section 5.6 and noise modeling in Section 5.8.* Sections 5.1 through 5.5 provide a review of background material on probability and random variables. Section 5.1 discusses basic concepts of probability: the most important of these for our purpose are the concepts of conditional probability and Bayes' rule. Sections 5.2 and 5.4 discuss random variables and functions of random variables. Multiple random variables, or random vectors, are discussed in Section 5.3. Section 5.5 discusses various statistical averages and their computation. The material which is not part of the assumed background starts with Section 5.6; this section goes in depth into Gaussian random variables and vectors, which play a critical role in the mathematical modeling of communication systems. Section 5.7 introduces random processes in sufficient depth that we can describe, and perform elementary computations with, the classical white-Gaussian-noise (WGN) model in Section 5.8. At this point, zealous followers of a "just in time" philosophy can move on to the discussion of optimal receiver design in Chapter 6. However, many others might wish to go through one more section, namely Section 5.9, which provides a more general treatment of the effect of linear operations on random processes. The results in this section allow us, for example, to model noise correlations and to compute quantities such as the signal-to-noise ratio (SNR). The material which we do not build on in later chapters, but which may be of interest to some readers, is placed in the appendices: this includes limit theorems, qualitative discussion of noise mechanisms, discussion of the structure of passband random processes, and quantification, via SNR computations, of the effect of noise on analog modulation.

5.1 Probability basics

In this section, we remind ourselves of some important definitions and properties.

Sample space The starting point in probability is the notion of an experiment whose outcome is not deterministic. The set of all possible outcomes from the experiment is termed the sample space Ω. For example, the sample space corresponding to the throwing of a six-sided die is $\Omega = \{1, 2, 3, 4, 5, 6\}$. An analogous example that is well-suited to our purpose is the sequence of bits sent by the transmitter in a digital communication system, which is modeled probabilistically by the receiver. For example, suppose that the transmitter can send a sequence of seven bits, each taking the value 0 or 1. Then our sample space consists of the $2^7 = 128$ possible bit sequences.

Event Events are sets of possible outcomes to which we can assign a probability. That is, an event is a subset of the sample space. For example, for a six-sided die, the event $\{1, 3, 5\}$ is the set of odd-numbered outcomes.

We are often interested in probabilities of events obtained from other events by basic set operations such as complementation, unions, and intersections; see Figure 5.1.

Complement of an event ("NOT") For an event A, the complement ("not A"), denoted by A^c, is the set of outcomes that do not belong to A.

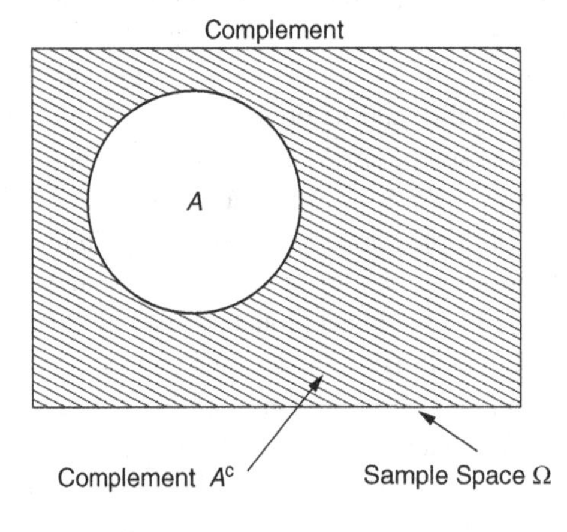

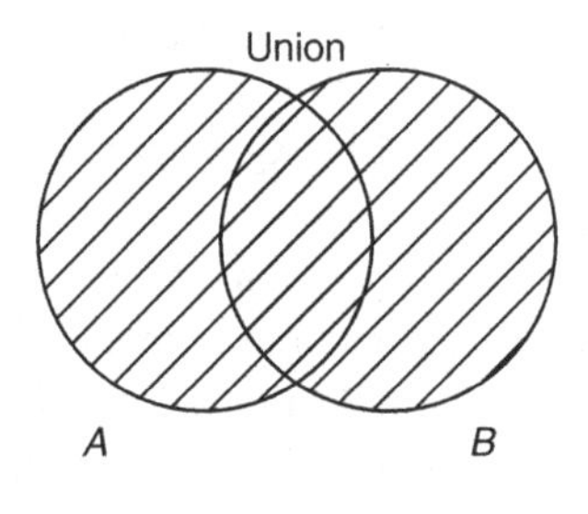

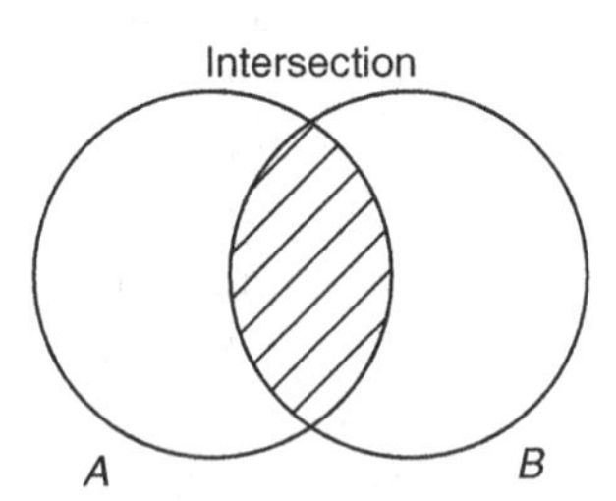

Figure 5.1 Basic set operations.

Union of events ("OR") The union of two events A and B, denoted by $A \cup B$, is the set of all outcomes that belong to either A or B. The term "or" always refers to the *inclusive or*, unless we specify otherwise. Thus, outcomes belonging to both events are included in the union.

Intersection of events ("AND") The intersection of two events A and B, denoted by $A \cap B$, is the set of all outcomes that belong to both A *and* B.

Mutually exclusive, or disjoint, events Events A and B are mutually exclusive, or disjoint, if their intersection is empty: $A \cap B = \emptyset$.

Difference of events The difference $A \setminus B$ is the set of all outcomes that belong to A but not to B. In other words, $A \setminus B = A \cap B^c$.

Probability Measure A probability measure is a function that assigns probability to events. Some properties are as follows.

Range of probability For any event A, we have $0 \leq P[A] \leq 1$. The probability of the empty set is zero: $P[\emptyset] = 0$. The probabilty of the entire sample space is one: $P[\Omega] = 1$.

Probabilities of disjoint events add up If two events A and B are mutually exclusive, then the probability of their union equals the sum of their probabilities:

$$P[A \cup B] = P[A] + P[B] \quad \text{if } A \cap B = \emptyset \tag{5.1}$$

By mathematical induction, we can infer that the probability of the union of a finite number of pairwise disjoint events also adds up. It is useful to review the principle of mathematical induction via this example. Specifically, suppose that we are given pairwise disjoint events $A_1, A_2, A_3, \ldots$. We wish to prove that, for any $n \geq 2$.

$$P[A_1 \cup A_2 \cup \cdots \cup A_n] = P[A_1] + \cdots + P[A_n] \quad \text{if } A_i \cap A_j = \emptyset \text{ for all } i \neq j \tag{5.2}$$

Mathematical induction consists of the following steps:

(a) verify that the result is true for the initial value of n, which in our case is $n = 2$;
(b) assume that the result is true for an arbitrary value of $n = k$;
(c) use (a) and (b) to prove that the result is true for $n = k + 1$.

In our case, step (a) does not require any work; it holds by virtue of our assumption of (5.1). Now, assume that (5.2) holds for $n = k$. Then we have

$$A_1 \cup A_2 \cup \cdots \cup A_k \cup A_{k+1} = B \cup A_{k+1}$$

where

$$B = A_1 \cup A_2 \cup \cdots \cup A_k$$

and A_{k+1} are disjoint. We can therefore conclude, using step (a), that

$$P[B \cup A_{k+1}] = P[B] + P[A_{k+1}]$$

But, using step (b), we know that

$$P[B] = P[A_1 \cup A_2 \cup \cdots \cup A_k] = P[A_1] + \cdots + P[A_k]$$

We can now conclude that

$$P[A_1 \cup A_2 \cup \cdots \cup A_k \cup A_{k+1}] = P[A_1] + \cdots + P[A_{k+1}]$$

thus accomplishing step (c).

The preceding properties are typically stated as *axioms,* which provide the starting point from which other properties, some of which are stated below, can be derived.

Probability of the complement of an event The probabilities of an event and its complement sum to one. By definition, A and A^c are disjoint, and $A \cup A^c = \Omega$. Since $P[\Omega] = 1$, we can now apply (5.1) to infer that

$$P[A] + P[A^c] = 1 \tag{5.3}$$

Probabilities of unions and intersections We can use the property (5.1) to infer the following property regarding the union and intersection of arbitrary events:

$$P[A_1 \cup A_2] = P[A_1] + P[A_2] - P[A_1 \cap A_2] \tag{5.4}$$

Let us get a feel for how to use the probability axioms by proving this. We break $A_1 \cup A_2$ into disjoint events as follows:

$$A_1 \cup A_2 = A_2 \cup (A_1 \setminus A_2)$$

Upon applying (5.1), we have

$$P[A_1 \cup A_2] = P[A_2] + P[A_1 \setminus A_2] \tag{5.5}$$

Furthermore, since A_1 can be written as the disjoint union $A_1 = (A_1 \cap A_2) \cup (A_1 \setminus A_2)$, we have $P[A_1] = P[A_1 \cap A_2] + P[A_1 \setminus A_2]$, or $P[A_1 \setminus A_2] = P[A_1] - P[A_1 \cap A_2]$. On plugging this into (5.5), we obtain (5.4).

Conditional probability The conditional probability of A given B is the probability of A assuming that we already know that the outcome of the experiment is in B. Outcomes corresponding to this probability must therefore belong to the intersection $A \cap B$. We therefore define the conditional probability as

$$P[A|B] = \frac{P[A \cap B]}{P[B]} \tag{5.6}$$

(We assume that $P[B] > 0$, otherwise the condition we are assuming cannot occur.)

Conditional probabilities behave just the same as regular probabilities, since all we are doing is restricting the sample space to the event being conditioned on. Thus, we still have $P[A|B] = 1 - P[A^c|B]$ and

$$P[A_1 \cup A_2|B] = P[A_1|B] + P[A_2|B] - P[A_1 \cap A_2|B]$$

Conditioning is a crucial concept in models for digital communication systems. A typical application is to condition on which of a number of possible transmitted signals is sent, in order to describe the statistical behavior of the communication medium. Such statistical models then form the basis for receiver design and performance analysis.

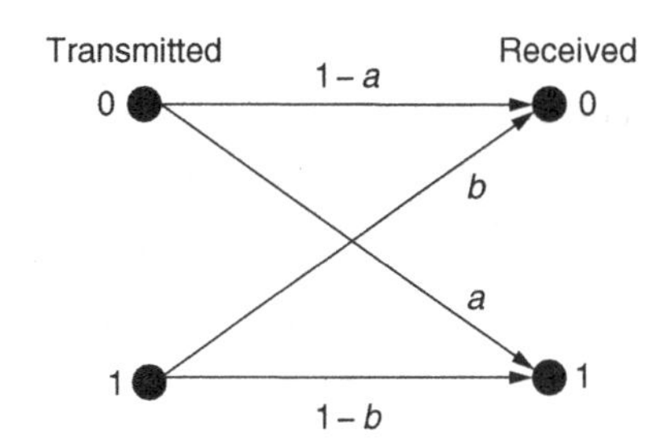

Figure 5.2 Conditional probabilities modeling a binary channel.

Example 5.1.1 (a binary channel) Figure 5.2 depicts the conditional probabilities for a noisy binary channel. On the left side are the two possible values of the bit sent, and on the right are the two possible values of the bit received. The label on a given arrow is the conditional probability of the received bit, given the transmitted bit. Thus, the binary channel is defined by means of the following conditional probabilities:

$$P[0 \text{ received}|0 \text{ transmitted}] = 1 - a, \; P[1 \text{ received}|0 \text{ transmitted}] = a;$$
$$P[0 \text{ received}|1 \text{ transmitted}] = b, \; P[1 \text{ received}|1 \text{ transmitted}] = 1 - b$$

These conditional probabilities are often termed the *channel transition probabilities.* The probabilities a and b are called the *crossover probabilities.* When $a = b$, we obtain the *binary symmetric channel.*

Law of total probability For events A and B, we have

$$P[A] = P[A \cap B] + P\big[A \cap B^c\big] = P[A|B]P[B] + P[A|B^c]P[B^c] \tag{5.7}$$

In the above, we have decomposed an event of interest, A, into a disjoint union of two events, $A \cap B$ and $A \cap B^c$, so that (5.1) applies. The sets B and B^c form a *partition* of the entire sample space; that is, they are disjoint, and their union equals Ω. This generalizes to any partition of the sample space; that is, if $B_1, B_2, \ldots$ are mutually exclusive events such that their union covers the sample space (actually, it suffices that the union contains A), then

$$P[A] = \sum_i P[A \cap B_i] = \sum_i P[A|B_i]P[B_i] \tag{5.8}$$

Example 5.1.2 (Applying the law of total probability to the binary channel) For the channel in Figure 5.2, set $a = 0.1$ and $b = 0.25$, and suppose that the probability of transmitting 0 is 0.6. This is called the *prior*, or *a priori,* probability of transmitting 0, because it is the statistical information that the receiver has *before* it sees the received bit. Using (5.3), the prior probability of 1 being transmitted is

$$P[0 \text{ transmitted}] = 0.6 = 1 - P[1 \text{ transmitted}]$$

(since our only options for this particular channel model are sending 0 or 1, the two events are complements of each other). We can now compute the probability that 0 is received using the law of total probability, as follows:

$$\begin{aligned} P[0 \text{ received}] &= P[0 \text{ received}|0 \text{ transmitted}]P[0 \text{ transmitted}] \\ &\quad + P[0 \text{ received}|1 \text{ transmitted}]P[1 \text{ transmitted}] \\ &= 0.9 \times 0.6 + 0.25 \times 0.4 = 0.64 \end{aligned}$$

We can also compute the probability that 1 is received using the same technique, but it is easier to infer this from (5.3) as follows:

$$P[1 \text{ received}] = 1 - P[0 \text{ received}] = 0.36$$

Bayes' rule Given $P[A|B]$, we compute $P[B|A]$ as follows:

$$P[B|A] = \frac{P[A|B]P[B]}{P[A]} = \frac{P[A|B]P[B]}{P[A|B]P[B] + P[A|B^c]P[B^c]} \tag{5.9}$$

where we have used (5.7). Similarly, in the setting of (5.8), we can compute $P[B_j|A]$ as follows:

$$P[B_j|A] = \frac{P[A|B_j]P[B_j]}{P[A]} = \frac{P[A|B_j]P[B_j]}{\sum_i P[A|B_i]P[B_i]} \tag{5.10}$$

Bayes' rule is typically used as follows in digital communication. The event B might correspond to which transmitted signal was sent. The event A may describe the received signal, so that $P[A|B]$ can be computed from our model for the statistics of the received signal, given the transmitted signal. Bayes' rule can then be used to compute the conditional probability $P[B|A]$ of a given signal having been transmitted, given information about the received signal, as illustrated in the example below.

Example 5.1.3 (Applying Bayes' rule to the binary channel) Continuing with the binary channel of Figure 5.2 with $a = 0.1$ and $b = 0.25$, let us find the probability that 0 was transmitted, given that 0 is received. This is called the *posterior*, or *a posteriori*, probability of 0 being transmitted, because it is the statistical model that the receiver infers after it sees the received bit. As in Example 5.1.2, we assume that the prior probability of 0 being transmitted is 0.6. We now apply Bayes' rule as follows:

$$\begin{aligned} P[0 \text{ transmitted}|0 \text{ received}] &= \frac{P[0 \text{ received}|0 \text{ transmitted}]P[0 \text{ transmitted}]}{P[0 \text{ received}]} \\ &= \frac{0.9 \times 0.6}{0.64} = \frac{27}{32} \end{aligned}$$

where we have used the computation from Example 5.1.2, which is based on the law of total probability, for the denominator. We can also compute the posterior probability of the complementary event as follows:

$$P[1 \text{ transmitted}|0 \text{ received}] = 1 - P[0 \text{ transmitted}|0 \text{ received}] = \frac{5}{32}$$

These results make sense. Since the binary channel in Figure 5.2 has a small probability of error, it is much more likely that 0 was transmitted than that 1 was transmitted when we receive 0. The situation would be reversed if 1 were received. The computation of the corresponding posterior probabilities is left as an exercise. Note that, for this example, the numerical values for the posterior probabilities may be different when we condition on 1 being received, since the channel transition probabilities and prior probabilities are not symmetric with respect to exchanging the roles of 0 and 1.

Two other concepts that we use routinely are independence and conditional independence.

Independence Events A_1 and A_2 are *independent* if

$$P[A_1 \cap A_2] = P[A_1]P[A_2] \tag{5.11}$$

Example 5.1.4 (Independent bits) Suppose that we transmit three bits. Each time, the probability of sending a 0 is 0.6. Assuming that the bits to be sent are selected independently each of these three times, we can compute the probability of sending any given three-bit sequence using (5.11):

$$\begin{aligned} P[000 \text{ transmitted}] &= P[\text{first bit } = 0, \text{ second bit } = 0, \text{ third bit } = 0] \\ &= P[\text{first bit } = 0]P[\text{second bit } = 0]P[\text{third bit } = 0] \\ &= 0.6^3 = 0.216 \end{aligned}$$

Let us do a few other computations similarly, where we now use the shorthand $P[x_1x_2x_3]$ to denote that $x_1x_2x_3$ is the sequence of three bits transmitted:

$$P[101] = 0.4 \times 0.6 \times 0.4 = 0.096$$

and

$$P[\text{two ones transmitted}] = P[110] + P[101] + P[011] = 3 \times (0.4)^2 \times 0.6 = 0.288$$

The number of ones is actually a binomial random variable (this is reviewed in Section 5.2).

Conditional independence Events A_1 and A_2 are *conditionally independent* given B if

$$P[A_1 \cap A_2|B] = P[A_1|B]P[A_2|B] \tag{5.12}$$

Example 5.1.5 (Independent channel uses) Now, suppose that we transmit three bits, with each bit seeing the binary channel depicted in Figure 5.2. We say that the channel is *memoryless* when the value of the received bit corresponding to a given channel use is conditionally independent of the other received bits, given the transmitted bits. For the setting of Example 5.1.4, where we choose the transmitted bits independently, the following example illustrates the computation of conditional probabilities for the received bits:

$$
\begin{aligned}
&P[100 \text{ received}|010 \text{ transmitted}] \\
&\quad = P[1 \text{ received}|0 \text{ transmitted}]P[0 \text{ received}|1 \text{ transmitted}]P[0 \text{ received}|0 \text{ transmitted}] \\
&\quad = 0.1 \times 0.25 \times 0.9 = 0.0225
\end{aligned}
$$

We end this section with a mention of two useful bounding techniques.

Union bound The probability of a union of events is upper bounded by the sum of the probabilities of the events:

$$P[A_1 \cup A_2] \leq P[A_1] + P[A_2] \tag{5.13}$$

This follows from (5.4) by noting that $P[A_1 \cap A_2] \geq 0$. This property generalizes to a union of a collection of events by mathematical induction:

$$P\left[\bigcup_{i=1}^{n} A_i\right] \leq \sum_{i=1}^{n} P[A_i] \tag{5.14}$$

If A implies B, then $P[A] \leq P[B]$ An event A implies an event B (denoted by $A \longrightarrow B$) if and only if A is contained in B (i.e., $A \subseteq B$). In this case, we can write B as a disjoint union as follows: $B = A \cup (B \setminus A)$. This means that $P[B] = P[A] + P[B \setminus A] \geq P[A]$, since $P[B \setminus A] \geq 0$.

5.2 Random variables

A random variable assigns a number to each outcome of a random experiment. That is, a random variable is a mapping from the sample space Ω to the set of real numbers, as shown in Figure 5.3. The underlying experiment that leads to the outcomes in the sample space can be quite complicated (e.g., generation of a noise sample in a communication system may involve the random movement of a large number of charge carriers, as well as the filtering operation performed by the receiver). However, we do not need to account for

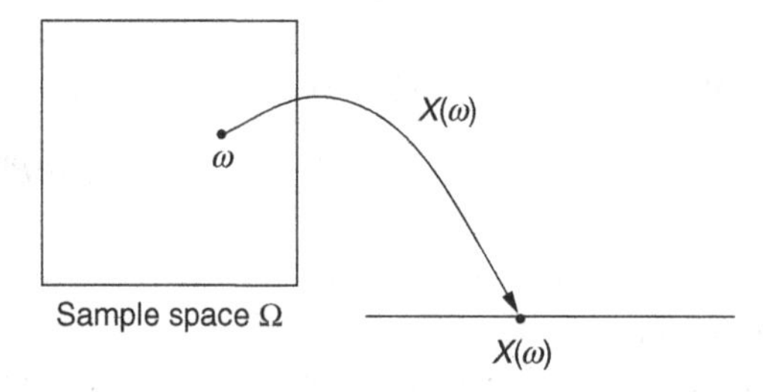

Figure 5.3 A random variable is a mapping from the sample space to the real line.

these underlying physical phenomena in order to specify the probabilistic description of the random variable. All we need to do is to describe how to compute the probabilities of the random variable taking on a particular set of values. In other words, we need to specify its *probability distribution,* or *probability law.* Consider, for example, the Bernoulli random variable, which may be used to model random bits sent by a transmitter, or to indicate errors in these bits at the receiver.

Bernoulli random variable X is a Bernoulli random variable if it takes values 0 or 1. The probability distribution is specified if we know $P[X = 0]$ and $P[X = 1]$. Since X can take only one of these two values, the events $\{X = 0\}$ and $\{X = 1\}$ constitute a partition of the sample space, so that $P[X = 0] + P[X = 1] = 1$. We therefore can characterize the Bernoulli distribution by a parameter $p \in [0, 1]$, where $p = P[X = 1] = 1 - P[X = 0]$. We denote this distribution as Bernoulli(p).

In general, if a random variable takes only a discrete set of values, then its distribution can be specified simply by specifying the probabilities that it takes each of these values.

Discrete random variable, probability mass function X is a discrete random variable if it takes a finite, or countably infinite, number of values. If X takes values $x_1, x_2, \ldots$, then its probability distribution is characterized by its *probability mass function (PMF),* or the probabilities $p_i = P[X = x_i]$, $i = 1, 2, \ldots$. These probabilities must add up to one, $\sum_i p_i = 1$, since the events $\{X = x_i\}, i = 1, 2, \ldots$ provide a partition of the sample space.

For random variables that take values in a continuum, the probability of taking any particular value is zero. Rather, we seek to specify the probability that the value taken by the random variable falls in a given set of interest. By choosing these sets to be intervals whose size shrinks to zero, we arrive at the notion of a probability density function, as follows.

Continuous random variable, probability density function X is a continuous random variable if the probability $P[X = x]$ is zero for each x. In this case, we define the *probability density function (PDF)* as follows:

$$p_X(x) = \lim_{\Delta x \to 0} \frac{P[x \leq X \leq x + \Delta x]}{\Delta x} \tag{5.15}$$

In other words, for small intervals, we have the approximate relationship

$$P[x \leq X \leq x + \Delta x] \approx p_X(x)\Delta x$$

Expressing an event of interest as a disjoint union of such small intervals, the probability of the event is the sum of the probabilities of these intervals; as we let the length of the intervals shrink, the sum becomes an integral (with Δx replaced by dx). Thus, the probability of X taking values in a set A can be computed by integrating its PDF over A, as follows:

$$P[X \in A] = \int_A p_X(x)dx \tag{5.16}$$

The PDF must integrate to one over the real line, since any value taken by X falls within this interval:

$$\int_{-\infty}^{\infty} p_X(x)dx = 1$$

Notation We use the notation $p_X(x)$ to denote the density of a random variable X, evaluated at the point x. Thus, the argument of the density is a dummy variable, and could be denoted by some other letter: for example, we could use the notation $p_X(u)$ as notation for the density of X, evaluated at the point u. Once we have firmly established these concepts, however, we plan to allow ourselves to get sloppy. As discussed in the note at the end of Section 5.3, if there is no scope for confusion, we plan to use the dummy variable to also denote the random variable we are talking about. For example, we use $p(x)$ as the notation for $p_X(x)$ and $p(y)$ as the notation for $p_Y(y)$. But, for now, we retain the subscripts in the introductory material in Sections 5.2 and 5.3.

Density We use the generic term "density" to refer to both the PDF and the PMF (but more often the PDF), relying on the context to clarify what we mean by the term.

The PMF or PDF cannot be used to describe *mixed* random variables that are neither discrete nor continuous. We can get around this problem by allowing PDFs to contain impulses, but a general description of the probability distribution of any random variable, irrespective of whether it is discrete, continuous, or mixed, can be provided in terms of its cumulative distribution function, defined below.

Cumulative distribution function (CDF) The CDF of a random variable X is defined as

$$F_X(x) = P[X \leq x]$$

and has the following general properties.

(1) $F_X(x)$ is nondecreasing in x. This is because, for $x_1 \leq x_2$, we have $\{X \leq x_1\} \subseteq \{X \leq x_2\}$, so that $P[X \leq x_1] \leq P[X \leq x_2]$.
(2) $F_X(-\infty) = 0$ and $F_X(\infty) = 1$. The event $\{X \leq -\infty\}$ contains no allowable values for X, and is therefore the empty set, which has probability zero. The event $\{X \leq \infty\}$ contains all allowable values for X, and is therefore the entire sample space, which has probability one.
(3) $F_X(x)$ is right-continuous: $F_X(x) = \lim_{\delta \to 0, \delta > 0} F_X(x + \delta)$. Denoting this right limit as $F_X(x^+)$, we can state the property compactly as $F_X(x) = F_X(x^+)$. The proof is omitted, since it requires going into probability theory to a depth that is unnecessary for our purpose.

Any function that satisfies (1)–(3) is a valid CDF. The CDFs for discrete and mixed random variables exhibit jumps. At each of these jumps, the left limit $F(x^-)$ is strictly smaller than the right limit $F_X(x^+) = F_X(x)$. Given that

$$P[X = x] = P[X \leq x] - P[X < x] = F_X(x) - F_X(x^-) \tag{5.17}$$

we note that the jumps correspond to the discrete set of points where nonzero probability mass is assigned. For a discrete random variable, the CDF remains constant between these jumps. The PMF is given by applying (5.17) for $x = x_i$, $i = 1, 2, \ldots$, where $\{x_i\}$ is the set of values taken by X.

For a continuous random variable, there are no jumps in the CDF, since $P[X = x] = 0$ for all x. That is, a continuous random variable can be defined as one whose CDF is a continuous function. From the definition (5.15) of PDF, it is clear that the PDF of a continuous random variable is the derivative of the CDF; that is,

$$p_X(x) = F'_X(x) \tag{5.18}$$

Actually, it is possible that the derivative of the CDF for a continuous random variable does not exist at certain points (i.e., when the slopes of $F_X(x)$ approaching from the left and the right are different). The PDF at these points can be defined as either the left or the right slope; it does not make a difference in our probability computations, which involve integrating the PDF (thus washing away the effect of individual points). We therefore do not worry any further about this technicality.

We obtain the CDF from the PDF by integrating the relationship (5.18):

$$F_X(x) = \int_{-\infty}^{x} p_X(z)dz \tag{5.19}$$

It is also useful to define the complementary CDF.

Complementary cumulative distribution function (CCDF) The CCDF of a random variable X is defined as

$$F^c_X(x) = P[X > x] = 1 - F_X(x)$$

The CCDF is often useful in talking about tail probabilities (e.g., the probability that a noise sample takes a large value, causing an error at the receiver). For a continuous random variable with PDF $p_X(x)$, the CCDF is given by

$$F_X(x) = \int_{x}^{\infty} p_X(z)dz \tag{5.20}$$

We now list a few more commonly encountered random variables.

Exponential random variable The random variable X has an exponential distribution with parameter λ, which we denote as $X \sim \text{Exp}(\lambda)$, if its PDF is given by

$$p_X(x) = \begin{cases} \lambda e^{-\lambda x}, & x \geq 0 \\ 0, & x < 0 \end{cases}$$

See Figure 5.4 for an example PDF. We can write this more compactly using the indicator function:

$$p_X(x) = \lambda e^{-\lambda x} I_{[0,\infty)}(x)$$

The CDF is given by

$$F_X(x) = (1 - e^{-\lambda x}) I_{[0,\infty)}(x)$$

For $x \geq 0$, the CCDF is given by

$$F^c_X(x) = P[X > x] = e^{-\lambda x}$$

That is, the tail of an exponential distribution decays (as befits its name) exponentially.

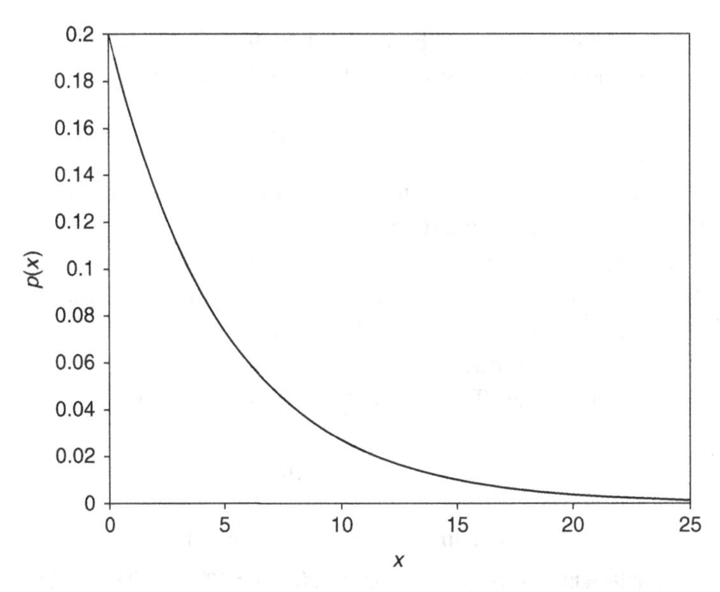

Figure 5.4 The PDF of an exponential random variable with parameter $\lambda = 1/5$ (or mean $1/\lambda = 5$).

Gaussian (or normal) random variable The random variable X has a Gaussian distribution with parameters m and v^2 if its PDF is given by

$$p_X(x) = \frac{1}{\sqrt{2\pi v^2}} \exp\left(-\frac{(x-m)^2}{2v^2}\right) \tag{5.21}$$

See Figure 5.5 for an example PDF. As we show in Section 5.5, m is the mean of X and v^2 is its variance. The PDF of a Gaussian has a well-known bell shape, as shown in Figure 5.5. The Gaussian random variable plays a very important role in communication system design, hence we discuss it in far more detail in Section 5.6, as a prerequisite for the receiver design principles to be developed in Chapter 6.

Example 5.2.1 (Recognizing a Gaussian density) Suppose that a random variable X has PDF

$$p_X(x) = ce^{-2x^2+x}$$

where c is an unknown constant, and x ranges over the real line. Specify the distribution of X and write down its PDF.

Solution

Any PDF with an exponential dependence on a quadratic can be put in the form (5.21) by completing squares in the exponent:

$$-2x^2 + x = -2(x^2 - x/2) = -2\left(\left(x - \frac{1}{4}\right)^2 - \frac{1}{16}\right)$$

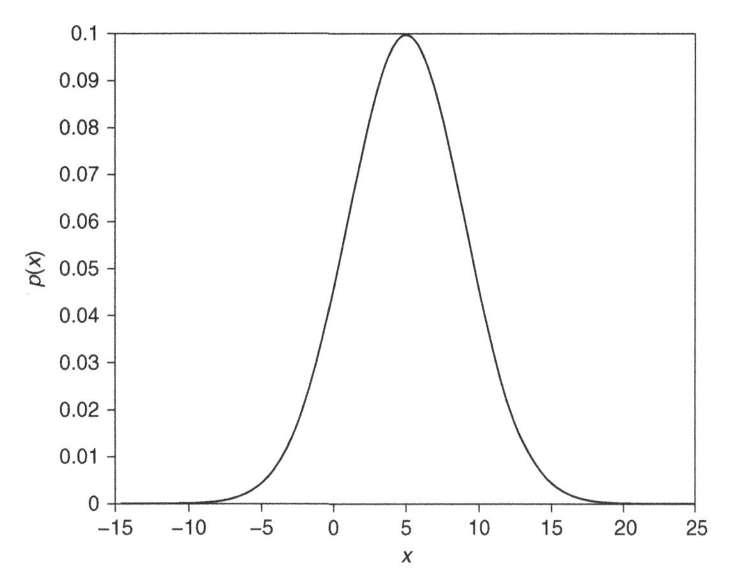

Figure 5.5 The PDF of a Gaussian random variable with parameters $m = 5$ and $v^2 = 16$. Note the bell shape for the Gaussian density, with a peak around its mean $m = 5$.

On comparing this with (5.21), we see that the PDF can be written as an $N(m, v^2)$ PDF with $m = \frac{1}{4}$ and $1/(2v^2) = 2$, so that $v^2 = \frac{1}{4}$. Thus, $X \sim N(\frac{1}{4}, \frac{1}{4})$ and its PDF is given by specializing (5.21):

$$p_X(x) = \frac{1}{\sqrt{2\pi/4}} e^{-2(x-\frac{1}{4})^2} = \sqrt{2/\pi}\, e^{-2x^2+x-\frac{1}{8}}$$

We usually do not really care about going back and specifying the constant c, since we already know the form of the density. But it is easy to check that $c = \sqrt{2/\pi}\, e^{-\frac{1}{8}}$.

Binomial random variable We say that a random variable Y has a binomial distribution with parameters n and p, and denote this by $Y \sim \text{Bin}(n, p)$, if Y takes integer values $0, 1, \ldots, n$, with probability mass function

$$p_k = P[Y = k] = \binom{n}{k} p^k (1-p)^{n-k}, \qquad k = 0, 1, \ldots, n$$

Recall that "n choose k" (the number of ways in which we can choose k items out of n identical items, is given by the expression

$$\binom{n}{k} = \frac{n!}{k!\,(n-k)!}$$

with $k! = 1 \times 2 \times \cdots \times k$ denoting the factorial operation. The binomial distribution can be thought of as a discrete-time analog of the Gaussian distribution; as can be seen in Figure 5.6, the PMF has a bell shape. We comment in more detail on this when we discuss the central limit theorem in Appendix 5.B.

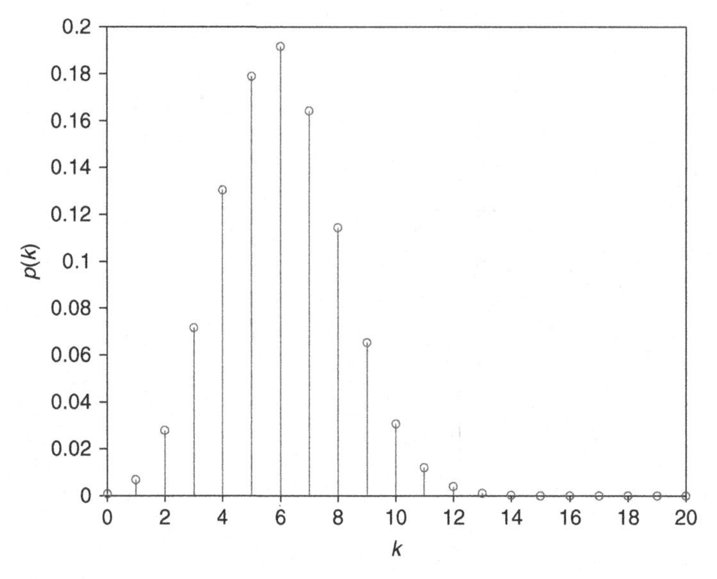

Figure 5.6 The PMF of a binomial random variable with $n = 20$ and $p = 0.3$.

Poisson random variable X is a Poisson random variable with parameter $\lambda > 0$ if it takes values from the nonnegative integers, with PMF given by

$$P[X = k] = \frac{\lambda^k}{k!}e^{-\lambda}, \quad k = 0, 1, 2, \ldots$$

As will be shown later, the parameter λ equals the mean of the Poisson random variable.

5.3 Multiple random variables, or random vectors

We are often interested in more than one random variable when modeling a particular scenario of interest. For example, a model of a received sample in a communication link may involve a randomly chosen transmitted bit, a random channel gain, and a random noise sample. In general, we are interested in multiple random variables defined on a "common probability space," where the latter phrase means simply that we can, in principle, compute the probability of events involving all of these random variables. Technically, multiple random variables on a common probability space are simply different mappings from the sample space Ω to the real line, as depicted in Figure 5.7. However, in practice, we do not usually worry about the underlying sample space (which can be very complicated), and simply specify the *joint distribution* of these random variables, which provides information sufficient to compute the probabilities of events involving these random variables.

In the following, suppose that $X_1, \ldots, X_n$ are random variables defined on a common probability space; we can also represent them as an n-dimensional *random vector* $\mathbf{X} = (X_1, \ldots, X_n)^T$.

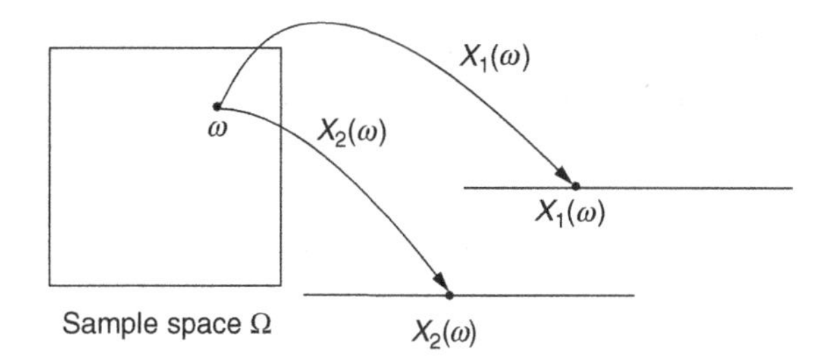

Figure 5.7 Multiple random variables defined on a common probability space.

Joint cumulative distribution function The joint CDF is defined as

$$F_{\mathbf{X}}(\mathbf{x}) = F_{X_1,\ldots,X_n}(x_1,\ldots,x_n) = P[X_1 \leq x_1,\ldots,X_n \leq x_n]$$

Joint probability density function When the joint CDF is continuous, we can define the joint PDF as follows:

$$p_{\mathbf{X}}(\mathbf{x}) = p_{X_1,\ldots,X_n}(x_1,\ldots,x_n) = \frac{\partial}{\partial x_1}\cdots\frac{\partial}{\partial x_n}F_{X_1,\ldots,X_n}(x_1,\ldots,x_n)$$

We can recover the joint CDF from the joint PDF by integrating:

$$F_{\mathbf{X}}(\mathbf{x}) = F_{X_1,\ldots,X_n}(x_1,\ldots,x_n) = \int_{-\infty}^{x_1}\cdots\int_{-\infty}^{x_n} p_{X_1,\ldots,X_n}(u_1,\ldots,u_n)du_1\ldots du_n$$

The joint PDF must be nonnegative and must integrate to one over n-dimensional space. The probability of a particular subset of n-dimensional space is obtained by integrating the joint PDF over the subset.

Joint probability mass function (PMF) For discrete random variables, the joint PMF is defined as

$$p_{\mathbf{X}}(\mathbf{x}) = p_{X_1,\ldots,X_n}(x_1,\ldots,x_n) = P[X_1 = x_1,\ldots,X_n = x_n]$$

Marginal distributions The marginal distribution for a given random variable (or set of random variables) can be obtained by integrating or summing over all possible values of the random variables that we are not interested in. For CDFs, this simply corresponds to setting the appropriate arguments in the joint CDF to infinity. For example,

$$F_X(x) = P[X \leq x] = P[X \leq x, Y \leq \infty] = F_{X,Y}(x,\infty)$$

For continuous random variables, the marginal PDF is obtained from the joint PDF by "integrating out" the undesired random variable:

$$p_X(x) = \int_{-\infty}^{\infty} p_{X,Y}(x,y)dy, \quad -\infty < x < \infty$$

For discrete random variables, we sum over the possible values of the undesired random variable:

$$p_X(x) = \sum_{y \in \mathcal{Y}} p_{X,Y}(x,y), \quad x \in \mathcal{X}$$

where $\mathcal{X}$ and $\mathcal{Y}$ denote the sets of possible values taken by X and Y, respectively.

Example 5.3.1 (Joint and marginal densities) Random variables X and Y have a joint density given by

$$p_{X,Y}(x,y) = \begin{cases} cxy, & 0 \leq x, y \leq 1 \\ 2cxy, & -1 \leq x, y \leq 0 \\ 0, & \text{else} \end{cases}$$

where the constant c is not specified.

(a) Find the value of c.
(b) Find $P[X + Y < 1]$.
(c) Specify the marginal distribution of X.

Solution

(a) We find the constant using the observation that the joint density must integrate to one:

$$\begin{aligned} 1 &= \int\int p_{X,Y}(x,y) dx\, dy \\ &= c\int_0^1 \int_0^1 xy\, dx\, dy + 2c \int_{-1}^0 \int_{-1}^0 xy\, dx\, dy \\ &= c\frac{x^2}{2}\Big|_0^1 \frac{y^2}{2}\Big|_0^1 + 2c\, \frac{x^2}{2}\Big|_{-1}^0 \frac{y^2}{2}\Big|_{-1}^0 = 3c/4 \end{aligned}$$

Thus, $c = 4/3$.

(b) The required probability is obtained by integrating the joint density over the shaded area in Figure 5.8. We obtain

$$\begin{aligned} P[X + Y < 1] &= \int_{y=0}^1 \int_{x=0}^{1-y} cxy\, dx\, dy + \int_{y=-1}^0 \int_{x=-1}^0 2c\, dx\, dy \\ &= c\int_{y=0}^1 \left(\frac{x^2}{2}\Big|_0^{1-y}\right) y\, dy + 2c\frac{x^2}{2}\Big|_{-1}^0 \frac{y^2}{2}\Big|_{-1}^0 \\ &= c\int_{y=0}^1 y\frac{(1-y)^2}{2}\, dy + 2c/4 = c/24 + c/2 = 13c/24 \\ &= 13/18 \end{aligned}$$

We could have computed this probability more quickly in this example by integrating the joint density over the unshaded area to find $P[X + Y \geq 1]$, since this area has a simpler shape:

$$\begin{aligned} P[X + Y \geq 1] &= \int_{y=0}^1 \int_{x=1-y}^1 cxy\, dx\, dy = c\int_{y=0}^1 \left(\frac{x^2}{2}\Big|_{1-y}^1\right) y\, dy \\ &= (c/2)\int_{y=0}^1 y(2y - y^2) dy = 5c/24 = 5/18 \end{aligned}$$

from which we get that $P[X + Y < 1] = 1 - P[X + Y \geq 1] = 13/18$.

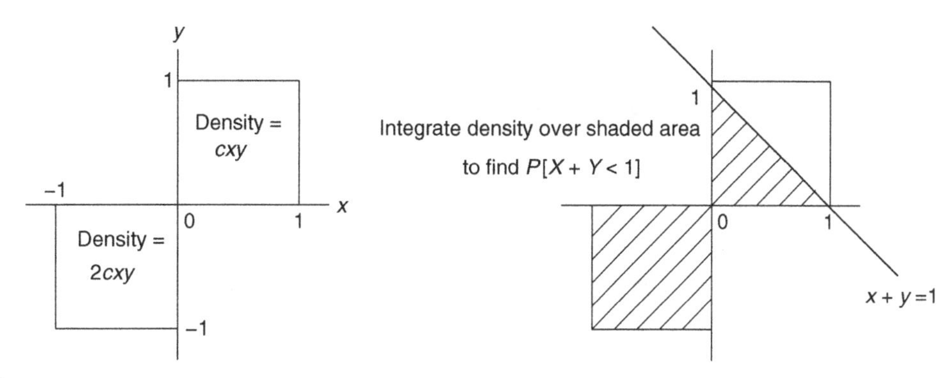

Figure 5.8 The joint density in Example 5.3.1.

(c) The marginal density of X is found by integrating the joint density over all possible values of Y. For $0 \leq x \leq 1$, we obtain

$$p_X(x) = \int_0^1 cxy\, dy = cx\frac{y^2}{2}\Big|_{y=0}^{1} = cx/2 = 2x/3 \qquad (5.22)$$

For $-1 \leq x \leq 0$, we have

$$p_X(x) = \int_0^1 2cxy\, dy = 2cx\frac{y^2}{2}\Big|_{y=0}^{1} = -cx = -4x/3 \qquad (5.23)$$

Conditional density The conditional density of Y given X is defined as

$$p_{Y|X}(y|x) = \frac{p_{X,Y}(x,y)}{p_X(x)} \qquad (5.24)$$

where the definition applies both for PDFs and for PMFs, and where we are interested in values of x such that $p_X(x) > 0$. For jointly continuous X and Y, the conditional density $p(y|x)$ has the interpretation

$$p_{Y|X}(y|x)\Delta y \approx P\left[Y \in [y, y+\Delta y] \middle| X \in [x, x+\Delta x]\right] \qquad (5.25)$$

for Δx and Δy small. For discrete random variables, the conditional PMF is simply the following conditional probability:

$$p_{Y|X}(y|x) = P[Y = y|X = x] \qquad (5.26)$$

Example 5.3.2 Continuing with Example 5.3.1, let us find the conditional density of Y given X. For $X = x \in [0, 1]$, we have $p_{X,Y}(x, y) = cxy$, with $0 \leq y \leq 1$ (the joint density is zero for other values of y, under this conditioning on X). By applying (5.24), and substituting (5.22) into it, we obtain

$$p_{Y|X}(y|x) = \frac{p_{X,Y}(x,y)}{p_X(x)} = \frac{cxy}{cx/2} = 2y,\ 0 \leq y \leq 1 \text{ (for } 0 \leq x \leq 1)$$

Similarly, for $X = x \in [-1, 0]$, we obtain, using (5.23), that

$$p_{Y|X}(y|x) = \frac{p_{X,Y}(x,y)}{p_X(x)} = \frac{2cxy}{-cx} = -2y, \quad -1 \leq y \leq 0 \text{ (for } -1 \leq x \leq 0)$$

We can now compute conditional probabilities using the preceding conditional densities. For example,

$$P[Y < -0.5|X = -0.5] = \int_{-1}^{-0.5} (-2y)dy = -y^2\Big|_{-1}^{-0.5} = 3/4$$

whereas $P[Y < 0.5|X = -0.5] = 1$ (why?).

Bayes' rule for conditional densities Given the conditional density of Y given X, the conditional density for X given Y is given by

$$\begin{aligned} p_{X|Y}(x|y) &= \frac{p_{Y|X}(y|x)p_X(x)}{p_Y(y)} \\ &= \frac{p_{Y|X}(y|x)p_X(x)}{\int p_{Y|X}(y|x)p_X(x)dx}, \quad \textbf{continuous random variables} \\ p_{X|Y}(x|y) &= \frac{p_{Y|X}(y|x)p_X(x)}{p_Y(y)} \\ &= \frac{p_{Y|X}(y|x)p_X(x)}{\sum_x p_{Y|X}(y|x)p_X(x)}, \quad \textbf{discrete random variables} \end{aligned}$$

We can also mix discrete and continuous random variables in applying Bayes' rule, as illustrated in the following example.

Example 5.3.3 (Conditional probability and Bayes' rule with discrete and continuous random variables) A bit sent by a transmitter is modeled as a random variable X taking values 0 and 1 with equal probability. The corresponding observation at the receiver is modeled by a real-valued random variable Y. The conditional distribution of Y given $X = 0$ is $N(0, 4)$. The conditional distribution of Y given $X = 1$ is $N(10, 4)$. This might happen, for example, with on–off signaling, where we send a signal to send 1, and send nothing when we want to send 0. The receiver therefore sees signal plus noise if 1 is sent; and sees only noise if 0 is sent; and the observation Y, presumably obtained by processing the received signal, has zero mean if 0 is sent, and nonzero mean if 1 is sent.

(a) Write down the conditional densities of Y given $X = 0$ and $X = 1$, respectively.
(b) Find $P[Y = 7|X = 0]$, $P[Y = 7|X = 1]$, and $P[Y = 7]$.
(c) Find $P[Y \geq 7|X = 0]$.
(d) Find $P[Y \geq 7|X = 1]$.
(e) Find $P[X = 0|Y = 7]$.

Solution *to (a)*. We simply plug in numbers into the expression (5.21) for the Gaussian density to obtain

$$p(y|x=0) = \frac{1}{\sqrt{8\pi}} e^{-y^2/8}, \quad p(y|x=1)dy = \frac{1}{\sqrt{8\pi}} e^{-(y-10)^2/8}$$

Solution *to (b)*. Conditioned on $X = 0$, Y is a continuous random variable, so the probability of taking a particular value is zero. Thus, $P[Y = 7|X = 0] = 0$. By the same reasoning, $P[Y = 7|X = 1] = 0$. The unconditional probability is given by the law of total probability:

$$P[Y=7] = P[Y=7|X=0]P[X=0] + P[Y=7|X=1]P[X=1] = 0$$

Solution *to (c)*. Finding the probability of Y lying in a region, conditioned on $X = 0$, simply involves integrating the conditional density over that region. We therefore have

$$P[Y \geq 7|X=0] = \int_7^\infty p(y|x=0)dy = \int_7^\infty \frac{1}{\sqrt{8\pi}} e^{-y^2/8}\, dy$$

We shall see in Section 5.6 how to express such probabilities involving Gaussian densities in compact form using standard functions (which can be evaluated using built-in functions in MATLAB), but, for now, we leave the desired probability in terms of the integral given above.

Solution *to (d)*. This is analogous to (c), except that we integrate the conditional probability of Y given $X = 1$:

$$P[Y \geq 7|X=1] = \int_7^\infty p(y|x=1)dy = \int_7^\infty \frac{1}{\sqrt{8\pi}} e^{-(y-10)^2/8}\, dy$$

Solution *to (e)*. Now we want to apply Bayes' rule to find $P[X = 0|Y = 7]$. But we know from (b) that the event $\{Y = 7\}$ has zero probability. How do we condition on an event that never happens? The answer is that we define $P[X = 0|Y = 7]$ to be the limit of $P[X = 0|Y \in (7-\epsilon, 7+\epsilon)]$ as $\epsilon \to 0$. For any $\epsilon > 0$, the event that we are conditioning on is $\{Y \in (7-\epsilon, 7+\epsilon)\}$, and we can show by methods beyond our present scope that one does get a well-defined limit as ϵ tends to zero. However, we do not need to worry about such technicalities when computing this conditional probability: we can simply compute it (for an arbitrary value of $Y = y$) as

$$P[X=0|Y=y] = \frac{p_{Y|X}(y|0)P[X=0]}{p_Y(y)} = \frac{p_{Y|X}(y|0)P[X=0]}{p_{Y|X}(y|0)P[X=0] + p_{Y|X}(y|1)P[X=1]}$$

On substituting the conditional densities from (a) and setting $P[X = 0] = P[X = 1] = 1/2$, we obtain

$$P[X=0|Y=y] = \frac{\frac{1}{2}e^{-y^2/8}}{\frac{1}{2}e^{-y^2/8} + \frac{1}{2}e^{-(y-10)^2/8}} = \frac{1}{1+e^{5(y-5)/2}}$$

By plugging $y = 7$ into this, we obtain

$$P[X=0|Y=7] = 0.0067$$

which of course implies that

$$P[X = 1|Y = 7] = 1 - P[X = 0|Y = 7] = 0.9933$$

Before seeing Y, we knew only that either 0 or 1 was sent with equal probability. After seeing $Y = 7$, however, our model tells us that 1 was far more likely to have been sent. This is of course what we want in a reliable communication system: we begin by not knowing the transmitted information at the receiver (otherwise there would be no point in sending it), but, after seeing the received signal, we can infer it with high probability. We shall see many more such computations in the next chapter: conditional distributions and probabilities are fundamental to principled receiver design.

Independent random variables Random variables $X_1, \ldots, X_n$ are independent if

$$P[X_1 \in A_1, \ldots, X_n \in A_n] = P[X_1 \in A_1] \ldots P[X_n \in A_n]$$

for any subsets $A_1, \ldots, A_n$. That is, events defined in terms of values taken by these random variables are independent of each other. This implies, for example, that the conditional probability of an event defined in terms of one of these random variables, conditioned on events defined in terms of the other random variables, equals the unconditional probability:

$$P[X_1 \in A_1 | X_2 \in A_2, \ldots, X_n \in A_n] = P[X_1 \in A_1]$$

In terms of distributions and densities, independence means that joint distributions are products of marginal distributions, and joint densities are products of marginal densities.

The joint distribution is the product of the marginals for independent random variables If $X_1, \ldots, X_n$ are independent, then their joint CDF is a product of the marginal CDFs:

$$F_{X_1,\ldots,X_n}(x_1, \ldots, x_n) = F_{X_1}(x_1) \ldots F_{X_n}(x_n)$$

and their joint density (PDF or PMF) is a product of the marginal densities:

$$p_{X_1,\ldots,X_n}(x_1, \ldots, x_n) = p_{X_1}(x_1) \ldots p_{X_n}(x_n)$$

Independent and identically distributed (i.i.d.) random variables We are often interested in collections of independent random variables in which each random variable has the same marginal distribution. We call such random variables independent and identically distributed.

Example 5.3.4 (A sum of i.i.d. Bernoulli random variables is a binomial random variable) Let $X_1, \ldots, X_n$ denote i.i.d. Bernoulli random variables with $P[X_1 = 1] = 1 - P[X_1 = 0] = p$, and let $Y = X_1 + \cdots + X_n$ denote their sum. We could think of X_i denoting whether the ith coin flip (of a possibly biased coin, if $p \neq \frac{1}{2}$) yields heads, where successive flips have independent outcomes, so that Y is the number of heads obtained in n flips. For communications applications, X_i could denote whether the ith bit in a sequence of n bits is incorrectly received, with successive bit errors modeled as independent, so that Y is the total number of bit errors. The random variable Y takes discrete values in $\{0, 1, \ldots, n\}$. Its PMF is given by

$$P[Y = k] = \binom{n}{k} p^k (1-p)^{n-k}, \qquad k = 0, 1, \ldots, n$$

That is, $Y \sim \text{Bin}(n, p)$. To see why, note that $Y = k$ requires that exactly k of the $\{X_i\}$ take the value 1, with the remaining $n - k$ taking the value 0. Let us compute the probability of one such outcome, $\{X_1 = 1, \ldots, X_k = 1, X_{k+1} = 0, \ldots, X_n = 0\}$:

$$\begin{aligned} P[X_1 = 1, \ldots, X_k = 1, X_{k+1} = 0, \ldots, X_n = 0] \\ = P[X_1 = 1] \ldots P[X_k = 1] P[X_{k+1} = 0] \ldots P[X_n = 0] \\ = p^k (1-p)^{n-k} \end{aligned}$$

Clearly, any other outcome with exactly k ones has the same probability, given the i.i.d. nature of the $\{X_i\}$. We can now sum over the probabilities of these mutually exclusive events, noting that there are exactly "n choose k" such outcomes (the number of ways in which we can choose the k random variables $\{X_i\}$ which take the value one) to obtain the desired PMF.

Density of sum of independent random variables Suppose that X_1 and X_2 are independent continuous random variables, and let $Y = X_1 + X_2$. Then the PDF of Y is a convolution of the PDFs of X_1 and X_2:

$$p_Y(y) = (p_{X_1} * p_{X_2})(y) = \int_{-\infty}^{\infty} p_{X_1}(x_1) p_{X_2}(y - x_1) dx_1$$

For discrete random variables, the same result holds, except that the PMF is given by a discrete-time convolution of the PMFs of X_1 and X_2.

Example 5.3.5 (Sum of two uniform random variables) Suppose that X_1 is uniformly distributed over $[0, 1]$, and X_2 is uniformly distributed over $[-1, 1]$. Then $Y = X_1 + X_2$ takes values in the interval $[-1, 2]$, and its density is the convolution shown in Figure 5.9.

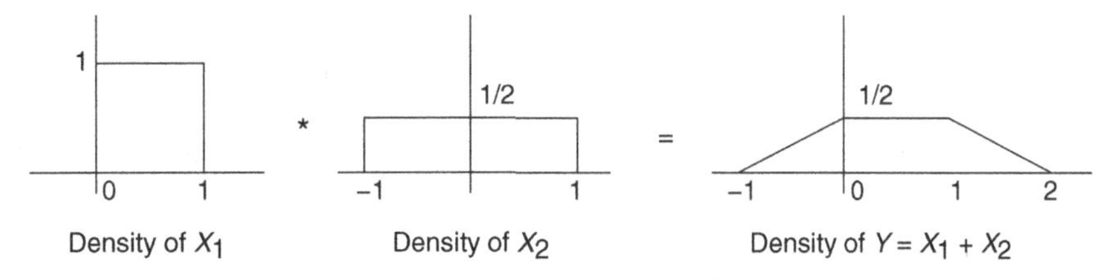

Figure 5.9 The sum of two independent uniform random variables has a PDF with trapezoidal shape, obtained by convolving two boxcar-shaped PDFs.

Of particular interest to us are jointly Gaussian random variables, which we discuss in more detail in Section 5.6.

Notational simplification In the preceding definitions, we have distinguished between different random variables by using subscripts. For example, the joint density of X and Y is denoted by $p_{X,Y}(x,y)$, where X, Y denote the random variables, and x, y are dummy variables that we might, for example, integrate over when evaluating a probability. We could easily use some other notation for the dummy variables, e.g., the joint density could be denoted as $p_{X,Y}(u,v)$. After all, we know that we are talking about the joint density of X and Y because of the subscripts. However, carrying around the subscripts is cumbersome. Therefore, from now on, when there is no scope for confusion, we drop the subscripts and use the dummy variables to also denote the random variables we are talking about. For example, we now use $p(x,y)$ as shorthand for $p_{X,Y}(x,y)$, choosing the dummy variables to be lower-case versions of the random variables they are associated with. Similarly, we use $p(x)$ to denote the density of X, $p(y)$ to denote the density of Y, and $p(y|x)$ to denote the conditional density of Y given X. Of course, we revert to the subscript-based notation whenever there is any possibility of confusion.

5.4 Functions of random variables

We review here methods of determining the distributions of functions of random variables. If $X = X(\omega)$ is a random variable, so is $Y(\omega) = g(X(\omega))$, since it is a mapping from the sample space to the real line which is a composition of the original mapping X and the function g, as shown in Figure 5.10.

Method 1 (Find the CDF first) We proceed from the definition to find the CDF of $Y = g(X)$ as follows:

$$F_Y(y) = P[Y \leq y] = P[g(X) \leq y] = P[X \in A(y)]$$

where $A(y) = \{x : g(x) \leq y\}$. We can now use the CDF or density of X to evaluate the extreme right-hand side. Once we have found the CDF of Y, we can find the PMF or PDF as usual.

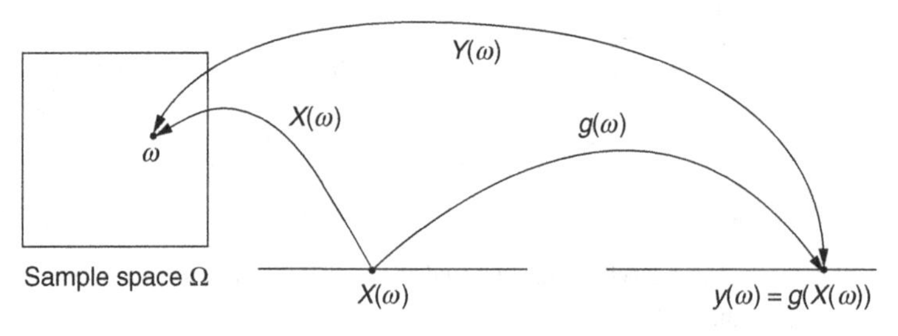

Figure 5.10 A function of a random variable is also a random variable.

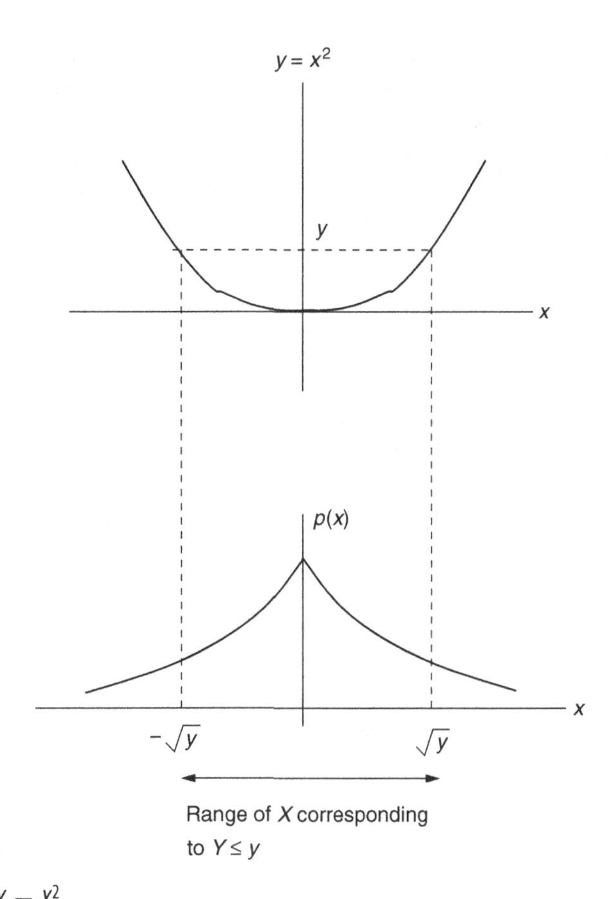

Figure 5.11 Finding the CDF of $Y = X^2$.

Example 5.4.1 (Application of Method 1) Suppose that X is a Laplacian random variable with density

$$p_X(x) = \frac{1}{2}e^{-|x|}$$

Find the CDF and PDF of $Y = X^2$.

In Method 1, we find the CDF of Y first, and then differentiate to find the PDF. First, note that Y takes only nonnegative values, so that $F_Y(y) = 0$ for $y < 0$. For $y \geq 0$, we have

$$\begin{aligned} F_Y(y) &= P[Y \leq y] = P[X^2 \leq y] = P[-\sqrt{y} \leq X \leq \sqrt{y}] \\ &= \int_{-\sqrt{y}}^{\sqrt{y}} p_X(x)dx = \int_{-\sqrt{y}}^{\sqrt{y}} \frac{1}{2}e^{-|x|}\,dx = \int_0^{\sqrt{y}} e^{-x}\,dx \\ &= 1 - e^{-\sqrt{y}}, \quad y \geq 0 \end{aligned}$$

We can now differentiate the CDF to obtain the PDF of Y:

$$p_Y(y) = \frac{d}{dy}F_Y(y) = \frac{e^{-\sqrt{y}}}{2\sqrt{y}}, \quad y \geq 0$$

(The CDF and PDF are zero for $y < 0$, since Y takes only nonnegative values.) See Figure 5.11.

Method 2 (Find the PDF directly) For differentiable $g(x)$ and continuous random variables, we can compute the PDF directly. Suppose that $g(x) = y$ is satisfied for $x = x_1, \ldots, x_m$. We can then express x_i as a function of y: $x_i = h_i(y)$ For $g(x) = x^2$, this corresponds to $x_1 = \sqrt{y}$ and $x_2 = -\sqrt{y}$. The probability of X lying in a small interval $[x_i, x_i + \Delta x]$ is approximately $p_X(x_i)\Delta x$, where we take the increment $\Delta x > 0$. For smooth g, this corresponds to Y lying in a small interval around y, where we need to sum up the probabilities corresponding to all possible values of x that get us near the desired value of y. We therefore get

$$p_Y(y)|\Delta y| = \sum_{i=1}^{m} p_X(x_i)\Delta x$$

where we take the magnitude of the Y increment Δy because a positive increment in x can cause a positive or negative increment in $g(x)$, depending on the slope at that point. We therefore obtain

$$p_Y(y) = \sum_{i=1}^{m} \frac{p_X(x_i)}{|dy/dx|}\Big|_{x_i = h_i(y)} \tag{5.27}$$

We now redo Example 5.4.1 using Method 2.

Example 5.4.2 (Application of Method 2) For the setting of Example 5.4.1, we wish to find the PDF using Method 2. For $y = g(x) = x^2$, we have $x = \pm\sqrt{y}$ (we consider only $y \geq 0$, since the PDF is zero for $y < 0$), with derivative $dy/dx = 2x$. We can now apply (5.27) to obtain

$$p_Y(y) = \frac{p_X(\sqrt{y})}{|2\sqrt{y}|} + \frac{p_X(-\sqrt{y})}{|-2\sqrt{y}|} = \frac{e^{-\sqrt{y}}}{2\sqrt{y}}, \quad y \geq 0$$

as before.

Since Method 1 starts from the definition of the CDF, it generalizes to multiple random variables (i.e., random vectors) in a straightforward manner, at least in principle. For example, suppose that $Y_1 = g_1(X_1, X_2)$ and $Y_2 = g_2(X_1, X_2)$. Then the joint CDF of Y_1 and Y_2 is given by

$$\begin{aligned} F_{Y_1,Y_2}(y_1, y_2) &= P[Y_1 \leq y_1, Y_2 \leq y_2] \\ &= P[g_1(X_1, X_2) \leq y_1, g_2(X_1, X_2) \leq y_2] \\ &= P[(X_1, X_2) \in A(y_1, y_2)] \end{aligned}$$

where $A(y_1, y_2) = \{(x_1, x_2) : g_1(x_1, x_2) \leq y_1, g_2(x_1, x_2) \leq y_2\}$. In principle, we can now use the joint distribution to compute the preceding probability for each possible value of (y_1, y_2). In general, Method 1 works for $\mathbf{Y} = \mathbf{g}(\mathbf{X})$, where $\mathbf{Y}$ is an n-dimensional random vector that is a function of an m-dimensional random vector $\mathbf{X}$ (in the preceding, we considered $m = n = 2$). However, evaluating probabilities involving m-dimensional random vectors can get pretty complicated even for $m = 2$. A generalization of Method 2 is often preferred as a way of directly obtaining PDFs when the functions involved are smooth enough, and when $m = n$. We review this next.

Method 2 for random vectors Suppose that $\mathbf{Y} = (Y_1, \ldots, Y_n)^T$ is an $n \times 1$ random vector that is a function of another $n \times 1$ vector, $\mathbf{X} = (X_1, \ldots, X_n)^T$. That is, $\mathbf{Y} = \mathbf{g}(\mathbf{X})$, or $Y_k = g_k(X_1, \ldots, X_n)$, $k = 1, .., n$. As before, suppose that $\mathbf{y} = g(\mathbf{x})$ has m solutions, $\mathbf{x}_1, \ldots, \mathbf{x}_m$, with the ith solution written in terms of $\mathbf{y}$ as $\mathbf{x}_i = \mathbf{h}_i(\mathbf{y})$. The probability of $\mathbf{Y}$ lying in an infinitesimal volume is now given by

$$p_{\mathbf{Y}}(\mathbf{y})|d\mathbf{y}| = \sum_{i=1}^{m} p_{\mathbf{X}}(\mathbf{x}_i)|d\mathbf{x}|$$

In order to relate the lengths of the *vector* increments $|d\mathbf{y}|$ and $|d\mathbf{x}|$, it no longer suffices to consider a scalar derivative. We now need the *Jacobian matrix* of partial derivatives of $\mathbf{y} = \mathbf{g}(\mathbf{x})$ with respect to $\mathbf{x}$, which is defined as

$$\mathbf{J}(\mathbf{y}; \mathbf{x}) = \begin{pmatrix} \partial y_1/\partial x_1 & \cdots & \partial y_1/\partial x_n \\ \vdots & & \vdots \\ \partial y_n/\partial x_1 & \cdots & \partial y_n/\partial x_n \end{pmatrix} \tag{5.28}$$

The lengths of the vector increments are related by

$$|d\mathbf{y}| = |\det(\mathbf{J}(\mathbf{y}; \mathbf{x}))||d\mathbf{x}|$$

where $\det(\mathbf{M})$ denotes the determinant of a square matrix $\mathbf{M}$. Thus, if $\mathbf{y} = g(\mathbf{x})$ has m solutions, $\mathbf{x}_1, \ldots, \mathbf{x}_m$, with the ith solution written in terms of $\mathbf{y}$ as $\mathbf{x}_i = \mathbf{h}_i(\mathbf{y})$, then the density at $\mathbf{y}$ is given by

$$p_{\mathbf{Y}}(\mathbf{y}) = \sum_{i=1}^{m} \frac{p_{\mathbf{X}}(\mathbf{x}_i)}{|\det(\mathbf{J}(\mathbf{y}; \mathbf{x}))|}\Bigg|_{\mathbf{x}_i = \mathbf{h}_i(\mathbf{y})} \tag{5.29}$$

Depending on how the functional relationship between $\mathbf{X}$ and $\mathbf{Y}$ is specified, it might sometimes be more convenient to find the Jacobian of $\mathbf{x}$ with respect to $\mathbf{y}$:

$$\mathbf{J}(\mathbf{x}; \mathbf{y}) = \begin{pmatrix} \partial x_1/\partial y_1 & \cdots & \partial x_1/\partial y_n \\ \vdots & & \vdots \\ \partial x_n/\partial y_1 & \cdots & \partial x_n/\partial y_n \end{pmatrix} \tag{5.30}$$

We can use this in (5.29) by noting that the two Jacobian matrices for a given pair of values $(\mathbf{x}, \mathbf{y})$ are inverses of each other:

$$\mathbf{J}(\mathbf{x}; \mathbf{y}) = (\mathbf{J}(\mathbf{y}; \mathbf{x}))^{-1}$$

This implies that their determinants are reciprocals of each other:

$$\det(\mathbf{J}(\mathbf{x}; \mathbf{y})) = \frac{1}{\det(\mathbf{J}(\mathbf{y}; \mathbf{x}))}$$

We can therefore rewrite (5.29) as follows:

$$p_{\mathbf{Y}}(\mathbf{y}) = \sum_{i=1}^{m} p_{\mathbf{X}}(\mathbf{x}_i)|\det(\mathbf{J}(\mathbf{x}; \mathbf{y}))|\Bigg|_{\mathbf{x}_i = \mathbf{h}_i(\mathbf{y})} \tag{5.31}$$

Example 5.4.3 (Rectangular to polar transformation) For random variables X_1 and X_2 with joint density p_{X_1,X_2}, think of (X_1, X_2) as a point in two-dimensional space in Cartesian coordinates. The corresponding polar coordinates are given by

$$R = \sqrt{X_1^2 + X_2^2}, \quad \Phi = \tan^{-1}\left(\frac{X_2}{X_1}\right) \tag{5.32}$$

(a) Find the general expression for the joint density $p_{R,\Phi}$.
(b) Specialize to a situation in which X_1 and X_2 are i.i.d. $N(0, 1)$ random variables.

Solution, *part (a)*. Finding the Jacobian involves taking partial derivatives in (5.32). However, in this setting, taking the Jacobian the other way around, as in (5.30), is simpler:

$$x_1 = r\cos\phi, \quad x_2 = r\sin\phi$$

so that

$$\mathbf{J}(\text{rect; polar}) = \begin{pmatrix} \partial x_1/\partial r & \partial x_1/\partial\phi \\ \partial x_2/\partial r & \partial x_2/\partial\phi \end{pmatrix} = \begin{pmatrix} \cos\phi & -r\sin\phi \\ \sin\phi & r\cos\phi \end{pmatrix}$$

We see that

$$\det(\mathbf{J}(\text{rect; polar})) = r\left(\cos^2\phi + \sin^2\phi\right) = r$$

On noting that the rectangular–polar transformation is one-to-one, we have from (5.31) that

$$\begin{aligned} p_{R,\Phi}(r,\phi) &= p_{X_1,X_2}(x_1,x_2)|\det(\mathbf{J}(\text{rect; polar}))|\Big|_{x_1=r\cos\phi, x_2=r\sin\phi} \\ &= r p_{X_1,X_2}(r\cos\phi, r\sin\phi), \quad r \geq 0, 0 \leq \phi \leq 2\pi \end{aligned} \tag{5.33}$$

Solution, *part (b)*. For X_1, X_2 i.i.d. $N(0, 1)$, we have

$$p_{X_1,X_2}(x_1,x_2) = p_{X_1}(x_1)p_{X_2}(x_2) = \frac{1}{\sqrt{2\pi}}e^{-x_1^2/2}\frac{1}{\sqrt{2\pi}}e^{-x_2^2/2}$$

By plugging the above into (5.33) and simplifying, we obtain

$$p_{R,\Phi}(r,\phi) = \frac{r}{2\pi}e^{-r^2/2}, \quad r \geq 0, 0 \leq \phi \leq 2\pi$$

We can find the marginal densities of R and Φ by integrating out the other variable, but, in this case, we can find them by inspection, since the joint density clearly decomposes into a product of functions of r and ϕ alone. With appropriate normalization, each of these functions is a marginal density. We can now infer that R and Φ are independent, with

$$p_R(r) = re^{-r^2/2}, \quad r \geq 0$$

and

$$p_\Phi(\phi) = \frac{1}{2\pi}, \quad 0 \leq \phi \leq 2\pi$$

The amplitude R in this case follows a Rayleigh distribution, while the phase Φ is uniformly distributed over $[0, 2\pi]$.

5.5 Expectation

We now discuss computation of statistical averages, which are often the performance measures on the basis of which a system design is evaluated.

Expectation The expectation, or statistical average, of a function of a random variable X is defined as

$$\begin{aligned} \mathbb{E}[g(X)] &= \int g(x)p(x)dx, \quad \textbf{continuous random variable} \\ \mathbb{E}[g(X)] &= \sum g(x)p(x), \quad \textbf{discrete random variable} \end{aligned} \tag{5.34}$$

Note that the expectation of a deterministic constant, therefore, is simply the constant itself.

Expectation is a linear operator We have

$$\mathbb{E}[a_1X_1 + a_2X_2 + b] = a_1\mathbb{E}[X_1] + a_2\mathbb{E}[X_2] + b$$

where a_1, a_2, and b are any constants.

Mean The mean of a random variable X is $\mathbb{E}[X]$.

Variance The variance of a random variable X is a measure of how much it fluctuates around its mean:

$$\text{var}(X) = \mathbb{E}\left[(X - \mathbb{E}[X])^2\right] \tag{5.35}$$

On expanding out the square, we have

$$\text{var}(X) = \mathbb{E}\left[X^2 - 2X\mathbb{E}[X] + (\mathbb{E}[X])^2\right]$$

Using the linearity of expectation, we can simplify to obtain the following alternative formula for the variance:

$$\text{var}(X) = \mathbb{E}[X^2] - (\mathbb{E}[X])^2 \tag{5.36}$$

The square root of the variance is called the *standard deviation.*

Effect of scaling and translation For $Y = aX + b$, it is left as an exercise to show that

$$\begin{aligned} \mathbb{E}[Y] &= \mathbb{E}[aX + b] = a\mathbb{E}[X] + b \\ \text{var}(Y) &= a^2\text{var}(X) \end{aligned} \tag{5.37}$$

Normalizing to zero mean and unit variance We can specialize (5.37) to $Y = X - \mathbb{E}[X]/\sqrt{\text{var}(X)}$, to see that $\mathbb{E}[Y] = 0$ and $\text{var}(Y) = 1$.

Example 5.5.1 (PDF after scaling and translation) If X has density $p_X(x)$, then $Y = (X - a)/b$ has density

$$p_Y(y) = |b| p_X(by + a) \tag{5.38}$$

This follows from a straightforward application of Method 2 in Section 5.4. Specializing to a Gaussian random variable $X \sim N(m, v^2)$ with mean m and variance v^2 (we review mean and variance later), consider a normalized version $Y = (X - m)/v$. On applying (5.38) to the Gaussian density, we obtain

$$p_Y(y) = \frac{1}{\sqrt{2\pi}} e^{-y^2/2}$$

which can be recognized as an $N(0, 1)$ density. Thus, if $X \sim N(m, v^2)$, then $Y = (X - m)/v \sim N(0, 1)$ is a standard Gaussian random variable. This enables us to express probabilities involving Gaussian random variables compactly in terms of the CDF and CCDF of a standard Gaussian random variable, as we see later when we deal extensively with Gaussian random variables when modeling digital communication systems.

Moments The nth moment of a random variable X is defined as $\mathbb{E}[X^n]$. From (5.36), we see that specifying the mean and variance is equivalent to specifying the first and second moments. Indeed, it is worth rewriting (5.36) as an explicit reminder that the second moment is the sum of the mean and variance:

$$\mathbb{E}[X^2] = (\mathbb{E}[X])^2 + \text{var}(X) \tag{5.39}$$

Example 5.5.2 (Moments of an exponential random variable) Suppose that $X \sim Exp(\lambda)$. We compute its mean using integration by parts, as follows:

$$\begin{aligned} \mathbb{E}[X] &= \int_0^\infty x\lambda e^{-\lambda x}\, dx = -xe^{-\lambda x}\Big|_0^\infty + \int_0^\infty \frac{d}{dx} x e^{-\lambda x}\, dx \\ &= \int_0^\infty e^{-\lambda x}\, dx = -\frac{e^{-\lambda x}}{\lambda}\Big|_0^\infty \\ &= 1/\lambda \end{aligned} \tag{5.40}$$

Similarly, using integration by parts twice, we can show that

$$\mathbb{E}[X^2] = \frac{2}{\lambda^2}$$

Using (5.36), we obtain

$$\text{var}(X) = \mathbb{E}[X^2] - (\mathbb{E}[X])^2 = \frac{1}{\lambda^2} \tag{5.41}$$

In general, we can use repeated integration by parts to evaluate higher moments of the exponential random variable to obtain

$$\mathbb{E}[X^n] = \int_0^\infty x^n \lambda e^{-\lambda x}\, dx = \frac{n!}{\lambda^n}, \quad n = 1, 2, 3, \ldots$$

(A proof of the preceding formula using mathematical induction is left as an exercise.)

As a natural follow-up to the computations in the preceding example, let us introduce the gamma function, which is useful for evaluating integrals associated with expectation computations for several important random variables.

Gamma function The gamma function, $\Gamma(x)$, is defined as

$$\Gamma(x) = \int_0^\infty t^{x-1} e^{-t}\, dt, \quad x > 0$$

In general, integration by parts can be used to show that

$$\Gamma(x+1) = x\Gamma(x), \quad x > 0 \tag{5.42}$$

Noting that $\Gamma(1) = 1$, we can now use induction to specify the gamma function for integer arguments:

$$\Gamma(n) = (n-1)!, \quad n = 1, 2, 3, \ldots \tag{5.43}$$

This is exactly the same computation as we did in Example 5.5.2: $\Gamma(n)$ equals the $(n-1)$th moment of an exponential random variable with $\lambda = 1$ (and hence mean $1/\lambda = 1$).

The gamma function can also be computed for non-integer arguments. Just as integer arguments of the gamma function are useful for exponential random variables, "integer-plus-half" arguments are useful for evaluating the moments of Gaussian random variables. We can evaluate these using (5.42) given the value of the gamma function at $x = 1/2$:

$$\Gamma(1/2) = \int_0^\infty t^{-1/2} e^{-t}\, dt = \sqrt{\pi} \tag{5.44}$$

For example, we can infer that

$$\Gamma(5/2) = (3/2)(1/2)\Gamma(1/2) = \frac{3}{4}\sqrt{\pi}$$

Example 5.5.3 (Mean and variance of a Gaussian random variable) We now show that $X \sim N(m, v^2)$ has mean m and variance v^2. The mean of X is given by the following expression:

$$\mathbb{E}[X] = \int_{-\infty}^\infty x \frac{1}{\sqrt{2\pi v^2}} e^{-(x-m)^2/(2v^2)}\, dx$$

Let us first consider the change of variables $t = (x-m)/v$, so that $dx = v\, dt$. Then

$$\mathbb{E}[X] = \int_{-\infty}^\infty (tv + m) \frac{1}{\sqrt{2\pi v^2}} e^{-t^2/2} v\, dt$$

Note that $te^{-t^2/2}$ is an odd function, and therefore integrates out to zero over the real line. We therefore obtain

$$\mathbb{E}[X] = \int_{-\infty}^{\infty} m \frac{1}{\sqrt{2\pi v^2}} e^{-t^2/2} v\, dt = m \int_{-\infty}^{\infty} \frac{1}{\sqrt{2\pi}} e^{-t^2/2}\, dt = m$$

recognizing that the integral on the extreme right-hand side is the $N(0, 1)$ PDF, which must integrate to one. The variance is given by

$$\mathrm{var}(X) = \mathbb{E}[(X - m)^2] = \int_{-\infty}^{\infty} (x - m)^2 \frac{1}{\sqrt{2\pi v^2}} e^{-(x-m)^2/(2v^2)}\, dx$$

With a change of variables $t = (x - m)/v$ as before, we obtain

$$\mathrm{var}(X) = v^2 \int_{-\infty}^{\infty} t^2 \frac{1}{\sqrt{2\pi}} e^{-t^2/2}\, dt = 2v^2 \int_{0}^{\infty} t^2 \frac{1}{\sqrt{2\pi}} e^{-t^2/2}\, dt$$

since the integrand is an even function of t. On substituting $z = t^2/2$, so that $dz = t\, dt = \sqrt{2z}\, dt$, we obtain

$$\mathrm{var}(X) = 2v^2 \int_{0}^{\infty} 2z \frac{1}{\sqrt{2\pi}} e^{-z} \frac{dz}{\sqrt{2z}} = 2v^2 \frac{1}{\sqrt{\pi}} \int_{0}^{\infty} z^{1/2} e^{-z}\, dz$$
$$= 2v^2 (1/\sqrt{\pi}) \Gamma(3/2) = v^2$$

since $\Gamma(3/2) = (1/2)\Gamma(1/2) = \sqrt{\pi}/2$.

The change of variables in the computations in the preceding example is actually equivalent to transforming the random variable $N(m, v^2)$ that we started with to a standard Gaussian random variable $N(0, 1)$ as in Example 5.5.1. As we mentioned earlier (this is important enough to be worth repeating), when we handle Gaussian random variables more extensively in later chapters, we prefer making the transformation up front when computing probabilities, rather than changing variables inside integrals.

As a final example, we show that the mean of a Poisson random variable with parameter λ is equal to λ.

Example 5.5.4 (Mean of a Poisson random variable) The mean is given by

$$\mathbb{E}[X] = \sum_{k=0}^{\infty} k P[X = k] = \sum_{k=1}^{\infty} k \frac{\lambda^k}{k!} e^{-\lambda}$$

where we have dropped the $k = 0$ term from the extreme right-hand side, since it does not contribute to the mean. On noting that $k/k! = 1/((k - 1)!)$, we have

$$\mathbb{E}[X] = \sum_{k=1}^{\infty} \frac{\lambda^k}{(k-1)!} e^{-\lambda} = \lambda e^{-\lambda} \sum_{k=1}^{\infty} \frac{\lambda^{k-1}}{(k-1)!} = \lambda$$

since

$$\sum_{k=1}^{\infty} \frac{\lambda^{k-1}}{(k-1)!} = \sum_{l=0}^{\infty} \frac{\lambda^l}{l!} = e^{\lambda}$$

where we set $l = k - 1$ to get an easily recognized form for the series expansion of an exponential.

5.5.1 Expectation for random vectors

So far, we have talked about expectations involving a single random variable. Expectations with multiple random variables are defined in exactly the same way as in (5.34), but replacing the scalar random variable and the corresponding dummy variable for summation or integration by a vector:

$$\begin{aligned}
\mathbb{E}[g(\mathbf{X})] &= \mathbb{E}[g(X_1,\ldots,X_n)] = \int g(\mathbf{x})p(\mathbf{x})d\mathbf{x}, \;\; \textbf{jointly continuous random variables} \\
&= \int_{x_1=-\infty}^{\infty} \cdots \int_{x_n=-\infty}^{\infty} g(x_1,\ldots,x_n)p(x_1,\ldots,x_n)\,dx_1\ldots dx_n \\
\mathbb{E}[g(\mathbf{X})] &= \mathbb{E}[g(X_1,\ldots,X_n)] = \sum_{\mathbf{x}} g(\mathbf{x})p(\mathbf{x}), \;\; \textbf{discrete random variables} \\
&= \sum_{x_1} \cdots \sum_{x_n} g(x_1,\ldots,x_n)p(x_1,\ldots,x_n)
\end{aligned} \tag{5.45}$$

Product of expectations for independent random variables When the random variables involved are independent, and the function whose expectation is to be evaluated decomposes into a product of functions of each individual random variable, then the preceding computation involves a product of expectations, each involving only one random variable:

$$\mathbb{E}[g_1(X_1)\ldots g_n(X_n)] = \mathbb{E}[g_1(X_1)]\ldots\mathbb{E}[g_n(X_n)], \;\; X_1,\ldots,X_n \;\textbf{independent} \tag{5.46}$$

Example 5.5.5 (Computing an expectation involving independent random variables) Suppose that $X_1 \sim N(1,1)$ and $X_2 \sim N(-3,4)$ are independent. Find $\mathbb{E}[(X_1+X_2)^2]$.

Solution

We have

$$\mathbb{E}[(X_1+X_2)^2] = \mathbb{E}[X_1^2 + X_2^2 + 2X_1X_2]$$

We can now use linearity to compute the expectations of each of the three terms on the right-hand side separately. We obtain $\mathbb{E}[X_1^2] = (\mathbb{E}[X_1])^2 + \text{var}(X_1) = 1^2 + 1 = 2$, $\mathbb{E}[X_2^2] = (\mathbb{E}[X_2])^2 + \text{var}(X_2) = (-3)^2 + 4 = 13$, and $\mathbb{E}[2X_1X_2] = 2\mathbb{E}[X_1]\mathbb{E}[X_2] = 2(1)(-3) = -6$, so

$$\mathbb{E}[(X_1+X_2)^2] = 2 + 13 - 6 = 9$$

The variance is a measure of how a random variable fluctuates around its means. The covariance, defined next, is a measure of how the fluctuations of two random variables around their means are correlated.

Covariance The covariance of X_1 and X_2 is defined as

$$\text{cov}(X_1,X_2) = \mathbb{E}[(X_1 - \mathbb{E}[X_1])(X_2 - \mathbb{E}[X_2])] \tag{5.47}$$

As with the variance, we can also obtain the following alternative formula:

$$\mathrm{cov}(X_1, X_2) = \mathbb{E}[X_1 X_2] - \mathbb{E}[X_1]\mathbb{E}[X_2] \tag{5.48}$$

The variance is the covariance of a random variable with itself It is immediately evident from the definition that

$$\mathrm{var}(X) = \mathrm{cov}(X, X)$$

Uncorrelated random variables Random variables X_1 and X_2 are said to be uncorrelated if $\mathrm{cov}(X_1, X_2) = 0$.

Independent random variables are uncorrelated If X_1 and X_2 are independent, then they are uncorrelated. This is easy to see from (5.48), since $\mathbb{E}[X_1 X_2] = \mathbb{E}[X_1]\mathbb{E}[X_2]$ using (5.46).

Uncorrelated random variables need not be independent Consider $X \sim N(0, 1)$ and $Y = X^2$. We see that $\mathbb{E}[XY] = \mathbb{E}[X^3] = 0$ by virtue of the symmetry of the $N(0, 1)$ density around the origin, so

$$\mathrm{cov}(X, Y) = \mathbb{E}[XY] - \mathbb{E}[X]\mathbb{E}[Y] = 0$$

Clearly, X and Y are not independent, since knowing the value of X determines the value of Y.

As we discuss in the next section, uncorrelated *jointly Gaussian* random variables are indeed independent. The joint distribution of such random variables is determined by means and covariances, hence we also postpone a more detailed discussion of covariance computation until our study of joint Gaussianity.

5.6 Gaussian random variables

We begin by repeating the definition of a Gaussian random variable.

Gaussian random variable The random variable X is said to follow a *Gaussian* or *normal*, distribution if its density is of the form

$$p(x) = \frac{1}{\sqrt{2\pi v^2}} \exp\left(-\frac{(x-m)^2}{2v^2}\right), \quad -\infty < x < \infty \tag{5.49}$$

where $m = \mathbb{E}[X]$ is the mean of X and $v^2 = \mathrm{var}(X)$ is the variance of X. The Gaussian density is therefore completely characterized by its mean and variance.

Notation for Gaussian distribution We use $N(m, v^2)$ to denote a Gaussian distribution with mean m and variance v^2, and use the shorthand $X \sim N(m, v^2)$ to denote that a random variable X follows this distribution.

We have already noted the characteristic bell shape of the Gaussian PDF in the example plotted in Figure 5.5: the bell is centered around the mean, and its width is determined by the variance. We now develop a detailed framework for efficient computations involving Gaussian random variables.

Standard Gaussian random variable A zero-mean, unit-variance Gaussian random variable, $X \sim N(0,1)$, is termed a standard Gaussian random variable.

An important property of Gaussian random variables is that they remain Gaussian under scaling and translation. Suppose that $X \sim N(m, v^2)$. Define $Y = aX + b$, where a and b are constants (assume $a \neq 0$ to avoid triviality). The density of Y can be found as follows:

$$p(y) = \left. \frac{p(x)}{|dy/dx|} \right|_{x=(y-b)/a}$$

On noting that $dy/dx = a$, and plugging (5.49) into this, we obtain

$$\begin{aligned} p(y) &= \frac{1}{|a|} \frac{1}{\sqrt{2\pi v^2}} \exp\left(\frac{-((y-b)/a - m)^2}{2v^2} \right) \\ &= \frac{1}{\sqrt{2\pi a^2 v^2}} \exp\left(\frac{-(y-(am+b))^2}{2a^2v^2} \right) \end{aligned}$$

On comparing this with (5.49), we can see that Y is also Gaussian, with mean $m_Y = am + b$ and variance $v_Y^2 = a^2v^2$. This is important enough to summarize and restate.

Gaussianity is preserved under scaling and translation

If $X \sim N(m, v^2)$, then $Y = aX + b \sim N(am + b, a^2v^2)$.

As a consequence of the preceding result, any Gaussian random variable can be scaled and translated to obtain a "standard" Gaussian random variable with zero mean and unit variance. For $X \sim N(m, v^2)$, $Y = aX + b \sim N(0,1)$ if $am + b = 0$ and $a^2v^2 = 1$ to have $a = v$ and $b = -vm$. That is, $Y = (X - m)/v \sim N(0,1)$.

Standard Gaussian random variable

A standard Gaussian random variable $N(0,1)$ has mean zero and variance one.

Conversion of a Gaussian random variable into standard form

If $X \sim N(m, v^2)$, then $(X - m)/v \sim N(0,1)$.

As the following example illustrates, this enables us to express probabilities involving any Gaussian random variable as probabilities involving a standard Gaussian random variable.

Example 5.6.1 Suppose that $X \sim N(5, 9)$. Then $(X - 5)/\sqrt{9} = (X - 5)/3 \sim N(0, 1)$. Any probability involving X can now be expressed as a probability involving a standard Gaussian random variable. For example,

$$P[X > 11] = P[(X - 5)/3 > (11 - 5)/3] = P[N(0, 1) > 2]$$

We therefore set aside special notation for the cumulative distribution function (CDF) $\Phi(x)$ and complementary cumulative distribution function (CCDF) $Q(x)$ of a standard Gaussian random variable. By virtue of the standard form conversion, we can now express probabilities involving any Gaussian random variable in terms of the Φ or Q functions. The definitions of these functions are illustrated in Figure 5.12, and the corresponding formulas are specified below.

$$\Phi(x) = P[N(0, 1) \leq x] = \int_{-\infty}^{x} \frac{1}{\sqrt{2\pi}} \exp\left(-\frac{t^2}{2}\right) dt \tag{5.50}$$

$$Q(x) = P[N(0, 1) > x] = \int_{x}^{\infty} \frac{1}{\sqrt{2\pi}} \exp\left(-\frac{t^2}{2}\right) dt \tag{5.51}$$

See Figure 5.13 for a plot of these functions. By definition, $\Phi(x) + Q(x) = 1$. Furthermore, by virtue of the symmetry of the Gaussian density around zero, $Q(-x) = \Phi(x)$. On combining these observations, we note that $Q(-x) = 1 - Q(x)$, so it suffices to consider only positive arguments for the Q function in order to compute probabilities of interest.

Let us now consider a few more Gaussian probability computations.

Example 5.6.2 X is a Gaussian random variable with mean $m = -5$ and variance $v^2 = 4$. Find expressions in terms of the Q function with positive arguments for the following probabilities: $P[X > 3]$, $P[X < -8]$, $P[X < -1]$, $P[3 < X < 6]$, and $P[X^2 - 2X > 15]$.

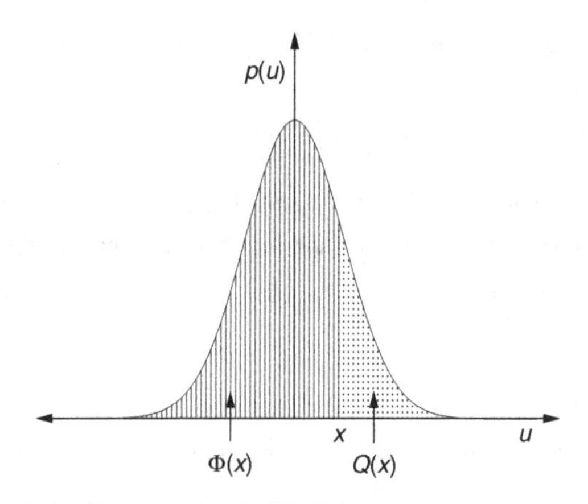

Figure 5.12 The Φ and Q functions are obtained by integrating the $N(0, 1)$ density over appropriate intervals.

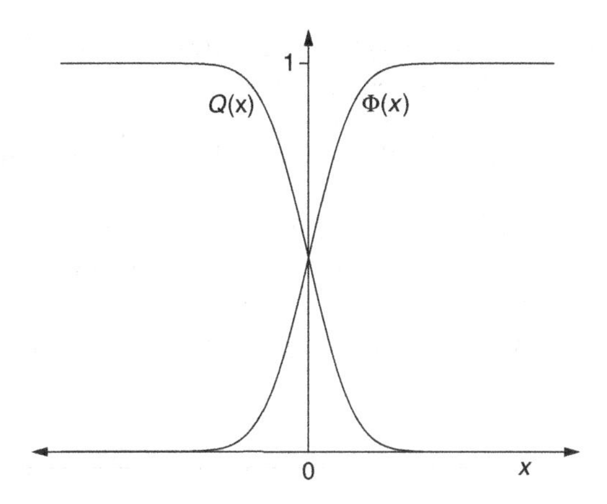

Figure 5.13 The Φ and Q functions.

Solution

We solve this problem by normalizing X to a standard Gaussian random variable $(X-m)/v = (X+5)/2$:

$$P[X > 3] = P\left[\frac{X+5}{2} > \frac{3+5}{2} = 4\right] = Q(4)$$
$$P[X < -8] = P\left[\frac{X+5}{2} < \frac{-8+5}{2} = -1.5\right] = \Phi(-1.5) = Q(1.5)$$
$$P[X < -1] = P\left[\frac{X+5}{2} < \frac{-1+5}{2} = 2\right] = \Phi(2) = 1 - Q(2)$$
$$P[3 < X < 6] = P\left[4 = \frac{3+5}{2} < \frac{X+5}{2} < \frac{6+5}{2} = 5.5\right]$$
$$= \Phi(5.5) - \Phi(4) = ((1 - Q(5.5)) - (1 - Q(4))) = Q(4) - Q(5.5)$$

Computation of the probability that $X^2 - 2X > 15$ requires that we express this event in terms of simpler events by factorization:

$$X^2 - 2X - 15 = X^2 - 5X + 3X - 15 = (X-5)(X+3)$$

This shows that $X^2 - 2X > 15$, or $X^2 - 2X - 15 > 0$, if and only if $X - 5 > 0$ and $X + 3 > 0$, or $X - 5 < 0$ and $X + 3 < 0$. The first event simplifies to $X > 5$, and the second to $X < -3$, so the desired probability is a union of two mutually exclusive events. We therefore have

$$P[X^2 - 2X > 15] = P[X > 5] + P[X < -3] = Q\left(\frac{5+5}{2}\right) + \Phi\left(\frac{-3+5}{2}\right)$$
$$= Q(5) + \Phi(1) = Q(5) + 1 - Q(1)$$

Interpreting the transformation to standard Gaussian For $X \sim N(m, v^2)$, the transformation to standard Gaussian tells us that

$$P[X > m + \alpha v] = P\left[\frac{X - m}{v} > \alpha\right] = Q(\alpha)$$

That is, the tail probability of a Gaussian random probability depends only on the number of standard deviations α away from the mean. More generally, the transformation is equivalent to the observation that the probability of an infinitesimal interval $[x, x + \Delta x]$ depends only on its normalized distance from the mean, $(x - m)/v$, and its normalized length $\Delta x/v$:

$$P[x \leq X \leq x + \Delta x] \approx p(x)\Delta x = \frac{1}{\sqrt{2\pi}} \exp\left(-\left(\frac{x - m}{v}\right)^2 \Big/ 2\right) \frac{\Delta x}{v}$$

Relating the Q function to the error function Mathematical software packages such as MATLAB often list the error function and the complementary error function, which is defined for $x \geq 0$ by

$$\text{erf}(x) = \frac{2}{\sqrt{\pi}} \int_0^x e^{-t^2}\, dt$$

$$\text{erfc}(x) = 1 - \text{erf}(x) = \frac{2}{\sqrt{\pi}} \int_x^{\infty} e^{-t^2}\, dt$$

On recognizing the form of the $N(0, \frac{1}{2})$ density, given by $(1/\sqrt{\pi})e^{-t^2}$, we see that

$$\text{erf}(x) = 2P[0 \leq X \leq x], \qquad \text{erfc}(x) = 2P[X > x]$$

where $X \sim N(0, \frac{1}{2})$. On transforming to standard Gaussian as usual, we see that

$$\text{erfc}(x) = 2P[X > x] = 2P\left[\frac{X - 0}{\sqrt{1/2}} > \frac{x - 0}{\sqrt{1/2}}\right] = 2Q\left(x\sqrt{2}\right)$$

We can invert this to compute the Q function for positive arguments in terms of the complementary error function, as follows:

$$Q(x) = \frac{1}{2}\,\text{erfc}\left(\frac{x}{\sqrt{2}}\right), \quad x \geq 0 \tag{5.52}$$

For $x < 0$, we can compute $Q(x) = 1 - Q(-x)$ using the preceding equation to evaluate the right-hand side. While the Communications System Toolbox in MATLAB has the Q function built in as *qfunc*$(\cdot)$, we provide a MATLAB code fragment for computing the Q function that is based on the complementary error function (which is available without subscription to separate toolboxes) below.

Code Fragment 5.6.1 (Computing the Q function)

```
%Q function computed using erfc (works for vector inputs)
function z = qfunction(x)
     b = (x >= 0);
     y1 = b.* x; %select the positive entries of x
     y2 = (1 - b).* (-x); %select, and flip the sign of, negative entries in x
     z1 = (0.5 * erfc(y1./sqrt(2))).* b; %Q(x) for positive entries in x
```

```
z2 = (1 - 0.5 * erfc(y2./sqrt(2))).* (1-b);
%Q(x) = 1 - Q(-x) for negative entries in x
z = z1 + z2; %final answer (works for x with positive or negative entries)
```

Example 5.6.3 (Binary on–off keying in Gaussian noise) A received sample Y in a communication system is modeled as follows: $Y = m + N$ if 1 is sent, and $Y = N$ if 0 is sent, where $N \sim N(0, v^2)$ is the contribution of the receiver noise to the sample and $|m|$ is a measure of the signal strength. Assuming that $m > 0$, suppose that we use the simple decision rule that splits the difference between the average values of the observation under the two scenarios: say that 1 is sent if $Y > m/2$ and 0 is sent if $Y \leq m/2$. Assuming that 0 and 1 are equally likely to be sent, the signal power is $(1/2)m^2 + (1/2)0^2 = m^2/2$. The noise power is $\mathbb{E}[N^2] = v^2$. Thus, $\text{SNR} = m^2/(2v^2)$.

(a) What is the conditional probability of error, conditioned on 0 being sent?
(b) What is the conditional probability of error, conditioned on 1 being sent?
(c) What is the (unconditional) probability of error if 0 and 1 are equally likely to have been sent?
(d) What is the error probability for SNR of 13 dB?

Solution

(a) Since $Y \sim N(0, v^2)$ given that 0 is sent, the conditional probability of error is given by

$$P_{e|0} = P[\text{say } 1|0 \text{ sent}] = P[Y > m/2|0 \text{ sent}] = Q\left(\frac{m/2 - 0}{v}\right) = Q\left(\frac{m}{2v}\right)$$

(b) Since $Y \sim N(m, v^2)$ given that 1 is sent, the conditional probability of error is given by

$$P_{e|1} = P[\text{say } 0|1 \text{ sent}] = P[Y \leq m/2|1 \text{ sent}] = \Phi\left(\frac{0 - m/2}{v}\right) = \Phi\left(-\frac{m}{2v}\right) = Q\left(\frac{m}{2v}\right)$$

(c) If π_0 is the probability of sending 0, then the unconditional error probability is given by

$$P_e = \pi_0 P_{e|0} + (1 - \pi_0)P_{e|1} = Q\left(\frac{m}{2v}\right) = Q\left(\sqrt{\text{SNR}/2}\right)$$

regardless of π_0 for this particular decision rule.

(d) For SNR of 13 dB, we have $\text{SNR(raw)} = 10^{\text{SNR(db)}/10} = 10^{1.3} \approx 20$, so the error probability evaluates to $P_e = Q(\sqrt{10}) = 7.8 \times 10^{-4}$.

Figure 5.14 shows the probability of error on a log scale, plotted against the SNR in dB. This is the first example of the many error probability plots that we will see in this chapter.

A MATLAB code fragment (with cosmetic touches omitted) for generating Figure 5.14 in Example 5.6.3 is as below.

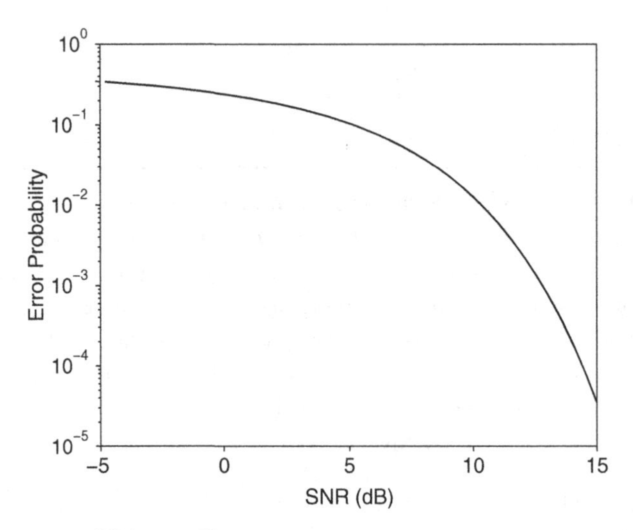

Figure 5.14 The probability of error versus SNR for on–off keying.

Code Fragment 5.6.2 (Error probability computation and plotting)

```
%Plot of error probability versus SNR for on-off keying
snrdb = -5:0.1:15; %vector of SNRs (in dB) for which to evaluate error probability
snr = 10.^(snrdb/10); %vector of raw SNRs
pe = qfunction(sqrt(snr/2)); %vector of error probabilities
%plot error probability on log scale versus SNR in dB
semilogy(snrdb,pe);
ylabel('Error Probability');
xlabel('SNR (dB)');
```

The preceding example illustrates a more general observation for signaling in AWGN: the probability of error involves terms such as $Q(\sqrt{a \times \text{SNR}})$, where the scale factor a depends on properties of the signal constellation, and SNR is the signal-to-noise ratio. It is therefore of interest to understand how the error probability decays with the SNR. As shown in Appendix 5.A, there are tight analytical bounds for the Q function, which can be used to deduce that it decays exponentially with its argument, as stated in the following.

Asymptotics of *Q(x)* for large arguments For large $x > 0$, the exponential decay of the Q function dominates. We denote this by

$$Q(x) \doteq e^{-x^2/2}, \quad x \to \infty \tag{5.53}$$

which is shorthand for the following limiting result:

$$\lim_{x \to \infty} \frac{\log Q(x)}{-x^2/2} = 1 \tag{5.54}$$

These asymptotics play a key role in the design of communication systems. Since events that cause bit errors have probabilities involving terms such as $Q(\sqrt{a \times \text{SNR}}) \doteq e^{-a \times \text{SNR}/2}$, when there are multiple events that can cause bit errors, the ones with the smallest rates of decay a dominate performance. We can therefore focus on these worst-case events in our designs for moderate and high SNR. This simplistic view does not quite hold in heavily coded systems operating at low SNR, but is still an excellent perspective for arriving at a coarse link design.

5.6.1 Joint Gaussianity

Often, we need to deal with multiple Gaussian random variables defined on the same probability space. These might arise, for example, when we sample filtered WGN. In many situations of interest, such random variables are not only individually Gaussian, but also satisfy a stronger *joint* Gaussianity property. Just as a Gaussian random variable is characterized by its mean and variance, jointly Gaussian random variables are characterized by means and covariances. We are also interested in what happens to these random variables under linear operations corresponding, for example, to filtering. Hence, we first review the mean and the covariance, and their evolution under linear operations and translations, for arbitrary random variables defined on the same probability space.

Covariance The covariance of random variables X_1 and X_2 measures the correlation between how they vary around their means, and is given by

$$\text{cov}(X_1, X_2) = \mathbb{E}[(X_1 - \mathbb{E}[X_1])(X_2 - \mathbb{E}[X_2])] = \mathbb{E}[X_1 X_2] - \mathbb{E}[X_1]\mathbb{E}[X_2]$$

The second formula is obtained from the first by multiplying out and simplifying:

$$\begin{aligned}\mathbb{E}\left[(X_1 - \mathbb{E}[X_1])(X_2 - \mathbb{E}[X_2])\right] &= \mathbb{E}[X_1 X_2 - \mathbb{E}[X_1]X_2 + \mathbb{E}[X_1]\mathbb{E}[X_2] - X_1\mathbb{E}[X_2]] \\ &= \mathbb{E}[X_1 X_2] - \mathbb{E}[X_1]\mathbb{E}[X_2] + \mathbb{E}[X_1]\mathbb{E}[X_2] - \mathbb{E}[X_1]\mathbb{E}[X_2] \\ &= \mathbb{E}[X_1 X_2] - \mathbb{E}[X_1]\mathbb{E}[X_2]\end{aligned}$$

where we use the linearity of the expectation operator to pull out constants.

Uncorrelatedness X_1 and X_2 are said to be uncorrelated if $\text{cov}(X_1, X_2) = 0$.

Independent random variables are uncorrelated If X_1 and X_2 are independent, then

$$\text{cov}(X_1, X_2) = \mathbb{E}[X_1 X_2] - \mathbb{E}[X_1]\mathbb{E}[X_2] = \mathbb{E}[X_1]\mathbb{E}[X_2] - \mathbb{E}[X_1]\mathbb{E}[X_2] = 0$$

The converse is not true in general; that is, uncorrelated random variables need not be independent. However, we shall see that *jointly Gaussian* uncorrelated random variables are indeed independent.

Variance Note that the variance of a random variable is its covariance with itself:

$$\text{var}(X) = \text{cov}(X, X) = \mathbb{E}\left[(X - \mathbb{E}[X])^2\right] = \mathbb{E}[X^2] - (\mathbb{E}[X])^2$$

The use of matrices and vectors provides a compact way of representing and manipulating means and covariances, especially using software programs such as MATLAB. Thus, for random variables $X_1, \ldots, X_m$, we define the random vector $\mathbf{X} = (X_1, \ldots, X_m)^{\mathrm{T}}$, and arrange the means and pairwise covariances in a vector and matrix, respectively, as follows.

Mean vector and covariance matrix Consider an arbitrary m-dimensional random vector $\mathbf{X} = (X_1, \ldots, X_m)^{\mathrm{T}}$. The $m \times 1$ mean vector of $\mathbf{X}$ is defined as $\mathbf{m}_X = \mathbb{E}[\mathbf{X}] =$

$(\mathbb{E}[X_1], \ldots, \mathbb{E}[X_m])^{\mathrm{T}}$. The $m \times m$ covariance matrix $\mathbf{C}_X$ has (i,j)th entry given by the covariance between the ith and jth random variables:

$$\mathbf{C}_X(i,j) = \mathrm{cov}(X_i, X_j) = \mathbb{E}\left[(X_i - \mathbb{E}[X_i])(X_j - \mathbb{E}[X_j])\right] = \mathbb{E}\left[X_i X_j\right] - \mathbb{E}[X_i]\mathbb{E}[X_j]$$

More compactly,

$$\mathbf{C}_X = \mathbb{E}[(\mathbf{X} - \mathbb{E}[\mathbf{X}])(\mathbf{X} - \mathbb{E}[\mathbf{X}])^{\mathrm{T}}] = \mathbb{E}[\mathbf{X}\mathbf{X}^{\mathrm{T}}] - \mathbb{E}[\mathbf{X}](\mathbb{E}[\mathbf{X}])^{\mathrm{T}}$$

Notes on covariance computation Computations of variance and covariance come up often when we deal with Gaussian random variables, hence it is useful to note the following properties of the covariance.

Property 1. The covariance is unaffected by adding constants:

$$\mathrm{cov}(X + a, Y + b) = \mathrm{cov}(X, Y) \quad \text{for any constants } a, b$$

Covariance provides a measure of the correlation between random variables after subtracting out their means, hence adding constants to the random variables (which just translates their means) does not affect covariance.

Property 2. The covariance is a bilinear function (i.e., it is linear in both its arguments):

$$\begin{aligned} \mathrm{cov}(a_1X_1 + a_2X_2, a_3X_3 + a_4X_4) = {} & a_1a_3\,\mathrm{cov}(X_1, X_3) + a_1a_4\,\mathrm{cov}(X_1, X_4) \\ & + a_2a_3\,\mathrm{cov}(X_2, X_3) + a_2a_4\,\mathrm{cov}(X_2, X_4) \end{aligned}$$

By virtue of Property 1, it is clear that we can always consider zero-mean, or centered, versions of random variables when computing the covariance. An example that frequently arises in performance analysis of communication systems is a random variable that is a sum of a deterministic term (e.g., due to a signal), and a zero-mean random term (e.g. due to noise). In this case, dropping the signal term is often convenient when computing the variance or covariance.

Affine transformations For a random vector $\mathbf{X}$, the analog of scaling and translating a random variable is a linear transformation using a matrix, together with a translation. Such a transformation is called an *affine* transformation. That is, $\mathbf{Y} = \mathbf{A}\mathbf{X} + \mathbf{b}$ is an affine transformation of $\mathbf{X}$, where $\mathbf{A}$ is a deterministic matrix and $\mathbf{b}$ a deterministic vector.

Example 5.6.4 (Mean and variance after an affine transformation) Let $Y = X_1 - 2X_2 + 4$, where X_1 has mean -1 and variance 4, X_2 has mean 2 and variance 9, and the covariance $\mathrm{cov}(X_1, X_2) = -3$. Find the mean and variance of Y.

Solution

The mean is given by

$$\mathbb{E}[Y] = \mathbb{E}[X_1] - 2\mathbb{E}[X_2] + 4 = -1 - 2(2) + 4 = -1$$

The variance is computed as

$$\begin{aligned} \mathrm{var}(Y) &= \mathrm{cov}(Y, Y) = \mathrm{cov}(X_1 - 2X_2 + 4, X_1 - 2X_2 + 4) \\ &= \mathrm{cov}(X_1, X_1) - 2\,\mathrm{cov}(X_1, X_2) - 2\,\mathrm{cov}(X_2, X_1) + 4\,\mathrm{cov}(X_2, X_2) \end{aligned}$$

where the constant drops out because of Property 1. We therefore obtain that

$$\text{var}(Y) = \text{cov}(X_1, X_1) - 4\,\text{cov}(X_1, X_2) + 4\,\text{cov}(X_2, X_2) = 4 - 4(-3) + 4(9) = 52$$

Computations such as those in the preceding example can be compactly represented in terms of matrices and vectors, which is particularly useful for computations for random vectors. In general, an affine transformation maps one random vector into another (of possibly different dimension), and the mean vector and covariance matrix evolve as follows.

Mean and covariance evolution under affine transformation

If $\mathbf{X}$ has mean $\mathbf{m}$ and covariance $\mathbf{C}$, and $\mathbf{Y} = \mathbf{AX} + \mathbf{b}$,
then $\mathbf{Y}$ has mean $\mathbf{m}_Y = \mathbf{Am} + \mathbf{b}$ and covariance $\mathbf{C}_Y = \mathbf{ACA}^T$.

To see this, first compute the mean vector of $\mathbf{Y}$ using the linearity of the expectation operator:

$$\mathbf{m}_Y = \mathbb{E}[\mathbf{Y}] = \mathbb{E}[\mathbf{AX} + \mathbf{b}] = \mathbf{A}\mathbb{E}[\mathbf{X}] + \mathbf{b} = \mathbf{Am} + \mathbf{b} \tag{5.55}$$

This also implies that the "zero mean" version of $\mathbf{Y}$ is given by

$$\mathbf{Y} - \mathbb{E}[\mathbf{Y}] = (\mathbf{AX} + \mathbf{b}) - (\mathbf{Am}_X + \mathbf{b}) = \mathbf{A}(\mathbf{X} - \mathbf{m}_X)$$

so the covariance matrix of $\mathbf{Y}$ is given by

$$\mathbf{C}_Y = \mathbb{E}[(\mathbf{Y} - \mathbb{E}[\mathbf{Y}])(\mathbf{Y} - \mathbb{E}[\mathbf{Y}])^T] = \mathbb{E}[\mathbf{A}(\mathbf{X} - \mathbf{m})(\mathbf{X} - \mathbf{m})^T\mathbf{A}^T] = \mathbf{ACA}^T \tag{5.56}$$

Note that the dimensions of $\mathbf{X}$ and $\mathbf{Y}$ can be different: $\mathbf{X}$ can be $m \times 1$, $\mathbf{A}$ can be $n \times m$, and $\mathbf{Y}$ and $\mathbf{b}$ can be $n \times 1$, where m and n are arbitrary. We also note below that the mean and covariance evolve separately under such transformations.

Mean and covariance evolve separately under affine transformations The mean of $\mathbf{Y}$ depends only on the mean of $\mathbf{X}$, and the covariance of $\mathbf{Y}$ depends only on the covariance of $\mathbf{X}$. Furthermore, the additive constant $\mathbf{b}$ in the transformation does not affect the covariance, since it influences only the mean of $\mathbf{Y}$.

Example 5.6.4 redone We can check that we get the same result as before by setting

$$\mathbf{m}_X = \begin{pmatrix} -1 \\ 2 \end{pmatrix}, \quad \mathbf{C}_X = \begin{pmatrix} 4 & -3 \\ -3 & 9 \end{pmatrix}, \mathbf{A} = (1 \ \ -2), \quad \mathbf{b} = 4 \tag{5.57}$$

and applying (5.55) and (5.56).

Jointly Gaussian random variables, or Gaussian random vectors Random variables $X_1, \ldots, X_m$ defined on a common probability space are said to be *jointly Gaussian,* or the $m \times 1$ random vector $\mathbf{X} = (X_1, \ldots, X_m)^T$ is termed a *Gaussian random vector*, if any linear combination of these random variables is a Gaussian random variable. That is, for any scalar constants $a_1, \ldots, a_m$, the random variable $a_1X_1 + \cdots + a_mX_m$ is Gaussian.

A Gaussian random vector is completely characterized by its mean vector and covariance matrix This is a generalization of the observation that a Gaussian random variable is completely characterized by its mean and variance. We derive this in Problem 5.47, but provide an intuitive argument here. The definition of joint Gaussianity merely requires us to characterize the distribution of an arbitrarily chosen linear combination of $X_1, \ldots, X_m$. For a Gaussian random vector $\mathbf{X} = (X_1, \ldots, X_m)^{\mathrm{T}}$, consider $Y = a_1X_1 + \cdots + a_mX_m$, where $a_1, \ldots, a_m$ can be any scalar constants. By definition, Y is a Gaussian random variable, and is completely characterized by its mean and variance. We can compute these in terms of $\mathbf{m}_X$ and $\mathbf{C}_X$ using (5.55) and (5.56) by noting that $Y = \mathbf{a}^{\mathrm{T}}\mathbf{X}$, where $\mathbf{a} = (a_1, \ldots, a_m)^{\mathrm{T}}$. Thus,

$$m_Y = \mathbf{a}^{\mathrm{T}}\mathbf{m}_X$$

and

$$C_Y = \mathrm{var}(Y) = \mathbf{a}^{\mathrm{T}}\mathbf{C}_X\mathbf{a}$$

We have therefore shown that we can characterize the mean and variance, and hence the density, of an arbitrarily chosen linear combination Y if and only if we know the mean vector $\mathbf{m}_X$ and the covariance matrix $\mathbf{C}_X$. As we shall see in Problem 5.47, this is the basis for the desired result that the distribution of a Gaussian random vector $\mathbf{X}$ is completely characterized by $\mathbf{m}_X$ and $\mathbf{C}_X$.

Notation for joint Gaussianity We use the notation $\mathbf{X} \sim N(\mathbf{m}, \mathbf{C})$ to denote a Gaussian random vector $\mathbf{X}$ with mean vector $\mathbf{m}$ and covariance matrix $\mathbf{C}$.

The preceding definitions and observations regarding joint Gaussianity apply even when the random variables involved do not have a joint density. For example, it is easy to check that, according to this definition, X_1 and $X_2 = 4X_1 - 1$ are jointly Gaussian. However, the joint density of X_1 and X_2 is not well defined (unless we allow delta functions), since all of the probability mass in the two-dimensional (x_1, x_2) plane is collapsed onto the line $x_2 = 4x_1 - 1$. Of course, since X_2 is completely determined by X_1, any probability involving X_1 and X_2 can be expressed in terms of X_1 alone. In general, when the m-dimensional joint density does not exist, probabilities involving $X_1, \ldots, X_m$ can be expressed in terms of a smaller number of random variables, and can be evaluated using a joint density over a lower-dimensional space. A necessary and sufficient condition for the joint density to exist is that the covariance matrix is invertible.

Joint Gaussian density exists if and only if the covariance matrix is invertible We do not prove this result, but discuss it in the context of the two-dimensional density in Example 5.6.5.

Joint Gaussian density For $\mathbf{X} = (X_1, \ldots, X_m) \sim N(\mathbf{m}, \mathbf{C})$, if $\mathbf{C}$ is invertible, the joint density exists and takes the following form (we skip the derivation, but see Problem 5.47):

$$p(x_1, \ldots, x_m) = p(\mathbf{x}) = \frac{1}{\sqrt{(2\pi)^m|\mathbf{C}|}} \exp\left(-\frac{1}{2}(\mathbf{x} - \mathbf{m})^{\mathrm{T}}\mathbf{C}^{-1}(\mathbf{x} - \mathbf{m})\right) \tag{5.58}$$

where $|\mathbf{C}|$ denotes the determinant of $\mathbf{C}$.

Example 5.6.5 (Two-dimensional joint Gaussian density) In order to visualize the joint Gaussian density (this is not needed for the remainder of the development, hence this example can be skipped), let us consider two jointly Gaussian random variables X and Y. In this case, it is convenient to define the *normalized correlation* between X and Y as

$$\rho(X, Y) = \frac{\mathrm{cov}(X, Y)}{\sqrt{\mathrm{var}(X)\mathrm{var}(Y)}} \tag{5.59}$$

Thus, $\mathrm{cov}(X, Y) = \rho\sigma_X\sigma_Y$, where $\mathrm{var}(X) = \sigma_X^2$, $\mathrm{var}(Y) = \sigma_Y^2$, and the covariance matrix for the random vector $(X, Y)^{\mathrm{T}}$ is given by

$$\mathbf{C} = \begin{pmatrix} \sigma_X^2 & \rho\sigma_X\sigma_Y \\ \rho\sigma_X\sigma_Y & \sigma_Y^2 \end{pmatrix} \tag{5.60}$$

It is shown in Problem 5.46 that $|\rho| \leq 1$. For $|\rho| = 1$, it is easy to check that the covariance matrix has determinant zero, hence the joint density formula (5.58) cannot be applied. As shown in Problem 5.46, this has a simple geometric interpretation: $|\rho| = 1$ corresponds to a situation when X and Y are affine functions of each other, so that all of the probability mass is concentrated on a line; hence a two-dimensional density does not exist. Thus, we need the strict inequality $|\rho| < 1$ for the covariance matrix to be invertible. Assuming that $|\rho| < 1$, we plug (5.60) into (5.58), setting the mean vector to zero without loss of generality (a nonzero mean vector simply shifts the density). We get the joint density shown in Figure 5.15 for $\sigma_X^2 = 1$, $\sigma_Y^2 = 4$, and $\rho = -0.5$. Since Y has larger variance, the density decays more slowly in Y than in X. The negative normalized correlation leads to contour plots given by tilted ellipses, corresponding to setting quadratic function $\mathbf{x}^{\mathrm{T}}\mathbf{C}^{-1}\mathbf{x}$ in the exponent of the density to different constants.

Exercise Show that the ellipses shown in Figure 5.15(b) can be described as

$$x^2 + ay^2 + bxy = c$$

specifying the values of a and b.

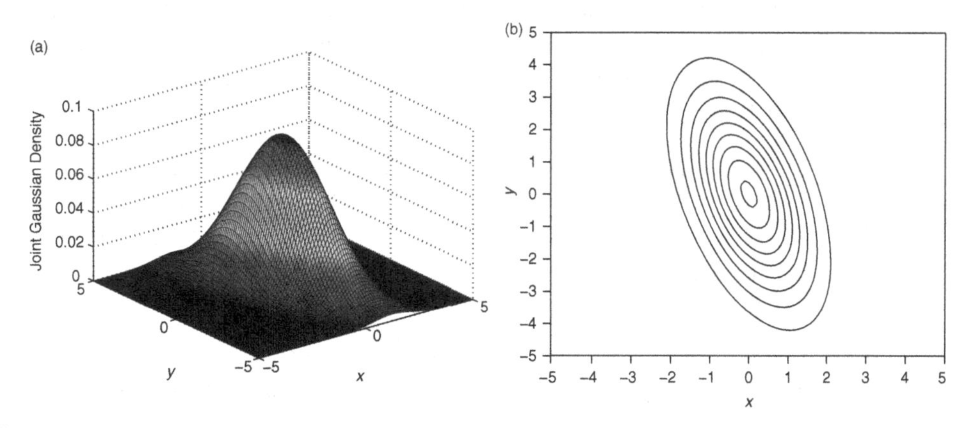

Figure 5.15 The joint Gaussian density (a) and its contours (b) for $\sigma_X^2 = 1, \sigma_Y^2 = 4$, and $\rho = -0.5$.

While we hardly ever integrate the joint Gaussian density to compute probabilities, we use its form to derive many important results. One such result is stated below.

Uncorrelated jointly Gaussian random variables are independent This follows from the form of the joint Gaussian density (5.58). If $X_1, \ldots, X_m$ are pairwise uncorrelated, then the off-diagonal entries of the covariance matrix $\mathbf{C}$ are zero: $\mathbf{C}(i,j) = 0$ for $i \neq j$. Thus, $\mathbf{C}$ and $\mathbf{C}^{-1}$ are both diagonal matrices, with diagonal entries given by $\mathbf{C}(i,i) = v_i^2$, $\mathbf{C}^{-1}(i,i) = 1/v_i^2$, $i = 1, \ldots, m$, and determinant $|\mathbf{C}| = v_1^2 \ldots v_m^2$. In this case, we see that the joint density (5.58) decomposes into a product of marginal densities:

$$p(x_1, \ldots, x_m) = \frac{1}{\sqrt{2\pi v_1^2}} e^{-(x_1-m_1)^2/(2v_1^2)} \ldots \frac{1}{\sqrt{2\pi v_m^2}} e^{-(x_1-m_m)^2/(2v_m^2)} = p(x_1) \ldots p(x_m)$$

so that $X_1, \ldots, X_m$ are independent.

Recall that, while independent random variables are uncorrelated, the converse need not be true. However, when we impose the additional restriction of joint Gaussianity, uncorrelatedness does imply independence.

We can now characterize the distribution of affine transformations of jointly Gaussian random variables. If $\mathbf{X}$ is a Gaussian random vector, then $\mathbf{Y} = \mathbf{AX} + \mathbf{b}$ is also Gaussian. To see this, note that any linear combination of $Y_1, \ldots, Y_n$ equals a linear combination of $X_1, \ldots, X_m$ (plus a constant), which is a Gaussian random variable by virtue of the Gaussianity of $\mathbf{X}$. Since $\mathbf{Y}$ is Gaussian, its distribution is completely characterized by its mean vector and covariance matrix, which we have just computed. We can now state the following result.

Joint Gaussianity is preserved under affine transformations

$$\text{If } \mathbf{X} \sim N(\mathbf{m}, \mathbf{C}), \text{ then } \mathbf{AX} + \mathbf{b} \sim N(\mathbf{Am} + \mathbf{b}, \mathbf{ACA}^T) \qquad (5.61)$$

Example 5.6.6 (Computations with jointly Gaussian random variables) As in Example 5.6.4, consider two random variables X_1 and X_2 such that X_1 has mean -1 and variance 4, X_2 has mean 2 and variance 9, and $\text{cov}(X_1, X_2) = -3$. Now assume in addition that these random variables are jointly Gaussian.

(a) Write down the mean vector and covariance matrix for the random vector $\mathbf{Y} = (Y_1, Y_2)^T$, where $Y_1 = 3X_1 - 2X_2 + 3$ and $Y_2 = X_1 + X_2 - 2$.

(b) Evaluate the probability $P[3X_1 - 2X_2 < 5]$ in terms of the Q function with positive arguments.

(c) Suppose that $Z = aX_1 + X_2$. Find the constant a such that Z is independent of $X_1 + X_2$.

Solution *to (a).* We have already found the mean and covariance of $\mathbf{X}$ in Example 5.6.4; they are given by (5.57). Now, $\mathbf{Y} = \mathbf{AX} + \mathbf{b}$, where

$$\mathbf{A} = \begin{pmatrix} 3 & -2 \\ 1 & 1 \end{pmatrix}, \qquad \mathbf{b} = \begin{pmatrix} 3 \\ -2 \end{pmatrix}$$

We can now apply (5.61) to obtain the mean vector and covariance matrix for $\mathbf{Y}$:

$$\mathbf{m}_Y = \mathbf{A}\mathbf{m}_X + \mathbf{b} = \begin{pmatrix} -4 \\ -1 \end{pmatrix}$$

and

$$\mathbf{C}_Y = \mathbf{A}\mathbf{C}_X\mathbf{A}^{\mathrm{T}} = \begin{pmatrix} 108 & -9 \\ -9 & 7 \end{pmatrix}$$

Solution *to (b).* Since $Y_1 = 3X_1 - 2X_2 + 3 \sim N(-4, 108)$, the required probability can be written as

$$P[3X_1 - 2X_2 < 5] = P[Y_1 < 8] = \Phi\left(\frac{8-(-4)}{\sqrt{108}}\right) = \Phi\left(2/\sqrt{3}\right) = 1 - Q\left(2/\sqrt{3}\right)$$

Solution *to (c).* Since $Z = aX_1 + X_2$ and X_1 are jointly Gaussian, they are independent if they are uncorrelated. The covariance is given by

$$\mathrm{cov}(Z, X_1) = \mathrm{cov}(aX_1 + X_2, X_1) = a\,\mathrm{cov}(X_1, X_1) + \mathrm{cov}(X_2, X_1) = 4a - 3$$

so that we need $a = 3/4$ for Z and X_1 to be independent.

Discrete time WGN The noise model $\mathbf{N} \sim N(0, \sigma^2\mathbf{I})$ is called discrete-time white Gaussian noise (WGN). The term *white* refers to the noise samples being uncorrelated and having equal variance. We will see how such discrete-time WGN arises from continuous-time WGN, which we discuss during our coverage of random processes later in this chapter.

Example 5.6.7 (Binary on–off keying in discrete time WGN) Let us now revisit on–off keying, explored for scalar observations in Example 5.6.3, for vector observations. The receiver processes a vector $\mathbf{Y} = (Y_1, \ldots, Y_n)^{\mathrm{T}}$ of samples modeled as follows: $\mathbf{Y} = \mathbf{s} + \mathbf{N}$ if 1 is sent, and $\mathbf{Y} = \mathbf{N}$ is 0 is sent, where $\mathbf{s} = (s_1, \ldots, s_n)^{\mathrm{T}}$ is the signal, and the noise $\mathbf{N} = (N_1, \ldots, N_n)^{\mathrm{T}} \sim N(0, \sigma^2\mathbf{I})$. That is, the noise samples $N_1, \ldots, N_n$ are i.i.d. $N(0, \sigma^2)$ random variables. Suppose we use the following correlator-based decision statistic:

$$Z = \mathbf{s}^{\mathrm{T}}\mathbf{Y} = \sum_{k=1}^{n} s_k Y_k$$

Thus, we have reduced the vector observation to a single number on the basis of which we will make our decision. The hypothesis framework developed in Chapter 6 will be used to show that this decision statistic is optimal, in a well-defined sense. For now, we simply accept it as given.

(a) Find the conditional distribution of Z given that 0 is sent.
(b) Find the conditional distribution of Z given that 1 is sent.
(c) Observe from (a) and (b) that we are now back to the setting of Example 5.6.3, with Z now playing the role of Y. Specify the values of m and v^2, and the SNR $= m^2/(2v^2)$, in terms of $\mathbf{s}$ and σ^2.
(d) As in Example 5.6.3, consider the simple decision rule that 1 is sent if $Z > m/2$, and say that 0 is sent if $Z \leq m/2$. Find the error probability (in terms of the Q function) as a function of $\mathbf{s}$ and σ^2.
(e) Evaluate the error probability for $\mathbf{s} = (-2, 2, 1)^{\mathrm{T}}$ and $\sigma^2 = 1/4$.

Solution

(a) If 0 is sent, then $\mathbf{Y} = \mathbf{N} \sim N(0, \sigma^2\mathbf{I})$. By applying (5.61) with $\mathbf{m} = 0$, $\mathbf{A} = \mathbf{s}^{\mathrm{T}}$, and $\mathbf{C} = \sigma^2\mathbf{I}$, we obtain $Z = \mathbf{s}^{\mathrm{T}}\mathbf{Y} \sim N(0, \sigma^2||\mathbf{s}||^2)$.
(b) If 1 is sent, then $\mathbf{Y} = \mathbf{s} + \mathbf{N} \sim N(\mathbf{s}, \sigma^2\mathbf{I})$. By applying (5.61) with $\mathbf{m} = \mathbf{s}$, $\mathbf{A} = \mathbf{s}^{\mathrm{T}}$, and $\mathbf{C} = \sigma^2\mathbf{I}$, we obtain $Z = \mathbf{s}^{\mathrm{T}}\mathbf{Y} \sim N(||\mathbf{s}||^2, \sigma^2||\mathbf{s}||^2)$. Alternatively, $\mathbf{s}^{\mathrm{T}}\mathbf{Y} = \mathbf{s}^{\mathrm{T}}(\mathbf{s} + \mathbf{N}) = ||\mathbf{s}||^2 + \mathbf{s}^{\mathrm{T}}\mathbf{N}$. Since $\mathbf{s}^{\mathrm{T}}\mathbf{N} \sim N(0, \sigma^2||\mathbf{s}||^2)$ from (a), we simply translate the mean by $||\mathbf{s}||^2$.
(c) Comparing with Example 5.6.3, we see that $m = ||\mathbf{s}||^2$, $v^2 = \sigma^2||\mathbf{s}||^2$, and SNR $= m^2/(2v^2) = ||\mathbf{s}||^2/(2\sigma^2)$.
(d) From Example 5.6.3, we know that the decision rule that splits the difference between the means has error probability

$$P_{\mathrm{e}} = P_{\mathrm{e}|0} = P_{\mathrm{e}|1} = Q\left(\frac{m}{2v}\right) = Q\left(\frac{||\mathbf{s}||}{2\sigma}\right)$$

plugging in the expressions for m and v^2 from (c).
(e) We have $||\mathbf{s}||^2 = 9$. Using (d), we obtain $P_{\mathrm{e}} = Q(3) = 0.0013$.

Noise is termed *colored* when it is not white; that is, when the noise samples are correlated and/or have different variances. We will see later how colored noise arises from linear transformations on white noise. Let us continue our sequence of examples regarding on–off keying, but now with colored noise.

Example 5.6.8 (Binary on–off keying in discrete-time colored Gaussian noise) As in the previous example, we have a vector observation $\mathbf{Y} = (Y_1, \ldots, Y_n)^{\mathrm{T}}$, with $\mathbf{Y} = \mathbf{s} + \mathbf{N}$ if 1 is sent, and $\mathbf{Y} = \mathbf{N}$ if 0 is sent, where $\mathbf{s} = (s_1, \ldots, s_n)^{\mathrm{T}}$ is the signal. However, we now allow the noise covariance matrix to be arbitrary: $\mathbf{N} = (N_1, \ldots, N_n)^{\mathrm{T}} \sim N(0, \mathbf{C}_N)$.

(a) Consider the decision statistic $Z_1 = \mathbf{s}^{\mathrm{T}}\mathbf{Y}$. Find the conditional distributions of Z_1 given 0 sent, and given 1 sent.

(b) Show that Z_1 follows the scalar on–off keying model Example 5.6.3, specifying the parameters m_1 and v_1^2, and $\mathrm{SNR}_1 = m_1^2/(2v_1^2)$, in terms of $\mathbf{s}$ and $\mathbf{C}_N$.
(c) Find the error probability of the simple decision rule comparing Z_1 with the threshold $m_1/2$.
(d) Repeat (a)–(c) for a decision statistic $Z_2 = \mathbf{s}^{\mathrm{T}}\mathbf{C}_N^{-1}\mathbf{Y}$ (use the notation m_2, v_2^2, and SNR_2 to denote the quantities analogous to those in (b)).
(e) Apply the preceding to the following example: a two-dimensional observation $\mathbf{Y} = (Y_1, Y_2)$ with $\mathbf{s} = (4, -2)^{\mathrm{T}}$ and

$$\mathbf{C}_N = \begin{pmatrix} 1 & -1 \\ -1 & 4 \end{pmatrix}$$

Find explicit expressions for Z_1 and Z_2 in terms of Y_1 and Y_2. Compute and compare the SNRs and error probabilities obtained with the two decision statistics.

Solution

We proceed similarly to Example 5.6.7.

(a) If 0 is sent, then $\mathbf{Y} = \mathbf{N} \sim N(0, \mathbf{C}_N)$. By applying (5.61) with $\mathbf{m} = 0$, $\mathbf{A} = \mathbf{s}^{\mathrm{T}}$, and $\mathbf{C} = \mathbf{C}_N$, we obtain $Z_1 = \mathbf{s}^{\mathrm{T}}\mathbf{Y} \sim N(0, \mathbf{s}^{\mathrm{T}}\mathbf{C}_N\mathbf{s})$.
If 1 is sent, then $\mathbf{Y} = \mathbf{s}+\mathbf{N} \sim N(\mathbf{s}, \mathbf{C}_N)$. By applying (5.61) with $\mathbf{m} = \mathbf{s}$, $\mathbf{A} = \mathbf{s}^{\mathrm{T}}$, and $\mathbf{C} = \mathbf{C}_N$, we obtain $Z_1 = \mathbf{s}^{\mathrm{T}}\mathbf{Y} \sim N(||\mathbf{s}||^2, \mathbf{s}^{\mathrm{T}}\mathbf{C}_N\mathbf{s})$. Alternatively, $\mathbf{s}^{\mathrm{T}}\mathbf{Y} = \mathbf{s}^{\mathrm{T}}(\mathbf{s}+\mathbf{N}) = ||\mathbf{s}||^2 + \mathbf{s}^{\mathrm{T}}\mathbf{N}$. Since $\mathbf{s}^{\mathrm{T}}\mathbf{N} \sim N(0, \mathbf{s}^{\mathrm{T}}\mathbf{C}_N\mathbf{s})$ from (a), we simply translate the mean by $||\mathbf{s}||^2$.
(b) Comparing with Example 5.6.3, we see that $m_1 = ||\mathbf{s}||^2$, $v_1^2 = \mathbf{s}^{\mathrm{T}}\mathbf{C}_N\mathbf{s}$, and $\mathrm{SNR}_1 = m_1^2/(2v_1^2) = ||\mathbf{s}||^2/(2\mathbf{s}^T\mathbf{C}_N\mathbf{s})$.
(c) From Example 5.6.3, we know that the decision rule that splits the difference between the means has error probability

$$P_{e1} = Q\left(\frac{m_1}{2v_1}\right) = Q\left(\frac{||\mathbf{s}||^2}{2\sqrt{\mathbf{s}^{\mathrm{T}}\mathbf{C}_N\mathbf{s}}}\right)$$

plugging in the expressions for m_1 and v_1^2 from (b).
(d) We now have $Z_2 = \mathbf{s}^{\mathrm{T}}\mathbf{C}_N^{-1}\mathbf{Y}$. If 0 is sent, $\mathbf{Y} = \mathbf{N} \sim N(0, \mathbf{C}_N)$. By applying (5.61) with $\mathbf{m} = 0$, $\mathbf{A} = \mathbf{s}^{\mathrm{T}}\mathbf{C}_N^{-1}$, and $\mathbf{C} = \mathbf{C}_N$, we obtain $Z_2 = \mathbf{s}^{\mathrm{T}}\mathbf{Y} \sim N(0, \mathbf{s}^{\mathrm{T}}\mathbf{C}_N^{-1}\mathbf{s})$.
If 1 is sent, then $\mathbf{Y} = \mathbf{s} + \mathbf{N} \sim N(\mathbf{s}, \mathbf{C}_N)$. By applying (5.61) with $\mathbf{m} = \mathbf{s}$, $\mathbf{A} = \mathbf{s}^{\mathrm{T}}\mathbf{C}_N^{-1}$, and $\mathbf{C} = \mathbf{C}_N$, we obtain $Z_2 = \mathbf{s}^{\mathrm{T}}\mathbf{C}_N^{-1}\mathbf{Y} \sim N(\mathbf{s}^{\mathrm{T}}\mathbf{C}_N^{-1}\mathbf{s}, \mathbf{s}^{\mathrm{T}}\mathbf{C}_N^{-1}\mathbf{s})$. That is, $m_2 = \mathbf{s}^{\mathrm{T}}\mathbf{C}_N^{-1}\mathbf{s}$, $v_2^2 = \mathbf{s}^{\mathrm{T}}\mathbf{C}_N^{-1}\mathbf{s}$, and $\mathrm{SNR}_2 = m_2^2/(2v_2^2) = \mathbf{s}^{\mathrm{T}}\mathbf{C}_N^{-1}\mathbf{s}/2$. The corresponding error probability is

$$P_{e2} = Q\left(\frac{m_2}{2v_2}\right) = Q\left(\frac{\sqrt{\mathbf{s}^{\mathrm{T}}\mathbf{C}_N^{-1}\mathbf{s}}}{2}\right)$$

(e) For the given example, we find $Z_1 = \mathbf{s}^{\mathrm{T}}\mathbf{Y} = 4Y_1 - 2Y_2$ and $Z_2 = \mathbf{s}^{\mathrm{T}}\mathbf{C}_N^{-1}\mathbf{Y} = \frac{2}{3}(7Y_1 + Y_2)$. We can see that the relative weights of the two observations are quite different in the two cases. Numerical computations using the MATLAB script below yield SNRs of 6.2 dB and 9.4 dB, and error probabilities of 0.07 and 0.02 in the two cases, so Z_2 provides better performance than Z_1. It can be shown, using the methods of Chapter 6, that Z_2 is actually the optimal decision statistic, both in terms of maximizing SNR and in terms of minimizing the error probability.

A MATLAB code fragment for generating the numerical results in Example 5.6.8(e) is given below.

Code Fragment 5.6.3 (Performance of on–off keying in colored Gaussian noise)

```
%%OOK with colored noise: N(s,C_N) versus N(0,C_N)
s = [4;-2]; %signal
Cn = [1 -1;-1 4]; %noise covariance matrix
%%decision statistic Z1 = s^T Y
m1 = s' * s; %mean if 1 sent
variance1 = s' * Cn * s; %variance under each hypothesis
v1 = sqrt(variance1); %standard deviation
SNR1 = m1^2/(2 * variance1); %SNR
Pe1 = qfunction(m1/(2 * v1)); %error probability for "split the
  difference" rule using Z1
%%decision statistic Z2 = s^T Cn^{-1} Y
m2 = s' * inv(Cn) * s; %mean if 1 sent
variance2 = s' * inv(Cn) * s; %variance = mean in this case
v2 = sqrt(variance2); %standard deviation
SNR2 = m2^2/(2 * variance2); %reduces to SNR2 = m2/2 in this case
Pe2 = qfunction(m2/(2 * v2)); %error probability for "split the
  difference" rule using Z2
%Compare performance of the two rules
10 * log10([SNR1 SNR2]) %SNRs in dB
[Pe1 Pe2] %error probabilities
```

5.7 Random processes

A key limitation on the performance of communication systems comes from receiver noise, which is an unavoidable physical phenomenon (see Appendix 5.C). Noise cannot be modeled as a deterministic waveform (i.e., we do not know what noise waveform we will observe at any given point of time). Indeed, neither can the desired signals in a communication system, even though we have sometimes pretended otherwise in prior chapters. Information-bearing signals such as speech, audio, and video are best modeled as being randomly chosen from a vast ensemble of possibilities. Similarly, the bit stream being transmitted in a digital communication system can be arbitrary, and can therefore be thought of as being randomly chosen from a large number of possible bit streams. It is time, therefore, to learn how to deal with *random processes,* which is the technical term we use for signals that are chosen randomly from an *ensemble,* or collection, of possible signals. A detailed investigation of random processes is well beyond our scope, and our goal here is limited to developing a working understanding of concepts critical to our study of communication systems. We shall see that this goal can be achieved using elementary extensions of the probability concepts covered earlier in this chapter.

5.7.1 Running example: a sinusoid with random amplitude and phase

Let us work through a simple example before we embark on a systematic development. Suppose that X_1 and X_2 are i.i.d. $N(0, 1)$ random variables, and define

$$X(t) = X_1 \cos(2\pi f_c t) - X_2 \sin(2\pi f_c t) \tag{5.62}$$

where $f_c > 0$ is a fixed frequency. The waveform $X(t)$ is not a deterministic signal, since X_1 and X_2 can take random values on the real line. Indeed, for each time t, $X(t)$ is a random variable, since it is a linear combination of two random variables X_1 and X_2 defined on a common probability space. Moreover, if we pick a number of times $t_1, t_2, \ldots$, then the corresponding samples $X(t_1), X(t_2), \ldots$ are random variables on a common probability space.

Another interpretation of $X(t)$ is obtained by converting (X_1, X_2) to polar form:

$$X_1 = A \cos \Theta, \qquad X_2 = A \sin \Theta$$

For X_1, X_2 i.i.d. $N(0, 1)$, we know from Problem 5.21 that A is Rayleigh, Θ is uniform over $[0, 2\pi]$, and A and Θ are independent. The random process $X(t)$ can be rewritten as

$$X(t) = A \cos \Theta \cos(2\pi f_c t) - A \sin \Theta \sin(2\pi f_c t) = A \cos(2\pi f_c t + \Theta) \tag{5.63}$$

Thus, $X(t)$ is a sinusoid with random amplitude and phase.

For a given time t, what is the distribution of $X(t)$? Since $X(t)$ is a linear combination of i.i.d. Gaussian, hence jointly Gaussian, random variables X_1 and X_2, we infer that it is a Gaussian random variable. Its distribution is therefore specified by computing its mean and variance, as follows:

$$\mathbb{E}[X(t)] = \mathbb{E}[X_1]\cos(2\pi f_c t) - \mathbb{E}[X_2]\sin(2\pi f_c t) = 0 \tag{5.64}$$

and

$$\begin{aligned} \mathrm{var}(X(t)) &= \mathrm{cov}(X_1 \cos(2\pi f_c t) - X_2 \sin(2\pi f_c t), X_1 \cos(2\pi f_c t) - X_2 \sin(2\pi f_c t)) \\ &= \mathrm{cov}(X_1, X_1)\cos^2(2\pi f_c t) + \mathrm{cov}(X_2, X_2)\sin^2(2\pi f_c t) \\ &\quad - 2\,\mathrm{cov}(X_1, X_2)\cos(2\pi f_c t)\sin(2\pi f_c t) \\ &= \cos^2(2\pi f_c t) + \sin^2(2\pi f_c t) = 1 \end{aligned} \tag{5.65}$$

using $\mathrm{cov}(X_i, X_i) = \mathrm{var}(X_i) = 1$, $i = 1, 2$, and $\mathrm{cov}(X_1, X_2) = 0$ (since X_1 and X_2 are independent). Thus, we have $X(t) \sim N(0, 1)$ for any t.

In this particular example, we can also easily specify the joint distribution of any set of n samples, $X(t_1), \ldots, X(t_n)$, where n can be arbitrarily chosen. The samples are jointly Gaussian, since they are linear combinations of the jointly Gaussian random variables X_1 and X_2. Thus, we need only to specify their means and pairwise covariances. We have just shown that the means are zero, and that the diagonal entries of the covariance matrix are one. More generally, the covariance of any two samples can be computed as follows:

$$\begin{aligned}\text{cov}\left(X(t_i), X(t_j)\right) &= \text{cov}\left(X_1 \cos(2\pi f_c t_i) - X_2 \sin(2\pi f_c t_i), X_1 \cos(2\pi f_c t_j) - X_2 \sin(2\pi f_c t_j)\right)\\ &= \text{cov}(X_1, X_1)\cos(2\pi f_c t_i)\cos(2\pi f_c t_j) + \text{cov}(X_2, X_2)\sin(2\pi f_c t_i)\sin(2\pi f_c t_j)\\ &\quad - 2\,\text{cov}(X_1, X_2)\cos(2\pi f_c t_i)\sin(2\pi f_c t_j)\\ &= \cos(2\pi f_c t_i)\cos(2\pi f_c t_j) + \sin(2\pi f_c t_i)\sin(2\pi f_c t_j)\\ &= \cos(2\pi f_c(t_i - t_j))\end{aligned} \tag{5.66}$$

While we have so far discussed the random process $X(t)$ from a statistical point of view, for fixed values of X_1 and X_2, we see that $X(t)$ is actually a deterministic signal. Specifically, if the random vector (X_1, X_2) is defined over a probability space Ω, a particular outcome $\omega \in \Omega$ maps to a particular realization $(X_1(\omega), X_2(\omega))$. This in turn maps to a deterministic "realization," or "sample path," of $X(t)$, which we denote as $X(t, \omega)$:

$$X(t, \omega) = X_1(\omega)\cos(2\pi f_c t) - X_2(\omega)\sin(2\pi f_c t)$$

To see what these sample paths look like, it is easiest to refer to the polar form (5.63):

$$X(t, \omega) = A(\omega)\cos(2\pi f_c t + \Theta(\omega))$$

Thus, as shown in Figure 5.16, different sample paths have different amplitudes, drawn from a Rayleigh distribution, along with phase shifts drawn from a uniform distribution.

5.7.2 Basic definitions

As we have seen earlier, a random vector $\mathbf{X} = (X_1, \ldots, X_n)^{\mathrm{T}}$ is a finite collection of random variables defined on a common probability space, as depicted in Figure 5.7. A random

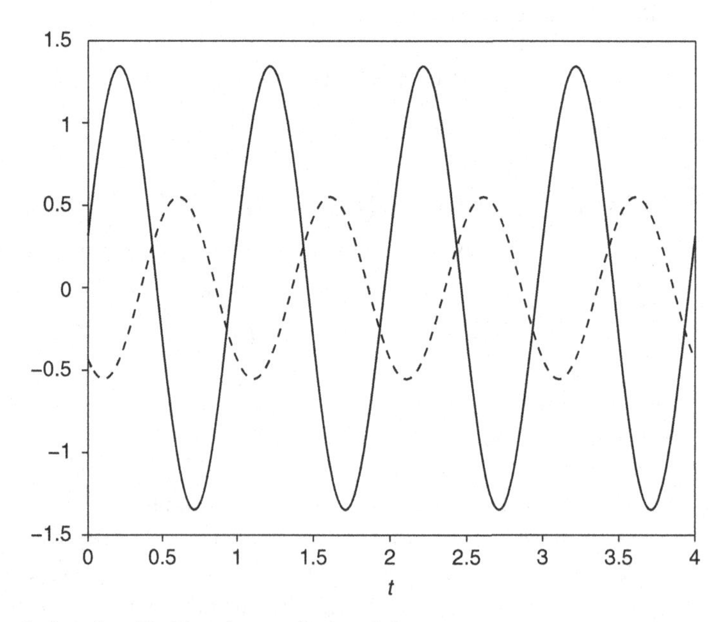

Figure 5.16 Two sample paths for a sinusoid with random amplitude and phase.

process is simply a generalization of this concept, where the number of such random variables can be infinite.

Random process A random process X is a collection of random variables $\{X(t), t \in \mathcal{T}\}$, where the index set $\mathcal{T}$ can be finite, countably infinite, or uncountably infinite. When we interpret the index set as denoting time, as we often do for the scenarios of interest to us, a countable index set corresponds to a *discrete-time* random process, and an uncountable index set corresponds to a *continuous-time* random process. We denote by $X(t, \omega)$ the value taken by the random variable $X(t)$ for any given outcome ω in the sample space.

For the sinusoid with random amplitude and phase, the sample space need only be rich enough to support the two random variables X_1 and X_2 (or A and Θ), from which we can create a continuum of random variables $X(t, \omega)$, $-\infty < t < \infty$:

$$\omega \rightarrow (X_1(\omega), X_2(\omega)) \rightarrow X(t, \omega)$$

In general, however, the source of randomness can be much richer. Noise in a receiver circuit is caused by random motion of a large number of charge carriers. A digitally modulated waveform depends on a sequence of randomly chosen bits. The preceding conceptual framework is general enough to cover all such scenarios.

Sample paths We can also interpret a random process as a signal drawn at random from an *ensemble,* or collection, of possible signals. The signal we get at a particular random draw is called a *sample path,* or *realization,* of the random process. Once we have fixed a sample path, it can be treated like a deterministic signal. Specifically, for each fixed outcome $\omega \in \Omega$, the sample path is $X(t, \omega)$, which varies only with t. We have already seen examples of sample paths for our running example in Figure 5.16.

Finite-dimensional distributions As indicated in Figure 5.17, the samples $X(t_1), \ldots, X(t_n)$ from a random process X are mappings from a common sample space to the real line, with $X(t_i, \omega)$ denoting the value of the random variable $X(t_i)$ for outcome $\omega \in \Omega$. The joint distribution of these random variables depends on the underlying probability measure on the sample space Ω. We say that we "know" the statistics of a random process if we know the joint statistics of an arbitrarily chosen finite collection of samples. That is, we know the

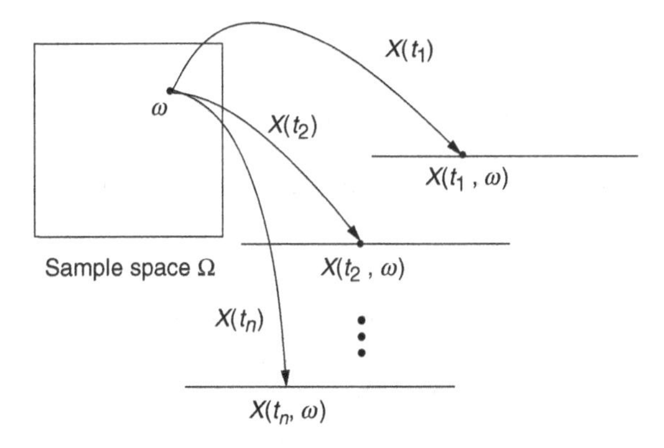

Figure 5.17 Samples of a random process are random variables defined on a common probability space.

joint distribution of the samples $X(t_1), \ldots, X(t_n)$, regardless of the number of samples n, and the sampling times $t_1, \ldots, t_n$. These joint distributions are called the *finite-dimensional distributions* of the random process, with the joint distribution of n samples called an nth-order distribution. Thus, while a random process may be comprised of infinitely many random variables, when we specify its statistics, we focus on a finite subset of these random variables.

For our running example (5.62), we observed that the samples are jointly Gaussian, and specified the joint distribution by computing the means and covariances. This is a special case of a broader class of Gaussian random processes (to be defined shortly) for which it is possible to characterize finite-dimensional distributions compactly in this fashion. Often, however, it is not possible to explicitly specify such distributions, but we can still compute useful quantities averaged across sample paths.

Ensemble averages Knowing the finite-dimensional distributions enables us to compute statistical averages across the collection, or *ensemble,* of sample paths. Such averages are called *ensemble averages.* We will be mainly interested in "second-order" statistics (involving expectations of products of at most two random variables), such as means and covariances. We define these quantities in sufficient generality that they apply to complex-valued random processes, but specialize to real-valued random processes in most of our computations.

5.7.3 Second-order statistics

Mean, autocorrelation, and autocovariance functions (ensemble averages) For a random process $X(t)$, the mean function is defined as

$$m_X(t) = \mathbb{E}[X(t)] \tag{5.67}$$

and the autocorrelation function as

$$R_X(t_1, t_2) = \mathbb{E}[X(t_1)X^*(t_2)] \tag{5.68}$$

Note that $R_X(t,t) = \mathbb{E}[|X(t)|^2]$ is the instantaneous power at time t. The autocovariance function of X is the autocorrelation function of the zero-mean version of X, and is given by

$$C_X(t_1, t_2) = \mathbb{E}[(X(t_1) - \mathbb{E}[X(t_1)])(X(t_2) - \mathbb{E}[X(t_2)])^*] = R_X(t_1, t_2) - m_X(t_1)m_X^*(t_2) \tag{5.69}$$

Second-order statistics for the running example We have from (5.64) and (5.66) that

$$m_X(t) \equiv 0, \quad C_X(t_1, t_2) = R_X(t_1, t_2) = \cos(2\pi f_c(t_1 - t_2)) \tag{5.70}$$

It is interesting to note that the mean function does not depend on t, and that the autocorrelation and autocovariance functions depend only on the difference of the times $t_1 - t_2$. This implies that, if we shift $X(t)$ by some time delay d, the shifted process $\tilde{X}(t) = X(t-d)$ would have the same mean and autocorrelation functions. Such translation invariance of statistics is interesting and important enough to merit a formal definition, which we provide next.

5.7.4 Wide-sense stationarity and stationarity

Wide-sense stationary (WSS) random process A random process X is said to be WSS if

$$m_X(t) \equiv m_X(0) \quad \text{for all } t$$

and

$$R_X(t_1, t_2) = R_X(t_1 - t_2, 0) \quad \text{for all } t_1, t_2$$

In this case, we change the notation, dropping the time dependence in the notation for the mean m_X, and expressing the autocorrelation function as a function of $\tau = t_1 - t_2$ alone. Thus, for a WSS process, we can define the autocorrelation function as

$$R_X(\tau) = E[X(t)X^*(t - \tau)] \quad \text{for } X \text{ WSS} \tag{5.71}$$

with the understanding that the expectation is independent of t. Since the mean is independent of time and the autocorrelation depends only on time differences, the autocovariance also depends only on time differences, and is given by

$$C_X(\tau) = R_X(\tau) - |m_X|^2 \quad \text{for } X \text{ WSS} \tag{5.72}$$

Second-order statistics for running example (new notation) With this new notation, we have

$$m_X \equiv 0, \quad R_X(\tau) = C_X(\tau) = \cos(2\pi f_c \tau) \tag{5.73}$$

A WSS random process has shift-invariant second-order statistics. An even stronger notion of shift-invariance is stationarity.

Stationary random process A random process $X(t)$ is said to be stationary if it is statistically indistinguishable from a delayed version of itself. That is, $X(t)$ and $X(t - d)$ have the same statistics for any delay $d \in (-\infty, \infty)$.

Running example The sinusoid with random amplitude and phase in our running example is stationary. To see this, it is convenient to consider the polar form in (5.63), namely $X(t) = A\cos(2\pi f_c t + \Theta)$, where Θ is uniformly distributed over $[0, 2\pi]$. Note that

$$Y(t) = X(t - d) = A\cos(2\pi f_c(t - d) + \Theta) = A\cos(2\pi f_c t + \Theta')$$

where $\Theta' = \Theta - 2\pi f_c d$ modulo 2π is uniformly distributed over $[0, 2\pi]$. Thus, X and Y are statistically indistinguishable.

Stationarity implies wide-sense stationarity For a stationary random process X, the mean function satisfies

$$m_X(t) = m_X(t - d)$$

for any t, regardless of the value of d. Choosing $d = t$, we infer that

$$m_X(t) = m_X(0) \tag{5.74}$$

That is, the mean function is a constant. Similarly, the autocorrelation function satisfies

$$R_X(t_1, t_2) = R_X(t_1 - d, t_2 - d)$$

for any t_1 and t_2, regardless of the value of d. On setting $d = t_2$, we have that

$$R_X(t_1, t_2) = R_X(t_1 - t_2, 0) \tag{5.75}$$

Thus, a stationary process is also WSS.

While our running example was easy to analyze, in general, stationarity is a stringent requirement that is not easy to verify. For our needs, the weaker concept of wide-sense stationarity typically suffices. Further, we are often interested in Gaussian random processes (which will be defined shortly), for which wide-sense stationarity actually implies stationarity.

5.7.5 Power spectral density

For deterministic finite-energy signals, we introduced the concept of the energy spectral density, which specifies how the energy in a signal is distributed in different frequency bands, in Chapter 2. Similarly, we defined the power spectral density (PSD) for finite-power deterministic signals in Chapter 4, "just in time" to characterize the spectral occupancy of digital communication signals. This deterministic framework directly applies to a given sample path of a random process, and, indeed, this is what we did when we computed the PSD of linearly modulated signals in Chapter 4. While we did not mention the term "random process" then (for the good reason that we had not introduced it yet), if we model the information encoded into a digitally modulated signal as random, then the latter is indeed a random process. Let us now begin by restating the definition of PSD in Chapter 4.

Power spectral density The power spectral density (PSD), $S_x(f)$, for a finite-power signal $x(t)$, which we can now think of as a sample path of a random process, is defined through the conceptual measurement depicted in Figure 5.18. Pass $x(t)$ through an ideal narrowband filter with transfer function

$$H_\nu(f) = \begin{cases} 1, & \nu - \Delta f/2 < f < \nu + \Delta f/2 \\ 0, & \text{else} \end{cases}$$

The PSD evaluated at ν, $S_x(\nu)$, is defined as the measured power at the filter output, divided by the filter width Δf (in the limit as $\Delta f \to 0$).

The power meter in Figure 5.18 is averaging over time to estimate the power in a frequency slice of a particular sample path. Let us review how this is done before discussing how to average across sample paths to define the PSD in terms of an ensemble average.

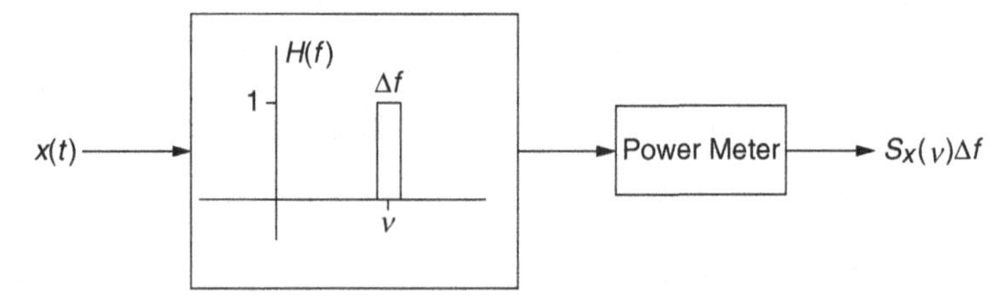

Figure 5.18 The operational definition of PSD for a sample path $x(t)$.

Periodogram-based PSD estimation The PSD can be estimated by computing the Fourier transform over a finite observation interval, and dividing its magnitude squared (which is the energy spectral density) by the length of the observation interval. The time-windowed version of x is defined as

$$x_{T_o}(t) = x(t)I_{[-T_o/2, T_o/2]}(t) \tag{5.76}$$

where T_o is the length of the observation interval. The Fourier transform of $x_{T_o}(t)$ is denoted as

$$X_{T_o}(f) = \mathcal{F}(x_{T_o})$$

The energy spectral density of x_{T_o} is therefore $|X_{T_o}(f)|^2$, and the PSD estimate is given by

$$\hat{S}_x(f) = \frac{|X_{T_o}(f)|^2}{T_o} \tag{5.77}$$

PSD for a sample path Formally, we define the PSD for a sample path in the limit of large time windows as follows:

$$S_x(f) = \lim_{T_o \to \infty} \frac{|X_{T_o}(f)|^2}{T_o} \quad \textbf{PSD for sample path} \tag{5.78}$$

The preceding definition involves time averaging across a sample path, and can be related to the time-averaged autocorrelation function, defined as follows.

The time-averaged autocorrelation function for a sample path For a sample path $x(t)$, we define the time-averaged autocorrelation function as

$$R_x(\tau) = \overline{x(t)x^*(t-\tau)} = \lim_{T_o \to \infty} \frac{1}{T_o} \int_{-T_o/2}^{T_o/2} x(t)x^*(t-\tau)dt$$

We now state the following important result.

The time-averaged PSD and the autocorrelation function form a Fourier transform pair We have

$$S_x(f) \leftrightarrow R_x(\tau) \tag{5.79}$$

We omit the proof, but the result can be derived using the techniques of Chapter 2.

The time-averaged PSD and the autocorrelation function for the running example For our random sinusoid (5.63), the time-averaged autocorrelation function is given by

$$\begin{aligned} R_x(\tau) &= \overline{A\cos(2\pi f_c t + \Theta) A\cos(2\pi f_c(t-\tau) + \Theta)} \\ &= (A^2/2)\overline{\cos(2\pi f_c \tau) + \cos(4\pi f_c t - 2\pi f_c \tau + 2\Theta)} \\ &= (A^2/2)\cos(2\pi f_c \tau) \end{aligned} \tag{5.80}$$

The time-averaged PSD is given by

$$S_x(f) = \frac{A^2}{4}\delta(f - f_c) + \frac{A^2}{4}\delta(f + f_c) \tag{5.81}$$

We now extend the concept of the PSD to a *statistical* average as follows.

The ensemble-averaged PSD The ensemble-averaged PSD for a random process is defined as follows:

$$S_X(f) = \lim_{T_o \to \infty} \mathbb{E}\left[\frac{|X_{T_o}(f)|^2}{T_o}\right] \quad \textbf{ensemble-averaged PSD} \tag{5.82}$$

That is, we take the expectations of the PSD estimates computed over an observation interval, and then let the observation interval get large.

Potential notational confusion We use capital letters (e.g., $X(t)$) to denote a random process and small letters (e.g., $x(t)$) to denote sample paths. However, we also use capital letters to denote the Fourier transform of a time-domain signal (e.g., $s(t) \leftrightarrow S(f)$), as introduced in Chapter 2. Rather than introducing additional notation to resolve this potential ambiguity, we rely on context to clarify the situation. In particular, (5.82) illustrates this potential problem. On the left-hand side, we use X to denote the random process whose PSD $S_X(f)$ we are interested in. On the right-hand side, we use $X_{T_o}(f)$ to denote the Fourier transform of a windowed sample path $x_{T_o}(t)$. Such opportunities for confusion arise seldom enough that it is not worth complicating our notation to avoid them.

A result analogous to (5.79) holds for ensemble-averaged quantities as well.

The ensemble-averaged PSD and the autocorrelation function for WSS processes form a Fourier-transform pair (the Wiener–Khintchine theorem) For a WSS process X with autocorrelation function $R_X(\tau)$, the ensemble-averaged PSD is the Fourier transform of the ensemble-averaged autocorrelation function:

$$S_X(f) = \mathcal{F}(R_X(\tau)) = \int_{-\infty}^{\infty} R_X(\tau) e^{-j2\pi f\tau}\, d\tau \tag{5.83}$$

This result is called the Wiener–Khintchine theorem, and can be proved under mild conditions on the autocorrelation function (the area under $|R_X(\tau)|$ must be finite and its Fourier transform must exist). The proof requires advanced probability concepts beyond our scope here, and is omitted.

The ensemble-averaged PSD for the running example For our running example, the PSD is obtained by taking the Fourier transform of (5.73):

$$S_X(f) = \frac{1}{2}\delta(f - f_c) + \frac{1}{2}\delta(f + f_c) \tag{5.84}$$

That is, the power in X is concentrated at $\pm f_c$, as we would expect for a sinusoidal signal at frequency f_c.

Power It follows from the Wiener–Khintchine theorem that the power of X can be obtained either by integrating the PSD or evaluating the autocorrelation function at $\tau = 0$:

$$P_X = \mathbb{E}\left[|X(t)|^2\right] = R_X(0) = \int_{-\infty}^{\infty} S_X(f) df \tag{5.85}$$

For our running example, we obtain from (5.73) or (5.84) that $P_X = 1$.

Ensemble versus time averages For our running example, we computed the ensemble-averaged autocorrelation function $R_X(\tau)$ and then used the Wiener–Khintchine theorem to compute the PSD by taking the Fourier transform. At other times, it is convenient to apply the operational definition depicted in Figure 5.18, which involves averaging across time for a given sample path. If the two approaches give the same answer, then the random process is said to be *ergodic* in PSD. In practical terms, ergodicity means that designs based on statistical averages across sample paths can be expected to apply to individual sample paths, and that measurements carried out on a particular sample path can serve as a proxy for statistical averaging across multiple realizations.

On comparing (5.81) and (5.84), we see that our running example is actually *not ergodic* in PSD. For any sample path $x(t) = A\cos(2\pi f_c t + \theta)$, it is quite easy to show that

$$S_x(f) = \frac{A^2}{2}\delta(f - f_c) + \frac{A^2}{2}\delta(f + f_c) \tag{5.86}$$

On comparing this with (5.84), we see that the time-averaged PSD varies across sample paths due to amplitude variations, with A^2 replaced by its expectation in the ensemble-averaged PSD.

Intuitively speaking, ergodicity requires sufficient richness of variation across time and sample paths. While this is not present in our simple running example (a randomly chosen amplitude that is fixed across the entire sample path is the culprit), it is often present in the more complicated random processes of interest to us, including receiver noise and digitally modulated signals (under appropriate conditions on the transmitted symbol sequences). When ergodicity holds, we have a choice of using either time averaging or ensemble averaging for computations, depending on which is most convenient or insightful.

The autocorrelation function and the PSD must satisfy the following structural properties (these apply to ensemble averages for WSS processes, as well as to time averages, although our notation corresponds to ensemble averages).

Structural properties of PSD and autocorrelation function

(P1) $S_X(f) \geq 0$ for all f. This follows from the sample-path-based definition in Figure 5.18, since the output of the power meter is always nonnegative. Averaging across sample paths preserves this property.

(P2a) The autocorrelation function is conjugate symmetric: $R_X(\tau) = R_X^*(-\tau)$. This follows quite easily from the definition (5.71). On setting $t = u + \tau$, we have

$$R_X(\tau) = \mathbb{E}[X(u+\tau)X^*(u)] = \left(\mathbb{E}[X(u)X^*(u+\tau)]\right)^* = R_X^*(-\tau)$$

(P2b) For real-valued X, both the autocorrelation function and the PSD are symmetric and real-valued. $S_X(f) = S_X(-f)$ and $R_X(\tau) = R_X(-\tau)$. (The proof of this is left as an exercise.)

Any function $g(\tau) \leftrightarrow G(f)$ must satisfy these properties in order to be a valid autocorrelation function/PSD.

Example 5.7.1 (Which function is an autocorrelation?) For each of the following functions, determine whether it is a valid autocorrelation function: (a) $g_1(\tau) = \sin(\tau)$, (b) $g_2(\tau) = I_{[-1,1]}(\tau)$, and (c) $g_3(\tau) = e^{-|\tau|}$.

Solution

(a) This is not a valid autocorrelation function, since it is not symmetric and violates property (P2b).
(b) This satisfies property (P2b). However, $I_{[-1,1]}(\tau) \leftrightarrow 2\,\mathrm{sinc}(2f)$, so property (P1) is violated, since the sinc function can take negative values. Hence, the boxcar function cannot be a valid autocorrelation function. This example shows that the nonnegativity property (P1) places a stronger constraint on the validity of a proposed function as an autocorrelation function than does the symmetry property (P2).
(c) The function $g_3(\tau)$ is symmetric and satisfies property (P2b). It is left as an exercise to check that $G_3(f) \geq 0$, and hence property (P1) is also satisfied.

Units for the PSD Power per unit frequency has the same units as power multiplied by time, or energy. Thus, the PSD is expressed in units of watts/hertz, or joules.

The one-sided PSD The PSD that we have talked about so far is the *two-sided* PSD, which spans both positive and negative frequencies. For a real-valued X, we can restrict attention to positive frequencies alone in defining the PSD, by virtue of property (P2b). This yields the *one-sided* PSD $S_X^+(f)$, defined as

$$S_X^+(f) = S_X(f) + S_X(-f) = 2S_X(f), \quad f \geq 0 \quad (\boldsymbol{X(t)} \ \textbf{real}) \tag{5.87}$$

It is useful to interpret this in terms of the sample-path-based operational definition shown in Figure 5.19. The signal is passed through a physically realizable filter (i.e., with real-valued impulse response) of bandwidth Δf, centered around ν. The filter transfer function must be conjugate symmetric, hence

$$H_\nu(f) = \begin{cases} 1, & \nu - \Delta f/2 < f < \nu + \Delta f/2 \\ 1, & -\nu - \Delta f/2 < f < -\nu + \Delta f/2 \\ 0, & \text{else} \end{cases}$$

The one-sided PSD is defined as the limit of the power of the filter output, divided by Δf, as $\Delta f \to 0$. Comparing Figures 5.18 and 5.19, we have that the sample-path-based one-sided PSD is simply twice the two-sided PSD: $S_x^+(f) = (S_x(f) + S_x(-f))\, I_{\{f \geq 0\}} = 2S_x(f) I_{\{f \geq 0\}}$.

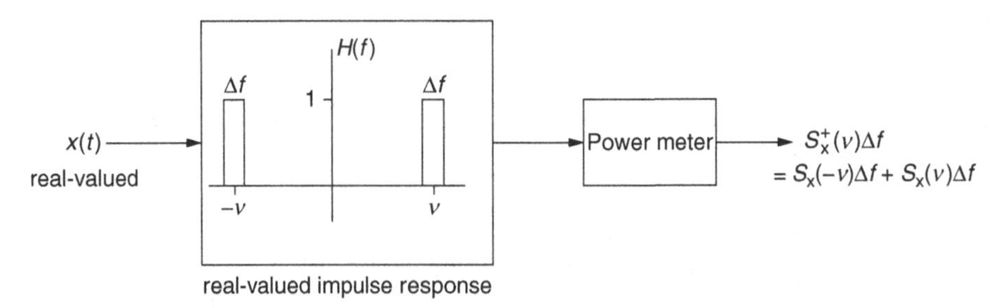

Figure 5.19 The operational definition of the one-sided PSD.

The one-sided PSD for the running example From (5.84), we obtain that

$$S_X^+(f) = \delta(f - f_c) \tag{5.88}$$

with all the power concentrated at f_c, as expected.

Power in terms of PSD We can express the power of a real-valued random process in terms of either the one-sided or the two-sided PSD:

$$\mathbb{E}[X^2(t)] = R_X(0) = \int_{-\infty}^{\infty} S_X(f)df = \textbf{(for } \boldsymbol{X} \textbf{ real)} \int_0^{\infty} S_X^+(f)df \tag{5.89}$$

Baseband and passband random processes A random process X is baseband if its PSD is baseband, and is passband if its PSD is passband. Thinking in terms of time-averaged PSDs, which are based on the Fourier transform of time-windowed sample paths, we see that a random process is baseband if its sample paths, time-windowed over a large enough observation interval, are (approximately) baseband. Similarly, a random process is passband if its sample paths, time-windowed over a large enough observation interval, are (approximately) passband. The caveat of a "large enough observation interval" is inserted because of the following consideration: timelimited signals cannot be strictly bandlimited, but as long as the observation interval is large enough, the time windowing (which corresponds to convolving the spectrum with a sinc function) does not spread out the spectrum of the signal significantly. Thus, the PSD (which is obtained taking the limit of large observation intervals) also defines the frequency occupancy of the sample paths over large enough observation intervals. Note that these intuitions, while based on time-averaged PSDs, also apply when the bandwidth occupancy is defined in terms of ensemble-averaged PSDs, as long as the random process is ergodic in PSD.

Example (PSD of a modulated passband signal) Consider a passband signal $u_p(t) = m(t)\cos(2\pi f_0 t)$, where $m(t)$ is a message modeled as a baseband random process with PSD $S_m(f)$ and power P_m. On timelimiting to an interval of length T_o and going to the frequency domain, we have

$$U_{p,T_o}(f) = \frac{1}{2}\left(M_{T_o}(f - f_0) + M_{T_o}(f - f_0)\right) \tag{5.90}$$

On taking the magnitude squared, dividing by T_o, and letting T_o get large, we obtain

$$S_{u_p}(f) = \frac{1}{4}(S_m(f - f_0) + S_m(f + f_0)) \tag{5.91}$$

An example is shown in Figure 5.20.

Thus, we start with the formula (5.90) relating the Fourier transform for a given sample path, which is identical to what we had in Chapter 2 (except that we now need to timelimit the finite-power message in order to obtain a finite-energy signal), and obtain the relation

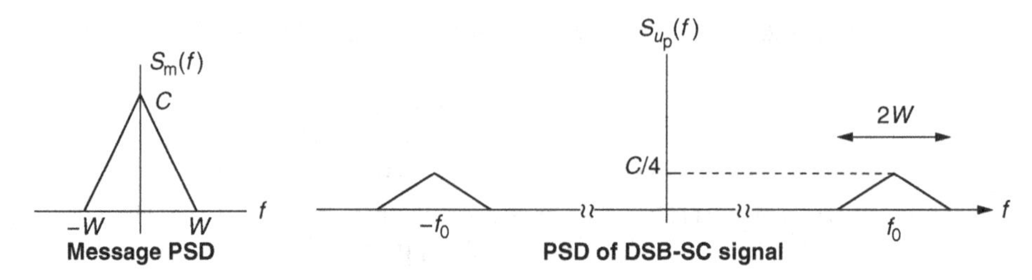

Figure 5.20 The relation between the PSDs of a message and the corresponding DSB-SC signal.

(5.91) relating the PSDs. An example is shown in Figure 5.20. We can now integrate the PSDs to get

$$P_{\mathrm{u}} = \frac{1}{4}(P_{\mathrm{m}} + P_{\mathrm{m}}) = \frac{P_{\mathrm{m}}}{2}$$

5.7.6 Gaussian random processes

Gaussian random processes are just generalizations of Gaussian random vectors to an arbitrary number of components (countable or uncountable).

Gaussian random process A random process $X = \{X(t), t \in T\}$ is said to be Gaussian if any linear combination of samples is a Gaussian random variable. That is, for any number n of samples, any sampling times $t_1, \ldots, t_n$, and any scalar constants $a_1, \ldots, a_n$, the linear combination $a_1 X(t_1) + \cdots + a_n X(t_n)$ is a Gaussian random variable. Equivalently, the samples $X(t_1), \ldots, X(t_n)$ are jointly Gaussian.

Our running example (5.62) is a Gaussian random process, since any linear combination of samples is a linear combination of the jointly Gaussian random variables X_1 and X_2, and is therefore a Gaussian random variable.

A linear combination of samples from a Gaussian random process is completely characterized by its mean and variance. To compute the latter quantities for an arbitrary linear combination, we can show, as we did for random vectors, that we need to know just the mean function (analogous to the mean vector) and the autocovariance function (analogous to the covariance matrix) of the random process. These functions therefore provide a complete statistical characterization of a Gaussian random process, since the definition of a Gaussian random process requires only that we be able to characterize the distribution of an arbitrary linear combination of samples.

Characterizing a Gaussian random process The statistics of a Gaussian random process are completely specified by its mean function $m_X(t) = \mathbb{E}[X(t)]$ and its autocovariance function $C_X(t_1, t_2) = \mathbb{E}[X(t_1)X(t_2)]$. Given the mean function, the autocorrelation function $R_X(t_1, t_2) = \mathbb{E}[X(t_1)X(t_2)]$ can be computed from $C_X(t_1, t_2)$, and vice versa, using the following relation:

$$R_X(t_1, t_2) = C_X(t_1, t_1) + m_X(t_1) m_X(t_2) \tag{5.92}$$

It therefore also follows that a Gaussian random process is completely specified by its mean and autocorrelation functions.

WSS Gaussian random processes are stationary We know that a stationary random process is WSS. The converse is not true in general, but Gaussian WSS processes are indeed stationary. This is because the statistics of a Gaussian random process are characterized by its first- and second-order statistics, and, if these are shift-invariant (as they are for WSS processes), the random process is statistically indistinguishable under a time shift.

Example 5.7.2 Suppose that Y is a Gaussian random process with mean function $m_Y(t) = 3t$ and autocorrelation function $R_Y(t_1, t_2) = 4e^{-|t_1 - t_2|} + 9t_1 t_2$.

(a) Find the probability that $Y(2)$ is bigger than 10.
(b) Specify the joint distribution of $Y(2)$ and $Y(3)$.
(c) **(True or false?)** Y is stationary.
(d) **(True or false?)** The random process $Z(t) = Y(t) - 3t$ is stationary.

Solution

(a) Since Y is a Gaussian random process, the sample $Y(2)$ is a Gaussian random variable with mean $m_Y(2) = 6$ and variance $C_Y(2, 2) = R_Y(2, 2) - (m_Y(2))^2 = 4$. More generally, note that the autocovariance function of Y is given by

$$C_Y(t_1, t_2) = R_Y(t_1, t_2) - m_Y(t_1)m_Y(t_2) = 4e^{-|t_1 - t_2|} + 9t_1 t_2 - (3t_1)(3t_2) = 4e^{-|t_1 - t_2|}$$

so that $\text{var}(Y(t)) = C_Y(t, t) = 4$ for any sampling time t. We have shown that $Y(2) \sim N(6, 4)$, so

$$P[Y(2) > 10] = Q\left(\frac{10 - 6}{\sqrt{4}}\right) = Q(2)$$

(b) Since Y is a Gaussian random process, $Y(2)$ and $Y(3)$ are jointly Gaussian, with distribution specified by the mean vector and covariance matrix given by

$$\mathbf{m} = \begin{pmatrix} m_Y(2) \\ m_Y(3) \end{pmatrix} = \begin{pmatrix} 6 \\ 9 \end{pmatrix}$$

and

$$\mathbf{C} = \begin{pmatrix} C_Y(2, 2) & C_Y(2, 3) \\ C_Y(3, 2) & C_Y(3, 3) \end{pmatrix} = \begin{pmatrix} 4 & 4e^{-1} \\ 4e^{-1} & 4 \end{pmatrix}$$

(c) Y has time-varying mean, and hence is not WSS. This implies that it is not stationary. The statement is therefore **false**.

(d) $Z(t) = Y(t) - 3t = Y(t) - m_Y(t)$ is the zero-mean version of Y. It inherits the Gaussianity of Y. The mean function $m_Z(t) \equiv 0$, and the autocorrelation function, given by

$$R_Z(t_1, t_2) = \mathbb{E}[(Y(t_1) - m_Y(t_1))(Y(t_2) - m_Y(t_2))] = C_Y(t_1, t_2) = 4e^{-|t_1 - t_2|}$$

depends on the time difference $t_1 - t_2$ alone. Thus, Z is WSS. Since it also Gaussian, this implies that Z is stationary. The statement is therefore **true**.

5.8 Noise modeling

We now have the background required in order to discuss mathematical modeling of noise in communication systems. A generic model for receiver noise is that it is a random process with zero DC value, and with a PSD that is flat, or *white,* over a band of interest. The key noise mechanisms in a communication receiver, thermal and shot noise, are both white, as discussed in Appendix 5.C. For example, Figure 5.21 shows the two-sided PSD of *passband* white noise $n_{\mathrm{p}}(t)$, which is given by

$$S_{n_{\mathrm{p}}}(f) = \begin{cases} N_0/2, & |f - f_{\mathrm{c}}| \leq B/2 \\ N_0/2, & |f + f_{\mathrm{c}}| \leq B/2 \\ 0, & \text{else} \end{cases}$$

Since $n_{\mathrm{p}}(t)$ is real-valued, we can also define the one-sided PSD as follows:

$$S^{+}_{n_{\mathrm{p}}}(f) = \begin{cases} N_0, & |f - f_{\mathrm{c}}| \leq B/2 \\ 0, & \text{else} \end{cases}$$

That is, white noise has a two-sided PSD $N_0/2$, and a one-sided PSD N_0, over the band of interest. The power of the white noise is given by

$$P_{n_{\mathrm{p}}} = \overline{n^2_{\mathrm{p}}} = \int_{-\infty}^{\infty} S_{n_{\mathrm{p}}}(f)df = (N_0/2)2B = N_0 B$$

The PSD N_0 is in units of watts/hertz, or joules.

Similarly, Figure 5.22 shows the one-sided and two-sided PSDs for *real-valued* white noise in a physical baseband system with bandwidth B. The power of this baseband white noise is again $N_0 B$. As we discuss in Appendix 5.D, as with deterministic passband signals, passband random processes can also be represented in terms of I and Q components. We note in Appendix 5.D that the I and Q components of passband white noise are baseband white-noise processes, and that the corresponding complex envelope is *complex-valued* white noise.

The noise figure The value of N_0 summarizes the net effects of white noise arising from various devices in the receiver. By comparing the noise power $N_0 B$ with the nominal figure

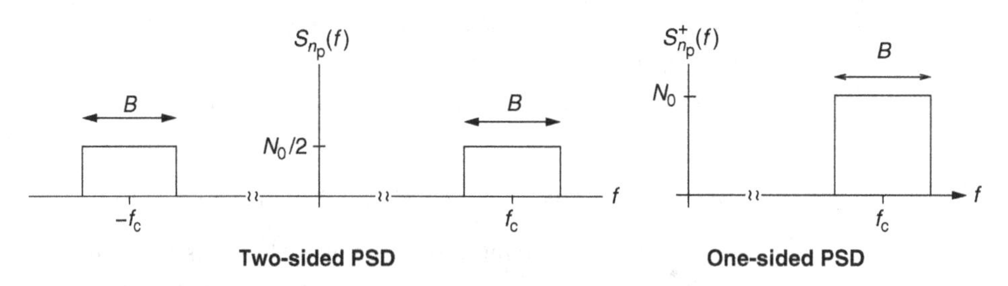

Figure 5.21 The PSD of passband white noise is flat over the band of interest.

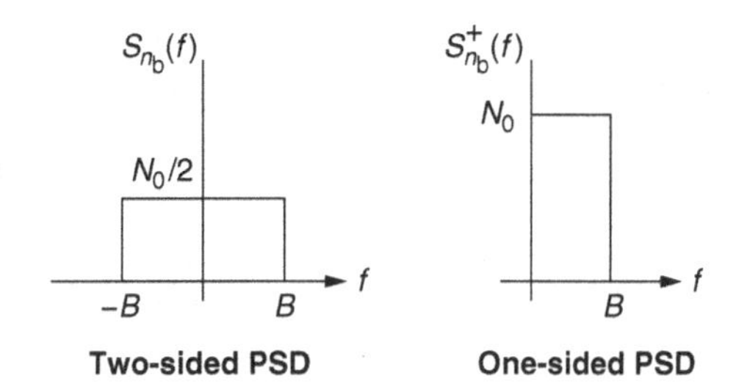

Figure 5.22 The PSD of baseband white noise.

of $k_B TB$ for the thermal noise of a resistor with matched impedance, we define the *noise figure* as

$$F = \frac{N_0}{k_B T_{\text{room}}}$$

where $k_B = 1.38 \times 10^{-23}$ J/K is Boltzmann's constant, and the nominal "room temperature" is taken by convention to be $T_{\text{room}} = 290$ K (the product $k_B T_{\text{room}} \approx 4 \times 10^{-21}$ J, so the numbers work out well for this slightly chilly choice of room temperature at 62.6 °F). The noise figure is usually expressed in dB.

The noise power for a bandwidth B is given by

$$P_n = N_0 B = k_B T_{\text{room}} 10^{F(\text{dB})/10} B$$

dBW and dBm It is customary to express power on the decibel (dB) scale:

$$\text{Power (dBW)} = 10 \log_{10}(\text{Power (watts)})$$

or

$$\text{Power (dBm)} = 10 \log_{10}(\text{Power (milliwatts)})$$

On the dB scale, the noise power over 1 Hz is therefore given by

$$\text{Noise power over 1 Hz} = -174 + F \text{ dBm} \tag{5.93}$$

Thus, the noise power in dBm over a bandwidth of B Hz is given by

$$P_n(\text{dBm}) = -174 + F + 10 \log_{10} B \text{ dBm} \tag{5.94}$$

Example 5.8.1 (Noise-power computation) A 5-GHz wireless local-area network (WLAN) link has a receiver bandwidth B of 20 MHz. If the receiver has a noise figure of 6 dB, what is the receiver noise power P_n?

Solution

The noise power

$$\begin{aligned} P_n &= N_0 B = k_B T_0 \times 10^{F/10} B = (1.38 \times 10^{-23})(290)(10^{6/10})(20 \times 10^6) \\ &= 3.2 \times 10^{-13} \text{ W} = 3.2 \times 10^{-10} \text{ mW} \end{aligned}$$

The noise power is often expressed in dBm, which is obtained by converting the raw number in milliwatts (mW) into dB. We therefore get

$$P_{\mathrm{n,dBm}} = 10\log_{10}[P_{\mathrm{n}}\ (\mathrm{mW})] = -95\,\mathrm{dBm}$$

Let us now redo this computation in the "dB domain," where the contributions to the noise power due to the various system parameters simply add up. Using (5.93), the noise power in our system can be calculated as follows:

$$P_{\mathrm{n}}\ (\mathrm{dBm}) = -174 + \text{Noise figure (dB)} + 10\log_{10}[\text{Bandwidth (Hz)}] \tag{5.95}$$

In our current example, we obtain P_{n} (dBm) $= -174 + 6 + 73 = -95$ dBm, as before.

We now add two more features to our noise model that greatly simplify computations. First, we assume that the noise is a Gaussian random process. The physical basis for this is that noise arises due to the random motion of a large number of charge carriers, which leads to Gaussian statistics based on the central limit theorem (see Appendix 5.B). The mathematical consequence of Gaussianity is that we can compute probabilities merely from knowledge of second-order statistics. Second, we remove band limitation, implicitly assuming that it will be imposed later by filtering at the receiver. That is, we model noise $n(t)$ (where n can be real-valued passband or baseband white noise) as a zero-mean WSS random process with PSD flat over the entire real line, $S_{\mathrm{n}}(f) \equiv N_0/2$. The corresponding autocorrelation function is $R_{\mathrm{n}}(\tau) = (N_0/2)\delta(\tau)$. This model is clearly physically unrealizable, since the noise power is infinite. However, since receiver processing in bandlimited systems always involves filtering, we can assume that the receiver noise prior to filtering is not bandlimited and still get the right answer. Figure 5.23 shows the steps we use to go from receiver noise in bandlimited systems to infinite-power white Gaussian noise (WGN), which we formally define below.

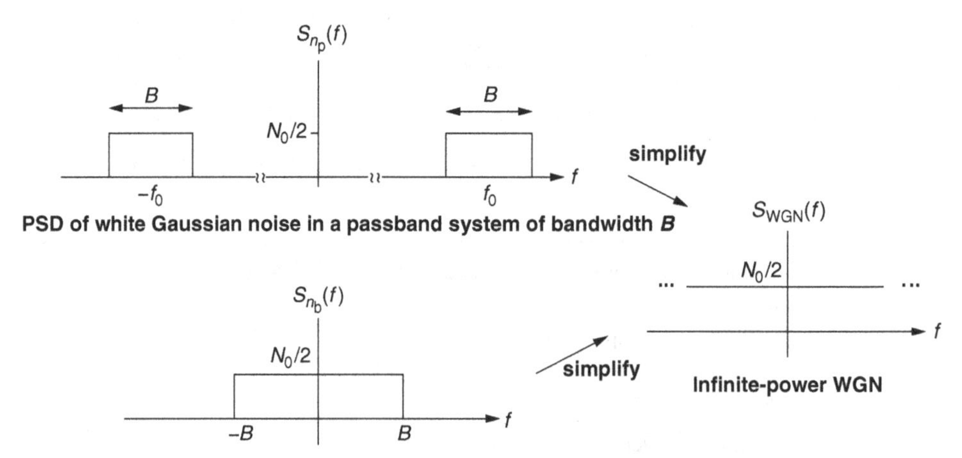

Figure 5.23 Since receiver processing always involves some form of band limitation, it is not necessary to impose band limitation on the WGN model.

White Gaussian noise Real-valued WGN $n(t)$ is a zero-mean, WSS, Gaussian random process with $S_n(f) \equiv N_0/2 = \sigma^2$. Equivalently, $R_n(\tau) = (N_0/2)\delta(\tau) = \sigma^2\delta(\tau)$. The quantity $N_0/2 = \sigma^2$ is often termed the two-sided PSD of WGN, since we must integrate over both positive and negative frequencies in order to compute power using this PSD. The quantity N_0 is therefore referred to as the one-sided PSD, and has the dimension of watts/hertz, or joules.

The following example provides a preview of typical computations for signaling in WGN, and illustrates why the model is so convenient.

Example 5.8.2 (On–off keying in continuous time) A receiver in an on–off keyed system receives the signal $y(t) = s(t) + n(t)$ if 1 is sent, and receives $y(t) = n(t)$ if 0 is sent, where $n(t)$ is WGN with PSD $\sigma^2 = N_0/2$. The receiver computes the following decision statistic:

$$Y = \int y(t)s(t)dt$$

(We shall soon show that this is actually the best thing to do.)

(a) Find the conditional distribution of Y if 0 is sent.
(b) Find the conditional distribution of Y if 1 is sent.
(c) Compare this case with the on–off keying model in Example 5.6.3.

Solution

(a) Conditioned on 0 being sent, $y(t) = n(t)$ and hence $Y = \int n(t)s(t)dt$. Since n is Gaussian, and Y is obtained from it by linear processing, Y is a Gaussian random variable (conditioned on 0 being sent). Thus, the conditional distribution of Y is completely characterized by its mean and variance, which we now compute:

$$\mathbb{E}[Y] = \mathbb{E}\left[Y = \int n(t)s(t)dt\right] = \int s(t)\mathbb{E}[n(t)]dt = 0$$

Here we can interchange expectation and integration because both are linear operations. Actually, there are some mathematical conditions (beyond our scope here) that need to be satisfied for such "natural" interchanges to be permitted, but these conditions are met for all the examples that we consider in this text. Since the mean is zero, the variance is given by

$$\text{var}(Y) = \mathbb{E}[Y^2] = \mathbb{E}\left[\int n(t)s(t)dt \int n(u)s(u)du\right]$$

Notice that we have written out $Y^2 = Y \times Y$ as the product of two identical integrals, but with the "dummy" variables of integration chosen to be different. This is because we need to consider all possible cross terms that could result from multiplying the integral by itself. We now interchange expectation and integration again, noting that all random quantities must be grouped inside the expectation. This gives us

$$\text{var}(Y) = \int \int \mathbb{E}[n(t)n(u)]s(t)s(u)dt\, du \tag{5.96}$$

Now this is where the WGN model makes our life simple. The autocorrelation function

$$\mathbb{E}[n(t)n(u)] = \sigma^2 \delta(t-u)$$

On plugging this into (5.96), the delta function collapses the two integrals into one, and we obtain

$$\text{var}(Y) = \sigma^2 \int \int \delta(t-u)s(t)s(u)dt\, du = \sigma^2 \int s^2(t)dt = \sigma^2 ||s||^2$$

We have therefore shown that $Y \sim N(0, \sigma^2 ||s||^2)$ conditioned on 0 being sent.

(b) Suppose that 1 is sent. Then $y(t) = s(t) + n(t)$ and

$$Y = \int (s(t) + n(t))\, s(t)dt = \int s^2(t)dt + \int n(t)s(t)dt = ||s||^2 + \int n(t)s(t)dt$$

We already know that the second term on the extreme right-hand side has distribution $N(0, \sigma^2||s||^2)$. The distribution remains Gaussian when we add a constant to it, with the mean being translated by this constant. We therefore conclude that $Y \sim N(||s||^2, \sigma^2||s||^2)$, conditioned on 1 being sent.

(c) The decision statistic Y obeys exactly the same model as in Example 5.6.3, with $m = ||s||^2$ and $v^2 = \sigma^2||s||^2$. Applying the intuitive decision rule in that example, we guess that 1 is sent if $Y > ||s||^2/2$, and that 0 is sent otherwise. The probability of error for that decision rule equals

$$P_{\text{e}|0} = P_{\text{e}|1} = P_{\text{e}} = Q\left(\frac{m}{2v}\right) = Q\left(\frac{||s||^2}{2\sigma||s||}\right) = Q\left(\frac{||s||}{2\sigma}\right)$$

Remark The preceding example illustrates that, for linear processing of a received signal corrupted by WGN, the signal term contributes to the mean, and the noise term to the variance, of the resulting decision statistic. The resulting Gaussian distribution is a *conditional* distribution, because it is conditioned on *which* signal is actually sent (or, for on–off keying, *whether* a signal is sent).

Complex-baseband WGN From the definition of a complex envelope that we have used so far (in Chapters 2 through 4), the complex envelope has twice the energy/power of the corresponding passband signal (which may be a sample path of a passband random process). In order to get a unified description of WGN, however, let us now divide the complex envelope of both signal and noise by $1/\sqrt{2}$. This cannot change the performance of the system, but leads to the complex envelope now having the same energy/power as the corresponding passband signal. Effectively, we are switching from defining the complex envelope via $u_{\text{p}}(t) = \text{Re}\left(u(t)e^{j2\pi f_c t}\right)$, to defining it via $u_{\text{p}}(t) = \text{Re}\left(\sqrt{2}u(t)e^{j2\pi f_c t}\right)$. This convention reduces the PSDs of the I and Q component by a factor of two: we now model them as

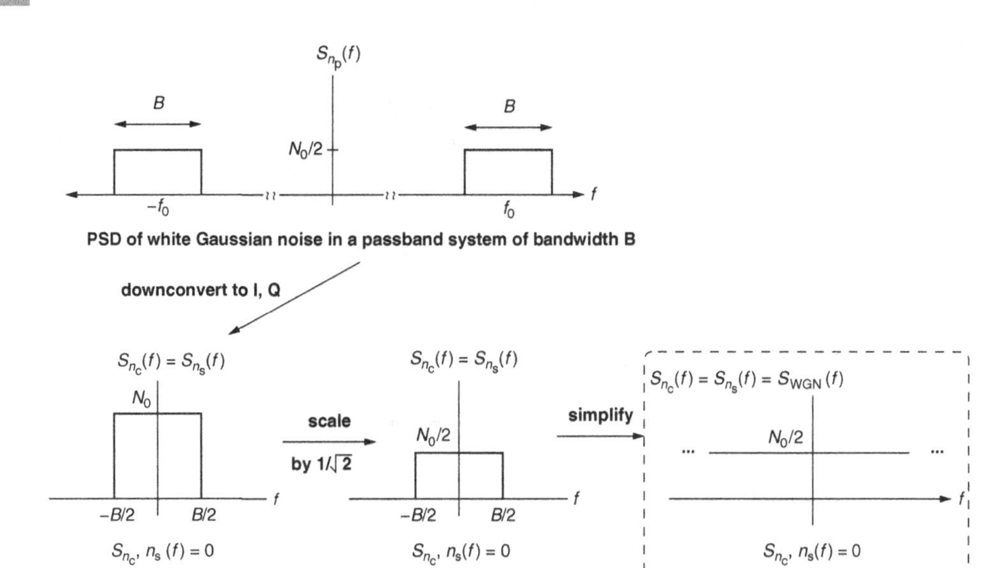

Figure 5.24 We scale the complex envelope for both signal and noise by $1/\sqrt{2}$, so that the I and Q components of passband WGN can be modeled as independent WGN processes with PSD $N_0/2$.

independent real WGN processes, with $S_{n_c}(f) = S_{n_s}(f) \equiv N_0/2 = \sigma^2$. The steps involved in establishing this model are shown in Figure 5.24.

We now have the noise modeling background needed for Chapter 6, where we develop a framework for optimal reception that is based on design criteria such as the error probability. The next section discusses linear processing of random processes, which is useful background for our modeling the effect of filtering on noise, as well as for computing quantities such as the signal-to-noise ratio (SNR). It can be skipped by readers anxious to get to Chapter 6, since the latter includes a self-contained exposition of the effects of the relevant receiver operations on WGN.

5.9 Linear operations on random processes

We now wish to understand what happens when we perform linear operations such as filtering and correlation on a random process. We have already seen an example of this in Example 5.8.2, where WGN was correlated against a deterministic signal. We now develop a more general framework.

It is useful to state up front the following result.

Gaussianity is preserved under linear operations Thus, if the input to a filter is a Gaussian random process, so is the output.

This is because any set of output samples can be expressed as a linear combination of input samples, or the limit of such linear combinations (an integral for computing, for example, a convolution, is the limit of a sum).

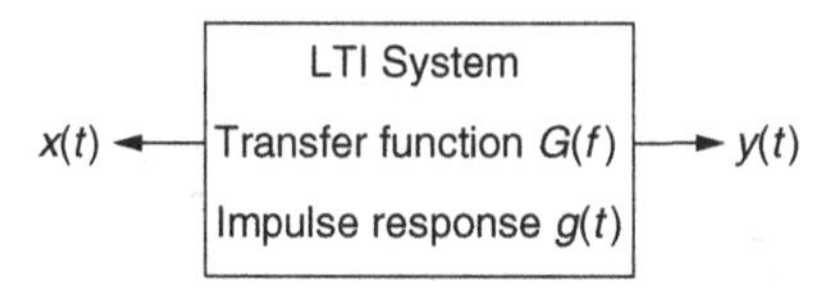

Figure 5.25 A random process through an LTI system.

In the remainder of this section, we discussion the evolution of second-order statistics under linear operations. Of course, for Gaussian random processes, this suffices to provide a complete statistical description of the output of a linear operation.

5.9.1 Filtering

Suppose that a random process $x(t)$ is passed through a filter, or an LTI system, with transfer function $G(f)$ and impulse response $g(t)$, as shown in Figure 5.25.

The PSD of the output $y(t)$ is related to that of the input as follows:

$$S_y(f) = S_x(f)|G(f)|^2 \tag{5.97}$$

This follows immediately from the operational definition of PSD in Figure 5.18, since the power gain due to the filter at frequency f is $|G(f)|^2$. Now,

$$|G(f)|^2 = G(f)G^*(f) \leftrightarrow (g * g_{\mathrm{MF}})(t)$$

where $g_{\mathrm{MF}}(t) = g^*(-t)$. Thus, on taking the inverse Fourier transform on both sides of (5.97), we obtain the following relation between the input and output autocorrelation functions:

$$R_y(\tau) = (R_x * g * g_{\mathrm{MF}})(\tau) \tag{5.98}$$

Let us now derive analogous results for ensemble averages for filtered WSS processes.

Filtered WSS random processes

Suppose that a WSS random process X is passed through an LTI system with impulse response $g(t)$ (which we allow to be complex-valued) to obtain an output $Y(t) = (X * g)(t)$. We wish to characterize the joint second-order statistics of X and Y.

On defining the *cross-correlation function* of Y and X as

$$R_{YX}(t+\tau, t) = \mathbb{E}[Y(t+\tau)X^*(t)]$$

we have

$$R_{YX}(t+\tau, t) = \mathbb{E}\left[\left(\int X(t+\tau-u)g(u)du\right)X^*(t)\right] = \int R_X(\tau-u)g(u)du \tag{5.99}$$

interchanging expectation and integration. Thus, $R_{YX}(t+\tau, t)$ depends only on the time difference τ. We therefore denote it by $R_{YX}(\tau)$. From (5.99), we see that

$$R_{YX}(\tau) = (R_X * g)(\tau)$$

The autocorrelation function of Y is given by

$$R_Y(t+\tau,t) = \mathbb{E}\left[Y(t+\tau)Y^*(t)\right] = \mathbb{E}\left[Y(t+\tau)\left(\int X(t-u)g(u)du\right)^*\right]$$
$$= \int \mathbb{E}[Y(t+\tau)X^*(t-u)]g^*(u)du = \int R_{YX}(\tau+u)g^*(u)du \qquad (5.100)$$

Thus, $R_Y(t+\tau,t)$ depends only on the time difference τ, and we denote it by $R_Y(\tau)$. On recalling that the matched filter $g_{MF}(u) = g^*(-u)$ and replacing u by $-u$ in the integral at the end of (5.100), we obtain that

$$R_Y(\tau) = (R_{YX} * g_{MF})(\tau) = (R_X * g * g_{MF})(\tau)$$

Finally, we note that the mean function of Y is a constant given by

$$m_Y = m_X * g = m_X \int g(u)du = m_X G(0)$$

Thus, X and Y are jointly WSS: X is WSS, Y is WSS, and their cross-correlation function depends on the time difference. The formulas for the second-order statistics, including the corresponding power spectral densities obtained by taking Fourier transforms, are collected below:

$$R_{YX}(\tau) = (R_X * g)(\tau), \qquad S_{YX}(f) = S_X(f)G(f)$$
$$R_Y(\tau) = (R_{YX} * g_{MF})(\tau) = (R_X * g * g_{MF})(\tau), \quad S_Y(f) = S_{YX}(f)G^*(f) = S_X(f)|G(f)|^2 \qquad (5.101)$$

Let us apply these results to infinite-power white noise (we do not need to invoke Gaussianity to compute second-order statistics). While the input has infinite power, as shown in the example below, if the filter impulse response is square integrable, then the output has finite power, and is equal to what we would have obtained if we had assumed that the noise was bandlimited to start with.

Example 5.9.1 (White noise through an LTI system – general formulas) White noise with PSD $S_n(f) \equiv N_0/2$ is passed through an LTI system with impulse response $g(t)$. We wish to find the PSD, autocorrelation function, and power of the output $y(t) = (n * g)(t)$. The PSD is given by

$$S_y(f) = S_n(f)|G(f)|^2 = \frac{N_0}{2}|G(f)|^2 \qquad (5.102)$$

We can compute the autocorrelation function directly or take the inverse Fourier transform of the PSD to obtain

$$R_y(\tau) = (R_n * g * g_{MF})(\tau) = \frac{N_0}{2}(g * g_{MF})(\tau) = \frac{N_0}{2}\int_{-\infty}^{\infty} g(s)g^*(s-\tau)ds \qquad (5.103)$$

The output power is given by

$$\overline{y^2} = \int_{-\infty}^{\infty} S_y(f)df = \frac{N_0}{2}\int_{-\infty}^{\infty}|G(f)|^2\,df = \frac{N_0}{2}\int_{-\infty}^{\infty}|g(t)|^2\,dt = \frac{N_0}{2}||g||^2 \qquad (5.104)$$

where the time-domain expression follows from Parseval's identity, or from setting $\tau = 0$ in (5.103). Thus, the output noise power equals the noise PSD times the energy of the filter impulse response. It is worth noting that the PSD of y is the same as what we would have obtained if the input had been bandlimited white noise, as long as the band is large enough to encompass frequencies where $G(f)$ is nonzero. Even if $G(f)$ is not strictly bandlimited, we get approximately the right answer if the input noise bandwidth is large enough that most of the energy in $G(f)$ falls within it.

When the input random process is Gaussian as well as WSS, the output is also WSS and Gaussian, and the preceding computations of second-order statistics provide a complete statistical characterization of the output process. This is illustrated by the following example, in which WGN is passed through a filter.

Example 5.9.2 (WGN through a boxcar impulse response) Suppose that WGN $n(t)$ with PSD $\sigma^2 = N_0/2 = \frac{1}{4}$ is passed through an LTI system with impulse response $g(t) = I_{[0,2]}(t)$ to obtain the output $y(t) = (n * g)(t)$.

(a) Find the autocorrelation function and PSD of y.
(b) Find $\mathbb{E}[y^2(100)]$.
(c) **(True or false?)** y is a stationary random process.
(d) **(True or false?)** $y(100)$ and $y(102)$ are independent random variables.
(e) **(True or false?)** $y(100)$ and $y(101)$ are independent random variables.
(f) Compute the probability $P[y(100) - y(101) + y(102) > 5]$.
(g) Which of the preceding results rely on the Gaussianity of n?

Solution

(a) Since n is WSS, so is y. The filter matched to g is a boxcar as well: $g_{\mathrm{MF}}(t) = I_{[-2,0]}(t)$. Their convolution is a triangular pulse centered at the origin: $(g * g_{\mathrm{MF}})(\tau) = 2\,(1 - |\tau|/2)\, I_{[-2,2]}(\tau)$. We therefore have

$$R_y(\tau) = \frac{N_0}{2}(g * g_{\mathrm{MF}})(\tau) = \frac{1}{2}\left(1 - \frac{|\tau|}{2}\right) I_{[-2,2]}(\tau) = C_y(\tau)$$

(since y is zero mean). The PSD is given by

$$S_y(f) = \frac{N_0}{2}|G(f)|^2 = \mathrm{sinc}^2(2f)$$

since $|G(f)| = |2\,\mathrm{sinc}(2f)|$. Note that these results do not rely on Gaussianity.

(b) The power $\mathbb{E}[y^2(100)] = R_y(0) = \frac{1}{2}$.

(c) The output y is a Gaussian random process, since it is obtained by a linear transformation of the Gaussian random process n. Since y is WSS and Gaussian, it is stationary. **True**.

(d) The random variables $y(100)$ and $y(102)$ are jointly Gaussian with zero mean and covariance $\text{cov}(y(100), y(102)) = C_y(2) = R_y(2) = 0$. Since they are jointly Gaussian and uncorrelated, they are independent. **True**.

(e) In this case, $\text{cov}(y(100), y(101)) = C_y(1) = R_y(1) = \frac{1}{4} \neq 0$, so $y(100)$ and $y(101)$ are not independent. **False**.

(f) The random variable $Z = y(100) - 2y(101) + 3y(102)$ is zero mean and Gaussian, with

$$\begin{aligned}\text{var}(Z) &= \text{cov}(y(100) - 2y(101) + 3y(102), y(100) - 2y(101) + 3y(102)) \\ &= \text{cov}(y(100), y(100)) + 4\,\text{cov}(y(101), y(101)) + 9\,\text{cov}(y(102), y(102)) \\ &\quad - 4\,\text{cov}(y(100), y(101)) + 6\,\text{cov}(y(100), y(102)) - 12\,\text{cov}(y(100), y(101)) \\ &= C_y(0) + 4C_y(0) + 9C_y(0) - 4C_y(1) + 6C_y(2) - 12C_y(1) \\ &= 14C_y(0) - 16C_y(1) + 6C_y(2) = 3\end{aligned}$$

substituting $C_y(0) = \frac{1}{2}$, $C_y(1) = \frac{1}{4}$, and $C_y(2) = 0$. Thus, $Z \sim N(0, 3)$, and the required probability can be evaluated as

$$P[Z > 5] = Q\left(\frac{5-0}{\sqrt{3}}\right) = 0.0019$$

(g) We invoke Gaussianity in (c), (d), and (f).

5.9.2 Correlation

As we shall see in Chapter 6, a typical operation in a digital communication receiver is to *correlate* a noisy received waveform against one or more noiseless templates. Specifically, the *correlation* of $y(t)$ (e.g., a received signal) against $g(t)$ (e.g., a noiseless template at the receiver) is defined as the inner product between y and g, given by

$$\langle y, g \rangle = \int_{-\infty}^{\infty} y(t) g^*(t) dt \qquad (5.105)$$

(We restrict our attention to real-valued signals in example computations provided here, but the preceding notation is general enough to include complex-valued signals.)

Signal-to-noise ratio and its maximization

If $y(t)$ is a random process, we can compute the mean and variance of $\langle y, g \rangle$ given the second-order statistics (i.e., mean function and autocorrelation function) of y, as shown in Problem 5.50. However, let us consider here a special case of particular interest in the study of communication systems:

$$y(t) = s(t) + n(t)$$

where we now restrict our attention to real-valued signals for simplicity, with $s(t)$ denoting a deterministic signal (e.g., corresponding to a specific choice of transmitted symbols) and

$n(t)$ zero-mean white noise with PSD $S_n(f) \equiv N_0/2$. The output of correlating y against g is given by

$$Z = \langle y, g \rangle = \langle s, g \rangle + \langle n, g \rangle = \int_{-\infty}^{\infty} s(t)g(t)dt + \int_{-\infty}^{\infty} n(t)g(t)dt$$

Since both the signal term and the noise term scale up by identical factors if we scale up g, a performance metric of interest is the *ratio* of the signal power to the noise power at the output of the correlator, defined as follows:

$$\text{SNR} = \frac{|\langle s, g \rangle|^2}{\mathbb{E}[|\langle n, g \rangle|^2]}$$

How should we choose g in order to maximize the SNR? In order to answer this, we need to compute the noise power in the denominator. We can rewrite it as

$$\mathbb{E}[|\langle n, g \rangle|^2] = \mathbb{E}\left[\int n(t)g(t)dt \int n(s)g(s)ds\right]$$

where we need to use two different dummy variables of integration to make sure we capture all the cross terms in the two integrals. Now, we take the expectation inside the integrals, grouping all random terms together inside the expectation:

$$\mathbb{E}[|\langle n, g \rangle|^2] = \int\int \mathbb{E}[n(t)n(s)]g(t)g(s)dt\,ds = \int\int R_n(t-s)g(t)g(s)dt\,ds$$

This is where the infinite-power white-noise model becomes useful: on plugging in $R_n(t-s) = (N_0/2)\delta(t-s)$, we find that the two integrals collapse into one, and obtain that

$$\mathbb{E}[|\langle n, g \rangle|^2] = \frac{N_0}{2}\int\int \delta(t-s)g(t)g(s)dt\,ds = \frac{N_0}{2}\int |g(t)|^2\,dt = \frac{N_0}{2}||g||^2 \qquad (5.106)$$

Thus, the SNR can be rewritten as

$$\text{SNR} = \frac{|\langle s, g \rangle|^2}{(N_0/2)||g||^2} = \frac{2}{N_0}|\langle s, g/||g|| \rangle|^2$$

Drawing on the analogy between signals and vectors, note that $g/||g||$ is the "unit vector" pointing along g. We wish to choose g such that the size of the projection of the signal s along this unit vector is maximized. Clearly, this is accomplished by choosing the unit vector along the direction of s. (A formal proof using the Cauchy–Schwarz inequality is provided in Problem 5.49.) That is, we must choose g to be a scalar multiple of s (any scalar multiple will do, since the SNR is a scale-invariant quantity). In general, for complex-valued signals in complex-valued white noise (useful for modeling in complex baseband), it can be shown that g must be a scalar multiple of $s^*(t)$. When we plug this in, the maximum SNR we obtain is $2||s||^2/N_0$. These results are important enough to state formally, and we do this below.

Theorem 5.9.1 *For linear processing of a signal $s(t)$ corrupted by white noise, the output SNR is maximized by correlating against $s(t)$. The resulting SNR is given by*

$$\text{SNR}_{\max} = \frac{2||s||^2}{N_0} \qquad (5.107)$$

The expression (5.106) for the noise power at the output of a correlator is analogous to the expression (5.104) (Example 5.9.1) for the power of white noise through a filter. This is no coincidence. Any correlation operation can be implemented using a filter and sampler, as we discuss next.

Matched filter

Correlation with a waveform $g(t)$ can be achieved using a filter $h(t) = g^*(-t)$ and sampling at time $t = 0$. To see this, note that

$$z(0) = (y * h)(0) = \int_{-\infty}^{\infty} y(\tau)h(-\tau)d\tau = \int_{-\infty}^{\infty} y(\tau)g^*(\tau)d\tau$$

On comparing this with the correlator output (5.105), we see that $Z = z(0)$. Now, applying Theorem 5.9.1, we see that the SNR is maximized by choosing the filter impulse response as $s^*(-t)$. As we know, this is called the *matched filter* for s, and we denote its impulse response as $s_{MF}(t) = s^*(-t)$. We can now restate Theorem 5.9.1 as follows.

Theorem 5.9.2 *For linear processing of a signal $s(t)$ corrupted by white noise, the output SNR is maximized by employing a matched filter with impulse response $s_{MF}(t) = s^*(-t)$, sampled at time $t = 0$.*

The statistics of the *noise* contribution to the matched filter output do not depend on the sampling time (WSS noise into an LTI system yields a WSS random process), hence the optimum sampling time is determined by the peak of the *signal* contribution to the matched filter output. The signal contribution to the output of the matched filter at time t is given by

$$z(t) = \int s(\tau)s_{MF}(t-\tau)d\tau = \int s(\tau)s^*(\tau - t)d\tau$$

This is simply the correlation of the signal with itself at delay t. Thus, the matched filter enables us to implement an infinite bank of correlators, each corresponding to a version of our signal template at a different delay. Figure 5.26 shows a rectangular pulse passed through its matched filter. For received signal $y(t) = s(t) + n(t)$, we have observed that the optimum sampling time (i.e., the correlator choice maximizing the SNR) is $t = 0$. More generally, when the received signal is given by $y(t) = s(t - t_0) + n(t)$, the peak of the signal contribution to the matched filter shifts to $t = t_0$, which now becomes the optimum sampling time.

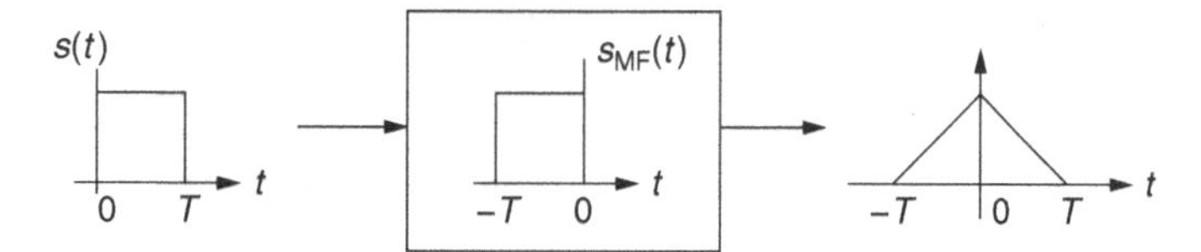

Figure 5.26 A signal passed through its matched filter gives a peak at time $t = 0$. When the signal is delayed by t_0, the peak occurs at $t = t_0$.

While the preceding computations rely only on second-order statistics, once we invoke the Gaussianity of the noise, as we do in Chapter 6, we will be able to compute probabilities (a preview of such computations is provided by Examples 5.8.2 and 5.9.2(f)). This will enable us to develop a framework for receiver design for minimizing the probability of error.

5.10 Concept summary

We do not summarize here the review of probability and random variables, but note that the key concepts relevant for communication systems modeling are conditional probabilities and densities, and associated results such as the law of total probability and Bayes' rule. As we see in much greater detail in Chapter 6, conditional probabilities and densities are used for statistical characterization of the received signal, given the transmitted signal, while Bayes' rule can be used to infer which signal was transmitted, given the received signal.

Gaussian random variables

- A Gaussian random variable $X \sim N(m, v^2)$ is characterized by its mean m and variance v^2.
- Gaussianity is preserved under translation and scaling. Particularly useful is the transformation to a standard ($N(0,1)$) Gaussian random variable: if $X \sim N(m, v^2)$, then $(X - m)/v \sim N(0,1)$. This allows probabilities involving any Gaussian random variable to be expressed in terms of the CDF $\Phi(x)$ and CCDF $Q(x)$ for a standard Gaussian random variable.
- Random variables $X_1, \ldots, X_n$ are jointly Gaussian, or $\mathbf{X} = (X_1, \ldots, X_n)^{\mathrm{T}}$ is a Gaussian random vector, if any linear combination $\mathbf{a}^{\mathrm{T}}\mathbf{X} = a_1X_1 + \cdots + a_nX_n$ is a Gaussian random variable.
- A Gaussian random vector $\mathbf{X} \sim N(\mathbf{m}, \mathbf{C})$ is completely characterized by its mean vector $\mathbf{m}$ and covariance matrix $\mathbf{C}$.
- Uncorrelated and jointly Gaussian random variables are independent.
- The joint density for $\mathbf{X} \sim N(\mathbf{m}, \mathbf{C})$ exists if and only if $\mathbf{C}$ is invertible.
- The mean vector and covariance matrix evolve separately under affine transformations: for $\mathbf{Y} = \mathbf{AX} + \mathbf{b}$, $\mathbf{m}_Y = \mathbf{Am}_X + \mathbf{b}$ and $\mathbf{C}_Y = \mathbf{AC}_X\mathbf{A}^{\mathrm{T}}$.
- Joint Gaussianity is preserved under affine transformations: if $\mathbf{X} \sim N(\mathbf{m}, \mathbf{C})$ and $\mathbf{Y} = \mathbf{AX} + \mathbf{b}$, then $\mathbf{Y} \sim N(\mathbf{Am} + \mathbf{b}, \mathbf{ACA}^{\mathrm{T}})$.

Random processes

- A random process is a generalization of the concept of a random vector; it is a collection of random variables on a common probability space.

- While statistical characterization of a random process requires specification of the finite-dimensional distributions, coarser characterization via its second-order statistics (the mean and autocorrelation functions) is often employed.
- A random process X is stationary if its statistics are shift-invariant; it is WSS if its second-order statistics are shift-invariant.
- A random process is Gaussian if any collection of samples is a Gaussian random vector, or equivalently, if any linear combination of any collection of samples is a Gaussian random variable.
- A Gaussian random process is completely characterized by its mean and autocorrelation (or mean and autocovariance) functions.
- A stationary process is WSS. A WSS Gaussian random process is stationary.
- The autocorrelation function and the power spectral density form a Fourier transform pair. (This observation applies both to time averages and to ensemble averages for WSS processes.)
- The most common model for noise in communication systems is WGN. WGN $n(t)$ is zero mean, WSS, Gaussian with a flat PSD $S_n(f) = \sigma^2 = N_0/2 \leftrightarrow R_n(\tau) = \sigma^2 \delta(\tau)$. While physically unrealizable (it has infinite power), it is a useful mathematical abstraction for modeling the flatness of the noise PSD over the band of interest. In complex baseband, noise is modeled as I and Q components that are independent real-valued WGN processes.
- A WSS random process X through an LTI system with impulse response $g(t)$ yields a WSS random process Y. X and Y are also jointly WSS. We have $S_Y(f) = S_X(f)|G(f)|^2 \leftrightarrow R_Y(\tau) = (R_X * g * g_{MF})(\tau)$.
- The statistics of WGN after linear operations such as correlation and filtering are easy to compute because of its impulsive autocorrelation function.
- When the received signal equals signal plus WGN, the SNR is maximized by matched filtering against the signal.

5.11 Notes

There are some textbooks on probability and random processes for engineers that can be used to supplement the brief communications-centric exposition here, including Yates and Goodman [25], Woods and Stark [26], Leon-Garcia [27], and Papoulis and Pillai [28].

A more detailed treatment of the noise analysis for analog modulation provided in Appendix 5.E can be found in a number of communication theory texts, with Ziemer and Tranter [4] providing a sound exposition.

As a historical note, thermal noise, which plays such a crucial role in communications systems design, was first experimentally characterized in 1928 by Johnson [29]. Johnson discussed his results with Nyquist, who quickly came up with a theoretical characterization [30]. See [31] for a modern re-derivation of Nyquist's formula, and [32] for a discussion of noise in transistors. These papers and the references therein are good resources for further exploration into the physical basis for noise, which we can only hint at here in Appendix

5.C. Of course, as discussed in Section 5.8, from a communication systems designer's point of view, it typically suffices to abstract away from such physical considerations, using the noise figure as a single number summarizing the effect of receiver circuit noise.

5.12 Problems

Conditional probabilities, law of total probability, and Bayes' rule

Problem 5.1 You are given a pair of dice (each with six sides). One is fair, the other is unfair. The probability of rolling 6 with the unfair die is $1/2$, while the probability of rolling 1 through 5 is $1/10$. You now pick one of the dice at random and begin rolling. Conditioned on the die picked, successive rolls are independent.

(a) Conditioned on picking the unfair die, what is the probability of the sum of the numbers in the first two rolls being equal to 10?
(b) Conditioned on getting a sum of 10 in your first two throws, what is the probability that you picked the unfair die?

Problem 5.2 A student who studies for an exam has a 90% chance of passing. A student who does not study for the exam has a 90% chance of failing. Suppose that 70% of the students studied for the exam.

(a) What is the probability that a student fails the exam?
(b) What is the probability that a student who fails studied for the exam?
(c) What is the probability that a student who fails did not study for the exam?
(d) Would you expect the probabilities in (b) and (c) to add up to one?

Problem 5.3 A receiver decision statistic Y in a communication system is modeled as exponential with mean 1 if 0 is sent, and as exponential with mean 10 if 1 is sent. Assume that we send 0 with probability 0.6.

(a) Find the conditional probability that $Y > 5$, given that 0 is sent.
(b) Find the conditional probability that $Y > 5$, given that 1 is sent.
(c) Find the unconditional probability that $Y > 5$.
(d) Given that $Y > 5$, what is the probability that 0 is sent?
(e) Given that $Y = 5$, what is the probability that 0 is sent?

Problem 5.4 Channel codes are constructed by introducing redundancy in a structured fashion. A canonical means of doing this is by introducing *parity checks*. In this problem, we see how one can make inferences based on three bits, b_1, b_2, and b_3 which satisfy a parity-check equation: $b_1 \oplus b_2 \oplus b_3 = 0$. Here $\oplus$ denotes an exclusive or (XOR) operation.

(a) Suppose that we know that $P[b_1 = 0] = 0.8$ and $P[b_2 = 1] = 0.9$, and model b_1 and b_2 as independent. Find the probability $P[b_3 = 0]$.

(b) Define the *log likelihood ratio (LLR)* for a bit b as $\text{LLR}(b) = \log(P[b = 0]/P[b = 1])$. Setting $L_i = \text{LLR}(b_i)$, $i = 1, 2, 3$, find an expression for L_3 in terms of L_1 and L_2, again modeling b_1 and b_2 as independent.

Problem 5.5 A bit $X \in \{0, 1\}$ is repeatedly transmitted using n independent uses of a binary symmetric channel (i.e., the binary channel in Figure 5.2 with $a = b$) with crossover probability $a = 0.1$. The receiver uses a majority rule to make a decision on the transmitted bit. Derive general expressions as a function of n (assume that n is odd, so there are no ties in the majority rule), and substitute $n = 5$ for numerical results and plots.

(a) Let Z denote the number of ones at the channel output. (Z takes values $0, 1, \ldots, n$.) Specify the probability mass function of Z, conditioned on $X = 0$.
(b) Conditioned on $X = 0$, what is the probability of deciding that one was sent (i.e., what is the probability of making an error)?
(c) Find the posterior probabilities $P[X = 0|Z = m]$, $m = 0, 1, \ldots, n$, assuming that 0 and 1 are equally likely to be sent. Make a stem plot against m.
(d) Repeat (c) assuming that the 0 is sent with probability 0.9.
(e) As an alternative visualization, plot the LLR $\log(P[X = 0|Z = m]/P[X = 1|Z = m])$ versus m for (c) and (d).

Problem 5.6 Consider the two-input, four-output channel with transition probabilities shown in Figure 5.27. In your numerical computations, take $p = 0.05$, $q = 0.1$, and $r = 0.3$. Denote the channel input by X and the channel output by Y.

(a) Assume that 0 and 1 are equally likely to be sent. Find the conditional probability of 0 being sent, given each possible value of the output. That is, compute $P[X = 0|Y = y]$ for each $y \in \{-3, -1, +1, +3\}$.
(b) Express the results in (a) as log likelihood ratios (LLRs). That is, compute $L(y) = \log(P[X = 0|Y = y]/P[X = 1|Y = y])$ for each $y \in \{-3, -1, +1, +3\}$.
(c) Assume that a bit X, chosen equiprobably from $\{0, 1\}$, is sent repeatedly, using three independent uses of the channel. The channel outputs can be represented as a vector $\mathbf{Y} = (Y_1, Y_2, Y_3)^T$. For channel outputs $\mathbf{y} = (+1, +3, -1)^T$, find the conditional probabilities $P[\mathbf{Y} = \mathbf{y}|X = 0]$ and $P[\mathbf{Y} = \mathbf{y}|X = 1]$.

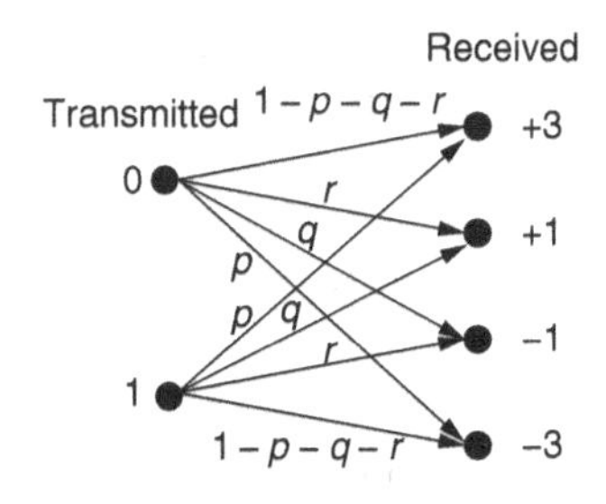

Figure 5.27 The two-input four-output channel for Problem 5.6.

(d) Use Bayes' rule and the result of (c) to find the posterior probability $P[X = 0|\mathbf{Y} = \mathbf{y}]$ for $\mathbf{y} = (+1, +3, -1)^{\mathrm{T}}$. Also compute the corresponding LLR $L(\mathbf{y}) = \log(P[X = 0|\mathbf{Y} = \mathbf{y}]/P[X = 1|\mathbf{Y} = \mathbf{y}])$.

(e) Would you decide that 0 or 1 was sent when you see the channel output $\mathbf{y} = (+1, +3, -1)^{\mathrm{T}}$?

Random variables

Problem 5.7 Let X denote an exponential random variable with mean 10.

(a) What is the probability that X is bigger than 20?

(b) What is the probability that X is smaller than 5?

(c) Suppose that we know that X is bigger than 10. What is the conditional probability that it is bigger than 20?

(d) Find $\mathbb{E}[e^{-X}]$.

(e) Find $\mathbb{E}[X^3]$.

Problem 5.8 Let $U_1, \ldots, U_n$ denote i.i.d. random variables with CDF $F_U(u)$.

(a) Let $X = \max(U_1, \ldots, U_n)$. Show that

$$P[X \leq x] = F_U^n(x)$$

(b) Let $Y = \min(U_1, \ldots, U_n)$. Show that

$$P[Y \leq y] = 1 - (1 - F_U(y))^n$$

(c) Suppose that $U_1, \ldots, U_n$ are uniform over $[0, 1]$. Plot the CDF of X for $n = 1$, $n = 5$, and $n = 10$, and comment on any trends that you notice.

(d) Repeat (c) for the CDF of Y.

Problem 5.9 **(True or false?)** The minimum of two independent exponential random variables is exponential.
(True or false?) The maximum of two independent exponential random variables is exponential.

Problem 5.10 Let U and V denote independent and identically distributed random variables, uniformly distributed over $[0, 1]$.

(a) Find and sketch the CDF of $X = \min(U, V)$.
Hint. It might be useful to consider the complementary CDF.

(b) Find and sketch the CDF of $Y = V/U$. Make sure you specify the range of values taken by Y.
Hint. It is helpful to draw pictures in the (u, v) plane when evaluating the probabilities of interest.

Problem 5.11 (The Relation between Gaussian and exponential) Suppose that X_1 and X_2 are i.i.d. $N(0, 1)$.

(a) Show that $Z = X_1^2 + X_2^2$ is exponential with mean 2.
(b) **(True or false?)** Z is independent of $\Theta = \tan^{-1}(X_2/X_1)$.

Hint. Use the results from Example 5.4.3, which tells us the joint distribution of $\sqrt{Z}$ and Θ.

Problem 5.12 (The role of the uniform random variable in simulations) Let U denote a uniform random variable that is uniformly distributed over $[0, 1]$.

(a) Let $F(x)$ denote an arbitrary CDF (assume for simplicity that it is continuous). Defining $X = F^{-1}(U)$, show that X has CDF $F(x)$.

Remark This gives us a way of generating random variables with arbitrary distributions, assuming that we have a random number generator for uniform random variables. The method works even if X is a discrete or mixed random variable, as long as F^{-1} is defined appropriately.

(b) Find a function g such that $Y = g(U)$ is exponential with mean 2, where U is uniform over $[0, 1]$.
(c) Use the result in (b) and MATLAB's rand() function to generate an i.i.d. sequence of 1000 exponential random variables with mean 2. Plot the histogram and verify that it has the right shape.

Problem 5.13 (Generating Gaussian random variables) Suppose that U_1 and U_2 are i.i.d. and uniform over $[0, 1]$.

(a) What is the joint distribution of $Z = -2 \ln U_1$ and $\Theta = 2\pi U_2$?
(b) Show that $X_1 = \sqrt{Z} \cos \Theta$ and $X_2 = \sqrt{Z} \sin \Theta$ are i.i.d. $N(0, 1)$ random variables. *Hint.* Use Example 5.4.3 and Problem 5.11.
(c) Use the result of (b) to generate 2000 i.i.d. $N(0, 1)$ random variables from 2000 i.i.d. random variables uniformly distributed over $[0, 1]$, using MATLAB's rand() function. Check that the histogram has the right shape.
(d) Use simulations to estimate $\mathbb{E}[X^2]$, where $X \sim N(0, 1)$, and compare your answer with the analytical result.
(e) Use simulations to estimate $P[X^3 + X > 3]$, where $X \sim N(0, 1)$.

Problem 5.14 (Generating discrete random variables) Let $U_1, \ldots, U_n$ denote i.i.d. random variables uniformly distributed over $[0, 1]$ (e.g., generated by the rand() function in MATLAB). Define, for $i = 1, \ldots, n$,

$$Y_i = \begin{cases} 1, & U_i > 0.7 \\ 0, & U_i \leq 0.7 \end{cases}$$

(a) Sketch the CDF of Y_1.
(b) Find (analytically) and plot the PMF of $Z = Y_1 + \cdots + Y_n$, for $n = 20$.

(c) Use simulation to estimate and plot the histogram of Z, and compare your result against the PMF in (b).
(d) Estimate $\mathbb{E}[Z]$ by simulation and compare your answer against the analytical result.
(e) Estimate $\mathbb{E}[Z^3]$ by simulation.

Gaussian random variables

Problem 5.15 Two random variables X and Y have joint density

$$p_{X,Y}(x,y) = \begin{cases} Ke^{-(2x^2+y^2)/2} & xy \geq 0 \\ 0 & xy < 0 \end{cases}$$

(a) Find K.
(b) Show that both X and Y are Gaussian random variables.
(c) Express the probability $P[X^2 + X > 2]$ in terms of the Q function.
(d) Are X and Y jointly Gaussian?
(e) Are X and Y independent?
(f) Are X and Y uncorrelated?
(g) Find the conditional density $p_{X|Y}(x|y)$. Is it Gaussian?

Problem 5.16 (Computations involving joint Gaussianity) The random vector $\mathbf{X} = (X_1 X_2)^T$ is Gaussian with mean vector $\mathbf{m} = (2, 1)^T$ and covariance matrix $\mathbf{C}$ given by

$$\mathbf{C} = \begin{pmatrix} 1 & -1 \\ -1 & 4 \end{pmatrix}$$

(a) Let $Y_1 = X_1 + 2X_2$ and $Y_2 = -X_1 + X_2$. Find $\text{cov}(Y_1, Y_2)$.
(b) Write down the joint density of Y_1 and Y_2.
(c) Express the probability $P[Y_1 > 2Y_2 + 1]$ in terms of the Q function.

Problem 5.17 (Computations involving joint Gaussianity) The random vector $\mathbf{X} = (X_1 X_2)^T$ is Gaussian with mean vector $\mathbf{m} = (-3, 2)^T$ and covariance matrix $\mathbf{C}$ given by

$$\mathbf{C} = \begin{pmatrix} 4 & -2 \\ -2 & 9 \end{pmatrix}$$

(a) Let $Y_1 = 2X_1 - X_2$ and $Y_2 = -X_1 + 3X_2$. Find $\text{cov}(Y_1, Y_2)$.
(b) Write down the joint density of Y_1 and Y_2.
(c) Express the probability $P[Y_2 > 2Y_1 - 1]$ in terms of the Q function with positive arguments.
(d) Express the probability $P[Y_1^2 > 3Y_1 + 10]$ in terms of the Q function with positive arguments.

Problem 5.18 (Plotting the joint Gaussian density) For jointly Gaussian random variables X and Y, plot the density and its contours as in Figure 5.15 for the following parameters in (a)–(c).

(a) $\sigma_X^2 = 1, \sigma_Y^2 = 1, \rho = 0$.
(b) $\sigma_X^2 = 1, \sigma_Y^2 = 1, \rho = 0.5$.
(c) $\sigma_X^2 = 4, \sigma_Y^2 = 1, \rho = 0.5$.
(d) Comment on the differences among the plots in the three cases.

Problem 5.19 (Computations involving joint Gaussianity) In each of the three cases in Problem 5.18,

(a) specify the distribution of $X - 2Y$; and
(b) determine whether $X - 2Y$ is independent of X?

Problem 5.20 (Computations involving joint Gaussianity) X and Y are jointly Gaussian, each with variance one, and with normalized correlation $-\frac{3}{4}$. The mean of X equals one, and the mean of Y equals two.

(a) Write down the covariance matrix.
(b) What is the distribution of $Z = 2X + 3Y$?
(c) Express the probability $P[Z^2 - Z > 6]$ in terms of Q function with positive arguments, and then evaluate it numerically.

Problem 5.21 (From Gaussian to Rayleigh, Rician, and exponential random variables) Let X_1 and X_2 be i.i.d. Gaussian random variables, each with mean zero and variance v^2. Define (R, Φ) as the polar representation of the point (X_1, X_2), i.e.,

$$X_1 = R \cos \Phi, \quad X_2 = R \sin \Phi$$

where $R \geq 0$ and $\Phi \in [0, 2\pi]$.

(a) Find the joint density of R and Φ.
(b) Observe from (a) that R and Φ are independent. Show that Φ is uniformly distributed in $[0, 2\pi]$, and find the marginal density of R.
(c) Find the marginal density of R^2.
(d) What is the probability that R^2 is at least 20 dB below its mean value? Does your answer depend on the value of v^2?

Remark The random variable R is said to have a Rayleigh distribution. Further, you should recognize that R^2 has an exponential distribution.

Random processes

Problem 5.22 Let $X(t) = 2\sin(20\pi t + \Theta)$, where Θ takes values with equal probability in the set $\{0, \pi/2, \pi, 3\pi/2\}$.

(a) Find the ensemble-averaged mean function and autocorrelation function of X.
(b) Is X WSS?
(c) Is X stationary?

(d) Find the time-averaged mean and autocorrelation function of X. Do these depend on the realization of Θ?
(e) Is X ergodic in mean and autocorrelation?

Problem 5.23 For each of the following functions, sketch it and state whether it can be a valid autocorrelation function. Give reasons for your answers.

(a) $f_1(\tau) = (1 - |\tau|)\, I_{[-1,1]}(\tau)$.
(b) $f_2(\tau) = f_1(\tau - 1)$.
(c) $f_3(\tau) = f_1(\tau) - \frac{1}{2}(f_1(\tau - 1) + f_1(\tau + 1))$.

Problem 5.24 Consider the random process $X_p(t) = X_c(t)\cos(2\pi f_c t) - X_s(t)\sin(2\pi f_c t)$, where X_c and X_s are random processes defined on a common probability space.

(a) Find conditions on X_c and X_s such that X_p is WSS.
(b) Specify the (ensemble-averaged) autocorrelation function and PSD of X_p under the conditions in (a).
(c) Assuming that the conditions in (a) hold, what are the additional conditions for X_p to be a passband random process?

Problem 5.25 Consider the square wave $x(t) = \sum_{n=-\infty}^{\infty}(-1)^n p(t - n)$, where $p(t) = I_{[-1/2,1/2]}(t)$.

(a) Find the time-averaged autocorrelation function of x by direct computation in the time domain.
Hint. The autocorrelation function of a periodic signal is periodic.
(b) Find the Fourier series for x, and use this to find the PSD of x.
(c) Are the answers in (a) and (b) consistent?

Problem 5.26 Consider again the square wave $x(t) = \sum_{n=-\infty}^{\infty}(-1)^n p(t - n)$, where $p(t) = I_{[-1/2,1/2]}(t)$. Define the random process $X(t) = x(t - D)$, where D is a random variable that is uniformly distributed over the interval $[0, 1]$.

(a) Find the ensemble-averaged autocorrelation function of X.
(b) Is X WSS?
(c) Is X stationary?
(d) Is X ergodic in mean and autocorrelation function?

Problem 5.27 Let $n(t)$ denote a zero mean baseband random process with PSD $S_n(f) = I_{[-1,1]}(f)$. Find and sketch the PSD of the following random processes.

(a) $x_1(t) = (dn/dt)(t)$.
(b) $x_2(t) = (n(t) - n(t - d))/d$, for $d = \frac{1}{2}$.
(c) Find the powers of x_1 and x_2.

Problem 5.28 Consider a WSS random process with autocorrelation function $R_X(\tau) = e^{-a|\tau|}$, where $a > 0$.

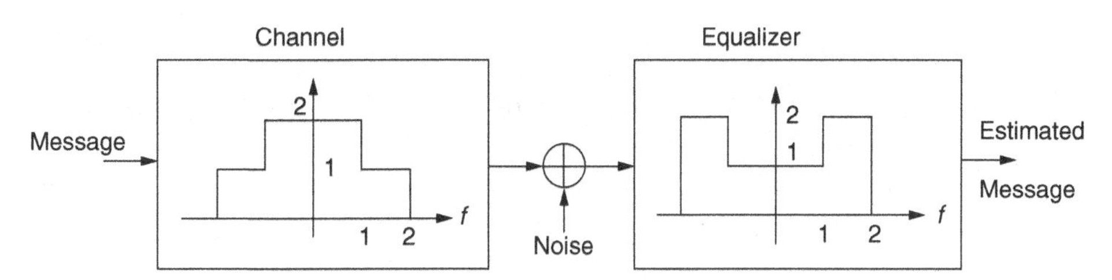

Figure 5.28 The baseband communication system in Problem 5.29.

(a) Find the output power when X is passed through an ideal LPF of bandwidth W.
(b) Find the 99% power containment bandwidth of X. How does it scale with the parameter a?

Problem 5.29 Consider the baseband communication system depicted in Figure 5.28, where the message is modeled as a random process with PSD $S_m(f) = 2(1 - |f|/2)\,I_{[-2,2]}(f)$. Receiver noise is modeled as bandlimited white noise with two-sided PSD $S_n(f) = \frac{1}{4}I_{[-3,3]}(f)$. The equalizer removes the signal distortion due to the channel.

(a) Find the signal power at the *channel input.*
(b) Find the signal power at the *channel output.*
(c) Find the SNR at the *equalizer input.*
(d) Find the SNR at the *equalizer output.*

Problem 5.30 A zero mean WSS random process X has power spectral density $S_X(f) = (1 - |f|)I_{[-1,1]}(f)$.

(a) Find $\mathbb{E}[X(100)X(100.5)]$, leaving your answer in as explicit a form as you can.
(b) Find the output power when X is passed through a filter with impulse response $h(t) = \operatorname{sinc} t$.

Problem 5.31 A signal $s(t)$ in a communication system is modeled as a zero mean random process with PSD $S_s(f) = (1 - |f|)I_{[-1,1]}(f)$. The received signal is given by $y(t) = s(t) + n(t)$, where n is WGN with PSD $S_n(f) \equiv 0.001$. The received signal is passed through an ideal lowpass filter with transfer function $H(f) = I_{[-B,B]}(f)$.

(a) Find the SNR (ratio of signal power to noise power) at the *filter input.*
(b) Is the SNR at the *filter output* better for $B = 1$ or $B = \frac{1}{2}$? Give a *quantitative* justification for your answer.

Problem 5.32 White noise n with PSD $N_0/2$ is passed through an RC filter with impulse response $h(t) = e^{-t/T_0}I_{[0,\infty)}(t)$, where T_0 is the RC time constant, to obtain the output $y = n * h$.

(a) Find the autocorrelation function, PSD, and power of y.
(b) Assuming now that the noise is a Gaussian random process, find a value of t_0 such that $y(t_0) - \frac{1}{2}y(0)$ is independent of $y(0)$, or say why such a t_0 cannot be found.

Problem 5.33 Find the noise power at the output of the filter for the following two scenarios:

(a) Baseband white noise with (two-sided) PSD $N_0/2$ is passed through a filter with impulse response $h(t) = \text{sinc}^2 t$.
(b) Passband white noise with (two-sided) PSD $N_0/2$ is passed through a filter with impulse response $h(t) = \text{sinc}^2 t \cos(100\pi t)$.

Problem 5.34 Suppose that WGN $n(t)$ with PSD $\sigma^2 = N_0/2 = 1$ is passed through a filter with impulse response $h(t) = I_{[-1,1]}(t)$ to obtain the output $y(t) = (n * h)(t)$.

(a) Find and sketch the output power spectral density $S_y(f)$, carefully labeling the axes.
(b) Specify the joint distribution of the three consecutive samples $y(1), y(2)$, and $y(3)$.
(c) Find the probability that $y(1) - 2y(2) + y(3)$ exceeds 10.

Problem 5.35 (Computations involving deterministic signal plus WGN) Consider the noisy received signal

$$y(t) = s(t) + n(t)$$

where $s(t) = I_{[0,3]}(t)$ and $n(t)$ is WGN with PSD $\sigma^2 = N_0/2 = 1/4$. The receiver computes the following statistics:

$$Y_1 = \int_0^2 y(t)dt, \quad Y_2 = \int_1^3 y(t)dt$$

(a) Specify the joint distribution of Y_1 and Y_2.
(b) Compute the probability $P[Y_1 + Y_2 < 2]$, expressing it in terms of the Q function with positive arguments.

Problem 5.36 (Filtered WGN) Let $n(t)$ denote WGN with PSD $S_n(f) \equiv \sigma^2$. We pass $n(t)$ through a filter with impulse response $h(t) = I_{[0,1]}(t) - I_{[1,2]}(t)$ to obtain $z(t) = (n * h)(t)$.

(a) Find and sketch the autocorrelation function of $z(t)$.
(b) Specify the joint distribution of $z(49)$ and $z(50)$.
(c) Specify the joint distribution of $z(49)$ and $z(52)$.
(d) Evaluate the probability $P[2z(50) > z(49) + z(51)]$. Assume $\sigma^2 = 1$.
(e) Evaluate the probability $P[2z(50) > z(49) + z(51) + 2]$. Assume $\sigma^2 = 1$.

Problem 5.37 (Filtered WGN) Let $n(t)$ denote WGN with PSD $S_n(f) \equiv \sigma^2$. We pass $n(t)$ through a filter with impulse response $h(t) = 2I_{[0,2]}(t) - I_{[1,2]}(t)$ to obtain $z(t) = (n * h)(t)$.

(a) Find and sketch the autocorrelation function of $z(t)$.
(b) Specify the joint distribution of $z(0), z(1)$, and $z(2)$.
(c) Compute the probability $P[z(0) - z(1) + z(2) > 4]$ (assume $\sigma^2 = 1$).

Problem 5.38 (Filtered and sampled WGN) Let $n(t)$ denote WGN with PSD $S_n(f) \equiv \sigma^2$. We pass $n(t)$ through a filter with impulse response $h(t)$ to obtain $z(t) = (n * h)(t)$, and then sample it at rate $1/T_s$ to obtain the sequence $z[n] = z(nT_s)$, where n takes integer values.

(a) Show that

$$\text{cov}(z[n], z[m]) = \mathbb{E}[z[n]z^*[m]] = \frac{N_0}{2} \int h(t)h^*(t - (n-m)T_s)dt$$

(We are interested in real-valued impulse responses, but we continue to develop a framework general enough to encompass complex-valued responses.)

(b) For $h(t) = I_{[0,1]}(t)$, specify the joint distribution of $(z[1], z[2], z[3])^T$ for a sampling rate of 2 ($T_s = \frac{1}{2}$).

(c) Repeat (b) for a sampling rate of 1.

(d) For a general h sampled at rate $1/T_s$, show that the noise samples are independent if $h(t)$ is square-root Nyquist at rate $1/T_s$.

Problem 5.39 Consider the signal $s(t) = I_{[0,2]}(t) - 2I_{[1,3]}(t)$.

(a) Find and sketch the impulse response $s_{MF}(t)$ of the matched filter for s.

(b) Find and sketch the output when $s(t)$ is passed through its matched filter.

(c) Suppose that, instead of the matched filter, all we have available is a filter with impulse response $h(t) = I_{[0,1]}(t)$. For an arbitrary input signal $x(t)$, show how $z(t) = (x * s_{MF})(t)$ can be synthesized from $y(t) = (x * h)(t)$.

Problem 5.40 (Correlation via filtering and sampling) A signal $x(t)$ is passed through a filter with impulse response $h(t) = I_{[0,2]}(t)$ to obtain an output $y(t) = (x * h)(t)$.

(a) Find and sketch a signal $g_1(t)$ such that

$$y(2) = \langle x, g_1 \rangle = \int x(t)g_1(t)dt$$

(b) Find and sketch a signal $g_2(t)$ such that

$$y(1) - 2y(2) = \langle x, g_2 \rangle = \int x(t)g_2(t)dt$$

Problem 5.41 (Correlation via filtering and sampling) Let us generalize the result we were hinting at in Problem 5.40. Suppose an arbitrary signal x is passed through an arbitrary filter $h(t)$ to obtain output $y(t) = (x * h)(t)$.

(a) Show that taking a linear combination of samples at the filter output is equivalent to a correlation operation on u. That is, show that

$$\sum_{i=1}^{n} \alpha_i y(t_i) = \langle x, g \rangle = \int x(t)g(t)dt$$

where

$$g(t) = \sum_{i=1}^{n} \alpha_i h(t_i - t) = \sum_{i=1}^{n} \alpha_i h_{MF}(t - t_i) \tag{5.108}$$

That is, taking a linear combination of samples is equivalent to correlating against a signal that is a linear combination of shifted versions of the matched filter for h.

(b) The preceding result can be applied to approximate a correlation operation by taking linear combinations at the output of a filter. Suppose that we wish to perform a correlation against a triangular pulse $g(t) = (1 - |t|)I_{[-1,1]}(t)$. How would you approximate this operation by taking a linear combination of samples at the output of a filter with impulse response $h(t) = I_{[0,1]}(t)$?

Problem 5.42 (Approximating a correlator by filtering and sampling) Consider the noisy signal

$$y(t) = s(t) + n(t)$$

where $s(t) = (1 - |t|)I_{[-1,1]}(t)$ and $n(t)$ is white noise with $S_n(f) \equiv 0.1$.

(a) Compute the SNR at the output of the integrator

$$Z = \int_{-1}^{1} y(t)dt$$

(b) Can you improve the SNR by modifying the integration in (a), while keeping the processing linear? If so, say how. If not, say why not.
(c) Now, suppose that $y(t)$ is passed through a filter with impulse response $h(t) = I_{[0,1]}(t)$ to obtain $z(t) = (y * h)(t)$. If you were to sample the filter output at a single time $t = t_0$, how would you choose t_0 so as to maximize the SNR?
(d) In the setting of (c), if you were now allowed to take two samples at times t_1, t_2, and t_3 and generate a linear combination $a_1 z(t_1) + a_2 z(t_2) + a_3 z(t_3)$, how would you choose $\{a_i\}$ and $\{t_i\}$, to improve the SNR relative to (c)? (We are looking for intuitively sensible answers rather than a provably optimal choice.)

Hint. See Problem 5.41. Taking linear combinations of samples at the output of a filter is equivalent to correlation with an appropriate waveform, which we can choose to approximate the optimal correlator.

Mathematical derivations

Problem 5.43 (Bounds on the *Q* function) We derive the bounds (5.117) and (5.116) for

$$Q(x) = \int_{x}^{\infty} \frac{1}{\sqrt{2\pi}} e^{-t^2/2}\, dt \tag{5.109}$$

(a) Show that, for $x \geq 0$, the following upper bound holds:

$$Q(x) \leq \frac{1}{2} e^{-x^2/2}$$

Hint. Try pulling out a factor of $e^{-x^2/2}$ from (5.109), and then bounding the resulting integrand. Observe that $t \geq x \geq 0$ in the integration interval.

(b) For $x \geq 0$, derive the following upper and lower bounds for the Q function:

$$\left(1 - \frac{1}{x^2}\right) \frac{e^{-x^2/2}}{\sqrt{2\pi}x} \leq Q(x) \leq \frac{e^{-x^2/2}}{\sqrt{2\pi}x}$$

Hint. Write the integrand in (5.109) as a product of $1/t$ and $te^{-t^2/2}$ and then integrate by parts to get the upper bound. Integrate by parts once more using a similar trick to get the lower bound. Note that you can keep integrating by parts to get increasingly refined upper and lower bounds.

Problem 5.44 (Geometric derivation of Q function bound) Let X_1 and X_2 denote independent standard Gaussian random variables.

(a) For $a > 0$, express $P[|X_1| > a, |X_2| > a]$ in terms of the Q function.
(b) Find $P[X_1^2 + X_2^2 > 2a^2]$.
Hint. Transform to polar coordinates. Or use the results of Problem 5.21.
(c) Sketch the regions in the (x_1, x_2) plane corresponding to the events considered in (a) and (b).
(d) Use (a)–(c) to obtain an alternative derivation of the bound $Q(x) \leq \frac{1}{2}e^{-x^2/2}$ for $x \geq 0$ (i.e., the bound in Problem 5.43(a)).

Problem 5.45 (Cauchy–Schwarz inequality for random variables) For random variables X and Y defined on a common probability space, define the *mean-squared error* in approximating X by a multiple of Y as

$$J(a) = \mathbb{E}\left[(X - aY)^2\right]$$

where a is a scalar. Assume that both random variables are nontrivial (i.e., neither of them is zero with probability one).

(a) Show that

$$J(a) = \mathbb{E}[X^2] + a^2\mathbb{E}[Y^2] - 2a\mathbb{E}[XY]$$

(b) Since $J(a)$ is quadratic in a, it has a global minimum (corresponding to the best approximation of X by a multiple of Y). Show that this is achieved for $a_{\text{opt}} = \mathbb{E}[XY]/\mathbb{E}[Y^2]$.
(c) Show that the mean-squared error in the best approximation found in (b) can be written as

$$J(a_{\text{opt}}) = \mathbb{E}[X^2] - \frac{(\mathbb{E}[XY])^2}{\mathbb{E}[Y^2]}$$

(d) Since the approximation error is nonnegative, conclude that

$$(\mathbb{E}[XY])^2 \leq \mathbb{E}[X^2]\mathbb{E}[Y^2] \tag{5.110}$$

This is the *Cauchy–Schwarz inequality* for random variables.
(e) Conclude also that equality is achieved in (5.110) if and only if X and Y are scalar multiples of each other.
Hint. Equality corresponds to $J(a_{\text{opt}}) = 0$.

Problem 5.46 (Normalized correlation)

(a) Apply the Cauchy–Schwarz inequality in the previous problem to "zero-mean" versions of the random variables, $X_1 = X - \mathbb{E}[X]$ and $Y_1 = Y - \mathbb{E}[Y]$ to obtain that

$$|\text{cov}(X, Y)| \leq \sqrt{\text{var}(X)\text{var}(Y)} \tag{5.111}$$

(b) Conclude that the normalized correlation $\rho(X, Y)$ defined in (5.59) lies in $[-1, 1]$.

(c) Show that $|\rho| = 1$ if and only if we can write $X = aY + b$. Specify the constants a and b in terms of the means and covariances associated with the two random variables.

Problem 5.47 (Characteristic function of a Gaussian random vector) Consider a Gaussian random vector $\mathbf{X} = (X_1, \ldots, X_m)^T \sim N(\mathbf{m}, \mathbf{C})$. The characteristic function of $\mathbf{X}$ is defined as follows:

$$\phi_{\mathbf{X}}(\mathbf{w}) = \mathbb{E}\left[e^{j\mathbf{w}^T\mathbf{X}}\right] = \mathbb{E}\left[e^{j(w_1X_1+\cdots+w_mX_m)}\right] \tag{5.112}$$

The characteristic function completely characterizes the distribution of a random vector, even if a density does not exist. If the density does exist, the characteristic function is a multidimensional inverse Fourier transform of it:

$$\phi_{\mathbf{X}}(\mathbf{w}) = \mathbb{E}\left[e^{j\mathbf{w}^T\mathbf{X}}\right] = \int e^{j\mathbf{w}^T\mathbf{x}} p_{\mathbf{X}}(\mathbf{x})d\mathbf{x}$$

The density is therefore given by the corresponding Fourier transform

$$p_{\mathbf{X}}(\mathbf{x}) = \frac{1}{(2\pi)^m} \int e^{-j\mathbf{w}^T\mathbf{x}} \phi_{\mathbf{X}}(\mathbf{w})d\mathbf{w} \tag{5.113}$$

(a) Show that $Y = \mathbf{w}^T\mathbf{X}$ is a Gaussian random variable with mean $\mu = \mathbf{w}^T\mathbf{m}$ and variance $v^2 = \mathbf{w}^T\mathbf{C}\mathbf{w}$.

(b) For $Y \sim N(\mu, v^2)$, show that

$$\mathbb{E}[e^{jY}] = e^{j\mu - v^2/2}$$

(c) Use the result of (b) to obtain that the characteristic function of $\mathbf{X}$ is given by

$$\phi_{\mathbf{X}}(\mathbf{w}) = e^{j\mathbf{w}^T\mathbf{m} - \frac{1}{2}\mathbf{w}^T\mathbf{C}\mathbf{w}} \tag{5.114}$$

which depends only on $\mathbf{m}$ and $\mathbf{C}$.

(d) Since the distribution of $\mathbf{X}$ is completely specified by its characteristic function, conclude that the distribution of a Gaussian random vector depends only on its mean vector and covariance matrix. When $\mathbf{C}$ is invertible, we can compute the density (5.58) by taking the Fourier transform of the characteristic function in (5.114), but we skip that derivation.

Problem 5.48 Consider a zero-mean WSS random process X with autocorrelation function $R_X(\tau)$. Let $Y_1(t) = (X * h_1)(t)$ and $Y_2(t) = (X * h_2)(t)$ denote random processes obtained by passing X through LTI systems with impulse responses h_1 and h_2, respectively.

(a) Find the cross-correlation function $R_{Y_1,Y_2}(t_1, t_2)$.
Hint. You can use the approach employed to obtain (5.101), first finding $R_{Y_1,X}$ and then R_{Y_1,Y_2}.

(b) Are Y_1 and Y_2 jointly WSS?

(c) Suppose that X is white noise with PSD $S_X(f) \equiv 1$, $h_1(t) = I_{[0,1]}(t)$, and $h_2(t) = e^{-t}I_{[0,\infty)}(t)$. Find $\mathbb{E}[Y_1(0)Y_2(0)]$ and $\mathbb{E}[Y_1(0)Y_2(1)]$.

Problem 5.49 (Cauchy–Schwarz inequality for signals) Consider two signals (assume that they are real-valued for simplicity, although the results we are about to derive apply for complex-valued signals as well) $u(t)$ and $v(t)$.

(a) We wish to approximate $u(t)$ by a scalar multiple of $v(t)$ so as to minimize the norm of the error. Specifically, we wish to minimize

$$J(a) = \int |u(t) - av(t)|^2 \, dt = ||u - av||^2 = \langle u - av, u - av \rangle$$

Show that

$$J(a) = ||u||^2 + a^2||v||^2 - 2a\langle u, v \rangle$$

(b) Show that the quadratic function $J(a)$ is minimized by choosing $a = a_{\text{opt}}$, given by

$$a_{\text{opt}} = \frac{\langle u, v \rangle}{||v||^2}$$

Show that the corresponding approximation $a_{\text{opt}}v$ can be written as a projection of u along a unit vector in the direction of v:

$$a_{\text{opt}}v = \langle u, v/||v|| \rangle \frac{v}{||v||}$$

(c) Show that the error due to the optimal setting is given by

$$J(a_{\text{opt}}) = ||u||^2 - \frac{|\langle u, v \rangle|^2}{||v||^2}$$

(d) Since the minimum error is nonnegative, conclude that

$$||u|| \, ||v|| \leq |\langle u, v \rangle| \tag{5.115}$$

This is the Cauchy–Schwarz inequality which applies to real- and complex-valued signals or vectors.

(e) Conclude also that equality in (5.115) occurs if and only if u is a scalar multiple of v or if v is a scalar multiple of u. (We need to say it both ways in case one of the signals is zero.)

Problem 5.50 Consider a random process X passed through a correlator g to obtain

$$Z = \int_{-\infty}^{\infty} X(t)g(t)dt$$

where $X(t)$ and $g(t)$ are real-valued.

(a) Show that the mean and variance of Z can be expressed in terms of the mean function and autocovariance function) of X as follows:

$$\mathbb{E}[Z] = \int m_X(t)g(t)dt = \langle m_X, g \rangle$$

$$\text{var}(Z) = \int \int C_X(t_1, t_2)g(t_1)g(t_2)dt_1 dt_2$$

(b) Suppose now that X is zero mean and WSS with autocorrelation $R_X(\tau)$. Show that the variance of the correlator output can be written as

$$\text{var}(Z) = \int R_X(\tau)R_g(\tau)d\tau = \langle R_X, R_g \rangle$$

where $R_g(\tau) = (g * g_{\text{MF}})(\tau) = \int g(t)g(t-\tau)dt$ is the "autocorrelation" of the waveform g.

Hint. An alternative to doing this from scratch is to use the equivalence of correlation and matched filtering. You can then employ (5.101), which gives the output autocorrelation function when a WSS process is sent through an LTI system, evaluate it at zero lag to find the power, and use the symmetry of autocorrelation functions.

Problems drawing on material from Chapter 3 and Appendix 5.E

These can be skipped by readers primarily interested in the digital communication material in the succeeding chapters.

Problem 5.51 Consider a noisy FM signal of the form

$$v(t) = 20\cos(2\pi f_c t + \phi_s(t)) + n(t)$$

where $n(t)$ is WGN with power spectral density $N_0/2 = 10^{-5}$ and $\phi_s(t)$ is the instantaneous phase deviation of the noiseless FM signal. Assume that the bandwidth of the noiseless FM signal is 100 kHz.

(a) The noisy signal $v(t)$ is passed through an ideal BPF that exactly spans the 100-kHz frequency band occupied by the noiseless signal. What is the SNR at the output of the BPF?
(b) The output of the BPF is passed through an ideal phase detector, followed by a differentiator that is normalized to give unity gain at 10 kHz, and an ideal (unity gain) LPF of bandwidth 10 kHz.

    (i) Sketch the noise PSD at the output of the differentiator.
    (ii) Find the noise power at the output of the LPF.

Problem 5.52 An FM signal of bandwidth 210 kHz is received at a power of −90 dBm, and is corrupted by bandpass AWGN with two-sided PSD 10^{-22} W/Hz. The message bandwidth is 5 kHz, and the peak-to-average power ratio for the message is 10 dB.

(a) What is the SNR (in dB) for the received FM signal? (Assume that the noise is bandlimited to the band occupied by the FM signal.)
(b) Estimate the peak frequency deviation.
(c) The noisy FM signal is passed through an ideal phase detector. Estimate and sketch the noise PSD at the output of the phase detector, carefully labeling the axes.
(d) The output of the phase detector is passed through a differentiator with transfer function $H(f) = jf$, and then an ideal lowpass filter of bandwidth 5 kHz. Estimate the SNR (in dB) at the output of the lowpass filter.

Problem 5.53 A message signal $m(t)$ is modeled as a zero-mean random process with PSD

$$S_m(f) = |f| I_{[-2,2]}(f)$$

We generate an SSB signal as follows:

$$u(t) = 20[m(t)\cos(200\pi t) - \check{m}(t)\sin(200\pi t)]$$

where $\check{m}$ denotes the Hilbert transform of m.

(a) Find the power of m and the power of u.
(b) The noisy received signal is given by $y(t) = u(t) + n(t)$, where n is passband AWGN with PSD $N_0/2 = 1$, and is independent of u. Draw the block diagram for an ideal synchronous demodulator for extracting the message m from y, specifying the carrier frequency as well as the bandwidth of the LPF, and find the SNR at the output of the demodulator.
(c) Find the signal-to-noise-plus-interference ratio if the local carrier for the synchronous demodulator has a phase error of $\pi/8$.

Appendix 5.A *Q* function bounds and asymptotics

The following upper and lower bounds on $Q(x)$ (derived in Problem 5.43) are asymptotically tight for large arguments; that is, the difference between the bounds tends to zero as x gets large.

Bounds on $Q(x)$, asymptotically tight for large arguments

$$\left(1 - \frac{1}{x^2}\right)\frac{e^{-x^2/2}}{x\sqrt{2\pi}} \le Q(x) \le \frac{e^{-x^2/2}}{x\sqrt{2\pi}}, \quad x \ge 0 \tag{5.116}$$

The asymptotic behavior (5.53) follows from these bounds. However, they do not work well for small x (the upper bound blows up to ∞, and the lower bound to $-\infty$, as $x \to 0$). The following upper bound is useful both for small and for large values of $x \ge 0$: it gives

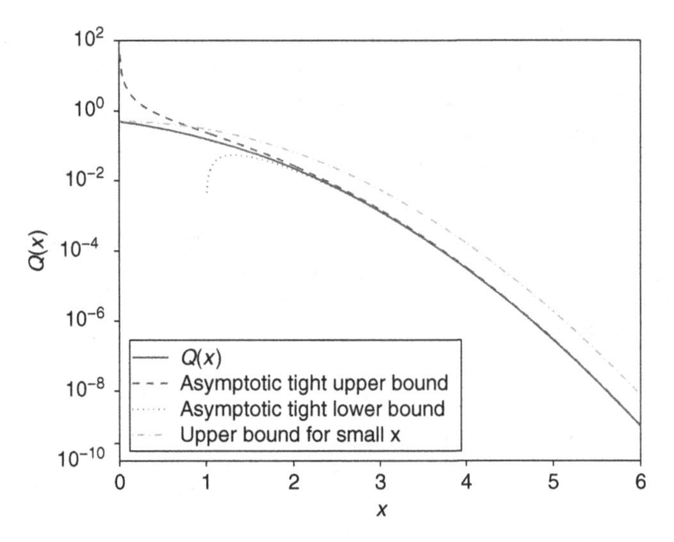

Figure 5.29 The Q function and bounds.

accurate results for small x, and, while it is not as tight as the bounds (5.116) for large x, it does give the correct exponent of decay.

Upper bound on $Q(x)$ useful both for small and for large arguments

$$Q(x) \leq \frac{1}{2} e^{-x^2/2}, \quad x \geq 0 \tag{5.117}$$

Figure 5.29 plots $Q(x)$ and its bounds for positive x. A logarithmic scale is used for the values of the function in order to demonstrate the rapid decay with x. The bounds (5.116) are seen to be tight even at moderate values of x (say $x \geq 2$), while the bound (5.117) shows the right rate of decay for large x, while also remaining useful for small x.

Appendix 5.B Approximations using limit theorems

We often deal with sums of independent (or approximately independent) random variables. Finding the exact distribution of such sums can be cumbersome. This is where limit theorems, which characterize what happens to these sums as the number of terms gets large, come in handy.

Law of large numbers (LLN) Suppose that $X_1, X_2, \ldots$ are i.i.d. random variables with finite mean m. Then their empirical average $(X_1 + \cdots + X_n)/n$ converges to their statistical average $\mathbb{E}[X_i] \equiv m$ as $n \to \infty$. (Let us not worry about exactly how convergence is defined for a sequence of random variables.)

When we do a simulation to estimate some quantity of interest by averaging over multiple runs, we are relying on the LLN. The LLN also underlies all of information theory, which is the basis for computing performance benchmarks for coded communication systems.

The LLN tells us that the empirical average of i.i.d. random variables tends to the statistical average. The central limit theorem characterizes the variation around the statistical average.

Central limit theorem (CLT) Suppose that $X_1, X_2, \ldots$ are i.i.d. random variables with finite mean m and variance v^2. Then the distribution of $Y_n = (X_1 + \cdots + X_n - nm)/\sqrt{nv^2}$ tends to that of a standard Gaussian random variable. Specifically,

$$\lim_{n \to \infty} P\left[\frac{X_1 + \cdots + X_n - nm}{\sqrt{nv^2}} \leq x\right] = \Phi(x) \tag{5.118}$$

Notice that the sum $S_n = X_1 + \cdots + X_n$ has mean nm and variance nv^2. Thus, the CLT is telling us that Y_n, a normalized, zero-mean, unit-variance version of S_n, has a distribution that tends to $N(0, 1)$ as n gets large. In practical terms, this translates to using the CLT to approximate S_n as a Gaussian random variable with mean nm and variance nv^2, for "large enough" n. In many scenarios, the CLT kicks in rather quickly, and the Gaussian approximation works well for values of n as small as 6–10.

Example 5.B.1 (Gaussian approximation for a binomial distribution) Consider a binomial random variable with parameters n and p. We know that we can write it as $S_n = X_1 + \cdots + X_n$, where X_i are i.i.d. Bernoulli(p). Note that $\mathbb{E}[X_i] = p$ and $\text{var}(X_i) = p(1-p)$, so that S_n has mean np and variance $np(1-p)$. We can therefore approximate Binomial(n, p) by $N(np, np(1-p))$ according to the CLT. The CLT tells us that we can approximate the CDF of a binomial by a Gaussian: thus, the *integral* of the Gaussian density from $(-\infty, k]$ should approximate the *sum* of the binomial pmf from 0 to k. The plot in Figure 5.30 shows that the Gaussian density itself (with mean $np = 6$ and variance $np(1-p) = 4.2$) approximates the binomial pmf quite well around the mean, so that we do expect the corresponding CDFs to be close.

Appendix 5.C Noise mechanisms

We have discussed mathematical models for noise. We provide here some motivation and physical feel for how noise arises.

Thermal noise Even in a resistor that has no external voltage applied across it, the charge carriers exhibit random motion because of thermal agitation, just as the molecules in a gas do. The amount of motion depends on the temperature, and results in *thermal noise*. Since the charge carriers are equally likely to move in either direction, the voltages and

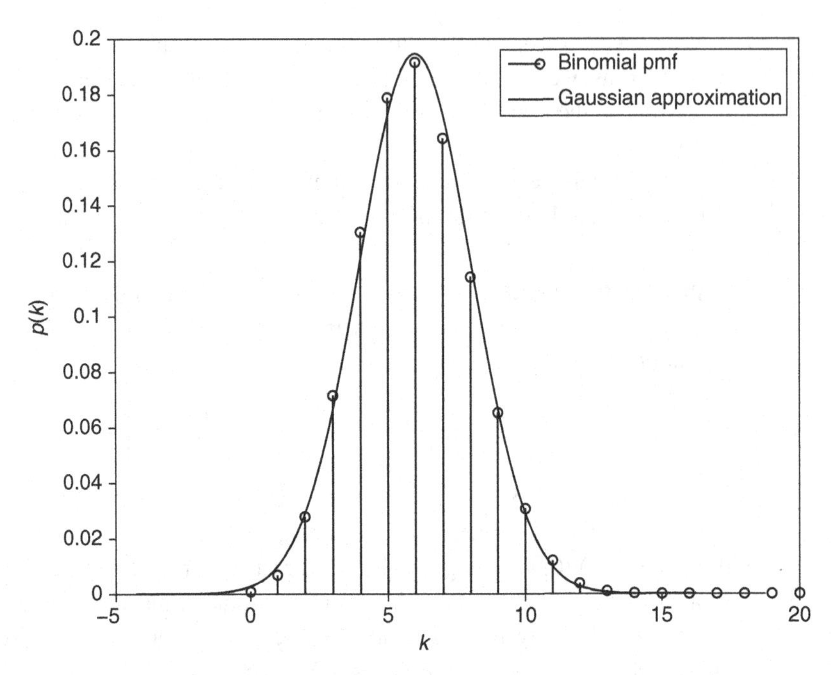

Figure 5.30 A binomial pmf with parameters $n = 20$ and $p = 0.3$, and its $N(6, 4.2)$ Gaussian approximation.

currents associated with thermal noise have zero DC value. We therefore quantify the noise power, or the average squared values of voltages and currents associated with the noise. These were first measured by Johnson, and then explained by Nyquist using statistical thermodynamics arguments, in the 1920s. As a result, thermal noise is often called *Johnson noise,* or *Johnson–Nyquist noise.*

Using arguments that we shall not go into, Nyquist concluded that the mean-squared value of the voltage associated with a resistor R, measured in a small frequency band $[f, f + \Delta f]$, is given by

$$\overline{v_n^2(f, \Delta f)} = 4Rk_B T \, \Delta f \tag{5.119}$$

where R is the resistance in ohms, $k_B = 1.38 \times 10^{-23}$ J/K is Boltzmann's constant, and T is the temperature in degrees Kelvin ($T_{\text{Kelvin}} = T_{\text{Centigrade}} + 273$). Notice that the mean-squared voltage depends only on the width of the frequency band, not its location; that is, thermal noise is *white.* Actually, a more accurate statistical mechanics argument does reveal a dependence on frequency, as follows:

$$\overline{v_n^2(f, \Delta f)} = \frac{4Rhf \, \Delta f}{e^{hf/(k_B T)} - 1} \tag{5.120}$$

where $h = 6.63 \times 10^{-34}$ J/Hz denotes Planck's constant, which relates the energy of a photon to the frequency of the corresponding electromagnetic wave (readers may recall the famous formula $E = h\nu$, where ν is the frequency of the photon). Now, $e^x \approx 1 + x$ for small x. Using this in (5.120), we obtain that it reduces to (5.119) for $hf/(k_B T) \ll 1$ or $f \ll k_B Th = f^*$. For $T = 290$K, we have $f^* \approx 6 \times 10^{12}$ Hz, or 6 THz. The practical operating range of communication frequencies today is much less than this (existing and emerging systems operate well below 100 GHz), so thermal noise is indeed very well modeled as white for current practice.

For bandwidth B, (5.119) yields the mean-squared voltage

$$\overline{v_{\mathrm{n}}^2} = 4Rk_{\mathrm{B}}TB$$

Now, if we connect the noise source to a matched load of impedance R, the mean-squared power delivered to the load is

$$\overline{P_{\mathrm{n}}^2} = \frac{\overline{(v_{\mathrm{n}}/2)^2}}{R} = k_{\mathrm{B}}TB \tag{5.121}$$

The preceding calculation provides a valuable benchmark, giving the communication link designer a ballpark estimate of how much noise power to expect in a receiver operating over a bandwidth B. Of course, the noise for a particular receiver is typically higher than this benchmark, and must be calculated by detailed modeling and simulation of internal and external noise sources, and the gains, input impedances, and output impedances for various circuit components. However, while the circuit designer must worry about these details, once the design is complete, he or she can supply the link designer with a single number for the noise power at the receiver output, referred to the benchmark (5.121).

Shot noise Shot noise occurs because of the discrete nature of the charge carriers. When a voltage applied across a device causes current to flow, if we could count the number of charge carriers going from one point in the device to the other (e.g., from the source to the drain of a transistor) over a time period τ, we would see a random number $N(\tau)$, which would vary independently across disjoint time periods. Under rather general assumptions, $N(\tau)$ is well modeled as a Poisson random variable with mean $\lambda\tau$, where λ scales with the direct current. The variance of a Poisson random variable equals its mean, so the variance of the rate of charge carrier flow equals

$$\mathrm{var}\left(\frac{N(\tau)}{\tau}\right) = \frac{1}{\tau^2}\,\mathrm{var}(N(\tau)) = \frac{\lambda}{\tau}$$

We can think of this as the power of the shot noise. Thus, increasing the observation interval τ smooths out the variations in charge-carrier flow, and reduces the shot-noise power. If we now think of the device being operated over a bandwidth B, we know that we are effectively observing the device at a temporal resolution $\tau \sim 1/B$. Thus, the shot-noise power scales linearly with B.

The preceding discussion indicates that both thermal noise and shot noise are white, in that their power scales linearly with the system bandwidth B, independently of the frequency band of operation. We can therefore model the aggregate system noise due to these two phenomena as a single white-noise process. Indeed, both phenomena involve random motions of a large number of charge carriers, and can be analyzed together in a statistical mechanics framework. This is well beyond our scope, but, for our purpose, we can simply model the aggregate system noise due to these phenomena as a single white-noise process.

Flicker noise Another commonly encountered form of noise is $1/f$ noise, also called flicker noise, whose power increases as the frequency of operation gets smaller. The sources of $1/f$

noise are poorly understood, and white noise dominates in the typical operating regimes for communication receivers. For example, in an RF system, the noise in the front end (antenna, low-noise amplifier, mixer) dominates the overall system noise, and $1/f$ noise is negligible at these frequencies. We therefore ignore $1/f$ noise in our noise modeling.

Appendix 5.D The structure of passband random processes

We discuss here the modeling of passband random processes, in particular, passband white noise, in more detail. These insights are useful for the analysis of the effect of noise in analog communication systems, as in Appendix 5.E.

We can define the complex envelope, and I and Q components, for a passband random process in exactly the same fashion as is done for deterministic signals in Chapter 2. For a passband random process, each sample path (observed over a large enough time window) has a Fourier transform restricted to passband. We can therefore define the complex envelope, I/Q components and envelope/phase as we do for deterministic signals. For any given reference frequency f_c in the band of interest, any sample path $x_p(t)$ for a passband random process can be written as

$$\begin{aligned} x_p(t) &= \text{Re}\left(x(t)e^{j2\pi f_c t}\right) \\ &= x_c(t)\cos(2\pi f_c t) - x_s(t)\sin(2\pi f_c t) \\ &= e(t)\cos(2\pi f_c t + \theta(t)) \end{aligned}$$

where $x(t) = x_c(t) + jx_s(t) = e(t)e^{j\theta(t)}$ is the complex envelope, $x_c(t)$ and $x_s(t)$ are the I and Q components, respectively, and $e(t)$ and $\theta(t)$ are the envelope and phase, respectively.

PSD of complex envelope Applying the standard frequency-domain relationship to the time-windowed sample paths, we have the frequency-domain relationship

$$X_{p,T_o}(f) = \frac{1}{2}X_{T_o}(f - f_c) + \frac{1}{2}X^*_{T_o}(-f - f_c)$$

We therefore have

$$|X_{p,T_o}(f)|^2 = \frac{1}{4}|X_{T_o}(f - f_c)|^2 + \frac{1}{4}|X^*_{T_o}(-f - f_c)|^2 = \frac{1}{4}|X_{T_o}(f - f_c)|^2 + \frac{1}{4}|X_{T_o}(-f - f_c)|^2$$

On dividing by T_o and letting $T_o \to \infty$, we obtain

$$S_{x_p}(f) = \frac{1}{4}S_x(f - f_c) + \frac{1}{4}S_x(-f - f_c) \tag{5.122}$$

where $S_x(f)$ is baseband. Using (5.87), the one-sided passband PSD is given by

$$S^+_{x_p}(f) = \frac{1}{2}S_x(f - f_c) \tag{5.123}$$

Similarly, we can go from passband to complex baseband using the formula

$$S_x(f) = 2S^+_{x_p}(f + f_c) \tag{5.124}$$

What about the I and Q components? Consider the complex envelope $x(t) = x_c(t) + jx_s(t)$. Its autocorrelation function is given by

$$R_x(\tau) = \overline{x(t)x^*(t-\tau)} = \overline{(x_c(t) + jx_s(t))(x_c(t-\tau) - jx_s(t-\tau))}$$

which yields

$$\begin{aligned} R_x(\tau) &= \left(R_{x_c}(\tau) + R_{x_s}(\tau)\right) + j\left(R_{x_s,x_c}(\tau) - R_{x_c,x_s}(\tau)\right) \\ &= \left(R_{x_c}(\tau) + R_{x_s}(\tau)\right) + j\left(R_{x_s,x_c}(\tau) - R_{x_s,x_c}(-\tau)\right) \end{aligned} \tag{5.125}$$

On taking the Fourier transform, we obtain

$$S_x(f) = S_{x_c}(f) + S_{x_s}(f) + j\left(S_{x_s,x_c}(f) - S^*_{x_s,x_c}(f)\right)$$

which simplifies to

$$S_x(f) = S_{x_c}(f) + S_{x_s}(f) - 2\,\mathrm{Im}\left(S_{x_s,x_c}(f)\right) \tag{5.126}$$

For simplicity, we henceforth consider situations in which $S_{x_s,x_c}(f) \equiv 0$ (i.e., the I and Q components are uncorrelated). Actually, for a given passband random process, even if the I and Q components for a given frequency reference are uncorrelated, we can make them correlated by shifting the frequency reference. However, such subtleties are not required for our purpose, which is to model digitally modulated signals and receiver noise.

5.D.1 Baseband representation of passband white noise

Consider passband white noise as shown in Figure 5.21. If we choose the reference frequency as the center of the band, then we get a simple model for the complex envelope and the I and Q components of the noise, as depicted in Figure 5.31. The complex envelope has PSD

$$S_n(f) = 2N_0, \quad |f| \le B/2$$

and the I and Q components have PSDs and cross-spectrum given by

$$\begin{aligned} S_{n_c}(f) &= S_{n_s}(f) = N_0, \quad |f| \le B/2 \\ S_{n_s,n_c}(f) &\equiv 0 \end{aligned}$$

Note that the power of the complex envelope is $2N_0B$, which is twice the power of the corresponding passband noise n_p. This is consistent with the convention in Chapter 2 for

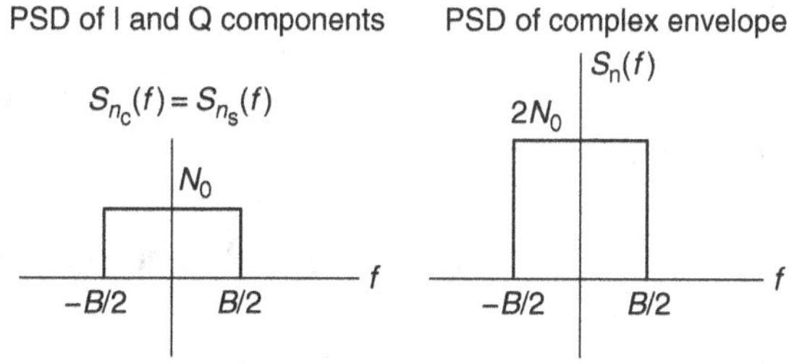

Figure 5.31 The PSDs of I and Q components, and the complex envelope, of passband white noise.

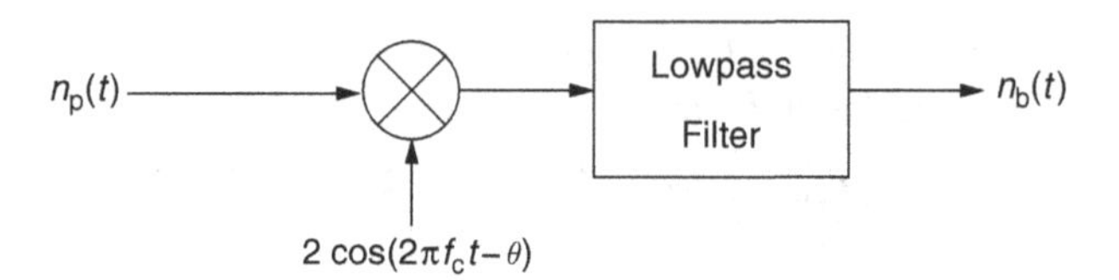

Figure 5.32 Circular symmetry implies that the PSD of the baseband noise $n_b(t)$ is independent of θ.

deterministic, finite-energy signals, where the complex envelope has twice the energy of the corresponding passband signal. Later, when we discuss digital communication receivers and their performance in Chapter 6, we find it convenient to scale signals and noise in complex baseband such that we get rid of this factor of two. In this case, we obtain that the PSDs of the I and Q components are given by $S_{n_c}(f) = S_{n_s}(f) = N_0/2$.

Passband white noise is circularly symmetric

An important property of passband white noise is its circular symmetry: the statistics of the I and Q components are unchanged if we change the phase reference. To understand what this means in practical terms, consider the downconversion operation shown in Figure 5.32, which yields a baseband random process $n_b(t)$. Circular symmetry corresponds to the assumption that the PSD of n_b does not depend on θ. Thus, it immediately implies that

$$S_{n_c}(f) = S_{n_s}(f) \leftrightarrow R_{n_c}(\tau) = R_{n_s}(\tau) \tag{5.127}$$

since $n_b = n_c$ for $\theta = 0$, and $n_b = n_s$, for $\theta = -\pi/2$, where n_c and n_s are the I and Q components, respectively, taking $f_c = f_0$ as a reference. Thus, changes in phase reference do not change the statistics of the I and Q components.

Appendix 5.E SNR computations for analog modulation

We now compute the SNR for the amplitude and angle modulation schemes discussed in Chapter 3. Since the format of the messages is not restricted in our analysis, the SNR computations apply to digital modulation (where the messages are analog waveforms associated with a particular sequence of bits being transmitted) as well as analog modulation (where the messages are typically "natural" audio or video waveforms beyond our control). However, such SNR computations are primarily of interest for analog modulation, since the performance measure of interest for digital communication systems is typically probability of error.

5.E.1 Noise model and SNR benchmark

For noise modeling, we consider passband, circularly symmetric, white noise $n_p(t)$ in a system of bandwidth B centered around f_c, with PSD as shown in Figure 5.21. As discussed in Section 5.D.1, we can write this in terms of its I and Q components with respect to the reference frequency f_c as

$$n_p(t) = n_c(t)\cos(2\pi f_c t) - n_s(t)\sin(2\pi f_c t)$$

where the relevant PSDs are given in Figure 5.31.

Baseband benchmark When evaluating the SNR for various passband analog modulation schemes, it is useful to consider a hypothetical baseband system as a benchmark. Suppose that a real-valued message of bandwidth B_m is sent over a baseband channel. The noise power over the baseband channel is given by $P_n = N_0 B_m$. If the received signal power is $P_r = P_m$, then the SNR benchmark for this baseband channel is given by

$$\mathrm{SNR_b} = \frac{P_r}{N_0 B_m} \tag{5.128}$$

5.E.2 SNR for amplitude modulation

We now quickly sketch SNR computations for some of the variants of AM. The signal and power computations are similar to earlier examples in this chapter, so we do not belabor the details.

SNR for DSB-SC For message bandwidth B_m, the bandwidth of the passband received signal is $B = 2B_m$. The received signal given by

$$y_p(t) = A_c m(t)\cos(2\pi f_c t + \theta_r) + n_p(t)$$

where θ_r is the phase offset between the incoming carrier and the LO. The received signal power is given by

$$P_r = \overline{(A_c m(t)\cos(2\pi f_c t + \theta_r))^2} = A_c^2 P_m/2$$

A coherent demodulator extracts the I component, which is given by

$$y_c(t) = A_c m(t)\cos\theta_r + n_c(t)$$

The signal power is

$$P_s = \overline{(A_c m(t)\cos\theta_r)^2} = A_c^2 P_m \cos^2\theta_r$$

while the noise power is

$$P_n = \overline{n_c^2(t)} = N_0 B = 2N_0 B_m$$

so that the SNR is

$$\mathrm{SNR_{DSB}} = \frac{A_c^2 P_m \cos^2\theta_r}{2N_0 B_m} = \frac{P_r}{N_0 B_m}\cos^2\theta_r = \mathrm{SNR_b}\cos^2\theta_r \tag{5.129}$$

which is the same as the baseband benchmark (5.128) For ideal coherent demodulation (i.e., $\theta_r = 0$), we obtain that the SNR for DSB equals the baseband benchmark $\mathrm{SNR_b}$ in (5.128).

SNR for SSB For message bandwidth B_m, the bandwidth of the passband received signal is $B = B_\mathrm{m}$. The received signal given by

$$y_\mathrm{p}(t) = A_\mathrm{c}m(t)\cos(2\pi f_\mathrm{c}t + \theta_\mathrm{r}) \pm A_\mathrm{c}\check{m}(t)\sin(2\pi f_\mathrm{c}t + \theta_\mathrm{r}) + n_\mathrm{p}(t)$$

where θ_r is the phase offset between the incoming carrier and the LO. The received signal power is given by

$$P_\mathrm{r} = \overline{(A_\mathrm{c}m(t)\cos(2\pi f_\mathrm{c}t + \theta_\mathrm{r}))^2} + \overline{(A_\mathrm{c}m(t)\sin(2\pi f_\mathrm{c}t + \theta_\mathrm{r}))^2} = A_\mathrm{c}^2 P_\mathrm{m}$$

A coherent demodulator extracts the I component, which is given by

$$y_\mathrm{c}(t) = A_\mathrm{c}m(t)\cos\theta_\mathrm{r} \mp A_\mathrm{c}\check{m}(t)\sin\theta_\mathrm{r} + n_\mathrm{c}(t)$$

The signal power is

$$P_\mathrm{s} = \overline{(A_\mathrm{c}m(t)\cos\theta_\mathrm{r})^2} = A_\mathrm{c}^2 P_\mathrm{m}\cos^2\theta_\mathrm{r}$$

while the noise plus interference power is

$$P_\mathrm{n} = \overline{n_\mathrm{c}^2(t)} + \overline{(A_\mathrm{c}\check{m}(t)\sin\theta_\mathrm{r})^2} = N_0 B + A_\mathrm{c}^2 P_\mathrm{m}\sin^2\theta_\mathrm{r} = N_0 B_\mathrm{m} + A_\mathrm{c}^2 P_\mathrm{m}\sin^2\theta_\mathrm{r}$$

so that the signal-to-interference-plus-noise ratio (SINR) is

$$\begin{aligned} \mathrm{SINR}_\mathrm{SSB} &= \frac{A_\mathrm{c}^2 P_\mathrm{m}\cos^2\theta_\mathrm{r}}{N_0 B_\mathrm{m} + A_\mathrm{c}^2 P_\mathrm{m}\sin^2\theta_\mathrm{r}} = \frac{P_\mathrm{r}\cos^2\theta_\mathrm{r}}{N_0 B_\mathrm{m} + P_\mathrm{r}\sin^2\theta_\mathrm{r}} \\ &= \frac{\mathrm{SNR}_\mathrm{b}\cos^2\theta_\mathrm{r}}{1 + \mathrm{SNR}_\mathrm{b}\sin^2\theta_\mathrm{r}} \end{aligned} \tag{5.130}$$

This coincides with the baseband benchmark (5.128) for ideal coherent demodulation (i.e., $\theta_\mathrm{r} = 0$). However, for $\theta_\mathrm{r} \neq 0$, even when the received signal power P_r gets arbitrarily large relative to the noise power, the SINR cannot be larger than $1/\tan^2\theta_\mathrm{r}$, which shows the importance of making the phase error as small as possible.

SNR for AM Now, consider conventional AM. While we would typically use envelope detection rather than coherent demodulation in this setting, it is instructive to compute SNR for both methods of demodulation. For message bandwidth B_m, the bandwidth of the passband received signal is $B = 2B_\mathrm{m}$. The received signal given by

$$y_\mathrm{p}(t) = A_\mathrm{c}\left(1 + a_\mathrm{mod}m_\mathrm{n}(t)\right)\cos(2\pi f_\mathrm{c}t + \theta_\mathrm{r}) + n_\mathrm{p}(t) \tag{5.131}$$

where $m_\mathrm{n}(t)$ is the normalized version of the message (with $\min_t m_\mathrm{n}(t) = -1$) and θ_r is the phase offset between the incoming carrier and the LO. The received signal power is given by

$$P_\mathrm{r} = \overline{(A_\mathrm{c}m(t)\cos(2\pi f_\mathrm{c}t + \theta_\mathrm{r}))^2} = A_\mathrm{c}^2(1 + a_\mathrm{mod}^2 P_{m_\mathrm{n}})/2 \tag{5.132}$$

where $P_{m_\mathrm{n}} = \overline{m_\mathrm{n}^2(t)}$ is the power of the normalized message. A coherent demodulator extracts the I component, which is given by

$$y_\mathrm{c}(t) = A_\mathrm{c} + A_\mathrm{c}a_\mathrm{mod}m_\mathrm{n}(t)\cos\theta_\mathrm{r} + n_\mathrm{c}(t)$$

The power of the information-bearing part of the signal (the DC term due to the carrier carries no information, and is typically rejected using AC coupling) is given by

$$P_s = \overline{(A_c a_{mod} m_n(t) \cos \theta_r)^2} = A_c^2 a_{mod}^2 P_{m_n} \cos^2\theta_r \qquad (5.133)$$

Recall that the AM power efficiency is defined as the power of the message-bearing part of the signal to the power of the overall signal (which includes an unmodulated carrier), and is given by

$$\eta_{AM} = \frac{a_{mod}^2 P_{m_n}}{1 + a_{mod}^2 P_{m_n}}$$

We can therefore write the signal power (5.133) at the output of the coherent demodulator in terms of the received power in (5.132) as

$$P_s = 2P_r \eta_{AM} \cos^2\theta_r$$

while the noise power is

$$P_n = \overline{n_c^2(t)} = N_0 B = 2N_0 B_m$$

Thus, the SNR is

$$\text{SNR}_{AM,coh} = \frac{P_s}{P_n} = \frac{2P_r \eta_{AM} \cos^2\theta_r}{2N_0 B_m} = \text{SNR}_b \eta_{AM} \cos^2\theta_r \qquad (5.134)$$

Thus, even with ideal coherent demodulation ($\theta_r = 0$), the SNR obtained in AM is less than that of the baseband benchmark, since $\eta_{AM} < 1$ (typically much smaller than one). Of course, the reason why we incur this power inefficiency is to simplify the receiver, by message recovery using an envelope detector. Let us now compute the SNR for the latter.

Upon expressing the passband noise in the received signal (5.131) with the incoming carrier as the reference, we have

$$y_p(t) = A_c\,(1 + a_{mod} m_n(t)) \cos(2\pi f_c t + \theta_r) + n_c(t)\cos(2\pi f_c t + \theta_r) - n_s(t)\sin(2\pi f_c t + \theta_r)$$

where, by virtue of circular symmetry, n_c and n_s have PSDs and cross-spectra as in Figure 5.31, regardless of θ_r. That is,

$$y_p(t) = y_c(t)\cos(2\pi f_c t + \theta_r) - y_s(t)\sin(2\pi f_c t + \theta_r)$$

where, as shown in Figure 5.33,

$$y_c(t) = A_c(1 + a_{mod} m_n(t)) + n_c(t), \quad y_s(t) = n_s(t)$$

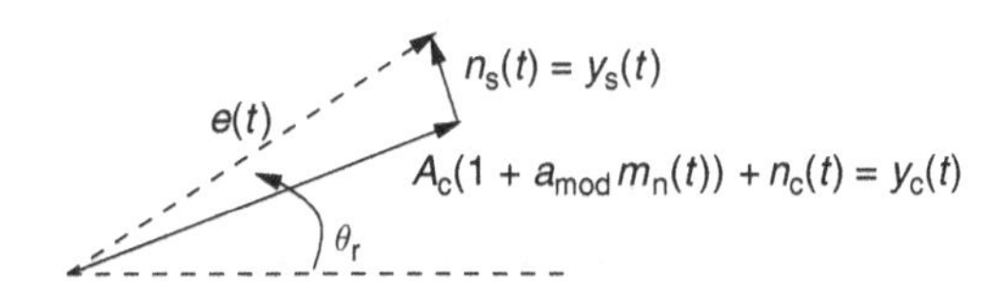

Figure 5.33 At high SNR, the envelope of an AM signal is approximately equal to its I component relative to the received carrier phase reference.

At high SNR, the signal term is dominant, so $y_c(t) \gg y_s(t)$. Furthermore, since the AM signal is positive (assuming $a_{mod} < 1$), $y_c > 0$ "most of the time," even though n_c can be negative. We therefore obtain that

$$e(t) = \sqrt{y_c^2(t) + y_s^2(t)} \approx |y_c(t)| \approx y_c(t)$$

That is, the output of the envelope detector is approximated, for high SNR, as

$$e(t) \approx A_c(1 + a_{mod} m_n(t)) + n_c(t)$$

The right-hand side is what we would get from ideal coherent detection. We can reuse our SNR computation for coherent detection to conclude that the SNR at the envelope detector output is given by

$$\mathrm{SNR}_{\mathrm{AM,envdet}} = \mathrm{SNR}_b \eta_{\mathrm{AM}} \tag{5.135}$$

Thus, for a properly designed ($a_{mod} < 1$) AM system operating at high SNR, the envelope detector approximates the performance of ideal coherent detection, without requiring carrier synchronization.

5.E.3 SNR for angle modulation

We have seen how to compute the SNR when white noise adds to a message encoded in the signal amplitude. Let us now see what happens when the message is encoded in the signal phase or frequency. The received signal is given by

$$y_p(t) = A_c \cos(2\pi f_c t + \theta(t)) + n_p(t) \tag{5.136}$$

where $n_p(t)$ is passband white noise with one-sided PSD N_0 over the signal band of interest, and where the message is encoded in the phase $\theta(t)$. For example,

$$\theta(t) = k_p m(t)$$

for phase modulation, and

$$\frac{1}{2\pi} \frac{d}{dt} \theta(t) = k_f m(t)$$

for frequency modulation. We wish to understand how the additive noise $n_p(t)$ perturbs the phase.

By decomposing the passband noise into I and Q components with respect to the phase of the noiseless angle-modulated signal, we can rewrite the received signal as follows:

$$\begin{aligned} y_p(t) &= A_c \cos(2\pi f_c t + \theta(t)) + n_c(t)\cos(2\pi f_c t + \theta(t)) - n_s(t)\sin(2\pi f_c t + \theta(t)) \\ &= (A_c + n_c(t))\cos(2\pi f_c t + \theta(t)) - n_s(t)\sin(2\pi f_c t + \theta(t)) \end{aligned} \tag{5.137}$$

where n_c and n_s have PSDs as in Figure 5.31 (with cross-spectrum $S_{n_s,n_c}(f) \equiv 0$), thanks to circular symmetry (we assume that it applies approximately even though the phase reference $\theta(t)$ is time-varying). The I and Q components with respect to this phase reference are shown in Figure 5.34, so the corresponding complex envelope can be written as

$$y(t) = e(t)e^{j\theta_n(t)}$$

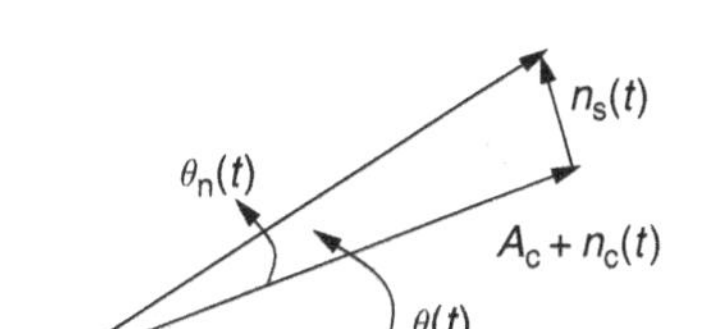

Figure 5.34 I and Q components of a noisy angle-modulated signal with the phase reference chosen as the phase of the noiseless signal.

where

$$e(t) = \sqrt{(A_c + n_c(t))^2 + n_s^2(t)} \tag{5.138}$$

and

$$\theta_n(t) = \tan^{-1}\left(\frac{n_s(t)}{A_c + n_c(t)}\right) \tag{5.139}$$

The passband signal in (5.137) can now be rewritten as

$$y_p(t) = \mathrm{Re}(y(t)e^{2\pi f_c t + \theta(t)}) = \mathrm{Re}\left(e(t)e^{j\theta_n(t)}e^{2\pi f_c t + \theta(t)}\right) = e(t)\cos(2\pi f_c t + \theta(t) + \theta_n(t))$$

At high SNR, $A_c \gg |n_c|$ and $A_c \gg |n_s|$. Thus,

$$\left|\frac{n_s(t)}{A_c + n_c(t)}\right| \ll 1$$

and

$$\frac{n_s(t)}{A_c + n_c(t)} \approx \frac{n_s(t)}{A_c}$$

For $|x|$ small, $\tan x \approx x$, and hence $x \approx \tan^{-1} x$. We therefore obtain the following high-SNR approximation for the phase perturbation due to the noise:

$$\theta_n(t) = \tan^{-1}\left(\frac{n_s(t)}{A_c + n_c(t)}\right) \approx \frac{n_s(t)}{A_c}, \quad \textbf{high-SNR approximation} \tag{5.140}$$

To summarize, we can model the received signal (5.136) as

$$y_p(t) \approx A_c \cos\left(2\pi f_c t + \theta(t) + \frac{n_s(t)}{A_c}\right), \quad \textbf{high-SNR approximation} \tag{5.141}$$

Thus, the Q component (relative to the desired signal's phase reference) of the passband white noise appears as phase noise, but is scaled down by the signal amplitude.

FM noise analysis

Let us apply the preceding discussion to develop an analysis of the effects of white noise on FM. It is helpful, but not essential, to have read Chapter 3 for this discussion. Suppose that we have an ideal detector for the phase of the noisy signal in (5.141), and that we differentiate it to recover a message encoded in the frequency. (For those who have read Chapter 3, we are talking about an ideal limiter–discriminator). The output is the instantaneous frequency deviation, which is given by

$$z(t) = \frac{1}{2\pi}\frac{d}{dt}(\theta(t) + \theta_{\rm n}(t)) \approx k_{\rm f}m(t) + \frac{n_{\rm s}'(t)}{2\pi A_{\rm c}} \tag{5.142}$$

using the high-SNR approximation (5.140).

We now analyze the performance of an FM system whose block diagram is shown in Figure 5.35. For wideband FM, the bandwidth $B_{\rm RF}$ of the received signal $y_{\rm p}(t)$ is significantly larger than the message bandwidth $B_{\rm m}$: $B_{\rm RF} \approx 2(\beta + 1)B_{\rm m}$ by Carson's formula, where $\beta > 1$. Thus, the RF front end in Figure 5.35 lets in passband white noise $n_{\rm p}(t)$ of bandwidth of the order of $B_{\rm RF}$, as shown in Figure 5.36. Figure 5.36 also shows the PSDs once we have passed the received signal through the limiter–discriminator. The estimated message at the output of the limiter–discriminator is a baseband signal that we can limit to the message bandwidth $B_{\rm m}$, which significantly reduces the noise that we see at the output of the limiter–discriminator. Let us now compute the output SNR. From (5.142), the signal power is given by

$$P_{\rm s} = \overline{(k_{\rm f}m(t))^2} = k_{\rm f}^2 P_{\rm m} \tag{5.143}$$

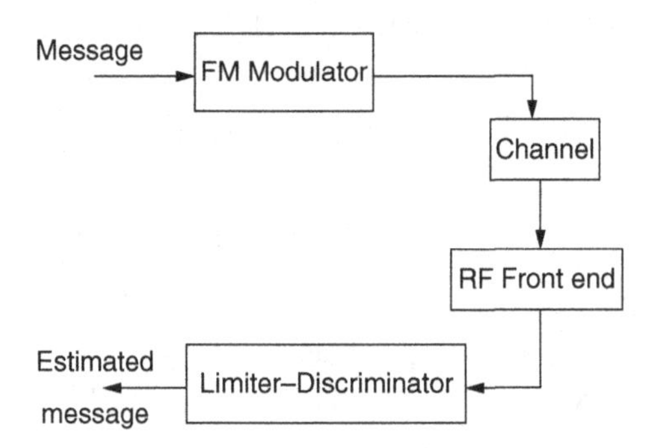

Figure 5.35 The block diagram for the FM system using limiter–discriminator demodulation.

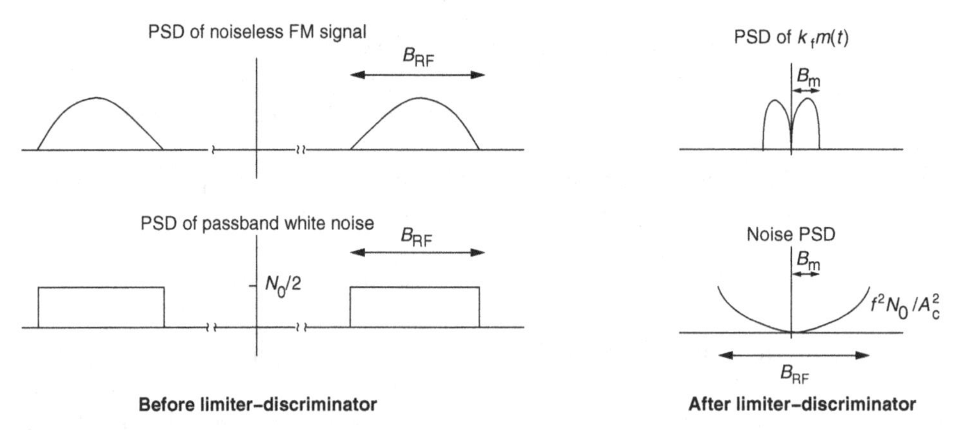

Figure 5.36 The PSDs of signal and noise before and after the limiter–discriminator.

The noise contribution at the output is given by

$$z_{\rm n}(t) = \frac{n'_{\rm s}(t)}{2\pi A_{\rm c}}$$

Since $d/dt \leftrightarrow j2\pi f$, $z_{\rm n}(t)$ is obtained by passing $n_{\rm s}(t)$ through an LTI system with $G(f) = j2\pi f/(2\pi A_{\rm c}) = jf/A_{\rm c}$. Thus, the noise PSD at the output of the limiter–discriminator is given by

$$S_{z_{\rm n}}(f) = |G(f)|^2 S_{n_{\rm s}}(f) = f^2 N_0/A_{\rm c}^2, \quad |f| \le B_{\rm RF}/2 \tag{5.144}$$

Once we limit the bandwidth to the message bandwidth $B_{\rm m}$ after the discriminator, the noise power is given by

$$P_{\rm n} = \int_{-B_{\rm m}}^{B_{\rm m}} S_{z_{\rm n}}(f)df = \int_{-B_{\rm m}}^{B_{\rm m}} \frac{f^2 N_0}{A_{\rm c}^2}\, df = \frac{2B_{\rm m}^3 N_0}{3A_{\rm c}^2} \tag{5.145}$$

From (5.143) and (5.145), we obtain that the SNR is given by

$$\mathrm{SNR}_{\rm FM} = \frac{P_{\rm s}}{P_{\rm n}} = \frac{3k_{\rm f}^2 P_{\rm m} A_{\rm c}^2}{2B_{\rm m}^3 N_0}$$

It is interesting to benchmark this against a baseband communication system in which the message is sent directly over the channel. To keep the comparison fair, we fix the received power to that of the passband system and the one-sided noise PSD to that of the passband white noise. Thus, the received signal power is $P_{\rm r} = A_{\rm c}^2/2$, and the noise power is $N_0 B_{\rm m}$, and the baseband benchmark SNR is given by

$$\mathrm{SNR}_{\rm b} = \frac{P_{\rm r}}{N_0 B_{\rm m}} = \frac{A_{\rm c}^2}{2N_0 B_{\rm m}}$$

We therefore obtain that

$$\mathrm{SNR}_{\rm FM} = \frac{3k_{\rm f}^2 P_{\rm m}}{B_{\rm m}^2} \mathrm{SNR}_{\rm b} \tag{5.146}$$

Let us now express this in terms of some interesting parameters. The maximum frequency deviation in the FM system is given by

$$\Delta f_{\rm max} = k_{\rm f} \max_t |m(t)|$$

and the modulation index is defined as the ratio between the maximum frequency deviation and the message bandwidth:

$$\beta = \frac{\Delta f_{\rm max}}{B_{\rm m}}$$

Thus, we have

$$\frac{k_{\rm f}^2 P_{\rm m}}{B_{\rm m}^2} = \frac{(\Delta f_{\rm max})^2}{B_{\rm m}^2} \frac{P_{\rm m}}{(\max_t |m(t)|)^2} = \beta^2/\mathrm{PAR}$$

defining the peak-to-average power ratio (PAR) of the message as

$$\mathrm{PAR} = \frac{(\max_t |m(t)|)^2}{\overline{m^2(t)}} = \frac{(\max_t |m(t)|)^2}{P_{\rm m}}$$

On substituting into (5.146), we obtain that

$$\mathrm{SNR}_{\mathrm{FM}} = \frac{3\beta^2}{\mathrm{PAR}} \mathrm{SNR}_{\mathrm{b}} \tag{5.147}$$

Thus, FM can improve upon the baseband benchmark by increasing the modulation index β. This is an example of a power–bandwidth tradeoff: by increasing the bandwidth beyond that strictly necessary for sending the message, we have managed to improve the SNR compared with the baseband benchmark. However, the quadratic power–bandwidth tradeoff offered by FM is highly suboptimal compared with the best possible tradeoffs in digital communication systems, where one can achieve exponential tradeoffs. Another drawback of the FM power–bandwidth tradeoff is that the amount of SNR improvement depends on the PAR of the message: messages with larger dynamic range, and hence larger PAR, will see less improvement. This is in contrast to digital communication, where message characteristics do not affect the power–bandwidth tradeoffs over the communication link, since messages are converted to bits via source coding before transmission. Of course, messages with larger dynamic range may well require more bits to represent them accurately, and hence a higher rate on the communication link, but such design choices are decoupled from the parameters governing reliable link operation.

Threshold effect It appears from (5.147) that the output SNR can be improved simply by increasing β. This is somewhat misleading. For a given message bandwidth B_{m}, increasing β corresponds to increasing the RF bandwidth: $B_{\mathrm{RF}} \approx 2(\beta + 1)B_{\mathrm{m}}$ by Carson's formula. Thus, an increase in β corresponds to an increase in the power of the passband white noise at the input of the limiter–discriminator, which is given by $N_0 B_{\mathrm{RF}} = 2N_0 B_{\mathrm{m}}(\beta + 1)$. Thus, if we increase β, the high-SNR approximation underlying (5.140), and hence the model (5.142) for the output of the limiter–discriminator, breaks down. It is easy to see this from Equation (5.139) for the phase perturbation due to noise: $\theta_{\mathrm{n}}(t) = \tan^{-1}(n_{\mathrm{s}}(t)/(A_{\mathrm{c}} + n_{\mathrm{c}}(t)))$. When A_{c} is small, variations in $n_{\mathrm{c}}(t)$ can change the sign of the denominator, which leads to phase changes of π, over a small time interval. This leads to impulses in the output of the discriminator. Indeed, as we start reducing the SNR at the input to the discriminator for FM audio below the threshold where the approximation (5.140) holds, we can actually hear these peaks as "clicks" in the audio output. As we reduce the SNR further, the clicks swamp out the desired signal. This is called the FM *threshold effect.*

To avoid this behavior, we must operate in the high-SNR regime where $A_{\mathrm{c}} \gg |n_{\mathrm{c}}|, |n_{\mathrm{s}}|$, so that the approximation (5.140) holds. In other words, the SNR for the passband signal at the *input* to the limiter–discriminator must be above a threshold, say γ (e.g., $\gamma = 10$ might be a good rule of thumb), for FM demodulation to work well. This condition can be expressed as follows:

$$\frac{P_{\mathrm{r}}}{N_0 B_{\mathrm{RF}}} \geq \gamma \tag{5.148}$$

Thus, in order to utilize a large RF bandwidth to improve the SNR at the output of the limiter–discriminator, the received signal power must also scale with the available bandwidth. Using Carson's formula, we can rewrite (5.148) in terms of the baseband benchmark as follows:

$$\text{SNR}_\text{b} = \frac{P_\text{r}}{N_0 B_\text{m}} \geq 2\gamma(\beta + 1), \quad \textbf{condition for operation above threshold} \tag{5.149}$$

To summarize, the power–bandwidth tradeoff (5.147) applies only when the received power (or equivalently, the baseband benchmark SNR) is above a threshold that scales with the bandwidth, as specified by (5.149).

Preemphasis and deemphasis

Since the noise at the limiter–discriminator output has a quadratic PSD (see (5.144) and Figure 5.36), higher frequencies in the message see more noise than lower frequencies. A commonly used approach to alleviate this problem is to boost the power of the higher message frequencies at the transmitter by using a highpass *preemphasis* filter. The distortion in the message due to preemphasis is undone at the receiver using a lowpass *deemphasis* filter, which attenuates the higher frequencies. The block diagram of an FM system using such an approach is shown in Figure 5.37.

A typical choice for the preemphasis filter is a highpass filter with a single zero, with a transfer function of the form

$$H_\text{PE}(f) = 1 + j2\pi f \tau_1$$

The corresponding deemphasis filter is a single-pole lowpass filter with transfer function

$$H_\text{DE}(f) = \frac{1}{1 + j2\pi f \tau_1}$$

For FM audio broadcast, τ_1 is chosen in the range 50–75 μs (e.g., 75 μs in the United States, 50 μs in Europe). The f^2 noise scaling at the output of the limiter–discriminator is compensated for by the (approximately) $1/f^2$ scaling provided by $|H_\text{DE}(f)|^2$ beyond the cutoff frequency $f_\text{pd} = 1/(2\pi\tau_1)$ (the subscript indicates the use of preemphasis and deemphasis), which evaluates to 2.1 kHz for $\tau_1 = 75$ μs.

Let us compute the SNR improvement obtained using this strategy. Assuming that the preemphasis and deemphasis filters compensate each other exactly, the signal contribution to the estimated message at the output of the deemphasis filter in Figure 5.37 is $k_\text{f} m(t)$,

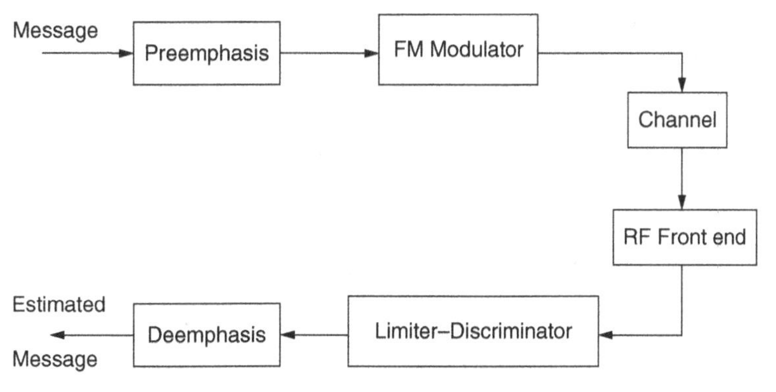

Figure 5.37 Preemphasis and deemphasis in FM systems.

which equals the signal contribution to the estimated message at the output of the limiter–discriminator in Figure 5.35, which shows a system not using preemphasis/deemphasis. Since the signal contributions in the estimated messages in both systems are the same, any improvement in SNR must come from a reduction in the output noise. Thus, we wish to characterize the noise PSD and power at the output of the deemphasis filter in Figure 5.37. To do this, note that the noise at the output of the limiter–discriminator is the same as before:

$$z_n(t) = \frac{n_s'(t)}{2\pi A_c}$$

with PSD

$$S_{z_n}(f) = |G(f)|^2 S_{n_s}(f) = f^2 N_0/A_c^2, \quad |f| \leq B_{RF}/2$$

The noise v_n obtained by passing z_n through the deemphasis filter has PSD

$$S_{v_n}(f) = |H_{DE}(f)|^2 S_{z_n}(f) = \frac{N_0}{A_c^2} \frac{f^2}{1+(f/f_{pd})^2} = \frac{N_0 f_{pd}^2}{A_c^2}\left(1 - \frac{1}{1+(f/f_{pd})^2}\right)$$

By integrating over the message bandwidth, we find that the noise power in the estimated message in Figure 5.37 is given by

$$P_n = \int_{-B_m}^{B_m} S_{v_n}(f)df = \frac{2N_0 f_{pd}^3}{A_c^2}\left(\frac{B_m}{f_{pd}} - \tan^{-1}\left(\frac{B_m}{f_{pd}}\right)\right) \tag{5.150}$$

where we have used the substitution $\tan x = f/f_{pd}$ to evaluate the integral. As we have already mentioned, the signal power is unchanged from the earlier analysis, so that the improvement in SNR is given by the reduction in noise power compared with (5.145), which gives

$$\begin{aligned} \text{SNR}_{\text{gain}} &= \frac{2B_m^3 N_0/(3A_c^2)}{\left(2N_0 f_{pd}^3/A_c^2\right)\left(B_m/f_{pd} - \tan^{-1}\left(B_m/f_{pd}\right)\right)} \\ &= \frac{1}{3}\frac{\left(B_m/f_{pd}\right)^3}{B_m/f_{pd} - \tan^{-1}\left(B_m/f_{pd}\right)} \end{aligned} \tag{5.151}$$

For $f_{pd} = 2.1$ kHz, corresponding to the United States guidelines for FM audio broadcast, and an audio bandwidth $B_m = 15$ kHz, the SNR gain in (5.151) evaluates to more than 13 dB.

For completeness, we give the formula for the SNR obtained using preemphasis and deemphasis as

$$\text{SNR}_{\text{FM,pd}} = \frac{\left(B_m/f_{pd}\right)^3}{B_m/f_{pd} - \tan^{-1}\left(B_m/f_{pd}\right)} \frac{\beta^2}{\text{PAR}} \text{SNR}_b \tag{5.152}$$

which is obtained by taking the product of the SNR gain (5.151) and the SNR without preemphasis/deemphasis given by (5.147).

6 Optimal demodulation

As we saw in Chapter 4, we can send bits over a channel by choosing one of a set of waveforms to send. For example, when sending a single 16QAM symbol, we are choosing one of 16 passband waveforms:

$$s_{b_c,b_s} = b_c p(t)\cos(2\pi f_c)t - b_s p(t)\sin(2\pi f_c t)$$

where b_c and b_s each take values in $\{\pm 1, \pm 3\}$. We are thus able to transmit $\log_2 16 = 4$ bits of information. In this chapter, we establish a framework for recovering these 4 bits when the received waveform is a noisy version of the transmitted waveform. More generally, we consider the fundamental problem of *M-ary signaling in additive white Gaussian noise (AWGN):* one of M signals, $s_1(t), \ldots, s_M(t)$ is sent, and the received signal equals the transmitted signal plus white Gaussian noise (WGN).

At the receiver, we are faced with a *hypothesis-testing* problem: we have M possible hypotheses about which signal was sent, and we have to make our "best" guess as to which one holds, given our observation of the received signal. We are interested in finding a guessing strategy, more formally termed a *decision rule,* which is the "best" according to some criterion. For communications applications, we are typically interested in finding a decision rule that minimizes the *probability of error* (i.e., the probability of making a wrong guess). We can now summarize the goals of this chapter as follows.

We wish to design optimal receivers when the received signal is modeled as follows:

$$H_i : y(t) = s_i(t) + n(t), \quad i = 1, \ldots, M$$

where H_i is the ith hypothesis, corresponding to signal $s_i(t)$ being transmitted, and $n(t)$ is white Gaussian noise. We then wish to analyze the performance of such receivers, to see how performance measures such as the probability of error depend on system parameters. It turns out that, for the preceding AWGN model, the performance depends only on the received signal-to-noise ratio (SNR) and on the "shape" of the signal constellation $\{s_1(t), \ldots, s_M(t)\}$. Underlying both the derivation of the optimal receiver and its analysis is a geometric view of signals and noise as vectors, which we term *signal-space concepts.* Once we have this background, we are in a position to discuss elementary power–bandwidth tradeoffs. For example, 16QAM has higher bandwidth efficiency than QPSK, so it makes sense that it has lower power efficiency; that is, it requires higher SNR, and hence higher transmit power, for the same probability of error. We will be able to quantify this intuition, previewed in Chapter 4, using the material in this chapter. We will also be able to perform link-budget calculations: for example, how much transmit power is needed to attain a given bit rate using a given constellation as a function of range, and transmit and receive antenna gains?

Chapter plan

The prerequisites for this chapter are Chapter 4 (digital modulation) and the material on Gaussian random variables (Section 5.6) and noise modeling (Section 5.8) in Chapter 5. We build up the remaining background required to attain our goals in this chapter in a step-by-step fashion, as follows.

Hypothesis testing. In Section 6.1, we establish the basic framework for hypothesis testing, derive the form of optimal decision rules, and illustrate the application of this framework for finite-dimensional observations.

Signal-space concepts. In Section 6.2, we show that continuous time M-ary signaling in AWGN can be reduced to an equivalent finite-dimensional system, in which transmitted signal vectors are corrupted by vector WGN. This is done by projecting the continuous-time signal into the finite-dimensional *signal space* spanned by the set of possible transmitted signals, $s_1, \ldots, s_M$. We apply the hypothesis-testing framework to derive the optimal receiver for the finite-dimensional system, and from this we infer the optimal receiver in continuous time.

Performance analysis. In Section 6.3, we analyze the performance of optimal reception. We show that the performance depends only on the SNR and the relative geometry of the signal constellation. We provide exact error probability expressions for binary signaling. While the probability of error for larger signal constellations must typically be computed by simulation or numerical integration, we obtain bounds and approximations, building on the analysis for binary signaling, that provide quick insight into power–bandwidth tradeoffs.

Link-budget analysis. In Section 6.5, we illustrate how performance analysis is applied to obtain the "link budget" for a typical radio link, which is the tool used to obtain coarse guidelines for the design of hardware, including transmit power, transmit and receive antennas, and receiver noise figure.

Software

Software Lab 6.1 in this chapter builds on Software Lab 4.1, providing a hands-on feel for Nyquist signaling over an AWGN channel. In turn, we build on this lab in Software Lab 8.1, which adds in channel dispersion to the model. Software Lab 6.2 illustrates the impact of fading on performance over a wireless mobile channel, using simulations based on a simple model for channel time variations.

Notational shortcut

In this chapter, we make extensive use of the notational simplification discussed at the end of Section 5.3. Given a random variable X, a common notation for the probability density function or probability mass function is $p_X(x)$, with X denoting the random variable, and x being a dummy variable that we might integrate out when computing probabilities. However, when there is no scope for confusion, we use the less cumbersome (albeit incomplete) notation $p(x)$, using the dummy variable x not only as the argument of the density, but also to indicate that the density corresponds to the random variable X. (Similarly, we would use $p(y)$ to denote the density for a random variable Y.) The same convention is used for joint and conditional densities as well. For random variables X and Y, we use the notation $p(x, y)$ instead of $p_{X,Y}(x, y)$, and $p(y|x)$ instead of $p_{Y|X}(y|x)$, to denote the joint and conditional densities, respectively.

6.1 Hypothesis testing

In Example 5.6.3, we considered a simple model for binary signaling, in which the receiver sees a single sample Y. If 0 is sent, the conditional distribution of Y is $N(0, v^2)$, whereas if 1 is sent, the conditional distribution is $N(m, v^2)$. We analyzed a simple decision rule in which we guess that 0 is sent if $Y \leq m/2$, and guess that 1 is sent otherwise. Thus, we wish to decide between two hypotheses (0 being sent or 1 being sent) on the basis of an observation (the received sample Y). The statistics of the observation depend on the hypothesis (this information is captured by the conditional distributions of Y given each of the hypotheses). We must now make a good guess as to which hypothesis is true, given the value of the observation. The guessing strategy is called the decision rule, which maps each possible value of Y to either 0 or 1.

The decision rule we have considered in Example 5.6.3 makes sense, splitting the difference between the conditional means of Y under the two hypotheses. But is this always the best thing to do? For example, if we know for sure that 0 is sent, then we should clearly always guess that 0 is sent, regardless of the value of Y that we see. As another example, if the noise variance is different under the two hypotheses, then it is no longer clear that splitting the difference between the means is the right thing to do. We therefore need a systematic framework for *hypothesis testing,* which allows us to derive good decision rules for a variety of statistical models.

In this section, we consider the general problem of M-ary hypothesis testing, in which we must decide which of M possible hypotheses, $H_0, \ldots, H_{M-1}$, "best explains" an observation Y. For our purpose, the observation Y can be a scalar or vector, and takes values in an observation space Γ. The link between the hypotheses and observation is statistical: for each hypothesis H_i, we know the conditional distribution of Y given H_i. We denote the conditional density of Y given H_i as $p(y|i)$, $i = 0, 1, \ldots, M-1$. We may also know the *prior probabilities* of the hypotheses (i.e., the probability of each hypothesis *prior* to seeing the observation), denoted by $\pi_i = P[H_i]$, $i = 0, 1, \ldots, M-1$, which satisfy $\sum_{i=0}^{M-1} \pi_i = 1$. The final ingredient of the hypothesis testing framework is the *decision rule:* for each possible value $Y = y$ of the observation, we must decide which of the M hypotheses we will bet on. Denoting this guess as $\delta(y)$, the decision rule $\delta(\cdot)$ is a mapping from the observation space Γ to $\{0, 1, \ldots, M-1\}$, where $\delta(y) = i$ means that we guess that H_i is true when we see $Y = y$. The decision rule partitions the observation space into decision regions, with Γ_i denoting the set of values of Y for which we guess H_i. That is, $\Gamma_i = \{y \in \Gamma : \delta(y) = i\}$, $i = 0, 1, \ldots, M-1$. We summarize these ingredients of the hypothesis testing framework as follows.

Ingredients of hypothesis-testing framework

- Hypotheses $H_0, H_1, \ldots, H_{M-1}$
- Observation $Y \in \Gamma$
- Conditional densities $p(y|i)$, for $i = 0, 1, \ldots, M-1$
- Prior probabilities $\pi_i = P[H_i], i = 0, 1, \ldots, M-1$, with $\sum_{i=0}^{M-1} \pi_i = 1$
- Decision rule $\delta : \Gamma \to \{0, 1, \ldots, M-1\}$
- Decision regions $\Gamma_i = \{y \in \Gamma : \delta(y) = i\}, i = 0, 1, \ldots, M-1$

To make the concepts concrete, let us quickly recall Example 5.6.3, where we have $M = 2$ hypotheses, with $H_0 : Y \sim N(0, v^2)$ and $H_1 : Y \sim N(m, v^2)$ with $m > 0$. The "sensible" decision rule in this example can be written as

$$\delta(y) = \begin{cases} 0, & y \leq m/2 \\ 1, & y > m/2 \end{cases}$$

so that $\Gamma_0 = (-\infty, m/2]$ and $\Gamma_1 = (m/2, \infty)$. Note that this decision rule need not be optimal if we know the prior probabilities. For example, if we know that $\pi_0 = 1$, we should say that H_0 is true, regardless of the value of Y: this would reduce the probability of error from $Q(m/(2v))$ (for the "sensible" rule) to zero!

6.1.1 Error probabilities

The performance measures of interest to us when choosing a decision rule are the conditional error probabilities and the average error probability. We have already seen these in Example 5.6.3 for binary on–off keying, but we now formally define them for a general M-ary hypothesis testing problem. For a fixed decision rule δ with corresponding decision regions $\{\Gamma_i\}$, we define the conditional probabilities of error as follows.

Conditional error probabilities The conditional error probability, conditioned on H_i, where $0 \leq i \leq M-1$, is defined as

$$P_{e|i} = P[\text{say } H_j \text{ for some } j \neq i | H_i \text{ is true}] = \sum_{j \neq i} P[Y \in \Gamma_j | H_i] = 1 - P[Y \in \Gamma_i | H_i] \quad (6.1)$$

Conditional probabilities of correct decision These are defined as

$$P_{c|i} = 1 - P_{e|i} = P[Y \in \Gamma_i | H_i] \tag{6.2}$$

Average error probability This is given by averaging the conditional error probabilities using the priors:

$$P_e = \sum_{i=1}^{M} \pi_i P_{e|i} \tag{6.3}$$

Average probability of correct decision This is given by

$$P_c = \sum_{i=1}^{M} \pi_i P_{c|i} = 1 - P_e \tag{6.4}$$

6.1.2 ML and MAP decision rules

For a general M-ary hypothesis-testing problem, an intuitively pleasing decision rule is the maximum likelihood rule, which, for a given observation $Y = y$, picks the hypothesis H_i for which the observed value $Y = y$ is most likely; that is, we pick i so as to maximize the conditional density $p(y|i)$.

Notation We denote by "arg max" the *argument* of the maximum. That is, if the maximum of a function $f(x)$ occurs at x_0, then x_0 is the argument of the maximum:

$$\max_x f(x) = f(x_0), \quad \arg\max_x f(x) = x_0$$

Note also that, while the *maximum value* of a function may change if we apply another function to it, if the second function is strictly increasing, the *argument of the maximum* remains the same. For example, when dealing with densities taking exponential forms (such as the Gaussian), it is useful to apply the logarithm (which is a strictly increasing function), as we note for the ML rule below.

Maximum likelihood (ML) decision rule

The ML decision rule is defined as

$$\delta_{ML}(y) = \arg\max_{0 \le i \le M-1} p(y|i) = \arg\max_{0 \le i \le M-1} \log p(y|i) \tag{6.5}$$

Another decision rule that "makes sense" is the maximum *a posteriori* probability (MAP) rule, where we pick the hypothesis which is most likely, conditioned on the value of the observation. The conditional probabilities $P[H_i|Y = y]$ are called the *a posteriori,* or *posterior,* probabilities, since they are probabilities that we can compute *after* we see the observation $Y = y$. Let us work through what this rule is actually doing. Using Bayes' rule, the posterior probabilities are given by

$$P[H_i|Y=y] = \frac{p(y|i)P[H_i]}{p(y)} = \frac{\pi_i p(y|i)}{p(y)}, \quad i = 0, 2, \ldots, M-1$$

Since we want to maximize this over i, the denominator $p(y)$, the unconditional density of Y, can be ignored in the maximization. We can also take the log as we did for the ML rule. The MAP rule can therefore be summarized as follows.

Maximum *a posteriori* probability (MAP) rule

The MAP decision rule is defined as

$$\begin{aligned}\delta_{\mathrm{MAP}}(y) &= \arg\max_{0 \le i \le M-1} P[H_i|Y=y] \\ &= \arg\max_{1 \le i \le M} \pi_i p(y|i) = \arg\max_{0 \le i \le M-1} \log \pi_i + \log p(y|i)\end{aligned} \tag{6.6}$$

Properties of the MAP rule

- The MAP rule reduces to the ML rule for equal priors.
- The MAP rule minimizes the probability of error. In other words, it is also the minimum probability of error (MPE) rule.

The first property follows from (6.6) on setting $\pi_i \equiv 1/M$: in this case π_i does not depend on i and can therefore be dropped when maximizing over i. The second property is important enough to restate and prove as a theorem.

Theorem 6.1.1 *The MAP rule (6.6) minimizes the probability of error.*

Proof of Theorem 6.1.1 We show that the MAP rule maximizes the probability of correct decision. To do this, consider an arbitrary decision rule δ, with corresponding decision regions $\{\Gamma_i\}$. The conditional probabilities of correct decision are given by

$$P_{\mathrm{c}|i} = P[Y \in \Gamma_i | H_i] = \int_{\Gamma_i} p(y|i)dy, \quad i = 0, 1, \ldots, M-1$$

so that the average probability of a correct decision is

$$P_{\mathrm{c}} = \sum_{i=0}^{M-1} \pi_i P_{\mathrm{c}|i} = \sum_{i=0}^{M-1} \pi_i \int_{\Gamma_i} p(y|i)dy$$

Any point $y \in \Gamma$ can belong in exactly one of the M decision regions. If we decide to put it in Γ_i, then the point contributes the term $\pi_i p(y|i)$ to the integrand. Since we wish to maximize the overall integral, we choose to put y in the decision region for which it makes the largest contribution to the integrand. Thus, we put it in Γ_i so as to maximize $\pi_i p(y|i)$, which is precisely the MAP rule (6.6). □

Example 6.1.1 (Hypothesis testing with exponentially distributed observations) A binary hypothesis problem is specified as follows:

$$H_0 : Y \sim \text{Exp}(1), \qquad H_1 : Y \sim \text{Exp}(1/4)$$

where $\text{Exp}(\mu)$ denotes an exponential distribution with density $\mu e^{-\mu y}$, CDF $1 - e^{-\mu y}$, and complementary CDF $e^{-\mu y}$, where $y \geq 0$ (all the probability mass falls on the nonnegative numbers). Note that the mean of an $\text{Exp}(\mu)$ random variable is $1/\mu$. Thus, in our case, the mean under H_0 is 1, while the mean under H_1 is 4.

(a) Find the ML rule and the corresponding conditional error probabilities.
(b) Find the MPE rule when the prior probability of H_1 is 1/5. Also find the conditional and average error probabilities.

Solution

(a) As shown in Figure 6.1, we have

$$p(y|0) = e^{-y} I_{\{y \geq 0\}}, \quad p(y|1) = (1/4)e^{-y/4} I_{\{y \geq 0\}}$$

The ML rule is given by

$$p(y|1) \underset{H_0}{\overset{H_1}{\gtrless}} p(y|0)$$

which reduces to

$$(1/4)e^{-y/4} \underset{H_0}{\overset{H_1}{\gtrless}} e^{-y} \qquad (y \geq 0)$$

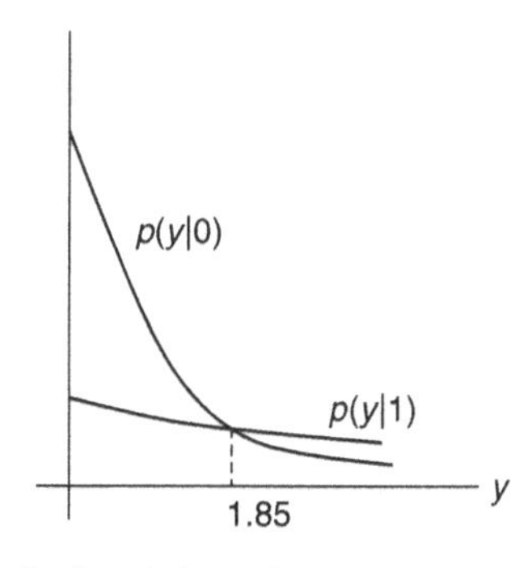

Figure 6.1 Hypothesis testing with exponentially distributed observations.

By taking logarithms on both sides and simplifying, we obtain that the ML rule is given by

$$y \underset{H_0}{\overset{H_1}{\gtrless}} (4/3)\log 4 = 1.8484$$

The conditional error probabilities are

$$\begin{aligned} P_{e|0} &= P[\text{say } H_1|H_0] = P[Y > (4/3)\log 4|H_0] \\ &= e^{-(4/3)\log 4} = (1/4)^{4/3} = 0.1575 \end{aligned}$$

$$\begin{aligned} P_{e|1} &= P[\text{say } H_0|H_1] = P[Y \leq (4/3)\log 4|H_1] \\ &= 1 - e^{-(1/3)\log 4} = 1 - (1/4)^{1/3} = 0.37 \end{aligned}$$

These conditional error probabilities are rather high, telling us that exponentially distributed observations with different means do not give us high-quality information about the hypotheses.

(b) The MPE rule is given by

$$\pi_1 p(y|1) \underset{H_0}{\overset{H_1}{\gtrless}} \pi_0 p(y|0)$$

which reduces to

$$(1/5)(1/4)e^{-y/4} \underset{H_0}{\overset{H_1}{\gtrless}} (4/5)e^{-y}$$

This gives

$$y \underset{H_0}{\overset{H_1}{\gtrless}} \frac{4}{3}\log 16 = 3.6968$$

Proceeding as in (a), we obtain

$$\begin{aligned} P_{e|0} &= e^{-(4/3)\log 16} = (1/16)^{4/3} = 0.0248 \\ P_{e|1} &= 1 - e^{-(1/3)\log 16} = 1 - (1/16)^{1/3} = 0.6031 \end{aligned}$$

with average error probability

$$P_e = \pi_0 P_{e|0} + \pi_1 P_{e|1} = (4/5) * 0.0248 + (1/5) * 0.6031 = 0.1405$$

Since the prior probability of H_1 is small, the MPE rule is biased towards guessing that H_0 is true. In this case, the decision rule is so skewed that the conditional probability of error under H_1 is actually worse than a random guess. Taking this one step further, if the prior probability of H_1 actually becomes zero, then the MPE rule would always guess that H_0 is true. In this case, the conditional probability of error under H_1 would be one! This shows that we must be careful about modeling when applying the MAP rule: if we are wrong about our prior probabilities, and H_1 does occur with nonzero probability, then our performance would be quite poor.

Both the ML rule and the MAP rule involve comparison of densities, and it is convenient to express them in terms of a ratio of densities, or *likelihood ratio,* as discussed next.

Binary hypothesis testing and the likelihood ratio For *binary hypothesis testing,* the ML rule (6.5) reduces to

$$p(y|1) \underset{H_0}{\overset{H_1}{\gtrless}} p(y|0), \quad \text{or} \quad \frac{p(y|1)}{p(y|0)} \underset{H_0}{\overset{H_1}{\gtrless}} 1 \tag{6.7}$$

The ratio of conditional densities appearing above is defined to be the *likelihood ratio (LR)* $L(y)$, a function of fundamental importance in hypothesis testing. Formally, we define the likelihood ratio as

$$L(y) = \frac{p(y|1)}{p(y|0)}, \quad y \in \Gamma \tag{6.8}$$

Likelihood ratio test A *likelihood ratio test (LRT)* is a decision rule in which we compare the likelihood ratio with a threshold:

$$L(y) \underset{H_0}{\overset{H_1}{\gtrless}} \gamma$$

where the choice of γ depends on our performance criterion. An equivalent form is the log likelihood ratio test (LLRT), where the log of the likelihood ratio is compared with a threshold.

We have already shown in (6.7) that the ML rule is an LRT with threshold $\gamma = 1$. From (6.6), we see that the MAP, or MPE, rule is also an LRT:

$$\pi_1 p(y|1) \underset{H_0}{\overset{H_1}{\gtrless}} \pi_0 p(y|0), \quad \text{or} \quad \frac{p(y|1)}{p(y|0)} \underset{H_0}{\overset{H_1}{\gtrless}} \frac{\pi_0}{\pi_1}$$

This is important enough to restate formally.

ML and MPE rules are likelihood ratio tests.

$$L(y) \underset{H_0}{\overset{H_1}{\gtrless}} 1 \quad \text{or} \quad \log L(y) \underset{H_0}{\overset{H_1}{\gtrless}} 0 \quad \textbf{ML rule} \tag{6.9}$$

$$L(y) \underset{H_0}{\overset{H_1}{\gtrless}} \frac{\pi_0}{\pi_1} \quad \text{or} \quad \log L(y) \underset{H_0}{\overset{H_1}{\gtrless}} \log\left(\frac{\pi_0}{\pi_1}\right) \quad \textbf{MAP/MPE rule} \tag{6.10}$$

We now specialize further to the setting of Example 5.6.3. The conditional densities are as shown in Figure 6.2. Since this example is fundamental to our understanding of signaling in AWGN, let us give it a name, the *basic Gaussian example,* and summarize the set-up in the language of hypothesis testing.

Basic Gaussian example

$$H_0 : Y \sim N(0, v^2), \qquad H_1 : Y \sim N(m, v^2),$$

or

$$p(y|0) = \frac{\exp(-y^2/(2v^2))}{\sqrt{2\pi v^2}}; \; p(y|1) = \frac{\exp(-(y-m)^2)/(2v^2))}{\sqrt{2\pi v^2}} \tag{6.11}$$

Likelihood ratio for basic Gaussian example By substituting (6.11) into (6.8) and simplifying (this is left as an exercise), we obtain that the likelihood ratio for the basic Gaussian example is

$$L(y) = \exp((1/v^2)(my - m^2/2))$$

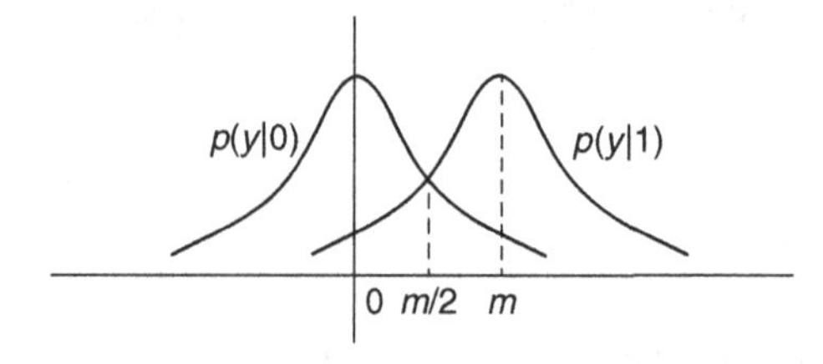

Figure 6.2 Conditional densities for the basic Gaussian example.

and

$$\log L(y) = (1/v^2)(my - m^2/2) \tag{6.12}$$

ML and MAP rules for basic Gaussian example Using (6.12) in (6.9), we leave it as an exercise to check that the ML rule reduces to

$$Y \underset{H_0}{\overset{H_1}{\gtrless}} m/2, \quad \textbf{ML rule } (m > 0) \tag{6.13}$$

(check that the inequalities get reversed for $m < 0$). This is exactly the "sensible" rule that we analyzed in Example 5.6.3. By using (6.12) in (6.10), we obtain the MAP rule:

$$Y \underset{H_0}{\overset{H_1}{\gtrless}} m/2 + \frac{v^2}{m} \log\left(\frac{\pi_0}{\pi_1}\right), \quad \textbf{MAP rule } (m > 0) \tag{6.14}$$

Example 6.1.2 (ML versus MAP for the basic Gaussian example) For the basic Gaussian example, we now know that the decision rule in Example 5.6.3 is the ML rule, and we showed in that example that the performance of this rule is given by

$$P_{e|0} = P_{e|1} = P_e = Q\left(\frac{m}{2v}\right) = Q\left(\sqrt{\text{SNR}/2}\right)$$

We also saw that, at 13 dB SNR, the error probability for the ML rule is

$$P_{e,\text{ML}} = 7.8 \times 10^{-4}$$

regardless of the prior probabilities. For equal priors, the ML rule is also MPE, and we cannot hope to do better than this. Let us now see what happens when the prior probability of H_0 is $\pi_0 = \frac{1}{3}$. The ML rule is no longer MPE, and we should be able to do better by using the MAP rule. We leave it as an exercise to show that the conditional error probabilities for the MAP rule are given by

$$P_{e|0} = Q\left(\frac{m}{2v} + \frac{v}{m}\log\left(\frac{\pi_0}{\pi_1}\right)\right), \qquad P_{e|1} = Q\left(\frac{m}{2v} - \frac{v}{m}\log\left(\frac{\pi_0}{\pi_1}\right)\right) \tag{6.15}$$

By plugging in the numbers for SNR of 13 dB and $\pi_0 = \frac{1}{3}$, we obtain

$$P_{e|0} = 1.1 \times 10^{-3}, \quad P_{e|1} = 5.34 \times 10^{-4}$$

which averages to

$$P_{e,\text{MAP}} = 7.3 \times 10^{-4}$$

which is a slight improvement on the error probability of the ML rule.

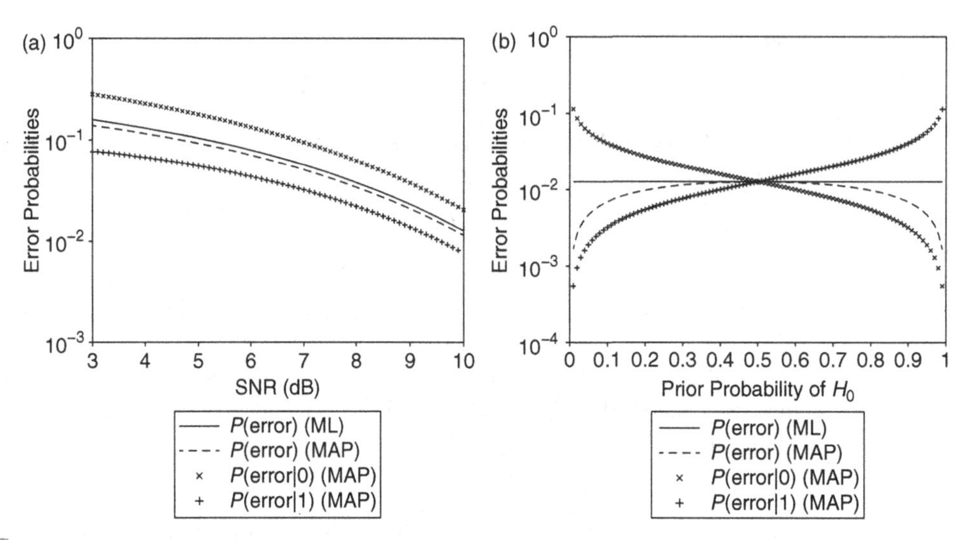

Figure 6.3 Conditional and average error probabilities for the MAP receiver compared with the error probability for the ML receiver. We consider the basic Gaussian example, fixing the priors ($\pi_0 = 0.3$) and varying the SNR in (a), and fixing the SNR (SNR $= 10$ dB) and varying the priors in (b). For the MAP rule, the conditional error probability given a hypothesis increases as the prior probability of the hypothesis decreases. The average error probability for the MAP rule is always smaller than the ML rule (which is the MAP rule for equal priors) when $\pi_0 \neq \frac{1}{2}$. The MAP error probability tends towards zero as $\pi_0 \to 0$ or $\pi_0 \to 1$.

Figure 6.3 shows the results of further numerical experiments (see the caption for a discussion).

6.1.3 Soft decisions

We have so far considered *hard* decision rules in which we must choose exactly one of the M hypotheses. In doing so, we are throwing away a lot of information in the observation. For example, suppose that we are testing $H_0 : Y \sim N(0,4)$ versus $H_1 : Y \sim N(10,4)$ with equal priors, so that the MPE rule is

$$Y \underset{H_0}{\overset{H_1}{\gtrless}} 5$$

We would guess H_1 if $Y = 5.1$ as well as if $Y = 10.3$, but we would be a lot more confident about our guess in the latter instance. Rather than throwing away this information, we can employ *soft* decisions that convey reliability information that could be used at a higher layer, for example, by a decoder that is processing a codeword consisting of many bits.

Actually, we already know how to compute soft decisions: the posterior probabilities $P[H_i|Y = y]$, $i = 0, 1, \ldots, M-1$, that appear in the MAP rule are actually the most information that we can hope to get about the hypotheses from the observation. For notational compactness, let us denote these by $\pi_i(y)$. The posterior probabilities can be computed using Bayes' rule as follows:

$$\pi_i(y) = P[H_i|Y = y] = \frac{\pi_i p(y|i)}{p(y)} = \frac{\pi_i p(y|i)}{\sum_{j=0}^{M-1} \pi_j p(y|j)} \tag{6.16}$$

In practice, we may settle for quantized soft decisions that convey less information than the posterior probabilities due to tradeoffs in precision or complexity versus performance.

Example 6.1.3 (Soft decisions for 4PAM in AWGN) Consider a 4-ary hypothesis-testing problem modeled as follows:

$$H_0 : Y \sim N(-3A, \sigma^2), \quad H_1 : Y \sim N(-A, \sigma^2), \; H_2 : Y \sim N(A, \sigma^2), \; H_3 : Y \sim N(3A, \sigma^2)$$

This is a model that arises for 4PAM signaling in AWGN, as we see later. For $\sigma^2 = 1$, $A = 1$, and $Y = -1.5$, find the posterior probabilities if $\pi_0 = 0.4$ and $\pi_1 = \pi_2 = \pi_3 = 0.2$.

Solution

The posterior probability for the ith hypothesis is of the form

$$\pi_i(y) = c\pi_i e^{-(y-m_i)^2/(2\sigma^2)}$$

where $m_i \in \{\pm A, \pm 3A\}$ is the conditional mean under H_i, and where c is a constant that does not depend on i. Since the posterior probabilities must sum to one, we have

$$\sum_{j=0}^{3} \pi_j(y) = c \sum_{j=0}^{3} \pi_j e^{-(y-m_j)^2/(2\sigma^2)} = 1$$

On solving for c, we obtain

$$\pi_i(y) = \frac{\pi_i e^{-(y-m_i)^2/(2\sigma^2)}}{\sum_{j=0}^{3} \pi_j e^{-(y-m_j)^2/(2\sigma^2)}}$$

By plugging in the numbers, we obtain

$$\pi_0(-1.5) = 0.4121, \qquad \pi_1(-1.5) = 0.5600,$$
$$\pi_2(-1.5) = 0.0279, \qquad \pi_3(-1.5) = 2.5 \times 10^{-5}$$

The MPE hard decision in this case is $\delta_{\text{MPE}}(-1.5) = 1$, but note that the posterior probability for H_0 is also quite high, which is information that would have been thrown away if just hard decisions were reported. However, if the noise strength is reduced, then the hard decision becomes more reliable. For example, for $\sigma^2 = 0.1$, we obtain

$$\pi_0(-1.5) = 9.08 \times 10^{-5}, \qquad \pi_1(-1.5) = 0.9999, \qquad \pi_2(-1.5) = 9.36 \times 10^{-14},$$
$$\pi_3(-1.5) = 3.72 \times 10^{-44}$$

where it is not wise to trust some of the smaller numbers. Thus, we can be quite confident about the hard decision from the MPE rule in this case.

For binary hypothesis testing, it suffices to output one of the two posterior probabilities, since they sum to one. However, it is often more convenient to output the log of the ratio of the posteriors, termed the log likelihood ratio (LLR):

$$\begin{aligned}\text{LLR}(y) &= \log\left(\frac{P[H_1|Y=y]}{P[H_0|Y=y]}\right) = \log\left(\frac{\pi_1 p(y|1)}{\pi_0 p(y|0)}\right)\\ &= \log\left(\frac{\pi_1}{\pi_0}\right) + \log\left(\frac{p(y|1)}{p(y|0)}\right)\end{aligned} \tag{6.17}$$

Notice how the information from the priors and the information from the observations, each of which also takes the form of an LLR, add up in the overall LLR. This simple additive combining of information is exploited in sophisticated decoding algorithms in which information from one part of the decoder provides priors for another part of the decoder. Note that the LLR contribution due to the priors is zero for equal priors.

Example 6.1.4 (LLRs for binary antipodal signaling) Consider $H_1 : Y \sim N(A, \sigma^2)$ versus $H_0 : Y \sim N(-A, \sigma^2)$. We shall see later how this model arises for binary antipodal signaling in AWGN. We leave it as an exercise to show that the LLR is given by

$$\text{LLR}(y) = \frac{2Ay}{\sigma^2}$$

for equal priors.

6.2 Signal-space concepts

We have seen in the previous section that the statistical relation between the hypotheses $\{H_i\}$ and the observation Y can be expressed in terms of the conditional densities $\{p(y|i)\}$. We are now interested in applying this framework to derive optimal decision rules (and the receiver structures required to implement them) for the problem of M-ary signaling in AWGN. In the language of hypothesis testing, the observation here is the received signal $y(t)$ modeled as follows:

$$H_i : y(t) = s_i(t) + n(t), \quad i = 0, 1, \ldots, M-1 \tag{6.18}$$

where $s_i(t)$ is the transmitted signal corresponding to hypothesis H_i, and $n(t)$ is WGN with PSD $\sigma^2 = N_0/2$. Before we can apply the framework of the previous section, however, we must figure out how to define conditional densities when the observation is a continuous-time signal. Here is how we do it.

- We first observe that, while the signals $s_i(t)$ live in an infinite-dimensional, continuous-time space, if we are interested only in the M signals that could be transmitted under each of the M hypotheses, then we can limit our attention to a finite-dimensional subspace of dimension at most M. We call this the *signal space*. We can then express the signals as vectors corresponding to an expansion with respect to an orthonormal basis for the subspace.

- The projection of WGN onto the signal space gives us a noise vector whose components are i.i.d. Gaussian. Furthermore, we observe that the component of the received signal orthogonal to the signal space is *irrelevant:* that is, we can throw it away without compromising performance.
- We can therefore restrict attention to projection of the received signal onto the signal space without loss of performance. This projection can be expressed as a finite-dimensional vector which is modeled as a discrete time analog of (6.18). We can now apply the hypothesis-testing framework of Section 6.1 to infer the optimal (ML and MPE) decision rules.
- We then translate the optimal decision rules back to continuous time to infer the structure of the optimal receiver.

6.2.1 Representing signals as vectors

Let us begin with an example illustrating how continuous-time signals can be represented as finite-dimensional vectors by projecting onto the signal space.

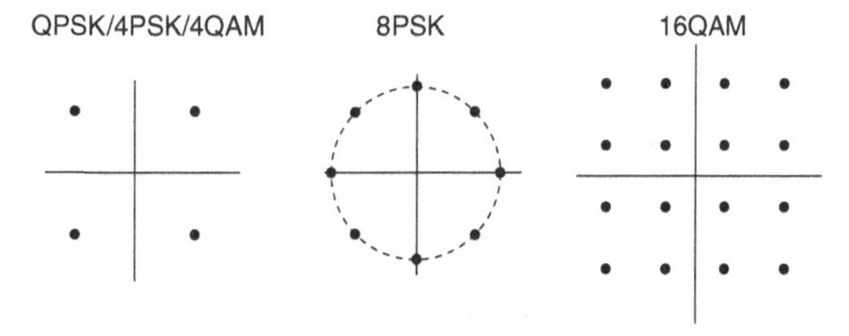

Figure 6.4 For linear modulation with no intersymbol interference, the complex symbols themselves provide a two-dimensional signal-space representation. Three different constellations are shown here.

Example 6.2.1 (Signal space for two-dimensional modulation) Consider a single complex-valued symbol $b = b_c + jb_s$ (assume that there is no intersymbol interference) sent using two-dimensional passband linear modulation. The set of possible transmitted signals is given by

$$s_{b_c,b_s}(t) = b_c p(t)\cos(2\pi f_c t) - b_s p(t)\sin(2\pi f_c t)$$

where (b_c, b_s) takes M possible values for an M-ary constellation (e.g., $M = 4$ for QPSK, $M = 16$ for 16QAM) and $p(t)$ is a baseband pulse of bandwidth smaller than the carrier frequency f_c. On setting $\phi_c(t) = p(t)\cos(2\pi f_c t)$ and $\phi_s(t) = -p(t)\sin(2\pi f_c t)$, we see that we can write the set of transmitted signals as a linear combination of these signals as follows:

$$s_{b_c,b_s}(t) = b_c\phi_c(t) + b_s\phi_s(t)$$

so that the signal space has dimension at most 2. From Chapter 2, we know that ϕ_c and ϕ_s are orthogonal (I–Q orthogonality), and hence linearly independent. Thus, the signal space has dimension exactly 2. By noting that $||\phi_c||^2 = ||\phi_s||^2 = \frac{1}{2}||p||^2$, we see that the normalized versions of ϕ_c and ϕ_s provide an orthonormal basis for the signal space:

$$\psi_c(t) = \frac{\phi_c(t)}{||\phi_c||}, \quad \psi_s(t) = \frac{\phi_s(t)}{||\phi_s||}$$

We can now write

$$s_{b_c,b_s}(t) = \frac{1}{\sqrt{2}}||p||b_c\psi_c(t) + \frac{1}{\sqrt{2}}||p||b_s\psi_s(t)$$

With respect to this basis, the signals can be represented as two-dimensional vectors:

$$s_{b_c,b_s}(t) \leftrightarrow \mathbf{s}_{b_c,b_s} = \frac{1}{\sqrt{2}}||p|| \begin{pmatrix} b_c \\ b_s \end{pmatrix}$$

That is, up to scaling, the signal-space representation for the transmitted signals are simply the two-dimensional symbols $(b_c, b_s)^T$. Indeed, while we have been careful about keeping track of the scaling factor in this example, we shall drop it henceforth, because, as we shall soon see, what matters in performance is the signal-to-noise *ratio,* rather than the absolute signal or noise strength.

Orthogonal modulation provides another example where an orthonormal basis for the signal space is immediately obvious. For example, if $s_1, \ldots, s_M$ are orthogonal signals with equal energy $||s||^2 \equiv E_s$, then $\psi_i(t) = s_i(t)/\sqrt{E_s}$ provide an orthonormal basis for the signal space, and the vector representation of the ith signal is the scaled unit vector $\sqrt{E_s}(0, \ldots, 0, 1 \text{ (in } i\text{th position)}, 0, \ldots, 0)^T$.

Yet another example where an orthonormal basis can be determined by inspection is shown in Figures 6.5 and 6.6, and discussed in Example 6.2.2.

Example 6.2.2 (Developing a signal-space representation for a 4-ary signal set) Consider the example depicted in Figure 6.5, where there are four possible transmitted signals, $s_0, \ldots, s_3$. It is clear from inspection that these span a three-dimensional signal space, with a convenient choice of basis signals

$$\psi_0(t) = I_{[0,1]}(t), \quad \psi_1(t) = I_{[1,2]}(t), \quad \psi_2(t) = I_{[2,3]}(t)$$

as shown in Figure 6.6. Let $\mathbf{s}_i = (s_i[1], s_i[2], s_i[3])^T$ denote the vector representation of the signal s_i with respect to the basis, for $i = 0, 1, 2, 3$. That is, the coefficients of the vector $\mathbf{s}_i$ are such that

$$s_i(t) = \sum_{k=0}^{2} s_i[k]\psi_k(t)$$

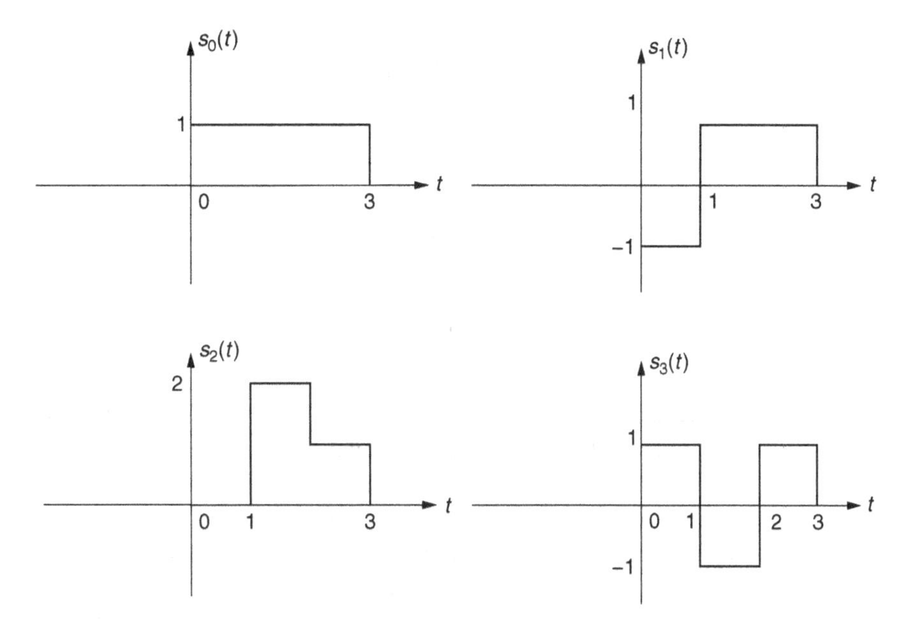

Figure 6.5 Four signals spanning a three-dimensional signal space.

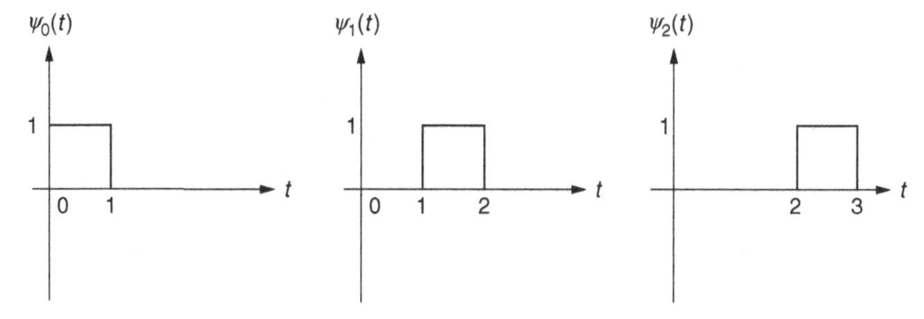

Figure 6.6 An orthonormal basis for the signal set in Figure 6.5, obtained by inspection.

We obtain, again by inspection, that

$$\mathbf{s}_0 = \begin{pmatrix} 1 \\ 1 \\ 1 \end{pmatrix}, \quad \mathbf{s}_1 = \begin{pmatrix} -1 \\ 1 \\ 1 \end{pmatrix}, \quad \mathbf{s}_2 = \begin{pmatrix} 0 \\ 2 \\ 1 \end{pmatrix}, \quad \mathbf{s}_3 = \begin{pmatrix} 1 \\ -1 \\ 1 \end{pmatrix}$$

Now that we have seen some examples, it is time to be more precise about what we mean by the "signal space." The signal space $\mathcal{S}$ is the finite-dimensional subspace (of dimension $n \leq M$) spanned by $s_0(t), \ldots, s_{M-1}(t)$. That is, $\mathcal{S}$ consists of all signals of the form $a_0 s_0(t) + \cdots + a_{M-1} s_{M-1}(t)$, where $a_0, \ldots, a_{M-1}$ are arbitrary scalars. Let $\psi_0(t), \ldots, \psi_{n-1}(t)$ denote an orthonormal basis for $\mathcal{S}$. We have seen in the preceding examples that such a basis can often be determined by inspection. In general, however, given an arbitrary set of signals, we can always construct an orthonormal basis using the Gram–Schmidt procedure described below. We do not need to use this procedure often – in most settings of interest, the way to go from continuous to discrete time is clear – but state it below for completeness.

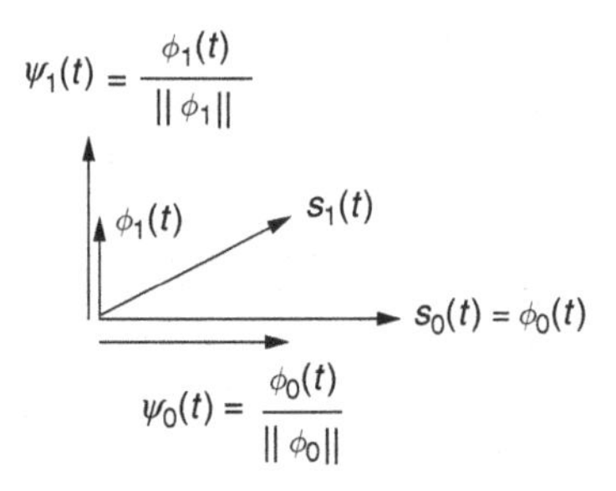

Figure 6.7 Illustrating Step 0 and Step 1 of the Gram–Schmidt procedure.

Gram–Schmidt orthogonalization The idea is to build up an orthonormal basis step by step, with the basis after the mth step spanning the first m signals. The first basis function is a scaled version of the first signal (assuming this is nonzero – otherwise we proceed to the second signal without adding a basis function). We then consider the component of the second signal orthogonal to the first basis function. This projection is nonzero if the second signal is linearly independent of the first; in this case, we introduce a basis function that is a scaled version of the projection. See Figure 6.7. This procedure goes on until we have covered all M signals. The number of basis functions n equals the dimension of the signal space, and satisfies $n \leq M$. We can summarize the procedure as follows.

Letting $\mathcal{S}_{k-1}$ denote the subspace spanned by $s_0, \ldots, s_{k-1}$, the Gram–Schmidt algorithm proceeds iteratively: given an orthonormal basis for $\mathcal{S}_{k-1}$, it finds an orthonormal basis for $\mathcal{S}_k$. The procedure stops when $k = M$. The method is identical to that used for finite-dimensional vectors, except that the definition of the inner product involves an integral, rather than a sum, for the continuous-time signals considered here.

Step 0 **(Initialization)** Let $\phi_0 = s_0$. If $\phi_0 \neq 0$, then set $\psi_0 = \phi_0/||\phi_0||$. Note that ψ_0 provides a basis function for $\mathcal{S}_0$.

Step k Suppose that we have constructed an orthonormal basis $\mathcal{B}_{k-1} = \{\psi_0, \ldots, \psi_{m-1}\}$ for the subspace $\mathcal{S}_{k-1}$ spanned by the first k signals, $s_0, \ldots, s_{k-1}$ (note that $m \leq k$). Define

$$\phi_k(t) = s_k(t) - \sum_{i=0}^{m-1} \langle s_k, \psi_i \rangle \psi_i(t)$$

The signal $\phi_k(t)$ is the component of $s_k(t)$ orthogonal to the subspace $\mathcal{S}_{k-1}$. If $\phi_k \neq 0$, define a new basis function $\psi_m(t) = \phi_k(t)/||\phi_k||$, and update the basis as $\mathcal{B}_k = \{\psi_1, \ldots, \psi_m, \psi_m\}$. If $\phi_k = 0$, then $s_k \in \mathcal{S}_{k-1}$, and it is not necessary to update the basis; in this case, we set $\mathcal{B}_k = \mathcal{B}_{k-1} = \{\psi_0, \ldots, \psi_{m-1}\}$.

The procedure terminates at step M, which yields a basis $\mathcal{B} = \{\psi_0, \ldots, \psi_{n-1}\}$ for the signal space $\mathcal{S} = \mathcal{S}_{M-1}$. The basis is not unique, and may depend (and typically does depend) on the order in which we go through the signals in the set. We use the Gram–Schmidt procedure here mainly as a conceptual tool, assuring us that there is indeed a finite-dimensional vector representation for a finite set of continuous-time signals.

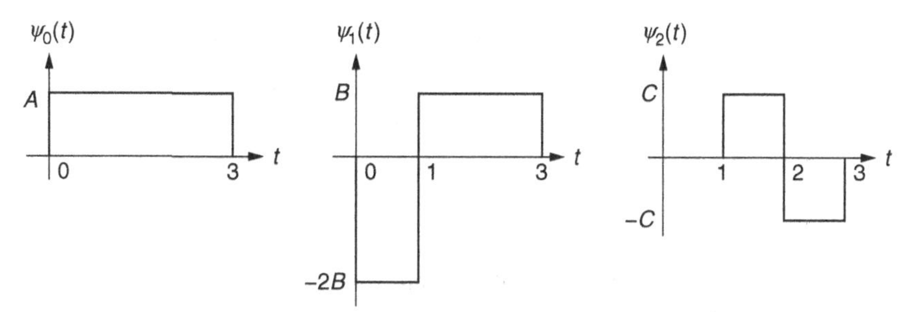

Figure 6.8 An orthonormal basis for the signal set in Figure 6.5, obtained by applying the Gram–Schmidt procedure. The unknowns A, B, and C are to be determined in Exercise 6.2.1.

Exercise 6.2.1 (Application of the Gram–Schmidt procedure) Apply the Gram–Schmidt procedure to the signal set in Figure 6.5. When the signals are considered in increasing order of index in the Gram–Schmidt procedure, verify that the basis signals are as in Figure 6.8, and fill in the missing numbers. While the basis thus obtained is not as "nice" as the one obtained by inspection in Figure 6.6, the Gram–Schmidt procedure has the advantage of general applicability.

Inner products are preserved We shall soon see that the performance of M-ary signaling in AWGN depends only on the inner products between the signals, if the noise PSD is fixed. Thus, an important observation when mapping the continuous-time hypothesis-testing problem to discrete time is to check that these inner products are preserved when projecting onto the signal space. Consider the continuous-time inner products

$$\langle s_i, s_j \rangle = \int s_i(t) s_j(t) dt, \; i,j = 0, 1, \ldots, M-1 \tag{6.19}$$

Now, expressing the signals in terms of their basis expansions, we have

$$s_i(t) = \sum_{k=0}^{n-1} s_i[k] \psi_k(t), \; i = 0, 1, \ldots, M-1$$

By plugging this into (6.19), we obtain

$$\langle s_i, s_j \rangle = \int \sum_{k=0}^{n-1} s_i[k] \psi_k(t) \sum_{l=0}^{n-1} s_j[l] \psi_l(t) dt$$

By interchanging integral and summations, we obtain

$$\langle s_i, s_j \rangle = \sum_{k=0}^{n-1} \sum_{l=0}^{n-1} s_i[k] s_j[l] \int \psi_k(t) \psi_l(t) dt$$

By virtue of the orthonormality of the basis functions $\{\psi_k\}$, we have

$$\langle \psi_k, \psi_l \rangle = \int \psi_k(t) \psi_l(t) dt = \delta_{kl} = \begin{cases} 1, & k = l \\ 0, & k \neq l \end{cases}$$

This collapses the two summations into one, so we obtain

$$\langle s_i, s_j \rangle = \int s_i(t) s_j(t) dt = \sum_{k=0}^{n-1} s_i[k] s_j[k] = \langle \mathbf{s}_i, \mathbf{s}_j \rangle \tag{6.20}$$

where the extreme right-hand side is the inner product of the signal vectors $\mathbf{s}_i = (s_i[0], \ldots, s_i[n-1])^{\mathrm{T}}$ and $\mathbf{s}_j = (s_j[0], \ldots, s_j[n-1])^{\mathrm{T}}$. This makes sense: the geometric relationship between signals (which is what the inner products capture) should not depend on the basis with respect to which they are expressed.

6.2.2 Modeling WGN in signal space

What happens to the noise when we project onto the signal space? Define the noise projection onto the ith basis function as

$$N[i] = \langle n, \psi_i \rangle = \int n(t) \psi_i(t) dt, \quad i = 0, 1, \ldots, n-1 \tag{6.21}$$

Then we can write the noise $n(t)$ as follows:

$$n(t) = \sum_{i=0}^{n-1} N[i] \psi_i(t) + n^{\perp}(t)$$

where $n^{\perp}(t)$ is the projection of the noise orthogonal to the signal space. Thus, we can decompose the noise into two parts: a noise vector $\mathbf{N} = (N[0], \ldots, N[n-1])^{\mathrm{T}}$ corresponding to the projection onto the signal space, and a component $n^{\perp}(t)$ orthogonal to the signal space. In order to characterize the statistics of these quantities, we need to consider random variables obtained by linear processing of WGN. Specifically, consider random variables generated by passing WGN through correlators:

$$Z_1 = \int_{-\infty}^{\infty} n(t) u_1(t) dt = \langle n, u_1 \rangle$$

$$Z_2 = \int_{-\infty}^{\infty} n(t) u_2(t) dt = \langle n, u_2 \rangle$$

where u_1 and u_2 are deterministic, finite-energy signals. We can now state the following result.

Theorem 6.2.1 (WGN through correlators) *The random variables $Z_1 = \langle n, u_1 \rangle$ and $Z_2 = \langle n, u_2 \rangle$ are zero mean, jointly Gaussian, with*

$$\operatorname{cov}(Z_1, Z_2) = \operatorname{cov}(\langle n, u_1 \rangle, \langle n, u_2 \rangle) = \sigma^2 \langle u_1, u_2 \rangle$$

On specializing to $u_1 = u_2 = u$, we obtain that

$$\operatorname{var}(\langle n, u \rangle) = \operatorname{cov}(\langle n, u \rangle, \langle n, u \rangle) = \sigma^2 ||u||^2$$

Thus, we obtain that $\mathbf{Z} = (Z_1, Z_2)^{\mathrm{T}} \sim N(0, \mathbf{C})$ *with covariance matrix*

$$\mathbf{C} = \begin{pmatrix} \sigma^2 ||u_1||^2 & \sigma^2 \langle u_1, u_2 \rangle \\ \sigma^2 \langle u_1, u_2 \rangle & \sigma^2 ||u_2||^2 \end{pmatrix}$$

Proof of Theorem 6.2.1 The random variables $Z_1 = \langle n, u_1 \rangle$ and $Z_2 = \langle n, u_2 \rangle$ are zero mean and jointly Gaussian, since n is zero mean and Gaussian. Their covariance is computed as

$$\begin{aligned} \mathrm{cov}(\langle n, u_1 \rangle, \langle n, u_2 \rangle) &= \mathbb{E}[\langle n, u_1 \rangle \langle n, u_2 \rangle] = \mathbb{E}\left[\int n(t) u_1(t) dt \int n(s) u_2(s) ds \right] \\ &= \int \int u_1(t) u_2(s) \mathbb{E}[n(t) n(s)] dt\, ds = \int \int u_1(t) u_2(s) \sigma^2 \delta(t - s) dt\, ds \\ &= \sigma^2 \int u_1(t) u_2(t) dt = \sigma^2 \langle u_1, u_2 \rangle \end{aligned}$$

The preceding computation is entirely analogous to the ones we did in Example 5.8.2 and in Section 5.10, but it is important enough that we repeat some points that we had mentioned then. First, we need to use two different variables of integration, t and s, in order to make sure we capture all the cross terms. Second, when we take the expectation inside the integrals, we must group all random terms inside it. Third, the two integrals collapse into one because the autocorrelation function of WGN is impulsive. Finally, specializing the covariance to get the variance leads to the remaining results stated in the theorem. □

We can now provide the following geometric interpretation of WGN.

Remark 6.2.1 (Geometric interpretation of WGN) Theorem 6.2.1 implies that the projection of WGN along any "direction" in the space of signals (i.e., the result of correlating WGN with a unit energy signal) has variance $\sigma^2 = N_0/2$. Also, its projections in *orthogonal* directions are jointly Gaussian and uncorrelated random variables, and are therefore independent.

Noise projection on the signal space is discrete-time WGN It follows from the preceding remark that the noise projections $N[i] = \langle n, \psi_i \rangle$ along the orthonormal basis functions $\{\psi_i\}$ for the signal space are i.i.d. $N(0, \sigma^2)$ random variables. In other words, the noise vector $\mathbf{N} = (N[0], \ldots, N[n-1])^{\mathrm{T}} \sim N\left(\mathbf{0}, \sigma^2 \mathbf{I}\right)$. In other words, the components of $\mathbf{N}$ constitute discrete-time white Gaussian noise ("white" in this case means uncorrelated and having equal variance across all components).

6.2.3 Hypothesis testing in signal space

Now that we have the signal and noise models, we can put them together in our hypothesis-testing framework. Let us condition on hypothesis H_i. The received signal is given by

$$y(t) = s_i(t) + n(t) \tag{6.22}$$

By projecting this onto the signal space by correlating against the orthonormal basis functions, we get

$$Y[k] = \langle y, \psi_k \rangle = \langle s_i + n, \psi_k \rangle = s_i[k] + N[k], \quad k = 0, 1, \ldots, n-1$$

On collecting these into an n-dimensional vector, we get the model

$$H_i : \mathbf{Y} = \mathbf{s}_i + \mathbf{N}$$

Note that the vector $\mathbf{Y} = (y[1], \ldots, y[n])^{\mathrm{T}}$ completely describes the component of the received signal $y(t)$ in the signal space, given by

$$y_{\mathcal{S}}(t) = \sum_{j=0}^{n-1} \langle y, \psi_j \rangle \psi_j(t) = \sum_{j=0}^{n-1} Y[j] \psi_j(t)$$

The component of the received signal orthogonal to the signal space is given by

$$y^{\perp}(t) = y(t) - y_{\mathcal{S}}(t)$$

It is shown in Appendix 6.A that this component is *irrelevant* to our decision. There are two reasons for this, as elaborated in the appendix: first, there is no signal contribution orthogonal to the signal space (by definition); second, for the WGN model, the noise component orthogonal to the signal space carries no information regarding the noise vector in the signal space. As illustrated in Figure 6.9, this enables us to reduce our infinite-dimensional problem to the following finite-dimensional vector model, *without loss of optimality.*

Model for received vector in signal space

$$H_i : \mathbf{Y} = \mathbf{s}_i + \mathbf{N}, \; i = 0, 1, \ldots, M-1 \tag{6.23}$$

where $\mathbf{N} \sim N(0, \sigma^2\mathbf{I})$.

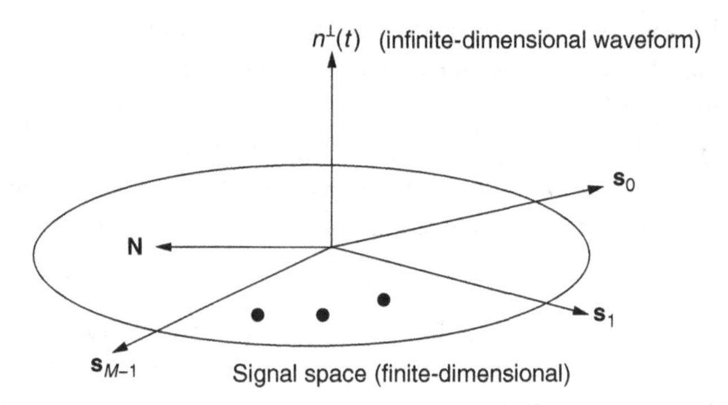

Figure 6.9 Illustration of signal-space concepts. The noise projection $n^{\perp}(t)$ orthogonal to the signal space is irrelevant. The relevant part of the received signal is the projection onto the signal space, which equals the vector $\mathbf{Y} = \mathbf{s}_i + \mathbf{N}$ under hypothesis H_i.

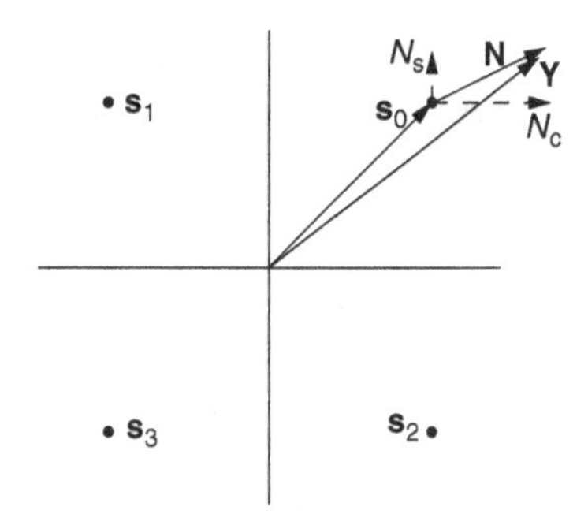

Figure 6.10 A signal-space view of QPSK. In the scenario shown, $\mathbf{s}_0$ is the transmitted vector, and $\mathbf{Y} = \mathbf{s}_0 + \mathbf{N}$ is the received vector after noise has been added. The noise components N_c and N_s are i.i.d. $N(0, \sigma^2)$ random variables.

Two-dimensional modulation (Example 6.2.1 revisited) For a single symbol sent using two-dimensional modulation, we have the hypotheses

$$H_{b_c,b_s} : y(t) = s_{b_c,b_s}(t) + n(t)$$

where

$$s_{b_c,b_s}(t) = b_c p(t)\cos(2\pi f_c t) - b_s p(t)\sin(2\pi f_c t)$$

On restricting our attention to the two-dimensional signal space identified in the example, we obtain the model

$$H_{b_c,b_s} : \mathbf{Y} = \begin{pmatrix} Y_c \\ Y_s \end{pmatrix} = \begin{pmatrix} b_c \\ b_s \end{pmatrix} + \begin{pmatrix} N_c \\ N_s \end{pmatrix}$$

where we have absorbed scale factors into the symbol (b_c, b_s), and where the I and Q noise components N_c and N_s are i.i.d. $N(0, \sigma^2)$. This is illustrated for QPSK in Figure 6.10. Thus, conditioned on H_{b_c,b_s}, $Y_c \sim N(b_c, \sigma^2)$ and $Y_s \sim N(b_s, \sigma^2)$, and Y_c and Y_s are conditionally independent. The conditional density of $\mathbf{Y} = (Y_c, Y_s)^T$ conditioned on H_{b_c,b_s} is therefore given by

$$p(y_c, y_s | b_c, b_s) = \frac{1}{2\sigma^2} e^{-(y_c - b_c)^2/(2\sigma^2)} \frac{1}{2\sigma^2} e^{-(y_s - b_s)^2/(2\sigma^2)}$$

We can now infer the ML and MPE rules using our hypothesis-testing framework. However, since the same reasoning applies to signal spaces of arbitrary dimensions, we provide a more general discussion in the next section, and then return to examples of two-dimensional modulation.

6.2.4 Optimal reception in AWGN

We begin by characterizing the optimal receiver when the received signal is a finite-dimensional vector. Using this, we infer the optimal receiver for continuous-time received signals.

Demodulation for M-ary signaling in discrete-time AWGN corresponds to solving an M-ary hypothesis-testing problem with observation model as follows:

$$H_i : \mathbf{Y} = \mathbf{s}_i + \mathbf{N}, \quad i = 0, 1, \ldots, M-1 \tag{6.24}$$

where $\mathbf{N} \sim N(0, \sigma^2 \mathbf{I})$ is discrete-time WGN. The ML and MPE rules for this problem are given as follows. As usual, we denote the prior probabilities required to specify the MPE rule by $\{\pi_i, i = 1, .., M\}$ $(\sum_{i=0}^{M-1} \pi_i = 1)$.

Optimal demodulation for signaling in discrete-time AWGN

ML rule

$$\begin{aligned}\delta_{ML}(\mathbf{y}) &= \arg\min_{0 \le i \le M-1} ||\mathbf{y} - \mathbf{s}_i||^2 \\ &= \arg\max_{0 \le i \le M-1} \langle \mathbf{y}, \mathbf{s}_i \rangle - ||\mathbf{s}_i||^2/2\end{aligned} \tag{6.25}$$

MPE rule

$$\begin{aligned}\delta_{MPE}(\mathbf{y}) &= \arg\min_{0 \le i \le M-1} ||\mathbf{y} - \mathbf{s}_i||^2 - 2\sigma^2 \log \pi_i \\ &= \arg\max_{0 \le i \le M-1} \langle \mathbf{y}, \mathbf{s}_i \rangle - ||\mathbf{s}_i||^2/2 + \sigma^2 \log \pi_i\end{aligned} \tag{6.26}$$

Interpretation of optimal decision rules The ML rule can be interpreted in two ways. The first is as a *minimum-distance* rule, choosing the transmitted signal which has minimum Euclidean distance to the noisy received signal. The second is as a "template matcher," choosing the transmitted signal with highest correlation with the noisy received signal, while adjusting for the fact that the energies of different transmitted signals may be different. The MPE rule adjusts the ML cost function to reflect prior information: the adjustment term depends on the noise level and the prior probabilities. The MPE cost functions decompose neatly into a sum of the ML cost function (which depends on the observation) and a term reflecting prior knowledge (which depends on the prior probabilities and the noise level). The latter term scales with the noise variance σ^2. Thus, we rely more on the observation at high SNR (small σ), and more on prior knowledge at low SNR (large σ).

Derivation of optimal receiver structures (6.25) and (6.26) Under hypothesis H_i, $\mathbf{Y}$ is a Gaussian random vector with mean $\mathbf{s}_i$ and covariance matrix $\sigma^2 \mathbf{I}$ (the translation of the noise vector $\mathbf{N}$ by the deterministic signal vector $\mathbf{s}_i$ does not change the covariance matrix), so that

$$p_{\mathbf{Y}|i}(\mathbf{y}|H_i) = \frac{1}{(2\pi\sigma^2)^{n/2}} \exp\left(-\frac{||\mathbf{y} - \mathbf{s}_i||^2}{2\sigma^2}\right) \tag{6.27}$$

By plugging (6.27) into the ML rule (6.5), we obtain the rule (6.25) upon simplification. Similarly, we obtain (6.26) by substituting (6.27) in the MPE rule (6.6). □

We now map the optimal decision rules in discrete time back to continuous time to obtain optimal detectors for the original continuous-time model (6.18), as follows.

Optimal demodulation for signaling in continuous-time AWGN

ML rule

$$\delta_{ML}(y) = \arg\max_{0 \le i \le M-1} \langle y, s_i \rangle - ||s_i||^2/2 \tag{6.28}$$

MPE rule

$$\delta_{MPE}(\mathbf{y}) = \arg\max_{0 \le i \le M-1} \langle y, s_i \rangle - ||s_i||^2/2 + \sigma^2 \log \pi_i \tag{6.29}$$

Derivation of optimal receiver structures (6.28) and (6.29) Owing to the irrelevance of $y^\perp$, the continuous-time model (6.18) reduces to the discrete-time model (6.24) on projection onto the signal space. It remains to map the optimal decision rules (6.25) and (6.26) for discrete-time observations back to continuous time. These rules involve correlation between the received and transmitted signals and the transmitted signal energies. It suffices to show that these quantities are the same for the continuous-time model and the equivalent discrete-time model. We know now that signal inner products are preserved, so that

$$||\mathbf{s}_i||^2 = ||s_i||^2$$

Further, the continuous-time correlator output can be written as

$$\begin{aligned}\langle y, s_i \rangle &= \langle y_S + y^\perp, s_i \rangle = \langle y_S, s_i \rangle + \langle y^\perp, s_i \rangle \\ &= \langle y_S, s_i \rangle = \langle \mathbf{y}, \mathbf{s}_i \rangle\end{aligned}$$

where the last equality follows because the inner product between the signals y_S and s_i (both of which lie in the signal space) is the same as the inner product between their vector representations. □

Why don't we have a "minimum distance" rule in continuous time? Notice that the optimal decision rules for the continuous-time model do not contain the continuous-time version of the minimum distance rule for discrete time. This is because of a technical subtlety. In continuous time, the squares of the distances would be

$$||y - s_i||^2 = ||y_S - s_i||^2 + ||y^\perp||^2 = ||y_S - s_i||^2 + ||n^\perp||^2$$

Under the AWGN model, the noise power orthogonal to the signal space is infinite; hence, from a purely mathematical point of view, the preceding quantities are infinite for each i (so we cannot minimize over i). Hence, it only makes sense to talk about the minimum-distance rule in a finite-dimensional space in which the noise power is finite. The correlator-based form of the optimal detector, on the other hand, automatically achieves the projection onto the finite-dimensional signal space, and hence does not suffer from this technical difficulty.

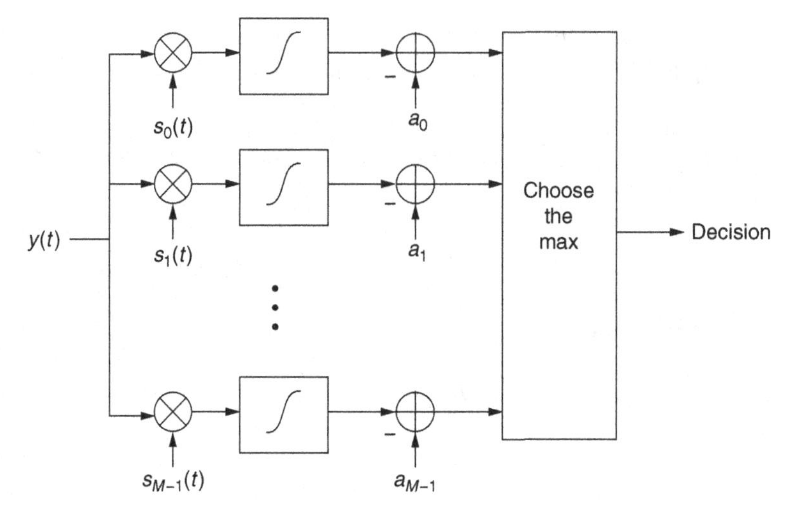

Figure 6.11 The optimal receiver for an AWGN channel can be implemented using a bank of correlators. For the ML rule, the constants are $a_i = ||s_i||^2/2$; for the MPE rule, $a_i = ||s_i||^2/2 - \sigma^2 \log \pi_i$.

Of course, in practice, even the continuous-time received signal may be limited to a finite-dimensional space by filtering and time-limiting, but correlator-based detection still has the practical advantage that only those components of the received signal which are truly useful appear in the decision statistics.

A bank of correlators or matched filters The optimal receiver involves computation of the decision statistics

$$\langle y, s_i \rangle = \int y(t) s_i(t) dt$$

and can therefore be implemented using a bank of correlators, as shown in Figure 6.11. Of course, any correlation operation can also be implemented using a matched filter, sampled at the appropriate time. On defining $s_{i,\mathrm{MF}}(t) = s_i(-t)$ as the impulse response of the filter matched to s_i, we have

$$\langle y, s_i \rangle = \int y(t) s_i(t) dt = \int y(t) s_{i,\mathrm{MF}}(-t) dt = (y * s_{i,\mathrm{MF}})(0)$$

Figure 6.12 shows an alternative implementation for the optimal receiver using a bank of matched filters.

Implementation in complex baseband We have developed the optimal receiver structures for real-valued signals, so that these apply to physical baseband and passband signals. However, recall from Chapter 2 that correlation and filtering in passband, which is what the optimal receiver does, can be implemented in complex baseband after downconversion. In particular, for passband signals $u_{\mathrm{p}}(t) = u_{\mathrm{c}}(t)\cos(2\pi f_{\mathrm{c}} t) - u_{\mathrm{s}}(t)\sin(2\pi f_{\mathrm{c}} t)$ and $v_{\mathrm{p}}(t) = v_{\mathrm{c}}(t)\cos(2\pi f_{\mathrm{c}} t) - v_{\mathrm{s}}(t)\sin(2\pi f_{\mathrm{c}} t)$, the inner product can be written as

$$\langle u_{\mathrm{p}}, v_{\mathrm{p}} \rangle = \frac{1}{2}(\langle u_{\mathrm{c}}, v_{\mathrm{c}} \rangle + \langle u_{\mathrm{s}}, v_{\mathrm{s}} \rangle) = \frac{1}{2} \operatorname{Re}\langle u, v \rangle \tag{6.30}$$

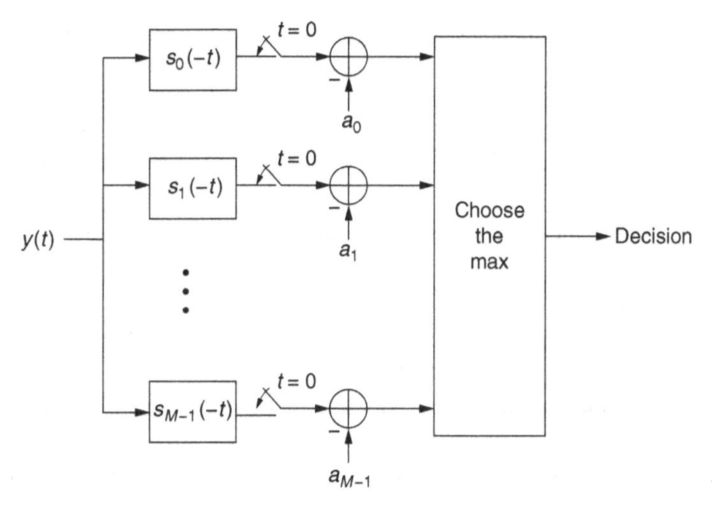

Figure 6.12 An alternative implementation for the optimal receiver using a bank of matched filters. For the ML rule, the constants are $a_i = ||s_i||^2/2$; for the MPE rule, $a_i = ||s_i||^2/2 - \sigma^2 \log \pi_i$.

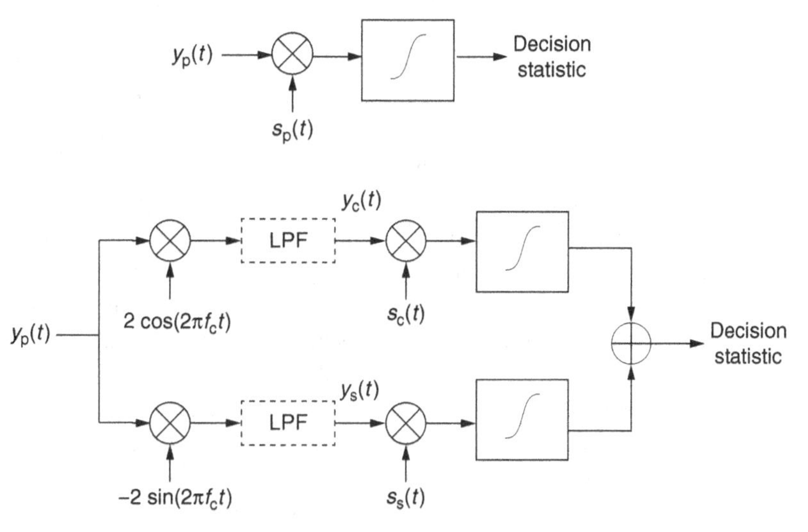

Figure 6.13 The passband correlations required by the optimal receiver can be implemented in complex baseband. Since the I and Q components are lowpass waveforms, correlation with them is an implicit form of lowpass filtering. Thus, the LPFs after the mixers could potentially be eliminated, which is why they are shown within dashed boxes.

where $u = u_c + ju_s$ and $v = v_c + jv_s$ are the corresponding complex envelopes. Figure 6.13 shows how a passband correlation can be implemented in complex baseband. Note that we correlate the I component with the I component, and the Q component with the Q component. This is because our optimal receiver is based on the assumption of *coherent* reception: our model assumes that the receiver has exact copies of the noiseless transmitted signals. Thus, ideal carrier synchronism is implicitly assumed in this model, so that the

I and Q components do not get mixed up as they would if the receiver's LO were not synchronized to the incoming carrier.

6.2.5 Geometry of the ML decision rule

The minimum-distance interpretation for the ML decision rule implies that the decision regions (in signal space) for M-ary signaling in AWGN are constructed as follows. Interpret the signal vectors $\{\mathbf{s}_i\}$, and the received vector $\mathbf{y}$, as points in n-dimensional Euclidean space. When deciding between any pair of signals $\mathbf{s}_i$ and $\mathbf{s}_j$ (which are points in n-dimensional space), we draw a line between these points. The decision boundary is the perpendicular bisector of this line, which is an $(n-1)$-dimensional hyperplane. This is illustrated in Figure 6.14, where, because we are constrained to draw on two-dimensional paper, the hyperplane reduces to a line. But we can visualize a plane containing the decision boundary coming out of the paper for a three-dimensional signal space. While it is hard to visualize signal spaces of more than three dimensions, the computation for deciding which side of the ML decision boundary the received vector $\mathbf{y}$ lies on is straightforward: simply compare the Euclidean distances $||\mathbf{y}-\mathbf{s}_i||$ and $||\mathbf{y}-\mathbf{s}_j||$.

The ML decision regions are constructed from drawing these pairwise decision regions. For any given i, draw a line between $\mathbf{s}_i$ and $\mathbf{s}_j$ for all $j \neq i$. The perpendicular bisector of the line between $\mathbf{s}_i$ and $\mathbf{s}_j$ defines two half spaces (half planes for $n=2$), one in which we choose $\mathbf{s}_i$ over $\mathbf{s}_j$, and the other in which we choose $\mathbf{s}_j$ over $\mathbf{s}_i$. The intersection of the half spaces in which $\mathbf{s}_i$ is chosen over $\mathbf{s}_j$, for $j \neq i$, defines the decision region Γ_i. This procedure is illustrated for a two-dimensional signal space in Figure 6.15. The line L_{1i} is the perpendicular bisector of the line between $\mathbf{s}_1$ and $\mathbf{s}_i$. The intersection of these lines defines Γ_1 as shown. Note that L_{16} plays no role in determining Γ_1, since signal $\mathbf{s}_6$ is "too far" from $\mathbf{s}_1$, in the following sense: if the received signal is closer to $\mathbf{s}_6$ than to $\mathbf{s}_1$, then it is also closer to $\mathbf{s}_i$ than to $\mathbf{s}_1$ for some $i = 2, 3, 4, 5$. This kind of observation plays an important role in the performance analysis of ML reception in Section 6.3.

The preceding procedure can now be applied to the simpler scenario of two-dimensional constellations to obtain ML decision regions as shown in Figure 6.16. For QPSK, the ML regions are simply the four quadrants. For 8PSK, the ML regions are sectors of a circle. For 16QAM, the ML regions take a rectangular form.

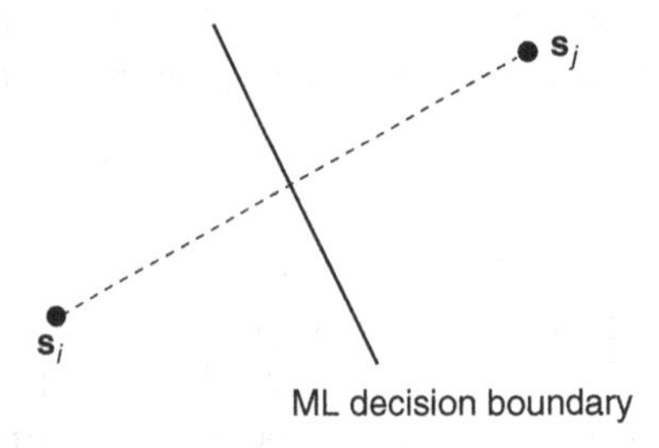

Figure 6.14 The ML decision boundary when testing between $\mathbf{s}_i$ and $\mathbf{s}_j$ is the perpendicular bisector of the line joining the signal points, which is an $(n-1)$-dimensional hyperplane for an n-dimensional signal space.

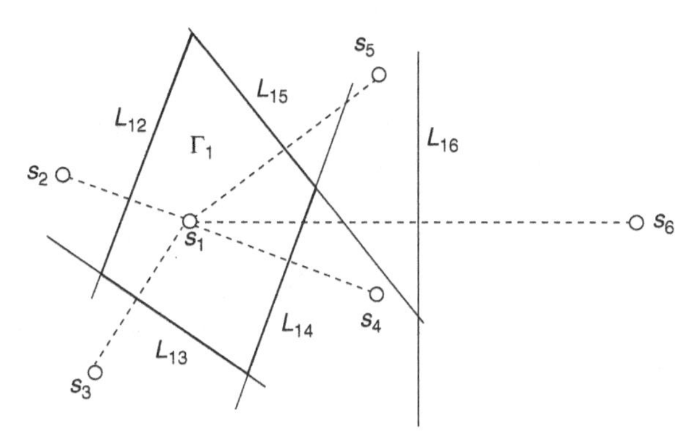

Figure 6.15 ML decision region Γ_1 for signal s_1.

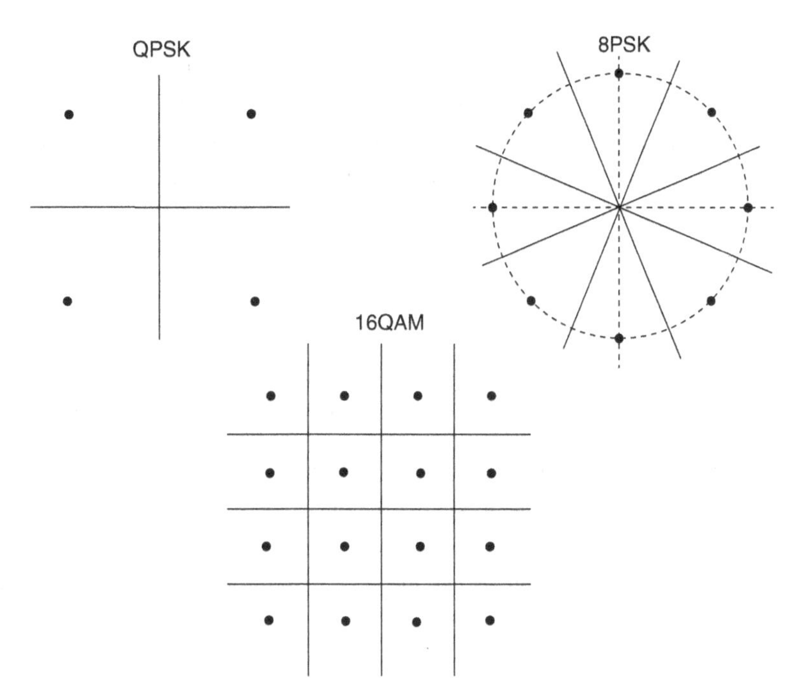

Figure 6.16 ML decision regions for some two-dimensional constellations.

6.3 Performance analysis of ML reception

We focus on performance analysis for the ML decision rule, assuming equal priors (for which the ML rule minimizes the error probability). The analysis for MPE reception with unequal priors is skipped, but it is a simple extension. We begin with a geometric picture of how errors are caused by WGN.

6.3.1 The geometry of errors

In Figure 6.17, suppose that signal $\mathbf{s}$ is sent, and we wish to compute the probability that the noise vector $\mathbf{N}$ causes the received vector to cross a given decision boundary. From the figure, it is clear that N_{perp}, the projection of the noise vector perpendicular to the decision boundary, is what determines whether or not we will cross the boundary. It does not matter what happens with the component $\mathbf{N}_{par}$ parallel to the boundary. While we draw the picture in two dimensions,the same conclusion holds in general for an n-dimensional signal space, where $\mathbf{s}$ and $\mathbf{N}$ have dimension n, $\mathbf{N}_{par}$ has dimension $n-1$, while N_{perp} is still a scalar. Since $N_{perp} \sim N(0, \sigma^2)$ (the projection of WGN in any direction has this distribution), we have

$$P[\text{cross a boundary at distance } D] = P[N_{perp} > D] = Q\left(\frac{D}{\sigma}\right) \tag{6.31}$$

Now, let us apply the same reasoning to the decision boundary corresponding to making an ML decision between two signals $\mathbf{s}_0$ and $\mathbf{s}_1$, as shown in Figure 6.18. Suppose that $\mathbf{s}_0$ is sent. What is the probability that the noise vector $\mathbf{N}$, when added to it, sends the received vector into the wrong region by crossing the decision boundary? We know from (6.31) that the answer is $Q(D/\sigma)$, where D is the distance between $\mathbf{s}_0$ and the decision boundary. For ML reception, the decision boundary is the plane which is the perpendicular bisector of the line between $\mathbf{s}_0$ and $\mathbf{s}_1$, whose length equals $d = ||\mathbf{s}_1 - \mathbf{s}_0||$, the Euclidean distance between the two signal vectors. Thus, $D = d/2 = ||\mathbf{s}_1 - \mathbf{s}_0||/2$, and the probability of crossing the ML decision boundary between the two signal vectors (starting from either of the two signal points) is

$$P[\text{cross ML boundary between } \mathbf{s}_0 \text{ and } \mathbf{s}_1] = Q\left(\frac{||\mathbf{s}_1 - \mathbf{s}_0||}{2\sigma}\right) = Q\left(\frac{||s_1 - s_0||}{2\sigma}\right) \tag{6.32}$$

where we note that the Euclidean distance between the signal vectors and the corresponding continuous time signals is the same.

Notation Now that we have established the equivalence between working with continuous-time signals and the vectors that represent their projections onto signal space,

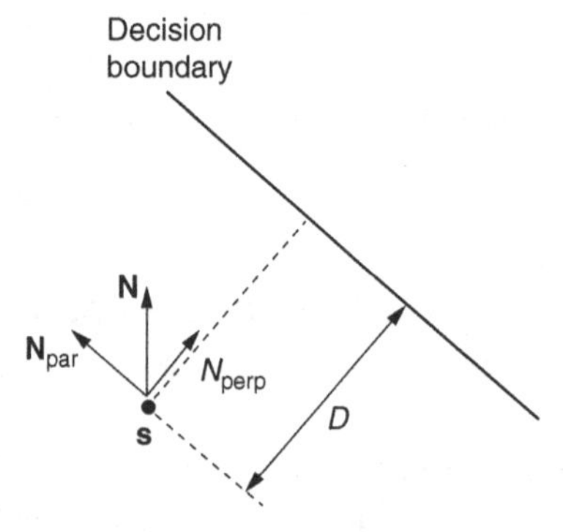

Figure 6.17 Only the component of noise perpendicular to the decision boundary, N_{perp}, can cause the received vector to cross the decision boundary, starting from the signal point $\mathbf{s}$.

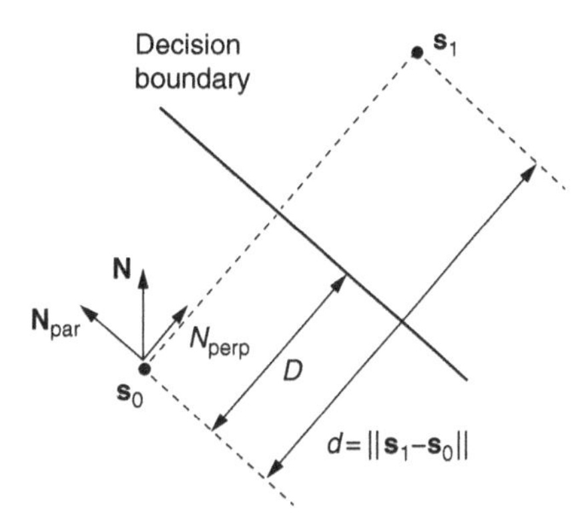

Figure 6.18 When making an ML decision between $\mathbf{s}_0$ and $\mathbf{s}_1$, the decision boundary is at distance $D = d/2$ from each signal point, where $d = ||\mathbf{s}_1 - \mathbf{s}_0||$ is the Euclidean distance between the two points.

we no longer need to be careful about distinguishing between them. Accordingly, we drop the use of boldface notation henceforth, using the notation y, s_i, and n to denote the received signal, the transmitted signal, and the noise, respectively, in both settings.

6.3.2 Performance with binary signaling

Consider binary signaling in AWGN, where the received signal is modeled using two hypotheses as follows:

$$\begin{aligned} H_1 &: y(t) = s_1(t) + n(t) \\ H_0 &: y(t) = s_0(t) + n(t) \end{aligned} \tag{6.33}$$

Geometric computation of error probability The ML decision boundary for this problem is as in Figure 6.18. The conditional error probability is simply the probability that, starting from one of the signal points, the noise makes us cross the boundary to the wrong side, the probability of which we have already computed in (6.32). Since the conditional error probabilities are equal, they also equal the average error probability regardless of the priors. We therefore obtain the following expression.

Error probability for binary signaling with ML reception

$$P_{e,ML} = P_{e|1} = P_{e|0} = Q\left(\frac{||s_1 - s_0||}{2\sigma}\right) = Q\left(\frac{d}{2\sigma}\right) \tag{6.34}$$

where $d = ||s_1 - s_0||$ is the distance between the two possible received signals.

Algebraic computation While this geometric computation is intuitively pleasing, it is important to also master algebraic approaches to computing the probabilities of errors due to WGN. It is easiest to first consider on–off keying:

$$H_1 : y(t) = s(t) + n(t)$$
$$H_0 : y(t) = n(t) \tag{6.35}$$

By applying (6.28), we find that the ML rule reduces to

$$\langle y, s \rangle \underset{H_0}{\overset{H_1}{\gtrless}} \frac{||s||^2}{2} \tag{6.36}$$

Setting $Z = \langle y, s \rangle$, we wish to compute the conditional error probabilities given by

$$P_{e|1} = P[Z < (||s||^2/2)|H_1], \qquad P_{e|0} = P[Z > (||s||^2/2)|H_0] \tag{6.37}$$

We have actually already done these computations in Example 5.8.2, but it pays to review them quickly. Note that, conditioned on either hypothesis, Z is a Gaussian random variable. The conditional mean and variance of Z under H_0 are given by

$$\mathbb{E}[Z|H_0] = \mathbb{E}[\langle n, s \rangle] = 0$$
$$\text{var}(Z|H_0) = \text{cov}(\langle n, s \rangle, \langle n, s \rangle) = \sigma^2 ||s||^2$$

where we have used Theorem 6.2.1 and the fact that $n(t)$ has zero mean. The corresponding computation under H_1 is as follows:

$$\mathbb{E}[Z|H_1] = \mathbb{E}[\langle s + n, s \rangle] = ||s||^2$$
$$\text{var}(Z|H_1) = \text{cov}(\langle s + n, s \rangle, \langle s + n, s \rangle) = \text{cov}(\langle n, s \rangle, \langle n, s \rangle) = \sigma^2 ||s||^2$$

noting that covariances do not change upon adding constants. Thus, $Z \sim N(0, v^2)$ under H_0 and $Z \sim N(m, v^2)$ under H_1, where $m = ||s||^2$ and $v^2 = \sigma^2 ||s||^2$. By substituting into (6.37), it is easy to check that

$$P_{e|1} = P_{e|0} = Q\left(\frac{||s||}{2\sigma}\right) \tag{6.38}$$

Going back to the more general binary signaling problem (6.33), the ML rule is given by (6.28) to be

$$\langle y, s_1 \rangle - \frac{||s_1||^2}{2} \underset{H_0}{\overset{H_1}{\gtrless}} \langle y, s_0 \rangle - \frac{||s_0||^2}{2}$$

We can analyze this system by considering the joint distribution of the correlator statistics $\langle y, s_1 \rangle$ and $\langle y, s_0 \rangle$, which are jointly Gaussian conditioned on each hypothesis. However, it is simpler and more illuminating to rewrite the ML decision rule as

$$\langle y, s_1 - s_0 \rangle \underset{H_0}{\overset{H_1}{\gtrless}} \frac{||s_1||^2}{2} - \frac{||s_0||^2}{2}$$

This is consistent with the geometry depicted in Figure 6.18: only the projection of the received signal along the line joining the signals matters in the decision, and hence only the noise along this direction can produce errors. The analysis now involves the conditional distributions of the single decision statistic $Z = \langle y, s_1 - s_0 \rangle$, which is conditionally Gaussian under either hypothesis. The computation of the conditional error probabilties is left as an exercise, but we already know that the answer should work out to (6.34).

A quicker approach is to consider a transformed system with received signal $\tilde{y}(t) = y(t) - s_0(t)$. Since this transformation is invertible, the performance of an optimal rule is unchanged under it. But the transformed received signal $\tilde{y}(t)$ falls under the on–off signaling model (6.35), with $s(t) = s_1(t) - s_0(t)$. The ML error probability formula (6.34) therefore follows from the formula (6.38).

Scale-invariance The formula (6.34) illustrates that the performance of the ML rule is scale-invariant: if we scale the signals and noise by the same factor α, the performance does not change, since both $||s_1 - s_0||$ and σ scale by α. Thus, the performance is determined by the *ratio* of signal and noise strengths, rather than individually on the signal and noise strengths. We now define some standard measures for these quantities, and then express the performance of some common binary signaling schemes in terms of them.

Energy per bit, E_b For binary signaling, this is given by

$$E_b = \frac{1}{2}(||s_0||^2 + ||s_1||^2)$$

assuming that 0 and 1 are equally likely to be sent.

Scale-invariant parameters If we scale up both s_1 and s_0 by a factor A, E_b scales up by a factor A^2, while the distance d scales up by a factor A. We can therefore define the *scale-invariant* parameter

$$\eta_P = \frac{d^2}{E_b} \tag{6.39}$$

Now, on substituting, $d = \sqrt{\eta_P E_b}$ and $\sigma = \sqrt{N_0/2}$ into (6.34), we obtain that the ML performance is given by

$$P_{e,ML} = Q\left(\sqrt{\frac{\eta_P E_b}{2N_0}}\right) = Q\left(\sqrt{\frac{d^2}{E_b}}\sqrt{\frac{E_b}{2N_0}}\right) \tag{6.40}$$

Two important observations follow.

Performance depends on signal-to-noise ratio We observe from (6.40) that the performance depends on the *ratio* E_b/N_0, rather than separately on the signal and noise strengths.

Power efficiency For fixed E_b/N_0, the performance is better for a signaling scheme that has a higher value of η_P. We therefore use the term *power efficiency* for $\eta_P = d^2/E_b$.

Let us now compute the performance of some common binary signaling schemes in terms of E_b/N_0, using (6.40). Since inner products (and hence energies and distances) are preserved in signal space, we can compute η_P for each scheme using the signal-space representations depicted in Figure 6.19. The absolute scale of the signals is irrelevant, since the performance depends on the signaling scheme only through the scale-invariant

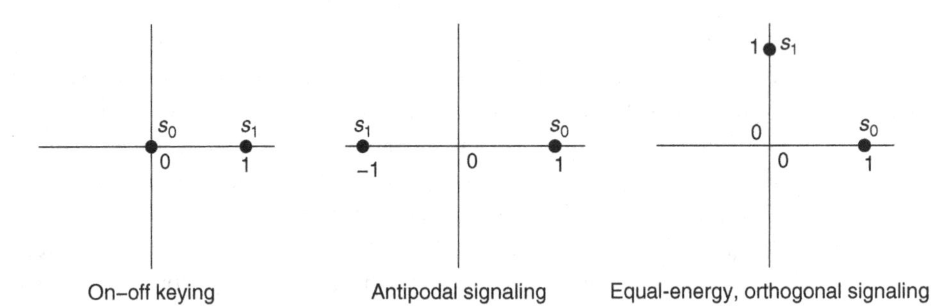

Figure 6.19 Signal-space representations with conveniently chosen scaling for three binary signaling schemes.

parameter $\eta_P = d^2/E_b$. We can therefore choose any convenient scaling for the signal-space representation for a modulation scheme.

On–off keying Here $s_1(t) = s(t)$ and $s_0(t) = 0$. As shown in Figure 6.19, the signal space is one-dimensional. For the scaling in the figure, we have $d = 1$ and $E_b = \frac{1}{2}(1^2 + 0^2) = \frac{1}{2}$, so that $\eta_P = d^2/E_b = 2$. By substituting into (6.40), we obtain $P_{e,ML} = Q(\sqrt{E_b/N_0})$.

Antipodal signaling: Here $s_1(t) = -s_0(t)$, leading again to a one-dimensional signal-space representation. One possible realization of antipodal signaling is BPSK, discussed in the previous chapter. For the scaling chosen, $d = 2$ and $E_b = \frac{1}{2}(1^2 + (-1)^2) = 1$, which gives $\eta_P = d^2/E_b = 4$. Substituting into (6.40), we obtain $P_{e,ML} = Q(\sqrt{2E_b/N_0})$.

Equal-energy, orthogonal signaling Here s_1 and s_0 are orthogonal, with $||s_1||^2 = ||s_0||^2$. This is a two-dimensional signal space. As discussed in the previous chapter, possible realizations of orthogonal signaling include FSK and Walsh–Hadamard codes. From Figure 6.19, we have $d = \sqrt{2}$ and $E_b = 1$, so $\eta_P = d^2/E_b = 2$. This gives $P_{e,ML} = Q\left(\sqrt{E_b/N_0}\right)$.

Thus, on–off keying (which is orthogonal signaling with unequal energies) and equal-energy orthogonal signaling have the same power efficiency, while the power efficiency of antipodal signaling is a factor of two (i.e., 3 dB) better.

In plots of error probability versus SNR, we typically express the error probability on a log scale (in order to capture its rapid decay with SNR) and express the SNR in decibels (in order to span a large range). We provide such a plot for antipodal and orthogonal signaling in Figure 6.20.

6.3.3 *M*-ary signaling: scale-invariance and SNR

We turn now to M-ary signaling with $M > 2$, modeled as the following hypothesis-testing problem:

$$H_i : y(t) = s_i(t) + n(t), \quad i = 0, 1, \ldots, M-1$$

for which the ML rule has been derived to be

$$\delta_{ML}(y) = \arg\max_{0 \le i \le M-1} Z_i$$

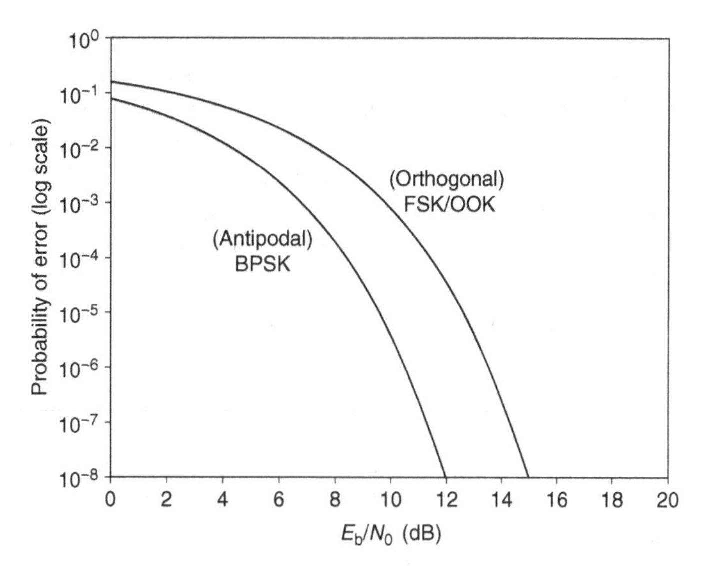

Figure 6.20 Error probability versus E_b/N_0 (dB) for binary antipodal and orthogonal signaling schemes.

with decision statistics

$$Z_i = \langle y, s_i \rangle - \frac{1}{2}||s_i||^2, \quad i = 0, 1, \ldots, M-1$$

and corresponding decision regions

$$\Gamma_i = \{y : Z_i > Z_j \text{ for all } j \neq i\}, \quad i = 0, 1, \ldots, M-1 \tag{6.41}$$

Before doing detailed computations, let us discuss some general properties that greatly simplify the framework for performance analysis.

Scale-invariance For binary signaling, we have observed through explicit computation of the error probability that performance depends only on the signal-to-noise ratio (E_b/N_0) and the geometry of the signal set (which determines the power efficiency d^2/E_b). Actually, we can make such statements in great generality for M-ary signaling without explicit computations. First, let us note that the performance of an optimal receiver does not change if we scale both signal and noise by the same factor. Specifically, optimal reception for the model

$$H_i : \tilde{y}(t) = As_i(t) + An(t), \quad i = 0, 1, \ldots, M-1 \tag{6.42}$$

does not depend on A. This is inferred from the following general observation: *the performance of an optimal receiver is unchanged when we pass the observation through an invertible transformation.* Specifically, suppose $z(t) = F(y(t))$ is obtained by passing $y(t)$ through an invertible transformation F. If the optimal receiver for z does better than the optimal receiver for y, then we could apply F to y to get z, then use optimal reception for z. This would give better performance than the optimal receiver for y, which is a contradiction. Similarly, if the optimal receiver for y does better than the optimal receiver for z, then we could apply F^{-1} to z to get y, and then use optimal reception for y to get a better performance than the optimal receiver for z, which is again a contradiction.

The preceding argument implies that performance depends only on the signal-to-noise ratio, once we have fixed the signal constellation. Let us now figure out what properties of the signal constellation are relevant in determining performance. For $M = 2$, we have seen that all that matters is the scale-invariant quantity d^2/E_b. What are the analogous quantities for $M > 2$? To determine these, let us consider the conditional error probabilities for the ML rule.

Conditional error probability The conditional error probability, conditioned on H_i, is given by

$$P_{e|i} = P[y \notin \Gamma_i | i \text{ sent}] = P[Z_i < Z_j \text{ for some } j \neq i | i \text{ sent}] \tag{6.43}$$

While computation of the conditional error probability in closed form is typically not feasible, we can actually get significant insight into what parameters it depends on by examining the conditional distributions of the decision statistics. Since $y = s_i + n$ conditioned on H_i, the decision statistics are given by

$$Z_j = \langle y, s_j \rangle - ||s_j||^2/2 = \langle s_i + n, s_j \rangle - ||s_j||^2/2 = \langle n, s_j \rangle + \langle s_i, s_j \rangle - ||s_j||^2/2, \quad 0 \leq j \leq M-1$$

By virtue of the Gaussianity of $n(t)$, the decision statistics $\{Z_j\}$ are jointly Gaussian (conditioned on H_i). Their joint distribution is therefore completely characterized by their means and covariances. Since the noise is zero mean, we obtain

$$\mathbb{E}[Z_j | H_i] = \langle s_i, s_j \rangle - ||s_j||^2/2$$

Using Theorem 6.2.1, and noting that covariance is unaffected by translation, we obtain that

$$\text{cov}(Z_j, Z_k | H_i) = \text{cov}(\langle n, s_j \rangle, \langle n, s_k \rangle) = \sigma^2 \langle s_j, s_k \rangle$$

Thus, conditioned on H_i, the joint distribution of $\{Z_j\}$ depends only on the noise variance σ^2 and the signal inner products $\{\langle s_i, s_j \rangle, 1 \leq i, j \leq M\}$. Now that we know the joint distribution, we can in principle compute the conditional error probabilities $P_{e|i}$. In practice, this is often difficult, and we may have to resort to Monte Carlo simulations. However, what we have found out about the joint distribution can now be used to refine our concepts of scale-invariance.

Performance depends only on normalized inner products Let us replace Z_j by Z_j/σ^2. Clearly, since we are simply picking the maximum among the decision statistics, scaling by a common factor does not change the decision (and hence the performance). However, we now obtain that

$$\mathbb{E}[Z_j/\sigma^2 | H_i] = \frac{\langle s_i, s_j \rangle}{\sigma^2} - \frac{||s_j||^2}{2\sigma^2}$$

and

$$\text{cov}\left(\frac{Z_j}{\sigma^2}, \frac{Z_k}{\sigma^2} | H_i\right) = \frac{1}{\sigma^4} \text{cov}(Z_j, Z_k | H_i) = \frac{\langle s_j, s_k \rangle}{\sigma^2}$$

Thus, the joint distribution of the normalized decision statistics $\{Z_j/\sigma^2\}$, conditioned on any of the hypotheses, depends only on the normalized inner products $\{\langle s_i, s_j\rangle/\sigma^2, 1 \leq i,j \leq M\}$. Of course, this means that the performance also depends only on these normalized inner products.

Let us now carry these arguments further, still without any explicit computations. We define the energy per symbol and energy per bit for M-ary signaling as follows.

Energy per symbol, E_s For M-ary signaling with equal priors, the energy per symbol E_s is given by

$$E_s = \frac{1}{M}\sum_{i=1}^{M} ||s_i||^2$$

Energy per bit, E_b Since M-ary signaling conveys $\log_2 M$ bits/symbol, the energy per bit is given by

$$E_b = \frac{E_s}{\log_2 M}$$

If all signals in an M-ary constellation are scaled up by a factor A, then E_s and E_b get scaled up by A^2, as do all inner products $\{\langle s_i, s_j\rangle\}$. Thus, we can define scale-invariant inner products $\{\langle s_i, s_j\rangle\}/E_b$ that depend only on the shape of the signal constellation. Indeed, we can *define* the shape of a constellation as these scale-invariant inner products. Setting $\sigma^2 = N_0/2$, we can now write the normalized inner products determining performance as follows:

$$\frac{\langle s_i, s_j\rangle}{\sigma^2} = \frac{\langle s_i, s_j\rangle}{E_b}\frac{2E_b}{N_0} \tag{6.44}$$

We can now make the following statement.

Performance depends only on E_b/N_0 and the constellation shape (as specified by the scale-invariant inner products) We have shown that the performance depends only on the normalized inner products $\{\langle s_i, s_j\rangle/\sigma^2\}$. From (6.44), we see that these in turn depend only on E_b/N_0 and the *scale-invariant* inner products $\left\{\langle s_i, s_j\rangle/E_b\right\}$. The latter depend only on the shape of the signal constellation, and are completely independent of the signal and noise strengths. What this means is that we can choose any convenient scaling that we want for the signal constellation when investigating its performance, as long as we keep track of the signal-to-noise ratio. We illustrate this via an example where we determine the error probability by simulation.

Example 6.3.1 (Using scale-invariance in error probability simulations) Suppose that we wish to estimate the error probability for 8PSK by simulation. The signal points lie in a two-dimensional space, and we can scale them to lie on a circle of unit radius, so that the constellation is given by $\mathcal{A} = \{(\cos\theta, \sin\theta)^T : \theta = k\pi/4, k = 0, 1, \ldots, 7\}$. The energy per symbol $E_s = 1$ for this scaling, so $E_b = E_s/\log_2 8 = 1/3$. We therefore have $E_b/N_0 = 1/(3N_0) = 1/(6\sigma^2)$, so the noise variance per dimension can be set to

$$\sigma^2 = \frac{1}{6(E_b/N_0)}$$

Typically, E_b/N_0 is specified in dB, so we need to convert it to the "raw" E_b/N_0. We now have a simulation consisting of the following steps, repeated over multiple symbol transmissions.

Step 1. Choose a symbol $\mathbf{s}$ at random from $\mathcal{A}$. For this symmetric constellation, we can actually keep sending the same symbol in order to compute the performance of the ML rule, since the conditional error probabilities are all equal. For example, set $\mathbf{s} = (1, 0)^T$.

Step 2. Generate two i.i.d. $N(0, 1)$ random variables U_c and U_s. The I and Q noises can now be set as $N_c = \sigma U_c$ and $N_s = \sigma U_s$, so that $\mathbf{N} = (N_c, N_s)^T$.

Step 3. Set the received vector $\mathbf{y} = \mathbf{s} + \mathbf{N}$.

Step 4. Compute the ML decision $\arg\max_i \langle \mathbf{y}, \mathbf{s}_i \rangle$ (the energy terms can be dropped, since the signals are of equal energy) or $\arg\min_i ||\mathbf{y} - \mathbf{s}_i||^2$.

Step 5. If there is an error, increment the error count.

The error probability is estimated as the error count, divided by the number of symbols transmitted. We repeat this simulation over a range of E_b/N_0, and typically plot the error probability on a log scale versus E_b/N_0 in dB.

These steps are carried out in the following code fragment, which generates Figure 6.21 comparing a simulation-based estimate of the error probability for 8PSK against the intelligent union bound, an analytical estimate that we develop shortly. The analytical

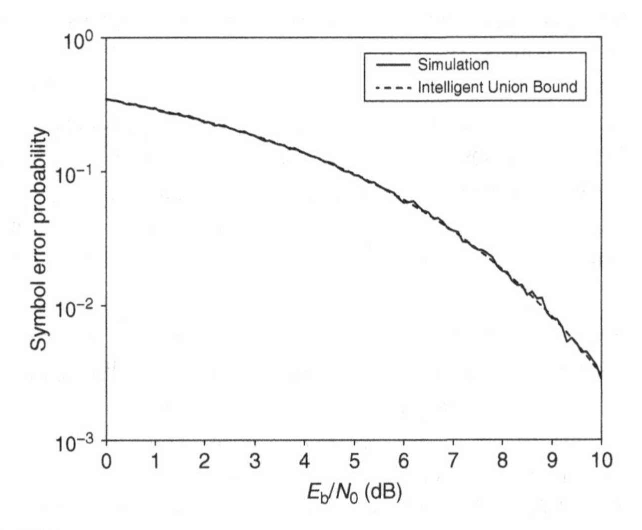

Figure 6.21 Error probability for 8PSK.

estimate requires very little computation (evaluation of a single Q function), but its agreement with simulations is excellent. As we shall see, developing such analytical estimates also gives us insight into how errors are most likely to occur for M-ary signaling in AWGN.

The code fragment is written for transparency rather than computational efficiency. The code contains an *outer for-loop* for varying SNR, and an *inner for-loop* for computing minimum distances for the symbols sent at each SNR. The inner loop can be avoided and the program sped up considerably by computing all minimum distances for all symbols at once using matrix operations (try it!). We use a less efficient program here to make the operations easy to understand.

Code Fragment 6.3.1 (Simulation of 8PSK performance in AWGN)

```
%generate 8PSK constellation as complex numbers
a = cumsum(ones(8,1)) - 1;
constellation = exp(i * 2 * pi.* a/8);
%number of symbols in simulation
nsymbols = 20000;
ebnodb = 0:0.1:10;
number_snrs = length(ebnodb);
perr_estimate = zeros(number_snrs,1);
for k = 1:number_snrs, %SNR for loop
ebnodb_now = ebnodb(k);
ebno = 10^(ebnodb_now/10);
sigma = sqrt(1/(6 * ebno));
%send first symbol without loss of generality, add two-dimensional Gaussian noise
received = 1 + sigma * randn(nsymbols,1) + j * sigma * randn(nsymbols,1);
decisions = zeros(nsymbols,1);
for n = 1:nsymbols, %Symbol for loop (can/should be avoided for fast implementation)
    distances = abs(received(n) - constellation);
    [min_dist,decisions(n)] = min(distances);
end
errors = (decisions ~= 1);
perr_estimate(k) = sum(errors)/nsymbols;
end
semilogy(ebnodb,perr_estimate);
hold on;
%COMPARE WITH INTELLIGENT UNION BOUND
etaP = 6 - 3 * sqrt(2); %power efficiency
Ndmin = 2;% number of nearest neighbors
ebno = 10.^(ebnodb/10);
perr_union = Ndmin * q_function(sqrt(etaP * ebno/2));
semilogy(ebnodb,perr_union,':r');
xlabel('Eb/N0 (dB)');
ylabel('Symbol error probability');
legend('Simulation','Intelligent Union Bound','Location','NorthEast');
```

6.3.4 Performance analysis for M-ary signaling

We begin by computing the error probability for QPSK, for which we can get simple expressions for the error probability in terms of the Q function. We then discuss why exact performance analysis can be more complicated in general, motivating the need for the bounds and approximations we develop in this section.

Exact analysis for QPSK Let us find $P_{e|0}$, the conditional error probability for the ML rule conditioned on s_0 being sent. For the scaling shown in Figure 6.22,

$$s_0 = \begin{pmatrix} d/2 \\ d/2 \end{pmatrix}$$

and the two-dimensional received vector is given by

$$y = s_0 + (N_c, N_s)^T = \begin{pmatrix} d/2 + N_c \\ d/2 + N_s \end{pmatrix}$$

where N_c and N_s are i.i.d. $N(0, \sigma^2)$ random variables, corresponding to the projections of WGN along the I and Q axes, respectively. An error occurs if the noise moves the observation out of the positive quadrant, which is the decision region for s_0. This happens if $N_c + d/2 < 0$ *or* $N_s + d/2 < 0$. We can therefore write

$$\begin{aligned} P_{e|0} &= P\left[N_c + \frac{d}{2} < 0 \text{ or } N_s + \frac{d}{2} < 0\right] \\ &= P\left[N_c + \frac{d}{2} < 0\right] + P\left[N_s + \frac{d}{2} < 0\right] - P\left[N_c + \frac{d}{2} < 0 \text{ and } N_s + \frac{d}{2} < 0\right] \end{aligned} \quad (6.45)$$

But

$$P\left[N_c + \frac{d}{2} < 0\right] = P\left[N_c < -\frac{d}{2}\right] = \Phi\left(-\frac{d}{2\sigma}\right) = Q\left(\frac{d}{2\sigma}\right)$$

This is also equal to $P[N_s + d/2 < 0]$, since N_c and N_s are identically distributed. Furthermore, since N_c and N_s are independent, we have

$$P\left[N_c + \frac{d}{2} < 0 \text{ and } N_s + \frac{d}{2} < 0\right] = P\left[N_c + \frac{d}{2} < 0\right] P\left[N_s + \frac{d}{2} < 0\right] = \left[Q\left(\frac{d}{2\sigma}\right)\right]^2$$

By substituting these expressions into (6.45), we obtain that

$$P_{e|1} = 2Q\left(\frac{d}{2\sigma}\right) - Q^2\left(\frac{d}{2\sigma}\right) \quad (6.46)$$

By symmetry, the conditional probabilities $P_{e|i}$ are equal for all i, which implies that the average error probability is also given by the expression above. We now express the

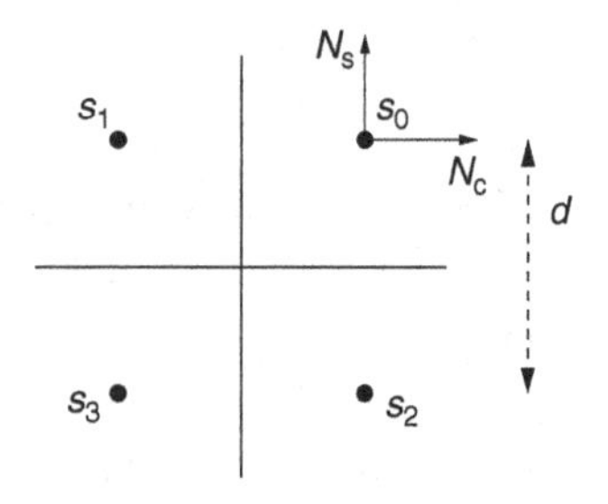

Figure 6.22 If s_0 is sent, an error occurs if N_c or N_s is negative enough to make the received vector fall out of the first quadrant.

error probability in terms of the scale-invariant parameter d^2/E_b and E_b/N_0, using the relation

$$\frac{d}{2\sigma} = \sqrt{\frac{d^2}{E_b}}\sqrt{\frac{E_b}{2N_0}}$$

The energy per symbol is given by

$$E_s = \frac{1}{M}\sum_{i=1}^{M} ||s_i||^2 = ||s_1||^2 = \left(\frac{d}{2}\right)^2 + \left(\frac{d}{2}\right)^2 = \frac{d^2}{2}$$

which implies that the energy per bit is

$$E_b = \frac{E_s}{\log_2 M} = \frac{E_s}{\log_2 4} = \frac{d^2}{4}$$

This yields $d^2/E_b = 4$, and hence $d/(2\sigma) = \sqrt{2E_b/N_0}$. By substituting into (6.46), we obtain

$$P_e = P_{e|1} = 2Q\left(\sqrt{\frac{2E_b}{N_0}}\right) - Q^2\left(\sqrt{\frac{2E_b}{N_0}}\right) \tag{6.47}$$

as the exact error probability for QPSK.

Why exact analysis can be difficult Let us first understand why we could find a simple expression for the error probability for QPSK. The decision regions are bounded by the I and Q axes. The noise random variable N_c can cause crossing of the Q axis, while N_s can cause crossing of the I axis. Since these two random variables are independent, the probability that at least one of these noise random variables causes a boundary crossing becomes easy to compute. Figure 6.23 shows an example where this is not possible. In the figure, we see that the decision region Γ_0 is bounded by three lines (in general, these would be $(n-1)$-dimensional hyperplanes in n-dimensional signal space). An error occurs if we cross any of these lines, starting from s_0. In order to cross the line between s_0 and s_i, the noise random variable N_i must be bigger than $||s_i - s_0||/2$, $i = 1, 2, 3$ (as we saw in Figures 6.17 and 6.18, only the noise component orthogonal to a hyperplane determines whether we cross it). Thus, the conditional error probability can be written as

$$P_{e|0} = P[N_1 > ||s_1 - s_0||/2 \text{ or } N_2 > ||s_2 - s_0||/2 \text{ or } N_3 > ||s_3 - s_0||/2] \tag{6.48}$$

The random variables N_1, N_2, and N_3 are, of course, jointly Gaussian, since each is a projection of WGN along a direction. Each of them is an $N(0, \sigma^2)$ random variable; that is, they are *identically distributed.* However, they are *not independent,* since they are projections of WGN along directions that are not orthogonal to each other. Thus, we cannot break down the preceding expression into probabilities in terms of the individual random variables N_1, N_2, and N_3, unlike what we did for QPSK (where N_c and N_s were independent). However, we can still find a simple upper bound on the conditional error probability using the union bound, as follows.

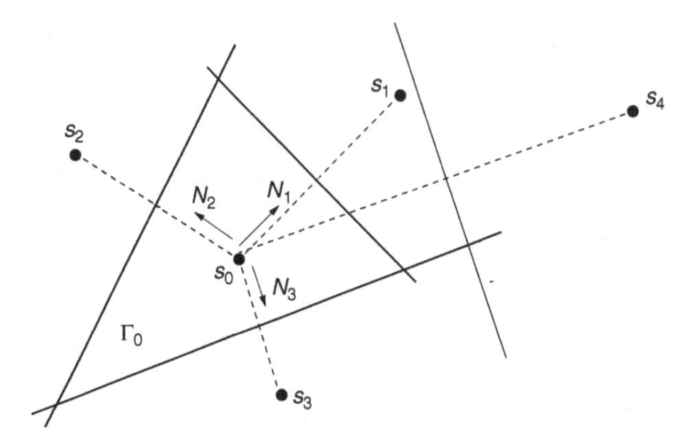

Figure 6.23 The noise random variables N_1, N_2, and N_3 which can drive the received vector outside the decision region Γ_0 are correlated, which makes it difficult to find an exact expression for $P_{e|0}$.

Union bound The probability of a union of events is upper bounded by the sum of the probabilities of the events:

$$P[A_1 \text{ or } A_2 \text{ or } \ldots \text{ or } A_n] = P[A_1 \cup A_2 \cup \ldots \cup A_n] \leq P[A_1] + P[A_2] + \cdots + P[A_n] \quad (6.49)$$

By applying (6.49) to (6.48), we obtain that, for the scenario depicted in Figure 6.23, the conditional error probability can be upper bounded as follows:

$$\begin{aligned} P_{e|0} &\leq P[N_1 > ||s_1 - s_0||/2] + P[N_2 > ||s_2 - s_0||/2] + P[N_3 > ||s_3 - s_0||/2] \\ &= Q\left(\frac{||s_1 - s_0||}{2\sigma}\right) + Q\left(\frac{||s_2 - s_0||}{2\sigma}\right) + Q\left(\frac{||s_3 - s_0||}{2\sigma}\right) \end{aligned} \quad (6.50)$$

Thus, the conditional error probability is upper bounded by a sum of probabilities, each of which corresponds to the error probability for a binary decision: s_0 versus s_1, s_0 versus s_2, and s_0 versus s_3. This approach applies in great generality, as we show next.

Union bound and variants Pictures such as the one in Figure 6.23 typically cannot be drawn when the signal-space dimension is high. However, we can still find union bounds on error probabilities, as long as we can enumerate all the signals in the constellation. To do this, let us rewrite (6.43), the conditional error probability, conditioned on H_i, as a union of $M - 1$ events as follows:

$$P_{e|i} = P[\cup_{j \neq i}\{Z_i < Z_j\}|i \text{ sent}]$$

where $\{Z_j\}$ are the decision statistics. Using the union bound (6.49), we obtain

$$P_{e|i} \leq \sum_{j \neq i} P[Z_i < Z_j|i \text{ sent}] \quad (6.51)$$

But the jth term on the right-hand side above is simply the error probability of ML reception for binary hypothesis testing between the signals s_i and s_j. From the results of Section 6.3.2, we therefore obtain the following *pairwise error probability:*

$$P[Z_i < Z_j | i \text{ sent}] = Q\left(\frac{||s_j - s_i||}{2\sigma}\right)$$

By substituting into (6.51), we obtain upper bounds on the conditional error probabilities and the average error probability as follows.

Union bound on conditional error probabilities

The conditional error probabilities for the ML rule are bounded as

$$P_{e|i} \leq \sum_{j \neq i} Q\left(\frac{||s_j - s_i||}{2\sigma}\right) = \sum_{j \neq i} Q\left(\frac{d_{ij}}{2\sigma}\right) \tag{6.52}$$

where $d_{ij} = ||s_i - s_j||$ is the distance between signals s_i and s_j.

Union bound on average error probability Averaging the conditional error using the prior probabilities gives an upper bound on the average error probability as follows:

$$P_e = \sum_i \pi_i P_{e|i} \leq \sum_i \pi_i \sum_{j \neq i} Q\left(\frac{||s_j - s_i||}{2\sigma}\right) = \sum_i \pi_i \sum_{j \neq i} Q\left(\frac{d_{ij}}{2\sigma}\right) \tag{6.53}$$

We can now rewrite the union bound in terms of E_b/N_0 and the scale-invariant squared distances d_{ij}^2/E_b as follows:

$$P_{e|i} \leq \sum_{j \neq i} Q\left(\sqrt{\frac{d_{ij}^2}{E_b}}\sqrt{\frac{E_b}{2N_0}}\right) \tag{6.54}$$

$$P_e = \sum_i \pi_i P_{e|i} \leq \sum_i \pi_i \sum_{j \neq i} Q\left(\sqrt{\frac{d_{ij}^2}{E_b}}\sqrt{\frac{E_b}{2N_0}}\right) \tag{6.55}$$

By applying the union bound to Figure 6.23, we obtain

$$P_{e|0} \leq Q\left(\frac{||s_1 - s_0||}{2\sigma}\right) + Q\left(\frac{||s_2 - s_0||}{2\sigma}\right) + Q\left(\frac{||s_3 - s_0||}{2\sigma}\right) + Q\left(\frac{||s_4 - s_0||}{2\sigma}\right)$$

Notice that this answer is different from the one we had in (6.50). This is because the fourth term corresponds to the signal s_4, which is "too far away" from s_0 to play a role in determining the decision region Γ_0. Thus, when we do have a more detailed geometric understanding of the decision regions, we can do better than the generic union bound (6.52) and get a tighter bound, as in (6.50). We term this the *intelligent union bound,* and give a general formulation in the following.

Denote by $N_{\text{ML}}(i)$ the indices of the set of neighbors of signal s_i (we exclude i from $N_{\text{ML}}(i)$ by definition) that characterize the ML decision region Γ_i. That is, the half planes that we intersect to obtain Γ_i correspond to the perpendicular bisectors of lines joining s_i

and s_j, $j \in N_{\text{ML}}(i)$. For example, in Figure 6.23, $N_{\text{ML}}(0) = \{1, 2, 3\}$; s_4 is excluded from this set, since it does not play a role in determining Γ_0. The decision region in (6.41) can now be expressed as

$$\Gamma_i = \{y : \delta_{\text{ML}}(y) = i\} = \{y : Z_i \geq Z_j \text{ for all } j \in N_{\text{ML}}(i)\} \tag{6.56}$$

We can now say the following: y falls outside Γ_i if and only if $Z_i < Z_j$ for some $j \in N_{\text{ML}}(i)$. We can therefore write

$$P_{\text{e}|i} = P[y \notin \Gamma_i | i \text{ sent}] = P[Z_i < Z_j \text{ for some } j \in N_{\text{ML}}(i) | i \text{ sent}] \tag{6.57}$$

and from there, following the same steps as in the union bound, get a tighter bound, which we express as follows.

Intelligent union bound

A better bound on $P_{\text{e}|i}$ is obtained by considering only the neighbors of s_i that determine its ML decision region, as follows:

$$P_{\text{e}|i} \leq \sum_{j \in N_{\text{ML}}(i)} Q\left(\frac{||s_j - s_i||}{2\sigma}\right) \tag{6.58}$$

In terms of E_b/N_0, we get

$$P_{\text{e}|i} \leq \sum_{j \in N_{\text{ML}}(i)} Q\left(\sqrt{\frac{d_{ij}^2}{E_\text{b}}}\sqrt{\frac{E_\text{b}}{2N_0}}\right) \tag{6.59}$$

(the bound on the average error probability P_e is computed as before by averaging the bounds on $P_{\text{e}|i}$ using the priors).

Union bound for QPSK For QPSK, we infer from Figure 6.22 that the union bound for $P_{\text{e}|1}$ is given by

$$P_\text{e} = P_{\text{e}|0} \leq Q\left(\frac{d_{01}}{2\sigma}\right) + Q\left(\frac{d_{02}}{2\sigma}\right) + Q\left(\frac{d_{03}}{2\sigma}\right) = 2Q\left(\frac{d}{2\sigma}\right) + Q\left(\frac{\sqrt{2}d}{2\sigma}\right)$$

Using $d^2/E_\text{b} = 4$, we obtain the union bound in terms of E_b/N_0 to be

$$P_\text{e} \leq 2Q\left(\sqrt{\frac{2E_\text{b}}{N_0}}\right) + Q\left(\sqrt{\frac{4E_\text{b}}{N_0}}\right) \quad \textbf{QPSK union bound} \tag{6.60}$$

For moderately large E_b/N_0, the dominant term in terms of the decay of the error probability is the first one, since $Q(x)$ falls off rapidly as x gets large. Thus, while the union bound (6.60) is larger than the exact error probability (6.47), as it must be, it gets the multiplicity and argument of the dominant term right. On tightening the analysis using the intelligent union bound, we get

$$P_{e|0} \leq Q\left(\frac{d_{01}}{2\sigma}\right) + Q\left(\frac{d_{02}}{2\sigma}\right) = 2Q\left(\sqrt{\frac{2E_b}{N_0}}\right) \quad \textbf{QPSK intelligent union bound} \tag{6.61}$$

since $N_{ML}(0) = \{1, 2\}$ (the decision region for s_0 is determined by the neighbors s_1 and s_2).

Another common approach for getting a better (and quicker to compute) estimate than the original union bound is the *nearest-neighbors approximation*. This is a loose term employed to describe a number of different methods for pruning the terms in the summation (6.52). Most commonly, it refers to regular signal sets in which each signal point has a number of nearest neighbors at distance d_{min} from it, where $d_{min} = \min_{i \neq j} ||s_i - s_j||$. Letting $N_{d_{min}}(i)$ denote the number of nearest neighbors of s_i, we obtain the following approximation.

Nearest-neighbors approximation

We have

$$P_{e|i} \approx N_{d_{min}}(i) Q\left(\frac{d_{min}}{2\sigma}\right) \tag{6.62}$$

By averaging over i, we obtain that

$$P_e \approx \bar{N}_{d_{min}} Q\left(\frac{d_{min}}{2\sigma}\right) \tag{6.63}$$

where $\bar{N}_{d_{min}}$ denotes the average number of nearest neighbors for a signal point. The rationale for the nearest-neighbors approximation is that, since $Q(x)$ decays rapidly, $Q(x) \sim e^{-x^2/2}$, as x gets large, the terms in the union bound corresponding to the smallest arguments for the Q function dominate at high SNR.

The corresponding formulas as a function of scale-invariant quantities and E_b/N_0 are

$$P_{e|i} \approx N_{d_{min}}(i) Q\left(\sqrt{\frac{d_{min}^2}{E_b}} \sqrt{\frac{E_b}{2N_0}}\right) \tag{6.64}$$

It is also worth explicitly writing down an expression for the average error probability, averaging the preceding over i:

$$P_e \approx \bar{N}_{d_{min}} Q\left(\sqrt{\frac{d_{min}^2}{E_b}} \sqrt{\frac{E_b}{2N_0}}\right) \tag{6.65}$$

where

$$\bar{N}_{d_{min}} = \frac{1}{M} \sum_{i=1}^{M} N_{d_{min}}(i)$$

is the *average* number of nearest neighbors for the signal points in the constellation.

For QPSK, we have from Figure 6.22 that

$$N_{d_{min}}(i) \equiv 2 = \bar{N}_{d_{min}}$$

and

$$\sqrt{\frac{d_{\min}^2}{E_b}} = \sqrt{\frac{d^2}{E_b}} = 4$$

yielding

$$P_e \approx 2Q\left(\sqrt{\frac{2E_b}{N_0}}\right)$$

In this case, the nearest-neighbors approximation coincides with the intelligent union bound (6.61). This happens because the ML decision region for each signal point is determined by its nearest neighbors for QPSK. Indeed, the latter property holds for many regular constellations, including all of the PSK and QAM constellations whose ML decision regions are depicted in Figure 6.16.

Power efficiency While exact performance analysis for M-ary signaling can be computationally demanding, we have now obtained simple enough estimates that we can define concepts such as power efficiency, in a manner analogous to the development for binary signaling. In particular, by comparing the nearest-neighbors approximation (6.63) with the error probability for binary signaling (6.40), we define in analogy the power efficiency of an M-ary signaling scheme as

$$\eta_P = \frac{d_{\min}^2}{E_b} \tag{6.66}$$

We can rewrite the nearest-neighbors approximation as

$$P_e \approx \bar{N}_{d_{\min}} Q\left(\sqrt{\frac{\eta_P E_b}{2N_0}}\right) \tag{6.67}$$

Since the argument of the Q function in (6.67) plays a bigger role than the multiplicity $\bar{N}_{d_{\min}}$ for moderately large SNR, η_P offers a means of quickly comparing the power efficiency of different signaling constellations, as well as for determining the dependence of performance on E_b/N_0.

Performance analysis for 16QAM We now apply the preceding performance analysis to the 16QAM constellation depicted in Figure 6.24, where we have chosen a convenient scale for

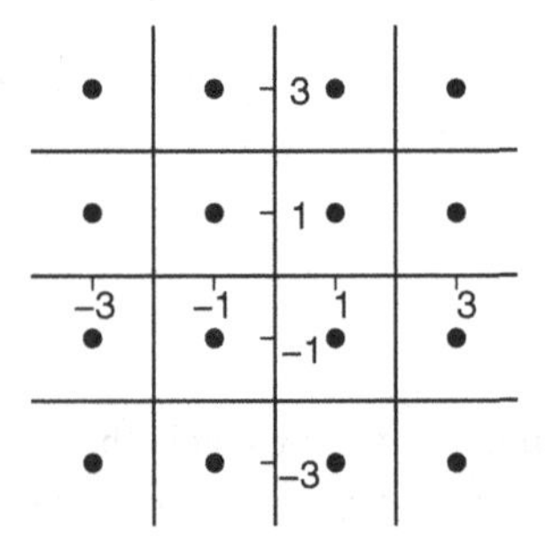

Figure 6.24 ML decision regions for 16QAM with scaling chosen for convenience in computing power efficiency.

the constellation. We compute the nearest-neighbors approximation, which coincides with the intelligent union bound, since the ML decision regions are determined by the nearest neighbors. On noting that the number of nearest neighbors is four for the four innermost signal points, two for the four outermost signal points, and three for the remaining eight signal points, we obtain upon averaging

$$\bar{N}_{d_{\min}} = 3 \tag{6.68}$$

It remains to compute the power efficiency η_{P} and apply (6.67). We did this in the preview in Chapter 4, but we repeat it here. For the scaling shown, we have $d_{\min} = 2$. The energy per symbol is obtained as follows:

$$\begin{aligned} E_{\mathrm{s}} &= \text{average energy of I component} + \text{average energy of Q component} \\ &= 2(\text{average energy of I component}) \end{aligned}$$

by symmetry. Since the I component is equally likely to take the four values ± 1 and ± 3, we have

$$\text{average energy of I component} = \frac{1}{2}(1^2 + 3^2) = 5$$

and

$$E_{\mathrm{s}} = 10$$

We therefore obtain

$$E_{\mathrm{b}} = \frac{E_{\mathrm{s}}}{\log_2 M} = \frac{10}{\log_2 16} = \frac{5}{2}$$

The power efficiency is therefore given by

$$\eta_{\mathrm{P}} = \frac{d_{\min}^2}{E_{\mathrm{b}}} = \frac{2^2}{\frac{5}{2}} = \frac{8}{5} \tag{6.69}$$

By substituting (6.68) and (6.69) into (6.67), we obtain that

$$P_{\mathrm{e}}(16\mathrm{QAM}) \approx 3Q\left(\sqrt{\frac{4E_{\mathrm{b}}}{5N_0}}\right) \tag{6.70}$$

as the nearest-neighbors approximation and intelligent union bound for 16QAM. The bandwidth efficiency for 16QAM is 4 bits/2 dimensions, which is twice that of QPSK, whose bandwidth efficiency is 2 bits/2 dimensions. It is not surprising, therefore, that the power efficiency of 16QAM ($\eta_{\mathrm{P}} = 1.6$) is smaller than that of QPSK ($\eta_{\mathrm{P}} = 4$). We often encounter such tradeoffs between power and bandwidth efficiency in the design of communication systems, including when the signaling waveforms considered are sophisticated codes that are constructed from multiple symbols drawn from constellations such as PSK and QAM.

Figure 6.25 shows the symbol error probabilities for QPSK and 16QAM, comparing the intelligent union bounds (which coincide with nearest-neighbors approximations) with exact results. The exact computations for 16QAM use the closed-form expression (6.70) derived in Problem 6.21. We see that the exact error probability and intelligent union bound

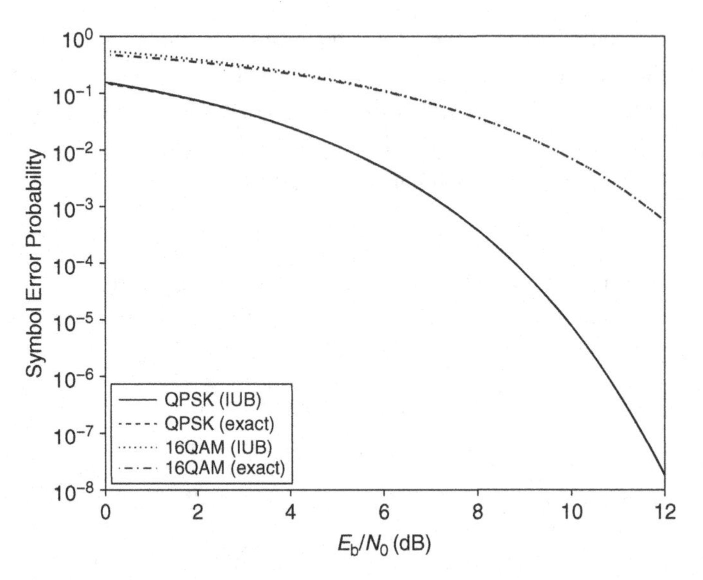

Figure 6.25 Symbol error probabilities for QPSK and 16QAM.

are virtually indistinguishable. The power efficiencies of the constellations (which depend on the argument of the Q function) accurately predict the distance between the curves: $\eta_P(\text{QPSK})/\eta_P(\text{16QAM}) = 4/1.6$, which equals about 4 dB. From Figure 6.25, we see that the distance between the QPSK and 16QAM curves at small error probabilities (high SNR) is indeed about 4 dB.

The performance analysis techniques developed here can also be applied to suboptimal receivers. Suppose, for example, that the receiver LO in a BPSK system is offset from the incoming carrier by a phase shift θ, but that the receiver uses decision regions corresponding to there being no phase offset. The signal-space picture is now as in Figure 6.26. The error probability is now given by

$$P_e = P_{e|0} = P_{e|1} = Q\left(\frac{D}{\sigma}\right) = Q\left(\sqrt{\frac{D^2}{E_b}\frac{2E_b}{N_0}}\right)$$

For the scaling shown, $D = \cos\theta$ and $E_b = 1$, which gives

$$P_e = Q\left(\sqrt{\frac{2E_b\cos^2\theta}{N_0}}\right)$$

so that there is a loss of $10\log_{10}\cos^2\theta$ dB in performance due to the phase offset (e.g. $\theta = 10°$ leads to a loss of 0.13 dB, while $\theta = 30°$ leads to a loss of 1.25 dB).

6.3.5 Performance analysis for *M*-ary orthogonal modulation

So far, our examples have focused on two-dimensional modulation, which is what we use when our primary concern is bandwidth efficiency. We now turn our attention to equal

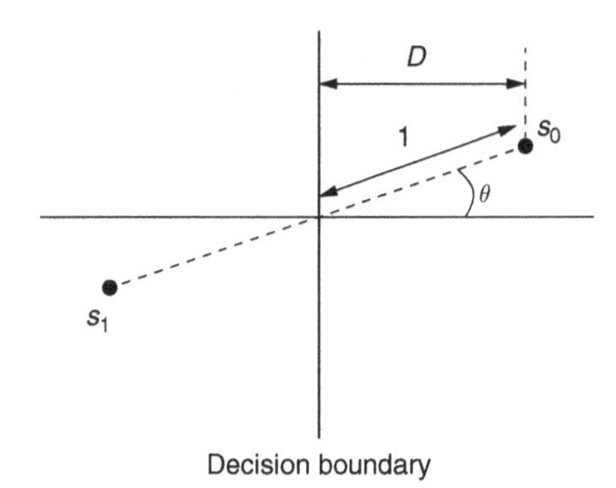

Figure 6.26 Performance analysis for BPSK with phase offset.

energy, M-ary orthogonal signaling, which, as we have mentioned before, lies at the other extreme of the power–bandwidth tradeoff space: as $M \to \infty$, the power efficiency reaches the highest possible value of any signaling scheme over the AWGN channel, while the bandwidth efficiency tends to zero. The signal space is M-dimensional in this case, but we can actually get expressions for the probability of error that involve a single integral rather than M-dimensional integrals, by exploiting the orthogonality of the signal constellation.

Let us first quickly derive the union bound. Without loss of generality, take the M orthogonal signals as unit vectors along the M axes in our signal space. With this scaling, we have $||s_i||^2 \equiv 1$, so that $E_s = 1$ and $E_b = 1/\log_2 M$. Since the signals are orthogonal, the squared distance between any two signals is

$$d_{ij}^2 = ||s_i - s_j||^2 = ||s_i||^2 + ||s_j||^2 - 2\langle s_i, s_j \rangle = 2E_s = 2, \quad i \neq j$$

Thus, $d_{\min} \equiv d_{ij}$ $(i \neq j)$ and the power efficiency

$$\eta_P = \frac{d_{\min}^2}{E_b} = 2\log_2 M$$

The union bound, intelligent union bound and nearest-neighbors approximation all coincide, and we get

$$P_e \equiv P_{e|i} \leq \sum_{j \neq i} Q\left(\frac{d_{ij}}{2\sigma}\right) = (M-1)Q\left(\sqrt{\frac{d_{\min}^2}{E_b}}\sqrt{\frac{E_b}{2N_0}}\right)$$

We now get the following expression in terms of E_b/N_0.

Union bound on error probability for M-ary orthogonal signaling

$$P_e \leq (M-1)Q\left(\sqrt{\frac{E_b \log_2 M}{N_0}}\right) \tag{6.71}$$

Exact expressions By symmetry, the error probability equals the conditional error probability, conditioned on any one of the hypotheses; similarly, the probability of correct decision equals the probability of correct decision given any of the hypotheses. Let us therefore condition on hypothesis H_0 (i.e., that s_0 is sent), so that the received signal $y = s_0 + n$. The decision statistics are

$$Z_i = \langle s_0 + n, s_i \rangle = E_s \delta_{0i} + N_i, \quad i = 0, 1, \ldots, M-1$$

where $\{N_i = \langle n, s_i \rangle\}$ are jointly Gaussian, zero mean, with

$$\text{cov}(N_i, N_j) = \sigma^2 \langle s_i, s_j \rangle = \sigma^2 E_s \delta_{ij}$$

Thus, $N_i \sim N(0, \sigma^2 E_s)$ are i.i.d. We therefore infer that, conditioned on s_0 sent, the $\{Z_i\}$ are conditionally independent, with $Z_0 \sim N(E_s, \sigma^2 E_s)$, and $Z_i \sim N(0, \sigma^2 E_s)$ for $i = 1, \ldots, M-1$.

Let us now express the decision statistics in scale-invariant terms, by replacing Z_i by $Z_i/(\sigma\sqrt{E_s})$. This gives $Z_0 \sim N(m, 1)$, $Z_1, \ldots, Z_{M-1} \sim N(0, 1)$, which are conditionally independent, where

$$m = \frac{E_s}{\sigma\sqrt{E_s}} = \sqrt{\frac{E_s}{\sigma^2}} = \sqrt{\frac{2E_s}{N_0}} = \sqrt{\frac{2E_b \log_2 M}{N_0}}$$

The conditional probability of *correct reception* is now given by

$$P_{c|0} = P[Z_1 \leq Z_0, \ldots, Z_{M-1} \leq Z_0 | H_0] = \int P[Z_1 \leq x, \ldots, Z_{M-1} \leq x | Z_0 = x, H_0] p_{Z_0|H_0}(x|H_0) dx$$

$$= \int P[Z_1 \leq x | H_0] \ldots P[Z_{M-1} \leq x | H_0] p_{Z_0|H_0}(x|H_0) dx$$

where we have used the conditional independence of the $\{Z_i\}$. By plugging in the conditional distributions, we get the following expression for the probability of correct reception.

Probability of correct reception for *M*-ary orthogonal signaling

$$P_c = P_{c|i} = \int_{-\infty}^{\infty} [\Phi(x)]^{M-1} \frac{1}{\sqrt{2\pi}} e^{-(x-m)^2/2} dx \tag{6.72}$$

where

$$m = \sqrt{\frac{2E_s}{N_0}} = \sqrt{\frac{2E_b \log_2 M}{N_0}}$$

The probability of *error* is, of course, one minus the preceding expression. But, for small error probabilities, the probability of correct reception is close to one, and it is difficult to get good estimates of the error probability using (6.72). We therefore develop an expression for the error probability that can be directly computed, as follows:

$$P_{e|0} = \sum_{j \neq 0} P[Z_j = \max_i Z_i | H_0] = (M-1)P[Z_1 = \max_i Z_i | H_0]$$

where we have used symmetry. Now,

$$\begin{aligned} P[Z_1 = \max_i Z_i | H_0] &= P[Z_0 \leq Z_1, Z_2 \leq Z_1, \ldots, Z_{M-1} \leq Z_1 | H_0] \\ &= \int P[Z_0 \leq x, Z_2 \leq x, \ldots, Z_{M-1} \leq x | Z_1 = x, H_0] p_{Z_1|H_0}(x|H_0) dx \\ &= \int P[Z_0 \leq x | H_0] P[Z_2 \leq x | H_0] \ldots P[Z_{M-1} \leq x | H_0] p_{Z_1|H_0}(x|H_0) dx \end{aligned}$$

Plugging in the conditional distributions, and multiplying by $M-1$, gives the following expression for the error probability.

Probability of error for M-ary orthogonal signaling

$$P_e = P_{e|i} = (M-1) \int_{-\infty}^{\infty} [\Phi(x)]^{M-2} \Phi(x-m) \frac{1}{\sqrt{2\pi}} e^{-x^2/2} \, dx \tag{6.73}$$

where

$$m = \sqrt{\frac{2E_s}{N_0}} = \sqrt{\frac{2E_b \log_2 M}{N_0}}$$

Asymptotics for large M The error probability for M-ary orthogonal signaling exhibits an interesting thresholding effect as M gets large:

$$\lim_{M \to \infty} P_e = \begin{cases} 0, & E_b/N_0 > \ln 2 \\ 1, & E_b/N_0 < \ln 2 \end{cases} \tag{6.74}$$

That is, by letting M get large, we can get arbitrarily reliable performance as long as E_b/N_0 exceeds -1.6 dB ($\ln 2$ expressed in dB). This result is derived in one of the problems. Actually, we can show using the tools of information theory that this is the best we can do over the AWGN channel in the limit of bandwidth efficiency tending to zero. That is, M-ary orthogonal signaling is asymptotically optimum in terms of power efficiency.

Figure 6.27 shows the probability of symbol error as a function of E_b/N_0 for several values of M. We see that the performance is quite far away from the asymptotic limit of -1.6 dB (also marked on the plot) for the moderate values of M considered. For example, the E_b/N_0 required for achieving an error probability of 10^{-6} for $M = 16$ is more than 9 dB away from the asymptotic limit.

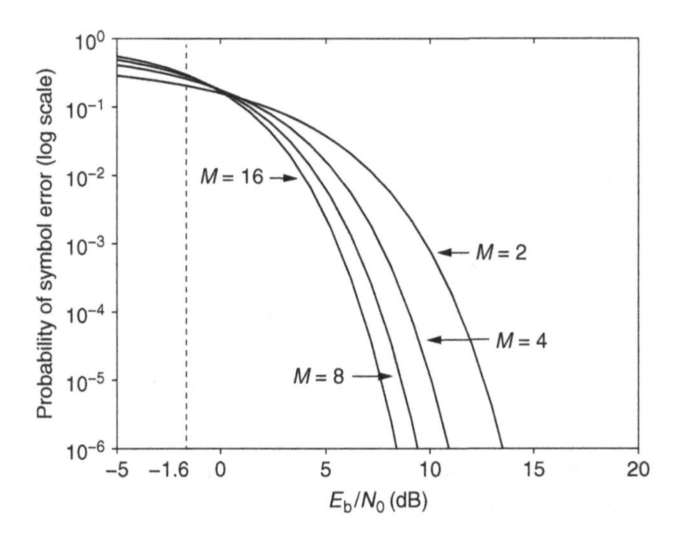

Figure 6.27 Symbol error probabilities for *M*-ary orthogonal signaling.

6.4 Bit error probability

We now know how to design rules for deciding which of M signals (or symbols) has been sent, and how to estimate the performance of these decision rules. Sending one of M signals conveys $m = \log_2 M$ bits, so that a hard decision on one of these signals actually corresponds to hard decisions on m bits. In this section, we discuss how to estimate the bit error probability, or the bit error rate (BER), as it is often called.

QPSK with Gray coding We begin with the example of QPSK, with the bit mapping shown in Figure 6.28. This bit mapping is an example of a Gray code, in which the bits corresponding to neighboring symbols differ by exactly one bit (since symbol errors are most likely going to be caused by decoding into neighboring decision regions, this reduces the number of bit errors). Let us denote the symbol labels as $b[1]b[2]$ for the transmitted symbol, where $b[1]$ and $b[2]$ each take values 0 and 1. Letting $\hat{b}[1]\hat{b}[2]$ denote the label for the ML symbol decision, the probabilities of bit error are given by $p_1 = P[\hat{b}[1] \neq b[1]]$ and $p_2 = P[\hat{b}[2] \neq b[2]]$. The average probability of bit error, which we wish to estimate, is given by $p_b = \frac{1}{2}(p_1 + p_2)$. Conditioned on 00 being sent, the probability of making an error on $b[1]$ is as follows:

$$P[\hat{b}[1] = 1|00 \text{ sent}] = P[\text{ML decision is 10 or 11}|00 \text{ sent}]$$
$$= P\left[N_c < -\frac{d}{2}\right] = Q\left(\frac{d}{2\sigma}\right) = Q\left(\sqrt{\frac{2E_b}{N_0}}\right)$$

where, as before, we have expressed the result in terms of E_b/N_0 using the power efficiency $d^2/E_b = 4$. We also note that, by virtue of the symmetry of the constellation and the bit map, the conditional probability of error of $b[1]$ is the same regardless of which symbol

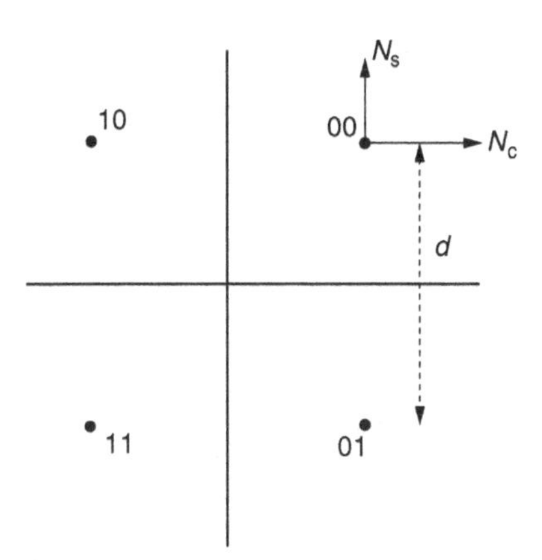

Figure 6.28 QPSK with Gray coding.

we condition on. Moreover, exactly the same analysis holds for $b[2]$, except that errors are caused by the noise random variable N_s. We therefore obtain that

$$p_b = p_1 = p_2 = Q\left(\sqrt{\frac{2E_b}{N_0}}\right) \tag{6.75}$$

The fact that this expression is identical to the bit error probability for binary antipodal signaling is not a coincidence. QPSK with Gray coding can be thought of as two independent BPSK systems, one signaling along the I component, and the other along the Q component.

Gray coding is particularly useful at low SNR (e.g., for heavily coded systems), where symbol errors happen more often. For example, in a coded system, we would pass up fewer bit errors to the decoder for the same number of symbol errors. We define it in general as follows.

Gray Coding Consider a 2^n-ary constellation in which each point is represented by a binary string $\mathbf{b} = (b_1, \ldots, b_n)$. The bit assigment is said to be *Gray coded* if, for any two constellation points $\mathbf{b}$ and $\mathbf{b}'$ that are nearest neighbors, the bit representations $\mathbf{b}$ and $\mathbf{b}'$ differ in exactly one bit location.

Nearest neighbors approximation for BER with Gray coded constellation Consider the ith bit b_i in an n-bit Gray code for a regular constellation with minimum distance $d_{\min}$. For a Gray code, there is at most one nearest neighbor that differs in the ith bit, and the pairwise error probability of decoding to that neighbor is $Q(d_{\min}/(2\sigma))$. We therefore have

$$P(\text{bit error}) \approx Q\left(\sqrt{\frac{\eta_P E_b}{2N_0}}\right) \quad \text{with Gray coding} \tag{6.76}$$

where $\eta_P = d_{\min}^2/E_b$ is the power efficiency.

Figure 6.29 shows the BER of 16QAM and 16PSK with Gray coding, comparing the nearest-neighbors approximation with exact results (obtained analytically for 16QAM, and by simulation for 16PSK). The slight pessimism and ease of computation of the nearest-neighbors approximation implies that it is an excellent tool for link design.

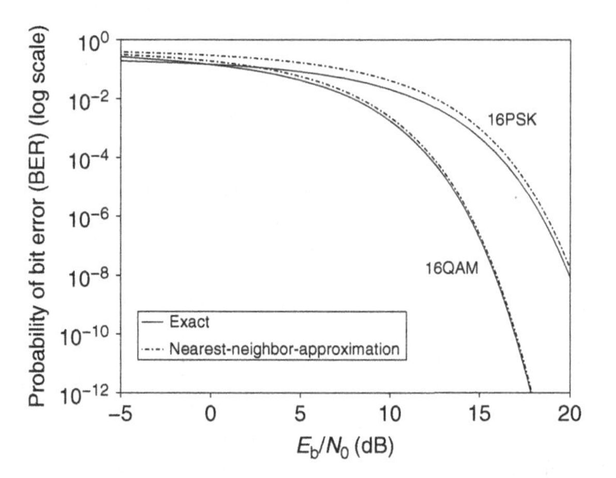

Figure 6.29 BER for 16QAM and 16PSK with Gray coding.

Gray coding might not always be possible. Indeed, for an arbitrary set of $M = 2^n$ signals, we might not understand the geometry well enough to assign a Gray code. In general, a necessary (but not sufficient) condition for an n-bit Gray code to exist is that the number of nearest neighbors for any signal point should be at most n.

BER for orthogonal modulation For $M = 2^m$-ary equal energy, orthogonal modulation, each of the m bits split the signal set into half. By the symmetric geometry of the signal set, any of the $M - 1$ wrong symbols are equally likely to be chosen, given a symbol error, and $M/2$ of these will correspond to error in a given bit. We therefore have

$$P(\text{bit error}) = \frac{M/2}{M-1} P(\text{symbol error}), \quad \textbf{BER for M-ary orthogonal signaling} \quad (6.77)$$

Note that Gray coding is out of the question here, since there are only m bits and $2^m - 1$ neighbors, all at the same distance.

6.5 Link-budget analysis

We have seen now that performance over the AWGN channel depends only on the constellation geometry and E_b/N_0. In order to design a communication link, however, we must relate E_b/N_0 to physical parameters such as the transmit power, transmit and receive antenna gains, range, and the quality of the receiver circuitry. Let us first take stock of what we know.

(a) Given the bit rate R_b and the signal constellation, we know the symbol rate (or more generally, the number of modulation degrees of freedom required per unit time), and hence the minimum Nyquist bandwidth $B_{\min}$. We can then factor in the excess bandwidth a dictated by implementation considerations to find the bandwidth

$B = (1 + a)B_{\min}$ required. (However, assuming optimal receiver processing, we show below that the excess bandwidth does not affect the link budget.)

(b) Given the constellation and a desired bit error probability, we can infer the E_b/N_0 at which we need to operate. Since the SNR satisfies $\mathrm{SNR} = E_b R_b/(N_0 B)$, we have

$$\mathrm{SNR}_{\mathrm{reqd}} = \left(\frac{E_b}{N_0}\right)_{\mathrm{reqd}} \frac{R_b}{B} \tag{6.78}$$

(c) Given the receiver noise figure F (dB), we can infer the noise power $P_n = N_0 B = N_{0,\mathrm{nom}} \cdot 10^{F/10} B$, and hence the minimum required received signal power is given by

$$P_{\mathrm{RX}}(\min) = \mathrm{SNR}_{\mathrm{reqd}} \cdot P_n = \left(\frac{E_b}{N_0}\right)_{\mathrm{reqd}} \frac{R_b}{B} N_0 B = \left(\frac{E_b}{N_0}\right)_{\mathrm{reqd}} R_b N_{0,\mathrm{nom}} \cdot 10^{F/10} \tag{6.79}$$

This is called the required *receiver sensitivity,* and is usually quoted in dBm, as $P_{\mathrm{RX,dBm}}(\min) = 10 \log_{10} P_{\mathrm{RX}}(\min)$ (mW). Using (5.93), we obtain that

$$P_{\mathrm{RX,dBm}}(\min) = \left(\frac{E_b}{N_0}\right)_{\mathrm{reqd,dB}} + 10 \log_{10} R_b - 174 + F \tag{6.80}$$

where R_b is in bits per second. Note that the dependence on the bandwidth B (and hence on excess bandwidth) cancels out in (6.79), so that the final expression for the receiver sensitivity depends only on the required E_b/N_0 (which depends on the signaling scheme and target BER), the bit rate R_b, and the noise figure F.

Once we know the receiver sensitivity, we need to determine the link parameters (e.g., transmitted power, choice of antennas, range) such that the receiver actually gets at least that much power, plus a link margin (typically expressed in dB). We illustrate such considerations via the Friis formula for propagation loss in free space, which we can think of as modeling a line-of-sight wireless link. While deriving this formula from basic electromagnetics is beyond our scope here, let us provide some intuition before stating it.

Suppose that a transmitter emits power P_{TX} that radiates uniformly in all directions. The power per unit area at a distance R from the transmitter is $P_{\mathrm{TX}}/(4\pi R^2)$, where we have divided by the area of a sphere of radius R. The receive antenna may be thought of as providing an effective area, termed the antenna *aperture,* for catching a portion of this power. (The aperture of an antenna is related to its size, but the relation is not usually straightforward.) If we denote the receive antenna aperture by A_{RX}, the received power is given by

$$P_{\mathrm{RX}} = \frac{P_{\mathrm{TX}}}{4\pi R^2} A_{\mathrm{RX}}$$

Now, if the transmitter can direct power selectively in the direction of the receiver rather than radiating it isotropically, we get

$$P_{\mathrm{RX}} = \frac{P_{\mathrm{TX}}}{4\pi R^2} G_{\mathrm{TX}} A_{\mathrm{RX}} \tag{6.81}$$

where G_{TX} is the transmit antenna's gain towards the receiver, relative to a hypothetical isotropic radiator. We now have a formula for received power in terms of transmitted power, which depends on the gain of the transmit antenna and the aperture of the receive antenna. We would like to express this formula solely in terms of antenna gains or antenna apertures. To do this, we need to relate the gain of an antenna to its aperture. To this end, we state without proof that the aperture of an isotropic antenna is given by $A = \lambda^2/(4\pi)$. Since the gain of an antenna is the ratio of its aperture to that of an isotropic antenna. This implies that the relation between gain and aperture can be written as

$$G = \frac{A}{\lambda^2/(4\pi)} = \frac{4\pi A}{\lambda^2} \tag{6.82}$$

Assuming that the aperture A scales up in some fashion with antenna size, this implies that, for a fixed form factor, we can get higher antenna gains as we decrease the carrier wavelength, or increase the carrier frequency.

Using (6.82) in (6.81), we get two versions of the Friis formula.

Friis formula for free-space propagation

$$P_{RX} = P_{TX} G_{TX} G_{RX} \frac{\lambda^2}{16\pi^2 R^2}, \quad \textbf{in terms of antenna gains} \tag{6.83}$$

$$P_{RX} = P_{TX} \frac{A_{TX} A_{RX}}{\lambda^2 R^2}, \quad \textbf{in terms of antenna apertures} \tag{6.84}$$

where

- G_{TX} and A_{TX} are the gain and aperture, respectively, of the transmit antenna;
- G_{RX} and A_{RX} are the gain and aperture, respectively, of the receive antenna;
- $\lambda = c/f_c$ is the carrier wavelength ($c = 3 \times 10^8$ m/s is the speed of light; f_c is the carrier frequency); and
- R is the range (line-of-sight distance between transmitter and receiver).

The first version (6.83) of the Friis formula tells us that, for antennas with fixed gain, we should try to use as low a carrier frequency (as large a wavelength) as possible. On the other hand, the second version tells us that, if we have antennas of a given form factor, then we can get better performance as we increase the carrier frequency (decrease the wavelength), assuming of course that we can "point" these antennas accurately at each other. Of course, higher carrier frequencies also have the disadvantage of incurring more attenuation from impairments such as obstacles, rain, and fog. Some of these tradeoffs are explored in the problems.

In order to apply the Friis formula (let us focus on version (6.83) for concreteness) to link-budget analysis, it is often convenient to take logarithms, thereby converting the multiplications into addition. On a logarithmic scale, antenna gains are expressed in dBi, where $G_{dBi} = 10\log_{10}G$ for an antenna with raw gain G. Expressing powers in dBm, we have

$$P_{\mathrm{RX,dBm}} = P_{\mathrm{TX,dBm}} + G_{\mathrm{TX,dBi}} + G_{\mathrm{RX,dBi}} + 10\log_{10}\left(\frac{\lambda^2}{16\pi^2 R^2}\right) \tag{6.85}$$

More generally, we have the link-budget equation

$$P_{\mathrm{RX,dBm}} = P_{\mathrm{TX,dBm}} + G_{\mathrm{TX,dBi}} + G_{\mathrm{RX,dBi}} - L_{\mathrm{path\ loss,dB}}(R) \tag{6.86}$$

where $L_{\mathrm{path\ loss,dB}}(R)$ is the path loss in dB. For free space propagation, we have from the Friis formula (6.85) that

$$L_{\mathrm{path\ loss,dB}}(R) = 10\log_{10}\left(\frac{16\pi^2 R^2}{\lambda^2}\right) \quad \text{path loss in dB for free-space propagation} \tag{6.87}$$

While the Friis formula is our starting point, the link-budget equation (6.86) applies more generally, in that we can substitute other expressions for path loss, depending on the propagation environment (e.g., for wireless communication in a cluttered environment, the signal power may decay as $1/R^4$ rather than the free-space decay of $1/R^2$). A mixture of empirical measurements and statistical modeling is typically used to characterize path loss as a function of range for the environments of interest. For example, the design of wireless cellular systems is accompanied by extensive "measurement campaigns" and modeling. Once we have decided on the path loss formula ($L_{\mathrm{path\ loss,dB}}(R)$) to be used in the design, the transmit power required to attain a given receiver sensitivity can be determined as a function of the range R. Such a path-loss formula typically characterizes an "average" operating environment, around which there might be significant statistical variations that are not captured by the model used to arrive at the receiver sensitivity. For example, the receiver sensitivity for a wireless link may be calculated from the AWGN channel model, whereas the link may exhibit rapid amplitude variations due to multipath fading, and slower variations due to shadowing (e.g., due to buildings and other obstacles). Even if fading/shadowing effects are factored into the channel model used to compute the BER, and the model for path loss, the actual environment encountered may be worse than that assumed in the model. In general, therefore, we add a link margin $L_{\mathrm{margin,dB}}$, again expressed in dB, in an attempt to budget for potential performance losses due to unmodeled or unforeseen impairments. The size of the link margin depends, of course, on the confidence of the system designer in the models used to arrive at the rest of the link budget.

Putting all this together, if $P_{\mathrm{RX,dBm}}(\mathrm{min})$ is the desired receiver sensitivity (i.e., the minimum required received power), then we compute the transmit power for the link to be as follows.

Required transmit power

$$P_{\mathrm{TX,dBm}} = P_{\mathrm{RX,dBm}}(\mathrm{min}) - G_{\mathrm{TX,dBi}} - G_{\mathrm{RX,dBi}} + L_{\mathrm{path\ loss,dB}}(R) + L_{\mathrm{margin,dB}} \tag{6.88}$$

Let us illustrate these concepts using some examples.

Example 6.5.1 Consider again the 5-GHz WLAN link of Example 5.8.1. We wish to utilize a 20-MHz channel, using Gray-coded QPSK and an excess bandwidth of 33%. The receiver has a noise figure of 6 dB.

(a) What is the bit rate?
(b) What is the receiver sensitivity required in order to achieve a BER of 10^{-6}?
(c) Assuming transmit and receive antenna gains of 2 dBi each, what is the range achieved for 100 mW transmit power, using a link margin of 20 dB? Use link-budget analysis that is based on the free-space path loss.

Solution

(a) For bandwidth B and fractional excess bandwidth a, the symbol rate

$$R_s = \frac{1}{T} = \frac{B}{1+a} = \frac{20}{1+0.33} = 15 \text{ Msymbols/s}$$

and the bit rate for an M-ary constellation is

$$R_b = R_s \log_2 M = 15 \text{ Msymbols/s} \times 2 \text{ bits/symbol} = 30 \text{ Mbits/s}$$

(b) BER for QPSK with Gray coding is $Q(\sqrt{2E_b/N_0})$. For a desired BER of 10^{-6}, we obtain that $(E_b/N_0)_{\text{reqd,dB}} \approx 10.2$. By plugging $R_b = 30$ Mbps and $F = 6$ dB into (6.80), we obtain that the required receiver sensitivity is $P_{\text{RX,dBm}}(\text{min}) = -83$ dBm.
(c) The transmit power is 100 mW, or 20 dBm. On rewriting (6.88), the allowed path loss to attain the desired sensitivity at the desired link margin is

$$\begin{aligned} L_{\text{pathloss,dB}}(R) &= P_{\text{TX,dBm}} - P_{\text{RX,dBm}}(\text{min}) + G_{\text{TX,dBi}} + G_{\text{RX,dBi}} - L_{\text{margin,dB}} \\ &= 20 - (-83) + 2 + 2 - 20 = 87 \text{ dB} \end{aligned} \tag{6.89}$$

We can now invert the formula for free-space loss, (6.87), noting that $f_c = 5$ GHz, which implies that $\lambda = c/f_c = 0.06$ m. We get a range R of 107 m, which is of the order of the advertised ranges for WLANs under nominal operating conditions. The range decreases, of course, for higher bit rates using larger constellations. What happens, for example, when we use 16QAM or 64QAM?

Example 6.5.2 Consider an indoor link at 10 meters range using unlicensed spectrum at 60 GHz. Suppose that both the transmitter and the receiver use antennas with horizontal beamwidths of 60° and vertical beamwidths of 30°. Use the following approximation to calculate the resulting antenna gains:

$$G \approx \frac{41{,}000}{\theta_{\text{horiz}}\theta_{\text{vert}}}$$

where G denotes the antenna gain (linear scale), θ_{horiz} and θ_{vert} denote horizontal and vertical beamwidths (in degrees). Set the noise figure to 8 dB, and assume a link margin of 10 dB at a BER of 10^{-6}.

(a) Calculate the bandwidth and transmit power required for a 2-Gbps link using Gray-coded QPSK and 50% excess bandwidth.

(b) How do your answers change if you change the signaling scheme to Gray-coded 16QAM, keeping the same *bit rate* as in (a)?

(c) If you now employ Gray-coded 16QAM keeping the same *symbol rate* as in (a), what is the bit rate attained and the transmit power required?

(d) How do the answers in the setting of (a) change if you increase the horizontal beamwidth to 120°, keeping all other parameters fixed?

Solution

(a) A 2-Gbps link using QPSK corresponds to a symbol rate of 1 Gsymbols/s. On factoring in the 50% excess bandwidth, the required bandwidth is $B = 1.5$ GHz. The target BER and constellation are as in the previous example, hence we still have $(E_b/N_0)_{\text{reqd,dB}} \approx 10.2$ dB. By plugging $R_b = 2$ Gbps and $F = 8$ dB into (6.80), we obtain that the required receiver sensitivity is $P_{\text{RX,dBm}}(\text{min}) = -62.8$ dBm.

The antenna gains at each end are given by

$$G \approx \frac{41,000}{60 \times 30} = 22.78$$

On converting to the dB scale, we obtain $G_{\text{TX,dBi}} = G_{\text{RX,dBi}} = 13.58$ dBi.

The transmit power for a range of 10 m can now be obtained using (6.88) to be 8.1 dBm.

(b) For the same bit rate of 2 Gbps, the symbol rate for 16QAM is 0.5 Gsymbols/s, so the bandwidth required is 0.75 GHz, factoring in 50% excess bandwidth. The nearest-neighbors approximation to BER for Gray coded 16QAM is given by $Q(\sqrt{4E_b/5N_0})$. Using this, we find that a target BER of 10^{-6} requires $(E_b/N_0)_{\text{reqd,dB}} \approx 14.54$ dB, an increase of 4.34 dB relative to (a). This leads to a corresponding increase in the receiver sensitivity to -58.45 dBm, which leads to the required transmit power increasing to 12.4 dBm.

(c) If we keep the symbol rate fixed at 1 Gsymbols/s, the bit rate with 16QAM is $R_b = 4$ Gbps. As in (b), $(E_b/N_0)_{\text{reqd,dB}} \approx 14.54$ dB. The receiver sensitivity is therefore given by -55.45 dBm, a 3-dB increase over (b), corresponding to the doubling of the bit rate. This translates directly to a 3-dB increase, relative to (b), in transmit power to 15.4 dBm, since the path loss, antenna gains, and link margin are as in (b).

(d) We now go back to the setting of (a), but with different antenna gains. The bandwidth is, of course, unchanged from (a). The new antenna gains are 3 dB smaller because of the doubling of the horizontal beamwidth. The receiver sensitivity, path loss and link margin are as in (a), thus the 3-dB reduction in antenna gain at each end must be compensated for by a 6-dB increase in transmit power relative to (a). Thus, the required transmit power is 14.1 dBm.

Discussion The parameter choices in the preceding examples illustrate how the physical characteristics of the medium change with the choice of carrier frequency, and affect system design tradeoffs. The 5-GHz system in Example 6.5.1 employs essentially omnidirectional antennas with small gains of 2 dBi, whereas it is possible to realize highly directional yet small antennas (e.g., using electronically steerable printed-circuit antenna arrays) for the 60-GHz system in Example 6.5.2 by virtue of the small (5 mm) wavelength. However, 60 GHz waves are easily blocked by walls, hence the range in Example 6.5.2 corresponds to in-room communication. We have also chosen parameters such that the transmit power required for 60 GHz is smaller than that at 5 GHz, since it is more difficult to produce power at higher radio frequencies. Finally, the link margin for 5 GHz is chosen to be higher than that for 60 GHz: propagation at 60 GHz is near line-of-sight, whereas fading due to multipath propagation at 5 GHz can be more significant, and hence may require a higher link margin relative to the AWGN benchmark which provides the basis for our link budget.

6.6 Concept summary

This chapter establishes a systematic hypothesis-testing-based framework for demodulation, develops tools for performance evaluation that enable exploration of the power–bandwidth tradeoffs exhibited by different signaling schemes, and relates these mathematical models to physical link parameters via the link budget. A summary of some key concepts and results is as follows.

Hypothesis testing

- The probability of error is minimized by choosing the hypothesis with the maximum *a posteriori* probability (i.e., the hypothesis that is most likely, conditioned on the observation). That is, the MPE rule is also the MAP rule:

$$\begin{aligned}\delta_{\text{MPE}}(y) = \delta_{\text{MAP}}(y) &= \arg\max_{1\le i\le M} P[H_i|Y=y] \\ &= \arg\max_{1\le i\le M} \pi_i p(y|i) = \arg\max_{1\le i\le M} \; \log\pi_i + \log p(y|i)\end{aligned}$$

  For equal priors, the MPE rule coincides with the ML rule:

$$\delta_{\text{ML}}(y) = \arg\max_{1\le i\le M} p(y|i) = \arg\max_{1\le i\le M} \; \log p(y|i)$$

- For binary hypothesis testing, ML and MPE rules can be written as likelihood, or log likelihood, ratio tests:

$$L(y) = \frac{p_1(y)}{p_0(y)} \underset{H_0}{\overset{H_1}{\gtrless}} 1 \quad \text{or} \quad \log L(y) \underset{H_0}{\overset{H_1}{\gtrless}} 0 \qquad \textbf{ML rule}$$

$$L(y) = \frac{p_1(y)}{p_0(y)} \underset{H_0}{\overset{H_1}{\gtrless}} \frac{\pi_0}{\pi_1} \quad \text{or} \quad \log L(y) \underset{H_0}{\overset{H_1}{\gtrless}} \frac{\pi_0}{\pi_1} \qquad \textbf{MPE/MAP rule}$$

Geometric view of signals

Continuous-time signals can be interpreted as vectors in Euclidean space, with inner product $\langle s_1, s_2 \rangle = \int s_1(t) s_2^*(t) dt$, norm $||s|| = \sqrt{\langle s, s \rangle}$, and energy $||s||^2 = \langle s, s \rangle$. Two signals are *orthogonal* if their inner product is zero.

Geometric view of WGN

- WGN $n(t)$ with PSD σ^2, when projected in any "direction" (i.e., correlated against any unit energy signal), yields an $N(0, \sigma^2)$ random variable.
- More generally, projections of the noise along any signals are jointly Gaussian, with zero mean and $\text{cov}(\langle n, u \rangle, \langle n, v \rangle) = \sigma^2 \langle v, u \rangle$.
- Noise projections along orthogonal signals are uncorrelated. Since they are jointly Gaussian, they are also independent.

Signal space

- M-ary signaling in AWGN in continuous time can be reduced, without loss of information, to M-ary signaling in finite-dimensional vector space with each dimension seeing i.i.d. $N(0, \sigma^2)$ noise, which corresponds to discrete-time WGN. This is accomplished by projecting the received signal onto the signal space spanned by the M possible signals.
- Decision rules derived using hypothesis testing in the finite-dimensional signal space map directly back to continuous time because of two key reasons: signal inner products are preserved, and the noise component orthogonal to the signal space is irrelevant. Because of this equivalence, we can stop making a distinction between continuous-time signals and finite-dimensional vector signals in our notation.

Optimal demodulation

- For the model $H_i = y = s_i + n$, $0 \leq i \leq M - 1$, optimum demodulation involves computation of the correlator outputs $Z_i = \langle y, s_i \rangle$. This can be accomplished by using a bank of correlators or matched filters, but any other receiver structure that yields the statistics $\{Z_i\}$ would also preserve all of the relevant information.
- The ML and MPE rules are given by

$$\delta_{\text{ML}}(y) = \arg\max_{0 \leq i \leq M-1} \langle y, s_i \rangle - \frac{||s_i||^2}{2}$$

$$\delta_{\text{MPE}}(y) = \arg\max_{0 \leq i \leq M-1} \langle y, s_i \rangle - \frac{||s_i||^2}{2} + \sigma^2 \log \pi_i$$

When the received signal lies in a finite-dimensional space in which the noise has finite energy, the ML rule can be written as a minimum-distance rule (and the MPE rule as a variant thereof) as follows:

$$\delta_{\mathrm{ML}}(\mathbf{y}) = \arg\min_{0\le i\le M-1} ||\mathbf{y}-\mathbf{s}_i||^2$$
$$\delta_{\mathrm{MPE}}(\mathbf{y}) = \arg\min_{0\le i\le M-1} ||\mathbf{y}-\mathbf{s}_i||^2 - 2\sigma^2\log\pi_i$$

Geometry of ML rule ML decision boundaries are formed from hyperplanes that bisect lines connecting signal points.

Performance analysis

- For binary signaling, the error probability for the ML rule is given by
$$P_{\mathrm{e}} = Q\left(\frac{d}{2\sigma}\right) = Q\left(\sqrt{\frac{d^2}{E_{\mathrm{b}}}}\sqrt{\frac{E_{\mathrm{b}}}{2N_0}}\right)$$
where $d = ||s_1 - s_0||$ is the Euclidean distance between the signals. The performance therefore depends on the power efficiency $\eta_{\mathrm{P}} = d^2/E_{\mathrm{b}}$ and the SNR E_{b}/N_0. Since the power efficiency is scale-invariant, we may choose any convenient scaling when computing it for a given constellation.
- For M-ary signaling, closed-form expressions for the error probability might not be available, but we know that the performance depends only on the scale-invariant inner products $\{\langle s_i, s_j\rangle/E_{\mathrm{b}}\}$, (which depend on the constellation "shape" alone) and on E_{b}/N_0.
- The conditional error probabilities for M-ary signaling can be bounded using the union bound (these can then be averaged to obtain an upper bound on the average error probability):
$$P_{\mathrm{e}|i} \le \sum_{j\ne i} Q\left(\frac{d_{ij}}{2\sigma}\right) = \sum_{j\ne i} Q\left(\sqrt{\frac{d_{ij}^2}{E_{\mathrm{b}}}}\sqrt{\frac{E_{\mathrm{b}}}{2N_0}}\right)$$
where $d_{ij} = ||s_i - s_j||$ are the pairwise distances between signal points.
- When we understand the shape of the decision regions, we can tighten the union bound into an intelligent union bound:
$$P_{\mathrm{e}|i} \le \sum_{j\in N_{\mathrm{ML}}(i)} Q\left(\frac{||d_{ij}||}{2\sigma}\right) = \sum_{j\in N_{\mathrm{ML}}(i)} Q\left(\sqrt{\frac{d_{ij}^2}{E_{\mathrm{b}}}}\sqrt{\frac{E_{\mathrm{b}}}{2N_0}}\right)$$
where $N_{\mathrm{ML}}(i)$ denotes the set of neighbors of s_i which defines the decision region Γ_i.
- For regular constellations, the nearest-neighbors approximation is given by
$$P_{\mathrm{e}|i} \approx N_{d_{\min}}(i) Q\left(\frac{d_{\min}}{2\sigma}\right) = N_{d_{\min}}(i) Q\left(\sqrt{\frac{d_{\min}^2}{E_{\mathrm{b}}}}\sqrt{\frac{E_{\mathrm{b}}}{2N_0}}\right)$$
$$P_e \approx \bar{N}_{d_{\min}} Q\left(\frac{d_{\min}}{2\sigma}\right) = \bar{N}_{d_{\min}} Q\left(\sqrt{\frac{d_{\min}^2}{E_{\mathrm{b}}}}\sqrt{\frac{E_{\mathrm{b}}}{2N_0}}\right)$$

with $\eta_P = d^2_{\min}/E_b$ providing a measure of power efficiency that can be used to compare across constellations.

- If Gray coding is possible, the bit error probability can be estimated as

$$P(\text{bit error}) \approx Q\left(\sqrt{\frac{\eta_P E_b}{2N_0}}\right)$$

Link budget This relates (e.g., using the Friis formula for free-space propagation) the performance of a communication link to physical parameters such as the transmit power, transmit and receive antenna gains, range, and receiver noise figure. A link margin is typically introduced to account for unmodeled impairments.

6.7 Notes

The geometric signal-space approach for deriving and analyzing is now standard in textbooks on communication theory, such as [7, 8]. It was first developed by Russian pioneer Vladimir Kotelnikov [33], and presented in a cohesive fashion in the classic textbook by Wozencraft and Jacobs [9].

Certain details of receiver design have been swept under the rug in this chapter. Our model for the received signal is that it equals the transmitted signal plus WGN. In practice, the transmitted signal can be significantly distorted by the channel (e.g., by scaling, delay, multipath propagation). However, the basic M-ary signaling model is still preserved: if M possible signals are sent, then, prior to the addition of noise, M possible signals are received after the deterministic (but *a priori* unknown) transformations due to channel impairments. The receiver can therefore estimate noiseless copies of the latter and then apply the optimum demodulation techniques developed here. This approach leads, for example, to the optimal equalization strategies developed by Forney [34] and Ungerboeck [35]; see Chapter 5 of [7] for a textbook exposition. Estimation of the noiseless received signals involves tasks such as carrier phase and frequency synchronization, timing synchronization, and estimation of the channel impulse response or transfer function. In modern digital communication transceivers, these operations are typically all performed using DSP on the complex-baseband received signal. Perhaps the best approach for exploring further is to acquire a basic understanding of the relevant estimation techniques, and to then go to technical papers of specific interest (e.g., IEEE conference and journal publications). Classic texts covering estimation theory include Kay [36], Poor [37], and Van Trees [38]. Several graduate texts on communications contain a brief discussion of the modern estimation-theoretic approach to synchronization that may provide a helpful orientation prior to going to the research literature; for example, see [7] (Chapter 4) and [11, 39] (Chapter 8 in both references).

6.8 Problems

Hypothesis testing

Problem 6.1 The received signal in a digital communication system is given by

$$y(t) = \begin{cases} s(t) + n(t) & 1 \text{ sent} \\ n(t) & 0 \text{ sent} \end{cases}$$

where n is AWGN with PSD $\sigma^2 = N_0/2$ and $s(t)$ is as shown in Figure 6.30. The received signal is passed through a filter, and the output is sampled to yield a decision statistic. An ML decision rule is employed based on the decision statistic. The set-up is shown in Figure 6.30.

(a) For $h(t) = s(-t)$, find the error probability as a function of E_b/N_0 if $t_0 = 1$.
(b) Can the error probability in (a) be improved by choosing the sampling time t_0 differently?
(c) Now, find the error probability as a function of E_b/N_0 for $h(t) = I_{[0,2]}$ and the best possible choice of sampling time.
(d) Finally, comment on whether you can improve the performance in (c) by using a linear combination of two samples as a decision statistic, rather than just using one sample.

Problem 6.2 Consider binary hypothesis testing that is based on the decision statistic Y, where $Y \sim N(2, 9)$ under H_1 and $Y \sim N(-2, 4)$ under H_0.

(a) Show that the optimal (ML or MPE) decision rule is equivalent to comparing a function of the form $ay^2 + by$ with a threshold.
(b) Specify the MPE rule explicitly (i.e., specify a, b, and the threshold) when $\pi_0 = \frac{1}{4}$.
(c) Express the conditional error probability $P_{e|0}$ for the decision rule in (b) in terms of the Q function with positive arguments. Also provide a numerical value for this probability.

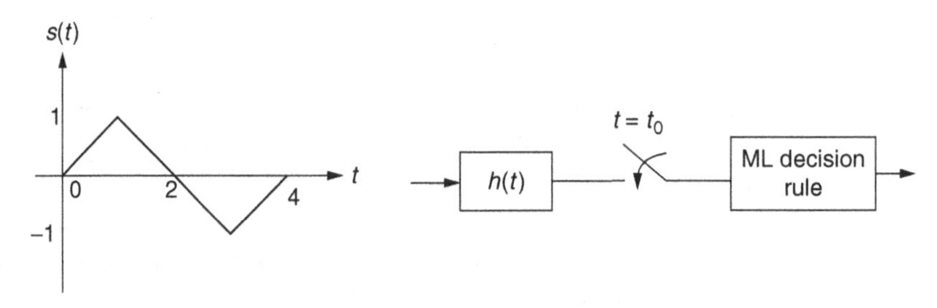

Figure 6.30 Set-up for Problem 6.1

Problem 6.3 Find and sketch the decision regions for a binary hypothesis-testing problem with observation Z, where the hypotheses are equally likely, and the conditional distributions are given by the following:

H_0: Z is uniform over $[-2, 2]$; and
H_1: Z is Gaussian with mean 0 and variance 1.

Problem 6.4 The receiver in a binary communication system employs a decision statistic Z that behaves as follows:

$Z = N$ if 0 is sent
$Z = 4 + N$ if 1 is sent

where N is modeled as Laplacian with density

$$p_N(x) = \frac{1}{2}e^{-|x|}, \quad -\infty < x < \infty$$

Note that parts (a) and (b) can be done independently.

(a) Find and sketch, as a function of z, the log likelihood ratio

$$K(z) = \log L(z) = \log\left(\frac{p(z|1)}{p(z|0)}\right)$$

where $p(z|i)$ denotes the conditional density of Z given that i is sent ($i = 0, 1$).

(b) Find $P_{e|1}$, the conditional error probability given that 1 is sent, for the decision rule

$$\delta(z) = \begin{cases} 0, & z < 1 \\ 1, & z \geq 1 \end{cases}$$

(c) Is the rule in (b) the MPE rule for any choice of prior probabilities? If so, specify the prior probability $\pi_0 = P[0 \text{ sent}]$ for which it is the MPE rule. If not, say why not.

Problem 6.5 Consider the MAP/MPE rule for the hypothesis-testing problem in Example 6.1.1.

(a) Show that the MAP rule always says H_1 if the prior probability of H_0 is smaller than some positive threshold. Specify this threshold.
(b) Compute and plot the conditional probabilities $P_{e|0}$ and $P_{e|1}$, and the average error probability P_e, versus π_0 as the latter varies in $[0, 1]$.
(c) Discuss any trends that you see from the plots in (b).

Problem 6.6 Consider a MAP receiver for the basic Gaussian example, as discussed in Example 6.1.2. Fix SNR at 13 dB. We wish to explore the effect of prior mismatch, by quantifying the performance degradation of a MAP receiver if the actual priors are different from the priors for which it has been designed.

(a) Plot the average error probability for a MAP receiver designed for $\pi_0 = 0.2$, as π_0 varies from 0 to 1. As usual, use a log scale for the probabilities. On the same plot, also plot the error probability of the ML receiver as a benchmark.

(b) From the plot in (a), comment on how much error you can tolerate in the prior probabilities before the performance of the MAP receiver designed for the given prior becomes unacceptable.
(c) Repeat (a) and (b) for a MAP receiver designed for $\pi_0 = 0.4$. Is the performance more or less sensitive to errors in the priors?

Problem 6.7 Consider binary hypothesis testing in which the observation Y is modeled as uniformly distributed over $[-2, 2]$ under H_0, and has conditional density $p(y|1) = c(1 - |y|/3)I_{[-3,3]}(y)$ under H_1, where $c > 0$ is a constant to be determined.

(a) Find c.
(b) Find and sketch the decision regions Γ_0 and Γ_1 corresponding to the ML decision rule.
(c) Find the conditional error probabilities.

Problem 6.8 Consider binary hypothesis testing with scalar observation Y. Under hypothesis H_0, Y is modeled as uniformly distributed over $[-5, 5]$. Under H_1, Y has conditional density $p(y|1) = \frac{1}{8}e^{-|y|/4}, \quad -\infty < y < \infty$.

(a) Specify the ML rule and clearly draw the decision regions Γ_0 and Γ_1 on the real line.
(b) Find the conditional probabilities of error for the ML rule under each hypothesis.

Problem 6.9 For the setting of Problem 6.8, suppose that the prior probability of H_0 is $1/3$.

(a) Specify the MPE rule and draw the decision regions.
(b) Find the conditional error probabilities and the average error probability. Compare your answers with the corresponding quantities for the ML rule considered in Problem 6.8.

Problem 6.10 The receiver output Z in an on–off-keyed optical communication system is modeled as a Poisson random variable with mean $m_0 = 1$ if 0 is sent, and mean $m_1 = 10$ if 1 is sent.

(a) Show that the ML rule consists of comparing Z with a threshold, and specify the numerical value of the threshold. Note that Z can take only nonnegative integer values.
(b) Compute the conditional error probabilities for the ML rule (compute numerical values in addition to deriving formulas).
(c) Find the MPE rule if the prior probability of sending 1 is 0.1.
(d) Compute the average error probability for the MPE rule.

Problem 6.11 The received sample Y in a binary communication system is modeled as follows: $Y = A + N$ if 0 is sent, and $Y = -A + N$ if 1 is sent, where N is *Laplacian* noise with density

$$p_N(x) = \frac{\lambda}{2}e^{-\lambda|x|}, \quad -\infty < x < \infty$$

(a) Find the ML decision rule. Simplify as much as possible.
(b) Find the conditional error probabilities for the ML rule.
(c) Now, suppose that the prior probability of sending 0 is 1/3. Find the MPE rule, simplifying as much as possible.
(d) In the setting of (c), find the LLR $\log(P[0|Y = A/2]/P[1|Y = A/2])$.

Problem 6.12 Consider binary hypothesis testing with scalar observation Y. Under hypothesis H_0, Y is modeled as an exponential random variable with mean 5. Under hypothesis H_1, Y is modeled as uniformly distributed over the interval $[0, 10]$.

(a) Specify the ML rule and clearly draw the decision regions Γ_0 and Γ_1 on the real line.
(b) Find the conditional probability of error for the ML rule, given that H_0 is true.
(c) Suppose that the prior probability of H_0 is 1/3. Compute the posterior probability of H_0 given that we observe $Y = 4$ (i.e., find $P[H_0|Y = 4]$).

Problem 6.13 Consider hypothesis testing in which the observation Y is given by the following model:

$$H_1 : Y = 6 + N$$
$$H_0 : Y = N$$

where the noise N has density

$$p_N(x) = \frac{1}{10}\left(1 - \frac{|x|}{10}\right) I_{[-10,10]}(x)$$

(a) Find the conditional error probability given H_1 for the following decision rule:

$$Y \underset{H_0}{\overset{H_1}{\gtrless}} 4$$

(b) Is there a set of prior probabilities for which the decision rule in (a) minimizes the error probability? If so, specify them. If not, say why not.

Receiver design and performance analysis for the AWGN channel

Problem 6.14 Consider binary signaling in AWGN, with $s_1(t) = (1 - |t|)I_{[-1,1]}(t)$ and $s_0(t) = -s_1(t)$. The received signal is given by $y(t) = s_i(t) + n(t)$, $i = 0, 1$, where the noise n has PSD $\sigma^2 = N_0/2 = 0.1$. For all of the error probabilities computed in this problem, specify your answers in terms of the Q function with positive arguments and also give numerical values.

(a) How would you implement the ML receiver using the received signal $y(t)$? What is its conditional error probability given that s_0 is sent?

Now, consider a suboptimal receiver, where the receiver generates the following decision statistics:

$$y_0 = \int_{-1}^{-0.5} y(t)dt, \qquad y_1 = \int_{-0.5}^{0} y(t)dt, \qquad y_2 = \int_{0}^{0.5} y(t)dt, \qquad y_0 = \int_{0.5}^{1} y(t)dt$$

(b) Specify the conditional distribution of $\mathbf{y} = (y_0, y_1, y_2, y_3)^{\mathrm{T}}$, conditioned on s_0 being sent.

(c) Specify the ML rule when the observation is $\mathbf{y}$. What is its conditional error probability given that s_0 is sent?

(d) Specify the ML rule when the observation is $y_0 + y_1 + y_2 + y_3$. What is its conditional error probability, given that s_0 is sent?

(e) Among the error probabilities in (a), (c), and (d), which is the smallest? Which is the biggest? Could you have rank ordered these error probabilities without actually computing them?

Problem 6.15 The received signal in an on–off-keyed digital communication system is given by

$$y(t) = \begin{cases} s(t) + n(t) & 1 \text{ sent} \\ n(t) & 0 \text{ sent} \end{cases}$$

where n is AWGN with PSD $\sigma^2 = N_0/2$, and $s(t) = A(1 - |t|)I_{[-1,1]}(t)$, where $A > 0$. The received signal is passed through a filter with impulse response $h(t) = I_{[0,1]}(t)$ to obtain $z(t) = (y * h)(t)$.

Remark *it would be helpful to draw a picture of the system before you start doing the calculations.*

(a) Consider the decision statistic $Z = z(0) + z(1)$. Specify the conditional distribution of Z given that 0 is sent, and the conditional distribution of Z given that 1 is sent.

(b) Assuming that the receiver must make its decision based on Z, specify the ML rule and its error probability in terms of E_b/N_0 (express your answer in terms of the Q function with positive arguments).

(c) Find the error probability (in terms of E_b/N_0) for ML decisions based on the decision statistic $Z_2 = z(0) + z(0.5) + z(1)$.

Problem 6.16 Consider binary signaling in AWGN using the signals depicted in Figure 6.31. The received signal is given by

$$y(t) = \begin{cases} s_1(t) + n(t), & 1 \text{ sent} \\ s_0(t) + n(t), & 0 \text{ sent} \end{cases}$$

where $n(t)$ is WGN with PSD $\sigma^2 = N_0/2$.

(a) Show that the ML decision rule can be implemented by comparing $Z = \int y(t)a(t)dt$ with a threshold γ. Sketch $a(t)$ and specify the corresponding value of γ.

(b) Specify the error probability of the ML rule as a function of E_b/N_0.

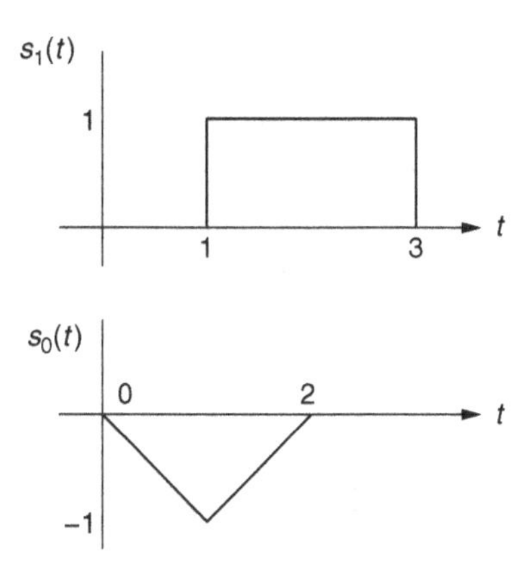

Figure 6.31 The signal set for Problem 6.16.

(c) Can the MPE rule, assuming that the prior probability of sending 0 is $1/3$, be implemented using the same receiver structure as in (a)? What would need to change? (Be specific.)

(d) Consider now a suboptimal receiver structure in which $y(t)$ is passed through a filter with impulse response $h(t) = I_{[0,1]}(t)$, and we take three samples: $Z_1 = (y * h)(1)$, $Z_2 = (y * h)(2)$, and $Z_3 = (y * h)(3)$. Specify the conditional distribution of $\mathbf{Z} = (Z_1, Z_2, Z_3)^{\mathrm{T}}$ given that 0 is sent.

(e) *(This is more challenging.)* Specify the ML rule based on $\mathbf{Z}$ and the corresponding error probability as a function of E_{b}/N_0.

Problem 6.17 Let $p_1(t) = I_{[0,1]}(t)$ denote a rectangular pulse of unit duration. Consider two 4-ary signal sets as follows:

signal set A: $s_i(t) = p_1(t-i)$, $i = 0, 1, 2, 3$; and
signal set B: $s_0(t) = p_1(t) + p_1(t-3)$, $s_1(t) = p_1(t-1) + p_1(t-2)$,
$s_2(t) = p_1(t) + p_1(t-2)$, and $s_3(t) = p_1(t-1) + p_1(t-3)$.

(a) Find signal-space representations for each signal set with respect to the orthonormal basis $\{p_1(t-i), i = 0, 1, 2, 3\}$.

(b) Find union bounds on the average error probabilities for both signal sets as a function of E_{b}/N_0. At high SNR, what is the penalty in dB for using signal set B?

(c) Find an exact expression for the average error probability for signal set B as a function of E_{b}/N_0.

Problem 6.18 Consider the 4-ary signaling set shown in Figure 6.32, to be used over an AWGN channel.

(a) Find a union bound, as a function of E_{b}/N_0, on the conditional probability of error given that $c(t)$ is sent.

(b) **(True or false?)** This constellation is more power efficient than QPSK. *Justify your answer.*

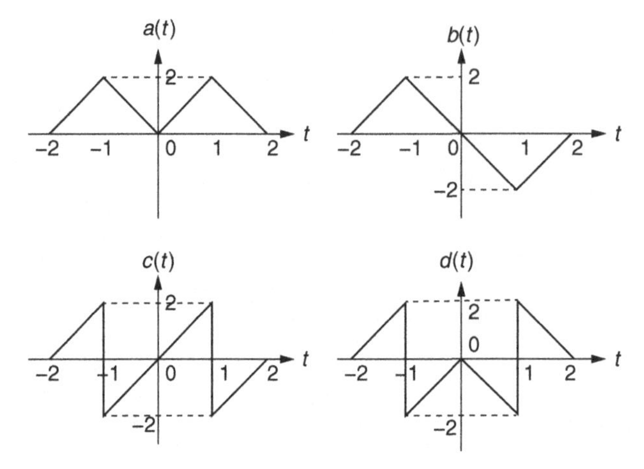

Figure 6.32 The signal set for Problem 6.18.

Problem 6.19 Three 8-ary signal constellations are shown in Figure 6.33.

(a) Express R and $d_{\min}^{(2)}$ in terms of $d_{\min}^{(1)}$ so that all three constellations have the same E_b.
(b) For a given E_b/N_0, which constellation do you expect to have the smallest bit error probability over a high-SNR AWGN channel?
(c) For each constellation, determine whether you can label signal points using three bits so that the label for nearest neighbors differs by at most one bit. If so, find such a labeling. If not, say why not and find some "good" labeling.
(d) For the labelings found in part (c), compute nearest-neighbors approximations for the average bit error probability as a function of E_b/N_0 for each constellation. Evaluate these approximations for $E_b/N_0 = 15$ dB.

Problem 6.20 Consider the signal constellation shown in Figure 6.34, which consists of two QPSK constellations of different radii, offset from each other by $\pi/4$. The constellation is to be used to communicate over a passband AWGN channel.

(a) Carefully redraw the constellation (roughly to scale, to the extent possible) for $r = 1$ and $R = \sqrt{2}$. Sketch the ML decision regions.

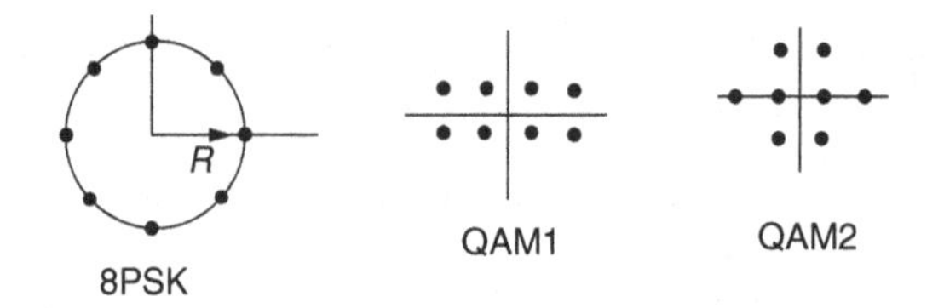

Figure 6.33 Signal constellations for Problem 6.19.

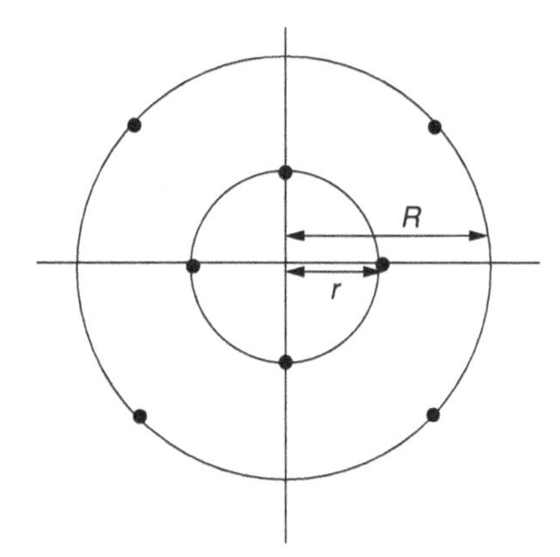

Figure 6.34 The constellation for Problem 6.20.

(b) For $r = 1$ and $R = \sqrt{2}$, find an intelligent union bound for the conditional error probability, given that a signal point from the inner circle is sent, as a function of E_b/N_0.
(c) How would you choose the parameters r and R so as to optimize the power efficiency of the constellation (at high SNR)?

Problem 6.21 (Exact symbol error probabilities for rectangular constellations) Assuming that each symbol is equally likely, derive the following expressions for the average error probability for 4PAM and 16QAM:

$$P_e = \frac{3}{2} Q\left(\sqrt{\frac{4E_b}{5N_0}}\right), \quad \textbf{symbol error probability for 4PAM} \tag{6.90}$$

$$P_e = 3Q\left(\sqrt{\frac{4E_b}{5N_0}}\right) - \frac{9}{4} Q^2\left(\sqrt{\frac{4E_b}{5N_0}}\right), \quad \textbf{symbol error probability for 16QAM} \tag{6.91}$$

(Assume 4PAM with equally spaced levels symmetric about the origin, and rectangular 16QAM equivalent to two 4PAM constellations independently modulating the I and Q components.)

Problem 6.22 The signal constellation shown in Figure 6.35 is obtained by moving the outer corner points in rectangular 16QAM to the I and Q axes.

(a) Sketch the ML decision regions.
(b) Is the constellation more or less power-efficient than rectangular 16QAM?

Problem 6.23 Consider a 16-ary signal constellation with four signals with coordinates $(\pm 1, \pm 1)$, four others with coordinates $(\pm 3, \pm 3)$, and two each having coordinates $(\pm 3, 0)$, $(\pm 5, 0)$, $(0, \pm 3)$, and $(0, \pm 5)$, respectively.

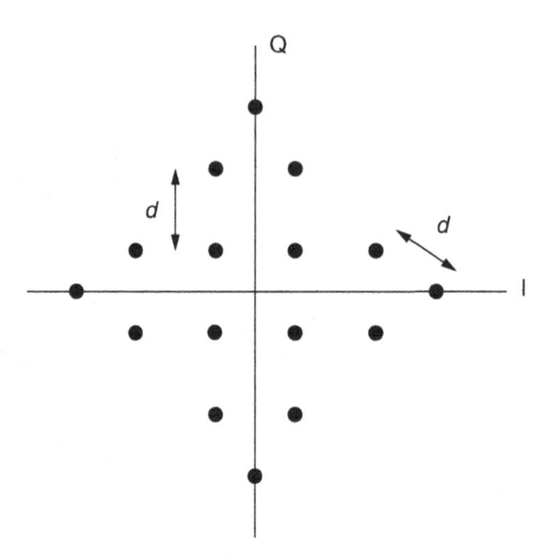

Figure 6.35 The constellation for Problem 6.22.

(a) Sketch the signal constellation and indicate the ML decision regions.
(b) Find an intelligent union bound on the average symbol error probability as a function of E_b/N_0.
(c) Find the nearest-neighbors approximation to the average symbol error probability as a function of E_b/N_0.
(d) Find the nearest-neighbors approximation to the average symbol error probability for 16QAM as a function of E_b/N_0.
(e) On comparing (c) and (d) (i.e., comparing the performance at high SNR), which signal set is more power efficient?

Problem 6.24 A QPSK demodulator is designed to put out an *erasure* when the decision is ambivalent. Thus, the decision regions are modified as shown in Figure 6.36, where the cross-hatched region corresponds to an erasure. Set $\alpha = d_1/d$, where $0 \leq \alpha \leq 1$.

(a) Use the intelligent union bound to find approximations to the probability p of symbol error and the probability q of symbol erasure in terms of E_b/N_0 and α.
(b) Find exact expressions for p and q as functions of E_b/N_0 and α.
(c) Using the approximations in (a), find an approximate value for α such that $q = 2p$ for $E_b/N_0 = 4$ dB.

Remark The motivation for (c) is that a typical error-correcting code can correct twice as many erasures as errors.

Problem 6.25 The constellation shown in Figure 6.37 consists of two QPSK constellations lying on concentric circles, with an inner circle of radius r and an outer circle of radius R.

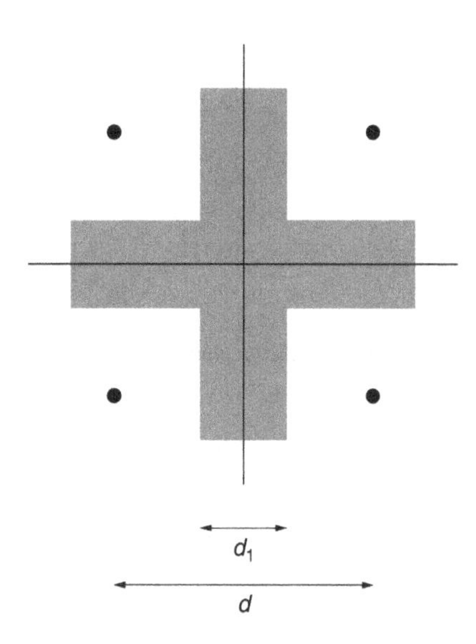

Figure 6.36 QPSK with erasures.

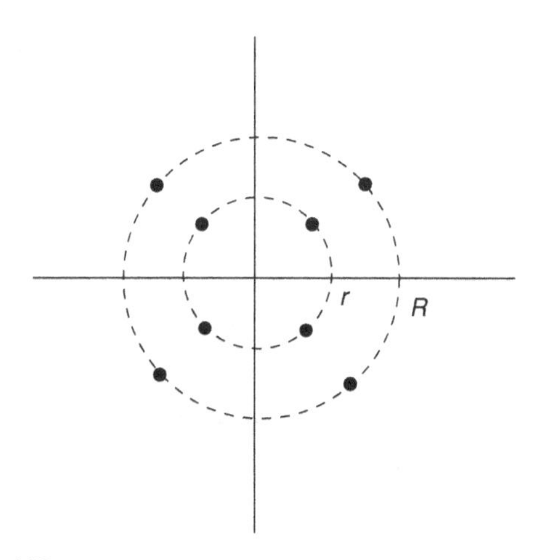

Figure 6.37 The constellation for Problem 6.25.

(a) For $r = 1$ and $R = 2$, redraw the constellation, and carefully sketch the ML decision regions.

(b) Still keeping $r = 1$ and $R = 2$, find an intelligent union bound for the symbol error probability as a function of E_b/N_0.

(c) For $r = 1$, find the best choice of R in terms of high-SNR performance. Compute the gain in power efficiency (in dB), if any, over the setting in (a) and (b).

Problem 6.26 Consider the constant-modulus constellation shown in Figure 6.38. where $\theta \leq \pi/4$. Each symbol is labeled by two bits (b_1, b_2) as shown. Assume that the constellation is used over a complex-baseband AWGN channel with noise power spectral

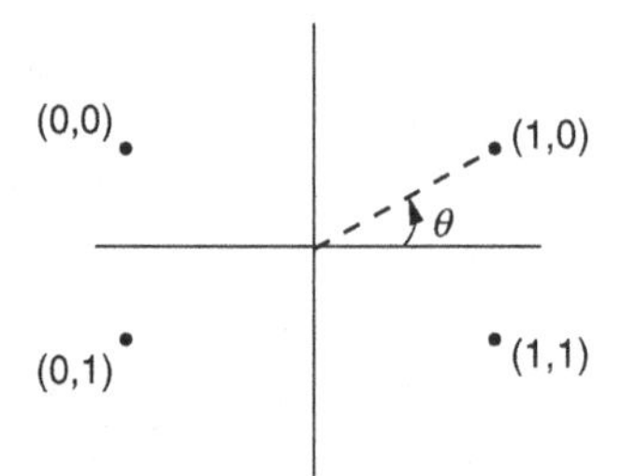

Figure 6.38 A signal constellation with unequal error protection (Problem 6.26).

density (PSD) $N_0/2$ in each dimension. Let $(\hat{b_1}, \hat{b_2})$ denote the maximum likelihood (ML) estimates of (b_1, b_2).

(a) Find $P_{e1} = P[\hat{b_1} \neq b_1]$ and $P_{e2} = P[\hat{b_2} \neq b_2]$ as functions of E_s/N_0, where E_s denotes the signal energy.
(b) Assume now that the transmitter is being heard by two receivers, R_1 and R_2, and that R_2 is twice as far away from the transmitter as R_1. Assume that the received signal energy falls off as $1/r^4$, where r is the distance from the transmitter, and that the noise PSD is identical for these two receivers. Suppose that R_1 can demodulate both bits b_1 and b_2 with error probability at least as good as 10^{-3}, i.e., so that $\max\{P_{e1}(R_1), P_{e2}(R_1)\} = 10^{-3}$. Design the signal constellation (i.e., specify θ) so that R_2 can demodulate at least one of the bits with the same error probability, i.e., such that $\min\{P_{e1}(R_2), P_{e2}(R_2)\} = 10^{-3}$.

Remark You have designed an unequal error protection scheme in which the receiver that sees a poorer channel can still extract part of the information sent.

Problem 6.27 The two-dimensional constellation shown in Figure 6.39 is to be used for signaling over an AWGN channel.

(a) Specify the ML decision if the observation is $(I, Q) = (1, -1)$.
(b) Carefully redraw the constellation and sketch the ML decision regions.
(c) Find an intelligent union bound for the symbol error probability conditioned on s_0 being sent, as a function of E_b/N_0.

Problem 6.28 (Demodulation with amplitude mismatch) Consider a 4PAM system using the constellation points $\{\pm 1, \pm 3\}$. The receiver has an accurate estimate of its noise level. An automatic gain control (AGC) circuit is supposed to scale the decision statistics so that the noiseless constellation points are in $\{\pm 1, \pm 3\}$. ML decision boundaries are set according to this nominal scaling.

(a) Suppose that the AGC scaling is faulty, and the *actual* noiseless signal points are at $\{\pm 0.9, \pm 2.7\}$. Sketch the points and the mismatched decision regions. Find an intelligent union bound for the symbol error probability in terms of the Q function and E_b/N_0.

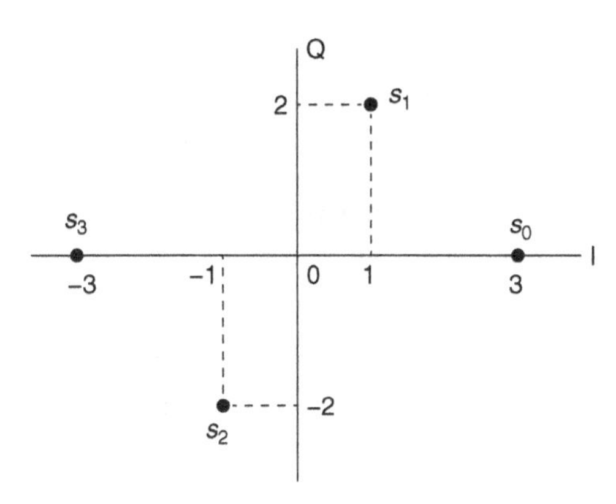

Figure 6.39 The constellation for Problem 6.27.

(b) Repeat (a), assuming that faulty AGC scaling puts the noiseless signal points at $\{\pm 1.1, \pm 3.3\}$.
(c) AGC circuits try to maintain a constant output power as the input power varies, and can be viewed as imposing on the input a scale factor inversely proportional to the square root of the input power. In (a), does the AGC circuit overestimate or underestimate the input power?

Problem 6.29 (Demodulation with phase mismatch) Consider a BPSK system in which the receiver's estimate of the carrier phase is off by θ.

(a) Sketch the I and Q components of the decision statistic, showing the noiseless signal points and the decision region.
(b) Derive the BER as a function of θ and E_b/N_0 (assume that $\theta < \pi/4$).
(c) Assuming now that θ is a random variable taking values uniformly in $[-\pi/4, \pi/4]$, numerically compute the BER averaged over θ, and plot it as a function of E_b/N_0. Plot the BER without phase mismatch as well, and estimate the dB degradation due to the phase mismatch.

Problem 6.30 (Simplex signaling set) Let $s_0(t), \ldots, s_{M-1}(t)$ denote a set of equal energy, orthogonal signals. Construct a new M-ary signal set from these as follows, by subtracting out the average of the M signals from each signal as follows:

$$u_k(t) = s_k(t) - \frac{1}{M} \sum_{j=0}^{M-1} s_j(t), \ k = 0, 1, \ldots, M-1$$

This is called the *simplex* signaling set.

(a) Find a union bound on the symbol error probability, as a function of E_b/N_0 and M, for signaling over the AWGN channel using the signal set $\{u_k(t), k = 0, 1, \ldots, M-1\}$.
(b) Compare the power efficiencies of the simplex and orthogonal signaling sets for a given M, and use these to estimate the performance difference in dB between these two signaling schemes, for $M = 4, 8, 16$, and 32. What happens as M gets large?

(c) Use computer simulations to plot, for $M = 4$, the error probability (on a log scale) versus E_b/N_0 (dB) of the simplex signaling set and the corresponding orthogonal signaling set. Are your results consistent with the prediction from (b)?

Problem 6.31 (Soft decisions for BPSK) Consider a BPSK system in which 0 and 1 are equally likely to be sent, with 0 mapped to +1 and 1 to -1 as usual. Thus, the decision statistic $Y = A + N$ if 0 is sent, and $Y = -A + N$ if 1 is sent, where $A > 0$ and $N \sim N(0, \sigma^2)$.

(a) Show that the LLR is conditionally Gaussian given the transmitted bit, and that the conditional distribution is scale-invariant, depending only on E_b/N_0.
(b) If the BER for hard decisions is 10%, specify the conditional distribution of the LLR, given that 0 is sent.

Problem 6.32 (Soft decisions for PAM) Consider soft decisions for 4PAM signaling as in Example 6.1.3. Assume that the signals have been scaled to $\pm 1, \pm 3$ (i.e., set $A = 1$ in Example 6.1.3). The system is operating at $E_b/N_0 = 6$ dB. Bits $b_1, b_2 \in \{0, 1\}$ are mapped to the symbols using Gray coding. Assume that $(b_1, b_2) = (0, 0)$ for symbol -3, and $(1, 0)$ for symbol +3.

(a) Sketch the constellation, along with the bit maps. Indicate the ML hard-decision boundaries.
(b) Find the posterior symbol probability $P[-3|y]$ as a function of the noisy observation y. Plot it as a function of y.
Hint. The noise variance σ^2 can be inferred from the signal levels and SNR.
(c) Find $P[b_1 = 1|y]$ and $P[b_2 = 1|y]$, and plot them as functions of y.

Remark The posterior probability of $b_1 = 1$ equals the sum of the posterior probabilities of all symbols that have $b_1 = 1$ in their labels.

(d) Display the results of part (c) in terms of LLRs.

$$\text{LLR}_1(y) = \log\left(\frac{P[b_1 = 0|y]}{P[b_1 = 1|y]}\right), \quad \text{LLR}_2(y) = \log\left(\frac{P[b_2 = 0|y]}{P[b_2 = 1|y]}\right)$$

Plot the LLRs as a function of y, saturating the values as ± 50.
(e) Try other values of E_b/N_0 (e.g., 0 dB and 10 dB). Comment on any trends you notice. How do the LLRs vary as a function of distance from the noiseless signal points? How do they vary as you change E_b/N_0?
(f) In order to characterize the conditional distribution of the LLRs, simulate the system over multiple symbols at E_b/N_0 such that the BER is about 5%. Plot the histograms of the LLRs for each of the two bits, and comment on whether they look Gaussian. What happens as you increase or decrease E_b/N_0?

Problem 6.33 (M-ary orthogonal signaling performance as $M \to \infty$) We wish to derive the result that

$$\lim_{M\to\infty} P(\text{correct}) = \begin{cases} 1, & E_b/N_0 > \ln 2 \\ 0, & E_b/N_0 < \ln 2 \end{cases} \tag{6.92}$$

(a) Show that

$$P(\text{correct}) = \int_{-\infty}^{\infty} \left[\Phi\left(x + \sqrt{\frac{2E_b \log_2 M}{N_0}} \right) \right]^{M-1} \frac{1}{\sqrt{2\pi}} e^{-x^2/2}\, dx$$

(b) Show that, for any x,

$$\lim_{M\to\infty} \left[\Phi\left(x + \sqrt{\frac{2E_b \log_2 M}{N_0}} \right) \right]^{M-1} = \begin{cases} 0, & E_b/N_0 < \ln 2 \\ 1, & E_b/N_0 > \ln 2 \end{cases}$$

Hint. Use l'Hôpital's rule on the log of the expression whose limit is to be evaluated.

(c) Substitute (b) into the integral in (a) to infer the desired result.

Problem 6.34 (Effect of Rayleigh fading) Constructive and destructive interference between multiple paths in wireless systems lead to large fluctuations in received amplitude, modeled as a Rayleigh random variable A (see Problem 5.21 for a definition). The energy per bit is therefore proportional to A^2, which, using Problem 5.21(c), is an exponential random variable. Thus, we can model E_b/N_0 as an exponential random variable with mean $\bar{E}_b/N_0$, where $\bar{E}_b$ is the *average* energy per bit. Simplify the notation by setting $E_b/N_0 = X$, and the mean $\bar{E}_b/N_0 = 1/\mu$, so that $X \sim \text{Exp}(\mu)$.

(a) Show that the average error probability for BPSK with Rayleigh fading can be written as

$$P_e = \int_0^{\infty} Q(\sqrt{2x})\mu e^{-\mu x}\, dx$$

Hint. The error probability for BPSK is given by $Q(\sqrt{2E_b/N_0})$, where E_b/N_0 is a random variable. We now find the expected error probability by averaging over the distribution of E_b/N_0.

(b) By integrating by parts and simplifying, show that the average error probability can be written as

$$P_e = \frac{1}{2}\left(1 - (1+\mu)^{-1/2}\right) = \frac{1}{2}\left(1 - \left(1 + \frac{N_0}{\bar{E}_b}\right)^{-1/2}\right)$$

Hint. $Q(x)$ is defined via an integral, so we can find its derivative (when integrating by parts) using the fundamental theorem of calculus.

(c) Using the approximation that $(1+a)^b \approx 1 + ba$ for $|a|$ small, show that

$$P_e \approx \frac{1}{4(\bar{E}_b/N_0)}$$

at high SNR. Comment on how this decay of error probability with the reciprocal of the SNR compares with the decay for the AWGN channel.

(d) Plot the error probability versus $\bar{E}_b/N_0$ for BPSK over the AWGN and Rayleigh fading channels (BER on a log scale, $\bar{E}_b/N_0$ in dB). Note that $\bar{E}_b = E_b$ for the AWGN channel. At a BER of 10^{-3}, what is the degradation in dB due to Rayleigh fading?

Link-budget analysis

Problem 6.35 You are given an AWGN channel of bandwidth 3 MHz. Assume that implementation constraints dictate an excess bandwidth of 50%. Find the achievable bit rate, the E_b/N_0 required for a BER of 10^{-8}, and the receiver sensitivity (assuming a receiver noise figure of 7 dB) for the following modulation schemes, assuming that the bit-to-symbol map is optimized to minimize the BER whenever possible: (a) QPSK, (b) 8PSK, (c) 64QAM, and (d) coherent 16-ary orthogonal signaling.

Remark Use nearest-neighbors approximations for the BER.

Problem 6.36 Consider the setting of Example 6.5.1.

(a) For all parameters remaining the same, find the range and bit rate when using a 64QAM constellation.

(b) Suppose now that the channel model is changed from AWGN to Rayleigh fading (see Problem 6.34). Find the receiver sensitivity required for QPSK at a BER of 10^{-5}. (In practice, we would shoot for a higher uncoded BER, and apply channel coding.) What is the range, assuming that all other parameters are as in Example 6.5.1? How does the range change if you reduce the link margin to 10 dB (now that fading is being accounted for, there are fewer remaining uncertainties).

Problem 6.37 Consider a line-of-sight communication link operating in the 60-GHz band (where large amounts of unlicensed bandwidth have been set aside by regulators). From version 1 of the Friis formula (6.83), we see that the received power scales as λ^2, and hence as the inverse square of the carrier frequency, so that 60-GHz links have much worse propagation than, say, 5 GHz links when antenna gains are fixed. However, from (6.82), we see that the we can get much better antenna gains at small carrier wavelengths for a fixed form factor, and version 2 of the Friis formula (6.84) shows that the received power scales as $1/\lambda^2$, which improves with increasing carrier frequency. Furthermore, *electronically steerable* antenna arrays with high gains can be implemented with a compact form factor (e.g., patterns of metal on a circuit board) at higher carrier frequencies such as 60 GHz. Suppose, now, that we wish to design a 2-Gbps link using QPSK with an excess bandwidth of 50%, operating at a BER of 10^{-6}. The receiver noise figure is 8 dB, and the desired link margin is 10 dB.

(a) What is the transmit power in dBm required to attain a range of 10 meters (e.g., for in-room communication), assuming that the transmit and receive antenna gains are each 10 dBi?

(b) For a transmit power of 20 dBm, what are the antenna gains required at the transmitter and receiver (assume that the gains at both ends are equal) to attain a range of 200 meters (e.g., for an outdoor last-hop link)?

(c) For the antenna gains found in (b), what happens to the attainable range if you account for additional path loss due to oxygen absorption (typical in the 60-GHz band) of 16 dB/km?

(d) In (c), what happens to the attainable range if there is a further path loss of 30 dB/km due to heavy rain (on top of the loss due to oxygen absorption)?

Problem 6.38 A 10-Mbps line-of-sight communication link operating at a carrier frequency of 1 GHz has a designed range of 5 km. The link employs 16QAM with an excess bandwidth of 25%, with a designed BER of 10^{-6} and a link margin of 10 dB. The receiver noise figure is 4 dB, and the transmit and receive antenna gains are 10 dBi each. *This is the baseline scenario against which each of the scenarios in (a)–(c) is to be compared.*

(a) Suppose that you change the carrier frequency to 5 GHz, keeping all other link parameters the same. What is the new range?

(b) Suppose that you change the carrier frequency to 5 GHz and increase the transmit and receive antenna gains by 3 dBi each, keeping all other link parameters the same. What is the new range?

(c) Suppose you change the carrier frequency to 5 GHz, increase the transmit and receive antenna directivities by 3 dBi each, and increase the data rate to 40 Mbps, still using 16QAM with excess bandwidth of 25%. All other link parameters are the same. What is the new range?

Software Lab 6.1: linear modulation with two-dimensional constellations

Lab objectives This is a follow-on to Software Lab 4.1, the code from which is our starting point here. The objective is to implement in complex baseband a linearly modulated system for a variety of signal constellations. We wish to estimate the performance of these schemes for an ideal channel via simulation, and to compare our estimate with analytical expressions. *As in Software Lab 4.1, we use a trivial channel filter in this lab.* The model is extended to dispersive channels in Software Lab 8.1 in Chapter 8.

Reading Basic modulation formats from Chapter 4, and error probability expressions from Chapter 6.

Laboratory assignment

(1.1) Use the code for Software Lab 4.1 as a starting point.

(1.2) Write a MATLAB function randbit that generates random bits taking values in $\{0, 1\}$ (not ± 1) with equal probability.

(1.3) Write the following functions mapping bits to symbols for different signal constellations. **Write the functions to allow for vector inputs and outputs.** The mapping is said to be a Gray code, or Gray labeling, if the bit maps for nearest neighbors in the constellation differ by exactly one bit. In all of the following, **choose the bit map to be a Gray code.**

(a) bpskmap: input a $0/1$ bit, output a ± 1 bit.

(b) qpskmap: input two $0/1$ bits, output a symbol taking one of four values in $\pm 1 \pm j$.

(c) fourpammap: input two $0/1$ bits, output a symbol taking one of four values in $\{\pm 1, \pm 3\}$.

(d) sixteenqammap: input four $0/1$ bits, output a symbol taking one of 16 values in $\{b_c + jb_s : b_c, b_s \in \{\pm 1, \pm 3\}\}$.

(e) eightpskmap: input three $0/1$ bits, output a symbol taking one of eight values in $e^{j2\pi i/8}, i = 0, 1, \ldots, 7$.

(1.4) *BPSK symbol generation.* Use (1.2) above to generate 12,000 $0/1$ bits. Map these to BPSK (± 1) bits using bpskmap. Pass these through the transmit and receive filter in Software Lab 4.1 to get noiseless received samples at rate $4/T$, as before.

(1.5) *Adding noise.* We consider discrete-time additive white Gaussian noise (AWGN). At the input to the receive filter, add independent and identically distributed (i.i.d.) complex Gaussian noise, such that the real and imaginary part of each sample are i.i.d. $N(0, \sigma^2)$ (you will choose $\sigma^2 = N_0/2$ corresponding to a specified value of E_b/N_0, as described in (1.6) below). Pass these (rate $4/T$) noise samples through the receive filter, and add the result to the output of (1.4).

Remark If the nth transmitted symbol is $b[n]$, the average received energy per symbol is $E_s = E[|b[n]|^2]||g_T * g_C||^2$. Divide that by the number of bits per symbol to get E_b. The noise variance per dimension is $\sigma^2 = N_0/2$. This enables you to compute E_b/N_0 for your simulation model. The signal-to-noise ratio E_b/N_0 is usually expressed in decibels (dB): E_b/N_0 (dB) $= 10 \log_{10} E_b/N_0$ (raw). Thus, if you fix the transmit and channel filter coefficients, then you can simulate any given value of E_b/N_0 in dB by varying the value of the noise variance σ^2.

(1.6) Plot the ideal bit error probability for BPSK, which is given by $Q(\sqrt{2E_b/N_0})$, on a log scale as a function of E_b/N_0 in dB over the range 0–10 dB. Find the value of E_b/N_0 which corresponds to an error probability of 10^{-2}.

(1.7) For the value of E_b/N_0 found in (1.6), choose the corresponding value of σ^2 in (1.2). Find the decision statistics corresponding to the transmitted symbols at the input and output of the receive filter, as in Software Lab 4.1 (parts (1.6) and (1.7)). Plot the imaginary versus the real parts of the decision statistics; you should see a noisy version of the constellation.

(1.8) Using an appropriate decision rule, make decisions on the 12,000 transmitted bits based on the 12,000 decision statistics, and measure the error probability obtained at the input and the output. Compare the results with the ideal error probability from (1.6). You should find that the error probability based on the receiver input samples is significantly worse than that based on the receiver output, and that the latter is a little worse than the ideal performance because of the ISI in the decision statistics.

(1.9) Now, map 12,000 0/1 bits into 6000 4PAM symbols using the function fourpammap (use as input two parallel vectors of 6000 bits). As shown in Chapter 6, a good approximation (the nearest neighbors approximation) to the ideal *bit* error probability for Gray-coded 4PAM is given by $Q(4E_b/(5N_0))$. As in (1.6), plot this on a log scale as a function of E_b/N_0 in dB over the range 0–10 dB. What is the value of E_b/N_0 (dB) corresponding to a bit error probability of 10^{-2}?

(1.10) Choose the value of the noise variance σ^2 corresponding to the E_b/N_0 found in (1.8). Now, find decision statistics for the 6000 transmitted symbols *based on the receive filter output only.*

(a) Plot the imaginary versus the real parts of the decision statistics, as before.
(b) Determine an appropriate decision rule for estimating the two parallel bit streams of 6000 bits from the 6000 complex decision statistics.
(c) Measure the *bit* error probability, and compare it with the ideal bit error probability.

(1.11) Repeat (1.9) and (1.10) for QPSK, the ideal bit error probability for which, as a function of E_b/N_0, is the same as for BPSK.

(1.12) Repeat (1.9) and (1.10) for 16QAM (four bit streams of length 3000 each), the ideal bit error probability for which, as a function of E_b/N_0, is the same as for 4PAM.

(1.13) Repeat (1.9) and (1.10) for 8PSK (three bit streams of length 4000 each). The ideal bit error probability for Gray-coded 8PSK is approximated by (using the nearest-neighbors approximation) $Q((6 - 3\sqrt{2})E_b/(2N_0))$.

(1.14) Since all your answers above will be off from the ideal answers because of some ISI, run a simulation with 12,000 bits sent using Gray-coded 16QAM with no ISI. To do this, generate the decision statistics by adding noise directly to the transmitted symbols, setting the noise variance appropriately to operate at the required E_b/N_0. Do this for two different values of E_b/N_0, the one in (1.12) and a value 3 dB higher. In each case, compare the nearest-neighbors approximation with the measured bit error probability, and plot the imaginary part versus the real part of the decision statistics.

Lab report Your lab report should document the results of the preceding steps in order. Describe the reasoning you used and the difficulties you encountered.

Tips. Vectorize as many of the functions as possible, including both the bit-to-symbol maps and the decision rules. Do BPSK and 4PAM first, where you will only use the real part of the complex decision statistics. Leverage this for QPSK and 16QAM, by replicating what you did for the imaginary part of the decision statistics as well. To avoid confusion, keep different MATLAB files for simulations regarding different signal constellations, and keep the analytical computations and plots separate from the simulations.

Software Lab 6.2: modeling and performance evaluation on a wireless fading channel

Lab objectives Introduction to statistical modeling and performance evaluation for signaling on wireless fading channels.

Laboratory assignment

Let us consider the following simple model of a wireless channel (obtained after filtering and sampling at the symbol rate, and assuming that there is no ISI). If $\{b[n]\}$ is the transmitted symbol sequence, then the complex-valued received sequence is given by

$$y[n] = h[n]b[n] + w[n] \tag{6.93}$$

where $\{w[n] = w_c[n] + jw_s[n]\}$ is an i.i.d. complex Gaussian noise sequence with $w_c[n]$ and $w_s[n]$ i.i.d. $N(0, \sigma^2 = N_0/2)$ random variables. We say that $w[n]$ has variance σ^2 per dimension. The channel sequence $\{h[n]\}$ is a time-varying sequence of complex gains.

Equation (6.93) models the channel at a given time as a simple scalar gain $h[n]$. On the other hand, as discussed in Example 2.5.6, a multipath wireless channel cannot be modeled as a simple scalar gain: it is dispersive in time, and exhibits frequency selectivity. However, it is shown in Chapter 8 that we can decompose complicated dispersive channels into scalar models by using frequency-domain modulation, or OFDM, which transmits data in parallel over narrow enough frequency slices that the channel over each slice can be modeled as a complex scalar. Equation (6.93) could therefore be interpreted as modeling time variations in such scalar gains.

Rayleigh fading The channel gain sequence is $\{h[n] = h_c[n] + jh_s[n]\}$, where $\{h_c[n]\}$ and $\{h_s[n]\}$ are zero-mean, independent, and identically distributed colored Gaussian random processes. The reason why this is called Rayleigh fading is that $|h[n]| = \sqrt{h_c^2[n] + h_s^2[n]}$ is a Rayleigh random variable.

Remark The Gaussianity arises because the overall channel gain results from a superposition of gains from multiple reflections off scatterers.

Simulation of Rayleigh fading We will use a simple model wherein the colored channel gain sequence $\{h[n]\}$ is obtained by passing white Gaussian noise through a first-order recursive filter, as follows:

$$\begin{aligned} h_c[n] &= \rho h_c[n-1] + u[n] \\ h_s[n] &= \rho h_s[n-1] + v[n] \end{aligned} \tag{6.94}$$

where $\{u[n]\}$ and $\{v[n]\}$ are independent real-valued white Gaussian sequences, with i.i.d. $N(0, \beta^2)$ elements. The parameter ρ $(0 < \rho < 1)$ determines how rapidly the channel varies. The models for I and Q gains in (6.94) are examples of *first-order autoregressive (AR(1))* random processes: they are autoregressive because future values depend on the past in a linear fashion, and of first order because only the immediately preceding value affects the current one.

Setting up the fading simulator

(2.1) Set up the AR(1) Rayleigh fading model in MATLAB, with ρ and β^2 as programmable parameters.

(2.2) Calculate $\mathbb{E}[|h[n]|^2] = 2\mathbb{E}\left[h_c^2[n]\right] = 2v^2$ analytically as a function of ρ and β^2. Use simulation to verify your results, setting $\rho = 0.99$ and $\beta = 0.01$. You may choose to initialize $h_c[0]$ and $h_s[0]$ as i.i.d. $N(0, v^2)$ in your simulation. Use at least 10,000 samples.

(2.3) Plot the instantaneous channel power relative to the average channel power, $|h[n]|^2/(2v^2)$ in dB as a function of n. Thus, 0 dB corresponds to the average value of $2v^2$. You will occasionally see sharp dips in the power, which are termed deep fades.

(2.4) Define the channel phase $\theta[n] = \text{angle}(h[n]) = \tan^{-1}(h_s[n]/h_c[n])$. Plot $\theta[n]$ versus n. Compare your result with the plot in (2.3); you should see sharp phase changes corresponding to deep fades.

QPSK in Rayleigh fading

Now, implement the model (6.93), where $\{b[n]\}$ correspond to Gray-coded QPSK, using an AR(1) simulation of Rayleigh fading as above. Assume that the receiver has perfect knowledge of the channel gains $\{h[n]\}$, and employs the decision statistic $Z[n] = h^*[n]y[n]$.

Remark In practice, the channel estimation required for implementing this is achieved by inserting pilot symbols periodically into the data stream. The performance will, of course, be worse than with the ideal channel estimates considered here.

(2.5) Make scatter plots of the two-dimensional received symbols $\{y[n]\}$ and of the decision statistics $\{Z[n]\}$. What does multiplying by $h^*[n]$ achieve?

(2.6) Implement a decision rule for the bits encoded in the QPSK symbols based on the statistics $\{Z[n]\}$. Estimate by simulation, and plot, the bit error probability (on a log scale) as a function of the average E_b/N_0 (dB), where E_b/N_0 ranges from 0 to 30 dB. Use at least 10,000 symbols for your estimate. On the same plot, also plot the analytical bit error probability as a function of E_b/N_0 when there is no fading. You should see a marked degradation due to fading. How do you think the error probability in fading varies with E_b/N_0?

Relating simulation parameters to E_b/N_0. The average symbol energy is $E_s = E[|b[n]|^2]E[|h[n]|^2]$, and $E_b = E_s/\log_2 M$. This is a function of the constellation scaling and the parameters β^2 and ρ in the fading simulator (see above). You can therefore fix E_s, and hence E_b, by fixing β and ρ (e.g., as in (2.2)), and fix the scaling of the $\{b[n]\}$ (e.g., keep the constellation points as $\pm 1 \pm j$). E_b/N_0 can now be varied by varying the variance σ^2 of the noise in (6.93).

Diversity

The severe degradation due to Rayleigh fading can be mitigated by using diversity: the probability that two paths are simultaneously in a deep fade is less likely than the probability that a single path is in a deep fade. Consider a receive antenna diversity system, where the received signals y_1 and y_2 at the two antennas are given by

$$\begin{aligned} y_1[n] &= h_1[n]b[n] + w_1[n] \\ y_2[n] &= h_2[n]b[n] + w_2[n] \end{aligned} \tag{6.95}$$

Thus, you get two looks at the data stream, through two different channels.

Implement the two-fold diversity system in (6.95) as you implemented (6.93), keeping the following in mind.

- The noises w_1 and w_2 are independent white-noise sequences with variance $\sigma^2 = N_0/2$ per dimension as before.
- The channels h_1 and h_2 are generated by passing independent white-noise streams through a first-order recursive filter. In relating the simulation parameters to E_b/N_0, keep in mind that the average symbol energy now is $E_s = E[|b[n]|^2]E[|h_1[n]|^2 + |h_2[n]|^2]$.
- Use the following *maximal ratio combining* rule to obtain the decision statistic:

$$Z_2[n] = h_1^*[n]y_1[n] + h_2^*[n]y_2[n]$$

The decision statistic above can be written as

$$Z_2[n] = (|h_1[n]|^2 + |h_2[n]|^2)b[n] + \tilde{w}[n]$$

where $\tilde{w}[n]$ is zero-mean complex Gaussian with variance $\sigma^2(|h_1[n]|^2 + |h_2[n]|^2)$ per dimension. Thus, the instantaneous SNR is given by

$$\text{SNR}[n] = \frac{\mathbb{E}\left[\left|(|h_1[n]|^2 + |h_2[n]|^2)b[n]\right|^2\right]}{\mathbb{E}[|\tilde{w}[n]|^2]} = \frac{|h_1[n]|^2 + |h_2[n]|^2\mathbb{E}[|b[n]|^2]}{2\sigma^2}$$

(2.7) Plot $|h_1[n]|^2 + |h_2[n]|^2$ in dB as a function of n, with 0 dB representing the average value as before. You should find that the fluctuations around the average are less than in (2.3).

(2.8) Implement a decision rule for the bits encoded in the QPSK symbols based on the statistics $\{Z_2[n]\}$. Estimate by simulation, and plot (on the same plot as in (2.5), the bit error probability (on a log scale) as a function of the average E_b/N_0 (dB), where E_b/N_0 ranges from 0 to 30 dB. Use at least 10,000 symbols for your estimate. You should see an improvement compared with the situation with no diversity.

Lab report Your lab report should document the results of the preceding steps in order. Describe the reasoning you used and the difficulties you encountered.

Bonus: a Glimpse of differential modulation and demodulation

Throughout this chapter, we have assumed that a noiseless "template" for the set of possible transmitted signals is available at the receiver. In the present context, this means assuming that estimates for the time-varying fading channel are available. But what if these estimates, which we used to generate the decision statistics earlier in this lab, are not available? One approach that avoids the need for explicit channel estimation is based on exploiting the fact that the channel does not change much from symbol to symbol. Let us illustrate this for the case of QPSK. The model is exactly as in (6.93) or (6.95), but the channel sequence(s) is(are) unknown *a priori*. This necessitates encoding the data in a different way. Specifically, let $d[n]$ be a Gray-coded QPSK *information* sequence, which contains information about the bits of interest. Instead of sending $d[n]$ directly, we generate the transmitted sequence $b[n]$ by *differential encoding* as follows:

$$b[n] = d[n]b[n-1], \quad n = 1, 2, 3, 4, \ldots$$

(You can initialize $b(0)$ as any element of the constellation, known by agreement to both transmitter and receiver. Or, just ignore the first information symbol in your demodulation.) At the receiver, use *differential demodulation* to generate the decision statistic for the information symbol $d[n]$ as follows:

$$Z^{\mathrm{nc}}[n] = y[n]y^*[n-1] \quad \text{single path}$$

$$Z_2^{\mathrm{nc}}[n] = y_1[n]y_1^*[n-1] + y_2[n]y_2^*[n-1] \quad \text{dual diversity}$$

where the superscript nc indicates *noncoherent* demodulation, i.e., demodulation that does not require an explicit channel estimate.

Lab report for bonus assignment. Estimate by simulation, and plot, the bit error probability of Gray-coded differentially encoded QPSK as a function of E_b/N_0 both for a single path and for dual diversity. Compare your results with the curves for coherent demodulation that you have obtained earlier. By how much (in dB) does the performance degrade? Document your results as in the earlier lab reports.

Appendix 6.A The irrelevance of the component orthogonal to the signal space

Conditioning on H_i, we have $y(t) = s_i(t) + n(t)$. The component of the received signal orthogonal to the signal space is given by

$$y^{\perp}(t) = y(t) - y_{\mathcal{S}}(t) = y(t) - \sum_{k=0}^{n-1} Y[k]\psi_k(t) = s_i(t) + n(t) - \sum_{k=0}^{n-1} (s_i[k] + N[k])\,\psi_k(t)$$

But the signal $s_i(t)$ lies in the signal space, so

$$s_i(t) - \sum_{k=0}^{n-1} s_i[k]\psi_k(t) = 0$$

That is, the signal contribution to $y^{\perp}$ is zero, and

$$y^{\perp}(t) = n(t) - \sum_{k=0}^{n-1} N[k]\psi_k(t) = n^{\perp}(t)$$

where $n^{\perp}$ denotes the noise projection orthogonal to the signal space.

We now show that $n^{\perp}(t)$ is independent of the signal-space noise vector $\mathbf{N}$. Since $n^{\perp}$ and $\mathbf{N}$ are jointly Gaussian, it suffices to show that they are uncorrelated. For any t and k, we have

$$\begin{aligned}\text{cov}(n^{\perp}(t), N[k]) &= \mathbb{E}[n^{\perp}(t)N[k]] = \mathbb{E}\left[\{n(t) - \sum\nolimits_{j=0}^{n-1} N[j]\psi_j(t)\}N[k]\right] \\ &= \mathbb{E}[n(t)N[k]] - \sum\nolimits_{j=0}^{n-1} \mathbb{E}[N[j]N[k]]\psi_j(t)\end{aligned} \tag{6.96}$$

The first term on the extreme right-hand side can be simplified as

$$\begin{aligned}\mathbb{E}[n(t)\langle n, \psi_k\rangle] &= \mathbb{E}\left[n(t)\int n(s)\psi_k(s)ds\right] \\ &= \int \mathbb{E}[n(t)n(s)]\psi_k(s)ds \\ &= \int \sigma^2\delta(s-t)\psi_k(s)ds = \sigma^2\psi_k(t)\end{aligned} \tag{6.97}$$

On plugging (6.97) into (6.96), and noting that $\mathbb{E}[N[j]N[k]] = \sigma^2\delta_{jk}$, we obtain that

$$\text{cov}(n^{\perp}(t), N[j]) = \sigma^2\psi_k(t) - \sigma^2\psi_k(t) = 0$$

What we have just shown is that the component of the received signal orthogonal to the signal space contains the noise component $n^{\perp}$ only, and thus does not depend on which signal is sent under a given hypothesis. Since $n^{\perp}$ is independent of $\mathbf{N}$, the noise vector in the signal space, knowing $n^{\perp}$ does not provide any information about $\mathbf{N}$. These two observations imply that $y^{\perp}$ is *irrelevant* for our hypothesis problem. The preceding discussion is illustrated in Figure 6.9, and enables us to reduce our infinite-dimensional problem to a finite-dimensional vector model restricted to the signal space.

Note that our irrelevance argument depends crucially on the property of WGN that its projections along orthogonal directions are independent. Even though $y^{\perp}$ does not contain any signal component (since these by definition fall into the signal space), if $n^{\perp}$ and $\mathbf{N}$ exhibited statistical dependence, one could hope to learn something about $\mathbf{N}$ from $n^{\perp}$, and thereby improve performance compared with a system in which $y^{\perp}$ is thrown away. However, since $n^{\perp}$ and $\mathbf{N}$ are independent for WGN, we can restrict our attention to the signal space for our hypothesis-testing problem.

7 Channel coding

We have seen in Chapter 6 that, for signaling over an AWGN channel, the error probability decays exponentially with the SNR, with the rate of decay determined by the power efficiency of the constellation. For example, for BPSK or Gray-coded QPSK, the error probability is given by $p = Q(\sqrt{2E_b/N_0})$. We have also seen in Chapter 6 how to engineer the link budget so as to guarantee a certain desired performance. So far, however, we have considered only *uncoded* systems, in which bits to be sent are directly mapped to symbols sent over the channel. We now indicate how it is possible to improve performance by *channel coding,* which corresponds to inserting redundancy strategically prior to transmission over the channel.

A bit of historical perspective is in order. As mentioned in Chapter 1, Shannon showed the optimality of separate source and channel coding back in 1948. Shannon also provided a theory for computing the limits of communication performance over any channel (given constraints such as power and bandwidth). He did not provide a constructive means of attaining these limits; his proofs employed randomized constructions. For reasons of computational complexity, it was assumed that such strategies could never be practical. Hence, for decades after Shannon's 1948 publication, researchers focused on algebraic constructions (for which decoding algorithms of reasonable complexity could be devised) to create powerful channel codes, but never quite succeeded in attaining Shannon's benchmarks. This changed with the invention of turbo codes by Berrou *et al.* in 1993: their conference paper laid out a simple coding strategy that got to within 1 dB of Shannon capacity. They took codes that were easy to encode, and used scramblers to make them random-like. Maximum likelihood decoding for such codes is too computationally complex, but Berrou *et al.* showed that suboptimal iterative decoding methods provide excellent performance with reasonable complexity. It was then realized that a different class of random-like codes, called low-density parity-check (LDPC) codes, along with an appropriate iterative decoding procedure, had actually been invented by Gallager in the 1960s. Since then, there has been a massive effort to devise and implement a wide variety of "turbo-like" codes (i.e., random-like codes amenable to iterative decoding), with the result that we can now approach Shannon's performance benchmarks over almost any channel.

In this chapter, we provide a glimpse of how Shannon's performance benchmarks are computed, how channel codes are constructed, and how iterative decoding works. A systematic and comprehensive treatment of information theory and channel coding would take up entire textbooks in itself, hence our goal is to provide just enough exposure to some of the key ideas to encourage further exploration.

Chapter plan

In Section 7.1, we discuss two extreme examples, uncoded transmission and repetition coding, in order to motivate the need for more sophisticated channel coding strategies. A generic model for channel coding is discussed in Section 7.2. Section 7.3 introduces Shannon's information-theoretic framework, which provides fundamental performance limits for any channel coding scheme, and discusses its practical implications. Linear codes, which are the most prevalent class of codes used in practice, are introduced in Section 7.4. Finally, we discuss belief-propagation decoding, which has been crucial for approaching Shannon performance limits in practice, in Section 7.5.

Software

Concepts in belief-propagation decoding are reinforced by Software Lab 7.1.

7.1 Motivation

Uncoded transmission First, let us consider what happens without channel coding. Suppose that we are sending a data block of 1500 bytes (i.e., $n = 12{,}000$ bits, since 1 byte comprises 8 bits) over a binary symmetric channel (see Chapter 5) with bit error probability p, where errors occur independently for each bit. Such a BSC could be induced, for example, by making hard decisions for Gray-coded QPSK over an AWGN channel; in this case, we have $p = Q(\sqrt{2E_{\rm b}/N_0})$. Let us now define *block error* as the event that one or more bits in the block are in error. The probability that *all* of the bits get through correctly is given by $(1 - p)^n$, so the probability of block error is given by

$$P_{\rm B} = 1 - (1 - p)^n$$

Figure 7.1 plots the probability of block error versus the probability of bit error on a log–log scale. Despite its simplicity, this computation leads to some useful observations.

(a) For $p > 10^{-4}$, the probability of block error is essentially unity. This is because the expected number of errors in the block is given by np, and, when this is of the order of unity, the probability of making at least one error is very close to unity, because of the law of large numbers. Using this reasoning, we see that it becomes harder and harder to guarantee reliability as the block size increases, since p must scale as $1/n$. Clearly, this is not a sustainable approach. For example, even the corruption of a single bit in a large computer file can cause chaos, so we must find more sophisticated means of protecting the data than just trying to drive the raw bit error probability to zero.

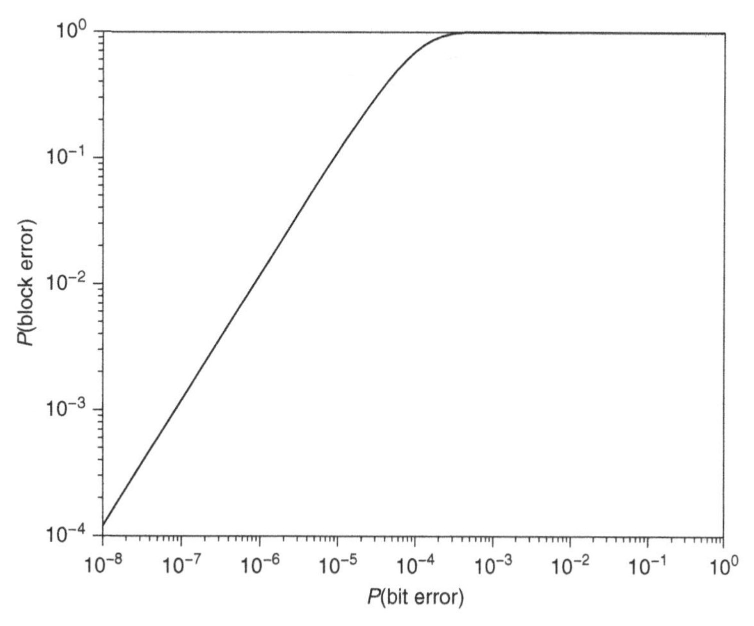

Figure 7.1 Block error probability versus bit error probability for uncoded transmission (the block size is 1500 bytes).

(b) It is often possible to efficiently *detect* block errors with very high probability. In practice, this might be achieved by using a cyclic redundancy check (CRC) code, but we do not discuss the specific error-detection mechanism here. If a block error is detected, then the receiver may ask the transmitter to retransmit the packet, if such retransmissions are supported by the underlying protocols. The link efficiency in this case becomes $1 - P_B$. Thus, if we can do retransmissions, uncoded transmission may actually not be a terrible idea. In our example, the link is 90% efficient ($P_B = 10^{-1}$) for a bit error probability p around 10^{-6}–10^{-5}, and 99% efficient ($P_B = 10^{-2}$) for p around 10^{-7}–10^{-6}.

(c) For Gray-coded QPSK, $p = Q(\sqrt{2E_b/N_0})$, so $p = 10^{-6}$ requires E_b/N_0 of about 10.55 dB. This is exactly the scenario in the link budget example modeling a 5-GHz WLAN link in Chapter 6. We see, therefore, that uncoded transmission, along with retransmissions, is a viable option in that setting.

Repetition coding Next, let us consider the other extreme, in which we send n copies of a single bit over a BSC with error probability p. That is, we send either a string of n zeros, or a string of n ones. The channel may flip some of these bits. Since the errors are independent, the number of errors is a binomial random variable, $\text{Bin}(n, p)$. For $p < \frac{1}{2}$, the average number of bits in error is $np < n/2$, hence a natural decoding rule is to employ *majority logic*: decide on 0 if the majority of received bits is zero, and on 1 otherwise. Taking n to be odd for simplicity (otherwise we need to specify a tiebreaker when there are equally many zeros and ones), a block error occurs if the number of errors is $\lceil n/2 \rceil$ or more. Using the binomial PMF, we have the following expression for the block error probability:

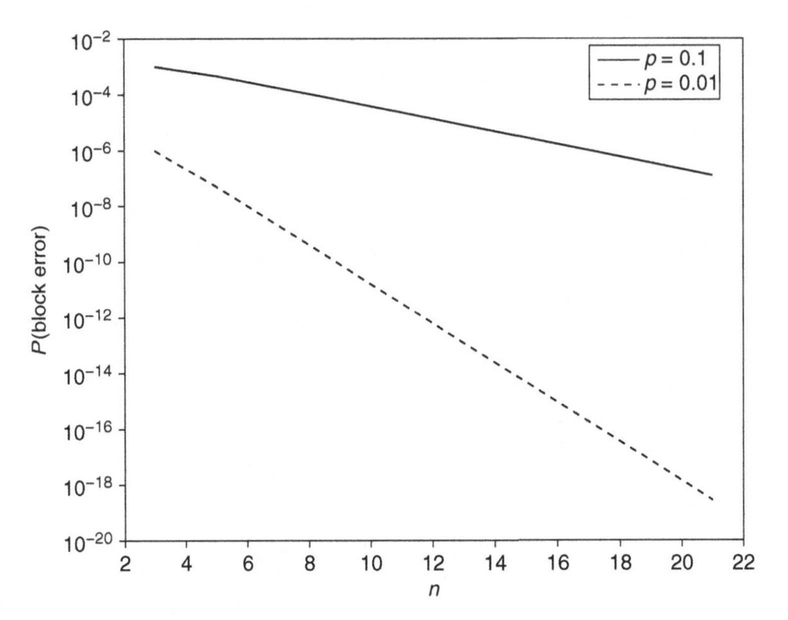

Figure 7.2 The error probability decays rapidly as a function of blocklength for a repetition code.

$$P_B = \sum_{m=\lceil n/2 \rceil}^{n} \binom{n}{m} p^m (1-p)^m$$

Figure 7.2 plots the probability of block error versus n for $p = 10^{-1}$ and $p = 10^{-2}$. Clearly, $P_B \to 0$ as $n \to \infty$, so we are doing well in terms of reliability. To see why, let us invoke the LLN again: the average number of errors is $np < \lceil n/2 \rceil$, so, as $n \to \infty$, the number of errors is smaller than $\lceil n/2 \rceil$ with probability one. However, we are sending only one bit of information for every n bits that we send over the channel, corresponding to a code rate of $1/n$ (one information bit for every n transmitted bits), which tends to zero as $n \to \infty$.

We have invoked the LLN to explain the performance of both uncoded transmission and repetition coding for large n, but neither of these approaches provides reliable performance at nonzero coding rates. As $n \to \infty$, the block error rate $P_B \to 1$ for uncoded transmission, while the code rate tends to zero for the repetition code. However, it is possible to design channel coding schemes between these two extremes that provide arbitrarily reliable communication ($P_B \to 0$ as $n \to \infty$) at non-vanishing code rates. The existence of such codes is guaranteed by LLN-style arguments. For example, for a BSC with crossover probability p, as n gets large, the number of errors clusters around np. Thus, the basic intuition is that, if we are able to insert enough redundancy to correct a number of errors of the order of np, then we should be able to approach zero block error probability. Giving precise form to such existence arguments is the realm of *information theory,* which can be used to establish fundamental performance limits for almost any reasonable channel model, while *coding theory* concerns itself with constructing practical coding schemes that approach these performance limits.

7.2 Model for channel coding

We introduce some basic terminology related to channel coding, and discuss where it fits within a communication link.

Binary code An (n, k) binary code maps k information bits to n transmitted bits, where $n \geq k$. Each of the k information bits can take any value in $\{0, 1\}$, hence the code $\mathcal{C}$ is a set of 2^k *codewords,* each a binary vector of length n. The code rate is defined as $R_c = k/n$.

Figure 7.3 provides a high-level view of how a binary channel code can be used over a communication link. The encoder maps the k-bit information word $\mathbf{u}$ to an n-bit codeword $\mathbf{x}$. As discussed shortly, the "channel" shown in Figure 7.3 is an abstraction that includes operations at the transmitter and the receiver, in addition to the physical channel. The output $\mathbf{y}$ of the channel is a length-n vector of hard decisions (bits) or soft decisions (real numbers) on the coded bits. These are then used by the decoder to provide an estimate $\hat{\mathbf{u}}$ of the information bits. We declare a block error if $\hat{\mathbf{u}} \neq \mathbf{u}$.

Figure 7.4 provides a specific example illustrating how the preceding abstraction connects to the transceiver design framework developed in earlier chapters. It shows a binary code used for signaling over an AWGN channel using Gray-coded 16QAM. We see that n coded bits are mapped to $n/4$ complex-valued symbols at the transmitter. Since channel codes are typically designed for random errors, we have inserted an *interleaver* between the channel encoder and the modulator in order to disperse potential correlations in errors among bits. The modulator could employ linear modulation as described in Chapter 4, with demodulation as in Chapter 6 for an ideal AWGN channel, or more sophisticated

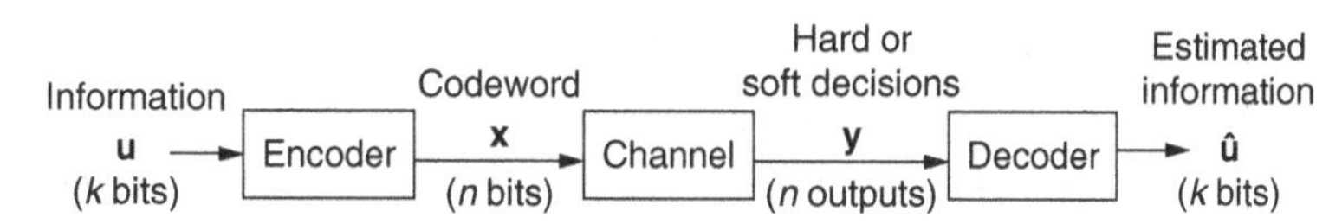

Figure 7.3 The high-level model for the coded system.

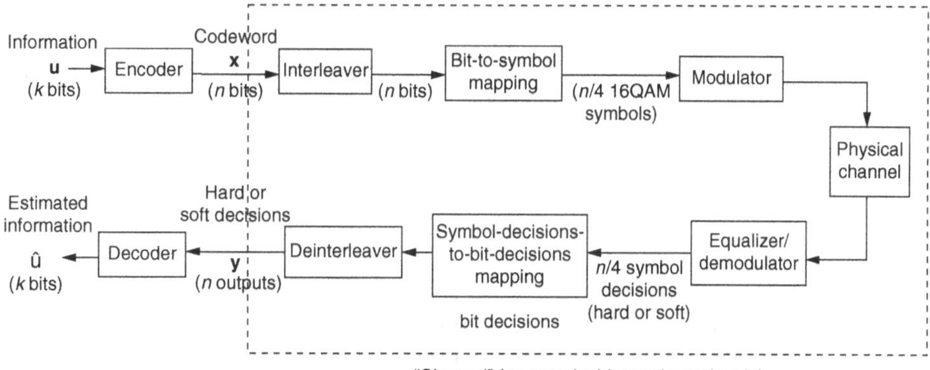

Figure 7.4 An example of bit-interleaved coded modulation.

equalization strategies for handling the intersymbol interference due to channel dispersion (see Chapter 8). An alternative frequency domain modulation strategy, termed orthogonal frequency-division multiplexing (OFDM), for handling channel dispersion is also discussed in Chapter 8. However, for our present purpose of discussing channel coding, we abstract all of these details away. Indeed, as shown in Figure 7.4, the "channel" from Figure 7.3 includes all of these operations, with the final output being the hard or soft decisions supplied to the decoder. Problem 7.4 explores the nature of this equivalent channel for an example constellation. Often, even if the physical channel has memory, the interleaving and deinterleaving operations allow us to model the equivalent channel as *memoryless:* the output y_i depends only on coded bit x_i, and the channel is completely characterized by the conditional density $p(y_i|x_i)$. For example, for hard decisions, we may model the equivalent channel as a binary symmetric channel with error probability p. For soft decisions, y_i may be a real number, or may comprise several bits, hence the channel model would be a little more complicated.

The preceding approach, which neatly separates out the binary channel code from the signal processing related to transmitting and receiving over a physical channel, is termed *bit-interleaved coded modulation (BICM).* If we use a binary code of rate $R_\mathrm{c} = k/n$ and a symbol alphabet of size M, then the overall rate of communication over the channel is given by $R_\mathrm{c} \log_2 M$ bits per symbol. From Chapter 4, we know that, using ideal Nyquist signaling, we can signal at rate W complex-valued symbols/s over a bandlimited passband channel of bandwidth W. Thus, the rate of communication in bits per second (bps) is given by $R_\mathrm{b} = R_\mathrm{c} W \log_2 M$. The bandwidth efficiency, or spectral efficiency, can now be defined as

$$r = \frac{R_\mathrm{b}}{W} = R_\mathrm{c} \log_2 M = \frac{k}{n} \log_2 M \tag{7.1}$$

in bps/Hz, or bits/symbol (for ideal Nyquist signaling). In comparison with Chapter 4 (where we termed this quantity η_W), what has changed is that we must now account for the rate of the binary code which we have wrapped around our communication link.

We also need to revisit our SNR concepts and carefully keep track of *information* bits, *coded* bits, and *modulated* symbols, when computing the signal power or energy. The quantity E_b refers to energy per *information* bit. When we encode these bits using a binary code of rate R_c, the energy per *coded* bit is $E_\mathrm{c} = R_\mathrm{c} E_\mathrm{b}$ (information bits per coded bit, times energy per information bit). When we then put the coded bits through a modulator that outputs M-ary symbols ($\log_2 M$ coded bits per symbol), we obtain that the energy per *modulated* symbol is given by

$$E_\mathrm{s} = E_\mathrm{c} \log_2 M = R_\mathrm{c} E_\mathrm{b} \log_2 M$$

In short, we have

$$E_\mathrm{s} = r E_\mathrm{b} \tag{7.2}$$

which makes sense: the energy per symbol equals the number of information bits per symbol, times the energy per information bit. While we have established (7.2) for BICM, it holds generally, since it is just a matter of energy bookkeeping.

BICM is a practical approach that applies to any physical communication channel, and the significant advances in channel coding over the past two decades ensure that there is little loss in optimality due to this decoupling of coding and modulation. In the preceding example, we have used it in conjunction with Nyquist sampling, which transforms the continuous-time channel into a discrete-time channel carrying complex-valued symbols. However, we can also view the Nyquist sampled channel in greater generality, in which the inputs to the effective channel are complex-valued symbols, and the outputs are the noisy received samples at the output of the equalizer/demodulator. A code of rate R bits/channel use over this channel is simply a collection of 2^{NR} discrete time complex-valued vectors of length N, where N is the number of symbols sent over the channel. In our BICM example with 16QAM, we have $R = R_c \log_2 M = 4R_c$ and $N = n/4$, but this framework also accommodates approaches that tie coding and modulation more closely together. The tools of information theory can be used to provide fundamental performance limits for any such coded modulation strategy. We provide a glimpse of such results in the next section.

7.3 Shannon's promise

Shannon established the field of information theory in the 1940s. Among its many consequences is the channel coding theorem, which states that, if we allow code block lengths to get large enough, then there is a well-defined quantity, called the *channel capacity,* which determines the maximum rate at which reliable communication can take place. A class of channel models of fundamental importance is the following.

Discrete memoryless channel (DMC) Inputs are fed to the channel in discrete time. If x is the channel input at a given time, then the output y at that time has conditional density $p(y|x)$. For multiple channel uses, the outputs are conditionally independent given the inputs, as follows:

$$p(y_1, \ldots, y_n|x_1, \ldots, x_n) = p(y_1|x_1) \ldots p(y_n|x_n)$$

The inputs may be constrained in some manner (e.g., to take values from a finite alphabet, or to be limited in average or peak power). A channel code of length n and rate R bits per channel use contains 2^{nR} codewords. That is, we employ M-ary signaling with $M = 2^{nR}$, where each signal, or codeword, is a vector of length n, with the jth codeword denoted by $\mathbf{X}^{(j)} = (X_1^{(j)}, \ldots, X_n^{(j)})^{\mathrm{T}}, j = 1, \ldots, 2^{nR}$.

Shannon's channel coding theorem gives us a compact characterization of the channel capacity C (in bits per channel use) for a DMC. It states that, for any code rate below capacity ($R < C$), and for a large enough block length n, there exist codes and decoding strategies such that the block error probability can be made arbitrarily small. The *converse* of this result also holds: for code rates above capacity ($R > C$), the block error probability is bounded away from zero for *any* coding strategy. The fundamental intuition is that, for large block lengths, events that cause errors cluster around some well-defined patterns with very high probability (because of the law of large numbers), hence it is possible to devise channel codes that can correct these patterns as long as we are not trying to fit in too many codewords.

Giving the expression for the Shannon capacity of a general DMC is beyond our scope, but we do provide intuitive derivations of the channel capacity for the two DMC models of greatest importance to us: the discrete-time AWGN channel and the BSC. We then discuss, via numerical examples, how these capacity computations can be used to establish design guidelines.

Discrete-time AWGN channel Let us consider the following real-valued discrete-time AWGN channel model, where we send a codeword consisting of a sequence of real numbers $\{X_i, i = 1, \ldots, n\}$, and obtain the noisy outputs

$$Y_i = X_i + N_i, \quad i = 1, \ldots, n \tag{7.3}$$

where $N_i \sim N(0, N)$ are i.i.d. Gaussian noise samples. We impose a power constraint $\mathbb{E}[X_i^2] \leq S$. This model is called the discrete-time AWGN channel. For Nyquist signaling over a continuous-time bandlimited AWGN channel with bandwidth W, we can signal at the rate of W complex-valued symbols per second, or $2W$ real-valued symbols per second. This can be interpreted as getting to use the discrete-time AWGN channel (7.3) $2W$ times per second. Thus, once we have figured out the capacity for the discrete-time AWGN channel in bits per channel use, we will be able to specify the maximum rate at which information can be transmitted reliably over a bandlimited continuous-time channel.

A channel code over the discrete-time channel (7.3) of rate R bits/channel contains 2^{nR} codewords, where the jth codeword $\mathbf{X}^{(j)} = (X_1^{(j)}, \ldots, X_n^{(j)})^{\mathrm{T}}$ satisfies the average power constraint if

$$||\mathbf{X}^{(j)}||^2 = \sum_{i=1}^{n} |X_i^{(j)}|^2 \leq nS$$

If the jth codeword is sent, the received vector is

$$\mathbf{Y} = \mathbf{X}^{(j)} + \mathbf{N}, \text{ codeword } j \text{ transmitted} \tag{7.4}$$

where $\mathbf{N} = (N_1, \ldots, N_n)^{\mathrm{T}}$ is the noise vector. The expected energy of the noise vector is given by

$$\mathbb{E}[||\mathbf{N}||^2] = \sum_{i=1}^{n} \mathbb{E}[N_i^2] = nN \tag{7.5}$$

The expected energy of the received vector is given by

$$\begin{aligned}\mathbb{E}\left[||\mathbf{Y}||^2\right] &= \sum_{i=1}^{n} \mathbb{E}[|X_i^{(j)} + N_i|^2] = \sum_{i=1}^{n} \left(|X_i^{(j)}|^2 + \mathbb{E}[N_i^2] + 2\mathbb{E}[X_i^{(j)} N_i]\right) \\ &= ||\mathbf{X}^{(j)}||^2 + nN \leq n(S+N)\end{aligned} \tag{7.6}$$

(The cross term involving signal and noise drops away, since they are independent and the noise is zero mean.)

We now provide a heuristic argument as to how reliable performance can be achieved by letting the code block length n get large. Invoking the law of large numbers, random quantities cluster around their averages with high probability, so the received vector $\mathbf{Y}$ lies inside an n-dimensional sphere of radius $\sqrt{n(S+N)}$, and the noise vector $\mathbf{N}$ lies inside

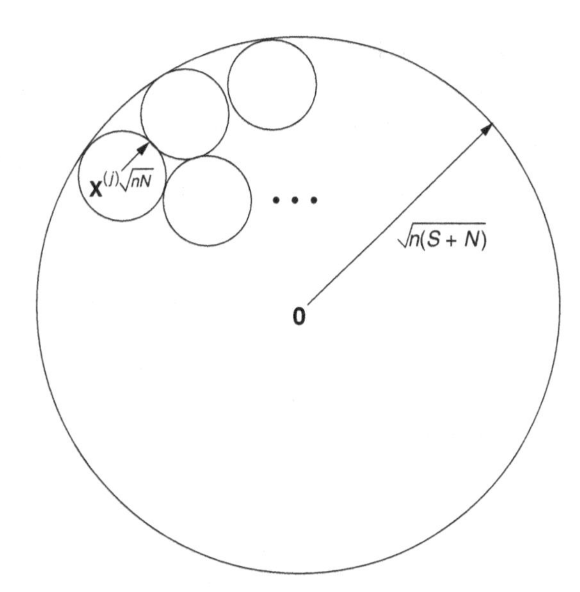

Figure 7.5 The sphere-packing argument for characterizing the rate of reliable communication.

an n-dimensional sphere of radius $\sqrt{nN}$. Consider now a "decoding sphere" around each codeword with radius just a little larger than $\sqrt{nN}$. Then we make correct decisions with high probability: if we send codeword j, the noise vector $\mathbf{N}$ is highly unlikely to push us outside the decoding sphere centered around $\mathbf{X}^{(j)}$. The question then is the following: what is the largest number of decoding spheres of radius $\sqrt{nN}$ that we can pack inside the n-dimensional sphere of radius $\sqrt{n(S+N)}$ in which the received vector $\mathbf{Y}$ lives? This sphere-packing argument, depicted in Figure 7.5, provides an estimate of the largest number of codewords 2^{nR} which we can accommodate while guaranteeing reliable communication.

We now invoke the result that the volume of an n-dimensional sphere of radius r is $K_n r^n$, where K_n is a constant depending on n whose explicit form we do not need. We can now estimate the maximum achievable rate R as follows:

$$2^{nR} \leq \frac{K_n \left(\sqrt{n(S+N)}\right)^n}{K_n \left(\sqrt{nN}\right)^n} = \left(1 + \frac{S}{N}\right)^{n/2}$$

from which we obtain

$$R \leq \frac{1}{2} \log_2 \left(1 + \frac{S}{N}\right)$$

While we have used heuristic arguments to arrive at this result, it can actually be rigorously demonstrated that the right-hand side of (7.7) is indeed the maximum possible rate of reliable communication over the discrete-time AWGN channel.

Capacity of the discrete-time AWGN channel We can now state that the capacity of the discrete-time AWGN channel (7.3) is given by

$$C_{\text{d-AWGN}} = \frac{1}{2} \log_2 \left(1 + \frac{S}{N}\right) \text{ bits/channel use} \tag{7.7}$$

Continuous-time bandlimited AWGN channel We now use this result to compute the maximum spectral efficiency attainable over a continuous-time bandlimited AWGN channel. The complex-baseband channel corresponding to a passband channel of physical bandwidth W spans $[-W/2, W/2]$ (taking the reference frequency at the center of the passband). Thus, Nyquist signaling over this channel corresponds to W complex-valued symbols per second, or $2W$ uses per second of the real-valued channel (7.3). Since each complex-valued symbol corresponds to two uses of the real discrete-time AWGN channel, the capacity of the bandlimited channel is given by $2WC_{\text{d-AWGN}}$ bits per second. We still need to specify S/N. For each complex-valued sample, the energy per symbol $E_s = rE_b$ (bits/symbol, times the energy per bit, gives the energy per complex-valued symbol). The noise variance per real dimension is $\sigma^2 = N_0/2$, hence the noise variance seen by a complex symbol is $2\sigma^2 = N_0$. We obtain

$$\frac{S}{N} = \frac{E_s}{N_0} \tag{7.8}$$

Putting these observations together, we can now state the following formula for the capacity of the bandlimited AWGN channel.

Capacity of the bandlimited AWGN channel We have

$$C_{BL}\left(W, \frac{E_s}{N_0}\right) = W \log_2\left(1 + \frac{E_s}{N_0}\right) \text{ bits per second} \tag{7.9}$$

It can be checked that we get exactly the same result for a physical (real-valued) baseband channel of physical bandwidth W. Such a channel spans $[-W, W]$, but the transmitted signal is constrained to be real-valued. Signals over such a channel can therefore be represented by $2W$ *real-valued* samples per second, which is the same as for a passband channel of bandwidth W.

For a system communicating reliably at a bit rate of R_b bps over such a bandlimited channel, we must have $R_b < C_{BL}$. Using (7.9), we see that the spectral efficiency $r = R_b/W$ in bps/Hz of the system must therefore satisfy

$$r < \log_2\left(1 + \frac{E_s}{N_0}\right) = \log_2\left(1 + r\frac{E_b}{N_0}\right) \text{ bps/Hz or bits/complex symbol} \tag{7.10}$$

where we have used (7.2).

The preceding argument defines the regime where reliable communication is possible. We can rewrite (7.10) to obtain the fundamental limits on the power–bandwidth tradeoffs achievable over the AWGN channel, as follows.

Fundamental power–bandwidth tradeoff for the bandlimited AWGN channel We have

$$\frac{E_b}{N_0} > \frac{2^r - 1}{r} \quad \text{regime where reliable communication is possible} \tag{7.11}$$

The quantity r is the bandwidth efficiency of a bandlimited AWGN channel. This fundamental power–bandwidth tradeoff is depicted in Figure 7.6, which plots the minimum required E_b/N_0 (dB) versus the bandwidth efficiency. Note that we cannot make the E_b/N_0 required for reliable communication arbitrarily small even as the bandwidth efficiency goes to zero. We leave it as an exercise (Problem 7.6) to show that the minimum possible

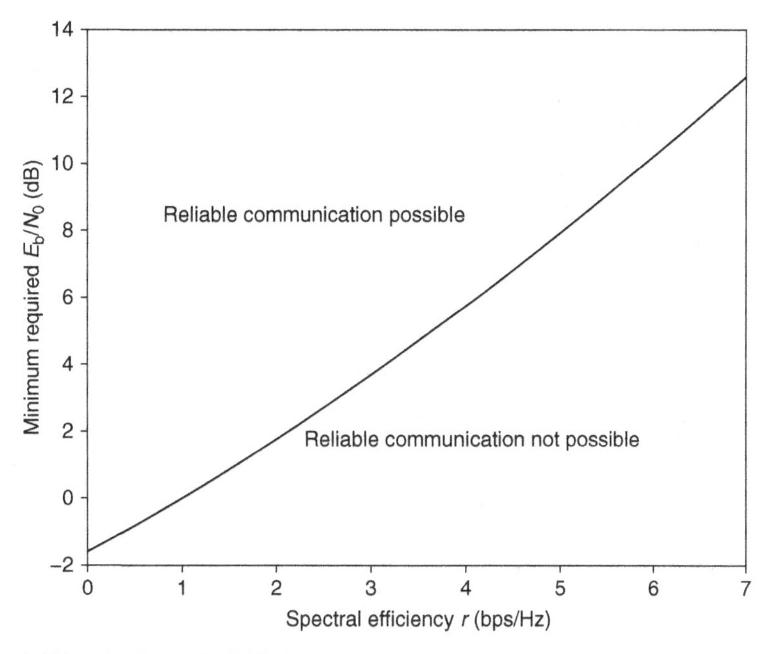

Figure 7.6 Power–bandwidth tradeoffs over the AWGN channel.

value of E_b/N_0 for reliable communication (corresponding to $r \to 0$) over the AWGN channel is -1.6 dB. Another point worth emphasizing is that these power–bandwidth tradeoffs assume powerful channel coding, and are different from those discussed for uncoded systems in Chapter 6: recall that the bandwidth efficiency for M-ary uncoded linear modulation was equal to $\log_2 M$, and that the power efficiency was defined as d_{min}^2/E_b. Numerical examples showing how channel coding fundamentally changes the achievable power–bandwidth tradeoffs are explored in more detail in Section 7.3.1. We provide here a quick example that illustrates how (7.11) relates to real-world scenarios.

Example 7.3.1 (Evaluating system feasibility using Shannon limits) A company claims to have developed a wireless modem with a receiver sensitivity of -82 dBm and a noise figure of 6 dB, operating at a rate of 100 Mbps over a bandwidth of 20 MHz. Do you believe their claim?

Shannon limit calculations. Modeling the channel as an ideal bandlimited AWGN channel, the proposed modem must satisfy (7.11). Assuming that there is no excess bandwidth, $r = 100$ Mbps/20 MHz $= 5$ bps/Hz or bits/symbol. From (7.11), we know that we must have $(E_b/N_0)_{\text{required}} > (2^5 - 1)/5 = 6.2$, or 7.9 dB. The noise PSD N_0 is given by $-174 + 6 = -168$ dBm over 1 Hz. The energy per bit equals the received power divided by the bit rate, so the actual E_b/N_0 for the advertised receiver sensitivity (i.e., the receive power at which the modem can operate) is given by $(E_b/N_0)_{\text{actual}} = -82 - 10\log_{10} 10^8 + 168 = 6$ dB. This is 1.9 dB short of the Shannon limit, hence our first instinct is not to believe them.

Tweaking the channel model. What if the channel were not a single AWGN channel, but two AWGN channels in parallel? As we shall see when we discuss multiple-antenna systems in Chapter 8, it is possible to use multiple antennas at the transmitter and receiver to obtain *spatial* degrees of freedom in addition to those in time and frequency. If there are indeed two spatial channels that are created using multiple antennas and we can model each of them as AWGN, then the spectral efficiency per channel is $5/2 = 2.5$ bps/Hz, and $(E_b/N_0)_{\text{required}} > (2^{2.5} - 1)/2.5 = 1.86$, or 2.7 dB. Since the actual E_b/N_0 is 6 dB, the system is operating more than 3 dB away from the Shannon limit. Since we do have practical channel codes that get to within 1 dB of Shannon capacity, or even closer to it, the claim now becomes believable.

Binary symmetric channel We now turn our attention to the BSC with crossover probability p shown in Figure 7.7, which might, for example, be induced by hard decisions on an AWGN channel. Note that we are interested only in $0 \leq p \leq \frac{1}{2}$. If $p > \frac{1}{2}$, then we can switch zeros and ones at the output of the channel to get back to a BSC with crossover probability $\tilde{p} = 1 - p < \frac{1}{2}$. The BSC can also be written as an additive noise channel, analogous to the discrete-time AWGN channel (7.3):

$$Y_i = X_i \oplus N_i, \quad i = 1, \ldots, n \tag{7.12}$$

where the *exclusive or (XOR)* symbol $\oplus$ corresponds to addition modulo 2, which follows the rules:

$$\begin{aligned} 1 \oplus 1 = 0 \oplus 0 = 0 \\ 1 \oplus 0 = 0 \oplus 1 = 1 \end{aligned} \tag{7.13}$$

Thus, we can flip a bit by adding (modulo 2) a 1 to it. The probability of a bit flip is p. Thus, the noise variables N_i are i.i.d. Bernoulli random variables with $P[N_i = 1] = p = 1 - P[N_i = 0]$.

Just as with the AWGN channel, we now develop a sphere-packing argument to provide an intuitive derivation of the BSC channel capacity. Of course, our concept of distance must be different from the Euclidean distance considered for the AWGN channel. Define the *Hamming distance* between two binary vectors of equal length to be the number of places in which they differ. For a codeword of length n, the average number of errors equals np. Assuming that the number of errors clusters around this average for large n, define a *decoding sphere* around a codeword as all sequences that are at Hamming distances of np or less from it. The number of such sequences is called the *volume* of the decoding sphere.

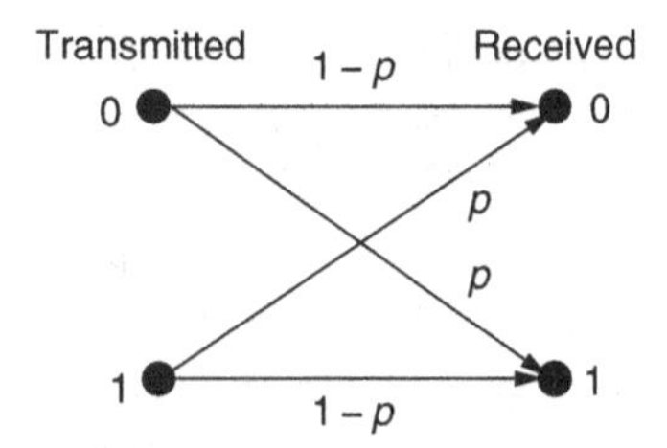

Figure 7.7 A binary symmetric channel with crossover p.

By virtue of (7.12) and (7.13), we see that this volume is exactly equal to the number of noise vectors $\mathbf{N} = (N_1, \ldots, N_n)^{\mathrm{T}}$ with np or fewer ones (the number of ones in a sequence is called its *weight*). The number of length-n vectors with weight m equals

$$\binom{n}{m},$$

hence the number of vectors with weight at most np is given by

$$V_n = \sum_{m=0}^{np} \binom{n}{m} \tag{7.14}$$

We state without proof the following asymptotic approximation for V_n for large n:

$$V_n \approx 2^{nH_{\mathrm{B}}(p)}, \quad 0 \leq p < \frac{1}{2} \tag{7.15}$$

where $H_{\mathrm{B}}(\cdot)$ is the *binary entropy function*, which is defined by

$$H_{\mathrm{B}}(p) = -p\log_2 p - (1-p)\log_2(1-p), \quad 0 \leq p \leq 1 \tag{7.16}$$

We plot the binary entropy function in Figure 7.8(a). Note the symmetry around $p = \frac{1}{2}$. This is because, as mentioned earlier, we can map $p > \frac{1}{2}$ to $1 - p < \frac{1}{2}$ by switching the roles of 0 and 1 at the output.

For a length-n code of rate R bits/channel use, the number of codewords equals 2^{nR}. The total number of binary sequences of length n, or the entire volume of the space we are working in, is 2^n. Thus, if we wish to put a decoding sphere of volume V_n around each codeword, the maximum number of codewords we can fit must satisfy

$$2^{nR} \leq \frac{2^n}{V_n} \approx \frac{2^n}{2^{nH_{\mathrm{B}}(p)}} = 2^{n(1-H_{\mathrm{B}}(p))}$$

which gives

$$R \leq 1 - H_{\mathrm{B}}(p)$$

It can be rigorously demonstrated that the right-hand side actually equals the capacity of the BSC. We therefore state this result formally.

Capacity of BSC The capacity of a BSC with crossover probability p is given by

$$C_{\mathrm{BSC}}(p) = 1 - H_{\mathrm{B}}(p) \text{ bits/channel use} \tag{7.17}$$

The capacity is plotted in Figure 7.8(b). We note the following points.

- For $p = \frac{1}{2}$, the channel is useless (its output does not depend on the input) and has capacity zero.
- For $p > \frac{1}{2}$, we switch zeros and ones at the output to obtain an effective BSC with crossover probability $1 - p$, hence $C_{\mathrm{BSC}}(p) = C_{\mathrm{BSC}}(1 - p)$.
- For $p = 0$ or $p = 1$, the channel is perfect, and the capacity attains its maximum value of 1 bit/channel use.

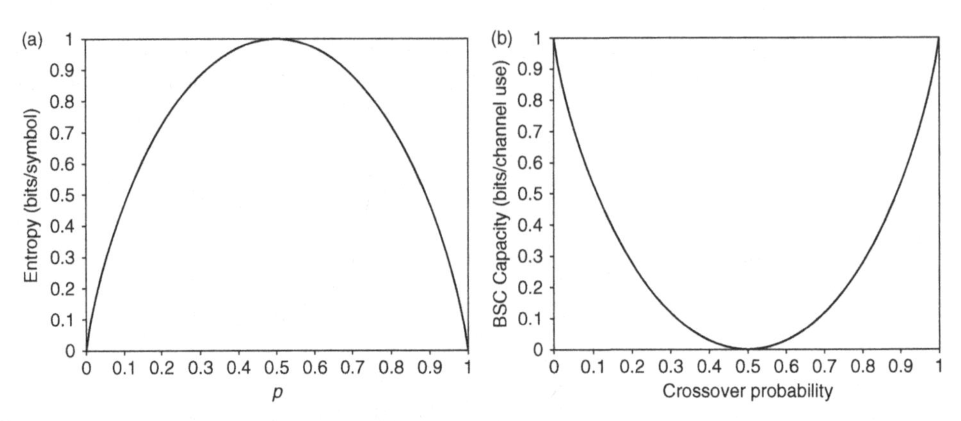

Figure 7.8 (a) The binary entropy function $H_B(p)$ and (b) the capacity of a BSC with crossover probability p, given by $1 - H_B(p)$.

7.3.1 Design implications of Shannon limits

Since the invention of turbo codes in 1993 and the subsequent rediscovery of LDPC codes, we now know how to construct random-looking codes that can be efficiently decoded (typically using iterative or message-passing methods) and come extremely close to Shannon limits. Such "turbo-like" coded modulation strategies have made, or are making, their way into almost every digital communication technology, including cellular, WiFi, digital video broadcast, optical communication, magnetic recording, and flash memory. Thus, we often summarize the performance of a practical coded modulation scheme by stating how far away it is from the Shannon limit. Let us discuss what this means via an example. A rate-$\frac{1}{2}$ binary code is employed, using bit-interleaved coded modulation with a 16QAM alphabet. We are told that it operates 2 dB away from the Shannon limit at a BER of 10^{-5}. What does this statement mean?

Since we can convey four coded bits every time we send a 16QAM symbol, the spectral efficiency $r = \frac{1}{2} \times 4 = 2$ bps/Hz, or information bits per symbol: the product of the binary code rate (number of information bits per coded bit) and the number of coded bits per symbol gives the number of information bits per symbol. From (7.11), the minimum possible E_b/N_0 is found to be about 1.8 dB. This is the minimum possible E_b/N_0 at which Shannon tells us that error-free operation (in the limit of large code blocklengths) is possible at the given spectral efficiency. Of course, any practical strategy at finite blocklength, no matter how large, will not give us error-free operation, hence we declare some value of error probability that we are satisfied with, and evaluate the E_b/N_0 for which that error probability is attained. Hence the statement that we started with says that our particular coded modulation strategy provides a BER of 10^{-5} at $E_b/N_0 = 1.8 + 2 = 3.8$ dB (2 dB higher than the Shannon limit).

How much gain does the preceding approach provide over uncoded communication? Let us compare it with uncoded QPSK, which has the same spectral efficiency of 2 bps/Hz. The BER is given by the expression $Q(\sqrt{2E_b/N_0})$, and we can check that a BER of 10^{-5} is attained at $E_b/N_0 = 9.6$ dB. Thus, we get a significant *coding gain* of $9.6 - 3.8 = 5.8$

dB from using a sophisticated coded modulation strategy, first expanding the constellation from QPSK to 16QAM so there is "room" to insert redundancy, and then using a powerful binary code.

The specific approach for applying a turbo-like coded modulation strategy depends on the system at hand. For systems with retransmissions (e.g., wireless data), we are often happy with block error rates of 1% or even higher, and may be able to use these relatively relaxed specifications to focus on reducing the computational complexity and coding delay (e.g., by considering simpler codes and smaller block lengths). For systems where there is no scope for retransmissions (e.g., storage or broadcast), we may use longer block lengths, and may even layer an outer code to clean up the residual errors from an inner turbo-like coded modulation scheme. Another common feature of many systems is the use of *adaptive* coded modulation, in which the spectral efficiency is varied as a function of the channel quality. BICM is particularly convenient for this purpose, since it allows us to mix and match a menu of well-optimized binary codes at different rates (e.g., ranging from $\frac{1}{4}$ to $15/16$) with a menu of standard constellations (e.g., QPSK, 8PSK, 16QAM, 64QAM) to provide a large number of options.

A detailed description of turbo-like codes is beyond our present scope, but we do provide a discussion of decoding via message passing after a basic exposition of linear codes.

7.4 Introducing linear codes

After decades of struggling to construct practical coding strategies that approach Shannon's performance limits, we can now essentially declare victory, with channel codes of reasonable blocklength coming within 1 dB of capacity. We refer the reader to more advanced texts and the research literature for details regarding such capacity-achieving codes, and limit ourselves here to establishing some basic terminology and concepts that provide a roadmap for further exploration. We restrict our attention to *linear* codes, which are by far the most prevalent class of codes in use today, and suffice to approach capacity. As we shall discuss shortly, a linear code is a subspace in a bigger vector space, but the arithmetic we use to define linearity is different from the real- and complex-valued arithmetic we are used to.

Finite fields We are used to doing calculations with real and complex numbers. The real numbers, together with the rules of arithmetic, comprise the real field, and the complex numbers, together with the rules of arithmetic, comprise the complex field. Each of these fields has infinitely many elements, forming a continuum. However, the basic rules of arithmetic (addition, multiplication, division by nonzero elements, and the associative, distributive, and commutative laws) can also be applied to fields with a finite number of discrete elements. Such fields are called *finite-fields*, or *Galois fields* (after the French mathematician who laid the foundations for finite-field theory). It turns out that, in order to be consistent with the basic rules of arithmetic, the number of elements in a finite field must be a power of a prime, and we denote a finite field with p^m elements, where p is a prime and m is a positive integer, as GF(p^m). The theory of finite fields, while outside our scope here, is essential for a variety of algebraic code constructions, and we provide some pointers for further study later in this chapter. Our own discussion here is restricted to codes over the binary field GF(2).

Binary arithmetic Binary arithmetic corresponds to operations with only two elements, 0 and 1, with addition modulo 2 as specified in (7.13). Binary subtraction is identical to binary addition. Multiplication and division (only division by nonzero elements is permitted) are trivial, since the only nonzero element is 1. The usual associative, distributive, and commutative laws apply. The elements $\{0, 1\}$, together with these rules of binary arithmetic, are said to comprise the binary field GF(2).

Linear binary code An (n, k) binary linear code $\mathcal{C}$ consists of 2^k possible codewords, each of length n, such that adding any two codewords in binary arithmetic yields another codeword. That is, $\mathcal{C}$ is closed under linear combinations (the coefficients of the linear combination can only take values 0 or 1, since we are working in binary arithmetic), and is therefore a k-dimensional *subspace* of the n-dimensional vector space of all length-n binary vectors, in a manner that is entirely analogous to the concept of a subspace in real-valued vector spaces. Pursuing this analogy further, we can specify a linear code $\mathcal{C}$ completely by defining a *basis* with k vectors, such that any vector in $\mathcal{C}$ (i.e., any codeword) can be expressed as a linear combination of the basis vectors.

Food for thought. The all-zero codeword is always part of any linear code. Why?

Notational convention. While we have preferred working with column vectors thus far, in deference to the convention in most coding-theory texts, we express codewords as row vectors. Letting $\mathbf{u}$ and $\mathbf{v}$ denote two binary vectors of the same length, we denote by $\mathbf{u} \oplus \mathbf{v}$ their component-by-component addition over the binary field. For example, $(00110) \oplus (10101) = (10011)$.

Example 7.4.1 (Repetition code) An $(n, 1)$ repetition code has only two codewords, the all-ones codeword $\mathbf{x}_1 = (1, \ldots, 1)$ and the all-zeros codeword $\mathbf{x}_0 = (0, \ldots, 0)$. We see that $\mathbf{x}_1 \oplus \mathbf{x}_1 = \mathbf{x}_0 \oplus \mathbf{x}_0 = \mathbf{x}_0$ and that $\mathbf{x}_1 \oplus \mathbf{x}_0 = \mathbf{x}_0 \oplus \mathbf{x}_1 = \mathbf{x}_1$, so this is indeed a linear code. There are only 2^1 codewords, so the dimension $k = 1$. Thus, the code, when viewed as a vector space over the binary field, is spanned by a single basis vector, $\mathbf{x}_1$. While the encoding operation is trivial (just repeat the information bit n times) for this code, let us write it in a manner that leads to a more general formalism. For example, for the $(5, 1)$ repetition code, the information bit $u \in \{0, 1\}$ is mapped to codeword $\mathbf{x}$ as follows:

$$\mathbf{x} = u\mathbf{G}$$

where

$$\mathbf{G} = (1 \ \ 1 \ \ 1 \ \ 1 \ \ 1) \tag{7.18}$$

is a matrix whose rows (just one row in this case) provide a basis for the code.

Example 7.4.2 (Single parity-check code) An $(n, n-1)$ single parity-check code takes as input $n-1$ unconstrained information bits $\mathbf{u} = (u_1, \ldots, u_{n-1})$, maps them unchanged to $n-1$ bits in the codeword, and adds a single parity-check bit to obtain a codeword $\mathbf{x} = (x_1, \ldots, x_{n-1}, x_n)$. For example, we can set the first $n-1$ code bits to the information bits ($x_1 = u_1, \ldots, x_{n-1} = u_{n-1}$) and append a parity check bit as follows:

$$x_n = x_1 \oplus x_2 \oplus \ldots \oplus x_{n-1}$$

Here, the code dimension $k = n-1$, so we can describe the code using $n-1$ linearly independent basis vectors. For example, for the $(5, 4)$ single parity-check code, we can make a particular choice of basis vectors, put as rows of a matrix as follows:

$$\mathbf{G} = \begin{pmatrix} 1 & 0 & 0 & 0 & 1 \\ 0 & 1 & 0 & 0 & 1 \\ 0 & 0 & 1 & 0 & 1 \\ 0 & 0 & 0 & 1 & 1 \end{pmatrix} \tag{7.19}$$

We can now check that any codeword can be written as

$$\mathbf{x} = \mathbf{u}\mathbf{G}$$

where $\mathbf{u} = (u_1, \ldots, u_4)$ is the information bit sequence.

Generator matrix While the preceding examples are very simple, they provide insight into the general structure of linear codes. An (n, k) linear code can be represented by a basis with k linearly independent vectors, each of length n. On putting these k basis vectors as the rows of a $k \times n$ matrix $\mathbf{G}$, we can then define a mapping from k information bits, represented as a $1 \times k$ row vector $\mathbf{u}$, to n code bits, represented as a $1 \times n$ row vector $\mathbf{x}$, as follows:

$$\mathbf{x} = \mathbf{u}\mathbf{G} \tag{7.20}$$

The matrix $\mathbf{G}$ is called the *generator matrix* for the code, since it can be used to generate all 2^k codewords by cycling through all possible values of the information vector $\mathbf{u}$.

Dual codes Drawing again on our experience with real-valued vector spaces, we know that, for any k-dimensional subspace $\mathcal{C}$ in an n-dimensional vector space, we can find an orthogonal $(n-k)$-dimensional subspace $\mathcal{C}^\perp$ such that every vector in $\mathbf{C}$ is orthogonal to every vector in $\mathcal{C}^\perp$. The subspace $\mathcal{C}^\perp$ is itself an $(n, n-k)$ code, and $\mathcal{C}$ and $\mathcal{C}^\perp$ are said to be duals of each other.

Example 7.4.3 (Duality of repetition and single parity-check codes) It can be checked that the $(5, 4)$ single parity-check code and $(5, 1)$ repetition codes are duals of each other. That is, each codeword in the $(5, 4)$ code is orthogonal to each codeword in the $(5, 1)$ code. Since codewords are linear combinations of rows of the generator matrix, it suffices to check that each row of a generator matrix for the (5, 4) code is orthogonal to each row of a generator matrix for the (5, 1) code. Specifically,

$$\mathbf{G}_{(5,1)}\mathbf{G}_{(5,4)}^{\mathrm{T}} = (1\;1\;1\;1\;1)\begin{pmatrix} 1 & 0 & 0 & 0 \\ 0 & 1 & 0 & 0 \\ 0 & 0 & 1 & 0 \\ 0 & 0 & 0 & 1 \\ 1 & 1 & 1 & 1 \end{pmatrix} = \mathbf{0}$$

Parity-check matrix The preceding discussion shows that we can describe an (n, k) linear code $\mathcal{C}$ by specifying its dual code $\mathcal{C}^{\perp}$. In particular, a generator matrix for the dual code serves as a parity-check matrix $\mathbf{H}$ for $\mathcal{C}$, in the sense that an n-dimensional binary vector $\mathbf{x}$ lies in $\mathcal{C}$ if and only if it is orthogonal to each row of $\mathbf{H}$. That is,

$$\mathbf{H}\mathbf{x}^{\mathrm{T}} = \mathbf{0} \text{ if and only if } \mathbf{x} \in \mathbf{C} \tag{7.21}$$

Each row of the parity-check matrix defines a parity-check equation. Thus, for a parity-check matrix $\mathbf{H}$ of dimension $(n-k) \times n$, each codeword must satisfy $n-k$ parity-check equations. Equivalently, if $\mathbf{G}$ is a generator matrix for $\mathcal{C}$, then it must satisfy

$$\mathbf{H}\mathbf{G}^{\mathrm{T}} = \mathbf{0} \tag{7.22}$$

In our examples, the generator matrix for the (5, 1) repetition code is a parity-check matrix for the (5, 4) code, and vice versa.

For an (n, k) code with large n and k, it is clearly difficult to check by brute-force search enumeration over 2^k codewords whether a particular n-dimensional vector $\mathbf{y}$ is a valid codeword. However, for a linear code, it becomes straightforward to verify this using only $n-k$ parity-check equations, as in (7.21). These parity-check equations, which provide the redundancy required to overcome channel errors, are not only important for verification of correct termination of decoding, but also play a crucial role during the decoding process, as we shall illustrate shortly.

Non-uniqueness An (n, k) linear code $\mathcal{C}$ is a unique subspace consisting of a set of 2^k codewords, and its dual $(n, n-k)$ code $\mathcal{C}^{\perp}$ is a unique subspace comprising 2^{n-k} codewords. However, in general, neither the generator nor the parity-check matrix for a code is unique, since the choice of basis for a nontrivial subspace is not unique. Thus, while the generator matrix for the (5, 1) code is unique because of its trivial nature (one dimension, binary field), the generator matrix for the (5, 4) code is not. For example, by taking linear combinations of the rows in (7.19), we obtain another linearly independent basis that provides an alternative generator matrix for the (5, 4) code:

$$\tilde{\mathbf{G}} = \begin{pmatrix} 1 & 1 & 0 & 0 & 0 \\ 0 & 1 & 1 & 0 & 0 \\ 1 & 1 & 1 & 0 & 1 \\ 0 & 1 & 1 & 1 & 1 \end{pmatrix} \tag{7.23}$$

From (7.20), we see that different choices of generator matrices correspond to different ways of encoding a k-dimensional information vector $\mathbf{u}$ into an n-dimensional codeword $\mathbf{x} \in \mathcal{C}$.

Systematic encoding A systematic encoding is one in which the information vector $\mathbf{u}$ appears directly in $\mathbf{x}$ (without loss of generality, we can take the bits of $\mathbf{u}$ to be the first k bits in $\mathbf{x}$), so that there is a clear separation between "information bits" and "parity-check" bits. In this case, the generator matrix can be written as

$$\mathbf{G} = [\mathbf{I}_k | \mathbf{P}] \quad \text{systematic encoding} \tag{7.24}$$

where $\mathbf{I}_k$ denotes the $k \times k$ identity matrix, and $\mathbf{P}$ is a $k \times (n-k)$ matrix specifying how the $n-k$ parity bits depend on the input. The identity matrix ensures that the k rows of $\mathbf{G}$ are linearly independent, so this does represent a valid generator matrix for an (n, k) code. The ith row of the generator matrix (7.24) corresponds to an information vector $\mathbf{u} = (u_1, \ldots, u_k)$ with $u_i = 1$ and $u_j = 0, j \neq i$. The generator matrices (7.18) and (7.19) for the $(5, 1)$ and $(5, 4)$ codes correspond to systematic encoding. The encoding of the $(5, 4)$ code corresponding to the generator matrix in (7.23) is not systematic.

Reading off a parity-check matrix from a systematic generator matrix If we are given a systematic encoding of the form (7.24), we can easily read off a parity-check matrix as follows:

$$\mathbf{H} = [-\mathbf{P}^{\mathrm{T}} | \mathbf{I}_{n-k}] \tag{7.25}$$

where the negative sign can be dropped for the binary field. The identity matrix ensures that $n-k$ rows of $\mathbf{H}$ are linearly independent, hence this is a valid parity-check matrix for an (n, k) linear code. We leave it as an exercise to verify, by directly substituting from (7.24) and (7.25), that $\mathbf{H}\mathbf{G}^{\mathrm{T}} = \mathbf{0}$.

Example 7.4.4 (Running example: a (5,2) linear code) Let us now construct a somewhat less trivial linear code that will serve as a running example for illustrating some basic concepts. Suppose that we have $k = 2$ information bits $u_1, u_2 \in \{0, 1\}$ that we wish to protect. We map this (using a systematic encoding) to a codeword of length 5 using a combination of repetition and parity check, as follows:

$$\mathbf{x} = (u_1, u_2, u_1, u_2, u_1 \oplus u_2) \tag{7.26}$$

A systematic generator matrix for this $(5, 2)$ code can be constructed by considering the two codewords corresponding to $\mathbf{u} = (1, 0)$ and $\mathbf{u} = (0, 1)$, respectively, which gives

$$\mathbf{G} = \begin{pmatrix} 1 & 0 & 1 & 0 & 1 \\ 0 & 1 & 0 & 1 & 1 \end{pmatrix} \tag{7.27}$$

We can read off a parity-check matrix using (7.24) and (7.25) to obtain

$$\mathbf{H} = \begin{pmatrix} 1 & 0 & 1 & 0 & 0 \\ 0 & 1 & 0 & 1 & 0 \\ 1 & 1 & 0 & 0 & 1 \end{pmatrix} \tag{7.28}$$

Any codeword $\mathbf{x} = (x_1, \ldots, x_5)$ must satisfy $\mathbf{H}\mathbf{x}^{\mathrm{T}} = \mathbf{0}$, which corresponds to the following parity-check equations:

$$x_1 \oplus x_3 = 0$$
$$x_2 \oplus x_4 = 0$$
$$x_1 \oplus x_2 \oplus x_3 = 0$$

Suppose, now, that we transmit the $(5, 2)$ code that we have just constructed over a BSC with crossover probability p. That is, we send a codeword $\mathbf{x} = (x_1, \ldots, x_5)$ using the channel $n = 5$ times. According to our discrete memoryless channel model, errors occur independently for each of the code symbols, and we get the output $\mathbf{y} = (y_1, \ldots, y_5)$, where $P[y_i|x_i] = x_i$ with probability $1 - p$, and $P[y_i|x_i] = x_i \oplus 1$ (i.e., the bit is flipped) with probability p. How should we try to decode this (i.e., estimate which codeword $\mathbf{x}$ was sent from the noisy output $\mathbf{y}$)? And how do we evaluate the performance of our decoding rule? In order to relate these issues to the structure of the code, it is useful to reiterate the notion of Hamming distance, and to introduce the concept of Hamming weight.

Hamming distance The Hamming distance $d_{\mathrm{H}}(\mathbf{u}, \mathbf{v})$ between two binary vectors $\mathbf{u}$ and $\mathbf{v}$ of equal length is the number of places in which they differ. For example, the Hamming distance between the two rows of the generator matrix $\mathbf{G}$ in (7.27) is given by $d_{\mathrm{H}}(\mathbf{g}_1, \mathbf{g}_2) = 4$.

Hamming weight The Hamming weight $w_{\mathrm{H}}(\mathbf{u})$ of a binary vector $\mathbf{u}$ equals the number of ones it contains. For example, the Hamming weight of each row of the generator matrix $\mathbf{G}$ in (7.27) is 3.

The Hamming distance between two vectors $\mathbf{u}$ and $\mathbf{v}$ is the Hamming weight of their binary sum:

$$d_{\mathrm{H}}(\mathbf{u}, \mathbf{v}) = w_{\mathrm{H}}(\mathbf{u} \oplus \mathbf{v}) \tag{7.29}$$

Structure of an (n, k) linear code Consider a specific codeword $\mathbf{x}_0$ in a linear code $\mathcal{C}$, and consider its Hamming distance from another codeword $\mathbf{x} \in \mathcal{C}$. We know that $d_{\mathrm{H}}(\mathbf{x}_0, \mathbf{x}) = w_{\mathrm{H}}(\mathbf{x}_0 \oplus \mathbf{x})$. By linearity, $\tilde{\mathbf{x}} = \mathbf{x}_0 \oplus \mathbf{x}$ is also a codeword in $\mathcal{C}$, and distinct choices of $\mathbf{x}$ give distinct codewords $\tilde{\mathbf{x}}$. Thus, as we run through all possible codewords $\mathbf{x} \in \mathcal{C}$, we obtain all possible codewords $\tilde{\mathbf{x}} \in \mathcal{C}$ (including $\tilde{\mathbf{x}} = \mathbf{0}$ for $\mathbf{x} = \mathbf{x}_0$). Thus, $d_{\mathrm{H}}(\mathbf{x}_0, \mathbf{x}) = w_{\mathrm{H}}(\tilde{\mathbf{x}})$, so that the set of Hamming distances between $\mathbf{x}_0$ and all codewords in $\mathcal{C}$ (running through all 2^k choices of $\mathbf{x}$) is precisely the set of weights that the codewords in $\mathcal{C}$ have (corresponding to the 2^k distinct vectors $\tilde{\mathbf{x}}$, one for each $\mathbf{x}$).

Minimum distance The minimum distance of a code is defined as

$$d_{\min} = \min_{\mathbf{x}_1, \mathbf{x}_2 \in \mathcal{C}, \mathbf{x}_1 \neq \mathbf{x}_2} d_{\mathrm{H}}(\mathbf{x}_1, \mathbf{x}_2)$$

Applying (7.29), and noting that, for a linear code, $\mathbf{x}_1 \oplus \mathbf{x}_2$ is a nonzero codeword in $\mathcal{C}$, we see that the minimum distance equals the minimum weight among all nonzero codewords. That is,

$$d_{\min} = w_{\min} = \min_{\mathbf{x} \in \mathcal{C}, \mathbf{x} \neq \mathbf{0}} w_{\mathrm{H}}(\mathbf{x}), \quad \text{for a linear code} \tag{7.30}$$

The $(5,2)$ code is small enough that we can simply list all four codewords: 00000, 10101, 01011, and 11110, from which we see that $w_{\min} = d_{\min} = 3$.

Guarantees on error correction A code is guaranteed to correct t errors if

$$2t + 1 \leq d_{\min} \tag{7.31}$$

It is quite easy to see why: we can set up non-overlapping "decoding spheres" of radius t around any codeword. The decoding sphere of radius t around a codeword $\mathbf{x}$ is defined as the set of vectors $\mathbf{y}$ within Hamming distance t of the codeword, as follows:

$$D_t(\mathbf{x}) = \{\mathbf{y} : d_{\mathrm{H}}(\mathbf{y}, \mathbf{x}) \leq t\}$$

The condition (7.31) guarantees that these decoding spheres do not overlap. Thus, if we make at most t errors, we are guaranteed that the received vector falls into the unique decoding sphere corresponding to the transmitted codeword.

Erasures There are some scenarios for which it is useful to introduce the concept of erasures, which correspond to assigning a "don't know" to a symbol rather than making a hard decision. Using a similar argument to the one used before, we can state that a code is guaranteed to correct t errors and e erasures if

$$2t + e + 1 \leq d_{\min} \tag{7.32}$$

Since it is "twice as easy" to correct erasures as it is to correct errors, we may choose to design a demodulator to output erasures in regions where we are uncertain about our hard decision. For a binary channel, this means that our input alphabet is $\{0, 1\}$ but our output alphabet is $\{0, 1, \epsilon\}$, where ϵ denotes erasure. As we see in Section 7.5, we can go further down this path in hedging our bets, with the decoder using soft decisions that take values in a real-valued output alphabet.

Running example Our $(5,2)$ code has $d_{\min} = 3$, and hence can correct one error or two erasures (but not both). Let us see how we would structure brute force decoding of a single error, by writing down which vectors fall within decoding spheres of unit radius around each codeword, and also pointing out which vectors are left over. This is done by writing all 2^5 possible binary vectors in what is termed a *standard array.*

Let us take advantage of this example to describe the general structure of a standard array for an (n, k) linear code. The array has 2^{n-k} rows and 2^k columns, and contains all possible binary vectors of length n. The first row of the array consists of the 2^k codewords, starting with the all-zero codeword. The first column consists of error patterns ordered by weight (ties broken arbitrarily), starting with no errors in the first row, $\mathbf{e}_1 = \mathbf{0}$. In general, denoting the first element of the ith row as the error pattern $\mathbf{e}_i$, the jth element in the ith row is $\mathbf{a}_{i,j} = \mathbf{e}_i + \mathbf{x}_j$, where $\mathbf{x}_j$ denotes the jth codeword, $j = 1, \ldots, 2^k$. That is, the (i,j)th element in the standard array is the jth codeword translated by the ith error pattern. For

Table 7.1 The standard array for the (5,2) code

00000	10101	01011	11110
10000	00101	11011	01110
01000	11101	00011	10110
00100	10001	01111	11010
00010	10111	01001	11100
00001	10100	01010	11111
11000	01101	10011	00110
01100	11001	00111	10010

the standard array in Table 7.1 for $(5, 2)$ code, the first row consists of the four codewords. We demarcate it from all the other entries in the table, which are not codewords, by a horizontal line. The next five rows correspond to the five possible one-bit error patterns, which we know can be corrected. Thus, for the jth column, the first six rows correspond to the decoding sphere of Hamming radius one around codeword $\mathbf{x}_j$. We demarcate this by drawing a double line under the sixth row. Beyond these, the first entries of the remaining row are arbitrarily set to be minimum-weight binary vectors that have not appeared yet. We cannot guarantee that we can correct these error patterns. For example, the first and fourth entries in the seventh and eighth rows are both equidistant from the first and fourth codewords, hence neither of these patterns can be mapped unambiguously to a decoding sphere.

Bounded-distance decoding For a code capable of correcting at least t errors, bounded-distance decoding at radius t corresponds to the following rule: decode a received word to the nearest codeword (in terms of Hamming distance), as long as the distance is at most t, and declare *decoding failure* if there is no such codeword. A conceptually simple, but computationally inefficient, way to think about this is in terms of the standard array. For our running example in Table 7.1, bounded distance decoding with $t = 1$ could be implemented by checking whether the received word is anywhere in the first six rows, and, if so, decoding it to the first element of the column it falls in. For example, the received word 10001 is in the fourth row and second column, and is therefore decoded to the second codeword 10101. If the received word is not in the first six rows, then we declare decoding failure. For example, the received word 01101 is in the seventh row and hence does not fall in the decoding sphere of radius one for any codeword, hence we would declare decoding failure if we received it.

Each row of the standard array is the translation of the code C by its first entry, $\mathbf{e}_i$, and is called a *coset* of the code. The first entry $\mathbf{e}_i$ is called the coset leader for the ith coset, $i = 1, \ldots, 2^{n-k}$. We now note that a coset can be described far more economically than by listing all its elements. By applying a parity-check matrix to the jth element of the ith coset, $\mathbf{H}(\mathbf{x}_j \oplus \mathbf{e}_i)^{\mathrm{T}} = \mathbf{H}\mathbf{e}_i^{\mathrm{T}}$, we get an answer that depends only on the coset leader, since $\mathbf{H}\mathbf{x}^{\mathrm{T}} = \mathbf{0}$ for any codeword $\mathbf{x}$. We therefore define the *syndrome* for the ith coset as $\mathbf{s}_i = \mathbf{H}\mathbf{e}_i^{\mathrm{T}}$. The syndrome is a binary vector of length $n - k$, and takes 2^{n-k} possible values. The coset

Table 7.2 Mapping between coset leaders and syndromes for the (5,2) code using (7.25)

Coset leader	Syndrome
00000	000
10000	101
01000	011
00100	100
00010	010
00001	001
11000	110
01100	111

leaders and syndromes corresponding to Table 7.1, using the parity-check matrix (7.25), are listed in Table 7.2.

Bounded-distance decoding using syndromes Consider a received word $\mathbf{y}$. Compute the syndrome $\mathbf{s} = \mathbf{H}\mathbf{y}^{\mathrm{T}}$. If the syndrome corresponds to a coset leader $\mathbf{e}$ that is within the decoding sphere of interest, then we estimate the transmitted codeword as $\hat{\mathbf{x}} = \mathbf{y} + \mathbf{e}$. Consider again the received word $\mathbf{y} = 10001$ and compute its syndrome $\mathbf{s} = \mathbf{H}\mathbf{y}^{\mathrm{T}} = 100$. This corresponds to the fourth row in Table 7.2, which we know is within a decoding sphere of radius one. The coset leader is $\mathbf{e} = 00100$. By adding this to the received word, we obtain $\hat{\mathbf{x}} = \mathbf{y} + \mathbf{e} = 10101$, which is the same result as we obtained by direct look-up in the standard array.

Performance of bounded-distance decoding Correct decoding occurs if the received word is mapped to the transmitted word. For bounded-distance decoding with $t = 1$ for the $(5, 2)$ code, this happens if and only if there is at most one error. Thus, when a codeword for the $(5, 2)$ code is sent over a BSC with crossover probability p, the probability of correct decoding is given by

$$P_{\mathrm{c}} = (1-p)^5 + \binom{5}{1} p(1-p)^4$$

If the decoding is not correct, let us term the event *incorrect decoding.* One of two things happens when the decoding is incorrect: the received word falls outside the decoding sphere of all codewords, hence we declare *decoding failure;* or the received word falls inside the decoding sphere of one of the incorrect codewords, and we have an *undetected error.* The sum of the probabilities of these two events is $P_{\mathrm{e}} = 1 - P_{\mathrm{c}}$. Since decoding failure (where we know something has gone wrong) is less damaging than decoding error (where we do not realize that we have made errors), we would like its probability P_{df} to be much larger than the probability P_{ue} of undetected error. For large block lengths n, we can typically design codes for which this is possible, hence we often take P_{e} as a proxy for decoding failure. For our simple running example, we compute the probabilities of decoding failure and decoding error in Problem 7.13. Exact computations of P_{df} and P_{ue} are difficult for more complex codes, hence we typically resort to bounds and simulations.

Even when we use syndromes to infer coset leaders rather than searching the entire standard array, look-up-based approaches to decoding do not scale well as we increase the code block length n and the decoding radius. A significant achievement of classical coding theory has been to construct codes whose algebraic structure can be exploited to devise efficient means of mapping syndromes to coset leaders for bounded distance decoding (such methods typically involve finding roots of polynomials over finite fields). However, much of the recent progress in coding has resulted from the development of iterative decoding algorithms that are based on message-passing architectures, which permit efficient decoding of very long, random-looking codes that can approach Shannon limits. We now provide a simple illustration of message passing via our running example of the $(5, 2)$ code.

Tanner graph A binary linear code with parity-check matrix $\mathbf{H}$ can be represented as a Tanner graph, with *variable nodes* representing the coded bits, and *check nodes* representing the parity-check equations. A variable node is connected to a parity-check node by an edge if it appears in that parity-check equation. A Tanner graph for our running example $(5, 2)$ code, which is based on the parity-check matrix (7.28), is shown in Figure 7.9. Check node c_1 corresponds to the parity-check equation specified by the first row of (7.28), $x_1 \oplus x_3 = 0$, and is therefore connected to x_1 and x_3. Check node c_2 corresponds to the second row, $x_2 \oplus x_4 = 0$, and is therefore connected to x_2 and x_4. Check node c_3 corresponds to the third row, $x_1 \oplus x_2 \oplus x_5 = 0$, and is connected to x_1, x_2, and x_5. The *degree* of a node is defined to be the number of edges incident on it. The variable nodes $x_1, \ldots, x_5$ have degrees 2, 2, 1, 1, and 1, respectively. The check nodes c_1, c_2, and c_3 have degrees 2, 2, and 3, respectively. The success of message passing on Tanner graphs is sensitive to these degrees, as we shall see shortly.

Bit-flipping-based decoding Let us now consider the following simple message passing algorithm, illustrated via the example in Figure 7.11. As shown in the example, each variable node maintains an estimate of the associated bit, initialized by what was received from the channel. In the particular example we consider, the received sequence is 10000. We know from Table 7.1 that a bounded distance decoder would map this to the codeword 00000. In message passing for bit flipping, each variable node sends out its current bit estimate on all outgoing edges. Each check node uses these incoming messages to generate new

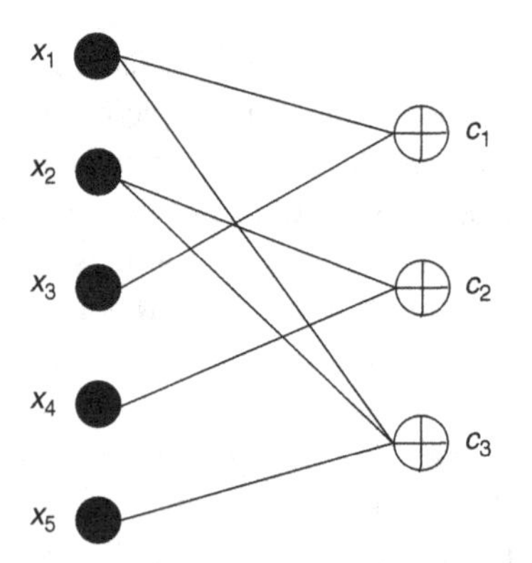

Figure 7.9 A Tanner graph for $(5, 2)$ code with the parity-check matrix given by (7.28).

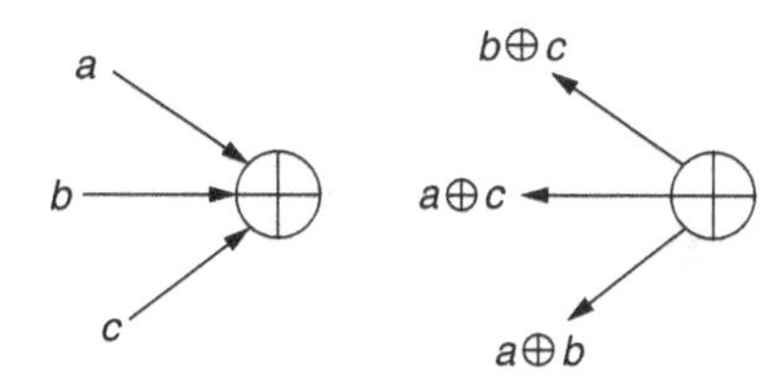

Figure 7.10 Incoming and outgoing messages for a check node.

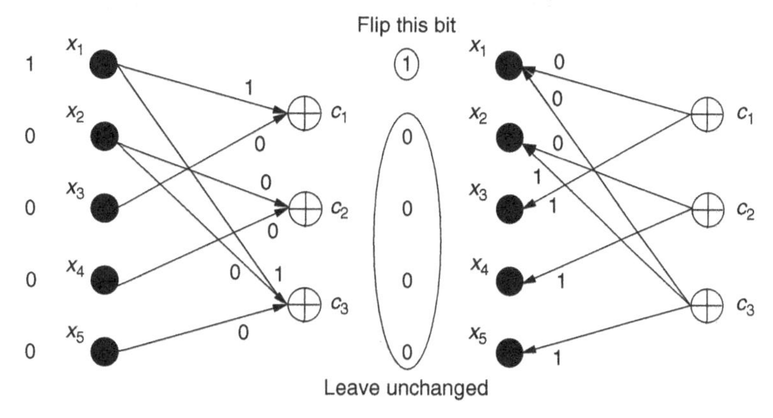

Figure 7.11 Bit-flipping-based decoding for the (5, 2) code is successful for this error pattern.

messages, which are sent back to the variable nodes, as illustrated in Figure 7.10, which shows a check node of degree 3. That is, the message sent back to a variable node is the value that bit should take in order to satisfy that particular parity check, assuming that the messages coming in from the other variable nodes are correct. When the variable nodes get these messages, they flip their bits if "enough" check-node messages tell them to. In our example of a $(5, 2)$ code, let us employ the following rule: a variable node flips its channel bit if (a) all the check messages coming into it tell it to, and (b) the number of check messages is more than one (so as to provide enough evidence to override the current estimate).

Figure 7.11 shows how bit flipping can be used to correct the one-bit error pattern 10000. Both check-node messages to variable node x_1 say that it should take value 0, and cause it to flip to the correct value. On the other hand, Figure 7.12 shows that bit flipping gets stuck for the one-bit error pattern 00001, because there is only one check message coming into variable node x_5, which is not enough to flip it. Note that both of these error patterns are correctable using bounded-distance decoding, using Table 7.1 or Table 7.2. This reveals an important property of iterative decoding on Tanner graphs: its success depends critically on the node degrees, which of course depend on the particular choice of parity-check matrix used to specify the Tanner graph.

Can we fix the problem revealed by the example in Figure 7.12? Perhaps we can choose a different parity-check matrix for which the Tanner graph has variable nodes of degree at least 2, so that bit flipping has a chance of working? For codes over large block lengths,

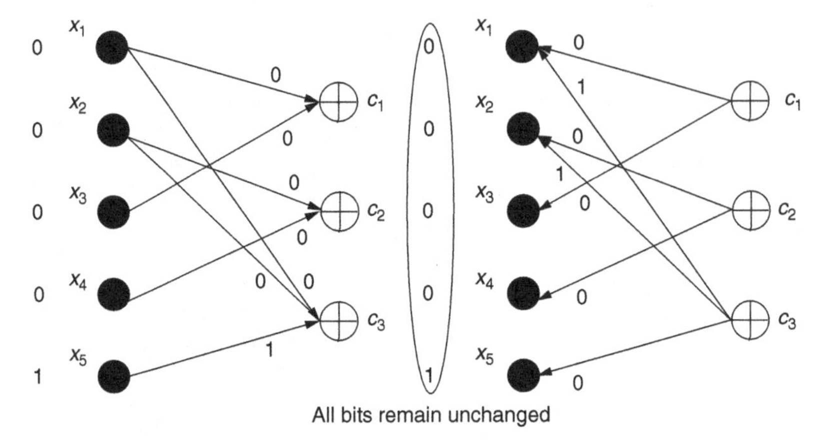

Figure 7.12 Bit-flipping-based decoding for the (5, 2) code is unsuccessful for this error pattern, even though it is correctable using bounded-distance decoding.

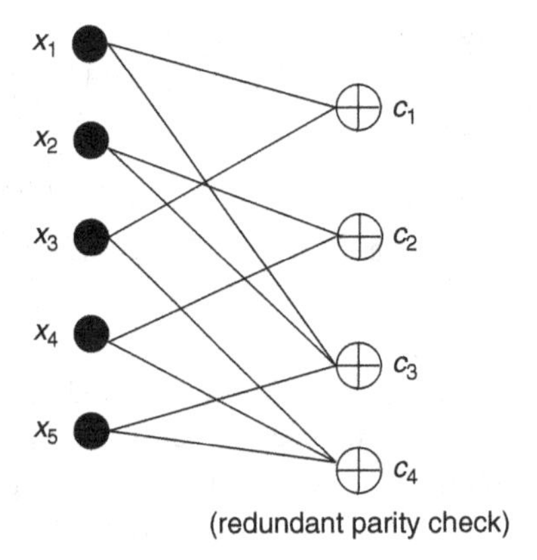

Figure 7.13 A Tanner graph for (5, 2) code with one redundant parity check.

it is actually possible to use a randomized approach for the design of parity-check matrices yielding desirable degree distributions, enabling spectacular performance approaching Shannon limits. In these regimes, iterative decoding goes well beyond the error-correction capability guarantees associated with the code's minimum distance. However, such results do not apply to the simple example we are considering, where iterative decoding is having trouble decoding even up to the guarantee of $t = 1$ associated with a minimum distance $d_{\min} = 3$. However, this gives us the opportunity to present a trick that can be useful even for large blocklengths: use redundant parity-check nodes, adding one or more rows to the parity-check matrix that are linearly dependent on other rows. Figure 7.13 shows a Tanner graph for the $(5, 2)$ code with a redundant check node c_4 corresponding to $x_3 \oplus x_4 \oplus x_5 = 0$. That is, we have added a fourth row 00111 to the parity-check matrix (7.25). This row is actually a sum of the first three rows, and hence would add no further information if we were just performing look-up-based bounded-distance decoding. However, on revisiting the troublesome error pattern 00001, we see that this redundant check makes all the

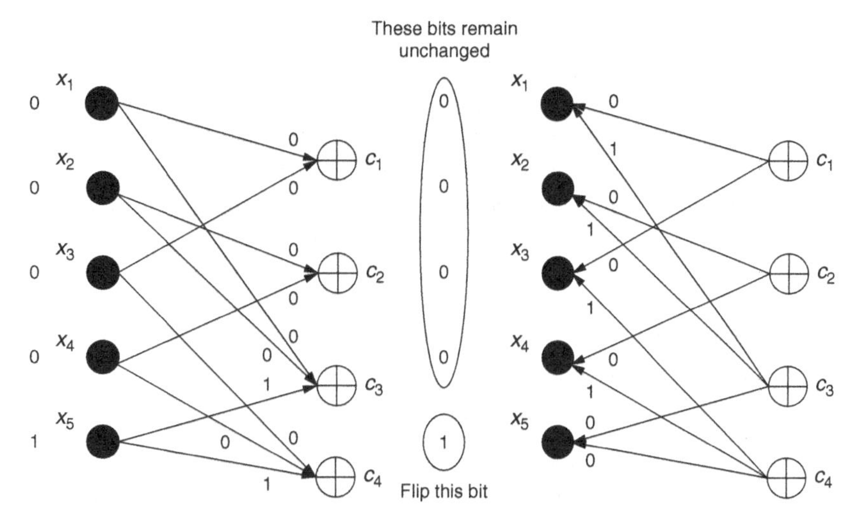

Figure 7.14 Bit-flipping-based decoding for the (5, 2) code using a redundant parity check is now successful for the 00001 error pattern.

difference in the performance of bit-flipping-based decoding; as Figure 7.14 shows, the pattern can now be corrected.

7.5 Soft decisions and belief propagation

We have discussed decoding of linear block codes based on *hard-decision* inputs, where the input to the decoder is a string of bits. However, these bits are sent over a channel using modulation techniques such as those discussed in Chapter 4; and, as discussed in Chapter 6, it is possible to extract *soft decisions* that capture more of the information we have about the physical channel. In this section, we discuss how soft decisions can be used in iterative decoding, illustrating the key concepts using our running example (5, 2) code. We restrict our attention to BPSK modulation over a discrete-time AWGN channel, but, as the discussion in Chapter 6 indicates, the concept of soft decisions is applicable to any signaling scheme.

A codeword $\mathbf{x} = (x[1], \ldots, x[n])$ with elements taking values in $\{0, 1\}$ can be mapped to a sequence of BPSK symbols using the transformation

$$b[m] = (-1)^{x[m]}, \quad m = 1, \ldots, n \tag{7.33}$$

The advantage of this map is that it transforms binary addition into real-valued multiplication. That is, $x[m_1] \oplus x[m_2]$ maps to $b[m_1]b[m_2]$. The BPSK symbols are transmitted over a discrete-time AWGN channel, with received symbols given by

$$y[m] = Ab[m] + w[m] = A(-1)^{x[m]} + w[m], \quad m = 1, \ldots, n \tag{7.34}$$

where the amplitude $A = \sqrt{E_s}$, where E_s denotes the energy/symbol, and $w[m] \sim N(0, \sigma^2)$ are i.i.d. discrete-time WGN samples. To simplify the notation, consider a single bit $x \in \{0, 1\}$, mapped to $b \in \{-1, +1\}$, with received sample $y = Ab + w$, $w \sim N(0, \sigma^2)$. Consider the posterior probabilities $P[x = 0|y]$ and $P[x = 1|y]$. Since $P[x = 0|y] + P[x = 1|y] = 1$, we can convey information regarding these probabilities in a number of ways. One particularly convenient format is the log likelihood ratio (LLR), defined as

$$L(x) = \log\left(\frac{P[x = 0]}{P[x = 1]}\right) = \log\left(\frac{P[b = +1]}{P[b = -1]}\right) \tag{7.35}$$

where we omit the conditioning on y to simplify the notation. We can go from LLRs to bit probabilities as follows:

$$P[x = 0] = \frac{e^{L(x)}}{e^{L(x)} + 1}, \qquad P[x = 1] = \frac{1}{e^{L(x)} + 1} \tag{7.36}$$

We can go from LLRs to hard decisions as follows:

$$\hat{b} = \text{sign}(L), \qquad \hat{x} = I_{\{L<0\}} = I_{\{\hat{b}<0\}} \tag{7.37}$$

where $\hat{b} \in \{-1, +1\}$ is the "BPSK" version of $\hat{x} \in \{0, 1\}$.

Suppose that the prior probability of bit x taking value 0 is $\pi_0(x)$. This notation implies that the prior probability could vary across bits: while we do not need this for the examples considered here, allowing this level of generality is useful for some decoder structures, such as for turbo codes, where the information about bit x supplied by a given decoder component may be interpreted as its prior probability by another decoder component. We can now apply Bayes' rule to show (see Problem 7.17) that the LLR decomposes as follows:

$$L(x) = L_{\text{prior}}(x) + L_{\text{channel}}(x) \tag{7.38}$$

where

$$L_{\text{prior}}(x) = \log\left(\frac{\pi_0(x)}{1 - \pi_0(x)}\right) \tag{7.39}$$

and

$$L_{\text{channel}}(x) = \frac{2Ay}{\sigma^2} \tag{7.40}$$

Thus, the use of the logarithm enables an additive decomposition of information from independent sources, which is both intuitively pleasing and computationally useful. For our present purpose, we can assume that x takes values from $\{0, 1\}$ with equal probability, so that $L_{\text{prior}} = 0$. The LLR $L(x)$ represents the strength of our *belief* in whether the bit is 0 or 1, and LLR-based message passing for iterative decoding is referred to as *belief propagation.*

Belief propagation We describe belief propagation over a Tanner graph for a linear block code by specifying message generation at a generic variable node and a generic check node. In belief propagation, the message going out on an edge is a function of the messages coming in on all of the *other* edges. At a variable node, all of the LLRs involved refer to a given bit, with information coming in from the channel and from check nodes.

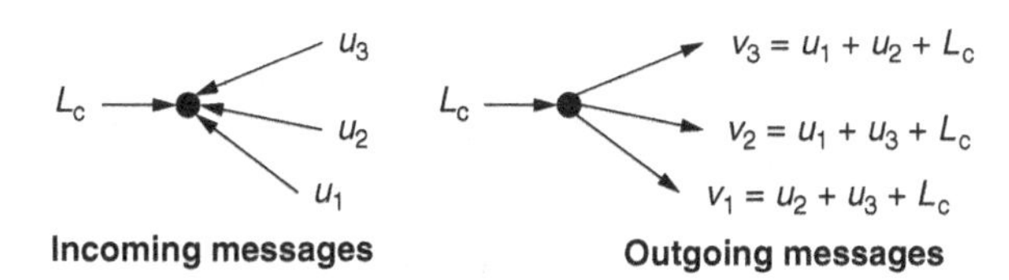

Figure 7.15 Variable node update.

A key approximation in belief propagation is to approximate all of these as independent sources of information, so that the corresponding LLRs add up; this an excellent approximation for large blocklengths that might not really apply to our small running example, but we will go ahead and use it anyway in our numerical examples. Figure 7.15 shows the generation of an outgoing message from a variable node: the outgoing message on an edge is the sum of the incoming message from all *other* edges (including from the channel as well as from the check nodes). Thus, the outgoing message on a given edge equals the sum of all incoming messages, *minus* the incoming message on that edge, and this is the way we implement it in the code fragment below. For simplicity, a node of degree 3 (not counting the edge coming from the channel) is shown in Figure 7.15, but the computation (and the code fragment implementing it) applies to variable nodes of arbitrary degrees.

Code Fragment 7.5.1 (Variable node update)

```
function Lout = variable_update(Lchannel,Lin )
%computes outgoing messages from a variable node
%Lchannel = LLR from channel for that variable
%Lin = vector of LLRs coming in from check nodes
%Lout = vector of LLRs going out to check nodes
%Note: dimension of Lin and Lout = variable node degree
%outgoing message on an edge = sum of incoming messages on all other edges
%(including LLR from channel)
%Efficient computation: sum over all edges and subtract incoming message for each edge
Lout = sum(Lin) + Lchannel - Lin; %vector of the same dimension as Lin
```

Exercise A variable node of degree 3 has channel LLR 0.25, and incoming LLR messages from check nodes $-1.5, 0.5$, and -2.

(a) If you had to make a hard decision on the variable using this information, what would it be?
(b) What are the outgoing messages back to the check nodes?

Answers.

(a) The hard decision would be $\hat{x} = 1$ ($\hat{b} = -1$).
(b) The outgoing messages to the check nodes are $-1.25, -3.25$, and -0.75.

Message generation for check nodes, depicted in Figure 7.16, is more complicated. Consider a check node of degree 3, corresponding to the parity-check equation $x_1 \oplus x_2 \oplus x_3 = 0$. Suppose that the incoming messages tells us that the LLRs for these three bits are

Figure 7.16 Check-node update. The outgoing messages are computed using the tanh rule: $\tanh(u_k/2) = \Pi_{i \neq k} \tanh(v_i/2)$.

$v_1 = L_{in}(x_1)$, $v_2 = L_{in}(x_2)$, and $v_3 = L_{in}(x_3)$. Let us compute the outgoing message $u_3 = L_{out}(x_3)$ on the edge corresponding to variable x_3. We have

$$\begin{aligned} P_{out}[x_3 = 0] &= P_{in}[x_1 = 0, x_2 = 0] + P_{in}[x_1 = 1, x_2 = 1] \\ &= P_{in}[x_1 = 0]P_{in}[x_2 = 0] + P_{in}[x_1 = 1]P_{in}[x_2 = 1] \end{aligned}$$

On plugging in from (7.36), we obtain that

$$\frac{e^{L_{out}(x_3)}}{e^{L_{out}(x_3)} + 1} = \frac{e^{L_{in}(x_1) + L_{in}(x_2)} + 1}{(e^{L_{in}(x_1)} + 1)(e^{L_{in}(x_2)} + 1)} \tag{7.41}$$

As shown in Problem 7.18, this simplifies to

$$\tanh(u_3/2) = \tanh(v_1/2)\tanh(v_2/2) \tag{7.42}$$

We can decompose the preceding into (intermediate) hard decisions and reliabilities as follows:

$$\text{sign}(u_3) = \text{sign}(v_1)\text{sign}(v_2) \tag{7.43}$$

$$\log|\tanh(u_3/2)| = \log|\tanh(v_1/2)| + \log|\tanh(v_2/2)| \tag{7.44}$$

Figure 7.16 illustrates the update for a check node of degree 3. However, these computations generalize to a check node of arbitrary degree, as implemented in the following code fragment.

Code Fragment 7.5.2 (Check-node update)

```
function Lout = check_update(Lin)
%computes messages going out from a check node
%Lin = vector of messages coming in from variable nodes
%Lout = vector of messages going out to variable nodes
%convert LLRs to reliabilities and signs
reliabilities_in = log(abs(tanh(Lin/2)));
signs_in = sign(Lin);
%compute check update
reliabilities_out = sum(reliabilities_in) - reliabilities_in;
sign_product = prod(signs_in);
signs_out = sign_product.* signs_in;
%convert reliabilities and signs back to LLRs
Lout = 2 * atanh(exp(reliabilities_out)).* signs_out;
```

Exercise A check node of degree 4 has incoming LLRs $-3.5, 2.2, 0.25$, and 1.3.

(a) Is the check satisfied by the incoming messages? That is, if we made hard decisions that were based on the incoming LLRs, would they satisfy the parity-check equation corresponding to this node?
(b) Use Code Fragment 7.5.2 to determine the corresponding outgoing LLRs. How are the signs and reliabilities of the outgoing LLRs related to those of the incoming messages?

Answers.

(a) No.
(b) The outgoing LLRs are $0.1139, -0.1340, -0.9217$, and -0.1880. The signs are flipped, and the larger reliabilities become smaller, while the smallest reliability increases. Why does this make sense?

Once we have defined the computations at the variable and check nodes, all that is needed to implement belief propagation is to route messages according to the edges defined by a given parity-check matrix (of course, the choice of code and parity-check matrix determines whether iterative decoding is effective). At any stage of iterative decoding, we can make hard decisions at a variable node using (7.37), where the LLR is the sum of all incoming LLRs, including the channel LLR. If the resulting estimated vector $\hat{\mathbf{x}}$ satisfies $\mathbf{H}\hat{\mathbf{x}} = \mathbf{0}$, then we know that we have obtained a valid codeword and we can terminate the decoding. Typically, if we do not obtain a valid codeword after a specified number of iterations, then we declare decoding failure. We implement belief-propagation-based iterative decoding in Software Lab 7.1; while we use two specific example codes, the software developed in this lab provides a generic implementation of belief propagation for any linear block code once the parity-check matrix has been specified.

7.6 Concept summary

This section provides a glimpse of channel coding concepts, including fundamental performance limits established by Shannon theory and constructive strategies for approaching these limits. Key points are summarized as follows.

- The need for nontrivial channel codes is motivated by examining two extremes when sending a block of bits over a binary symmetric channel: uncoded communication (the probability of packet error tends to one as the blocklength increases) and repetition coding (the code rate tends to zero as the blocklength increases).
- Channel coding consists of introducing structured redundancy into the transmitted bits/symbols. While there are many possible coded modulation strategies, we focus on BICM, a simple, flexible, and effective approach cascading a binary code and an interleaver, followed by mapping of bits to modulated symbols.

Shannon limits

- For a given channel (fixing parameters such as power and bandwidth), Shannon theory tells us that there is a well-defined maximum possible rate of reliable communication, termed the channel capacity. For a passband bandlimited AWGN channel with bandwidth W (Hz), the capacity is given by $W \log_2(1 + E_s/N_0)$, which translates to the following fundamental power–bandwidth tradeoff:

$$E_s/N_0 > 2^r - 1, \quad E_b/N_0 > \frac{2^r - 1}{r}$$

  where E_s is the energy per transmitted symbol, E_b is the energy per *information* bit, and r is the spectral efficiency (the information bit rate normalized by the bandwidth). These results were derived after first showing that the capacity of a discrete-time real AWGN channel is given by $\frac{1}{2} \log_2(1 + S/N)$ bits per channel use.
- The channel capacity for a BSC with crossover probability p is $1 - H_B(p) = 1 + p \log_2 p + (1 - p)\log_2(1 - p)$ bits per channel use. For BICM, such a channel is obtained, for example, by making hard decisions on Gray-coded constellations.
- Shannon limits can be used for guidelines for choosing system sizing: for example, the combination of code rate and constellation size that is appropriate for a given SNR.
- The performance of a given coded modulation strategy can be compared with fundamental limits by comparing the SNR at which it attains a certain performance (e.g., a BER of 10^{-5}) with the minimum SNR required for reliable communication at that spectral efficiency.

Linear codes

- Linear codes are a popular and well-understood design choice in modern communication systems. The 2^k codewords in an (n, k) binary linear code $\mathcal{C}$ form a k-dimensional subspace of the space of n-dimensional binary vectors, under addition and multiplication over the binary field. The dual code $\mathcal{C}^\perp$ is an $(n, n - k)$ linear code such that each codeword in $\mathcal{C}$ is orthogonal (under binary inner products) to each codeword in $\mathcal{C}^\perp$.
- A basis for an (n, k) linear code $\mathcal{C}$ can be used to form a generator matrix $\mathbf{G}$. A k-dimensional information vector $\mathbf{u}$ can be encoded into an n-dimensional codeword $\mathbf{x}$ using the generator matrix: $\mathbf{x} = \mathbf{uG}$.
- A basis for the dual code $\mathcal{C}^\perp$ can be used to form a parity-check matrix $\mathbf{H}$ satisfying $\mathbf{Hx}^T = \mathbf{0}$ for any $\mathbf{x} \in \mathcal{C}$.
- The choices for $\mathbf{G}$ and $\mathbf{H}$ are not unique, since the choice of basis for a linear vector space is not unique. Furthermore, we may add redundant rows to $\mathbf{H}$ to aid in decoding.
- The number of errors t that a code can be *guaranteed* to correct satisfies $2t + 1 \leq d_{min}$, where d_{min} is the minimum Hamming distance between codewords. For a linear code, d_{min} equals the minimum weight among nonzero codewords, since the all-zero vector is always a codeword, and since the difference between codewords is a codeword.

- The translation of codewords by error vectors can be enumerated in a standard array, whose rows correspond to translations of the entire code, termed *cosets,* by a given error pattern, termed the *coset leader.* A more compact representation lists only coset leaders and syndromes obtained by operating the parity-check matrix on a given received word. These can be used to carry out a look-up-based implementation of bounded-distance decoding.

Tanner graphs

- An (n, k) linear code with an $n \times r$ $(r \geq n - k)$ parity-check matrix $\mathbf{H}$ can be represented by a Tanner graph, with n variable nodes on one side and r check nodes on the other side, with an edge between the jth variable and ith check node if and only if $\mathbf{H}(i, j) = 1$.
- Message passing on the Tanner graph can be used for iterative decoding, which scales well to very large code blocklengths. One approach is to employ bit-flipping algorithms with hard-decision inputs and binary messages, but a more powerful approach is to use soft decisions and belief propagation.

Soft decisions and belief propagation

- The messages passed between the variable and check nodes are the bit LLRs. The message going out on an edge depends on the messages coming in on all the other edges.
- Outgoing messages from a variable node are generated simply by summing LLRs. Outgoing messages from a check node are more complicated, but can be viewed as a product of signs, and a sum of reliabilities.

7.7 Notes

The material in this chapter has been selected to make two points: (a) information theory provides fundamental performance benchmarks that can be used to guide parameter selection for communication links; and (b) coding theory provides constructive strategies for approaching these fundamental benchmarks. We now list some *keywords* associated with topics that a systematic exposition of information and coding theory might cover, and then provide some references for further study.

Keywords A systematic study of information theory, and its application to derive theorems in source and channel coding, includes concepts such as entropy, mutual information, divergence, and typicality. A systematic study of the structure of algebraic codes, such as BCH and RS codes, is required in order to understand their construction, their distance properties, and decoding algorithms such as the Berlekamp–Massey algorithm. A study of convolutional codes, their decoding using the Viterbi algorithm, and their performance analysis, is another important component of a study of channel coding. Tight integration of convolutional codes with modulation leads to trellis coded modulation. Suitably interleaving convolutional codes leads to turbo codes, which can be decoded iteratively using the forward–backward, or BCJR, algorithm. LDPC codes, which can be decoded iteratively by message passing over a Tanner graph (as described here and in Software Lab 7.1), are of course an indispensable component in modern communication design.

Further reading One level up from the glimpse provided here is a self-contained introduction to "just enough" information theory to compute performance benchmarks for communication systems, and a selection of constructive coding and decoding strategies (including convolutional, turbo, and LDPC codes), in the author's graduate text [7] (Chapters 6 and 7). The textbook by Cover and Thomas [40] is highly recommended for a systematic and lucid exposition of information theory. Shannon's beautifully written work [41] establishing the field is also highly recommended as a source of inspiration. The textbook by McEliece [42] is a good source for a first exposure to information theory and algebraic coding. A detailed treatment of algebraic coding is provided by the textbook by Blahut [43], while comprehensive treatments of channel coding, including both algebraic and turbo-like codes, are provided in the texts by Lin and Costello [44] and Moon [45].

7.8 Problems

Shannon limits

Problem 7.1 Consider a coded modulation strategy pairing a rate $\frac{1}{4}$ binary code with QPSK. Assuming that this scheme performs 1.5 dB away from the Shannon limit, what are the minimum values of E_s/N_0 (dB) and E_b/N_0 (dB) required in order for the scheme to work?

Problem 7.2 At a BER of 10^{-5}, how far away are the following uncoded constellations from the corresponding Shannon limits: QPSK, 8PSK, 16QAM, and 64QAM. Use the nearest-neighbors approximation for the BER of Gray-coded constellations in Section 6.4.

Problem 7.3 Consider Gray-coded QPSK, 8PSK, 16QAM, and 64QAM.

(a) Assuming that we make ML hard decisions, use the nearest-neighbors approximation for the BER of Gray-coded constellations in Section 6.4 to plot the BER as a function of E_s/N_0 (dB) for each of these constellations.
(b) The hard decisions induce a BSC with crossover probability given by the BERs computed in (a). Using the BSC capacity formula (7.17), plot the capacity in bits per symbol (i.e., the spectral efficiency) as a function of E_s/N_0 (dB) for each constellation. Also plot for comparison the capacity of the bandlimited AWGN channel given by (7.10). Comment on the penalty for hard decisions, as well as any other trends that you see.

Problem 7.4 Consider a BICM system employing a rate $\frac{2}{3}$ binary code with Gray-coded QPSK modulation.

(a) What is E_s/N_0 in terms of E_b/N_0?

(b) In terms of the AWGN capacity region (7.11), what is the Shannon limit for this system (i.e., the minimum required E_b/N_0 in dB)?

(c) Now, consider the suboptimal strategy of making hard decisions, thus inducing a BSC. What is the Shannon limit for the system? What is the degradation in dB due to making hard decisions?

Hint. Hard decisions on Gray-coded QPSK symbols induce a BSC with crossover probability $p = Q\left(\sqrt{2E_s/N_0}\right)$, whose capacity is given by (7.17). The Shannon limit is the minimum value of E_b/N_0 for the capacity to be larger than the code rate being used.

Problem 7.5 A rate $\frac{1}{2}$ binary code is employed using bit-interleaved coded modulation with QPSK, 16QAM, and 64QAM.

(a) What are the bit rates attained by these three schemes when operating over a passband channel of bandwidth 10 MHz (ignore excess bandwidth).

(b) Assuming that each coded modulation scheme operates 2 dB from the Shannon limit, what is the minimum value of E_s/N_0 (dB) required for each of the three schemes to provide reliable communication?

(c) Assume that these three schemes are employed in an adaptive modulation strategy that adapts the data rate as a function of the range. Suppose that the *largest* attainable range among the three schemes is 10 km. Assuming inverse square path loss, what are the ranges corresponding to the other two schemes?

(d) Now, if we add binary codes of rates $\frac{2}{3}$ and $\frac{3}{4}$, plot the attainable bit rate versus range for an adaptive modulation scheme allowing all possible pairings of code rates and constellations. Assume that each scheme is 2 dB away from the corresponding Shannon limit.

Problem 7.6

(a) Apply l'Hôpital's rule to evaluate the limit of the right-hand side of (7.11) as $r \to 0$. What is the minimum value of E_b/N_0 in dB at which reliable communication is possible over the AWGN channel?

(b) Re-plot the region for reliable communication shown in Figure 7.6, but this time with spectral efficiency r (bps/Hz) versus the SNR E_s/N_0 (dB). Is there any lower limit to E_s/N_0 below which reliable communication is not possible? If so, what is it? If not, why not?

Linear codes and bounded-distance decoding

Problem 7.7 A parity-check matrix for the (7, 4) Hamming code is given by

$$\mathbf{H} = \begin{pmatrix} 1 & 0 & 0 & 1 & 0 & 1 & 1 \\ 0 & 1 & 0 & 1 & 1 & 0 & 1 \\ 0 & 0 & 1 & 0 & 1 & 1 & 1 \end{pmatrix} \tag{7.45}$$

(a) Find a generator matrix for the code.

(b) Find the minimum distance of the code. How many errors can be corrected using bounded distance decoding?

Answer. $d_{\min} = 3$, hence a bounded distance decoder can correct one error.

(c) Write down the standard array. Comment on any structural differences you see between this and the standard array for the $(5, 2)$ code in Table 7.1.

Answer. Unlike in Table 7.1, no binary vectors are "left over" after running through the single-error patterns. The Hamming code is a *perfect* code: the decoding spheres of radius one cover the entire space of length-7 binary vectors. "Perfect" in this case just refers to how well decoding spheres can be packed into the available space; it definitely does not mean "good," since the Hamming code is a weak code.

(d) Write down the mapping between coset leaders and syndromes for the given parity-check matrix (as was done in Table 7.2 for the $(5, 2)$ code).

Problem 7.8 Suppose that the $(7, 4)$ Hamming code is used over a BSC with crossover probability $p = 0.01$. Assuming that bounded-distance decoding with decoding radius one is employed, find the probability of correct decoding, the probability of decoding failure, and the probability of undetected error.

Problem 7.9 Append a single parity check to the $(7, 4)$ Hamming code. That is, given a codeword $\mathbf{x} = (x_1, \ldots, x_7)$ for the $(7, 4)$ code, define a new codeword $\mathbf{z} = (x_1, \ldots, x_7, x_8)$ by appending a parity check to the existing code bits:

$$x_8 = x_1 \oplus x_2 \oplus \ldots \oplus x_7$$

This new code is called an *extended* Hamming code.

(a) What are n and k for the new code?

(b) What is the minimum distance for the new code?

Problem 7.10 Hamming codes of different lengths can be constructed using the following prescription: the parity-check matrix consists of all nonzero binary vectors of length m, where m is a positive integer.

(a) What is the value of m for the $(7, 4)$ Hamming code?

(b) For arbitrary m, what are the values of the code blocklength n and the number of information bits k as a function of m?

Hint. The code blocklength is the number of columns in the parity-check matrix. The dimension of the dual code is the rank of the parity-check matrix. Remember that the row rank equals the column rank. Which is easier to find in this case?

Problem 7.11 BCH codes (named after their discoverers, Bose, Ray-Chaudhuri, and Hocquenghem) are a popular class of linear codes with a well-defined algebraic structure and well-understood algorithms for bounded-distance coding. For a positive integer m, we can construct a binary BCH code that can correct at least t errors with the following parameters:

$$n = 2^m - 1, \qquad k \geq n - mt, \qquad d_{\min} \geq 2t + 1 \tag{7.46}$$

so that the code rate $R = k/n \geq 1 - mt/(2^m - 1)$, where the inequality for k is often tight for small values of t. For example, Hamming codes are actually $(2^m - 1, 2^m - 1 - m)$ BCH codes with $t = 1$.

Remark The price of increasing the blocklength of a BCH code is decoding complexity. Algebraic decoding of a code of length $n = 2^m - 1$ requires operations over GF(2^m).

(a) Consider a (1023, 923) BCH code. Assuming that the inequality for k is tight, how many errors can it correct?
Answer. $t = 10$.

(b) Assuming that the inequality for k in (7.46) is tight, what is the rate of a BCH code with $n = 511$ and $t = 10$?

Problem 7.12 Consider an (n, k) linear code used over a BSC channel with crossover probability p. The number of errors among n code bits is $X \sim \text{Bin}(n, p)$. A bounded-distance decoder of radius t is used to decode it (assume that the code is capable of correcting at least t errors). The probability of incorrect decoding is therefore given by

$$P_e = P[X > t] = \sum_{k=t+1}^{n} \binom{n}{k} p^k (1 - p)^{n-k} \tag{7.47}$$

The computation in (7.47) is straightforward, but, for large n, numerical problems can arise when evaluating the terms in the sum directly, because

$$\binom{n}{k}$$

can take very large values, and p^k can take very small values. One approach to alleviate this problem is to compute the binomial pmf recursively.

(a) Show that

$$P[X = k] = \frac{p}{1-p} \frac{n-k+1}{k} P[X = k - 1], \quad k = 1, \ldots, n \tag{7.48}$$

(b) Use the preceding, together with the initial condition $P[X = 0] = (1 - p)^n$, to write a MATLAB program to compute $P[X > t]$. Evaluate for the (1023, 923) BCH code in Problem 7.11(a), assuming that $P = 0.02$.

Problem 7.13 Use the standard array in Table 7.1 for an exact computation of the probabilities of decoding failure and decoding error for the (5, 2) code, for bounded-distance decoding with $t = 1$ over a BSC with crossover probability p. Plot these probabilities as a function of p on a log–log scale.
Hint. Assume that the all-zero codeword is sent. Find the number and weight of error patterns resulting in decoding failure and decoding error, respectively.

Problem 7.14 For the binomial tail probability (7.47) associated with the probability of incorrect decoding, we are often interested in large n and relatively small t; for example, consider the $(1023, 923)$ BCH code in Problem 7.11, for which $t = 10$. While recursive computations as in Problem 7.12 are relatively numerically stable, we are often interested in quick approximations that do not require the evaluation of a large summation with $n - t$ terms. In this problem, we discuss some simple approximations.

(a) We are interested in designing systems to obtain small values of P_e, which we hope are significantly smaller than the input BER p. Argue that $p \geq t/n$ is an uninteresting regime from this point of view. What is the uninteresting regime for the $(1023, 923)$ BCH code?
(b) For $p \ll t/n$, argue that the sum in (7.47) is well approximated by its first term.
(c) Since X is a sum of n i.i.d. Bernoulli random variables, show that the CLT can be used to approximate its distribution by a Gaussian: $X \sim N(np, np(1-p))$.
(d) For the $(1023, 923)$ BCH code, compute a numerical estimate of the probability of incorrect decoding for $t = 10$ and $p = 10^{-3}$ in three different ways: (i) direct computation, (ii) estimation by the first term as in (b), and (iii) estimation using the Gaussian approximation as in (c).
(e) Repeat (d) for $p = 10^{-4}$.
(f) Comment on the match (or otherwise) between the three estimates in (d) and (e). What happens with smaller p?

Problem 7.15 Here are the (n, k, t) parameters for some other binary BCH codes for which the computations of Problem 7.14 can be repeated: $(1023, 863, 16)$, $(511, 421, 10)$, and $(255, 215, 5)$.

Problem 7.16 Reed–Solomon (RS) codes are a widely used class of codes on non-binary alphabets. While we do not discuss the algebraic structure of any of the codes we have mentioned, we state in passing that RS codes can be viewed as a special class of BCH codes. The symbols in an RS code come from GF(2^m) (a finite field with 2^m elements, where m is a positive integer), hence each symbol can be represented by m bits. The code blocklength equals $n = 2^m - 1$. The minimum distance is given by

$$d_{\min} = n - k + 1 \tag{7.49}$$

This is actually the best possible minimum distance attainable for an (n, k) code. It is possible to extend the RS code by one symbol to obtain $n = 2^m$, and to shorten the code to obtain $n < 2^m - 1$, all the while maintaining the minimum-distance relationship (7.49). Bounded-distance decoding can be used to correct up to $\lfloor (d_{\min} - 1)/2 \rfloor = \lfloor (n-k)/2 \rfloor$ or $d_{\min} - 1 = n - k$ erasures, or any pattern of t errors and e erasures satisfying $2t + e + 1 \leq d_{\min} = n - k + 1$. One drawback of RS codes is that it is not possible to obtain code block lengths larger than 2^m, the alphabet size.

(a) What is the maximum number of symbol errors that a $(255, 235)$ RS code can correct? How many bits does each symbol represent? In the worst case, how many bits can the code correct? How about in the best case?

(b) The $(255, 235)$ RS code is used as an outer code in a system in which the inner code produces a BER of 10^{-3}. What is the symbol error probability, assuming that the bit errors are i.i.d.? Assuming bounded-distance decoding up to the maximum possible number of correctable errors, find the probability of incorrect decoding.
Note. The symbol error probability $p = 1 - (1 - p_b)^m$, where p_b is the BER and m the number of bits per symbol.

(c) What is the BER that the inner code must produce in order for the $(255, 235)$ RS code to attain a decoding failure probability of less than 10^{-12}?

(d) If the BER of the inner code is fixed at 10^{-3} and the blocklength and alphabet size of the RS code are as in (b) and (c), what is the value of k for which the decoding failure probability is less than 10^{-12}?

Remark While we consider random bit errors in this problem, inner decoders may often output a burst of errors, and this is where outer RS codes become truly valuable. For example, a burst of errors spanning 30 bits corresponds to at most five symbol errors in an RS code with 8-bit symbols. On the other hand, correcting up to 30 errors using, say, a binary BCH code would cost a lot in terms of redundancy.

LLR computations

Problem 7.17 Consider a BPSK system with a typical received sample given by

$$Y = A(-1)^x + N \tag{7.50}$$

where $A > 0$ is the amplitude, $x \in \{0, 1\}$ is the transmitted bit, and $N \sim N(0, \sigma^2)$ is the noise. Let $\pi_0 = P[x = 0]$ denote the prior probability that $x = 0$.

(a) Show that the LLR

$$L(x) = \log\left(\frac{P[x=0|y]}{P[x=1|y]}\right) = \log\left(\frac{\pi_0 p(y|0)}{(1-\pi_0)p(y|1)}\right)$$

Conclude that

$$L(x) = L_{\text{channel}}(x) + L_{\text{prior}}(x)$$

where $L_{\text{channel}}(x) = \log(p(y|0)/p(y|1))$ and $L_{\text{prior}}(x) = \log(\pi_0/(1-\pi_0))$.

(b) Write down the conditional densities $p(y|x = 0)$ and $p(y|x = 1)$.

(c) Show that the channel LLR $L_{\text{channel}}(x)$ is given by

$$L_{\text{channel}}(x) = \log\left(\frac{p(y|0)}{p(y|1)}\right) = \frac{2Ay}{\sigma^2}$$

(d) Specify (in terms of the parameters A and σ) the conditional distribution of L_{channel}, conditioned on $x = 0$ and $x = 1$.

(e) Suppose that the preceding is used to model either the I sample or the Q sample of a Gray-coded QPSK system. Express E_s/N_0 for the system in terms of A and σ.
Answer. $E_s/N_0 = A^2/\sigma^2$.

(f) Suppose that we use BICM using a binary code of rate R_{code} prior to QPSK modulation. Express E_s/N_0 for the QPSK symbols in terms of E_b/N_0.
Answer. $E_s/N_0 = 2R_{\text{code}}E_b/N_0$.

(e) For E_b/N_0 of 3 dB and a rate $\frac{2}{3}$ binary code, what is the value of A if the noise variance per dimension is scaled to $\sigma^2 = 1$.

(g) For the system parameters in (f), specify numerical values for the parameters governing the conditional distributions of the LLR found in (c).

(h) For the system parameters in (f), specify the probability of bit error for hard decisions that are based on Y.

Problem 7.18 In this problem, we derive the tanh rule (7.42) for the check update, in a way that we hope provides some insight into where the tanh comes from.

(a) For any bit x with LLR L, we have observed that $P[x = 0] = e^L/(e^L + 1)$. Now, show that

$$\delta = P[x = 0] - \frac{1}{2} = \frac{1}{2}\tanh(L/2) \tag{7.51}$$

Thus, the tanh provides a measure of how much the distribution of x deviates from an equiprobable distribution.

Now, suppose that $x_3 = x_1 \oplus x_2$, where x_1 and x_2 are modeled as independent for the purpose of belief propagation. Let L_i denote the LLR for x_i, and set $P[x_i = 0] - \frac{1}{2} = \delta_i$, $i = 1, 2, 3$. (Note that $P[x_i = 1] = \frac{1}{2} - \delta_i$.) Under our model,

$$P[x_3 = 0] = P[x_1 = 0]P[x_2 = 0] + P[x_1 = 1]P[x_2 = 1]$$

(b) Plug in expressions for these probabilities in terms of the δ_i and simplify to show that

$$\delta_3 = 2\delta_1\delta_2$$

(c) Use the result in (a) to infer the tanh rule

$$\tanh(L_3/2) = \tanh(L_1/2)\tanh(L_2/2)$$

Problem 7.19 Consider the Gray-coded 4PAM constellation depicted in Figure 7.17. Denote the label for each constellation point by x_1x_2, where $x_1, x_2 \in \{0, 1\}$. The received sample is given by

$$Y = s + N$$

where $s \in \{-3A, -A, A, 3A\}$ is the transmitted symbol, and $N \sim N(0, \sigma^2)$ is the noise.

(a) Find expressions for the channel LLRs for the two bits:

$$L_{\text{channel}}(x_1) = \log\left(\frac{p(y|x_1 = 0)}{p(y|x_1 = 1)}\right), \qquad L_{\text{channel}}(x_2) = \log\left(\frac{p(y|x_2 = 0)}{p(y|x_2 = 1)}\right)$$

Hint. Note that, assuming all four symbols are equiprobable,

$$p(y|x_1 = 0) = \frac{1}{2}p(y|x_1x_2 = 00) + \frac{1}{2}p(y|x_1x_2 = 01)$$

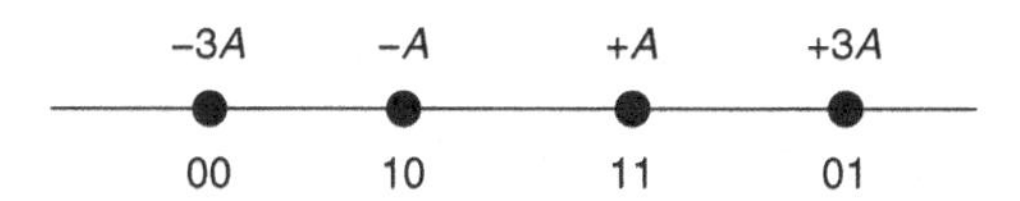

Figure 7.17 A Gray-coded 4PAM constellation.

(b) Simulate the system for $A = 2$, normalizing $\sigma = 1$, and choosing the bits x_1 and x_2 independently and with equal probability from $\{0, 1\}$. Plot the histogram for LLR_1 conditioned on $x_1 = 0$ and conditioned on $x_1 = 1$ on the same plot. Plot the histogram for LLR_2 conditioned on $x_2 = 0$ and conditioned on $x_2 = 1$ on the same plot. Are the conditional distributions in each case well separated?

(c) You wish to design a BICM system with a binary code of rate R_{code} to be used with 4PAM modulation with A and σ as in (b). Using the formula (7.7) for the discrete-time AWGN channel, estimate the code rate to be used, assuming that you can operate 3 dB from the Shannon limit.
Hint. Compute the SNR in terms of A and σ, but then reduce it by 3 dB before plugging into (7.7) to find the bits per channel use.

(d) Repeat (b) and (c) for $A = 1, \sigma = 1$.

Software Lab 7.1: belief propagation

Lab objectives The purpose of this lab is to provide hands-on experience with belief propagation (BP) for decoding. As a warm-up exercise, we first apply BP to our running example (5, 2) code, for which we can compare its performance against bounded-distance decoding. We then introduce *array codes,* a class of LDPC codes with a simple deterministic construction, which can provide excellent performance. While the performance of the array codes we consider here is inferior to that of the best available LDPC codes, the gap can be narrowed considerably by tweaking them (discussion of such modifications is beyond our scope).

Reading Sections 7.4 (linear codes) and 7.5 (belief propagation).

Lab assignment

(1.1) Write a function implementing belief propagation. The inputs are the parity-check matrix, the channel LLRs, and the maximum number of iterations. The outputs are a binary vector that is an estimate of the transmitted codeword, a bit indicating whether this binary vector is a valid codeword, and the number of iterations actually taken. To be concrete, we start defining the function below.

```
function [xhat,valid_codeword,iter] = belief_propagation(H,Lchannel,max_iter)
%%INPUTS
%H = parity-check matrix
%Lchannel = LLRs obtained from channel
```

```
%max_iter = maximum allowed number of iterations
%%OUTPUTS
%xhat = binary vector(estimate of transmitted codeword)
%valid_codeword = 1 if x is a codeword
%iter = number of iterations taken to decode

%%NEED TO FILL IN THE FUNCTION NOW
```

One possible approach to filling in the function is to take the following steps.

(a) *Build the Tanner graph.* Given the parity-check matrix **H**, find and store the neighbors for each variable node and each check node. This can be done using a cell array, as follows.

```
%determine number of nodes on each side of the Tanner graph
[number_check_nodes,n] = size(H);
%Store indices of edges from variable to check nodes
variables_edges_index = cell(n,1);
for j = 1:n
    variables_edges_index{j} = find(H(:,j)==1);
end
%Store indices of edges from check to variable nodes
check_node_edges_index = cell(number_check_nodes,1);
for i = 1:number_check_nodes
    check_node_edges_index{i} = find(H(i,:)==1)';
end
```

(b) *Build the message data structure.* We can maintain messages (LLRs) in a matrix of the same dimension as **H**, with nonzero entries only where **H** is nonzero. The jth variable node will read/write its messages from/to the jth column, while the ith check node will read/write its messages from/to the ith row. Initialize messages from variable nodes to the channel LLRs, and from check nodes to zeros. We maintain two matrices, one corresponding to messages from variable nodes, and one corresponding to messages from check nodes.

```
%messages from variable nodes
Lout_variables = H.* repmat(Lchannel',number_check_nodes,1);
%messages from check nodes
Lout_check_nodes = zeros(size(H));
```

(c) *Implement message passing.* We can now use the variable update and check update functions (Code Fragments 7.5.1 and 7.5.2, respectively), along with the preceding data structure, to implement message passing.

```
%initialize message passing
valid_codeword = 0; %indicates valid codeword found
iter = 0;
while(iter<max_iter && ~valid_codeword)
%loop over check nodes to generate messages
for i = 1:number_check_nodes
    Lout_check_nodes(i,check_node_edges_index{i})
    = check_update(Lout_variables(i,check_node_edges_index{i}));
end
%loop over variable nodes
for j = 1:n
    Lout_variables(variables_edges_index{j},j)
    = variable_update(Lchannel(j),Lout_check_nodes(variables_edges_index{j},j));
end
%check for valid codeword
```

```
bhat = sign(sum(Lout_check_nodes)' + Lchannel); %hard decisions +1,-1
x = (1 - bhat)/2; %convert hard decisions from {+1,-1} to {0,1}
if(mod(H * x,2)==zeros(number_check_nodes,1))
    valid_codeword = 1;
end
iter = iter +1;
end
```

Putting (a)–(c) together gives the desired function.

(1.2) Write a program to check that the preceding belief-propagation function works for our example (5, 2) code, using the parity-check matrix corresponding to the Tanner graph in Figure 7.13, and generating the channel LLRs as described in Problem 7.17. Specifically, consider Gray-coded QPSK modulation, where the I and Q components follow the BPSK model in Problem 7.17. Note that A and σ in Problem 7.17 must be chosen appropriately (fix one, say $\sigma = 1$, and scale the other) based on the spectral efficiency r ($r = 4/5$ for QPSK with the (5, 2) code) and E_b/N_0. Assume, without loss of generality, that the all-zero codeword is sent. Decoding error therefore occurs when the belief-propagation function returns a nonzero codeword, or reports that a valid codeword was not found after the maximum allowed number of iterations.

(1.3) Use simulations to estimate and plot the probability of decoding error (log scale) with BP as a function of E_b/N_0 (dB). On the same graph, also plot the probability of decoding error for bounded-distance decoding with hard decisions (this can be computed analytically, as described in Problem 7.12), and the probability of error for uncoded QPSK. Comment on the results. Does BP with soft decisions provide an improvement over bounded-distance decoding? Is the performance better than that of uncoded QPSK? For your reference, an example *unlabeled* plot is provided in Figure 7.18. Guess the labels for the three plots before verifying them using your own computations and simulations.

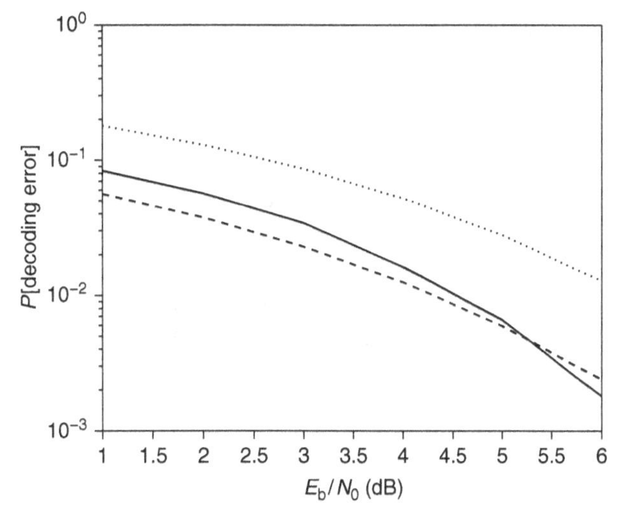

Figure 7.18 The performance of the (5, 2) code with QPSK modulation, comparing belief propagation with soft decisions against bounded-distance decoding with hard decisions. Also plotted for comparison is the performance of uncoded QPSK. Which curve is which?

Array codes We now introduce the class of array codes, whose parity-check matrix is characterized by three positive integers (p, J, L), and is of the following form:

$$\mathbf{H} = \begin{pmatrix} \mathbf{I} & \mathbf{I} & \mathbf{I} & \dots & \mathbf{I} \\ \mathbf{I} & \mathbf{P} & \mathbf{P}^2 & \dots & \mathbf{P}^{L-1} \\ \mathbf{I} & \mathbf{P}^2 & \mathbf{P}^4 & \dots & \mathbf{P}^{2(L-1)} \\ \dots & & & & \\ \mathbf{I} & \mathbf{P}^{J-1} & \mathbf{P}^{2(J-1)} & \dots & \mathbf{P}^{(J-1)(L-1)} \end{pmatrix} \tag{7.52}$$

where $\mathbf{I}$ denotes a $p \times p$ identity matrix, and p is a prime number. The matrix $\mathbf{P}$ is obtained by cyclically shifting the rows of $\mathbf{I}$ by one. Thus, for $p = 3$, we have

$$\mathbf{I} = \begin{pmatrix} 1 & 0 & 0 \\ 0 & 1 & 0 \\ 0 & 0 & 1 \end{pmatrix}, \quad \mathbf{P} = \begin{pmatrix} 0 & 1 & 0 \\ 0 & 0 & 1 \\ 1 & 0 & 0 \end{pmatrix}, \quad \text{for } p = 3$$

The matrix $\mathbf{P}$ is a permutation matrix, in the sense that for any $p \times 1$ vector $\mathbf{u} = (u_1, \dots, u_p)^T$, the vector $\mathbf{Pu}$ is a permutation of $\mathbf{u}$. For this choice of $\mathbf{P}$, the vector $\mathbf{Pu} = (u_p, u_1, \dots, u_{p-1})^T$ is a cyclic shift of $\mathbf{u}$ by one. Raising $\mathbf{P}$ to an integer power k simply corresponds to applying k successive cyclic shifts, so that $\mathbf{P}^k$ is a cyclic shift of the rows of $\mathbf{I}$ by k, and $\mathbf{P}^k\mathbf{u}$ is a cyclic shift of $\mathbf{u}$ by k.

The parity-check matrix $\mathbf{H}$ in (7.52) consists of JL $p \times p$ blocks, with the (j, l)th block being $\mathbf{P}^{(j-1)(l-1)}$, $1 \leq j \leq J$, $1 \leq l \leq L$. The code length $n = pL$, the number of columns (or variable nodes). The column weight equals J for each column (make sure you check this); that is, each variable node has degree J. The number of rows (or check nodes) equals pJ, but some of these rows may be redundant, so the dimension of the dual code $n - k \leq pJ$. In fact, it can be shown that exactly $J - 1$ rows are redundant. (To see why this might be true, add the first p rows, and then the next p rows. What answers do you get? What does this tell you about the number of linearly independent rows?) Thus, the rank of $\mathbf{H}$ equals $n - k = pJ - (J - 1)$, so that the code dimension $k = p(L - J) + J - 1$. We therefore summarize as follows:

$$n = pL, \qquad k = p(L - J) + J - 1 \quad \text{for a } (p, J, L) \text{ array code} \tag{7.53}$$

Popular choices of the variable node degree are $J = 3$ and $J = 4$. Analysis of code properties shows that we should impose the restriction $L \leq p$. We can, for example, use a large prime p and moderate-sized L, or set $L = p$ for a relatively small value of p.

(1.4) Write a function to generate the parity-check matrix of an array code whose inputs are p, J, and L and whose outputs are $\mathbf{H}, n$, and k.

```
function [H,n,k] = array_code(p,J,L)
%Generates the parity-check matrix for an array code
%%INPUTS
%p is a prime
%J = check node degree (column weight), usually set to 3 or 4
%L = parameter <= p that determines code length
%%OUTPUTS
%H = parity check matrix
```

```
%%n = pL (code length)
%%k = p(L - J)+J - 1 (number of info bits)

%p times p identity matrix
Iblock = eye(p);
% can use Matlab's circshift operation on Iblock to generate P and its powers
%for example, circshift(Iblock,[0 (j - 1) * (l - 1)]) generates
  the (j,l)th block of H
%%NOW FILL IN THE FUNCTION!%%%
```

(1.5) Consider an array code with $p = 11$, with $L = p$ and $J = 4$, used as before with Gray-coded QPSK and BICM. As before, use simulations to estimate and plot the probability of decoding error (log scale) with BP as a function of E_b/N_0 (dB) for a BICM system employing QPSK. Compare the performance (E_b/N_0 for a decoding error probability of 10^{-4}) with the Shannon limit for that spectral efficiency. To limit the simulation cost, you may wish to use a relatively small number of simulation runs to generate your plots, and to estimate the value of E_b/N_0 at which the probability of decoding error starts falling below, say, 10^{-2}, and then use a larger number of runs for a few carefully chosen values of E_b/N_0 to see when the decoding error probability hits 10^{-4}. How does this E_b/N_0 compare with that required for a BER of 10^{-4} with uncoded QPSK?

(1.6) Repeat (1.5) for larger values of the prime number p (still keeping $L = p$ and $J = 4$), within the limits of your computational infrastructure. For example, try $p = 47$.

(1.7) Repeat (1.5) with large p and relatively small L; for example, $p = 911$ and $L = 8$, still keeping $J = 4$. How do the code rate and spectral efficiency (with QPSK) compare with those in (1.5) and (1.6)?

Lab report

Your lab report should answer the preceding questions in order, and should document the reasoning you used and the difficulties you encountered. Comment on the decoding error probability trends as you vary the code parameters.

8 Dispersive channels and MIMO

From the material in Chapters 4–6, we now have an understanding of commonly used modulation formats, noise models, and optimum demodulation for the AWGN channel model. Chapter 7 discusses channel coding strategies for these idealized models. In this final chapter, we discuss more sophisticated channel models, and the corresponding signal processing schemes required at the demodulator.

We first consider the following basic model for a *dispersive* channel: the transmitted signal passes through a linear time-invariant system, and is then corrupted by white Gaussian noise. The LTI model is broadly applicable to wireline channels, including copper wires, cable and fiber-optic communication (at least over shorter distances, over which fiber nonlinearities can be neglected), as well as to wireless channels with quasi-stationary transmitters and receivers. For wireless mobile channels, the LTI model is a good approximation over durations that are small compared with the time constants of mobility, but still fairly long on an electronic timescale (e.g., of the order of milliseconds). Methods for compensating for the effects of a dispersive channel are generically termed *equalization*. We introduce two common design approaches for this purpose.

The first approach is *single-carrier* modulation, which refers to the linear modulation schemes discussed in Chapter 4, where the symbol sequence modulates a transmit pulse occupying the entire available bandwidth. We discuss *linear* zero forcing (ZF) and minimum mean-squared error (MMSE) equalization techniques, which are suboptimal from the point of view of minimizing error probability, but are intuitively appealing and less computationally complex than optimum equalization. (We refer the reader to more advanced texts for discussion of optimum equalization and its performance analysis.) We discuss adaptive implementation and geometric interpretation for linear equalizers.

The second approach to channel dispersion is orthogonal frequency-division multiplexing (OFDM), where linear modulation is applied in parallel to a number of *subcarriers,* each of which occupies a bandwidth that is small compared with the overall bandwidth. OFDM may be viewed as a mechanism for ISI *avoidance.* It is based on the observation that any complex exponential $e^{j2\pi f_0 t}$ passes through any LTI system with transfer function $H(f)$ unchanged except for multiplication by $H(f_0)$. Thus, we can send a number of complex exponentials $\{e^{j2\pi f_i t}\}$, termed *subcarriers,* in parallel through the channel, each multiplied by an information-bearing symbol, such that interference across subcarriers is avoided. The task of channel equalization therefore reduces to compensating separately for the channel gains $H(f_i)$ for each such subcarrier. Parallelizing the problem of equalization in this manner is particularly attractive when the underlying time-domain impulse response $h(t)$ is complicated (e.g., an indoor wireless channel where there are many paths with multiple bounces off walls and ceilings between transmitter and receiver). We discuss how this intuition is translated into practice using transceiver implementations employing digital signal processing (DSP).

Finally, we discuss multiple antenna communication, also popularly known as *multiple-input, multiple-output (MIMO)*, or *space–time, communication*. There is a great deal of commonality between signal processing for dispersive channels and for MIMO, which is why we treat these topics within the same chapter. Furthermore, the combination of OFDM with MIMO allows parallelization of transceiver signal processing for complicated channels, and has become the architecture of choice both for WiFi (the IEEE 802.11n standard) and for fourth-generation cellular systems (LTE, or long term evolution). Three key concepts for MIMO are covered: beamforming (directing energy towards a desired communication partner), diversity (combating fading by using multiple paths from transmitter to receiver), and spatial multiplexing (using multiple antennas to support parallel data streams).

Chapter plan

Compared with the earlier chapters, this chapter has a somewhat unusual organization. For dispersive channels, a key goal is to provide hands-on exposure via software labs. A model for single-carrier linear modulation over a dispersive channel, including code fragments for modeling the transmitter and the channel, is presented in Section 8.1. Linear equalization is discussed in Section 8.2. Sections 8.1 and 8.2.1 provide just enough background, including code fragments, for Software Lab 8.1 on adaptive implementation of linear equalization. Section 8.2.2 provides geometric insight into why the implementation in Software Lab 8.1 works, and provides a framework for analytical computations related to MMSE equalization and the closely related notion of zero-forcing (ZF) equalization. It is not required for actually doing Software Lab 8.1. The key concepts behind OFDM and its DSP-centric implementation are discussed in Section 8.3, whose entire focus is to provide background for developing a simplified simulation model for an OFDM link in Software Lab 8.2. Finally, MIMO is discussed in Section 8.4, with the signal processing concepts for MIMO communication reinforced by Software Lab 8.3. The problems at the end of this chapter focus on the linear equalization concepts discussed in Section 8.2.2, and on the performance evaluation of core MIMO techniques (beamsteering, diversity, and spatial multiplexing) discussed in Section 8.4.

Software

As has already been mentioned, this chapter is structured to give the reader some exposure to advanced concepts through the associated software labs: Software Lab 8.1 for single-carrier modulation over dispersive channels, Software Lab 8.2 for OFDM, and Software Lab 8.3 for MIMO signal processing.

8.1 The single-carrier system model

We first provide a system-level overview of single-carrier linear modulation over a dispersive channel. Figure 8.1 shows block diagrams corresponding to a typical DSP-centric realization of the transceiver. The DSP operations are performed on digital streams at an integer multiple of the symbol rate, denoted by m/T. For example, we might choose $m = 4$ for implementing the transmit and receive filters, but we might subsample the output of the receive filter down to $m = 2$ before implementing an equalizer. We model the core components of such a system using the complex-baseband representation, as shown in Figure 8.2. Given the equivalence of passband and complex baseband, we are only skipping modeling of finite-precision effects due to digital-to-analog conversion (DAC) and analog-to-digital conversion (ADC). These effects can easily be incorporated into models such as those we develop, but are beyond our current scope.

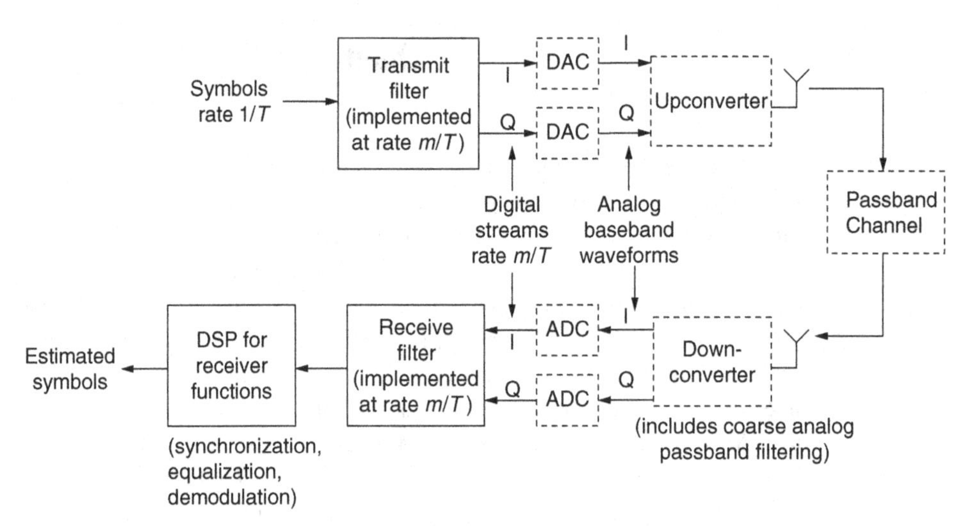

Figure 8.1 A typical DSP-centric transceiver realization. Our model does not include the blocks shown in dashed lines. Finite-precision effects due to digital-to-analog conversion (DAC) and analog-to-digital conversion (ADC) are not considered. The upconversion and downconversion operations are not modeled. The passband channel is modeled as an LTI system in complex baseband.

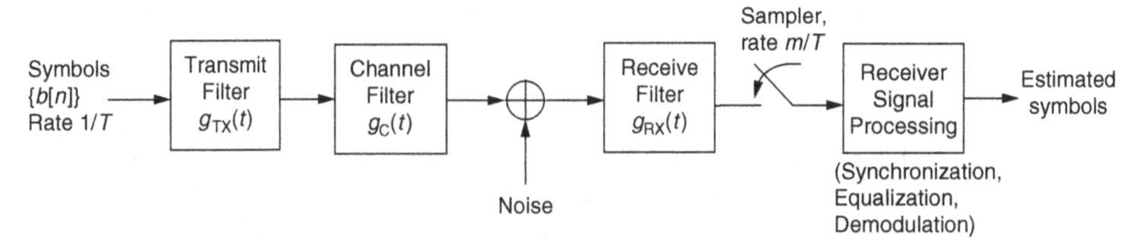

Figure 8.2 The block diagram of a linearly modulated system, modeled in complex baseband.

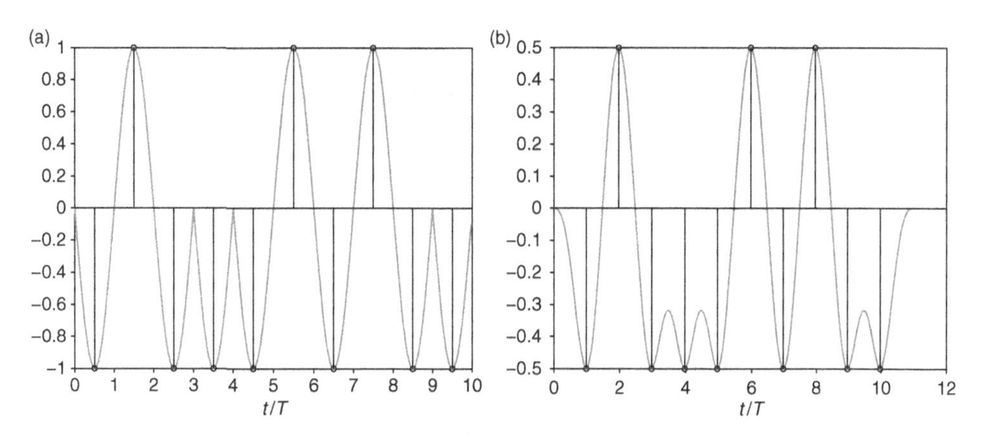

Figure 8.3 The outputs of (a) the transmit filter and (b) the receive filter without channel dispersion. The symbols can be read off from sampling each waveform at the times indicated by the stem plot.

We focus on a hands-on development of the key ideas using discrete-time simulation models, illustrated by code fragments.

8.1.1 The signal model

We begin with an example of linear modulation, to see how ISI arises and can be modeled. Consider linear modulation using BPSK with a sine pulse, which leads to the transmitted baseband waveform shown in Figure 8.3(a). The MATLAB code used for generating this plot is given below. We have sampled much faster than the symbol rate (at $32/T$) in order to obtain a smooth plot. In practice, we would typically sample at a smaller multiple of the symbol rate (e.g., at $4/T$) to generate the input to the DAC in Figure 8.1.

We provide MATLAB code fragments that convey the concepts underlying discrete-time modeling and implementation. The code fragments also show how some of the plots here are generated, with cosmetic touches omitted.

The following code fragment shows how to work with discrete time samples using oversampling at rate m/T, including how to generate the plot of the transmitted waveform in Figure 8.3(a).

Code Fragment 8.1.1 (Transmitted waveform)

```
%choose large oversampling factor for smooth plots
oversampling_factor = 32;
m = oversampling_factor; %for brevity
%generate sine pulse
time_over_symbol = cumsum(ones(m,1)) - 1;
transmit_filter = sin(time_over_symbol * pi/m);
%number of symbols
nsymbols = 10;
%BPSK symbol generation
symbols = sign(rand(nsymbols,1) - .5);
%express symbol sequence at oversampled rate using zero-padding,
%(starts and ends with nonzero symbols)
Lpadded = m * (nsymbols - 1) + 1; %length of zeropadded sequence
```

```
symbolspadded = zeros(Lpadded,1); %initialize
symbolspadded (1:m:Lpadded) = symbols; %fill in bit values every m entries
%now all convolutions can be performed in oversampled domain
transmit_output = conv(symbolspadded,transmit_filter);
%plot transmitted waveform and sampling times
t1 = (cumsum(ones(length(transmit_output))) - 1)/m;
figure;
plot(t1,transmit_output,'b');
xlabel('t/T');
hold on;
%choose sampling times in accordance with peak of transmit filter response
[maxval maxloc] = max(transmit_filter); %find peak location
sampling_times = maxloc:m:(nsymbols - 1) * m + maxloc;
sampled_outputs = transmit_output(sampling_times);
stem((sampling_times - 1)/m,sampled_outputs,'r')
hold off;
```

If this waveform now goes through an ideal channel, and we use a receive filter with impulse response matched to the transmitted pulse, then the waveform we obtain is as shown in Figure 8.3(b). The transmit filter impulse response is timelimited to length T and hence square-root Nyquist (see Chapter 4), hence the net response to a single symbol, which is a cascade of the transmit filter with its matched filter, is Nyquist. It follows that, by sampling at the right moments (as marked on the plot), we can recover the symbols exactly.

We now provide a code fragment to model the channel and receive filter; it can be employed for modeling both ideal and dispersive channels. Appending it to Code Fragment 8.1.1 generates and plots the noiseless received waveform.

Code Fragment 8.1.2 (Modeling the channel and receive filter)

```
dispersive = 0; %set this to 0 for ideal channel, and to 1 for dispersive channel
if dispersive == 0,
channel = 1;
else
channel = [0.8;zeros(m/2,1);-0.7;zeros(m,1);-0.6];
%(or substitute your favorite choice of dispersive channel)
end
%noiseless receiver input
receive_input = conv(transmit_output,channel);
t2 = (cumsum(ones(length(receive_input))) - 1)/m;
figure;
plot(t2,receive_input);
xlabel('t/T');
%receive filter matched to transmit filter
%(would also need to conjugate if complex-valued)
receive_filter = flipud(transmit_filter);
%receive filter output (normalized to account for oversampling)
receive_output = (1/m) * conv(receive_input,receive_filter);
t3 = (cumsum(ones(length(receive_output),1)) - 1)/m;
%plot receive filter output together with sample locations chosen on
%basis of peak of net response
figure;
plot(t3,receive_output,'b');
xlabel('t/T');
hold on;
%effective pulse at channel output
pulse = conv(transmit_filter,channel);
%effective pulse at receive filter output (normalized to account for oversampling)
```

```
rx_pulse = conv(pulse,receive_filter)/m;
[maxval maxloc] = max(rx_pulse);
rx_sampling_times = maxloc:m:(nsymbols - 1) * m + maxloc;
rx_sampled_outputs = receive_output(rx_sampling_times);
stem((rx_sampling_times - 1)/m,rx_sampled_outputs,'r');
hold off;
```

Figure 8.4 shows a dispersive channel and the corresponding noiseless receive filter output. The effective pulse given by the cascade of the transmit, channel, and receive filters is no longer Nyquist, hence we do not expect a symbol decision that is based on a single sample to be reliable. Figure 8.4(b) shows the severe distortion due to ISI with a "best effort" choice of sampling times (chosen based on the peak of the effective pulse). In particular, for the specific symbol sequence shown, one (out of ten) of the symbol estimates obtained by taking the signs of these samples is incorrect.

Eye diagrams A classical technique for visualizing the effect of ISI is the eye diagram. It is constructed by overlapping multiple segments of the received waveform over a fixed window, which tells us how different combinations of symbols could potentially create ISI. For an ideal channel and square-root Nyquist pulses at either end, the eye is *open,* as shown in Figure 8.5(a). However, for the dispersive channel in Figure 8.4(b), we see from Figure 8.5(b) that the eye is *closed.* An open eye implies that, by an appropriate choice of sampling times, we can make reliable single-sample symbol decisions, while a closed eye means that more sophisticated equalization techniques are needed for symbol recovery.

Physically, an eye diagram can be generated using an oscilloscope with the baseband modulated signal as the vertical input, with horizontal sweep triggered at the symbol rate. A code fragment for generating the eye pattern from discrete time samples at rate m/T is given below. (While MATLAB has its own eye-diagram routine, this code fragment is

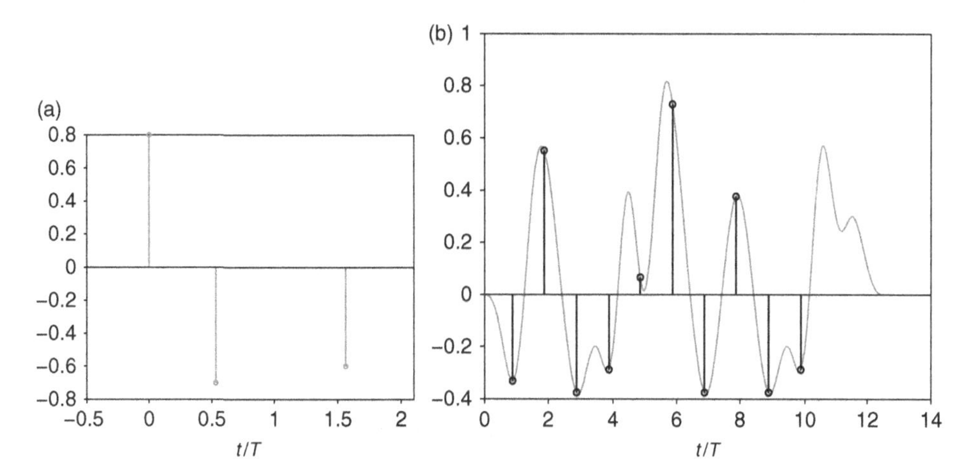

Figure 8.4 When the transmitted waveform passes through the dispersive channel shown in (a), we can no longer read off the symbols reliably by sampling the output of the receive filter shown in (b). For this particular set of symbols, one of the symbols is estimated incorrectly, even though there is no noise.

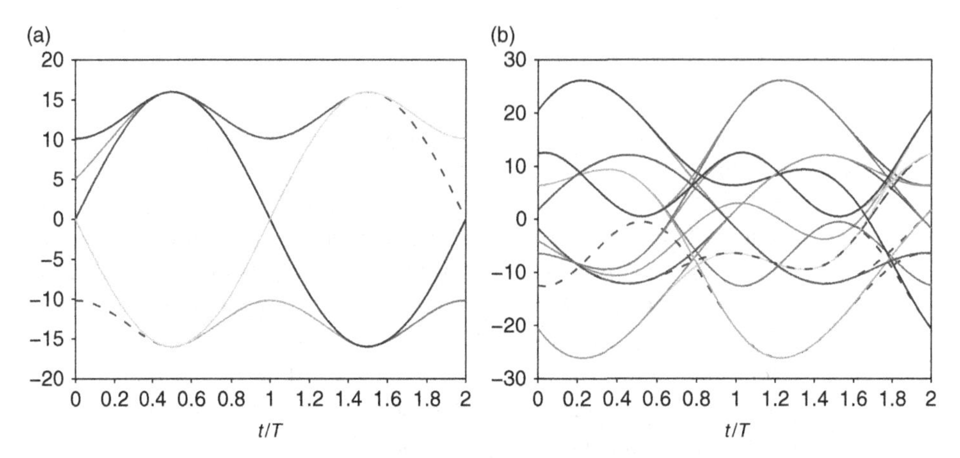

Figure 8.5 Eye diagrams with and without channel dispersion: (a) ideal channel and (b) dispersive channel. The eye is closed for the channel considered, which means that reliable symbol decisions are not possible without equalization.

provided in order to clearly convey the concept.) The output of the receive filter generated in Code Fragment 8.1.2 is the input to this fragment, but in general, we could plot an eye diagram based on the baseband waveform at any stage in the system. For complex baseband signals, we would plot the eye diagrams for the I and Q components separately.

Code Fragment 8.1.3 (Eye diagram)

```
%remove edge effects before doing eye diagram
r1 = receive_out(m/2:m/2 + nsymbols * m - 1);
%horizontal display length in number of symbol intervals
K = 2;
%break into non-overlapping traces
R1 = reshape(r1,K * m,length(r1)/(K * m));
%now enforce continuity across traces
%(append to each trace the first element of the next trace)
row1 = R1(1,:);
L = length(row1);
row_pruned = row1(2:L);
R_pruned = R1(:,1:L - 1);
R2 = [R_pruned;row_pruned];
time = (0:K * m)/m; %time as a multiple of symbol interval
plot(time,R2);
xlabel('t/T');
```

8.1.2 The noise model and SNR

In continuous time, our model for the noisy input to the receive filter is

$$y(t) = \sum_n b[n]p(t - nT) + n(t) \tag{8.1}$$

where $p(t) = (g_{TX} * g_C)(t)$ is the "effective pulse" given by the cascade of the transmit pulse and the channel filter, $\{b[n]\}$ is the symbol sequence, which is in general complex-valued, and $n(t)$ is complex WGN with PSD $\sigma^2 = N_0/2$. We translate this model directly into discrete time by constraining $t = kT/m + \tau$, where m/T is the sampling rate (m is a positive integer) and τ is the sampling offset. The noise at the input to the receive filter is now modeled as discrete-time white Gaussian noise (WGN) with variance $\sigma^2 = N_0/2$ per dimension. As we well know from Chapter 6, the absolute value of the noise variance is meaningless unless we also specify the signal scaling, hence we fix either the signal or the noise strength, and set the other on the basis of SNR measures such as E_b/N_0 or E_s/N_0. Here $E_s = \mathbb{E}[|b[n]|^2]||p||^2$ for the model (8.1), and $E_b = E_s/\log_2 M$ as usual, where M is the constellation size. Inner products and norms are computed in discrete time.

Note that, with the preceding convention, the noise energy in a fixed time interval scales up with the sampling rate, and so does the signal energy (since we have more samples whose energies we are adding up), with the SNR converging to the continuous-time SNR as the sampling rate gets large. However, for a sampling rate that is a small multiple of the symbol rate, the SNR for the discrete-time system can, in general, be different from that in the original continuous-time system. We do not worry about this distinction here.

We now provide a code fragment which adds discrete-time WGN to the receive filter input, resulting in colored noise at the output. We add this to the signal component already computed in Code Fragment 8.1.2.

Code Fragment 8.1.4 (Noise modeling)

```
bn_energy = 1; % for BPSK with current normalization
Es = bn_energy * (pulse' * pulse); %pulse is cascade of transmit and channel filters
constellation_size = 2; %for BPSK
Eb = Es/log2(constellation_size);
%specify Eb/N0 in dB
ebnodb = 5;
ebnoraw = 10^(ebnodb/10); %raw Eb/N0
N0 = Eb/ebnoraw;
%noise standard deviation per dimension
sigma = sqrt(N0/2);
%noise at input to receive filter
noise_receive_input = sigma * randn(size(receive_input));
%(would also need to add an imaginary component for complex-valued signals)
%noise_receive_input = noise_receive_input + 1i * sigma * randn(size(receive_input));
%noise at output of receive filter
noise_receive_output = (1/m) * conv(noise_receive_input,receive_filter);
%noisy receive filter output
receive_output_noisy = receive_output + noise_receive_output;
```

8.2 Linear equalization

We have seen that single-sample symbol decisions are unreliable when the eye is closed. However, what if we are willing to use multiple samples for each symbol decision? Typically, the transmitter and receiver may implement fixed filters in DSP at a faster sampling rate than the sampling rate used eventually for equalization. Thus, suppose that we have samples at rate m/T from the output of the receive filter, but we now wish to use rate-q/T samples for equalization, where q divides m. For example, we may have $m = 4$ and $q = 2$. We subsample the output of the receive filter, taking one out of every m/q samples, and then use L consecutive samples, collected into a vector $\mathbf{r}[n]$, to make a decision on symbol $b[n]$. We would want to choose these samples so that the bulk of the response due to $b[n]$ falls within the *observation interval* over which we collect these samples. When we want to make a decision on the next symbol $b[n+1]$, we must slide this observation interval over by T in order to obtain the received vector $\mathbf{r}[n+1]$. Since our sampling rate is now q/T, this corresponds to an offset of q samples between successive observation intervals. Note that an observation interval typically spans multiple symbol intervals, so that successive observation intervals overlap. Figures 8.6 and 8.7 illustrate this concept for a channel of length $L = 4$ obtained by sampling at rate $2/T$, so that $q = 2$. The overlap between successive observation intervals equals $L - q = 2$ samples.

The transmit, channel, and receive filters are LTI systems and the noise is stationary, and successive symbols are input to the system spaced by time T. Since the discrete-time symbol sequence is stationary as well, the statistics of the signal at any stage of the system are invariant with respect to shifts by integer multiples of T. Such periodicity in the statistics is termed *cyclostationarity*. This implies that the statistics of the noise and ISI seen in different observation intervals are identical: the only change is in which symbol plays the role of the desired symbol. In particular, on comparing Figures 8.6 and 8.7, we see that the roles of desired and interfering symbols shifts by one as we go from the observation interval for $b[0]$ to that for $b[1]$. Thus, an appropriately designed strategy for handling ISI over a given observation interval should work for other observation intervals as well. This opens up the possibility of realizing *adaptive* equalizers that can learn enough about the statistics of the ISI and noise to compensate for them.

We focus here on *linear* equalization, which corresponds to using the decision statistic $\mathbf{c}^{\mathrm{T}}\mathbf{r}[n]$ to estimate $b[n]$, where $\mathbf{c}$ is an appropriately chosen correlator. The choice of $\mathbf{c}$ can be independent of n, by virtue of cyclostationarity. For BPSK signaling, for example, this leads to a decision rule

$$\hat{b}[n] = \mathrm{sign}\left(\mathbf{c}^{\mathrm{T}}\mathbf{r}[n]\right) \tag{8.2}$$

8.2.1 Adaptive MMSE equalization

While constraining ourselves to linear equalization is suboptimal (discussion of optimal equalization is beyond our present scope), we can try to optimize $\mathbf{c}$ to combat ISI and noise. In particular, the linear MMSE criterion corresponds to choosing $\mathbf{c}$ so as to minimize the mean-squared error (MSE) between the decision statistic and the desired symbol, defined as

$$MSE = J(\mathbf{c}) = \mathbb{E}\left[(\mathbf{c}^{\mathrm{T}}\mathbf{r}[n] - b[n])^2\right] \tag{8.3}$$

Minimizing the MSE in this fashion leads to minimizing the contribution due to ISI and noise at the correlator output, which is clearly a desirable outcome.

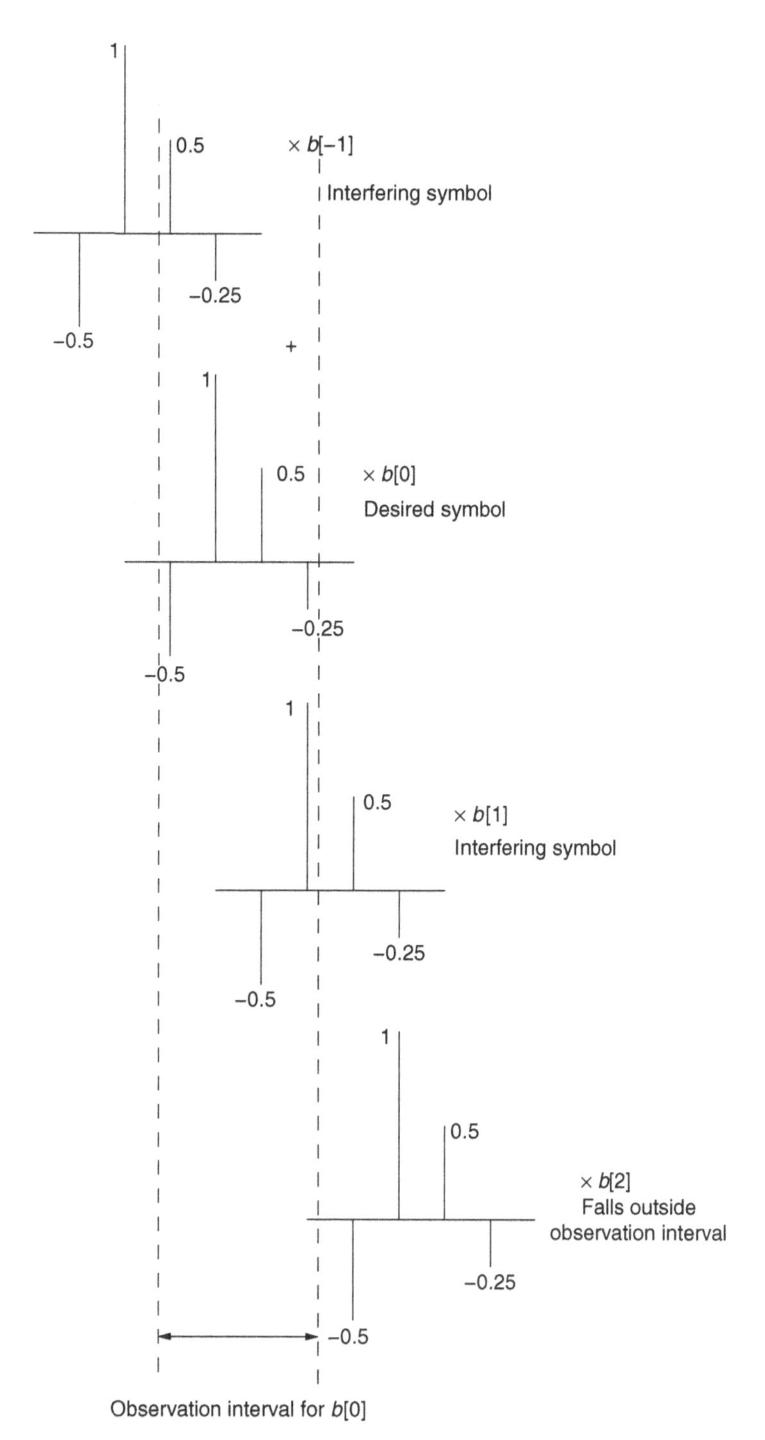

Figure 8.6 The observation interval used to make a decision on $b[0]$ sees contributions from the desired symbol $b[0]$ and interfering symbols $b[-1]$ and $b[1]$.

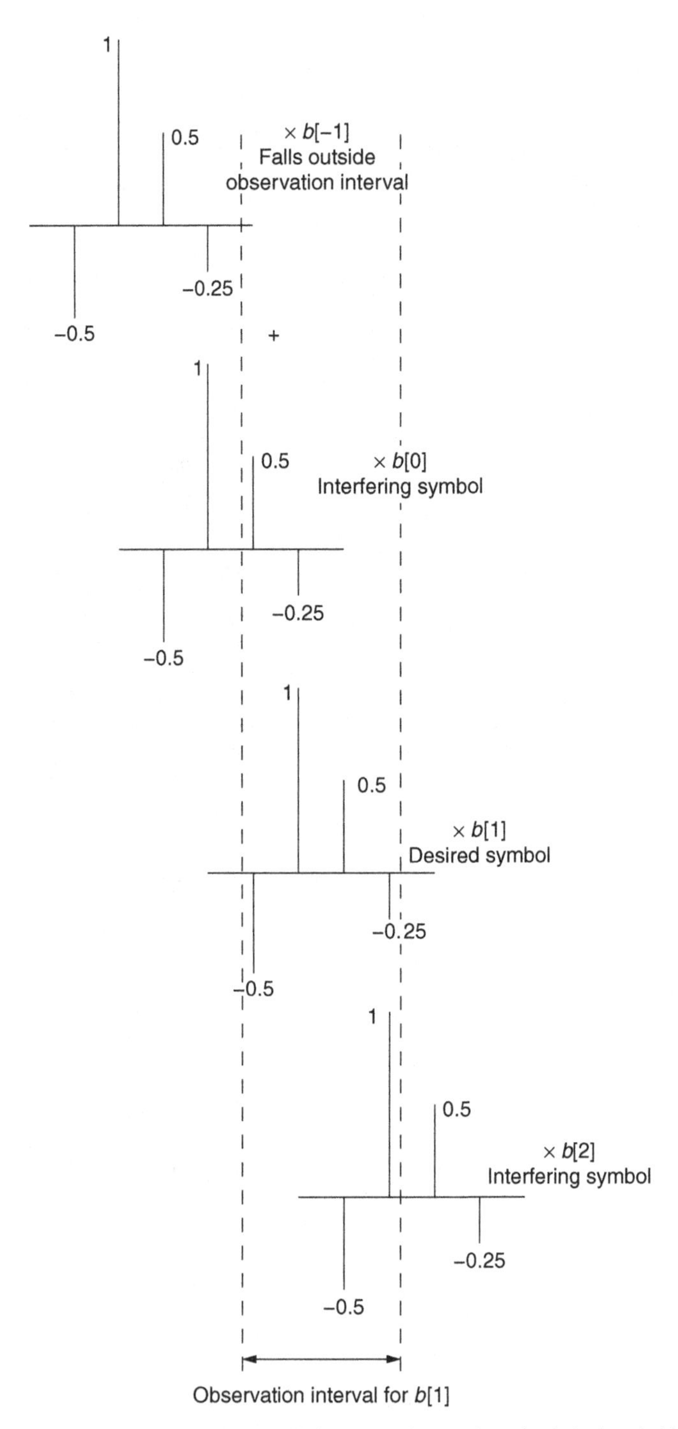

Figure 8.7 The observation interval used to make a decision on $b[1]$ sees contributions from the desired symbol $b[1]$ and interfering symbols $b[0]$ and $b[2]$. In comparison with Figure 8.6, the roles of the symbols have been shifted by one.

The MSE is a quadratic function of $\mathbf{c}$, and can therefore be minimized by setting its gradient with respect to $\mathbf{c}$ to zero. Owing to linearity, the gradient can be taken inside the expectation, and we obtain

$$\nabla_{\mathbf{c}} J(\mathbf{c}) = 2\mathbb{E}\left[\mathbf{r}[n](\mathbf{c}^{\mathrm{T}}\mathbf{r}[n] - b[n])\right] = 2\mathbb{E}\left[\mathbf{r}[n](\mathbf{r}^{\mathrm{T}}[n]\mathbf{c} - b[n])\right]$$

On defining

$$\mathbf{R} = \mathbb{E}\left[\mathbf{r}[n]\mathbf{r}^{\mathrm{T}}[n]\right], \quad \mathbf{p} = \mathbb{E}[b[n]\mathbf{r}[n]] \tag{8.4}$$

we can rewrite the gradient of the MSE as

$$\nabla_{\mathbf{c}} J(\mathbf{c}) = 2(\mathbf{R}\mathbf{c} - \mathbf{p}) \tag{8.5}$$

Setting the gradient to zero yields the following expression for the MMSE correlator:

$$\mathbf{c}_{\mathrm{MMSE}} = \mathbf{R}^{-1}\mathbf{p} \tag{8.6}$$

In order to compute this, we must know, or be able to estimate, the expectations in (8.4). If we know the transmit filter, the channel filter, the receive filter, the sampling times, and the noise PSD, we can compute these expectations using a model such as (8.13). However, we often do not have explicit knowledge of one or more of these quantities. Thus, an attractive approach in practice is to exploit the stationarity of the model as we vary n to estimate expectations using their *empirical averages*. These expectations involve the received vectors $\mathbf{r}[n]$, which we of course have access to, and the symbols $b[n]$, which we assume we have access to over a *training* period during which a known sequence of symbols is transmitted. This approach leads to *adaptive* equalization techniques that do not require explicit knowledge or estimates of the model parameters.

Least-squares adaptation Assuming that the first n_{training} symbols are known, least-squares adaptation corresponds to replacing the expectations in (8.4) by their empirical averages as follows:

$$\hat{\mathbf{R}} = \frac{1}{n_{\mathrm{training}}} \sum_{n=1}^{n_{\mathrm{training}}} \mathbf{r}[n]\mathbf{r}^{\mathrm{T}}[n], \quad \hat{\mathbf{p}} = \frac{1}{n_{\mathrm{training}}} \sum_{n=1}^{n_{\mathrm{training}}} b[n]\mathbf{r}[n] \tag{8.7}$$

where the normalization by $1/n_{\mathrm{training}}$ is not needed, but is put in to make the averaging interpretation transparent. The MMSE correlator is now approximated by the least-squares solution:

$$\hat{\mathbf{c}}_{\mathrm{LS}} = (\hat{\mathbf{R}})^{-1}\hat{\mathbf{p}} \tag{8.8}$$

This correlator can now be used to make decisions on the unknown symbols following the training period. It can be checked that the preceding solution minimizes the empirical MSE over the training period:

$$\hat{\mathrm{MSE}} = \sum_{n=1}^{n_{\mathrm{training}}} \left(\mathbf{c}^{\mathrm{T}}\mathbf{r}[n] - b[n]\right)^2$$

Filter implementation of linear equalization For conceptual clarity, we have introduced linear equalization as a correlator operating on the received vectors $\{\mathbf{r}[n]\}$ obtained by windowing the samples at the output of the receive filter. However, an efficient technique for generating the decision statistics $\mathbf{c}^{\mathrm{T}}\mathbf{r}[n]$ is by passing the received samples through a discrete-time filter matched to $\mathbf{c}$, and then subsampling the output at the symbol rate with an appropriate delay.

The following code fragment implements and tests least-squares adaptation, comparing it with unequalized estimates obtained by sampling at the peaks of the net response to a symbol.

Code Fragment 8.2.1 (Least-squares adaptive equalization)

```
%Use Code Fragments 8.1.1 and 8.1.2 with a large value of nsymbols
%(first ntraining symbols assumed to be known)
%Insert noise using Code Fragment 8.1.4
%downsample to q/T to get input to equalizer
q = 2;
r = receive_output_noisy(1:m/q:length(receive_output_noisy));
%figure out net response to a single symbol at receive filter output
rx_pulse = (1/m) * conv(pulse,receive_filter);
%effective response after downsampling
h = rx_pulse(1:m/q:length(rx_pulse));
%set equalizer length
L = 6;
%choose how to align correlator (e.g., to maximize desired vector energy)
desired_energy = conv(h.^2,ones(L,1));
[max_energy loc_max_energy] = max(desired_energy);
%choose offset to align correlator with desired vector to maximize energy
offset = max(loc_max_energy - L,0)
%another option: set equalizer length equal to effective response
%L = length(h);
%offset = 0;
%initialize for least-squares adaptation
phat = zeros(L,1);
Rhat = zeros(L,L);
for n = 1:ntraining,
    rn = r(1 + q * (n - 1) + offset:L + q * (n - 1) + offset); %current received
        vector r[n]
    phat = phat + symbols(n) * rn;
    Rhat = Rhat + rn * rn';
end
%least-squares estimate of MMSE correlator
cLS = Rhat\phat; %often more stable computation than inv(Rhat) * phat
%implement equalizer as filter
h_equalizer = flipud(cLS); %would also need conjugation for complex signals
equalizer_output = conv(r,h_equalizer);
%sample filter output at symbol rate after appropriate delay
delay = length(h_equalizer) + offset;
%symbol decision statistics
decision_stats = equalizer_output(delay:q:delay + (nsymbols - 1) * q);
%payload = non-training symbols
payload = symbols(ntraining + 1:nsymbols);
%estimate of payload (for BPSK)
payload_estimate = sign(decision_stats(ntraining + 1:nsymbols));
%number of errors
nerrors = sum(ne(payload,payload_estimate))
%COMPARE WITH UNEQUALIZED ESTIMATES
%unequalized estimates obtained by sampling at peaks of effective response
[maxval maxloc] = max(h);
sampling_times = maxloc:q:(nsymbols - 1) * q + maxloc;
unequalized_decision_stats = r(sampling_times);
sampled_outputs = transmit_output(sampling_times);
%estimate of payload (for BPSK)
payload_estimate_unequalized = sign(unequalized_decision_stats(ntraining
    + 1:nsymbols));
%number of errors
nerrors_unequalized = sum(ne(payload,payload_estimate_unequalized))
```

On putting Code Fragments 8.1.1, 8.1.2, 8.1.4, and 8.2.1 together, we obtain a simulation model for adaptive linear equalization over a dispersive channel. As a quick example, for the dispersive channel considered, at E_b/N_0 of 7 dB, we estimate (using $n_{symbols} = 10{,}000$, and $n_{training} = 100$) the error probability after equalization at rate $2/T$ ($q = 2$) to be about 3.5×10^{-3} and the unequalized error probability to be about 0.16. Linear equalization is quite effective in this case, although it exhibits some degradation relative to the ideal BPSK error probability of 7.7×10^{-4}. We can now build on this code base to run a variety of experiments, as suggested in Software Lab 8.1: for example, evaluating the probability of error as a function of E_b/N_0 for different equalizer lengths, for different channel models, and for different choices of the transmit and receive filters. Our model extends easily to complex-valued constellations, as discussed below.

Extension to complex-valued signals All of the preceding development goes through for complex-valued constellations and signals, except that vector transposes $\mathbf{x}^T$ are replaced by conjugate transposes $\mathbf{x}^H$. Indeed, the MATLAB code fragments we provide here already include this level of generality, since we use the conjugate-transpose operation $\mathbf{x}'$ when computing the transpose for real-valued $\mathbf{x}$. All that is needed to employ these code fragments is to make the symbols complex-valued, and to add an imaginary component to the noise model in Code Fragment 8.1.4. We skip derivations, and state that the decision statistics are given by $\mathbf{c}^H\mathbf{r}[n]$, the MSE expression is

$$\text{MSE} = J(\mathbf{c}) = \mathbb{E}\left[|\mathbf{c}^H\mathbf{r}[n] - b[n]|^2\right]$$

and the MMSE solution is given by (8.6) as before, with

$$\mathbf{R} = \mathbb{E}[\mathbf{r}[n]\mathbf{r}^H[n]], \quad \mathbf{p} = \mathbb{E}[b^*[n]\mathbf{r}[n]] \tag{8.9}$$

As before, these statistical expectations can be replaced by empirical averages for a least-squares implementation.

We now have the background required for a hands-on exposure to equalization through Software Lab 8.1.

8.2.2 Geometric interpretation and analytical computations

Computer simulations using the code fragments in Sections 8.1 and 8.2 show that adaptive MMSE equalization works well, at least in the specific examples considered in these sections and in Software Lab 8.1. We now develop geometric insight into why linear equalization works well when it does, and when it might run into trouble. We stick with real-valued signals, but the results extend easily to complex-valued signals, as noted in the appropriate places. *This section is not required for doing Software Lab 8.1.*

Consider the example depicted in Figures 8.6 and 8.7, where the overall sampled response (at rate $2/T$) to a single symbol is assumed to be

$$\mathbf{h} = (\ldots, 0, -0.5, 1, 0.5, -0.25, 0, \ldots)$$

Consider an observation interval (i.e., equalizer length) of length $L = 4$, aligned with the response to the desired symbol as depicted in Figures 8.6 and 8.7. As shown in Code Fragment 8.2.1, we can also choose smaller or larger observation intervals, and optimize their alignment using some criterion (in the code fragment, the criterion is maximizing the energy of the desired response falling into the observation interval). In addition to the contribution to $\mathbf{r}[n]$ due to $b[n]$, we also have contributions from other symbols before and after it in the sequence, corresponding to parts of appropriately shifted versions of the response $\mathbf{h}$. For example, the response to $b[n+1]$ falling in the nth observation interval is obtained by shifting $\mathbf{h}$ by $q = 2$ and then windowing. The received vector $\mathbf{r}[n]$ can therefore be written as follows.

Model for $L = 4$ Two interfering symbols fall into the observation interval. The observation interval is large enough to accommodate the entire response due to the desired symbol:

$$\mathbf{r}[n] = b[n]\begin{pmatrix} -0.5 \\ 1 \\ 0.5 \\ -0.25 \end{pmatrix} + b[n+1]\begin{pmatrix} 0 \\ 0 \\ -0.5 \\ 1 \end{pmatrix} + b[n-1]\begin{pmatrix} 0.5 \\ -0.25 \\ 0 \\ 0 \end{pmatrix} + \mathbf{w}[n] \tag{8.10}$$

where our convention is that time progresses downward, and $\mathbf{w}[n]$ denotes noise. The vector multiplying $b[n]$ is the *desired* vector, while the others are *interference* vectors. Figure 8.6 corresponds to $n = 0$, while Figure 8.7 corresponds to $n = 1$.

In order to obtain the preceding model, the vector corresponding to a given symbol is obtained by appropriately shifting $\mathbf{h}$, and then windowing to the observation interval. In order to ensure that the modeling approach is clear, we also provide the model for $L = 3$, where the observation interval is lined up with the first three elements of the response to the desired symbol, and that for $L = 6$, where the observation interval contains two additional samples on either side of the response to the desired symbol.

Model for $L = 3$ The observation interval is smaller than the desired symbol response. Two interfering symbols fall in the interval:

$$\mathbf{r}[n] = b[n]\begin{pmatrix} -0.5 \\ 1 \\ 0.5 \end{pmatrix} + b[n+1]\begin{pmatrix} 0 \\ 0 \\ -0.5 \end{pmatrix} + b[n-1]\begin{pmatrix} 0.5 \\ -0.25 \\ 0 \end{pmatrix} + \mathbf{w}[n] \tag{8.11}$$

Model for $L = 6$ The observation interval is larger than the desired symbol response. Four interfering symbols fall in the interval:

$$\mathbf{r}[n] = b[n]\begin{pmatrix} 0 \\ -0.5 \\ 1 \\ 0.5 \\ -0.25 \\ 0 \end{pmatrix} + b[n+1]\begin{pmatrix} 0 \\ 0 \\ 0 \\ -0.5 \\ 1 \\ 0.5 \end{pmatrix} + b[n+2]\begin{pmatrix} 0 \\ 0 \\ 0 \\ 0 \\ -0.5 \end{pmatrix}$$

$$+ b[n-1]\begin{pmatrix} 1 \\ 0.5 \\ -0.25 \\ 0 \\ 0 \\ 0 \end{pmatrix} + b[n-2]\begin{pmatrix} -0.25 \\ 0 \\ 0 \\ 0 \\ 0 \\ 0 \end{pmatrix} + \mathbf{w}[n] \tag{8.12}$$

Vector model for ISI In general, we can write the received vector over observation interval n as follows:

$$\mathbf{r}[n] = b[n]\mathbf{u}_0 + \sum_{k \neq 0} b[n+k]\mathbf{u}_k + \mathbf{w}[n] \quad (8.13)$$

where $b[n]$ and $\mathbf{u}_0$ are the desired symbol and vector, respectively; $b[n+k]$ and $\mathbf{u}_k$ for $k \neq 0$ are interference symbols and vectors, respectively; and $\mathbf{w}[n] \sim N(0, \mathbf{C_w})$ denotes the vector of noise samples at the output of the receive filter, windowed to the current observation interval. For an equalizer working with rate-q/T samples, we have already noted that successive observation intervals are offset by q samples. Clearly, the structure of the ISI remains the same as we go from observation interval n to $n+1$, but the roles of the symbols are shifted by one: for the $(n+1)$st observation interval, $b[n+1]$ is the desired symbol multiplying $\mathbf{u}_0$, while $b[n+1+k]$ for $k \neq 0$ is the interfering symbol multiplying $\mathbf{u}_k$.

Modeling the output of a linear correlator A linear correlator $\mathbf{c}$ operating on the received vector produces the following output:

$$\mathbf{c}^T\mathbf{r}[n] = b[n]\mathbf{c}^T\mathbf{u}_0 + \sum_{k \neq 0} b[n+k]\mathbf{c}^T\mathbf{u}_k + \mathbf{c}^T\mathbf{w}[n] \quad (8.14)$$

where the first term is the desired term, the second term is the ISI at the correlator output, and the third term is the noise at the correlator output. While the ultimate performance metric of interest is the error probability, a convenient metric that is easy to compute is the signal-to-interference-plus-noise ratio (SINR) at the output of the linear correlator. This is defined as the ratio of the average energy of the desired term to those of the undesired terms:

$$\text{SINR} = \frac{\mathbb{E}[|b[n]\mathbf{c}^T\mathbf{u}_0|^2]}{\mathbb{E}\left[|\sum_{k \neq 0} b[n+k]\mathbf{c}^T\mathbf{u}_k + \mathbf{c}^T\mathbf{w}[n]|^2\right]} \quad (8.15)$$

Assuming that the symbols are uncorrelated with $\mathbb{E}[|b[n]|^2] \equiv \sigma_b^2$ and are independent of the noise, we obtain the following expression for the SINR:

$$\text{SINR} = \frac{\sigma_b^2|\mathbf{c}^T\mathbf{u}_0|^2}{\sigma_b^2 \sum_{k \neq 0} |\mathbf{c}^T\mathbf{u}_k|^2 + \mathbf{c}^T\mathbf{C_w}\mathbf{c}} \quad (8.16)$$

Choosing $\mathbf{c}$ to minimize the MSE (8.3) means that we would like to have $\mathbf{c}^T\mathbf{r}[n] \approx b[n]$. This means that, if the linear MMSE equalizer is working, then $\mathbf{c}^T\mathbf{u}_0 \approx 1$, and the ISI terms $\mathbf{c}^T\mathbf{u}_k$, $k \neq 0$, and the output noise variance $\mathbf{c}^T\mathbf{C_w}\mathbf{c}$, are small. The MMSE criterion represents a tradeoff between ISI and noise at the output. To see why, let us consider the closely related criterion of *zero-forcing* equalization. While the noise in the example considered in our code fragments is colored, let us first consider white noise for simplicity: $\mathbf{w}[n] \sim N(0, \sigma^2\mathbf{I})$, so that the output noise $\mathbf{c}^T\mathbf{w}[n] \sim N(0, \sigma^2||\mathbf{c}||^2)$.

The geometry of zero-forcing equalization The zero-forcing (ZF) equalizer is a linear equalizer chosen to set the ISI terms at the output exactly to zero:

$$\mathbf{c}^T\mathbf{u}_k = 0, \quad k \neq 0 \quad (8.17)$$

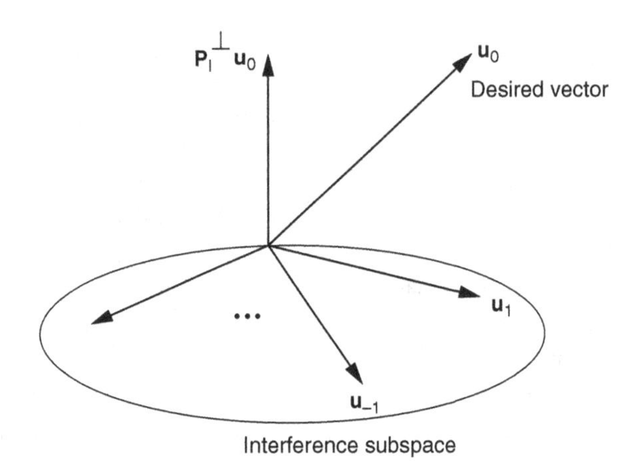

Figure 8.8 The zero-forcing correlator projects the received signal along $\mathbf{P}_I^{\perp}\mathbf{u}_0$, the projection of the desired signal vector orthogonal to the interference subspace.

while scaling the desired term to the right level:

$$\mathbf{c}^T\mathbf{u}_0 = 1 \tag{8.18}$$

The first condition (8.17) means that $\mathbf{c}$ must be orthogonal to the *interference subspace,* which is our term for the subspace spanned by the interference vectors $\{\mathbf{u}_k, k \neq 0\}$. If (8.17) is satisfied, then the second condition (8.18) can be satisfied only if the desired vector $\mathbf{u}_0$ does not lie in the interference subspace, otherwise we would have $\mathbf{c}^T\mathbf{u}_0 = 0$ (why?). Thus, the zero-forcing equalizer exists only if the desired vector $\mathbf{u}_0$ is linearly independent of the interference vectors $\{u_k, k \neq 0\}$, in which case it has a nonzero component $\mathbf{P}_I^{\perp}\mathbf{u}_0$ orthogonal to the interference subspace, as shown in Figure 8.8. In this case, if we choose $\mathbf{c}$ to be a scalar multiple of this orthogonal component, then (8.17) is satisfied by construction, and (8.18) can be satisfied by choosing the scale factor appropriately, as discussed shortly. Indeed, while the solution to (8.17) and (8.18) need not be unique, it can be shown (see Problem 8.5) that choosing $\mathbf{c}_{ZF} = \alpha\mathbf{P}_I^{\perp}\mathbf{u}_0$ is optimal (in terms of minimizing the error probability or maximizing the SNR) among all possible ZF solutions, assuming that the noise vector $\mathbf{w}[n]$ is white Gaussian. As we shall see, the performance of the ZF correlator depends on the length of this orthogonal projection $\mathbf{P}_I^{\perp}\mathbf{u}_0$ relative to that of the desired signal vector $\mathbf{u}_0$: the smaller this relative length $||\mathbf{P}_I^{\perp}\mathbf{u}_0||/||\mathbf{u}_0||$, the poorer the performance.

For the model (8.10) for an equalizer of length $L = 4$, the signal vectors live in a space of dimension 4, with two interference vectors. It is quite clear that the desired vector is indeed linearly independent of the interference vectors, and we expect the ZF correlator to exist. For the model (8.10) for $L = 3$, we again have two interference vectors, and it again appears that the ZF correlator should exist, although we would expect the performance to be poorer because the relative length of the orthogonal projection can be expected to be smaller. Of course, such intuition must be quantified by explicit computation of the ZF correlator and its performance, which we discuss next.

Computation of the ZF correlator Let us now obtain an explicit expression for the ZF correlator given the vector ISI model (8.13). Suppose that the signal vectors $\{\mathbf{u}_k\}$ are written as columns in a matrix $\mathbf{U}$ as follows:

$$\mathbf{U} = [\ldots\ \mathbf{u}_{-1}\ \mathbf{u}_0\ \mathbf{u}_1\ \ldots] \tag{8.19}$$

The ZF conditions (8.17) and (8.18) can be compactly written as

$$\mathbf{U}^{\mathrm{T}}\mathbf{c}_{ZF} = \mathbf{e} \tag{8.20}$$

where $\mathbf{e} = (\ldots,\ 0, 1, 0, \ldots)^{\mathrm{T}}$ is a unit vector with a one corresponding to the column $\mathbf{u}_0$ and zeros corresponding to columns $\mathbf{u}_k$, $k \neq 0$. Further, we can write the ZF correlator as a linear combination of the signal vectors (any component orthogonal to all of the $\{\mathbf{u}_k\}$ can only add noise):

$$\mathbf{c}_{ZF} = \mathbf{Ua} \tag{8.21}$$

By plugging this into (8.20), we obtain

$$\mathbf{U}^{\mathrm{T}}\mathbf{Ua} = \mathbf{e}$$

so

$$\mathbf{a} = \left(\mathbf{U}^{\mathrm{T}}\mathbf{U}\right)^{-1}\mathbf{e}$$

assuming invertibility, which in turn requires that the signal vectors $\{\mathbf{u}_k\}$ are linearly independent (see Problem 8.6). By substituting into (8.21), we obtain that

$$\mathbf{c}_{ZF} = \mathbf{U}\left(\mathbf{U}^{\mathrm{T}}\mathbf{U}\right)^{-1}\mathbf{e}, \quad \textbf{ZF correlator for white noise} \tag{8.22}$$

Noise enhancement By "looking" along the direction of the orthogonal component $\mathbf{P}_{\mathrm{I}}^{\perp}\mathbf{u}_0$ shown in Figure 8.8, the ZF equalizer nulls out the interference vectors. When we plug in (8.17)–(8.18), the output SINR expression in (8.16) reduces to the output SNR. On setting $\mathbf{C}_{\mathbf{w}} = \sigma^2\mathbf{I}$, we obtain

$$\mathrm{SNR}_{ZF} = \frac{\sigma_b^2}{\sigma^2||\mathbf{c}_{ZF}||^2}, \quad \textbf{ZF SNR for white noise} \tag{8.23}$$

On the other hand, if we ignore ISI, then we know from Chapter 6 that the optimal correlator in AWGN is a scalar multiple of the desired vector $\mathbf{u}_0$. Relative to this "matched filter" solution, the ZF correlator incurs loss in SNR, termed *noise enhancement.* The reason why we say the noise is getting enhanced (as opposed to the signal getting reduced) is that, if we scale the correlator to keep the desired contribution at the output constant as in (8.18), then the degradation in SNR corresponds to an increase in the noise variance. For an ideal system with no ISI, the received vector is given by $\mathbf{r}[n] = b[n]\mathbf{u}_0 + \mathbf{w}[n]$. Setting $\mathbf{c} = \mathbf{u}_0$, we have $\mathbf{c}^{\mathrm{T}}\mathbf{r}[n] = b[n]||\mathbf{u}_0||^2 + N(0, \sigma^2||\mathbf{u}_0||^2)$, from which it is easy to see that the output SNR is given by

$$\mathrm{SNR}_{MF} = \frac{\mathbb{E}[|b[n]|^2]||\mathbf{u}_0||^2}{\sigma^2} = \frac{\sigma_b^2||\mathbf{u}_0||^2}{\sigma^2}, \quad \textbf{matched-filter bound for white noise} \tag{8.24}$$

This is termed the *matched-filter (MF) bound* on the SNR, and is an unrealizable (because we have ignored ISI) benchmark that we can compare the performance of linear equalization strategies against. In particular, the *noise enhancement* ζ can be defined as the ratio by which the ZF SNR is smaller than the MF benchmark:

$$\zeta = \frac{\text{SNR}_{\text{MF}}}{\text{SNR}_{\text{ZF}}} = ||\mathbf{u}_0||^2||\mathbf{c}_{\text{ZF}}||^2, \quad \textbf{noise enhancement for white noise} \tag{8.25}$$

Let us first interpret this geometrically. Setting $\mathbf{c}_{\text{ZF}} = \alpha \mathbf{P}_{\text{I}}^{\perp}\mathbf{u}_0$, the condition (8.18) corresponds to

$$1 = \langle \mathbf{c}_{\text{ZF}}, \mathbf{u}_0 \rangle = \alpha \langle \mathbf{P}_{\text{I}}^{\perp}\mathbf{u}_0, \mathbf{u}_0 \rangle = \alpha ||\mathbf{P}_{\text{I}}^{\perp}\mathbf{u}_0||^2 \tag{8.26}$$

The last equality follows because $\mathbf{u}_0$ decomposes into its projection onto the interference subspace $\mathbf{P}_{\text{I}}\mathbf{u}_0$ and its orthogonal projection $\mathbf{P}_{\text{I}}^{\perp}\mathbf{u}_0$. Since these two components are orthogonal by definition, we have

$$\langle \mathbf{P}_{\text{I}}^{\perp}\mathbf{u}_0, \mathbf{u}_0 \rangle = \langle \mathbf{P}_{\text{I}}^{\perp}\mathbf{u}_0, \mathbf{P}_{\text{I}}\mathbf{u}_0 \rangle + \langle \mathbf{P}_{\text{I}}^{\perp}\mathbf{u}_0, \mathbf{P}_{\text{I}}^{\perp}\mathbf{u}_0 \rangle = 0 + ||\mathbf{P}_{\text{I}}^{\perp}\mathbf{u}_0||^2$$

We see from (8.26) that

$$\alpha = \frac{1}{||\mathbf{P}_{\text{I}}^{\perp}\mathbf{u}_0||^2}$$

In other words, a ZF correlator satisfying (8.17) and (8.18) can be written in terms of the projection of the desired vector orthogonal to the interference subspace as follows:

$$\mathbf{c}_{\text{ZF}} = \frac{\mathbf{P}_{\text{I}}^{\perp}\mathbf{u}_0}{||\mathbf{P}_{\text{I}}^{\perp}\mathbf{u}_0||^2} \tag{8.27}$$

from which it follows that

$$||\mathbf{c}_{\text{ZF}}||^2 = \frac{1}{||\mathbf{P}_{\text{I}}^{\perp}\mathbf{u}_0||^2} \tag{8.28}$$

Thus, the smaller the orthogonal projection $\mathbf{P}_{\text{I}}^{\perp}\mathbf{u}_0$, the more we must scale up the correlator in order to maintain the normalization (8.18) of the contribution of the desired symbol at the output. By plugging (8.28) into (8.25), we obtain the following geometric interpretation for the noise enhancement:

$$\zeta = \frac{\text{SNR}_{\text{MF}}}{\text{SNR}_{\text{ZF}}} = \frac{||\mathbf{u}_0||^2}{||\mathbf{P}_{\text{I}}^{\perp}\mathbf{u}_0||^2} \tag{8.29}$$

This is intuitively reasonable: the noise enhancement is the inverse of the factor by which the effective signal energy is reduced because of looking along the orthogonal projection $\mathbf{P}_{\text{I}}^{\perp}\mathbf{u}_0$, instead of along the desired vector $\mathbf{u}_0$.

The following code fragment computes the ZF correlator and the noise enhancement for the model (8.10). We find that the noise enhancement is 4.4 dB.

Code Fragment 8.2.2 (Computing the ZF solution and its noise enhancement)

```
%ZF example
%matrix with signal vectors as columns
U = transpose([0.5 -0.25 0 0;-0.5 1 0.5 -0.25;0 0 -0.5 1]);
%unit vector with one corresponding to u0
```

```
e = transpose([0 1 0]);
%coeffs of linear combination
a = (U' * U)\e;
%ZF correlator: linear comb of columns of U
czf = U * a;
%desired vector is second column
u0 = U(:,2);
%check that ZF equations are satisfied
U' * czf %should be equal to the vector e
%noise_enhancement
noise_enhancement = (u0' * u0) * (czf' * czf)
%in dB
noise_enhancement_db = 10 * log10(noise_enhancement)
```

While the matrix $\mathbf{U}$ is specified manually in the preceding code fragment, for longer channels, we would typically automate the generation of $\mathbf{U}$ given the channel impulse response $\mathbf{h}$, the equalizer length L, the oversampling factor q, and the specification of how the observation interval lines up with the response to the desired symbol (i.e., how to generate $\mathbf{u}_0$ from $\mathbf{h}$).

ZF correlator for colored noise Let us now discuss how to generalize the expressions for the ZF correlator and its noise enhancement for colored noise, where $\mathbf{w}[n]$ has covariance matrix $\mathbf{C_w}$ (which is assumed to be strictly positive definite, and hence invertible). We limit ourselves here to stating the results; guidance for deriving these results is provided in Problem 8.9. The optimal ZF solution, in terms of maximizing the output SNR while satisfying (8.17) and (8.18), is given by

$$\mathbf{c}_{\mathrm{ZF}} = \mathbf{C}_{\mathbf{w}}^{-1}\mathbf{U}\left(\mathbf{U}^{\mathrm{T}}\mathbf{C}_{\mathbf{w}}^{-1}\mathbf{U}\right)^{-1}\mathbf{e}, \quad \textbf{ZF correlator for colored noise} \tag{8.30}$$

The corresponding SNR is given by

$$\mathrm{SNR}_{\mathrm{ZF}} = \frac{\sigma_b^2}{\mathbf{c}_{\mathrm{ZF}}^{\mathrm{T}}\mathbf{C}_{\mathbf{w}}\mathbf{c}_{\mathrm{ZF}}}, \quad \textbf{ZF SNR for colored noise} \tag{8.31}$$

If there were no ISI, then the optimal correlator is given by the *whitened* matched filter $\mathbf{c} = \mathbf{C}_{\mathbf{w}}^{-1}\mathbf{u}_0$, and the corresponding matched filter bound on SNR is given by

$$\mathrm{SNR}_{\mathrm{MF}} = \sigma_b^2\mathbf{u}_0^{\mathrm{T}}\mathbf{C}_{\mathbf{w}}^{-1}\mathbf{u}_0, \quad \textbf{matched filter bound for colored noise} \tag{8.32}$$

Proceeding as before, the noise enhancement is given by

$$\zeta = \frac{\mathrm{SNR}_{\mathrm{MF}}}{\mathrm{SNR}_{\mathrm{ZF}}} = (\mathbf{u}_0^{\mathrm{T}}\mathbf{C}_{\mathbf{w}}^{-1}\mathbf{u}_0)(\mathbf{c}_{\mathrm{ZF}}^{\mathrm{T}}\mathbf{C}_{\mathbf{w}}\mathbf{c}_{\mathrm{ZF}}), \quad \textbf{noise enhancement for colored noise} \tag{8.33}$$

The reader is encouraged to check that, when we set $\mathbf{C}_{\mathbf{w}} = \sigma^2\mathbf{I}$ in the preceding expressions, we recover the expressions derived earlier for white noise.

MMSE correlator While we have seen how to adaptively implement the MMSE equalizer, if we are given the vector ISI model (8.13), then we can compute the MMSE solution analytically (see Problem 8.8 for the derivation) as follows:

$$\mathbf{c}_{\mathrm{MMSE}} = \mathbf{R}^{-1}\mathbf{p} \tag{8.34}$$

where

$$\mathbf{R} = \sigma_b^2 \mathbf{U}\mathbf{U}^{\mathrm{T}} + \mathbf{C}_{\mathbf{w}}, \quad \mathbf{p} = \sigma_b^2 \mathbf{u}_0$$

We state without proof the following results:

(1) Among the class of linear correlators, the MMSE correlator is optimal in terms of the SINR. Thus, it achieves the best tradeoff between the ISI and noise at the output, attaining an SINR that is better than the SNR attained by the ZF correlator (for the ZF correlator, the SINR equals the SNR, since there is no residual ISI at the output).
(2) The MMSE correlator tends to the ZF correlator (if the latter exists) as the noise variance, or more generally, the noise covariance matrix, tends to zero. This makes sense: if we can neglect noise, then the MSE $\mathbb{E}[|\mathbf{c}^{\mathrm{T}}\mathbf{r}[n] - b[n]|^2]$ can be driven to zero by forcing the ISI to zero as in (8.17) and by scaling the desired contribution according to (8.18), since we then obtain $\mathbf{c}^{\mathrm{T}}\mathbf{r}[n] = b[n]$.

We summarize as follows. The zero-forcing equalizer drives the ISI to zero, while the linear MMSE equalizer trades off ISI and noise at its output so as to maximize the SINR. For large SNR, the contribution of the ISI is dominant, and the MMSE equalizer tends in the limit to the zero-forcing equalizer (if it exists), and hence pays the same asymptotic penalty in terms of noise enhancement. In practice, the MMSE equalizer often performs significantly better than the ZF equalizer at moderate SNRs, but, in order to improve equalization performance at high SNR, one must look to nonlinear equalization strategies, which are beyond our present scope.

Extension to complex-valued signals All of the preceding development applies to complex-valued constellations and signals, except that vector transposes $\mathbf{x}^{\mathrm{T}}$ are replaced by conjugate transposes $\mathbf{x}^{\mathrm{H}}$, and the noise covariance matrix must include the effect both of the real part and of the imaginary part of the noise.

Noise model In order to model complex-valued WGN, we set $\mathbf{C}_{\mathbf{w}} = 2\sigma^2\mathbf{I}$. This can be generated by setting $\mathrm{Re}(\mathbf{w})$ and $\mathrm{Im}(\mathbf{w})$ to be i.i.d. $N(0, \sigma^2)$. More generally, we consider *circularly symmetric,* zero-mean, complex Gaussian noise vectors $\mathbf{w}$, which are completely characterized by their complex covariance matrix,

$$\mathbf{C}_{\mathbf{w}} = \mathbb{E}\left[(\mathbf{w} - \mathbb{E}[\mathbf{w}])(\mathbf{w} - \mathbb{E}[\mathbf{w}])^{\mathrm{H}}\right] = \mathbb{E}\left[\mathbf{w}\mathbf{w}^{\mathrm{H}}\right]$$

We use the notation $\mathbf{w} \sim CN(\mathbf{0}, \mathbf{C}_{\mathbf{w}})$. Detailed discussion of circularly symmetric Gaussian random vectors would distract us from our present purpose. Suffice it to say that circular symmetry and Gaussianity are preserved under linear transformations. The covariance matrix evolves as follows: if $\mathbf{w} = \mathbf{B}\tilde{\mathbf{w}}$, then $\mathbf{C}_{\mathbf{w}} = \mathbf{B}\mathbf{C}_{\tilde{\mathbf{w}}}\mathbf{B}^{\mathrm{H}}$. Thus, we can generate colored circularly symmetric Gaussian noise $\mathbf{w}$ by passing complex WGN through a linear transformation. Specifically, if we can write $\mathbf{C}_{\mathbf{w}} = \mathbf{B}\mathbf{B}^{\mathrm{H}}$ (this can always be done for a positive definite matrix), we can generate $\mathbf{w}$ as $\mathbf{w} = \mathbf{B}\tilde{\mathbf{w}}$, where $\tilde{\mathbf{w}} \sim CN(0, \mathbf{I})$.

The expressions for the ZF and MMSE correlators are as follows:

$$\mathbf{c}_{\text{ZF}} = \mathbf{C}_{\mathbf{w}}^{-1}\mathbf{U}\left(\mathbf{U}^{\text{H}}\mathbf{C}_{\mathbf{w}}^{-1}\mathbf{U}\right)^{-1}\mathbf{e}, \quad \textbf{ZF correlator for complex-valued signals} \tag{8.35}$$

$$\mathbf{c}_{\text{MMSE}} = \mathbf{R}^{-1}\mathbf{p}\,, \quad \textbf{MMSE correlator for complex-valued signals}$$

where

$$\mathbf{R} = \sigma_{\text{b}}^2\mathbf{U}\mathbf{U}^{\text{H}} + \mathbf{C}_{\mathbf{w}}, \qquad \mathbf{p} = \sigma_{\text{b}}^2\mathbf{u}_0, \qquad \sigma_{\text{b}}^2 = \mathbb{E}[|b[n]|^2]$$

MMSE and SINR While the SINR for any linear correlator can be computed as in (8.16), we can obtain particularly simple expressions for the MSE and SINR achieved by the MMSE correlator, as follows:

$$\begin{aligned} \text{MMSE} &= \sigma_{\text{b}}^2 - \mathbf{p}^{\text{H}}\mathbf{c}_{\text{MMSE}} = \sigma_{\text{b}}^2 - \mathbf{p}^{\text{H}}\mathbf{R}^{-1}\mathbf{p} \\ \text{SINR}_{\max} &= \sigma_{\text{b}}^2/\text{MMSE} - 1 \end{aligned} \tag{8.36}$$

8.3 Orthogonal frequency-division multiplexing

We now introduce an alternative approach to communication over dispersive channels whose goal is to isolate symbols from each other for any dispersive channel. The idea is to employ frequency-domain transmission, sending symbols $B[n]$ using complex exponentials $s_n(t) = e^{j2\pi f_n t}$, which have two key properties:

(P1) When $s_n(t)$ goes through an LTI system with impulse response $h(t)$ and transfer function $H(f)$, the output is a scalar multiple of $s_n(t)$. Specifically,

$$e^{j2\pi f_n t} * h(t) = H(f_n)e^{j2\pi f_n t}$$

(P2) Complex exponentials at different frequencies are orthogonal:

$$\langle s_n, s_m \rangle = \int s_n(t)s_m^*(t)dt = \int_{-\infty}^{\infty} e^{j2\pi(f_n - f_m)t}\,dt = \delta(f_m - f_n) = 0,\ f_n \neq f_m$$

This is analogous to the properties of eigenvectors of matrices. Thus, complex exponentials are *eigenfunctions* of any LTI system, as pointed out in Chapter 2.

Conceptual basis for OFDM For frequency-domain transmission with symbol $B[k]$ modulating the complex exponential $s_n(t) = e^{j2\pi f_n t}$, the transmitted signal is given by

$$u(t) = \sum_n B[n]e^{j2\pi f_n t}$$

When this goes through a dispersive channel $h(t)$, we obtain (ignoring noise)

$$(u * h)(t) = \sum_n B[n]H(f_n)e^{j2\pi f_n t}$$

Note that the symbols $\{B[n]\}$ do not interfere with each other after passing through the channels, since different complex exponentials are orthogonal. Furthermore, regardless of how complicated the time-domain channel $h(t)$ is, we have managed to parallelize the problem of equalization by going to the frequency domain. Thus, we need only estimate and compensate for the complex scalar $H(f_n)$ in demodulating the nth symbol. We now discuss how to translate this concept into practice.

Finite signaling interval The first step is to constrain the signaling interval, say to length T. The complex-baseband transmitted signal is therefore given by

$$u(t) = \sum_{n=0}^{N-1} B[n]e^{j2\pi f_n t} I_{[0,T]}(t) = \sum_{n=0}^{N-1} B[n]p_n(t) \tag{8.37}$$

where $B[n]$ is the symbol transmitted using the modulating signal $p_n(t) = e^{j2\pi f_n t} I_{[0,T]}$, using the nth subcarrier at frequency f_n. Let us now see how the properties P1 and P2 are affected by timelimiting. The timelimited tone $p_n(t)$ has Fourier transform $P_n(f) = T\,\mathrm{sinc}((f-f_n)T)e^{-\pi fT}$, which decays quickly as $|f-f_n|$ takes on values of the order of k/T. For a channel whose impulse response $h(t)$ is approximately timelimited to T_d (the channel delay spread), the transfer function is approximately constant over frequency intervals of length B_c roughly inversely proportional to $1/T_\mathrm{d}$ (the channel coherence bandwidth). If the signaling interval is large compared with the channel delay spread ($T \gg T_\mathrm{d}$), then $1/T$ is small compared with the channel coherence bandwidth ($1/T \ll B_\mathrm{c}$), so that the gain seen by $P_n(f)$ is roughly constant, and the eigenfunction property is roughly preserved. That is, when $P_n(f)$ goes through a channel with transfer function $H(f)$, the output

$$Q_n(f) = H(f)P_n(f) \approx H(f_n)P_n(f) \tag{8.38}$$

Regarding the orthogonality property P2, two complex exponentials that are constrained to an interval of length T are orthogonal if the frequency separation is an integer multiple of $1/T$:

$$\int_0^T e^{j2\pi f_n t} e^{-j2\pi f_m t}\, dt = \frac{e^{j2\pi(f_n - f_m)T} - 1}{j2\pi(f_n - f_m)} = 0, \quad \text{for } (f_n - f_m)T = \text{ nonzero integer} \tag{8.39}$$

Thus, if we wish to send N symbols in parallel using N *subcarriers* (the term used for each time-constrained complex exponential), we need a bandwidth of roughly N/T in order to preserve orthogonality among the timelimited tones. Of course, even if we enforce orthogonality in this fashion, the timelimited tones are not eigenfunctions of LTI systems, so the output corresponding to the nth timelimited tone is not just a scalar multiple of itself. However, using (8.38), we can approximate the channel output for the nth timelimited tone as $H(f_n)e^{j2\pi f_n t} I_{[0,T]}(t)$. Thus, the output corresponding to the transmitted signal (8.37) can be approximated as follows:

$$y(t) \approx \sum_{n=0}^{N-1} B[n]H(f_n)e^{j2\pi f_n t} I_{[0,T]}(t) + n(t) \tag{8.40}$$

To summarize, once we have limited the signaling duration to being finite, the ISI-avoidance property of OFDM is approximate rather than exact. However, as we now discuss, orthogonality between subcarriers can be restored *exactly* in digital implementations of OFDM. Before discussing such implementations, we provide some background on discrete-time signal processing.

8.3.1 DSP-centric implementation

The proliferation of OFDM in commercial systems (including wireline DSL, wireless local area networks, and wireless cellular systems) has been enabled by the implementation of its transceiver functionalities in DSP, which leverages the economies of scale of digital computation (Moore's "law"). For T large enough, the bandwidth of the OFDM signal u is approximately N/T, where N denotes the number of subcarriers. Thus, we can represent $u(t)$ accurately by sampling at rate $1/T_s = N/T$, where T_s denotes the sampling interval. From (8.37), the samples are given by

$$u(kT_s) = \sum_{n=0}^{N-1} B[n] e^{j2\pi nk/N}$$

We can recognize this simply as the inverse DFT (IDFT) of the symbol sequence $\{B[n]\}$. We make this explicit in the notation as follows:

$$b[k] = u(kT_s) = \sum_{n=0}^{N-1} B[n] e^{j2\pi nk/N} \tag{8.41}$$

If N is a power of 2 (which can be achieved by zero-padding if necessary), the samples $\{b[k]\}$ can be efficiently generated from the symbols $\{B[n]\}$ using an inverse fast Fourier transform (IFFT). The complex-baseband waveform $u(t)$ can now be obtained from its samples by digital-to-analog (D/A) conversion. This implementation of an OFDM transmitter is as shown in Figure 8.9: the bits are mapped to symbols, the symbols are fed in parallel to the inverse FFT (IFFT) block, and the complex baseband signal is obtained by D/A conversion of the samples (after insertion of a cyclic prefix, to be discussed after we have motivated it in the context of receiver implementation). Typically, the D/A converter is an interpolating filter, so that its effect can be subsumed within the channel impulse response.

Note that the relation (8.41) can be inverted as follows:

$$B[n] = \frac{1}{N} \sum_{k=0}^{N-1} b[k] e^{-j2\pi nk/N} \tag{8.42}$$

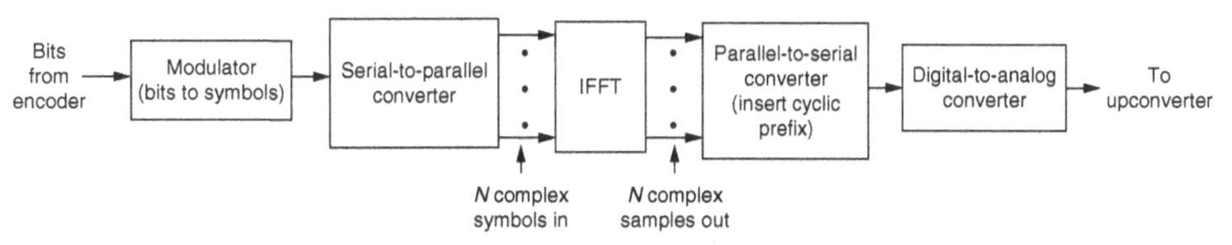

Figure 8.9 DSP-centric implementation of an OFDM transmitter.

This is exploited in the digital implementation of the OFDM receiver, which will be discussed next.

Remark on MATLAB FFT and IFFT conventions MATLAB puts a factor of $1/N$ in the IFFT rather than in the FFT as done in (8.41) and (8.42). In both cases, however, IFFT followed by FFT gives the identity. Note also that MATLAB numbers vector entries starting with one, so the FFT of $x[n], n = 1, \ldots, N$ is given by

$$X[k] = \sum_{n=1}^{N} x[n] e^{j2\pi(n-1)(k-1)/N}$$

The corresponding IFFT is given by

$$x[n] = \frac{1}{N} \sum_{k=1}^{N} X[k] e^{j2\pi(n-1)(k-1)/N}$$

We have observed that, once we have limited the signaling duration to being finite, the ISI avoidance property of OFDM is approximate rather than exact. However, as we now show, orthogonality between subcarriers can be restored *exactly* in discrete time by using a cyclic prefix, which allows efficient demodulation using an FFT. The noiseless received OFDM signal is modeled as

$$v(t) = \sum_{k=0}^{N-1} b[k] p(t - kT_s)$$

where the "effective" channel impulse response $p(t)$ includes the effect of the D/A converter at the transmitter, the physical channel, and the receive filter. When we sample this signal at rate $1/T_s$, we obtain the discrete-time model

$$v[m] = \sum_{k=0}^{N-1} b[k] h[m - k] \tag{8.43}$$

where $\{h[l] = p(lT_s)\}$ is the effective discrete-time channel of length L, assumed to be smaller than N. We assume, without loss of generality, that $h[l] = 0$ for $l < 0$ and $l \geq L$. We can rewrite (8.43) as

$$v[m] = \sum_{l=0}^{L-1} h[l] b[m - l] \tag{8.44}$$

Let H denote the N-point DFT of h, where $N > L$:

$$H[n] = \sum_{l=0}^{N-1} h[l] e^{-j2\pi nl/N} = \sum_{l=0}^{L-1} h[l] e^{-j2\pi nl/N} \tag{8.45}$$

As noted in (8.42), the DFT of $\{b[k]\}$ is the symbol sequence $B[n]$. (The normalization is chosen differently in (8.42) and (8.45) to simplify the forthcoming equations.) In order to parallelize equalization across the N subcarriers, we would like the noiseless signal to equal $V[n] = H[n]B[n]$. However, this is not quite satisfied in our setting. We now discuss why not, and how to modify the system so as to indeed enforce such a relationship. Before doing this, we need a brief discussion of the DFT and its dual operation, the *cyclic convolution.*

DFT multiplication and cyclic convolution The time-domain samples $\{b[k]\}$ defined via the IDFT in (8.41) have range $0 \leq k \leq N-1$. If we now plug in integer values of k outside this range, we simply get a periodic extension $b_N[k]$ of these samples with period N, satisfying $b_N[k+N] = b_N[k]$ for all k, with $b_N[k] = b[k]$, $0 \leq k \leq N-1$. Thus, the IDFT can be viewed as a discrete-time analog of a Fourier series for a periodic time-domain sample sequence $\{b_N[k]\}$. We know that, for the Fourier transform, "multiplication in the frequency domain corresponds to convolution in the time domain." We skipped the analogous result for Fourier series in Chapter 2 because we did not need it then. Now, however, we establish the appropriate result for the discrete-time Fourier series of interest here: for the DFT, if we multiply two sequences in the frequency domain, then it corresponds to a *cyclic,* or periodic, convolution in the time domain.

While the result we wish to establish is general, let us stick with the notation we have already established. Consider the "desired" sequence $\tilde{V}[n] = H[n]B[n]$, $n = 0, \ldots, N-1$, that we would like to get when we take the DFT of the output of the channel. What is the corresponding time domain sequence? To see this, take the IDFT:

$$\tilde{v}[m] = \sum_{n=0}^{N-1} H[n]B[n]e^{j2\pi mn/N}$$

On plugging in the expression (8.45) for the channel DFT coefficients, we obtain

$$\begin{aligned}\tilde{v}[m] &= \sum_{n=0}^{N-1}\sum_{l=0}^{L-1} h[l]e^{-j2\pi nl/N}B[n]e^{j2\pi mn/N} \\ &= \sum_{l=0}^{L-1} h[l] \sum_{n=0}^{N-1} B[n]e^{j2\pi(m-l)n/N}\end{aligned} \tag{8.46}$$

Now, the summation over n corresponds to an IDFT, and therefore gives us $b[m-l]$ as long as $0 \leq m-l \leq N-1$. Outside this range, it gives us the periodic extension $\{b_N[m-l]\}$:

$$\sum_{n=0}^{N-1} B[n]e^{j2\pi(m-l)n/N} = b_N[m-l]$$

Thus, we can write (8.46) as

$$\tilde{v}[m] = \sum_{l=0}^{L-1} h[l]b_N[m-l] = \sum_{l=0}^{L-1} h[l]b[(m-l)\text{mod } N] = (h \odot b)[m] \tag{8.47}$$

where we have introduced the notation $h \odot b$ to denote the *cyclic convolution* of h and b. While we have derived this result in our particular context, it is worth stating that it holds generally: the cyclic convolution *modulo* N between two sequences p and q, each of length at most N (it is often convenient to think of them as having length N, using zero-padding if necessary) is defined as the convolution over a period of length N of their periodic extensions with period N:

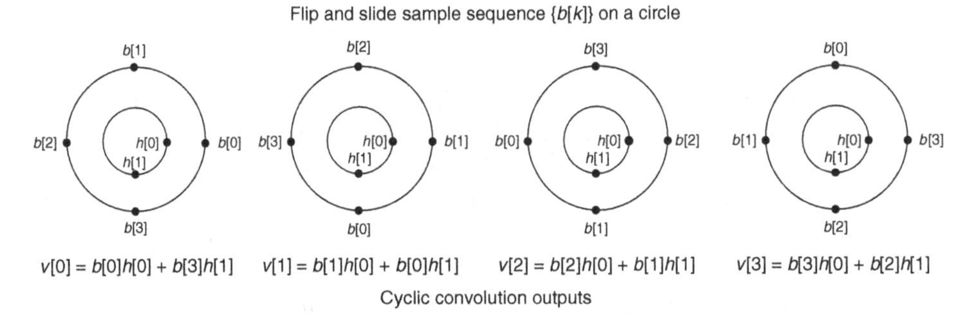

Figure 8.10 An example of cyclic convolution. Time progresses clockwise on the circle. The sequence $\{b[k]\}$ is flipped, and hence goes counter-clockwise. We then "slide" this flipped sequence clockwise in order to compute successive outputs. Clearly, the output is periodic with period $N = 4$.

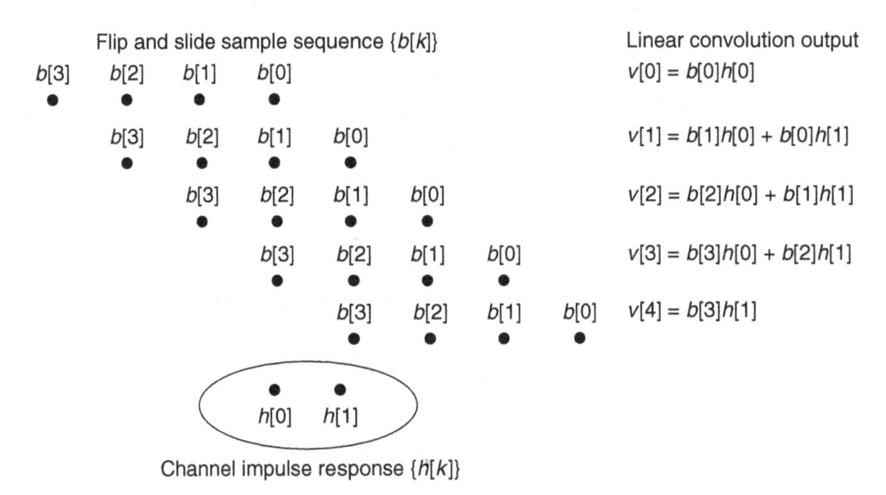

Figure 8.11 Linear convolution of the two sequences in Figure 8.10 leads to an aperiodic sequence of length $2 + 4 - 1 = 5$. Note that the outputs at times 1, 2, and 3 coincide with the outputs of the corresponding cyclic convolution.

$$(p \odot q)[m] = \sum_{l=0}^{N-1} p_N[l] q_N(m - l)$$

The N-point DFT of the cyclic convolution of these two sequences is the product of their DFTs.

Figure 8.10 illustrates cyclic convolution modulo $N = 4$ between a sample sequence $\{b[k]\}$ of length 4 and a channel impulse response of length 2, while Figure 8.11 illustrates the corresponding linear convolution.

Let us summarize where we now stand. In order to parallelize the channel in the DFT domain, we need a cyclic convolution in the time domain given by (8.47). However, what the physical channel actually gives us is the linear convolution of the form (8.44). In order to get the cyclic convolution we want, we simply need to send an appropriately large segment of a periodic extension of the time domain samples $\{b[k]\}$ through the channel.

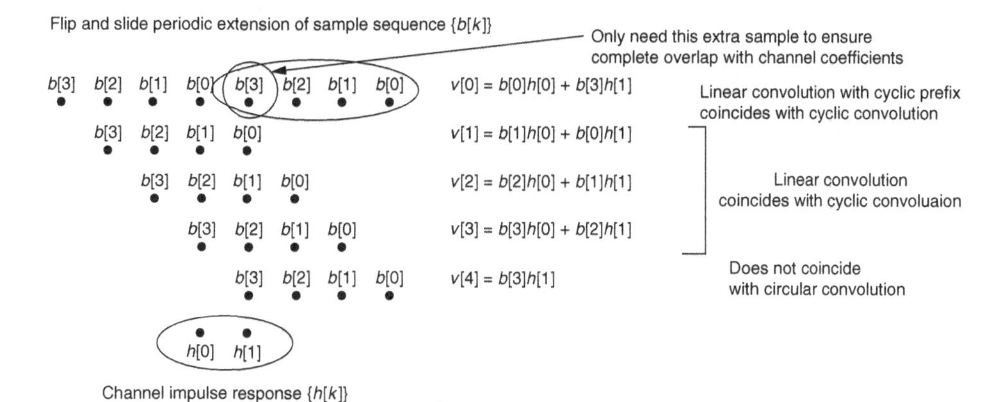

Figure 8.12 Using linear convolution to emulate a circular convolution.

Indeed, if we just want to get N outputs corresponding to a single period of the output of the circular convolution, then we do not need a full-fledged periodic extension. Figure 8.12 shows how to get the first $N = 4$ outputs of a linear convolution to be equal to a period of the cyclic convolution in Figure 8.10 by inserting a single sample. More generally, we need a *cyclic prefix* of length $L - 1$ for a channel of length L, as discussed below.

Since $L < N$, we can write the circular convolution (8.47) as

$$\tilde{v}[m] = \sum_{l=0}^{\min(L-1,m)} h[l]b[m-l] + \sum_{l=m+1}^{L-1} h[l]b[m-l+N] \tag{8.48}$$

Comparing the linear convolution (8.47) and the cyclic convolution (8.48), we see that they are identical *except* when the index $m - l$ takes negative values: in this case, $b[m - l] = 0$ in the linear convolution, while $b[(m - l) \bmod N] = b_N(m - l) = b[m - l + N]$ contributes to the circular convolution. Thus, we can emulate a cyclic convolution using the physical linear convolution by sending a *cyclic prefix*; that is, by sending

$$b[k] = b_N[k] = b[N + k], \quad k = -(L - 1), -(L - 2), \ldots, -1$$

before we send the samples $b[0], \ldots, b[N - 1]$. That is, we transmit the samples

$$b[N - L + 1], \ldots, b[N - 1], b[0], \ldots, b[N - 1]$$

incurring an overhead of $(L - 1)/N$, which can be made small by choosing N to be large.

In the example depicted in Figure 8.12, $N = 4$ and $L = 2$, and we insert the cyclic prefix $b[3]$, sending $b[3], b[0], b[1], b[2]$, and $b[3]$ (when this is flipped for the pictorial convolution in the figure, the extra sample $b[3]$ appears at the end).

At the receiver (see Figure 8.13), the complex-baseband signal is sampled at rate $1/T_s$ to obtain noisy versions of the samples $\{b[k]\}$. The FFT of these samples then yields the model

$$Y[n] = H[n]B[n] + N[n] \tag{8.49}$$

where the frequency-domain noise samples $N[n]$ are modeled as i.i.d. complex Gaussian, with $\text{Re}(N[n])$ and $\text{Im}(N[n])$ being i.i.d. $N(0, \sigma^2)$. If the receiver knows the channel, then

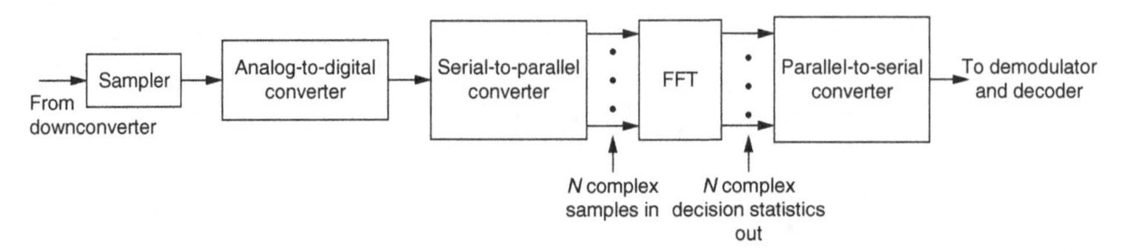

Figure 8.13 DSP-centric implementation of an OFDM receiver. Carrier and timing synchronization blocks are not shown.

it can implement ML reception based on the statistic $H^*[n]Y[n]$. Thus, the task of channel equalization has been reduced to compensating for scalar channel gains for each subcarrier. This makes OFDM extremely attractive for highly dispersive channels, for which time-domain single-carrier equalization strategies would be difficult to implement.

Channel estimation Channel estimation (along with timing and carrier synchronization, which are not considered here) is accomplished by sending pilot symbols. In Software Lab 8.2, we send an entire OFDM symbol as a pilot, followed by a succession of other OFDM symbols with payload.

8.4 MIMO

The term *multiple-input, multiple-output (MIMO)*, or *space–time communication*, refers to communication systems employing multiple antennas at the transmitter and receiver. We now provide a brief introduction to key concepts in MIMO systems, along with pointers for further exploration.

While much effort and expertise must go into the design of antennas and their interface to RF circuits, the following abstract view suffices for our purpose here: at the transmitter, an antenna transduces electrical signals at radio frequencies into electromagnetic waves at the same frequency that propagate in space; at the receiver, the antenna transduces electromagnetic waves in a certain frequency range into electrical signals at the same set of frequencies. Antennas that are insensitive to the direction of arrival/departure of the waves are termed *omnidirectional* or *isotropic* (while there is no such thing as an ideal isotropic antenna, it is a convenient conceptual building block). Antennas that are sensitive to the direction of arrival or departure are termed *directional.* It is possible to synthesize directional responses using an array of omnidirectional antenna elements, as we discuss next.

8.4.1 The linear array

Consider a plane wave impinging on the uniformly spaced linear array shown in Figure 8.14. We see that the wave sees slightly different path lengths, and hence different phase shifts, in reaching different antenna elements. The path-length difference between two successive elements is given by $\ell = d \sin\theta$, where d is the inter-element spacing and θ is the

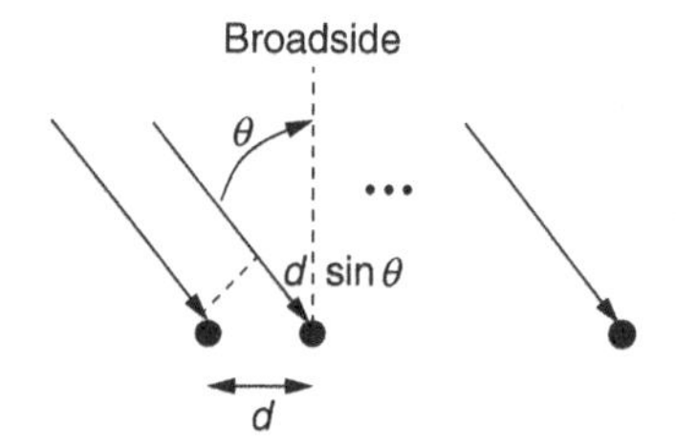

Figure 8.14 A plane wave impinging on a linear array.

angle of arrival (AoA) relative to the broadside. The corresponding phase shift across successive elements is given by $\phi = 2\pi \ell/\lambda = 2\pi d \sin\theta/\lambda$, where λ denotes the wavelength. Another way to get the same result is that the delay difference between successive elements is $\tau = \ell/c$, where c is the speed of wave propagation (equal to 3×10^8 m/s in free space). For carrier frequency f_c, the corresponding phase shift is $\phi = 2\pi f_c \tau = 2\pi f_c d \sin\theta/c$. The two expressions are equivalent, since $\lambda = c/f_c$.

The narrowband assumption What is the effect of the differences in delays seen by successive elements? Suppose that the wave impinging on element 1 is represented as

$$u_p(t) = u_c(t)\cos(2\pi f_c t) - u_s(t)\sin(2\pi f_c t) = \text{Re}(u(t)e^{j2\pi f_c t})$$

where $u(t) = u_c(t) + ju_s(t)$ is the complex envelope, assumed to be of bandwidth W. Suppose that the bandwidth $W \ll f_c$: this is the so-called "narrowband assumption," which typically holds in most practical settings. For the scenario shown in Figure 8.14, the wave arrives $\tau = \ell/c$ time units earlier at element 2. The wave impinging on element 2 can therefore be represented as

$$v_p(t) = u_c(t+\tau)\cos(2\pi f_c(t+\tau)) - u_s(t+\tau)\sin(2\pi f_c(t+\tau)) = \text{Re}(u(t+\tau)e^{j\phi}e^{j2\pi f_c t})$$

where $\phi = 2\pi f_c \tau$. Thus, the complex envelope of the wave at element 2 is $v(t) = u(t+\tau)e^{j\phi}$. The time shift τ has two effects on the complex envelope: a time shift in the baseband waveform u and a phase rotation ϕ due to the carrier. However, for most settings of interest, the time shift in the baseband waveform can be ignored. To see why, suppose that the array parameters are such that ϕ is of the order of 2π or less, in which case τ is of the order of $1/f_c$ or less. Under the narrowband assumption, the time shift τ produces little distortion in u. To see this, note that

$$u(t+\tau) \leftrightarrow U(f)e^{j2\pi f\tau}$$

As f varies over a range W, the frequency-dependent phase change produced by the time shift varies over a range $2\pi W\tau \sim 2\pi W/f_c \ll 2\pi$ for $W \ll f_c$. Thus, we can ignore the effect of the time shift on the complex envelope, and model the complex envelope at element 2 as $v(t) \approx u(t)e^{j\phi}$. Similarly, for element 3, the complex envelope is well approximated as $u(t)e^{j2\phi}$.

Array response and spatial frequency Under the narrowband assumption, if the complex envelope at element 1 is $u(t)$, then the complex envelopes at the various elements can be collected into a vector $u(t)\mathbf{a}$, where

$$\mathbf{a} = (1, e^{j\phi}, e^{j2\phi}, \ldots, e^{j(N-1)\phi})^{\mathrm{T}} \tag{8.50}$$

is the *array response* for a particular AoA. Making the dependence on the AoA θ explicit for the linear array, we have $\phi(\theta) = 2\pi d \sin\theta/\lambda$, which yields a corresponding array response $\mathbf{a}(\theta)$. The linear increase in phase across antenna elements (i.e., across space) is analogous to the linear increase of phase with time for a sinusoid. Thus, we call $\phi = \phi(\theta)$ the *spatial frequency* corresponding to AoA θ. The collection of array responses $\{\mathbf{a}(\theta), \theta \in [-\pi, \pi]\}$ as we vary the AoA is termed the *array manifold.*

Reciprocity While Figure 8.14 depicts an antenna array receiving a wave, exactly the same reasoning applies to an antenna array emitting a wave. In particular, the principle of reciprocity tells us that the propagation channel from transmitter to receiver is the same as that from receiver to transmitter. Thus, the array response of a linear array for angle of arrival θ is the same as the array response for angle of departure θ.

Signal processing architecture What the preceding complex-baseband model means physically is that, if we downconvert the RF signals at the outputs of the antenna elements (using the same LO frequency and phase, and filters with identical responses, in each such "RF chain"), then the complex envelopes corresponding to the different antenna elements will be related as described above. Once the I and Q components for these complex envelopes have been obtained, they would typically be sampled and quantized using analog-to-digital converters (ADCs), and then processed digitally. Such a DSP-centric signal processing architecture, depicted in Figure 8.15, allows the implementation of sophisticated MIMO algorithms in today's cellular and WiFi systems. While the figure depicts a receiver architecture, an entirely analogous block diagram can be drawn for a MIMO transmitter, simply by reversing the arrows and replacing downconverters by upconverters.

While the DSP-centric architecture depicted in Figure 8.15 has been key to enabling the widespread deployment of low-cost MIMO transceivers, it may need to be revisited as carrier frequencies, and the available signaling bandwidths, scale up. Both the cost and the power consumption of ADCs with adequate precision can be prohibitively large at very high sampling rates, hence alternative architectures with MIMO processing done, wholly or in part, *prior* to ADC may need to be considered. See the epilogue for further discussion.

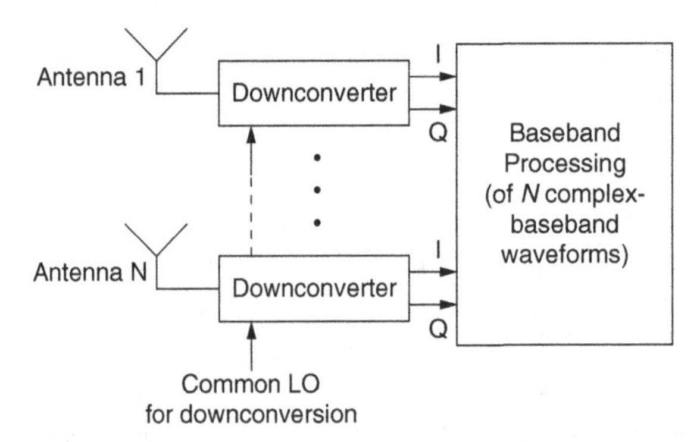

Figure 8.15 MIMO signal processing architecture. There is one "RF chain" per antenna, downconverting the signal received at that antenna to I and Q components.

8.4.2 Beamsteering

Once we know the array response for a given direction, we can maximize the received power (for a receive antenna array) or the transmitted power (for a transmit antenna array) in that direction by employing a *spatial matched filter* or *spatial correlator.* If the first antenna element receives a complex baseband waveform (after downconversion and sampling) $s[n]$ from AoA θ, then the output of the antenna array is modeled as a vector of complex-baseband discrete-time signals with kth component

$$y_k[n] = e^{j(k-1)\phi(\theta)} s(t) + w_k[n], \quad k = 1, 2, \ldots, N \tag{8.51}$$

where $\phi(\theta)$ is the spatial frequency corresponding to θ, and $w_k[n]$ are typically modeled as complex WGN, independent across space and time: $\mathrm{Re}(w_k[n])$ and $\mathrm{Im}(w_k[n])$ i.i.d. $N(0, \sigma^2)$ for all k, n. In vector notation, we can write

$$\mathbf{y}[n] = \mathbf{a}(\theta)s[n] + \mathbf{w}[n] \tag{8.52}$$

where $\mathbf{y}[n] = (y_1[n], \ldots, y_N[n])^{\mathrm{T}}$, $\mathbf{w}[n] = (w_1[n], \ldots, w_N[n])^{\mathrm{T}}$, and $\mathbf{a}(\theta)$ is the array response corresponding to direction θ. We have not discussed complex WGN in detail in this text, but, in analogy with the results in Chapters 5 and 6 for real WGN, it is possible to show that correlation against a noiseless signal template is the right thing to do. Thus, regardless of the value of the time-domain sample $s[n]$, the *spatial* processing that maximizes the SNR is to correlate against the noiseless template $\mathbf{a}(\theta)$. That is, we wish to compute the decision statistics

$$Z[n] = \langle \mathbf{y}[n], \mathbf{a}(\theta) \rangle = \mathbf{a}^{\mathrm{H}}(\theta)\mathbf{y}[n] \tag{8.53}$$

Correlating the spatial signal against the array response in this fashion is termed *beamforming.* The desired signal contribution to the decision statistic obtained from beamforming is $||\mathbf{a}(\theta)||^2 s[n] = Ns[n]$. Thus, the signal amplitude gets scaled by a factor of N, and hence the signal power gets scaled by a factor of N^2. It can be shown that the variance of the noise contribution to the decision statistic gets amplified by a factor of N. Thus, the SNR gets amplified by a factor of N by beamforming at the receiver. This is called the beamforming gain. Receive beamforming is also termed *maximal ratio combining,* because it combines the spatial signal in a manner that maximizes the signal-to-noise *ratio.*

Receive beamforming gathers energy coming from a given direction. Conversely, transmit beamforming can be used to direct energy in a given direction. For example, if a linear transmit antenna array seeks to direct energy towards an angle of departure θ, then, in order to send a time-domain sample $s[n]$, it should transmit the spatial vector $s[n]\mathbf{a}^{\mathrm{H}}(\theta)$. Since the spatial channel to the receiver is $\mathbf{a}(\theta)$, the signal received is given by $s[n]\mathbf{a}^{\mathrm{H}}(\theta)\mathbf{a}(\theta) = Ns[n]$. Thus, the received amplitude scales as N, and the received power as N^2. Since the noise at the receiver does not get the benefit of this transmit beamforming gain, transmit beamforming with N antennas leads to an SNR gain of N^2 relative to a single-antenna system, if we fix the per-antenna emitted power. The signal transmitted from antenna k is $s[n]e^{j(k-1)\phi(\theta)}$, which has power $|s[n]|^2$, and, since we have N antenna elements, we are transmitting at N times the power. The additional factor of N in received power comes from the fact that, by choosing the beamforming coefficients appropriately, we are ensuring that the signals from these N antenna elements add up in phase at the receiver, which leads to an N-fold gain.

Thus, both transmit and receive beamforming perform *spatial matched filtering*, leading to a *beamforming gain* of N. That is, the SNR is enhanced by a factor of N. In addition, if each element in a transmit antenna array transmits at a power equal to that of a reference single-element antenna, then we have an additional *power combining gain* of N for transmit beamforming, leading to a net SNR gain of N^2.

Beamforming directs energy in a given direction by ensuring that the radio waves emitted or received from that direction (or their complex envelopes) add constructively, or in phase. The radio waves in other directions may add constructively or destructively, depending on the array geometry. Thus, it is of interest to characterize the *beam pattern* corresponding to a particular set of beamforming coefficients. If we are beamforming in direction θ_0, then the gain in an arbitrary direction θ is given by

$$G(\theta;\theta_0) = |\langle \mathbf{a}(\theta), \mathbf{a}(\theta_0)\rangle| = \left|\mathbf{a}^{\mathrm{H}}(\theta_0)\mathbf{a}(\theta)\right|$$

The following code fragment computes and plots the beam pattern for a linear array.

Code Fragment 8.4.1 (Plotting beam patterns for a linear array)

```
d = 1/3; %normalized inter-element spacing
N = 10; %number of array elements
theta0_degrees = 0; %desired angle from broadside in degrees
theta0 = theta0_degrees * pi/180; %desired angle from broadside in radians
phi0 = 2 * pi * d * sin(theta0); %desired spatial frequency
a0 = exp(j * phi0 * [0:N - 1]); %array response in desired direction
theta_degrees = -90:1:90; %sweep of angles with respect to broadside
theta = theta_degrees * pi/180; %(angles in radians)
phi = 2 * pi * d * sin(theta); %spatial frequency as a function of angle w.r.t.
    broadside
%(as a row)
%array responses corresponding to the spatial frequencies as columns
array_responses = exp(j * transpose([0:N - 1]) * phi);
%inner product of desired array response with array responses in other directions
rho = conj(a0) * array_responses;
plot(theta_degrees,10 * log10(abs(rho))); %plot gain (dB) versus angle
hold on;
stem(theta0_degrees,10 * log10(N),'r'); %indicates desired direction
xlabel('Angle with respect to broadside');
ylabel('Gain (dB)');
```

Array spacing In the preceding code fragment, we have set the element spacing at $\lambda/3$. In Problem 8.10, we explore the effect of varying the element spacing, and, in particular, what happens as the element spacing exceeds $\lambda/2$.

Notational convention We say that we employ beamforming weights or coefficients $\mathbf{c} = (c_1,\ldots,c_N)^{\mathrm{T}}$ when we apply the coefficient c_i^* to the ith antenna element. For a receive beamformer, if the spatial signal being received is $\mathbf{y} = (y_1,\ldots,y_N)^{\mathrm{T}}$, then the use of beamforming weights $\mathbf{c}$ corresponds to computing the inner product $\langle \mathbf{y}, \mathbf{c}\rangle = \mathbf{c}^{\mathrm{H}}\mathbf{y} = \sum_{i=1}^{N} c_i^* y_i$. With this convention, the beamforming weights for directing a beam in direction θ are given by $\mathbf{c} = \mathbf{a}(\theta)$.

Steering nulls As we see from Figure 8.16, when we form a beam in a given direction, we maximize the beam pattern in that direction, creating a *main lobe* in the beam pattern, while

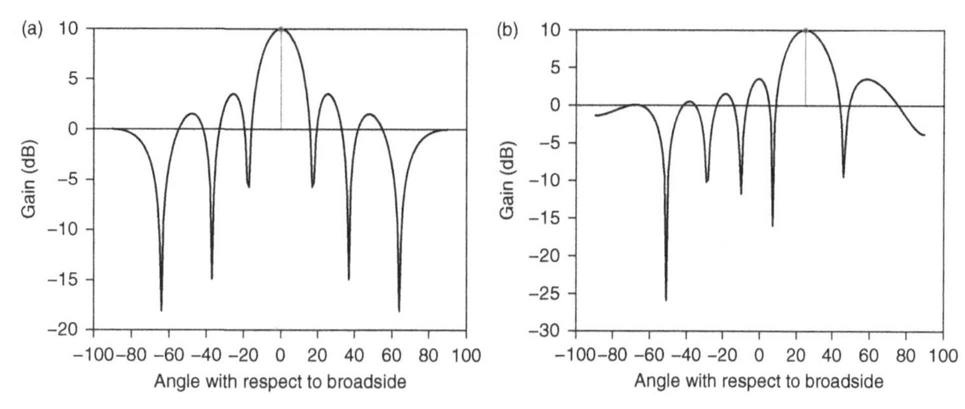

Figure 8.16 Example beam patterns with a linear array, generated using Code Fragment 8.4.1: (a) beam directed at 0° and (b) beam directed at 25°.

also generating other local maxima (typically of lower strength) in other directions. The latter are called *sidelobes,* and are often small enough compared with the main lobe that we do not need to worry about them. Sometimes, however, we want to be extra careful in guaranteeing that power is not accidentally steered in an undesired direction. For example, a cellular base station employing a beamforming array to receive a signal from mobile A may wish to null out interference from mobile B. We can use a ZF approach, analogous to the one discussed in detail in Section 8.2.2. If mobile A is in direction θ_A and mobile B in direction θ_B, then we wish to align $\mathbf{c}$ with $\mathbf{a}(\theta_A)$ as best we can, while keeping it orthogonal to $\mathbf{a}(\theta_B)$. Thus, we can choose the beamforming weights to be a scaled version of the projection of $\mathbf{a}(\theta_A)$ orthogonal to the interference subspace spanned by $\mathbf{a}(\theta_B)$, which is given by

$$\mathbf{c}_A = \mathbf{a}(\theta_A) - \langle \mathbf{a}(\theta_A), \mathbf{a}(\theta_B) \rangle \frac{\mathbf{a}(\theta_B)}{\langle \mathbf{a}(\theta_B), \mathbf{a}(\theta_B) \rangle} \tag{8.54}$$

While the ZF approach has the advantage of having a clear geometric interpretation, in practice, when implementing this at the receiver, we may often employ the MMSE criterion (see Sections 8.2 and 8.2.2), which lends itself to adaptive implementation.

We can combine beam and null steering in this fashion at the transmitter as well as the receiver. There are some additional issues when employing this approach at the transmitter. First, the transmitter must know the array responses corresponding to the different receivers it is steering beams or nulls towards, which requires either explicit feedback or implicit feedback derived from reciprocity. Second, we must scale the weights appropriately depending on constraints on transmit power: the average power scales with $||\mathbf{c}||^2$, while the peak power scales with $\max_i |c_i|^2$.

Space-division multiple access (SDMA) Beamforming and nullforming can enable a single receiver to receive from multiple transmitters, and conversely, a single transmitter to transmit separate messages to different receivers, using a common set of time–frequency resources. This is termed space-division multiple access (SDMA). For example, in order to send a message signal $s_A(t)$ to mobile A without interfering with mobile B, and message signal $s_B(t)$ to mobile B without interfering with mobile A, the transmitter sends the "space–time" signal

$$\mathbf{y}(t) = s_A(t)\mathbf{c}_A^* + s_B(t)\mathbf{c}_B^* \tag{8.55}$$

where $\mathbf{c}_A$ is the zero-forcing solution in (8.54), and $\mathbf{c}_B$ is a zero-forcing solution with the roles of A and B interchanged. That is, the signal transmitted from the ith antenna is a linear combination of the two message signals, $y_i(t) = s_A(t)c_{i,A}^* + s_B(t)c_{i,B}^*$, where the conjugation of the beamforming weights is in accordance with the convention discussed earlier. A receiver with an antenna array can use similar techniques to receive signals from multiple transmitters at the same time. SDMA is explored further in Problem 8.11.

8.4.3 Rich scattering and MIMO-OFDM

While Section 8.4.2 focuses on beamsteering and nullsteering along specific directions, the channel between transmitter and receiver may often be characterized by a large number of paths, possibly corresponding to different directions of arrival or departure. Indoor WiFi channels are one example of such "rich-scattering" channels. Figure 8.17 shows some of the paths obtained from two-dimensional ray tracing between a transmitter and a receiver in a rectangular room. These include all four first-order reflections (single bounces) and two second-order reflections (two bounces). Not all of these have equal attenuation (the attenuation of a path depends on its length, as well as the angles of incidence and the type of material at each surface from which it reflects), but we can see from the construction of the second-order reflections that the number of paths quickly becomes large as we start accounting for multiple bounces. Of course, the path strengths start dying out as the number of bounces increases, since there is a loss in strength for each bounce, but, for typical indoor

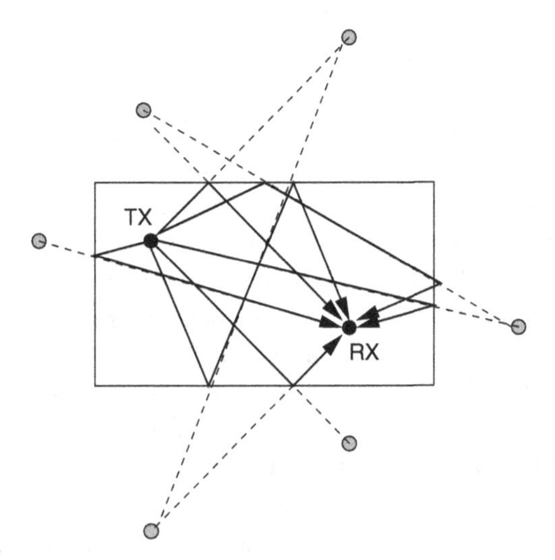

Figure 8.17 Ray tracing to determine paths between a transmitter and a receiver inside a "two-dimensional room." All first-order reflections, and two second-order reflections, are shown. The lightly shaded circles depict "virtual sources" employed to perform ray tracing.

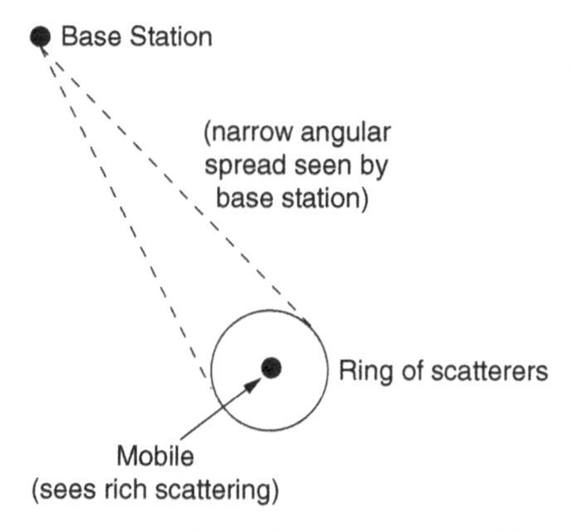

Figure 8.18 A typical propagation environment between an elevated base station and a mobile in urban clutter. The mobile sees a rich scattering environment locally, due to reflections from buildings and street surfaces. However, from the base station's viewpoint, the paths to the mobile fall within a narrow angular spread.

environments in the WiFi bands (2.4 and 5 GHz), there are many paths with nontrivial gains.

Exercise What is the total number of second-order reflections in the scenario depicted in Figure 8.17?

Even in outdoor settings, such as for cellular networks, mobiles in an urban environment may see rich scattering because of bounces from buildings around them. An elevated base station, however, may still see a relatively sparse scattering environment. Such a situation is depicted in Figure 8.18. Since the base station sees a narrow angular spread, it may be able to employ beamforming strategies effectively (e.g., forming a beam along the "mean" angle of arrival/departure). However, the mobile transceiver must account for the rich scattering environment that it sees.

At this point, the reader is encouraged to quickly review Section 2.9. As we noted there, a multipath channel has a transfer function that is "frequency-selective" (i.e., it varies with frequency). Now that we have multiple antennas, each antenna sees a frequency-selective channel, so the net array response is frequency-selective. However, we can model the array response as constant for a small enough frequency slice (smaller than the coherence bandwidth – see the discussion in Section 2.9). OFDM (see Section 8.3) naturally decomposes the channel into such slices, and each subcarrier in a MIMO-OFDM system may see a different array response. Thus, we can apply MIMO processing in parallel to each subcarrier after downconversion and OFDM processing, as shown in Figure 8.19.

Focusing on a single subcarrier in a MIMO-OFDM system (this model also applies to narrowband signaling with bandwidth smaller than the channel coherence bandwidth), consider a link with M transmit antennas and N receive antennas. Over a subcarrier, the channel from transmit element m to receive element n is a complex-valued scalar, which we denote by H_{nm}. If the transmitter sends a complex symbol $x[m]$ from antenna m, then the nth receive antenna sees the linear combination

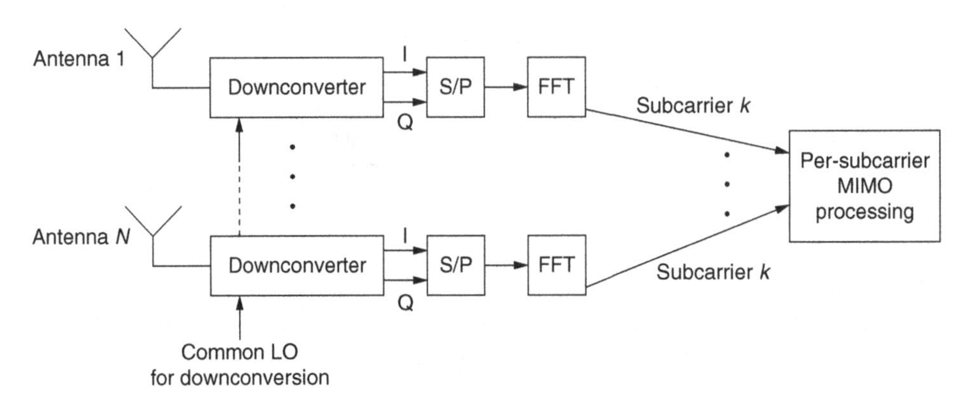

Figure 8.19 A typical MIMO-OFDM receiver architecture. After downconverting and sampling the received signal from each antenna, we apply OFDM processing to separate out the subcarriers. After the FFT, the samples for a given subcarrier, say k, from the different antennas are collected together for per-subcarrier MIMO processing. Thus, each subcarrier sees a different narrowband MIMO channel.

$$y_n = \sum_{m=1}^{M} H_{nm} x_m + w_n \tag{8.56}$$

where w_n is the complex-valued noise seen at the nth receive antenna. The preceding can be written in matrix–vector notation as

$$\mathbf{y} = \mathbf{H}\mathbf{x} + \mathbf{w} \tag{8.57}$$

where $\mathbf{y} = (y_1, \ldots, y_N)^{\mathrm{T}}$ is the received vector, $\mathbf{x} = (x_1, \ldots, x_M)^{\mathrm{T}}$ is the transmitted vector, and $\mathbf{H}$ is the $N \times M$ channel matrix, whose mth column is the receive array response seen by the mth transmit element.

Noise model The complex-valued noise w_n is typically modeled as follows: $\mathrm{Re}(w_n)$ and $\mathrm{Im}(w_n)$ are i.i.d. $N(0, \sigma^2)$, and are independent across receive antennas. The noise vector $\mathbf{w}$ is said to be a complex Gaussian random vector, which is completely characterized by its mean $\mathbb{E}[\mathbf{w}] = \mathbf{0}$ and covariance matrix $\mathbf{C_w} = \mathbb{E}\left[(\mathbf{w} - \mathbb{E}[\mathbf{w}])(\mathbf{w} - \mathbb{E}[\mathbf{w}])^{\mathrm{H}}\right] = 2\sigma^2\mathbf{I}$. Its distribution is denoted by $\mathbf{w} \sim CN(\mathbf{0}, 2\sigma^2\mathbf{I})$, and the distribution of any entry is specified as $w_n \sim CN(0, 2\sigma^2)$.

Remark on notation According to our convention (which is consistent with most literature in the field), for an $M \times N$ MIMO system (i.e., with M transmit antennas and N receive antennas), the channel matrix $\mathbf{H}$ is an $N \times M$ matrix. The reason for this choice of convention is that we like working with column vectors: $\mathbf{x}$ is the $M \times 1$ column vector of symbols transmitted from the different transmit antennas, the mth column of $\mathbf{H}$ is the receiver's spatial response to the mth transmit antenna, and $\mathbf{y}$ is the $N \times 1$ column vector of received samples.

Operations such as beamforming and nullforming can now be performed separately for each subcarrier. However, these operations are no longer associated with directing energy or nulls towards particular physical directions, since the spatial response in each subchannel is a linear combination of array responses associated with many directions. A particularly simple model for the resulting channel gains for a given subcarrier is described next.

The rich-scattering model The path gains $H(n, m)$ for a given subcarrier are a function of the channel impulse responses for each transmit/receive pair, but are often modeled statistically in order to provide quick insights into design tradeoffs in a manner that is independent of the specific propagation geometry. We now discuss a particularly simple model, motivated by "rich-scattering environments" in which there are many paths of roughly equal strength between the transmitter and the receiver. Let $h = H(m, n)$ denote the complex gain for a typical transmit/receive antenna pair. We can write

$$h = \sum_{i=1}^{L} A_i e^{j\theta_i}$$

where L is the number of paths, and where $A_i \geq 0$, $\theta_i \in [0, 2\pi]$ are the amplitude and phase of the complex-valued path gain for the given subcarrier. We therefore have

$$\text{Re}(h) = \sum_{i=1}^{L} A_i \cos\theta_i, \qquad \text{Im}(h) = \sum_{i=1}^{L} A_i \sin\theta_i$$

If the differences between the lengths of the different paths are comparable to, or larger than, a carrier wavelength (which is typically the case even for WiFi links indoors, and certainly for cellular links outdoors), then we can model the phases θ_i as i.i.d. uniform over $[0, 2\pi]$. Now, if the amplitudes for the different paths are roughly comparable, then we can apply the central limit theorem to approximate the joint distribution of Re(h) and Im(h) as i.i.d. $N(0, \sum_{i=1}^{L} A_i^2/2)$. Let us now normalize $\sum_{i=1}^{L} A_i^2 = 1$ without loss of generality; we can scale the noise variance to adjust the average SNR: $\overline{\text{SNR}} = \mathbb{E}[|h|^2]/(2\sigma^2) = 1/(2\sigma^2)$ for the model. We can therefore model h as a zero-mean complex Gaussian random variable: $h \sim CN(0, 1)$. Furthermore, for rich-scattering environments, it is assumed that the phases seen by different transmit/receive antenna pairs are sufficiently different that we can model the gains $H(n, m)$ as i.i.d. $CN(0, 1)$ for different transmit/receive antenna pairs (m, n).

8.4.4 Diversity

When the transmitter and receiver each have only one antenna ($M = N = 1$), under the rich-scattering model, the SNR is given by

$$\text{SNR} = \frac{|h|^2}{2\sigma^2} \tag{8.58}$$

Since h is a random variable, so is the SNR. In fact, since Re(h) and Im(h) are i.i.d. $N(0, \frac{1}{2})$, the sum of their squares is an exponential random variable (see Problems 5.11 and 5.21). Taking into account the scaling by $2\sigma^2$, we can show that the SNR is an exponential random variable with mean equal to the average SNR, $\overline{\text{SNR}} = 1/(2\sigma^2)$. If we now design our coded modulation strategy for a nominal SNR of SNR_0, we say that the system is in *outage* when the SNR is smaller than this value. The probability of outage is given by

$$P_{\text{out}} = P[\text{SNR} < \text{SNR}_0] = 1 - e^{-\text{SNR}_0/\overline{\text{SNR}}} \quad (8.59)$$

We would typically choose the nominal SNR, SNR_0, to be smaller than the average SNR, $\overline{\text{SNR}}$, by a link margin. For example, for a link margin of 10 dB, we have $\text{SNR}_0 = 0.1\overline{\text{SNR}}$, so $P_{\text{out}} = 1-e^{-0.1} \approx 0.1$ (for $|x|$ small, $e^x \approx 1+x$ for $|x|$ small, so $1-e^{-x} \approx x$). Thus, even after giving up 10 dB in link margin, we still get a relatively high outage rate of 10%. Of course, there is a nontrivial probability that the SNR with fading is *higher* than the nominal, hence we can have *negative* link margins if we are willing to live with large enough outage rates. For example, a link margin of -3 dB corresponds to $\text{SNR}_0 = 2\overline{\text{SNR}}$, with outage rate $P_{\text{out}} = 1 - e^{-2} = 0.865$ (which is too high for most practical applications).

In order to reduce the outage rate without increasing the link margin, we must employ *diversity,* which is a generic term used for any strategy that involves multiple, approximately independent, "looks" at a fading channel. We saw diversity in action for our simulation-based model in Software Lab 2.2. We now explore it for the rich-scattering model, skipping some details in the derivation in the interest of arriving quickly at the key insights.

Benchmark We continue to define our link margin relative to an unfaded single-input, single-output (SISO) system with average SNR of $\overline{\text{SNR}} = 1/(2\sigma^2)$.

Receive diversity Consider a receiver equipped with two antennas ($N = 2$). If they are spaced far enough apart in a rich-scattering environment, we can assume that the channel gains (for a given subcarrier) seen by the two antennas are i.i.d. $CN(0, 1)$ random variables. The received samples at the two antennas are modeled as

$$y_n = h_n x + w_n, \quad n = 1, 2$$

where x is the transmitted symbol, $h[1], h[2] \sim CN(0, 1)$ are i.i.d. (independent Rayleigh fading), and $w[1], w[2] \sim CN(0, 2\sigma^2)$ are i.i.d. (independent noise samples). The optimal decision statistic is obtained using receive beamforming, and is given by

$$Z = h_1^* y_1 + h_2^* y_2 \quad (8.60)$$

It can be shown that the SNR is now given by

$$\text{SNR} = G\overline{\text{SNR}} \quad (8.61)$$

where the gain relative to the benchmark SISO system is given by

$$G = |h_1|^2 + |h_2|^2 \quad (8.62)$$

We can break this up into two gains: $G_{\text{diversity}} = (|h_1|^2 + |h_2|^2)/2$ due to averaging channel fluctuations across antennas, and $G_{\text{coh}} = 2$ due to averaging noise across antennas (the signal terms are being combined coherently, so the phases line up, while the noise terms are being combined incoherently, across the two receive antennas). Thus,

$$G = G_{\text{diversity}} G_{\text{coh}} \quad (8.63)$$

Equations (8.61) and (8.63) generalize directly to N receive antennas:

$$G = |h_1|^2 + \cdots + |h_N|^2 = G_{\text{diversity}} G_{\text{coh}} \quad (8.64)$$

where $G_{\text{diversity}} = (|h_1|^2 + \cdots + |h_N|^2)/N$ and $G_{\text{coh}} = N$. As N gets large, the fluctuations due to fading get smoothed away, and $G_{\text{diversity}} \rightarrow 1$ by the law of large numbers. In practice, however, even small values of N (e.g., $N = 2, 4$) give significant performance gains.

Suppose now that we design our coding and modulation so as to provide reliable performance at a nominal SNR, say SNR_0, which is smaller than the SISO benchmark $\overline{\text{SNR}}$ by a link margin of L dB: $\text{SNR}_0\ (dB) = \overline{\text{SNR}}(\text{dB}) - L\ (\text{dB})$. The probability of outage is given by

$$P_{\text{out}} = P[\text{SNR} < \text{SNR}_0] = P[G < 10^{-L\ (\text{dB})/10}] \tag{8.65}$$

Figure 8.20 plots the outage probability as a function of the link margin for several different values of N. The plots are obtained using the procedure outlined in Problem 8.12.

Transmit diversity If the transmitter has multiple antennas, it can beamform towards the receiver if it has implicit or explicit feedback regarding the channel, as has already been noted. When such feedback is not available, we would like to use open-loop strategies that provide diversity. Consider a transmitter with two antennas communicating with a receiver with a single antenna ($M = 2, N = 1$). In a MIMO-OFDM system, for a given subcarrier, suppose that the transmit antenna 1 sends the sample x_1 and transmit antenna 2 sends the sample x_2. If the transmitter knows the channel coefficients h_1 and h_2 from the two transmit antennas to the receive antenna, then it could choose $x_1 = h_1^* x$ and $x_2 = h_2^* x$, where x is the symbol to be transmitted. What do we do when h_1 and h_2 are unknown? In general, if we send x_1 and x_2 from the two transmit elements, then the received sample is given by

$$y = h_1 x_1 + h_2 x_2 + w \tag{8.66}$$

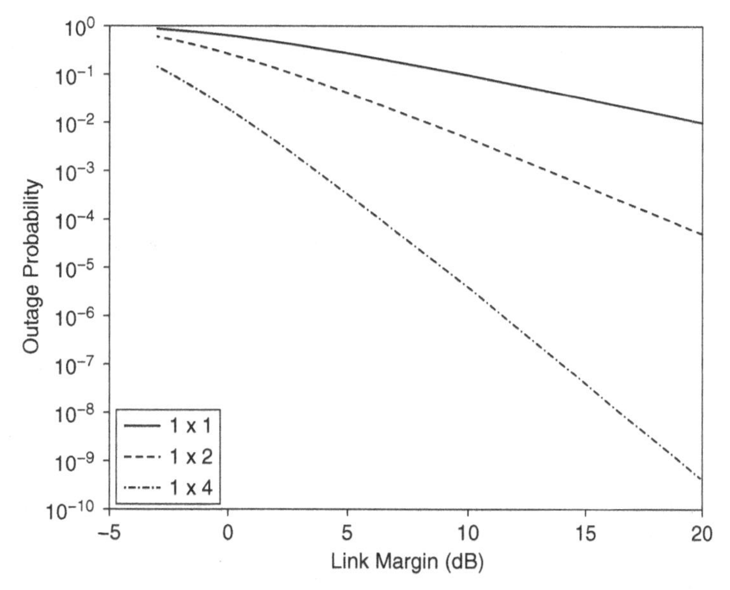

Figure 8.20 The probability of outage versus the link margin (dB) for receive diversity in $1 \times N$ MIMO systems.

where w is noise. Thus, if x_1 and x_2 are two independent symbols, then they interfere with each other at the receiver. On the other hand, if we set $x_1 = x_2 = x/\sqrt{2}$ (normalizing the transmit power across the two antennas to that of a transmitter with a single antenna), then we receive

$$y = \frac{h_1 + h_2}{\sqrt{2}} x + w$$

If h_1 and h_2 are i.i.d. $CN(0, 1)$, it is easy to show that the effective channel coefficient $h_{\text{eff}} = (h_1 + h_2)/\sqrt{2}$ is also $CN(0, 1)$. Thus, we still have Rayleigh fading, and have not made any progress relative to a single-antenna transmitter! An ingenious solution to this problem is the *Alamouti space–time code* (named after its inventor), which resolves the interference between the signals sent by the two transmit antennas over two time samples. Let $b[1]$ and $b[2]$ be two symbols to be transmitted. For a single-antenna transmitter, they would be transmitted in sequence. For the two-antenna transmitter now being considered, let us expand the signal-space dimension to two at the receiver by considering two successive time samples. This allows us to orthogonalize the contributions of these two symbols at the receiver. Denoting by $x_i[1]$ and $x_i[2]$, $i = 1, 2$, the samples transmitted from antenna i at two successive time intervals, we set

$$\begin{aligned} x_1[1] &= b[1]/\sqrt{2}, \qquad & x_2[1] &= b[2]/\sqrt{2} \\ x_1[2] &= -b^*[2]/\sqrt{2}, \qquad & x_2[2] &= b^*[1]/\sqrt{2} \end{aligned} \tag{8.67}$$

where we have again normalized the net transmit power to that of a single-antenna system. Figure 8.21 depicts the operation of the Alamouti space–time code, taking a sequence of symbols as input, and mapping them in groups of two to a sequence of samples at the output of each antenna.

The received samples in the two successive time intervals are given by

$$\begin{aligned} y[1] &= h_1 x_1[1] + h_2 x_2[1] + w[1] = (h_1/\sqrt{2})b[1] + (h_2/\sqrt{2})b[2] + w[1] \\ y[2] &= h_1 x_1[2] + h_2 x_2[2] + w[2] = -(h_1/\sqrt{2})b^*[2] + (h_2/\sqrt{2})b^*[1] + w[2] \end{aligned} \tag{8.68}$$

We assume that the receiver has estimates of the channel coefficients h_1 and h_2 (e.g., using known training signals). We would like to write the two observations as a received vector in which each symbol modulates a different signal vector. Since the symbols are conjugated when sent over the second time interval, we conjugate the second

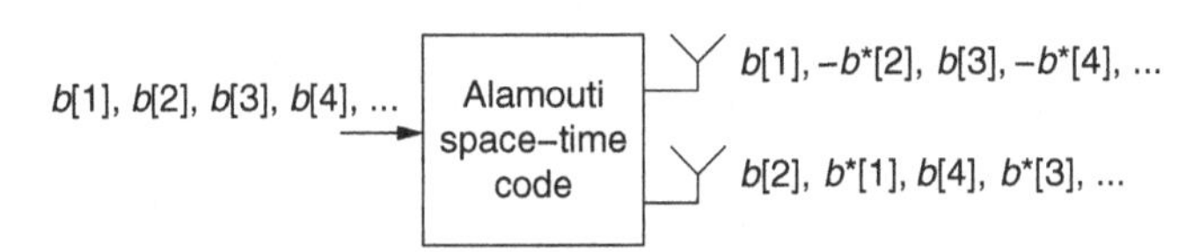

Figure 8.21 The transmitter in an Alamouti space–time code takes two symbols at a time, and maps them to two consecutive symbols to be sent from each of the two transmit antennas. The input to the space–time encoder is the sequence of symbols to be transmitted, $\{b[n]\}$, while the outputs are the sequences $\{x_i[n]\}$, $i = 1, 2$, to be transmitted from antenna i. The $1/\sqrt{2}$ factor for power normalization is omitted from the figure.

received sample when creating the received vector. This yields the following vector model:

$$\tilde{\mathbf{y}} = \begin{pmatrix} y[1] \\ y^*[2] \end{pmatrix} = b[1]\begin{pmatrix} h_1/\sqrt{2} \\ h_2^*/\sqrt{2} \end{pmatrix} + b[2]\begin{pmatrix} h_2/\sqrt{2} \\ -h_1^*/\sqrt{2} \end{pmatrix} + \begin{pmatrix} w[1] \\ w^*[2] \end{pmatrix}$$
$$= b[1]\mathbf{u}_1 + b[2]\mathbf{u}_2 + \mathbf{w} \tag{8.69}$$

The vectors $\mathbf{u}_1 = (1/\sqrt{2})(h_1, h_2^*)^{\mathrm{T}}$ and $\mathbf{u}_2 = (1/\sqrt{2})(h_2, -h_1^*)^{\mathrm{T}}$ are orthogonal (i.e., $\mathbf{u}_1^{\mathrm{H}}\mathbf{u}_2 = 0$), regardless of the values of the channel coefficients, hence the symbols $b[1]$ and $b[2]$ do not interfere with each other. The vector $\mathbf{w} \sim CN(0, 2\sigma^2\mathbf{I})$. The optimal decision statistic $Z[i]$ for symbol $b[i]$ is given by matched filtering against $\mathbf{u}_i$:

$$Z[i] = \mathbf{u}_i^{\mathrm{H}}\tilde{\mathbf{y}}, \quad i = 1, 2 \tag{8.70}$$

Exercise Write out these decision statistics explicitly in terms of $y[1]$, $y[2]$, h_1, and h_2.
Answer. $Z[1] = h_1^* y[1] + h_2 y^*[2]$, $Z[2] = h_2^* y[1] - h_1 y^*[2]$ (up to scale).

The SNR seen by each symbol is given by

$$\mathrm{SNR}_{\mathrm{Alamouti}} = \frac{||\mathbf{u}_i||^2}{2\sigma^2} = \frac{(|h_1|^2 + |h_2|^2)/2}{2\sigma^2} = G_{\mathrm{Alamouti}}\overline{\mathrm{SNR}} \tag{8.71}$$

where

$$G_{\mathrm{Alamouti}} = \frac{|h_1|^2 + |h_2|^2}{2} \tag{8.72}$$

On comparing with (8.64) for receive diversity, we see that the Alamouti scheme in a 2×1 system achieves the same diversity gain as the receive diversity in a 1×2 system, but does not provide the coherent gain obtained from averaging across receive antennas in the latter. Of course, as we shall see in Problem 8.13 and in Software Lab 8.3, the Alamouti scheme applies to $2 \times N$ MIMO systems for arbitrary N, so we can get noise averaging and receive diversity gains for $N > 1$. The outage rates computed in Problem 8.13 are plotted in Figure 8.22.

The simplicity of the Alamouti construction (and its optimality for 2×1 MIMO) has led to its adoption by a number of cellular and WiFi standards (just do an Internet search to see this). Unfortunately, the orthogonalization provided by the Alamouti space–time code does not scale to more than two transmit antennas. Several "quasi-orthogonal" constructions have been investigated, but, so far, these have had less impact on practice. Indeed, when there are many transmit antennas, the trend is to engineer the system so that the transmitter has enough information about the channel to perform some form of transmit beamforming (possibly using multiple beams).

8.4.5 Spatial multiplexing

We have already seen that a transceiver with multiple antennas can use SDMA to communicate with multiple nodes at different locations. For example, a cellular base station with multiple antennas can use the same time–frequency resources to send data streams in parallel to different mobile devices (even if such devices have only one antenna each).

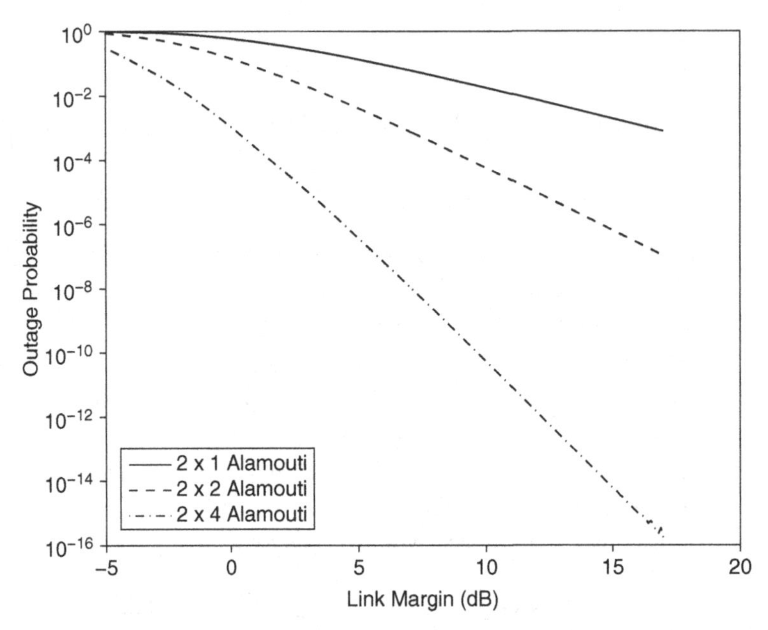

Figure 8.22 The probability of outage versus the link margin (dB) for Alamouti space–time coding for $2 \times N$ MIMO with rich scattering.

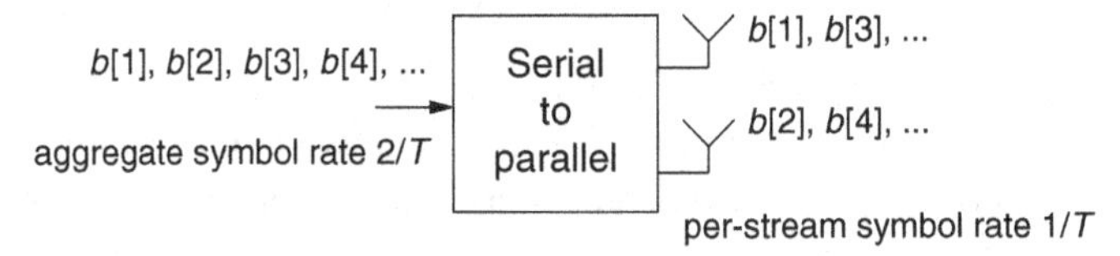

Figure 8.23 For spatial multiplexing, the transmitter may take a sequence of incoming symbols $\{b[n]\}$, and do a serial-to-parallel conversion to map them to subsequences to be transmitted from the different antennas. In the example shown, the odd symbols are transmitted from antenna 1, and the even symbols from antenna 2. The aggregate symbol rate is twice the per-stream symbol rate.

When both transmitter and receiver have multiple antennas, if the propagation environment is "rich enough," then multiple parallel data streams can be sent between transmitter and receiver. This is termed *spatial multiplexing.* Figure 8.23 depicts spatial multiplexing with two antennas, modeling, for example, one subcarrier in a MIMO-OFDM system. The per-stream symbol rate $1/T$ is the rate of sending symbols along a subcarrier, where T is the length of an OFDM symbol. With M-fold spatial multiplexing, the aggregate symbol rate for a given subcarrier is M/T. This should be scaled up by the number of subcarriers to get the overall symbol rate.

For example, suppose that the transmitter and receiver in a MIMO-OFDM system each have two antennas ($M = N = 2$). For a given subcarrier in an OFDM system, consider a particular time interval. Suppose that the transmitter sends x_1 from antenna 1 and x_2 from antenna 2 (referring to Figure 8.23, $x_1 = b[1]$ and $x_2 = b[2]$ in the first time interval). The samples at the two receive elements are given by

$$y_1 = H_{11}x_1 + H_{12}x_2 + w_1$$
$$y_2 = H_{21}x_1 + H_{22}x_2 + w_2$$

which we can write in vector form as

$$\mathbf{y} = \begin{pmatrix} y_1 \\ y_2 \end{pmatrix} = x_1 \begin{pmatrix} H_{11} \\ H_{21} \end{pmatrix} + x_2 \begin{pmatrix} H_{12} \\ H_{22} \end{pmatrix} + \begin{pmatrix} w_1 \\ w_2 \end{pmatrix} = x_1\mathbf{u}_1 + x_2\mathbf{u}_2 + \mathbf{w} \tag{8.73}$$

where $\mathbf{u}_1$ is the response seen by transmit element 1 at the two receive antennas, $\mathbf{u}_2$ is the response seen by transmit element 2 at the receive antennas, and $\mathbf{w} \sim CN(0, 2\sigma^2)$ is complex WGN. While we have considered a 2×2 MIMO system for illustration, the model is generally applicable for $2 \times N$ MIMO systems with $N \geq 2$, with $\mathbf{u}_1$ and $\mathbf{u}_2$ denoting the two columns of the channel matrix $\mathbf{H}$, corresponding to the received responses for each of the two transmit antennas, respectively.

In a MIMO-OFDM system, we have eliminated interference across subcarriers using OFDM, but we have introduced interference in space by sending multiple symbols from different antennas. The vector spatial interference model (8.73) for each subcarrier is analogous to the vector time-domain interference model for ISI in single-carrier systems discussed in Chapter 6. Just as we can compensate for ISI using a time-domain equalizer if the time-domain channel has appropriate characteristics, we can compensate for spatial interference using a spatial equalizer if the spatial channel has appropriate characteristics. For example, if $\mathbf{u}_1$ and $\mathbf{u}_2$ are linearly independent, then we can use linear ZF or MMSE techniques to demodulate the symbols x_1 and x_2. Thus, if there are M parallel data streams being sent from the transmit antennas, then we need at least M receive antennas in order to obtain a signal space of large enough dimension for the linear independence condition to be satisfied. Indeed, it can be shown more generally (without restricting ourselves to ZF or MMSE techniques) that, for rich-scattering models, the capacity of an $M \times N$ MIMO channel scales as $\min(M, N)$, the minimum of the number of transmit and receive antennas.

The ZF and MMSE receivers have been discussed in detail in Section 8.2.2. For our purpose, the relevant expressions are those for complex-valued signals in (8.35). We reproduce the expression for a ZF correlator here before adapting it to our present purpose:

$$\mathbf{c}_{ZF} = \mathbf{C}_\mathbf{w}^{-1}\mathbf{U}\left(\mathbf{U}^H\mathbf{C}_\mathbf{w}^{-1}\mathbf{U}\right)^{-1}\mathbf{e}$$

Recall that $\mathbf{U}$ is a matrix containing the signal vectors as columns, and that $\mathbf{e}$ is a unit vector with nonzero entry corresponding to the desired vector $\mathbf{u}_0$ in the ISI vector model (8.13). In our spatial multiplexing model (8.73), $\mathbf{U} = \mathbf{H}$ (the signal vectors are simply the columns of the channel matrix), the noise covariance $\mathbf{C}_\mathbf{w} = 2\sigma^2\mathbf{I}$, and we wish to demodulate the data corresponding to both of the signal vectors. Letting $\mathbf{e}_1 = (1, 0)^T$ and $\mathbf{e}_2 = (0, 1)^T$, the ZF correlators for the two streams can be written as (dropping scale factors corresponding to the noise variance)

$$\mathbf{c}_1 = \mathbf{H}\left(\mathbf{H}^H\mathbf{H}\right)^{-1}\mathbf{e}_1, \qquad \mathbf{c}_2 = \mathbf{H}\left(\mathbf{H}^H\mathbf{H}\right)^{-1}\mathbf{e}_2$$

We can represent this compactly as a single ZF matrix $\mathbf{C}_{ZF} = [\mathbf{c}_1\ \mathbf{c}_2]$ containing these correlators as the columns. On noting that $[\mathbf{e}_1\ \mathbf{e}_2] = \mathbf{I}$, we obtain that

$$\mathbf{C}_{ZF} = \mathbf{H}\left(\mathbf{H}^H\mathbf{H}\right)^{-1}, \quad \textbf{ZF matrix for spatial demultiplexing} \tag{8.74}$$

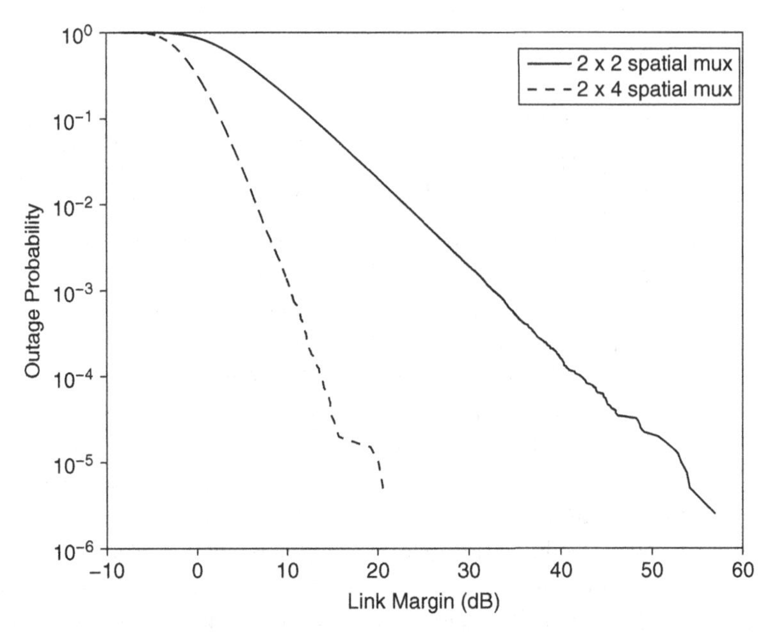

Figure 8.24 Outage rate versus link margin (with respect to the SISO benchmark) for $2 \times N$ spatial multiplexing (max; $N = 2, 4$) with zero-forcing reception.

The decision statistics for the multiplexed streams are given by

$$\mathbf{Z} = \mathbf{C}_{\mathrm{ZF}}^{\mathrm{H}} \mathbf{y} \tag{8.75}$$

While we have focused on the 2×2 example (8.73) in this derivation, it applies in general to M spatially multiplexed streams in an $M \times N$ MIMO system with $N \geq M$. The outage rate with zero-forcing reception for 2×2 and 2×4 is plotted in Figure 8.24, using software developed in Problem 8.14.

The MMSE receiver can be similarly derived to be

$$\mathbf{C}_{\mathrm{MMSE}} = \left(\mathbf{H}\mathbf{H}^{\mathrm{H}} + 2\sigma^2 \mathbf{I}\right)^{-1} \mathbf{H}, \quad \textbf{MMSE matrix for spatial demultiplexing} \tag{8.76}$$

where we have normalized the transmitted symbols to unit energy ($\mathbb{E}\left[|b[n]|^2\right] = \sigma_b^2 = 1$).

It is interesting to compare the spatial multiplexing model (8.73) with the diversity model (8.69) for the Alamouti space–time code. The Alamouti code does not rely on the receiver having multiple antennas, and therefore uses *time* to create enough dimensions for two symbols to be sent in parallel. Furthermore, the vectors $\mathbf{u}_1$ and $\mathbf{u}_2$ in the Alamouti model (8.69) are constructed such that they are orthogonal regardless of the propagation channel. In contrast, the spatial multiplexing model (8.73) relies on nature to provide vectors $\mathbf{u}_1$ and $\mathbf{u}_2$ that are "different enough" to support two parallel symbols. We explore these differences in Software Lab 8.3. The spectral efficiency of spatial multiplexing is twice that of the Alamouti code, but the diversity gain that it sees is smaller, as is evident from a comparison of the outage-rate versus link-margin curves in Figures 8.22 and 8.24. It is possible to systematically quantify the tradeoff between diversity and multiplexing, but this is beyond our scope here.

8.5 Concept summary

This chapter begins with an introduction to modeling and equalization for communication over dispersive channels, including single-carrier and OFDM modulation. All models and algorithms are developed in complex baseband, such that upconversion and downconversion are not explicitly modeled.

Modeling of single-carrier systems

- Symbols in a linearly modulated system pass through a cascade of the transmit, channel, and receive filters, where the cascade typically does not satisfy the Nyquist criterion for ISI avoidance. Any technique for handling the resulting ISI is termed *equalization.*
- Receiver noise is modeled as AWGN passed through the receive filter.
- Eye diagrams enable visualization of the effect of various ISI patterns, and equalization techniques are needed for reliable demodulation if the eye is closed.

Linear equalization

- The decision statistic for a given symbol is computed by a linear operation on a vector of received samples in an observation interval that is typically chosen to be large enough that it contains a significant contribution from the symbol of interest. Observation intervals for successive symbols are offset by the symbol time, so that the statistics of the desired symbol and the ISI are identical across observation intervals.
- The linear MMSE equalizer minimizes the MSE between the decision statistic and the desired symbol, and also maximizes the SINR over the class of linear equalizers.
- The MSE and the MMSE equalizer can be expressed in terms of statistical averages, hence the MMSE equalizer can be computed *adaptively* by replacing statistical averages by empirical averages. Such adaptive implementation requires transmission of a known training sequence.
- The received vector over an observation interval is the sum of the desired symbol modulating a desired vector, interfering symbols modulating interference vectors, and a noise vector. This vector ISI model can be characterized completely if we know the transmit filter, the channel filter, the receive filter, and the noise statistics at the input to the receive filter.
- Explicit analytical formulas can be given for the ZF and MMSE equalizer, and the associated SINRs, once the vector ISI model has been specified.
- At high SNR, the MMSE equalizer converges to the ZF equalizer, which (for white noise) can be interpreted geometrically as projecting the received vector orthogonal to the interference subspace spanned by the interference vectors, thus nulling out the ISI while incurring noise enhancement. The ZF equalizer exists only if the desired vector is linearly independent of the interference vectors.

- The geometric interpretation and analytical formulas for the ZF and MMSE equalizers developed for white noise can be extended to colored noise, with the derivation using the concept of noise whitening.

OFDM

- Since complex exponentials are eigenfunctions of any LTI system, multiple complex exponentials, each modulated by a complex-valued symbol, do not interfere with each other when transmitted through a dispersive channel. Each complex exponential simply gets scaled by the channel transfer function at that frequency. The task of equalization corresponds to undoing this complex gain in parallel for each complex exponential. This is the conceptual basis for OFDM, which enables parallelization of the task of equalization even for very complicated channels by transmitting along subcarriers.
- OFDM can be implemented efficiently in DSP using an IDFT at the transmitter (frequency-domain symbols to time-domain samples) and a DFT at the receiver (time-domain samples to frequency-domain observations, which are the symbols scaled by the channel gain and corrupted by noise).
- In order to maintain orthogonality across subcarriers (which is required for parallelization of the task of equalization) when we take the DFT at the receiver, the effect of the channel on the transmitted samples must be that of a circular convolution. Since the physical channel corresponds to linear convolution, OFDM systems emulate circular convolution by inserting a cyclic prefix into the transmitted time-domain samples.

The second part of the chapter provides an initial exposure to how multiple antennas at the transmitter and receiver (i.e., MIMO or space–time techniques) can be employed to enhance the performance of wireless systems. Three key techniques, which in practice are combined in various ways, are beamforming, diversity, and spatial multiplexing.

Beamforming and nullforming

- The array response for a linear array can be viewed as a mapping from the angle of arrival/departure to a spatial frequency.
- For an N-element array, a spatial matched filter, or beamforming, leads to a factor of N gain in the SNR. For transmit beamforming in which each antenna element is transmitting at a fixed power, we obtain an additional power combining gain of N.
- By forming a beam at a desired transceiver and nulls at other transceivers, an antenna array can support SDMA.

MIMO-OFDM abstraction

- Decomposing a time-domain channel into subcarriers using OFDM allows a simple model for MIMO systems, in which the channel between each pair of transmit and receive antennas is modeled as a single complex gain for each subcarrier.

- When the propagation environment is complex enough, the central limit theorem motivates modeling the channel gains between transmit/receive antenna pairs as i.i.d. zero-mean complex Gaussian random variables. We term this the *rich-scattering* model.
- Under the rich-scattering model, each transmit/receive antenna pair sees Rayleigh fading, but performance degradation due to fading can be alleviated using diversity.

Diversity

- Diversity strategies average over fades by exploiting roughly independent looks at the channel.
- Receive spatial diversity using spatial matched filtering provides a channel averaging gain (averaging the fading gains across antennas) and a noise averaging gain (due to coherent combining across antennas).
- Transmit spatial diversity provides channel averaging gains alone. It is trickier than receive diversity, since samples transmitted from different transmit antennas can interfere at the receiver. For two transmit antennas, the Alamouti space–time code is an optimal scheme for avoiding interference between different transmitted symbols, while providing channel averaging gains.

Spatial multiplexing

- Sending parallel data streams from different antennas increases the symbol rate in a manner proportional to the number of data streams, with space playing a role analogous to bandwidth.
- The parallel data streams interfere at the receiver, but can be demodulated using spatial equalization techniques analogous to the time-domain equalization techniques studied in Section 8.2 (e.g., suboptimal techniques such as ZF and MMSE).

8.6 Notes

While we have shown that ISI in single-carrier systems can be handled using linear equalization, significant performance improvements can be obtained using nonlinear strategies, including optimal maximum likelihood sequence estimation (MLSE), whose complexity is often prohibitive for long channels and/or large constellations, as well as suboptimal strategies such as decision feedback equalization (DFE), whose complexity is comparable to that of linear equalization. An introduction to such strategies, as well as pointers for further reading, can be found in more advanced communication theory texts such as [7, 8].

OFDM has now become ubiquitous in both wireless and wireline communication systems in recent years, because it provides a standardized mechanism for parallelizing equalization of arbitrarily complicated channels in a way that leverages the dropping cost and increasing speed of digital computation. For more detail than we have presented here, we refer the reader to the relevant chapters in the books on wireless communication by Goldsmith [46] and Tse and Viswanath [47]. These should provide the background required to access the huge research literature on OFDM, which focuses on issues such as channel estimation, synchronization and reduction of the PAPR.

There has been an explosion of research and development activity in MIMO, or space–time communication, starting from the 1990s: this is the decade in which the large capacity gains provided by spatial multiplexing were pointed out by Foschini [48] and Telatar [49], and the Alamouti space–time code was published by Alamouti [50]. MIMO techniques have been incorporated into 3G and 4G (WiMax and LTE) cellular standards, and WiFi (IEEE 802.11n) standards. An excellent reference for exploring MIMO-OFDM further is the textbook by Tse and Viswanath [47], while a brief introduction is provided in Chapter 8 of Madhow [7]. Other books devoted to MIMO include Paulraj *et al.* [51], Jafarkhani [52], and the compilation edited by Bolcskei *et al.* [53].

As discussed in the epilogue, a new frontier in MIMO is opening up with research and development for wireless communication systems at higher carrier frequencies, starting with the "millimeter-wave" band (i.e., carrier frequencies in the range 30–300 GHz, for which the wavelength is in the range 1–10 mm). Of particular interest is the 60-GHz band, where there is a huge amount (7 GHz!) of unlicensed spectrum, in contrast to the crowding in existing cellular and WiFi bands. While fundamental MIMO concepts such as beamforming, diversity, and spatial multiplexing still apply, the order-of-magnitude-smaller carrier wavelength and the order-of-magnitude-larger bandwidth require fundamentally rethinking many aspects of link and network design, as we briefly indicate in the epilogue.

8.7 Problems

ZF and MMSE equalization: modeling and numerical computations

Problem 8.1 (Noise-enhancement computations) Consider the ISI vector models in (8.10), (8.12), and (8.11).

(a) Compute the noise enhancements (dB) for the three equalizer lengths in these models, assuming white noise.

(b) Now, assume that the noise $\mathbf{w}[n]$ is colored, with $\mathbf{C_w}$ specified as follows:

$$\mathbf{C_w}(i,j) = \begin{cases} \sigma^2, & i=j \\ \sigma^2/2, & |i-j|=1 \\ 0, & \text{else} \end{cases} \tag{8.77}$$

Compute the noise enhancements for the three equalizer lengths considered, and compare the answers with your results in (a).

Problem 8.2 (Noise enhancement as a function of correlator length) Now, consider the discrete-time channel model leading to ISI vector models in (8.10), (8.12), and (8.11).

(a) Assuming white noise, compute and plot the noise enhancement (dB) as a function of the equalizer length, for L ranging from 4 to 16, increasing the observation interval by two by adding one sample to each side of the current observation interval, and starting from an observation interval of length $L = 4$ lined up with the impulse response for the desired symbol. Does the noise enhancement decrease monotonically? Does it plateau?

(b) Repeat for colored noise as in Problem 8.1(b).

Problem 8.3 (MMSE correlator and SINR computations) Consider the ISI vector model (8.10).

(a) Assume $\mathbf{C_w} = \sigma^2\mathbf{I}$, and define SNR $= ||\mathbf{u}_0||^2/\sigma^2$ as the MF bound on achievable SNR. For an SNR of 6 dB, compute the MMSE correlator and the corresponding SINR (dB), using (8.34) and (8.16). Check that the results match the alternative formula (8.36). Compare your result with the SNR achieved by the ZF correlator.

(b) Plot the SINR (dB) of the MMSE and ZF correlators as a function of the MF SNR (dB). Comment on any trends that you notice.

Problem 8.4 (From continuous-time to discrete-time vector ISI model) In this problem, we discuss an example of how to derive the vector ISI model (8.13) from a continuous-time model, using the system shown in Figure 8.25. The symbol rate $1/T = 1$, and the input to the transmit filter is $\sum_n b[n]\delta(t - nT)$, where $b[n] \in \{-1, 1\}$. Thus, the continuous-time noiseless signal at the output of the receive filter is $\sum_n b[n]q(t - nT)$, where $q(t) = (g_{TX} * g_C * g_{RX})(t)$ is the system response to a single symbol. The noise $n(t)$ at the input to the receive filter is WGN with PSD $\sigma^2 = N_0/2$, so the noise $w(t) = (n * g_{RX})(t)$ at the output of the receive filter, using the material in Section 5.9, is zero-mean, WSS, Gaussian with autocorrelation/autocovariance function $R_w(\tau) = C_w(\tau) = \sigma^2(g_{RX} * g_{RX,MF})(\tau)$.

(a) Sketch the end-to-end response $q(t)$. Compute the energy per bit $E_b = ||q||^2$.

Remark Note that $E_b/N_0 = ||q||^2/(2\sigma^2)$. If we fix the signal scaling, and hence $||q||^2$, then the value of σ^2 is fixed once we specify E_b/N_0.

(b) Assume that the sampler operates at rate $2/T = 2$, taking samples at times $t = m/2$ for integer m. Show that the discrete-time end-to-end response to a single symbol (i.e., the sampled version of $q(t)$) is

$$\mathbf{h} = (\ldots, 0, 1, 2, -\tfrac{1}{2}, -1, -\tfrac{1}{2}, 0, \ldots)$$

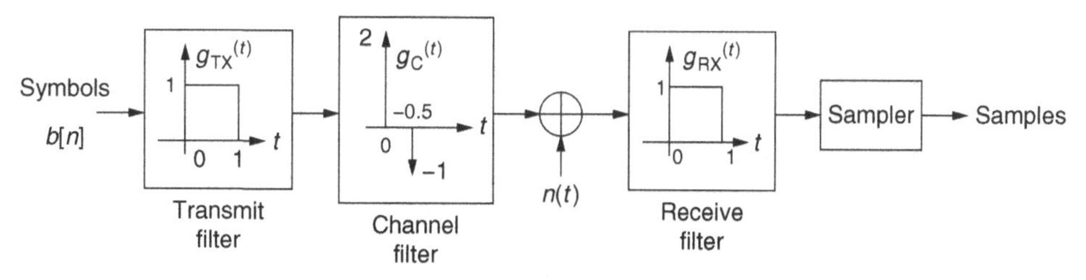

Figure 8.25 A continuous-time model for a link with ISI.

(c) For the given sampling rate, show that, for the vector ISI model (8.13), the noise covariance matrix satisfies (8.77).
(d) Specify the matrix $\mathbf{U}$ corresponding to the ISI model (8.13) for a linear equalizer of length 5, with its observation interval lined up with the channel response for the desired symbol.
(e) Taking into account the noise coloring, compute the optimal ZF correlator, and its noise enhancement relative to the matched filter bound.
(f) Compute the MMSE correlator for E_b/N_0 of 10 dB (see (a) and the associated remark). What is the output SINR, and how does it compare with the SNR of the ZF correlator in (e)?

ZF and MMSE equalization: theoretical derivations

Problem 8.5 (ZF geometry) For white noise, the output of a ZF correlator satisfying (8.17) and (8.18) is given by

$$\mathbf{c}^T \mathbf{r}[n] = b[n] + N(0, \sigma^2 ||\mathbf{c}||^2)$$

Since the signal scaling is fixed, the optimal ZF correlator is one that minimizes the noise variance $\sigma^2||\mathbf{c}||^2$. Thus, the optimal ZF correlator minimizes $||\mathbf{c}||^2$ subject to (8.17) and (8.18).

(a) Suppose a correlator $\mathbf{c}_1$ satisfies (8.17) and (8.18), and is a linear combination of the signal vectors $\{\mathbf{u}_k\}$. Now, suppose that we add a component $\Delta\mathbf{c}$ orthogonal to the space spanned by the signal vectors. Show that $\mathbf{c}_2 = \mathbf{c}_1 + \Delta\mathbf{c}$ is also a ZF correlator.
(b) How is the output noise variance for $\mathbf{c}_2$ related to that for $\mathbf{c}_1$?
(c) Conclude that, in order to be optimal, a ZF correlator must lie in the signal subspace spanned by $\{\mathbf{u}_k\}$.
(d) Observe that the condition (8.17) implies that $\mathbf{c}$ must be orthogonal to the interference subspace spanned by $\{\mathbf{u}_k, k \neq 0\}$. By combining this observation with (c), infer that $\mathbf{c}$ must be a scalar multiple of $\mathbf{P}_I^{\perp}\mathbf{u}_0$.

Problem 8.6 (Invertibility requirement for computing ZF correlator) The ZF correlator expression (8.22) requires inversion of the matrix $\mathbf{U}^T\mathbf{U}$ of correlations among the signal vectors, with (i,j)th entry $\mathbf{u}_i, \mathbf{u}_j\rangle = \mathbf{u}_i^T\mathbf{u}_j$.

(a) Show that

$$\mathbf{U}^T\mathbf{U}\mathbf{a} = 0 \tag{8.78}$$

if and only

$$\mathbf{U}\mathbf{a} = 0 \tag{8.79}$$

Hint. In order to show that (8.78) implies (8.79), suppose that (8.78) holds. Multiply by $\mathbf{a}^T$ and show that you get an expression of the form $\mathbf{x}^T\mathbf{x} = ||\mathbf{x}||^2 = 0$, which implies $\mathbf{x} = 0$.

(b) Use the result in (a) to infer that $\mathbf{U}^T\mathbf{U}$ is invertible if and only if the signal vectors $\{\mathbf{u}_k\}$ are linearly independent.

Problem 8.7 (Alternative computation of the ZF correlator) An alternative computation for the ZF correlator is by developing an expression for $\mathbf{P}_I^{\perp}\mathbf{u}_0$ in terms of $\mathbf{u}_0$ and $\mathbf{U}_I$, a matrix containing the interference vectors $\{\mathbf{u}_k, k \neq 0\}$ as columns. That is, $\mathbf{U}_I$ is obtained from the signal matrix $\mathbf{U}$ by deleting the column corresponding to the desired vector $\mathbf{u}_0$. Let us define the projection of $\mathbf{u}_0$ onto the interference subspace by $\mathbf{P}_I\mathbf{u}_0$. By definition, this is a linear combination of the interference vectors, and can be written as

$$\mathbf{P}_I\mathbf{u}_0 = \mathbf{U}_I\mathbf{a}_I \tag{8.80}$$

The orthogonal projection $\mathbf{P}_I^{\perp}\mathbf{u}_0$ is therefore given by

$$\mathbf{P}_I^{\perp}\mathbf{u}_0 = \mathbf{u}_0 - \mathbf{P}_I\mathbf{u}_0 = \mathbf{u}_0 - \mathbf{U}_I\mathbf{a}_I \tag{8.81}$$

(a) Note that $\mathbf{P}_I^{\perp}\mathbf{u}_0$ must be orthogonal to each of the interference vectors $\{\mathbf{u}_k, k \neq 0\}$, hence (going directly to the general complex-valued setting)

$$\mathbf{U}_I^H\mathbf{P}_I^{\perp}\mathbf{u}_0 = \mathbf{0}$$

(b) Infer from (a) that

$$\mathbf{a}_I = \left(\mathbf{U}_I^H\mathbf{U}_I\right)^{-1}\mathbf{U}_I^H\mathbf{u}_0$$

(c) Derive the following explicit expression for the orthogonal projection:

$$\mathbf{P}_I^{\perp}\mathbf{u}_0 = \mathbf{u}_0 - \mathbf{U}_I\left(\mathbf{U}_I^H\mathbf{U}_I\right)^{-1}\mathbf{U}_I^H\mathbf{u}_0 \tag{8.82}$$

(d) Derive the following explicit expression for the energies of the projection onto the interference subspace and the orthogonal projection:

$$\begin{aligned} ||\mathbf{P}_I\mathbf{u}_0||^2 &= \mathbf{u}_0^H\mathbf{U}_I\left(\mathbf{U}_I^H\mathbf{U}_I\right)^{-1}\mathbf{U}_I^H\mathbf{u}_0 \\ ||\mathbf{P}_I^{\perp}\mathbf{u}_0||^2 &= ||\mathbf{u}_0||^2 - \mathbf{u}_0^H\mathbf{U}_I\left(\mathbf{U}_I^H\mathbf{U}_I\right)^{-1}\mathbf{U}_I^H\mathbf{u}_0 = \mathbf{u}_0^H\left(\mathbf{I} - \mathbf{U}_I\left(\mathbf{U}_I^H\mathbf{U}_I\right)^{-1}\mathbf{U}_I^H\right)\mathbf{u}_0 \end{aligned} \tag{8.83}$$

(e) Note that (8.82) and (8.83), together with (8.27), give us an expression for a ZF correlator $\mathbf{c}_{ZF}$ scaled such that $\langle \mathbf{c}_{ZF}, \mathbf{u}_0 \rangle = 1$.

Problem 8.8 (Analytical expression for MMSE correlator) Let us derive the expression (8.34) for the MMSE correlator for the vector ISI model (8.13). We consider the general scenario of complex-valued symbols and signals. Suppose that the symbols $\{b[n]\}$ in the model are uncorrelated, satisfying

$$\mathbb{E}[b[n]b^*[m]] = \begin{cases} 0, & m \neq n \\ \sigma_b^2, & m = n \end{cases} \tag{8.84}$$

We have

$$\mathbf{R} = \mathbb{E}[\mathbf{r}[n]\mathbf{r}^H[n]] = \mathbb{E}\left[\left(\sum_k b[n+k]\mathbf{u}_k + \mathbf{w}[n]\right)\left(\sum_l b[n+l]\mathbf{u}_l + \mathbf{w}[n]\right)^H\right]$$

and

$$\mathbf{p} = \mathbb{E}[b^*[n]\mathbf{r}[n]] = \mathbb{E}\left[b^*[n]\left(\sum_k b[n+k]\mathbf{u}_k + \mathbf{w}[n]\right)\right]$$

Now use (8.84), and the independence of the symbols and the noise, to infer that

$$\mathbf{R} = \sigma_b^2 \sum_k \mathbf{u}_k\mathbf{u}_k^H + \mathbf{C}_w, \quad \mathbf{p} = \sigma_b^2\mathbf{u}_0$$

Problem 8.9 (ZF correlator for colored noise) Consider the model (8.13) where the noise covariance is a positive definite matrix $\mathbf{C_w}$. We now derive the formula (8.30) for the ZF correlator, by mapping, via a linear transformation, the system to a white-noise setting for which we have already derived the ZF correlator in (8.22). Specifically, suppose that we apply an invertible matrix $\mathbf{A}$ to the received vector $\mathbf{r}[n]$, then we obtain a *transformed* received vector

$$\tilde{\mathbf{r}}[n] = \mathbf{A}\mathbf{r}[n] = \sum_k b[n+k]\tilde{\mathbf{u}}_k + \tilde{\mathbf{w}}[n] \tag{8.85}$$

where

$$\tilde{\mathbf{u}}_k = \mathbf{A}\mathbf{u}_k \tag{8.86}$$

and

$$\tilde{\mathbf{w}}[n] = \mathbf{A}\mathbf{w}[n] \sim N(0, \mathbf{A}\mathbf{C_w}\mathbf{A}^T) \tag{8.87}$$

(a) Suppose that we find a linear correlator $\tilde{\mathbf{c}}$ for the transformed system (8.85), leading to a decision statistic $Z[n] = \tilde{\mathbf{c}}^T\tilde{\mathbf{r}}[n]$. Show that we can write the decision statistic $Z[n] = \mathbf{c}^T\mathbf{r}[n]$ (i.e., as the output of a linear correlator operating on the original received vector), where

$$\mathbf{c} = \mathbf{A}^T\tilde{\mathbf{c}} \tag{8.88}$$

(b) Suppose that we can find $\mathbf{A}$ such that the noise in the transformed system is white:

$$\mathbf{A}\mathbf{C_w}\mathbf{A}^T = \mathbf{I} \tag{8.89}$$

Show that

$$\mathbf{C_w}^{-1} = \mathbf{A}^T\mathbf{A} \tag{8.90}$$

(c) Show that the optimal ZF correlator $\tilde{\mathbf{c}}$ for the transformed system is given by

$$\tilde{\mathbf{c}}_{ZF} = \tilde{\mathbf{U}}\left(\tilde{\mathbf{U}}^T\tilde{\mathbf{U}}\right)^{-1}\mathbf{e} \tag{8.91}$$

(d) Show that the optimal ZF correlator $\mathbf{c}_{ZF}$ in the original system is given by (8.30). *Hint.* Use (8.86), (8.88) and (8.90).

(e) While we have used whitening only as an intermediate step to deriving the formula (8.30) for the ZF correlator in the original system, we note for completeness that a whitening matrix $\mathbf{A}$ satisfying (8.89) is guaranteed to exist for any positive definite $\mathbf{C_w}$, and spell out two possible choices for $\mathbf{A}$. For example, we can take $\mathbf{A} = \mathbf{B}^{-1}$, where $\mathbf{B}$ is the *square root* of $\mathbf{C_w}$, which is a symmetric matrix satisfying $\mathbf{C_w} = \mathbf{B}^2$. Another, often more numerically stable, choice is $\mathbf{A} = \mathbf{L}^{-1}$, where $\mathbf{L}$ is a lower-triangular matrix obtained by the *Cholesky* decomposition of $\mathbf{C_w}$. which satisfies $\mathbf{C_w} = \mathbf{L}\mathbf{L}^{\mathrm{T}}$. MATLAB functions implementing these are given below:

```
%square root of Cw
B = sqrtm(Cw); %symmetric matrix
%Cholesky decomposition of Cw
L = chol(Cw,'lower'); %lower-triangular matrix
```

Throughout the preceding problem, replacing transpose by conjugate transpose gives the corresponding results for the complex-valued setting. The MATLAB code segment above applies for both real- and complex-valued noise.

MIMO

Problem 8.10 (Effect of array spacing) Consider a regular linear array with N elements and inter-element spacing d.

(a) For $N = 8$, plot the beam pattern for a beam directed at 30° from broadside for $d = \lambda/4$.
(b) Repeat (a) for $d = 2\lambda$.
(c) Comment on any differences that you notice between the beamforming patterns in (a) and (b).
(d) For inter-element spacings of $\alpha\lambda$, the maximum of the beam pattern is not unique as α gets larger. That is, the beam pattern takes its maximum value not just in the desired direction, but also in a few other directions. These other maxima are called *grating lobes*. What is the value of α beyond which you expect to see grating lobes?

Problem 8.11 (SDMA) The base station in a cellular network wishes to simultaneously send different data streams to two different mobiles. Assume that it has a linear array with 16 elements uniformly spaced at $\lambda/3$. Mobile A is at angle 20° from broadside and mobile B is at angle $-30°$ from broadside.

(a) Compute the array responses corresponding to each mobile, and plot the beamforming patterns if the base station were communicating with only one mobile at a time.
(b) Now, suppose that the base station employs SDMA using zero-forcing interference suppression to send to both mobiles simultaneously. Plot the beam patterns used to send to mobile A and mobile B, respectively.
(c) What is the noise enhancement in (b) relative to (a)?
(d) Repeat (a)–(c) when mobile B is at angle 10° from broadside (i.e., closer to mobile A in angular spacing). You should notice a significant increase in noise enhancement.
(e) Try playing around with different values of angular spacing between mobiles to determine when the base station should attempt to use SDMA (e.g., what is the minimum angular spacing at which the noise enhancement is, say, less than 3 dB).

Problem 8.12 (Outage rates with receive diversity) Consider a $1 \times N$ MIMO system with receive diversity. The gain relative to a SISO system is given by (8.64):

$$G = |h_1|^2 + \cdots + |h_N|^2$$

For our rich-scattering model, $h_i \sim CN(0, 1)$ are i.i.d., hence $h_i|^2$ are i.i.d. exponential random variables, each with mean one. We state without proof that the sum of N such random variables is a gamma random variable with PDF and CDF given by

$$p_G(g) = \frac{g^{N-1}}{(N-1)!} e^{-g} I_{g \geq 0} \tag{8.92}$$

and

$$F_G(g) = P[G \leq g] = e^{-g} \sum_{k=N}^{\infty} \frac{g^k}{k!} = 1 - e^{-g} \sum_{k=0}^{N-1} \frac{g^k}{k!}, \quad g \geq 0 \tag{8.93}$$

(a) Use the preceding results to compute the probability of outage (on a log scale) for N-fold receive diversity versus the link margin (dB) relative to the SISO benchmark for $N = 1, 2$, and 4. That is, reproduce the results displayed in Figure 8.20.
(b) *Optional.* It may be an interesting exercise to use simulations to compute the empirical CDF of G, and to check that you get the same outage-rate curves as those in (a).

Problem 8.13 (Alamouti scheme with multiple receive antennas) Consider a $2 \times N$ MIMO system where the transmitter employs Alamouti space–time coding as in (8.67). Let $\mathbf{H} = (\mathbf{h}_1\ \mathbf{h}_2)$ denote the $N \times 2$ channel matrix, with $(N \times 1)$ columns $\mathbf{h}_1$ and $\mathbf{h}_2$.

(a) Show that the optimal receiver is given by (8.70), where $\mathbf{u}_1 = (1/\sqrt{2})(\mathbf{h}_1, \mathbf{h}_2^*)^{\mathrm{T}}$ and $\mathbf{u}_2 = (1/\sqrt{2})(\mathbf{h}_2, -\mathbf{h}_1^*)^{\mathrm{T}}$, and

$$\tilde{\mathbf{y}} = \begin{pmatrix} \mathbf{y}[1] \\ \mathbf{y}^*[2] \end{pmatrix}$$

(b) Show that the SNR gain relative to our unfaded SISO system with the same transmit power and constant channel gain of unity is given by

$$G = \frac{1}{2} \sum_{j=1}^{N} \sum_{k=1}^{2} |H(j, k)|^2$$

By comparing with the receive diversity gain (8.64) in a $1 \times N$ system, answer the following true/false questions *(give reasons for your answers)*.

(c) **(True or false?)** A 2×2 MIMO system with Alamouti space–time coding is 3 dB better than a 1×2 MIMO system with receive diversity.
(d) **(True or false?)** A 2×2 MIMO system with Alamouti space–time coding is 3 dB worse than a 1×4 MIMO system with receive diversity.
(e) Use the approach in Problem 8.12 to compute and plot the outage probability (log scale) versus the link margin (dB) relative to the unfaded SISO system for a $2 \times N$ MIMO system, $N = 1, 2, 4$. You should be able to reproduce the plots in Figure 8.22.

Problem 8.14 (Outage rates for spatial multiplexing with ZF reception) Consider two-fold spatial multiplexing in a $2 \times N$ MIMO system with $N \times 2$ channel matrix $\mathbf{H}$. Define the 2×2 matrix $\mathbf{R} = \mathbf{H}^{\mathrm{H}}\mathbf{H}$.

(a) Referring to the spatial multiplexing model (8.73), how do the entries of $\mathbf{R}$ relate to the signal vectors $\mathbf{u}_1$ and $\mathbf{u}_2$?

(b) Show that the energy of the projection of $\mathbf{u}_1$ orthogonal to the subspace spanned by $\mathbf{u}_2$ is given by

$$E_1 = \mathbf{R}(1,1) - \frac{|\mathbf{R}(1,2)|^2}{\mathbf{R}(2,2)}$$

where $\mathbf{R}(i,j)$ denotes the (i,j)th entry of $\mathbf{R}$, $i,j = 1,2$.

(c) If we fix the transmit power to that of the SISO benchmark (splitting it equally between the two data streams), show that the gain seen by the first data stream is given by

$$G_1 = \frac{E_1}{2} = \frac{1}{2}\left(\mathbf{R}(1,1) - \frac{|\mathbf{R}(1,2)|^2}{\mathbf{R}(2,2)}\right)$$

Similarly, the gain seen by the second data stream is given by

$$G_2 = \frac{1}{2}\left(\mathbf{R}(2,2) - \frac{|\mathbf{R}(1,2)|^2}{\mathbf{R}(1,1)}\right)$$

Note that, under our rich-scattering model, G_1 and G_2 are identically distributed random variables.

(d) Use computer simulations with the rich-scattering model to plot the outage rate versus link margin for 2×2 and 2×4 MIMO with two-fold spatial multiplexing and ZF reception. You should get a plot similar to Figure 8.24. Discuss how the performance compares with that of the Alamouti scheme.

Problem 8.15 (Outage rates for spatial multiplexing with ZF reception)

(a) Argue that $2 \times N$ Alamouti space–time coding is exactly 3 dB worse than $1 \times 2N$ receive diversity.

(b) For $2 \times N$ spatial multiplexing with ZF reception, approximate its performance as x dB worse than a $1 \times N'$ receive diversity. (Note that spatial multiplexing has twice the bandwidth efficiency of receive diversity, but it loses 3 dB of power up front due to splitting it between the two data streams.)
Answer. It is approximately 3 dB worse than $1 \times (N-1)$ receive diversity. That is, if the gain relative to the SISO benchmark for a $1 \times N$ receive diversity system is denoted as $G_{\text{RX-div}}(N)$, then the gain for a $2 \times N$ spatial multiplexed system is $G_{\text{smux}}(N) \approx \frac{1}{2}G_{\text{RX-div}}(N-1)/2$. Thus, the CDF, and hence outage rate, of the spatially multiplexed system is approximated as

$$P[G_{\text{smux}}(N) \leq x] \approx P[G_{\text{RX-div}}(N-1) \leq 2x]$$

(c) Use the results in (b), and the analytical framework in Problem 8.12, to obtain an analytical approximation for the simulation results in Problem 8.14(c).

Software Lab 8.1: introduction to equalization in single-carrier systems

Lab objectives To understand the need for equalization in communication systems, and to implement linear MMSE equalizers adaptively.

Reading Sections 8.1 and 8.2; Chapter 4 (linear modulation); and Section 5.6 (Gaussian random variables and the Q function). This lab can be completed without systematic coverage of Chapter 6; we state and use probability-of-error expressions from Chapter 6, but knowing how they are derived is not required for the lab.

Laboratory assignment

(1.1) Use as your transmit and receive filters the SRRC pulse employed in Software Labs 4.1 and 6.1. Putting the code for realizing these together with the code fragments developed in this chapter provides the code required for this lab. As in Software Labs 4.1 and 6.1, the transmit, channel, and receive filters are implemented at rate $4/T$. For simplicity, we consider BPSK signaling throughout this lab, and consider only real-valued signals. Generate $n_{\text{symbols}} = n_{\text{training}} + n_{\text{payload}}$ (numbers to be specified later) ± 1 BPSK symbols as in Lab 6.1, and pass them through the transmit, channel, and receive filters to get noiseless received samples at rate $4/T$.

(1.2) Let us start with a trivial channel filter as before. Set $n_{\text{symbols}} = 200$. The number of rate-$4/T$ samples at the output of the receive filter is therefore 800, plus tails at either end because the length of the effective pulse modulating each symbol extends over multiple symbol intervals. Plot an eye diagram (e.g., using Code Fragment 8.1.3) using, say, 400 samples in the middle. You should get an eye diagram that looks like Figure 8.26: the cascade of the transmit and receive filter is approximately Nyquist, and the eye is *open,* so we can find a sampling time such that we can distinguish between +1 and -1 well, despite the influence of neighboring symbols.

(1.3) Now introduce a nontrivial channel filter. In particular, consider a channel filter specified (at rate $4/T$) using the following MATLAB command:

$$\text{channel\_filter} = [-0.7, -0.3, 0.3, 0.5, 1, 0.9, 0.8, -0.7, -0.8, 0.7, 0.8, 0.6, 0.3]';$$

Generate an eye diagram again. You should get something that looks like Figure 8.27. Notice now that there is no sampling time at which you can clearly make out the difference between +1 and -1 symbols. The eye is now said to be *closed* due to ISI, so we cannot make symbol decisions just by passing appropriately timed received samples through a thresholding device.

(1.4) We are now going to evaluate the probability of error without and with equalization. First, let us generate the noisy output of the receive filter. We need to generate $n_{\text{symbols}} = n_{\text{training}} + n_{\text{payload}}$ (numbers to be specified later) ± 1 BPSK symbols as in Software Lab 6.1, and pass them through the transmit filter, the dispersive channel, and the receive filter to get noiseless received samples at rate $4/T$. Since we are

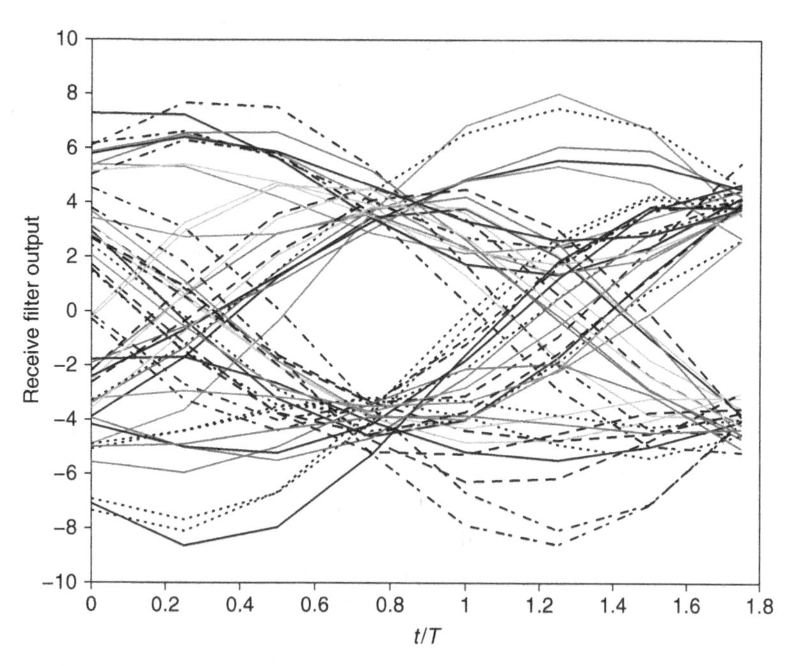

Figure 8.26 The eye diagram for a non-dispersive channel. The eye is open.

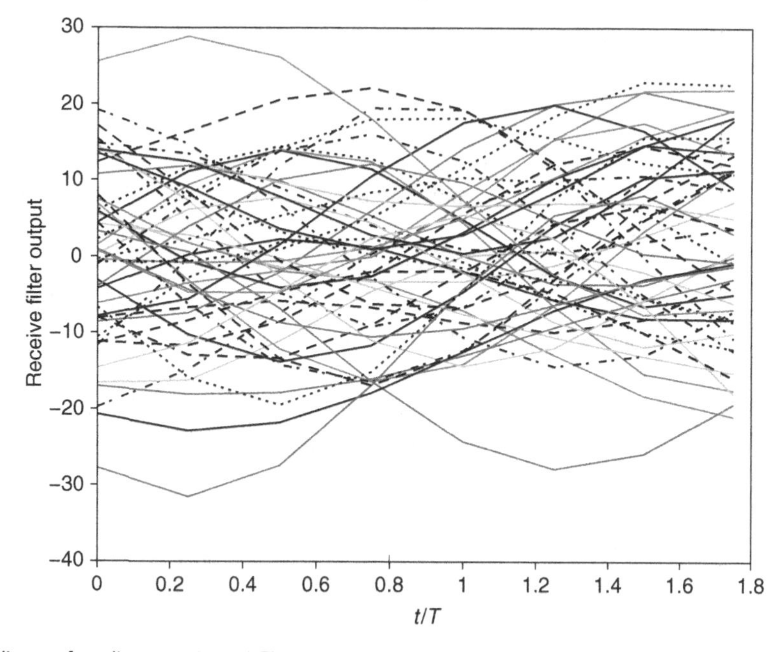

Figure 8.27 The eye diagram for a dispersive channel. The eye is closed.

signaling along the real axis only, at the *input* to the receive filter, add i.i.d. $N(0, \sigma^2)$ real-valued noise samples (as in Lab 6.1, choose $\sigma^2 = N_0/2$ corresponding to a specified value of E_b/N_0). Pass these (rate-$4/T$) noise samples through the receive filter, and add the result to the signal contribution at the receive filter output.

(1.5) Performance without equalization Let $\{r_k\}$ denote the output of the receive filter, and let $Z[n] = r_{d+4(n-1)}$, $n = 1, 2, \ldots, n_{\text{symbols}}$ denote the best symbol-rate decision statistics you can obtain by subsampling at rate $1/T$ the receive filter output. As in the solutions to earlier labs, choose the decision delay d equal to the location of the maximum of the overall response (which now includes the channel) to a single symbol. For $n_{\text{symbols}} = 10, 100$, compute the error probability of the decision rule $\hat{b}[n] = \text{sign}(Z[n])$ as a function of E_b/N_0, where the latter ranges from 5 to 20 dB. Compare with the ideal error probability curve for BPSK signaling for the same range of E_b/N_0. This establishes that a simple one-sample-per-symbol decision rule does not work well for non-ideal channels, and motivates the equalization schemes discussed below.

Linear equalization We now consider *linear equalization,* where the decision for symbol b_n is based on linear processing of a vector of samples $\mathbf{r}[n]$ of length $L = 2M + 1$, where the entries of $\mathbf{r}[n]$ are samples spaced by T/q, with the center sample being the same as the decision statistic in (1.4) above $q = 1$ corresponds to symbol-spaced sampling, and $q > 1$ corresponds to fractionally spaced sampling. We consider two cases: $q = 1$ and $q = 2$:

$$\mathbf{r}[n] = (r_{k+4(n-1)+d-(4/q)M}, r_{k+4(n-1)+d-(4/q)(M-1)}, \ldots,$$
$$r_{k+4(n-1)+d}, r_{k+4(n-1)+d+(4/q)}, \ldots, r_{k+4(n-1)+d+(4/q)M})^{\mathrm{T}}$$

The decision rule we use is

$$\hat{b} = \text{sign}(\mathbf{c}^{\mathrm{T}}\mathbf{r}[n]) \tag{8.94}$$

where $\mathbf{c}$ is a correlator whose choice is to be specified. Note that the decision rule in (1.4) above corresponds to the choice $\mathbf{c} = (0, .., 0, 1, 0, \ldots, 0)^{\mathrm{T}}$, since it uses only the center sample.

The vector of samples $\mathbf{r}[n]$ contains contributions both from the desired symbol b_n and from ISI due to $b_{n\pm1}$, $b_{n\pm2}$, etc. We implement the linear minimum mean-squared error (MMSE) equalizer using a least-squares adaptive implementation, as discussed in Section 8.2.1.

(1.6) For the least-squares implementation, assume that the first n_{training} symbols are known training symbols, $b_1, \ldots, b_{n_{\text{training}}}$. Define the $L \times L$ matrix

$$\hat{\mathbf{R}} = \frac{1}{n_{\text{training}}} \sum_{n=1}^{n_{\text{training}}} \mathbf{r}[n]\mathbf{r}^{\mathrm{T}}[n]$$

and the $L \times 1$ vector

$$\hat{\mathbf{p}} = \frac{1}{n_{\text{training}}} \sum_{n=1}^{n_{\text{training}}} b[n]\mathbf{r}[n]$$

The MMSE correlator is now approximated as

$$\hat{\mathbf{c}}_{\text{MMSE}} = (\hat{\mathbf{R}})^{-1}\hat{\mathbf{p}} \tag{8.95}$$

(1.7) Now, the correlator obtained via (8.8) is used to make decisions, using the decision rule (8.94), on the unknown symbols $n = n_{\text{training}} + 1, \ldots, n_{\text{symbols}}$.

(1.8) Fix $n_{\text{training}} = 100$ and $n_{\text{payload}} = 10,000$. For $L = 3, 5, 7, 9$ and $q = 1, 2$, implement linear MMSE equalizers, and plot their error probabilities (for the payload symbols) as a function of E_b/N_0, in the range 5–20 dB. Compare your results with the unequalized error probability and the ideal error probability found in (1.5).
Hint. An efficient way to generate the statistics $\mathbf{c}^T\mathbf{r}[n]$ is to pass an appropriate rate-$2/T$ subsequence of the receive filter output through a filter whose impulse response is the time reverse of $\mathbf{c}$, and to then appropriately subsample at rate $1/T$ the output of the equalizing filter. This is much faster than correlating $\mathbf{c}$ with $\mathbf{r}[n]$ for each n.

(1.9) Comment on the performance of symbol-spaced versus fractionally spaced equalization. Comment on the effect of the equalizer length on performance. What is the effect of increasing or decreasing the training period?

Lab report Your lab report should document the results of the preceding steps in order. Describe the reasoning you used and the difficulties you encountered.

Software Lab 8.2: simplified simulation model for an OFDM link

Lab objectives To develop a hands-on understanding of basic OFDM transmission and reception.

Reading Section 8.3 (OFDM) and Chapter 4 (linear modulation).

Laboratory assignment

We would like to leverage the code from Software Lab 8.1 as much as possible, so we set the DAC filter to be the transmit filter in that lab, and the receive filter to be its matched filter as before. The main difference is that the time-domain samples sent through the DAC filter are obtained by taking the inverse FFT of the frequency-domain symbols, and inserting a cyclic prefix. We fix the constellation as Gray-coded QPSK.

Step 1. Exploring time- and frequency-domain relationships in OFDM

Let us first discuss the structure of a single "OFDM symbol," which carries N complex-valued symbols in the frequency domain. Here N is the number of subcarriers, chosen to be a power of 2. Set L to be length of the cyclic prefix. Set $N = 256$ and $L = 20$ for these initial explorations, but keep the parameters programmable for later use.

(1a) Generate N Gray-coded QPSK symbols $\mathbf{B} = \{B[k], k = 1, \ldots, N\}$. (You can use the function qpskmap developed in Software Lab 6.1 for this purpose.) Take the inverse FFT to obtain time-domain samples $\mathbf{b} = \{b[n], n = 1, \ldots, N\}$.

(1b) Append the last L time-domain samples to the beginning, to get a length $N+L$ sequence of time-domain samples $\mathbf{b}' = \{b'[n], n = 1, \ldots, N+L\}$. That is, $b'[1] = b[N-L+1], \ldots, b'[L] = b[N], b'[L+1] = b[1], \ldots, b'[N+L] = b[N]$.

(1c) Take the first N symbols of $\mathbf{b}'$, say $\mathbf{r}_1 = \{b'[1], \ldots, b'[N]\}$. Show that the FFT output (say $\mathbf{R}_1 = \{R_1[k]\}$) is related to the original frequency-domain symbols $\mathbf{B}$ through a frequency-domain channel $\mathbf{H}$ as follows: $R_1[k] = H[k]B[k]$. Find and plot the amplitude $|H[k]|$ and phase $\arg(H[k])$ versus k.

(1d) Repeat Step (1c) for the time-domain samples $\{b'[3], \ldots, b'[N+3]\}$ (i.e., skip the first two samples of $\mathbf{b}'$. How are the frequency-domain channels in Steps (1c) and (1d) related? (What we are doing here is exploring what cyclic shifts in the time domain do in the frequency domain.)

Step 2. Generating multiple OFDM symbols

Now, we generate K frames, each carrying N Gray-coded QPSK symbols. Set $N = 256, L = 20$, and $K = 5$ for numerical results and plots in this step.

(2a) For each frame, generate time-domain samples and add a cyclic prefix, as in Steps (1a) and (1b). Then, append the time-domain samples for successive frames together. We now have a stream of $K(N+L)$ time-domain samples, analogous to the time-domain symbols sent in Lab 8.1.

(2c) Pass the time-domain symbols through the same transmit filter (this is the DAC in Figure 8.9) as in Software Labs 4.1, 6.1, and 8.1, again oversampling by a factor of 4. (That is, if the time-domain samples are at rate $1/T_s$, the filter is implemented as rate $4/T_s$. This gives us a rate-$4/T_s$ transmitted signal.)

(2d) Compute the peak-to-average power ratio (PAPR) in dB for this transmitted signal (OFDM is notorious for having a large PAPR). This is done by taking the ratio of the maximum to the average value of the magnitude squared of the time-domain samples.

(2e) Note that the original QPSK symbols in the frequency domain have a PAPR of one, but the time-domain samples are generated by mixing these together. The time-domain samples could be expected, therefore, to have a Gaussian distribution, invoking the central limit theorem. Plot a histogram of the I and Q components from the time-domain samples. Do they look Gaussian?

(2f) As in Lab 6.1, assume an ideal channel filter and pass the transmitted signal through a receive filter matched to the transmit filter. This gives a rate-$4/T_s$ noiseless received signal.

(2g) Subsample the received signal at rate $1/T_s$, starting with a delay of d samples (play around and see what choice of d works well – perhaps one based on the peak of the cascade of the transmit, channel, and receive filters). The first N samples correspond to the first frame. Take the FFT of these N samples to get $\{R_1[k]\}$. Now, estimate the frequency-domain channel coefficients $\{H[k]\}$ by using the known transmitted symbols $B_1[k]$ in the first frame as training. That is,

$$\hat{H}[k] = R_1[k]/B_1[k]$$

Plot the magnitude and phase of the channel estimates and comment on how it compares with what you saw in Steps (1c) and 1d).

(2h) Now, use the channel estimate from Step (2g) to demodulate the succeeding frames. If frame m uses time-domain samples over a window $[a, b]$, then frame $m + 1$ uses time-domain samples over a window $[a + (N + L)T_s, b + (N + L)T_s]$. Denoting the FFT of the time-domain samples for frame m as $R_m[k]$, the decision statistics for the frequency-domain symbols for the mth frame are given by

$$\hat{B}_m[k] = \hat{H}^*[k]R_m[k]$$

You can now decode the bits and check that you get a BER of zero (there is no noise so far). Also, display scatter plots of the decision statistics to verify that you are indeed seeing a QPSK constellation after compensating for the channel.

Step 3. Channel compensation

We now introduce a nontrivial channel (still no noise). Increase the cyclic prefix length if needed (it should be long enough to cover the cascade of the transmit, channel, and receive filters. But remember that the cyclic prefix is at rate $1/T_s$, whereas the filter cascade is at rate $4/T_s$.

(3a) Repeat Step (2f), but now with a nontrivial channel filter modeled at rate $4/T_s$. Use the channels you have tried out in Lab 8.1 (still no noise). For example,

$$\text{channel\_filter} = [-0.7, -0.3, 0.3, 0.5, 1, 0.9, 0.8, -0.7, -0.8, 0.7, 0.8, 0.6, 0.3]';$$

(3b) Repeat Step (2g). Comment on how the magnitude and phase of the frequency-domain channel differ from what you saw in Steps (2g), (1c), and (1d).

(3c) Repeat Step (2h). Check that you get a BER of zero, and that your decision statistics give nice QPSK scatter plots.

(3d) Check that everything still works out as you vary the number of subcarriers N (e.g., $N = 512, 1024, 2048$), the cyclic prefix length L, and the number of frames K.

Step 4. Effect of noise

Now, add noise as in Software Labs 6.1 and 8.1. Specifically, at the input to the receive filter, add independent and identically distributed (i.i.d.) complex Gaussian noise, such that the real part and the imaginary part of each sample are i.i.d. $N(0, \sigma^2)$ (we choose $\sigma^2 = N_0/2$ corresponding to a specified value of E_b/N_0). Let us fix $N = 1024$ for concreteness, and set the cyclic prefix to just a little longer than the minimum required for the channel you are considering. Set the number of frames to $K = 10$. Try a couple of values of E_b/N_0 of 5 dB and 8 dB.

(4a) While you can estimate E_b analytically, estimate it by taking the energy of the transmitted signal in (3a), and dividing it by the number of bits in the payload (i.e., excluding the first frame). Use this to set the value of N_0 for generating the noise samples.

(4b) Pass the (rate-$4/T_s$) noise samples through the receive filter, and add the result to the output of Step (3a).

(4c) Consider first a noiseless channel estimate, in which you carry out Step (3b) (estimating the channel based on frame 1) *before* you add noise to the output of Step (3a). Now add the noise and carry out Step (3c) (demodulating the other frames). Estimate the BER and compare it with the analytical value for ideal QPSK. Show the scatter plots of the decision statistics.

(4d) Repeat Step (4c), except that you now estimate the channel based on frame 1 *after* adding noise. Discuss how the BER degrades. Compare the channel estimates from parts Steps (4c) and (4d) on the same plot.

Note. You may notice a significant BER degradation, but that is because the channel estimation technique is naive (the channel coefficients for neighboring subcarriers are highly correlated, but our estimate is not exploiting this property). Exploring better channel estimation techniques is beyond the scope of this lab, but you are encouraged to browse the literature on OFDM channel estimation to dig deeper.

Step 5. Consolidation

Once you are happy with your code, plot the BER (log scale) as a function of E_b/N_0 (dB) for the channel in Lab 3. Plot three curves: ideal QPSK, OFDM with noiseless channel estimation, and OFDM with noisy channel estimation. Comment on the relation between the curves.

Lab report Your lab report should document the results of the preceding steps in order. Describe the reasoning you used and the difficulties you encountered.

Software Lab 8.3: MIMO signal processing

Lab objectives To gain hands-on exposure to basic MIMO signal processing at the transmitter and receiver.

Reading Section 8.4.

Laboratory assignment

Background Consider the rich-scattering model for a single subcarrier in a MIMO-OFDM system with M transmit and N receive antennas. The $N \times M$ channel matrix $\mathbf{H}$ is modeled as having i.i.d. $CN(0, 1)$ entries.

Code Fragment 8.7.1 (MIMO matrix with i.i.d. complex Gaussian entries)

```
%M,N specified earlier
%MIMO matrix with i.i.d. CN(0,1) entries
H = (randn(N,M) + j * randn(N,M))/sqrt(2);
```

Let T denote the number of time-domain samples for our system. Let $x_i[t]$ denote the sample transmitted from transmit antenna i at time t, where $1 \leq i \leq M$ and $1 \leq t \leq T$. Let $\mathbf{x}[t] = (x_1[t], \ldots, x_M[t])^T$ denote the $M \times 1$ vector of samples transmitted at time t, and let $\mathbf{X} = (\mathbf{x}[1], \ldots, \mathbf{x}[T])$ denote the $M \times T$ matrix containing all the transmitted samples. Our convention is to normalize the net transmit power to one, so that $\overline{|x_i[t]|^2} = 1/M$. For a single-input, single-output (SISO) system, this would lead to an average received SNR of $\overline{\text{SNR}} = 1/(2\sigma^2)$, since the magnitude squared of the channel gain is normalized to one, and the noise per receive antenna is modeled as $CN(0, 2\sigma^2)$, and we vary this hypothetical SISO system SNR when evaluating performance.

The $N \times T$ received matrix $\mathbf{Y}$, with $\mathbf{y}_j[t]$, $1 \leq j \leq N$, denoting the spatial vector of received samples at time t, is then modeled as

$$\mathbf{Y} = \mathbf{HX} + \mathbf{N}$$

where $\mathbf{N}$ is an $N \times T$ matrix with i.i.d. $CN(0, 2\sigma^2)$ entries. This model is implemented in the following code fragment.

Code Fragment 8.7.2 (Received signal in MIMO system)

```
%snrbardb specified earlier
%express snrbar in linear scale
snrbar = 10^(snrbardb/10);
%find noise variance, assuming TX power = 1
%(snrbar = 1/(2 * sigma^2))
sigma = sqrt(1/(2 * snr));
%x = M x T vector of symbols, already specified earlier
%(normalized to unit power per time)
%%RECEIVED SIGNAL MODEL: N x T matrix
y = H * x + sigma * randn(N,T) + j * sigma * randn(N,T);
```

In order to use the preceding generic code fragments for a particular MIMO scheme, we must (a) map the transmitted symbols into the matrix $\mathbf{X}$ of transmitted samples, and (b) process the matrix $\mathbf{Y}$ of received samples appropriately.

Alamouti space–time code We first consider the Alamouti space–time code for a 2×1 MIMO system. The transmitted samples can be generated using the following code fragment.

Code Fragment 8.7.3 (Transmitted samples for Alamouti space–time code)

```
%assume number of time samples T has been specified
%QPSK symbols normalized to unit power per symbol
symbols = (sign(rand(1,T) - 0.5) + j * sign(rand(1,T) - 0.5))/sqrt(2);
%Alamouti space--time code mapping
X = zeros(2,T); %M = 2
X(1,1:2:T) = symbols(1:2:T); %odd samples from antenna 1
X(2,1:2:T) = symbols(2:2:T); %odd samples from antenna 2
X(1,2:2:T) = -conj(symbols(2:2:T)); %even samples from antenna 1
X(2,2:2:T) = conj(symbols(1:2:T)); %even samples from antenna 2
```

Step 1. Consider a 2×1 MIMO system. Setting $M = 2$, $N = 1$, and $\overline{\text{SNR}}$ at 10 dB, put Code Fragments 8.7.1, 8.7.3, and 8.7.2 together to model the transmitted and received matrices $\mathbf{X}$ and $\mathbf{Y}$. Setting $T = 100$, make a scatter plot of the real and imaginary parts of the received samples. The received samples should be smeared out over the complex plane, since the signals from the two transmit antennas interfere with each other at the receive antenna.

Step 2. Compute the decision statistics (8.70) on the basis of the received matrix **Y**. You may use the following code fragment, but you must explain what it is doing. Make a scatter plot of the decision statistics. You should recover the noisy QPSK constellation.

Code Fragment 8.7.4 (Receiver processing for Alamouti space–time code for a 2 × 1 MIMO system)

```
Ytilde = zeros(2,T/2); %assume T even
Ytilde(1,:) = Y(1,1:2:T);
Ytilde(2,:) = conj(Y(1,2:2:T));
%u1 = [H(1,1);conj(H(1,2))]; u2 = [H(1,2);-conj(H(1,1))];
u1 = [H(1,1);conj(H(1,2))]; u2 = [H(1,2);-conj(H(1,1))];
Z(1:2:T) = u1' * Ytilde;
Z(2:2:T) = u2' * Ytilde;
```

Step 3. Repeat Steps 1 and 2 for a few different realizations of the channel matrix. The quality of the scatter plot in Step 2 should depend on $G = (|H(1,1)|^2 + |H(1,2)|^2)/2$.

Step 4. Now, suppose that we use the Alamouti space–time code for a $2 \times N$ MIMO system, where N may be larger than one. Show that only the receiver processing Code Fragment 8.7.4 needs to be modified (other than changing the value of N in the other code fragments), with $\tilde{\mathbf{Y}}$ having dimension $2N \times T/2$ and $\mathbf{u}_1$ and $\mathbf{u}_2$ each having dimension $2N \times 1$. Implement these modifications, and make a scatter plot of the decision statistics for $N = 2$ and $N = 4$, fixing the equivalent SISO $\overline{\text{SNR}}$ to 10 dB. You should notice a qualitative improvement with increasing N as you run several channel matrices, although the plots depend on the channel realization.
Hint. See Problem 8.13.

We now consider spatial multiplexing in a $2 \times N$ MIMO system. We can now send $2T$ symbols over T time intervals, as in the following code fragment.

Code Fragment 8.7.5 (Transmitted samples for two-fold spatial multiplexing)

```
%QPSK symbols normalized to unit power
symbols = (sign(rand(1,2 * T) - 0.5) + j * sign(rand(1,2 * T) - 0.5))/sqrt(2);
x = zeros(M,T);
%normalize samples so as to emit unit power per unit time
x(1,:) = symbols(1:2:2 * T)/sqrt(2);
x(2,:) = symbols(2:2:2 * T)/sqrt(2);
```

Step 5. Setting $M = 2$, $N = 4$ and $\overline{\text{SNR}}$ at 10 dB, put Code Fragments 8.7.1, 8.7.5, and 8.7.2 together to model the transmitted and received matrices **X** and **Y**. Setting $T = 100$, again make a scatter plot of the real and imaginary parts of the received samples for each received antenna. As before, the received samples should be smeared out over the complex plane, since the signals from the two transmit antennas interfere with each other at the receive antennas.

Step 6. Now, apply a ZF correlator as in (8.74) to **Y** to separate the two data streams. Make scatter plots of the two estimated data streams. You should recover noisy QPSK constellations.

Step 7. Fixing $\overline{\text{SNR}}$ at 10 dB and fixing a 4×2 channel matrix, compare the scatter plots of the decision statistics for the Alamouti scheme with those for two-fold spatial multiplexing. Which ones appear to be cleaner?

Lab report Your lab report should document the results of the preceding steps in order. Describe the reasoning you used and the difficulties you encountered.

Epilogue

We conclude with a brief discussion of research and development frontiers in communication systems. This discussion is speculative by its very nature (it is difficult to predict progress in science and technology) and is significantly biased by the author's own research experience. There is no attempt to be comprehensive. The goal is to highlight a few of the exciting challenges in communication systems in order to stimulate the reader to explore further.

The continuing wireless story

The growth of content on the Internet continues unabated, driven by applications such as video on demand, online social networks, and online learning. At the same time, there have been significant advances in the sophistication of mobile devices such as smart phones and tablet computers, which greatly enhance the quality of the content these devices can support (e.g., smart phones today provide high-quality displays for video on demand). As a result, users increasingly expect Internet content to be ubiquitously and seamlessly available on their mobile device. This means that, even after the runaway growth of cellular and WiFi starting in the 1990s, wireless remains the big technology story. Mobile operators today face the daunting task of evolving networks originally designed to support voice into *broadband* networks supplying data rates of the order of tens of Mbps or more to their users. By some estimates, this requires a 1000-fold increase in cellular network capacity in urban areas! On the other hand, since charging by the byte is not an option, this growth in capacity must be accomplished in an extremely cost-effective manner, which demands significant technological breakthroughs.

At the other end of the economic spectrum, cellular connectivity has reached the remotest corners of this planet, with even basic voice and text messaging transforming lives in developing nations by providing access to critical information (e.g., enabling farmers to obtain timely information on market prices and weather). The availability of more sophisticated mobile devices implies that ongoing revolutionary developments in online education and healthcare can reach underserved populations everywhere, as long as there is adequate connectivity to the Internet. The lack of such connectivity is commonly referred to as the *digital divide.*

Wireless researchers now face the challenge of building on the great expectations created by the success of the technologies they have created. At one end, how can we scale cellular network capacity by several orders of magnitude, in order to address the exponential growth in demand for wireless data created by smart mobile devices? At the other extreme, how do we close the digital divide, ensuring that even the most remote regions of our planet gain access to the wealth of information available online? In addition, there are some specialized applications of wireless that may assume significant importance as time evolves.

We summarize some key concepts driving this continuing technology story in the following.

Small cells There are two fundamental approaches to scaling up data rates: increasing spatial reuse (i.e., using the same time–bandwidth resources at locations that are far enough apart), and increasing communication bandwidth. Decreasing cell sizes from *macrocells* with diameters of the order of kilometers to *picocells* with diameters of the order of 100–200 meters increases spatial reuse, and hence potentially the network capacity, by two orders of magnitude. Picocellular base stations may be opportunistically deployed on lampposts or rooftops, and will see a very different propagation and interference environment from that for macrocellular base stations carefully placed at elevated locations. Interference among adjacent picocells becomes a major bottleneck, as does the problem of handing off rapidly moving users as they cross cell boundaries (indeed, cell boundaries are difficult to even define in picocellular networks due to the complexity of below-rooftop propagation). Thus, it is important to rethink the design philosophy of tightly controlled deployment and operation in today's macrocellular networks. The scaling and organic growth of picocellular networks is expected to require a significantly greater measure of decentralized self-organization, including, for example, auto-configuration for plug-and-play deployment, decentralized coordination for interference and mobility management, and automatic fault detection and self-healing. Another critical issue with small cells is backhaul (i.e., connecting each base station to the wired Internet): pulling optical fiber to every lamppost on which a picocellular base station is deployed might not be feasible. Finally, we can go to even smaller cells called *femtocells,* with base stations typically deployed indoors, in individual homes or businesses, and using the last-mile broadband technology already deployed in such places for backhaul. Both for picocells and for femtocells, it is important to devise efficient techniques for sharing spectrum, and managing potential interference, with the macrocellular network. In essence, we would like to be able to opportunistically deploy base stations as we do WiFi access points, but coordinate just enough to avoid the tragedy of the commons resulting from purely selfish behavior in unmanaged WiFi networks. Of course, as we learn more about how to scale such self-organized cellular networks, we might be able to apply some of the ideas to promote peaceful co-existence in densely deployed and independently operated WiFi networks using unlicensed spectrum. In short, it is fair to say that there is a clear opportunity and great need for significant innovations in overall design approach as well as specific technological breakthroughs, in order to truly attain the potential of "small cells."

Millimeter-wave communication While commercial wireless networks deployed today employ bands well below 10 GHz, there is significant interest in exploring higher carrier frequencies, where there are vast amounts of spectrum. Of particular interest are the *millimeter (mm)-wave* frequencies from 30 to 300 GHz, corresponding to wavelengths from 10 mm down to 1 mm. Historically, RF front-end technology for these bands has been expensive and bulky, hence there was limited commercial interest in using them. This has changed in recent years, with the growing availability of low-cost silicon radio-frequency integrated circuits (RFICs) in these bands. The particular slice of spectrum that has received the most attention is the 60-GHz band (57–64 GHz). Most of this band is unlicensed worldwide. The availability of 7 GHz of unlicensed spectrum (vastly more than the bandwidth in current cellular and WiFi systems) opens up the possibility for another revolution in wireless communication, with links operating at several gigabits per second (Gbps). Potential applications of 60 GHz in particular, and mm wave in general, include order-of-magnitude increases in the data rates for indoor wireless networks, multiGbps wireless backhaul networks, and base-station-to-mobile links in picocells, and even wireless data centers. However, realizing the vision of multiGbps wireless everywhere is going to take some work. While we can draw upon the existing toolkit of ideas developed for wireless communication to some extent, we may have to rethink many of these ideas because of the unique characteristics of mm-wave communication. The latter largely follow from the order-of-magnitude-smaller carrier wavelength relative to existing wireless systems.

At the most fundamental level, consider propagation loss. As discussed in Section 6.5 (see also Problems 6.37 and 6.38), the propagation loss for omnidirectional transmission scales with the square of the carrier frequency, but, for the same antenna aperture, the antenna directivity scales with the square of the carrier frequency. Thus, given that generating RF power at high carrier frequencies is difficult, we anticipate that mm-wave communication systems will employ antenna directivity at both ends. Since the inter-element spacing scales with the carrier wavelength, it becomes possible to accommodate a large number of antenna elements in a small area (e.g., a 1000-element antenna array at 60 GHz is palm-sized!), and to use electronic beamsteering to realize pencil beams at the transmitter and receiver. Of course, this is easier said than done. Hardware realization of such large arrays remains a challenge. On the algorithmic side, since building a separate upconverter or downconverter for every antenna element becomes infeasible as we scale up the array, it is essential to devise signal processing algorithms that do not assume the availability of the separate complex-baseband signals for each antenna element. The nature of diversity and spatial multiplexing also fundamentally changes at tiny wavelengths: due to the directionality of mm-wave links, there are only a few dominant propagation paths, so designs for rich-scattering models no longer apply. For indoor environments, blockage by humans and furniture becomes inevitable, since the ability of electromagnetic waves to diffract around obstacles depends on how large they are relative to the wavelength (i.e., obstacles "look bigger" at tiny wavelengths). For outdoor environments, performance is limited by the oxygen absorption loss (about 16 dB/km) in the 60-GHz band, and rain loss for mm-wave communication in general (e.g., as high as 30 dB/km in heavy rain). While link ranges of hundreds of meters can be achieved with reasonable margins to account for these effects, longer ranges than these would be fighting physics, hence multihop networks become interesting. Of course, once we start forming pencil beams, networking protocols that rely on the broadcast nature of the wireless medium no longer apply. These are just a

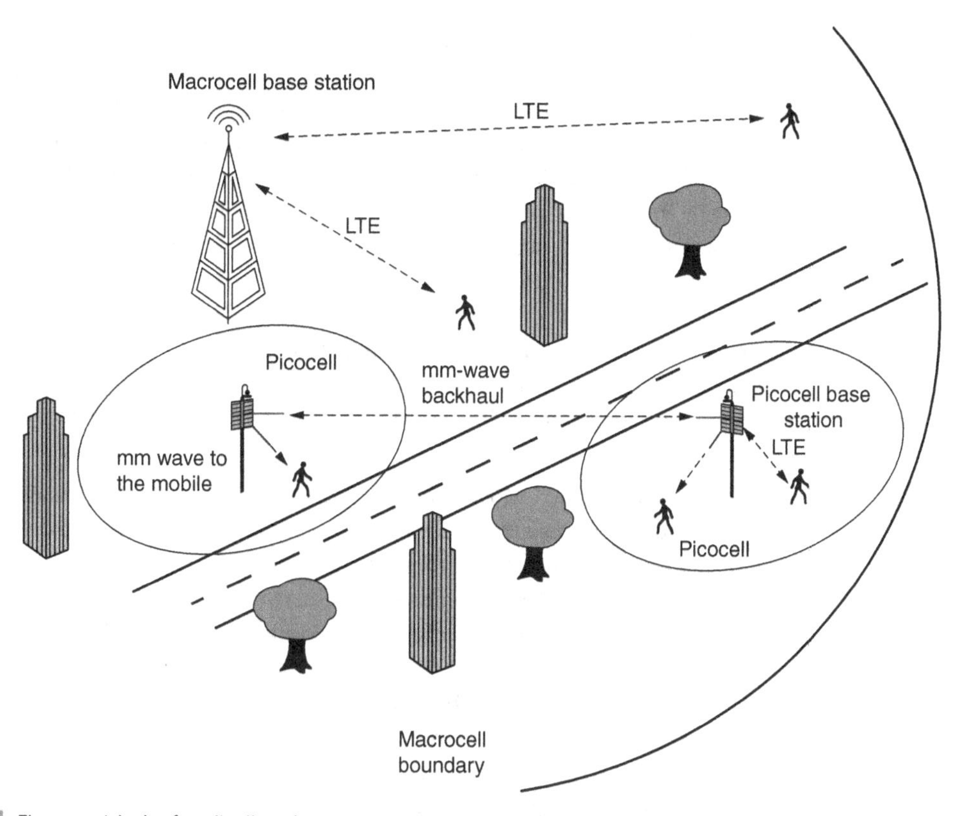

Figure E.1 The potential role of small cells and mm-wave communication in future cellular systems (figure courtesy of Dinesh Ramasamy).

few of the issues that are probably going to take significant research and development to iron out, which bodes well for aspiring communication engineers.

Figure E.1 depicts how picocells and mm-wave communication might come together to address the cellular capacity crisis. A large macrocellular base station provides default connectivity via Long Term Evolution (LTE), a fourth-generation cellular technology standardized relatively recently; despite its name, it might not suffice for the long term because of exponentially increasing demand. Picocellular base stations are deployed opportunistically on lampposts, and may be connected via a mm-wave backhaul network. Users in a picocell could talk to the base stations using LTE, or perhaps even mm wave. Users not covered by picocells talk to the macrocell using LTE.

Cooperative communication While we have restricted our attention in this book to the study of communication between a single transmitter and a single receiver, this provides a building block for emerging ideas in cooperative communication. For example, neighboring nodes could form a virtual antenna array, forming distributed MIMO (DMIMO) systems with significantly improved power–bandwidth tradeoffs. This allows us, for example, to bring the benefits of MIMO to systems with low carrier frequencies, which propagate well over large distances, but are not compatible with centralized antenna arrays because of the large carrier wavelength. For example, the wavelength at 50 MHz is 6 meters, hence conventional antenna arrays would be extremely bulky, but neighboring nodes naturally spaced by tens of meters could form a DMIMO array. DMIMO at low carrier frequencies is a promising approach for bridging the digital divide in cost-effective fashion, providing interference suppression and multiplexing capabilities as in MIMO, along with link ranges of tens of kilometers. Another promising example of cooperative communication is interference alignment, in which multiple transmitters, each of which is sending to a different receiver, coordinate so as to ensure that the interference they generate for each other is limited in time–frequency space. Of course, realizing the benefits of cooperative communication will require fundamental advances in distributed synchronization and channel estimation, along with new network protocols that support these innovations. More good news for the next generation of communication engineers!

Full-duplex communication Most communication transceivers cannot transmit and receive at the same time on the same frequency band (or even closely spaced bands). This is because even a small amount of leakage from the transmit chain can swamp the received signal, which is much weaker. Thus, communication networks typically operate in *time-division duplexed (TDD)* mode (which is also more loosely termed *half-duplex* mode), in which the transmitter and receiver use the same band, but are not active at the same time, or in *frequency-division duplexed (FDD)* mode, in which the transmitter and receiver may be simultaneously active, but in different, typically widely separated, bands. There has been promising progress recently, however, on relatively low-cost approaches to canceling out interference from the transmit chain, seeking to make full duplex operation (i.e., sending and receiving at the same time in the same band) feasible. If these techniques turn out to be robust and practical, then they could lead to significant performance enhancements in wireless networks. Of course, networking with full duplex links will require revisiting current protocols, which are based on either TDD or FDD.

Challenging channels While we have discussed issues related to significant improvements in wireless data rates and ranges relative to large-scale commercial wireless networks today, there are important applications where simply forming and maintaining a viable link is a challenge. Examples include underwater acoustic networks (for sensing and exploration in oceans, rivers and lakes) and body area networks (for continuous health monitoring).

Wireless-enabled multi-agent systems Wireless is at the heart of many emerging systems that require communication and coordination between a variety of "agents" (these may be machines or humans). Examples include asset tracking and inventory management using radio-frequency identification (RFID) tags; sensor networks for automation in manufacturing, environmental monitoring, healthcare and assisted living; vehicular communication; the smart grid; and nascent concepts such as autonomous robot swarms. Such "multi-agent" systems rely on wireless to provide tetherless connectivity among agents, as well as to possibly provide radar-style measurements, hence characterization and optimization of the wireless network in each specific context is essential for sound system design.

Scaling mostly digital transceivers

As discussed in Chapter 1, a key technology story that has driven the growth of communication systems is Moore's law, which allows us to inexpensively implement sophisticated DSP algorithms in communication transceivers. A modern "mostly digital" receiver typically has analog-to-digital converters (ADCs) representing each I and Q sample with 8–12 bits of precision. As communication data rates and bandwidths increase, Moore's law will probably be able to keep up for a while longer, but the ADC becomes a bottleneck as signal bandwidths, and hence the required sampling rates, scale to GHz and beyond. High-speed, high-precision ADCs are power-hungry, occupy large chip areas (and are therefore expensive), and are difficult to build. Thus, a major open question in communication systems is whether we can continue to enjoy the economies of scale provided by "mostly digital" architectures as communication bandwidths increase.

We have already discussed the potential for mm-wave communication and its unique challenges. The ADC bottleneck is one more challenge we must add to the list, but this challenge applies to any communication system that seeks to employ DSP while scaling up bandwidth. An important example is fiber-optic communication, where the bandwidths involved are huge. For the longest time, these systems have operated using elementary signaling schemes such as on–off keying, with mostly analog processing at the receiver. However, researchers are now seeking to bring the sophistication of wireless transceivers to optical communication. By making optical communication more spectrally efficient, we can increase data rates on fibers that are already buried in the ground, simply by replacing the transceivers at each end. Sophisticated algorithms are critical for achieving this, and these are best implemented in DSP. Furthermore, by making optical transceivers mostly digital, we could obtain the economies of scale required for high-volume applications such as very-short-range chip-to-chip, or intra-chip, communication. Compared with wireless communication, optical communication represents special challenges due to fiber nonlinearities and because of its higher bandwidth, while not facing the difficulties arising from mobility.

Yet another area where we seek increased speed and sophistication is wired backplane communication for interconnecting hardware modules on a circuit board (e.g., inputs and outputs for a high-speed router, or processor and memory modules in a computer), and "networks on a chip" for communicating between modules on a single integrated circuit (e.g., for a "multi-core" processor chip with multiple processor and memory modules).

How does one overcome the ADC bottleneck? We do not have answers yet, but there are some natural ideas to try. One possibility is to try and get by with fewer bits of precision per sample. Severe quantization introduces a significant nonlinearity, but it is possible that we could still extract enough information for reliable communication if the dynamic range of the analog signal being quantized is not too large. Of course, the algorithms that we have seen in this textbook (e.g., for demodulation, linear equalization, and MIMO processing) all rely on the linearity of the channel not being disturbed by the ADC, which is an excellent approximation for high-precision ADC. This assumption now needs to be thrown out: in essence, we must "redo" DSP for communication if we are going to live with low-precision ADC at the receiver. Another possibility is to parallelize: we could implement a high-speed ADC by running lower-speed ADCs in parallel, or we could decompose the communication signal in the frequency domain, such that relatively low-speed ADCs with high precision can be used in parallel for different subbands. These are areas of active research.

Beyond Moore's law

Moore's law has been working because the semiconductor industry keeps managing to shrink feature sizes on integrated circuits, making transistors (which are then used as building blocks for digital logic) tinier and tinier. Many in the industry now say that the time is approaching when shrinking transistors in this fashion will make their behavior non-deterministic (i.e., their output can have errors, just like the output of a demodulator in a communication system). Doing deterministic logic computations with non-deterministic units is a serious challenge, which looks to be more difficult than reliable communication over a noisy channel. However, it is intriguing to ask whether it is possible to use ideas similar to those in digital communication to evolve new paradigms for reliable computation with unreliable units. This is a grand challenge to which experts in communication systems might be able to make significant contributions.

Parting thoughts

The introductory treatment in this textbook is intended to serve as a gateway to an exciting future in communications research and technology development. We hope that this discussion gives the reader the motivation for further study in this area, using, for example, more advanced textbooks and the research literature. We do not provide specific references for the topics mentioned in this epilogue, because research in many of these areas is evolving too rapidly for a few books or papers to do it justice. Of course, the discussion does provide plenty of keywords for online searches, which should bring up interesting material to follow up on.

References

[1] S. Haykin, *Communications Systems*. Wiley, 2000.
[2] J. G. Proakis and M. Salehi, *Fundamentals of Communication Systems*. Prentice Hall, 2004.
[3] M. B. Pursley, *Introduction to Digital Communications*. Prentice Hall, 2003.
[4] R. E. Ziemer and W. H. Tranter, *Principles of Communication: Systems, Modulation and Noise*. Wiley, 2001.
[5] J. R. Barry, E. A. Lee, and D. G. Messerschmitt, *Digital Communication*. Kluwer Academic Publishers, 2004.
[6] S. Benedetto and E. Biglieri, *Principles of Digital Transmission; with Wireless Applications*. Springer, 1999.
[7] U. Madhow, *Fundamentals of Digital Communication*. Cambridge University Press, 2008.
[8] J. G. Proakis and M. Salehi, *Digital Communications*. McGraw-Hill, 2007.
[9] J. M. Wozencraft and I. M. Jacobs, *Principles of Communication Engineering*. Wiley, 1965; reissued by Waveland Press in 1990.
[10] A. J. Viterbi and J. K. Omura, *Principles of Digital Communication and Coding*. McGraw-Hill, 1979.
[11] R. E. Blahut, *Digital Transmission of Information*. Addison-Wesley, 1990.
[12] J. D. Gibson, ed., *The Mobile Communications Handbook*. CRC Press, 2012.
[13] K. Sayood, *Introduction to Data Compression*. Morgan Kaufmann, 2005.
[14] D. P. Bertsekas and R. G. Gallager, *Data Networks*. Prentice Hall, 1991.
[15] A. Kumar, D. Manjunath, and J. Kuri, *Communication Networking: An Analytical Approach*. Morgan Kaufmann, 2004.
[16] J. Walrand and P. Varaiya, *High Performance Communication Networks*. Morgan Kaufmann, 2000.
[17] A. V. Oppenheim, A. S. Willsky, and S. H. Nawab, *Signals and Systems*. Prentice Hall, 1996.
[18] B. P. Lathi, *Linear Systems and Signals*. Oxford University Press, 2004.
[19] A. V. Oppenheim and R. W. Schafer, *Discrete-Time Signal Processing*. Prentice Hall, 2009.
[20] S. K. Mitra, *Digital Signal Processing: A Computer-based Approach*. McGraw-Hill, 2010.
[21] F. M. Gardner, *Phaselock techniques*. Wiley, 2005.
[22] A. J. Viterbi, *Principles of Coherent Communication*. McGraw-Hill, 1966.
[23] R. Best, *Phase Locked Loops: Design, Simulation, and Applications*. McGraw-Hill, 2007.
[24] B. Razavi, *Phase Locking in High-Performance Systems: From Devices to Architectures*. Wiley-IEEE Press, 2003.
[25] R. D. Yates and D. J. Goodman, *Probability and Stochastic Processes: A Friendly Introduction for Electrical and Computer Engineers*. Wiley, 2004.
[26] J. W. Woods and H. Stark, *Probability and Random Processes with Applications to Signal Processing*. Prentice Hall, 2001.

[27] A. Leon-Garcia, *Probability and Random Processes for Electrical Engineering*. Prentice Hall, 1993.
[28] A. Papoulis and S. U. Pillai, *Probability, Random Variables and Stochastic Processes*. McGraw-Hill, 2002.
[29] J. B. Johnson, "Thermal agitation of electricity in conductors," *Physical Review*, vol. 32, pp. 97–109, 1928.
[30] H. Nyquist, "Thermal agitation of electric charge in conductors," *Physical Review*, vol. 32, pp. 110–113, 1928.
[31] D. Abbott, B. Davis, N. Phillips, and K. Eshraghian, "Simple derivation of the thermal noise formula using window-limited Fourier transforms and other conundrums," *IEEE Transactions on Education*, vol. 39, pp. 1–13, 1996.
[32] R. Sarpeshkar, T. Delbruck, and C. Mead, "White noise in MOS transistors and resistors," *IEEE Circuits and Devices Magazine,* vol. 9, pp. 23–29, 1993.
[33] V. A. Kotelnikov, *The Theory of Optimum Noise Immunity*. McGraw-Hill, 1959.
[34] G. D. Forney, "Maximum-likelihood sequence estimation of digital sequences in the presence of intersymbol interference," *IEEE Transactions on Information Theory*, vol. 18, pp. 363–378, 1972.
[35] G. Ungerboeck, "Adaptive maximum-likelihood receiver for carrier-modulated data-transmission systems," *IEEE Transactions on Communications*, vol. 22, pp. 624–636, 1974.
[36] S. Kay, *Fundamentals of Statistical Signal Processing, Volume I: Estimation Theory*. Prentice Hall, 1993.
[37] H. V. Poor, *An Introduction to Signal Detection and Estimation*. Springer, 2005.
[38] H. L. V. Trees, *Detection, Estimation, and Modulation Theory, Part I*. Wiley, 2001.
[39] R. E. Blahut, *Modem Theory: An Introduction to Telecommunications*. Cambridge University Press, 2009.
[40] T. M. Cover and J. A. Thomas, *Elements of Information Theory*. Wiley, 2006.
[41] C. E. Shannon, "A mathematical theory of communication," *Bell Systems Technical Journal*, vol. 27, pp. 379–423, 623–656, 1948.
[42] R. J. McEliece, *The Theory of Information and Coding*. Cambridge University Press, 2002.
[43] R. E. Blahut, *Algebraic Codes for Data Transmission*. Cambridge University Press, 2003.
[44] S. Lin and D. J. Costello, *Error Control Coding*. Prentice Hall, 2004.
[45] T. K. Moon, *Error Correction Coding: Mathematical Methods and Algorithms*. Wiley, 2005.
[46] A. Goldsmith, *Wireless Communications*. Cambridge University Press, 2005.
[47] D. Tse and P. Viswanath, *Fundamentals of Wireless Communication*. Cambridge University Press, 2005.
[48] G. Foschini, "Layered space–time architecture for wireless communication in a fading environment when using multi-element antennas," *Bell-Labs Technical Journal*, vol. 1, no. 2, pp. 41–59, 1996.
[49] E. Telatar, "Capacity of multi-antenna Gaussian channels," AT&T Bell Labs Internal Technical Memo # BL0112170-950615-07TM, June 1995.
[50] S. Alamouti, "A simple transmit diversity technique for wireless communications," *IEEE Transactions on Selected Areas in Communications*, vol. 16, pp. 1451–1458, 1998.
[51] A. Paulraj, R. Nabar, and D. Gore, *Introduction to Space-Time Wireless Communications*. Cambridge University Press, 2003.

[52] H. Jafarkhani, *Space-Time Coding: Theory and Practice*. Cambridge University Press, 2003.

[53] H. Bolcskei, D. Gesbert, C. B. Papadias, and A. J. van der Veen, eds., *Space-Time Wireless Systems: From Array Processing to MIMO Communications*. Cambridge University Press, 2006.

Index

Printed in the United States
by Baker & Taylor Publisher Services